Methods in Enzymology

Volume 319
SINGLET OXYGEN, UV-A, AND OZONE

METHODS IN ENZYMOLOGY

EDITORS-IN-CHIEF

John N. Abelson   Melvin I. Simon

DIVISION OF BIOLOGY
CALIFORNIA INSTITUTE OF TECHNOLOGY
PASADENA, CALIFORNIA

FOUNDING EDITORS

Sidney P. Colowick and Nathan O. Kaplan

Methods in Enzymology

Volume 319

Singlet Oxygen, UV-A, and Ozone

EDITED BY

Lester Packer

UNIVERSITY OF CALIFORNIA
BERKELEY, CALIFORNIA

Helmut Sies

HENRICH-HEINE-UNIVERSITÄT
DÜSSELDORF, GERMANY

Editorial Advisory Board

ACADEMIC PRESS

San Diego   London   Boston   New York   Sydney   Tokyo   Toronto

Academic Press
A Harcourt Science and Technology Company
525 B Street, Suite 1900, San Diego, California 92101-4495, USA

http://www.academicpress.com

Academic Press Limited
24-28 Oval Road, London NW1 7DX, UK

International Standard Book Number: 0-12-182220-6

PRINTED IN THE UNITED STATES OF AMERICA
00 01 02 03 04 05 06 MM 9 8 7 6 5 4 3 2 1

Table of Contents

CONTRIBUTORS TO VOLUME 319 xi

PREFACE xvii

VOLUMES IN SERIES xix

Section I. Singlet Oxygen

A. Generation and Detection: Chemical Systems

1. Naphthalene Endoperoxides as Generators of Sin- CHRISTEL PIERLOT,
   glet Oxygen in Biological Media   JEAN-MARIE AUBRY,
   KARLIS BRIVIBA,
   HELMUT SIES, AND
   PAOLO DI MASCIO   3

2. Photosensitized Production of Singlet Oxygen   IRENE E. KOCHEVAR AND
   ROBERT W. REDMOND   20

3. Singlet Oxygen from Irradiated Titanium Dioxide YORIHIRO YAMAMOTO,
   and Zinc Oxide   NORITAKA IMAI,
   RYUICHI MASHIMA,
   RYUSEI KONAKA,
   MASAYASU INOUE, AND
   WALTER C. DUNLAP   29

4. Time-Resolved Singlet Oxygen Detection   SANTI NONELL AND
   SILVIA E. BRASLAVSKY   37

5. Measurement of Photogenerated Singlet Oxygen in VÉRONIQUE NARDELLO AND
   Aqueous Media   JEAN-MARIE AUBRY   50

B. Generation and Detection: Biological Systems

6. Assay for Singlet-Oxygen Generation by Peroxi- JEFFREY R. KANOFSKY   59
   dases Using 1270-nm Chemiluminescence

7. Formation of Electronically Excited States during ENRIQUE CADENAS AND
   the Oxidation of Arachidonic Acid by Prostaglan- HELMUT SIES   67
   din Endoperoxide Synthase

8. Singlet Oxygen Detection with Sterically Hindered ÉVA HIDEG,
   Amine Derivatives in Plants under Light Stress IMRE VASS,
   TAMÁS KÁLAI, AND
   KÁLMÁN HIDEG   77

9. Cholesterol as a Singlet Oxygen Detector in Biologi-   ALBERT W. GIROTTI AND
   cal Systems   WITOLD KORYTOWSKI   85

10. Singlet Oxygen Scavenging in Phospholipid Mem-   KENJI FUKUZAWA   101
    branes

11. Catalase Modification as a Marker for Singlet   FERNANDO LLEDIAS AND
    Oxygen   WILHELM HANSBERG   110

C. Biological Effects: Role in Signaling

12. Nuclear Factor-κB Activation by Singlet Oxygen   JEAN-YVES MATROULE AND
    Produced during Photosensitization   JACQUES PIETTE   119

13. Mitogen-Activated Protein Kinase Activation by   LARS-OLIVER KLOTZ,
    Singlet Oxygen and Ultraviolet A   KARLIS BRIVIBA, AND
                                                         HELMUT SIES   130

D. Toxic Effects

14. Singlet Oxygen DNA Damage Products: Formation   JEAN CADET,
    and Measurement   THIERRY DOUKI,
                                                         JEAN-PIERRE POUGET, AND
                                                         JEAN-LUC RAVANET   143

15. Ultraviolet A- and Singlet Oxygen-Induced Muta-   ANNE STARY AND
    tion Spectra   ALAIN SARASIN   153

16. Damage to DNA by Long-Range Charge Transport   MEGAN E. NÚÑEZ,
                                                         SCOTT R. RAJSKI, AND
                                                         JACQUELINE K. BARTON   165

17. Cholesterol Photodynamic Oxidation by Ultraviolet   KYOICHI OSADA AND
    Irradiation and Cholesterol Ozonization by   ALEX SEVANIAN   188
    Ozone Exposure

18. Bactericidal and Virucidal Activities of Singlet Oxy-   CORINNE PELLIEUX,
    gen Generated by Thermolysis of Naphthalene   ANNY DEWILDE,
    Endoperoxides   CHRISTEL PIERLOT, AND
                                                         JEAN-MARIE AUBRY   197

19. Inactivation of Viruses in Human Plasma   Harald MOHR   207

20. 3-(4'-Methyl-1'-naphthyl)propionic   Acid,   1',4'-   MINORU NAKANO,
    Endoperoxide for Dioxygenation of Squalene   YASUHIRO KAMBAYASHI, AND
    and for Bacterial Killing   HIDETAKA TATSUZAWA   216

E. Protection: Singlet Oxygen Quenchers

21. Biological Singlet Oxygen Quenchers Assessed by   KARLIS BRIVIBA AND
    Monomol Light Emission   HELMUT SIES   222

22. Synthetic Singlet Oxygen Quenchers   STEFAN BEUTNER,
BRITTA BLOEDORN,
THOMAS HOFFMANN, AND
HANS-DIETER MARTIN   226

Section II. Ultraviolet A

A. Dosimetry

23. Dosimetry of Ultraviolet A Radiation   BRIAN L. DIFFEY   245

B. Biological Responses: Signaling

24. Radiation-Induced Signal Transduction   AXEL KNEBEL,
FRANK D. BÖHMER, AND
PETER HERRLICH   255

25. Signaling Pathways Leading to Nuclear Factor-κB   NANXIN LI AND
Activation   MICHAEL KARIN   273

26. Gene Regulation by Ultraviolet A Radiation and   SUSANNE GRETHER-BECK AND
Singlet Oxygen   JEAN KRUTMANN   280

27. Role for Singlet Oxygen in Biological Effects of   REX M. TYRRELL   290
Ultraviolet A Radiation

28. Ultraviolet A-1 Irradiation as a Tool to Study the   JEAN KRUTMANN   296
Pathogenesis of Atopic Dermatitis

29. Ultraviolet A Radiation-Induced Apoptosis   AKIMICHI MORITA AND
JEAN KRUTMANN   302

30. Singlet Oxygen-Triggered Immediate Prepro-   DIANNE E. GODAR   309
grammed Apoptosis

31. Determination of DNA Damage, Peroxide Gen-   SAEKO TADA-OIKAWA,
eration, Mitochondrial Membrane Potential,   SHINJI OIKAWA, AND
and Caspase-3 Activity during Ultraviolet A-   SHOSUKE KAWANISHI   331
Induced Apoptosis

32. Mechanism of Photodynamic Therapy-Induced   NIHAL AHMAD AND
Cell Death   HASAN MUKHTAR   342

C. Biological Responses: Photocarcinogenesis, Photoaging, and Photoallergic Reactions

33. Photocarcinogenesis: UVA vs UVB   FRANK R. DE GRUIJL   359

34. Photoaging-Associated Large-Scale Deletions of   MARK BERNEBURG AND
Mitochondrial DNA   JEAN KRUTMANN   366

35. Role of Activated Oxygen Species in Photody-
namic Therapy — WESLEY M. SHARMAN, CYNTHIA M. ALLEN, AND JOHAN E. VAN LIER   376

36. Gas Chromatography–Mass Spectrometry Analysis of DNA: Optimization of Protocols for Isolation and Analysis of DNA from Human Blood — ALMAS REHMAN, ANDREW JENNER, AND BARRY HALLIWELL   401

D. Oxidative Damage Markers

37. Sequence Specificity of Ultraviolet A-Induced DNA Damage in the Presence of Photosensitizer — KIMIKO ITO AND SHOSUKE KAWANISHI   417

38. Protein Oxidative Damage — EMILY SHACTER   428

39. DNA Damage Induced by Ultraviolet and Visible Light and Its Wavelength Dependence — CHRISTOPHER KIELBASSA AND BERND EPE   436

40. Photoprotection of Skin against Ultraviolet A Damage — HANS SCHAEFER, ALAIN CHARDON, AND DOMINIQUE MOYAL   445

E. Protection

41. Topically Applied Antioxidants in Skin Protection — FRANZ STÄB, RAINER WOLBER, THOMAS BLATT, REZA KEYHANI, AND GERHARD SAUERMANN   465

42. Erythropoietic Protoporphyria: Treatment with Antioxidants and Potential Cure with Gene Therapy — MICHELINE M. MATHEWS-ROTH   479

43. Porphyrias: Photosensitivity and Phototherapy — MAUREEN B. POH-FITZPATRICK   485

44. Carotenoids in Human Skin: Noninvasive Measurement and Identification of Dermal Carotenoids and Carotenol Esters — WILHELM STAHL, ULRIKE HEINRICH, HOLGER JUNGMANN, HAGEN TRONNIER, AND HELMUT SIES   494

Section III. Ozone

45. Reactive Absorption of Ozone: An Assay for Reaction Rates of Ozone with Sulfhydryl Compounds and with Other Biological Molecules — JEFFREY R. KANOFSKY AND PAUL D. SIMA   505

46. Assay for Singlet Oxygen Generation by Plant Leaves Exposed to Ozone — JEFFERY R. KANOFSKY AND PAUL D. SIMA   512

47. Ozone Effects on Plant Defense — CHRISTIAN LANGEBARTELS, DIETER ERNST, JAAKKO KANGASJÄRVI, AND HEINRICH SANDERMANN, JR.   520

48. High-Pressure Liquid Chromatography Analysis of STEFAN U. WEBER,
Ozone-Induced Depletion of Hydrophilic and Li- SUMANA JOTHI, AND
pophilic Antioxidants in Murine Skin JENS J. THIELE 536

49. Reactions of Vitamin E with Ozone DANIEL C. LIEBLER 546

50. Induction of Nuclear Factor-κB by Exposure to Kian Fan CHUNG AND
Ozone and Inhibition by Glucocorticoids Ian M. ADCOCK 551

51. Detection of 4-Hydroxy-2-nonenol Adducts Fol- LUKE I. SZWEDA,
lowing Lipid Peroxidation from Ozone Exposure PAMELA A. SZWEDA, AND
ANDRIJ HOLIAN 562

52. Synthesis of Inflammatory Signal Transduction Spe- GIUSEPPE L. SQUADRITO,
cies Formed during Ozonation and/or Peroxida- MARIA G. SALGO,
tion of Tissue Lipids FRANK R. FRONCZEK, AND
WILLIAM A. PRYOR 570

Section IV. General Methods

53. Assay for Redox-Sensitive Transcription Factor MADAN M. CHATURVEDI,
ASOK MUKHOPADHYAY, AND
BHARAT B. AGGARWAL 585

54. Fluorescent Fatty Acid to Monitor Reactive Oxygen Eward H. W. PAP,
in Single Cells G. P. C. DRUMMEN,
J. A. POST,
P. J. RIJKEN, AND
K. W. A. WIRTZ 603

55. Noninvasive Techniques for Measuring Oxidation DANIEL MAES,
Products on the Surface of Human Skin TOM MOMMONE,
MARYANN MCKEEVER,
ED PELLE,
CHRISTINA FTHENAKIS,
LIEVE DECLERCQ,
PAOLO U. GIACOMONI, AND
KEN MARENUS 612

AUTHOR INDEX 623

SUBJECT INDEX 661

Contributors to Volume 319

Article numbers are in parentheses following the names of contributors.
Affiliations listed are current.

IAN M. ADCOCK (50), *National Heart and Lung Institute, Imperial College School of Medicine, London, United Kingdom*

BHARAT B. AGGARWAL (53), *Cytokine Research Section, Department of Molecular Oncology, The University of Texas MD Anderson Cancer Center, Houston, Texas, 77030*

NIHAL AHMAD (32), *Department of Dermatology, Case Western Reserve University, Cleveland, Ohio 44106*

CYNTHIA M. ALLEN (35), *MRC Group in the Radiation Sciences, Department of Nuclear Medicine and Radiobiology, Université de Sherbrooke, Sherbrooke, Quebéc, Canada JIH 5N4*

JEAN-MARIE AUBRY (1, 5, 18), *Equipe de Recherche, Oxydation et Formulation, F-59652 Villeneuve d'Ascq Cedex, France*

JACQUELINE K. BARTON (16), *Division of Chemistry and Chemical Engineering, California Institute of Technology, Pasadena, California 91125*

MARK BERNEBURG (34), *Clinical and Experimental Photodermatology, Department of Dermatology, Heinrich-Heine-University, D-40225 Düsseldorf, Germany*

STEFAN BEUTNER (22), *Institute of Organic Chemistry and Macromolecular Chemistry, University of Düsseldorf, D-40225 Düsseldorf, Germany*

THOMAS BLATT (41), *Department 4212, Beiersdorf AG, D-20245 Hamburg, Germany*

BRITTA BLOEDORN (22), *Institute of Organic Chemistry and Macromolecular Chemistry, University of Düsseldorf, D-40225 Düsseldorf, Germany*

FRANK D. BÖHMER (24), *Arbeitsgruppe Molekulare Zellbiologie, Klinikum der Friedrich-Schiller-Universität, D-07747 Jena, Germany*

SILVIA E. BRASLAVSKY (4), *Max-Planck-Institut für Strahlenchemie, D-45413 Mülheim an der Ruhr, Germany*

KARLIS BRIVIBA (1, 13, 21), *Institut für Ernährungsphysiologie, Bundesforschungsanstalt für Ernährung, 76131 Karlsruhe, Germany*

ENRIQUE CADENAS (7), *Department of Molecular Pharmacology and Toxicology, School of Pharmacy, University of Southern California, Los Angeles, California 90089-9121*

JEAN CADET (14), *Laboratoire "Lésions des Acides Nucléiques" Service de Chimie Inorganique et Biologique, Département de Recherche Fondamentale sur la Matière Condensée, F-38054 Grenoble Cedex 9, France*

ALAIN CHARDON (40), *L'Oreal Cosmetometrie, 92M7 Clichy Cedex, France*

MADAN M. CHATURVEDI (53), *Biochemistry and Molecular Biology Laboratory, Center for Advanced Study in Zoology, Banaras Hindu University, Varanasi 221005, India*

KIAN FAN CHUNG (50), *National Heart and Lung Institute, Imperial College School of Medicine, London, United Kingdom*

LIEVE DECLERCQ (55), *Estée Lauder Laboratories, Oevel 2260, Belgium*

FRANK R. DE GRUIJL (33), *Department of Dermatology, University Medical Center, NL-3508 GA Utrecht, The Netherlands*

ANNY DEWILDE (18), *Laboratoire de Virologie, Institut Gernez-Rieux, F-59037 Lille Cedex, France*

BRIAN L. DIFFEY (23), *Regional Medical Physics Department, Newcastle General Hospital, Newcastle NE4 GBE, United Kingdom*

PAOLO DI MASCIO (1), *Departamento de Bio-*

quimica, Instituto de Quimica, Universidade de São Paulo, CEP 05599-970, São Paulo, Brazil

THIERRY DOUKI (14), Laboratoire "Lésions des Acides Nucléiques" Service de Chimie Inorganique et Biologique, Département de Recherche Fondamentale sur la Matière Condensée, F-38054 Grenoble Cedex 9, France

G. P. C. DRUMMEN (54), Department of Biochemistry of Lipids, Centre for Biomembranes and Lipid Enzymology, Utrecht University, NL-3584CH Utrecht, The Netherlands

WALTER C. DUNLAP (3), Australian Institute of Marine Science, Townsville, Queensland 4810, Australia

BERND EPE (39), Institute of Pharmacy, University of Mainz, D-55099 Mainz, Germany

DIETER ERNST (47), GSF-Forschungszentrum für Umwelt und Gesundheit, GmbH, Institut für Biochemische Pflanzenpathologie, D-85764 Oberschleißheim, Germany

FRANK R. FRONCZEK (52), Department of Chemistry, Louisiana State University, Baton Rouge, Louisiana 70803-1800

CHRISTINA FTHENAKIS (55), Estée Lauder Laboratories, Melville, New York 11747

KENJI FUKUZAWA (10), Faculty of Pharmaceutical Sciences, Tokushima University, Tokushima 770-8505, Japan

PAOLO U. GIACOMONI (55), Estée Lauder Laboratories, Melville, New York 11747

ALBERT W. GIROTTI (9), Department of Biochemistry, Medical College of Wisconsin, Milwaukee, Wisconsin 53226

DIANNE E. GODAR (30), Food and Drug Administration, Center for Devices and Radiological Health, Rockville, Maryland 20852

SUSANNE GRETHER-BECK (26), Clinical and Experimental Photodermatology, Department of Dermatology, Heinrich-Heine-University, D-40225 Düsseldorf, Germany

BARRY HALLIWELL (36), Department of Biochemistry, National University of Singapore, Kent Ridge Crescent, 119260 Singapore

WILHELM HANSBERG (11), Departamento de Bioquímica, Instituto de Fisiologia Celúlar, Universidad Nacional Autónoma de México, 04510 México, D.F.

ULRIKE HEINRICH (44), Institut für Experimentelle Dermatologie, Universität Witten-Herdecke, D-58453 Witten, Germany

PETER HERRLICH (24), Forschungszentrum Karlsrube Institute of Toxicology and Genetics, 76021 Karlsrube, Germany

ÉVA HIDEG (8), Institute of Plant Biology, Biological Research Center, H-6701 Szeged, Hungary

KÁLMÁN HIDEG (8), Institute of Organic and Medicinal Chemistry, University of Pécs, H-7643 Pécs, Hungary

THOMAS HOFFMANN (22), Institute of Organic Chemistry and Macromolecular Chemistry, University of Düsseldorf, D-40225 Düsseldorf, Germany

ANDRIJ HOLIAN (51), Department of Pharmaceutical Sciences, University of Montana, Missoula, Montana 59812-1552

NORITAKA IMAI (3), Research Center for Advanced Science and Technology, University of Tokyo, Tokyo 153-8904, Japan

MASAYASU INOUE (3), Department of Biochemistry, Osaka City University Medical School, Osaka 545-8585, Japan

KIMITO ITO (37), Department of Public Health, Kyoto University, Kyoto 606-8501, Japan

ANDREW JENNER (36), Department of Pharmacology, King's College, London SW3 6LX, United Kingdom

SUMANA JOTHI (48), Department of Molecular and Cellular Biology, University of California, Berkeley, California 94720-3200

HOLGER JUNGMANN (44), Krebsforschung Herdecke e.V., D-58313 Herdecke, Germany

TAMÁS KÁLAI (8), Institute of Organic and Medicinal Chemistry, University of Pécs, H-7643 Pécs, Hungary

YASUHIRO KAMBAYASHI (20), Department of Photon and Free Radical Research, Japan

Immunoresearch Laboratories, Takasaki 370-0021, Japan

JAAKKO KANGASJÄRVI (47), *Institute of Biotechnology, Division of Genetics, Department of Biological Sciences, University of Helsinki, Helsinki 00014 Finland*

JEFFREY R. KANOFSKY (6, 45, 46), *Medical Service, Department of Veterans Affairs Hospital, Hines, Illinois 60141, and Departments of Medicine and Molecular and Cellular Biochemistry, Loyola University, Stritch School of Medicine, Maywood, Illinois 60153*

MICHAEL KARIN (25), *Department of Pharmacology, University of California, San Diego, La Jolla, California 92093-0636*

SHOSUKE KAWANISHI (31, 37), *Department of Hygiene, Mie University School of Medicine, Mie 514-8507 Japan*

REZA KEYHANI (41), *Department 4212, Beiersdorf AG, D-20245 Hamburg, Germany*

CHRISTOPHER KIELBASSA (39), *Institute of Pharmacy, University of Mainz, D-55099 Mainz, Germany*

LARS-OLIVER KLOTZ (13), *Institut für Physiologische Chemie I, Heinrich-Heine-Universität, D-40001 Düsseldorf, Germany*

AXEL KNEBEL (24), *Department of Biochemistry, MRC Protein Phosphorylation Unit, University of Dundee, Dundee, DD1-5EH, United Kingdom*

IRENE E. KOCHEVAR (2), *Wellman Laboratories of Photomedicine, Department of Dermatology, Massachusetts General Hospital, Harvard Medical School, Boston, Massachusetts 02114-2696*

RYUSEI KONAKA (3), *Department of Biochemistry, Osaka City University Medical School, Osaka 545-8585, Japan*

WITOLD KORYTOWSKI (9), *Institute of Molecular Biology, Jagiellonian University, Krakow, Poland*

JEAN KRUTMANN (26, 28, 29, 34), *Clinical and Experimental Photodermatology, Department of Dermatology, Heinrich-Heine-University, D-40225 Düsseldorf, Germany*

CHRISTIAN LANGEBARTELS (47), *GSF-Forschungszentrum für Umwelt und Gesundheit, GmbH, Institut für Biochemische Pflanzenpathologie, D-85764 Oberschleißheim, Germany*

NANXIN LI (25), *Department of Pharmacology, University of California, San Diego, La Jolla, California 92093-0636*

DANIEL C. LIEBLER (49), *Department of Pharmacology and Toxicology, College of Pharmacy, University of Arizona, Tucson, Arizona 85721-0207*

FERNANDO LLEDIAS (11), *Departamento de Bioquímica, Instituto de Fisiologia Celúlar, Universidad Nacional Autónoma de México, 04510 México, D.F.*

DANIEL MAES (55), *Estée Lauder Laboratories, Melville, New York 11747*

KEN MARENUS (55), *Estée Lauder Laboratories, Melville, New York 11747*

HANS-DIETER MARTIN (22), *Institute of Organic Chemistry and Macromolecular Chemistry, University of Düsseldorf, D-40225 Düsseldorf, Germany*

RYUICHI MASHIMA (3), *Research Center for Advanced Science and Technology, University of Tokyo, Tokyo 153-8904, Japan*

MICHELINE M. MATHEWS-ROTH (42), *Department of Medicine, Channing Laboratory, Brigham & Women's Hospital, Harvard Medical School, Boston, Massachusetts 02115*

JEAN-YVES MATROULE (12), *Department of Microbiology, Laboratory of Virology and Immunology, University of Liège, B-4000 Liège, Belgium*

MARYANN MCKEEVER (55), *Estée Lauder Laboratories, Melville, New York 11747*

HARALD MOHR (19), *Blood Center of the German Red Cross, Institute Springe, 31832 Springe, Germany*

TOM MOMMONE (55), *Estée Lauder Laboratories, Melville, New York 11747*

AKIMICHI MORITA (29), *Department of Dermatology, Nagoya City University Medical School, Nagoya 467-8601, Japan*

DOMINIQUE MOYAL (40), L'Oreal Cosmetometrie, 92M7 Clichy Cedex, France

ASOK MUKHOPADHYAY (53), Department of
Molecular Oncology, The University of
Texas MD Anderson Cancer Center, Hous-
ton, Texas, 77030

HASAN MUKHTAR (32), Department of Dermatology, Case Western Reserve University,
Cleveland, Ohio 44106

MINORU NAKANO (20), Department of Photon
and Free Radical Research, Japan Immunoresearch Laboratories, Takasaki 370-
0021, Japan

VÉRONIQUE NARDELLO (5), Equipe de Recherche, Oxydation et Formulation, F-
59652 Villeneuve d'Ascq Cedex, France

SANTI NONELL (4), Institut Químic de Sarrià,
Universitat Ramon Llull, E-08017 Barce-
lona, Spain

MEGAN E. NÚÑEZ (16), Division of Chemistry
and Chemical Engineering, California Institute of Technology, Pasadena, California
91125

SHINJI OIKAWA (31), Department of Hygiene,
Mie University School of Medicine, Mie
514-8507 Japan

KYOICHI OSADA (17), Hirosaki University,
Hirosaki 036, Japan

EWARD H. W. PAP (54), Department of Biochemistry of Lipids, Centre for Biomembranes and Lipid Enzymology, Utrecht
University, NL-3584CH Utrecht, The Netherlands

ED PELLE (55), Estée Lauder Laboratories,
Melville, New York 11747

CORINNE PELLIEUX (18), Division of Hyper-
tension, University of Lausanne Medical
School, CH-1011 Lausanne, Switzerland

CHRISTEL PIERLOT (1, 18), Equipe de Recherche, Oxydation et Formulation, F-59652
Villeneuve d'Ascq Cedex, France

JACQUES PIETTE (12), Department of Microbiology, Laboratory of Virology and Immunology, University of Liège, B-4000
Liège, Belgium

MAUREEN B. POH-FITZPATRICK (43), University of Tennessee College of Medicine,
Memphis, Tennessee 38103

J. A. POST (54), Department of Molecular Cell
Biology, Centre for Biomembranes and
Lipid Enzymology, Utrecht University, NL-
3584CH Utrecht, The Netherlands

JEAN-PIERRE POUGET (14), Laboratoire "Lésions des Acides Nucléiques" Service de
Chimie Inorganique et Biologique, Département de Recherche Fondamentale sur
la Matière Condensée, F-38054 Grenoble
Cedex 9, France

WILLIAM A. PRYOR (52), Biodynamics Institute, Louisiana State University, Baton
Rouge, Louisiana 70803-1800

SCOTT R. RAJSKI (16), Division of Chemistry
and Chemical Engineering, California Institute of Technology, Pasadena, California
91125

JEAN-LUC RAVANAT (14), Laboratoire "Lésions des Acides Nucléiques" Service de
Chimie Inorganique et Biologique, Département de Recherche Fondamentale sur
la Matière Condensée, F-38054 Grenoble
Cedex 9, France

ROBERT W. REDMOND (2), Wellman Laboratories of Photomedicine, Department of
Dermatology, Massachusetts General Hospital, Harvard Medical School, Boston,
Massachusetts 02114-2696

ALMAS REHMAN (36), Department of Pharmacology, King's College, London SW3
6LX, United Kingdom

P. J. RIJKEN (54), Unilever Research Vlaardingen, 3130AC Vlaardingen, The Netherlands

MARIA G. SALGO (52), Department of Biomedical Sciences and Biotechnology, University of Cagliari, 09100 Cagliari, Italy

HEINRICH SANDERMANN, JR. (47), GSF-
Forschungszentrum für Umwelt und Gesundheit, GmbH, Institut für Biochemische Pflanzenpathologie, D-85764
Oberschleißheim, Germany

ALAIN SARASIN (15), UPR 2169—CNRS,
Laboratory of Molecular Genetics, 94800
Villejuif, France

GERHARD SAUERMANN (41), Department of

Skin Physiology, Beiersdorf AG, D-20245 Hamburg, Germany

HANS SCHAEFER (40), *L'Oreal, Centre de Recherche Charles Zviak, 92583 Clichy Cedex, France*

ALEX SEVANIAN (17), *School of Pharmacy, University of Southern California, Los Angeles, California, 90033*

EMILY SHACTER (38), *Food and Drug Administration, Center for Biologics Evaluation and Research, Division of Therapeutic Proteins, Laboratory of Immunology, Bethesda, Maryland 20892*

WESLEY M. SHARMAN (35), *MRC Group in the Radiation Sciences, Department of Nuclear Medicine and Radiobiology, Université de Sherbrooke, Sherbrooke, Quebéc, Canada JIH 5N4*

HELMUT SIES (1, 7, 13, 21, 44), *Institut für Physiologische Chemie I and Biologisch-Medizinisches, Forschungszentrum, Heinrich-Heine-Universität Düsseldorf, D-40001 Düsseldorf, Germany*

PAUL D. SIMA (45, 46), *Research Service, Department of Veterans Affairs Hospital, Hines, Illinois 60141*

GIUSEPPE L. SQUADRITO (52), *Biodynamics Institute and Department of Chemistry, Louisiana State University, Baton Rouge, Louisiana 70803-1800*

FRANZ STÄB (41), *Department of Skin Physiology, Beiersdorf AG, D-20245 Hamburg, Germany*

WILHELM STAHL (44), *Institut für Physiologische Chemie I and Biologisch-Medizinisches, Forschungszentrum, Heinrich-Heine-Universität Düsseldorf, D-40001 Düsseldorf, Germany*

ANNE STARY (15), *UPR 2169—CNRS, Laboratory of Molecular Genetics, 94800 Villejuif, France*

LUKE I. SZWEDA (51), *Department of Physiol-ogy and Biophysics, Case Western Reserve University, Cleveland, Ohio 44106-4970*

PAMELA A. SZWEDA (51), *Department of Physiology and Biophysics, Case Western Reserve University, Cleveland, Ohio 44106-4970*

SAEKO TADA-OIKAWA (31), *Department of Hygiene, Mie University School of Medicine, Mie 514-8507 Japan*

HIDETAKA TATSUZAWA (20), *Shimizu Laboratories, Marine Biology Institute, Shimizu 424-0037, Japan*

JENS J. THIELE (48), *Department of Dermatology, Freidrich Schiller University, Jena, D-07740 Jena, Germany*

HAGEN TRONNIER (44), *Institut für Experimentelle Dermatologie, Universität Witten-Herdecke, D-58453 Witten, Germany*

REX M. TYRRELL (27), *Department of Pharmacy and Pharmacology, University of Bath, Bath BA2 7AY, United Kingdom*

JOHAN E. VAN LIER (35), *MRC Group in the Radiation Sciences, Department of Nuclear Medicine and Radiobiology, Faculty of Medicine, Université de Sherbrooke, Sherbrooke, Quebéc, Canada JIH 5N4*

IMRE VASS (8), *Institute of Plant Biology, Biological Research Center, H-6701 Szeged, Hungary*

STEFAN U. WEBER (48), *Department of Molecular and Cellular Biology, University of California, Berkeley, California 94720-3200*

K. W. A. WIRTZ (54), *Department of Biochemistry of Lipids, Centre for Biomembranes and Lipid Enzymology, Utrecht University, NL-3584CH Utrecht, The Netherlands*

RAINER WOLBER (41), *Department of Skin Physiology, Beiersdorf AG, D-20245 Hamburg, Germany*

YORIHIRO YAMAMOTO (3), *Research Center for Advanced Science and Technology, University of Tokyo, Tokyo 153-8904, Japan*

Preface

Reactive oxygen species and oxygen-derived free radicals are generated in biological systems by metabolism or by exposure to environmental stressors. Thus UV-A irradiation generates singlet oxygen on exposure to the skin or other organs.

Ozone exposure as well as singlet oxygen, both of which are strong oxidants, can generate free radicals and oxidative reactions with biomolecules. Reactive oxygen species at low levels regulate signal transduction and gene expression. Indeed hydrogen peroxide (H_2O_2) and nitric oxide (NO) may act as second messengers in cells. At higher concentrations, these species can be highly toxic and lead to molecular damage.

Relatively recently the methods for the study of UV-A, singlet oxygen, and ozone in biological systems have improved in accuracy and specificity. Hence a volume of *Methods in Enzymology* on the latest methods and techniques for studying these species and evaluating their actions in biological systems seemed warranted at this time.

LESTER PACKER
HELMUT SIES

METHODS IN ENZYMOLOGY

VOLUME I. Preparation and Assay of Enzymes
Edited by SIDNEY P. COLOWICK AND NATIIAN O. KAPLAN

VOLUME II. Preparation and Assay of Enzymes
Edited by SIDNEY P. COLOWICK AND NATHAN O. KAPLAN

VOLUME III. Preparation and Assay of Substrates
Edited by SIDNEY P. COLOWICK AND NATHAN O. KAPLAN

VOLUME IV. Special Techniques for the Enzymologist
Edited by SIDNEY P. COLOWICK AND NATHAN O. KAPLAN

VOLUME V. Preparation and Assay of Enzymes
Edited by SIDNEY P. COLOWICK AND NATHAN O. KAPLAN

VOLUME VI. Preparation and Assay of Enzymes (*Continued*)
Preparation and Assay of Substrates
Special Techniques
Edited by SIDNEY P. COLOWICK AND NATHAN O. KAPLAN

VOLUME VII. Cumulative Subject Index
Edited by SIDNEY P. COLOWICK AND NATHAN O. KAPLAN

VOLUME VIII. Complex Carbohydrates
Edited by ELIZABETH F. NEUFELD AND VICTOR GINSBURG

VOLUME IX. Carbohydrate Metabolism
Edited by WILLIS A. WOOD

VOLUME X. Oxidation and Phosphorylation
Edited by RONALD W. ESTABROOK AND MAYNARD E. PULLMAN

VOLUME XI. Enzyme Structure
Edited by C. H. W. HIRS

VOLUME XII. Nucleic Acids (Parts A and B)
Edited by LAWRENCE GROSSMAN AND KIVIE MOLDAVE

VOLUME XIII. Citric Acid Cycle
Edited by J. M. LOWENSTEIN

VOLUME XIV. Lipids
Edited by J. M. LOWENSTEIN

VOLUME XV. Steroids and Terpenoids
Edited by RAYMOND B. CLAYTON

VOLUME XVI. Fast Reactions
Edited by KENNETH KUSTIN

VOLUME XVII. Metabolism of Amino Acids and Amines (Parts A and B)
Edited by HERBERT TABOR AND CELIA WHITE TABOR

VOLUME XVIII. Vitamins and Coenzymes (Parts A, B, and C)
Edited by DONALD B. MCCORMICK AND LEMUEL D. WRIGHT

VOLUME XIX. Proteolytic Enzymes
Edited by GERTRUDE E. PERLMANN AND LASZLO LORAND

VOLUME XX. Nucleic Acids and Protein Synthesis (Part C)
Edited by KIVIE MOLDAVE AND LAWRENCE GROSSMAN

VOLUME XXI. Nucleic Acids (Part D)
Edited by LAWRENCE GROSSMAN AND KIVIE MOLDAVE

VOLUME XXII. Enzyme Purification and Related Techniques
Edited by WILLIAM B. JAKOBY

VOLUME XXIII. Photosynthesis (Part A)
Edited by ANTHONY SAN PIETRO

VOLUME XXIV. Photosynthesis and Nitrogen Fixation (Part B)
Edited by ANTHONY SAN PIETRO

VOLUME XXV. Enzyme Structure (Part B)
Edited by C. H. W. HIRS AND SERGE N. TIMASHEFF

VOLUME XXVI. Enzyme Structure (Part C)
Edited by C. H. W. HIRS AND SERGE N. TIMASHEFF

VOLUME XXVII. Enzyme Structure (Part D)
Edited by C. H. W. HIRS AND SERGE N. TIMASHEFF

VOLUME XXVIII. Complex Carbohydrates (Part B)
Edited by VICTOR GINSBURG

VOLUME XXIX. Nucleic Acids and Protein Synthesis (Part E)
Edited by LAWRENCE GROSSMAN AND KIVIE MOLDAVE

VOLUME XXX. Nucleic Acids and Protein Synthesis (Part F)
Edited by KIVIE MOLDAVE AND LAWRENCE GROSSMAN

VOLUME XXXI. Biomembranes (Part A)
Edited by SIDNEY FLEISCHER AND LESTER PACKER

VOLUME XXXII. Biomembranes (Part B)
Edited by SIDNEY FLEISCHER AND LESTER PACKER

VOLUME XXXIII. Cumulative Subject Index Volumes I–XXX
Edited by MARTHA G. DENNIS AND EDWARD A. DENNIS

VOLUME XXXIV. Affinity Techniques (Enzyme Purification: Part B)
Edited by WILLIAM B. JAKOBY AND MEIR WILCHEK

VOLUME XXXV. Lipids (Part B)
Edited by JOHN M. LOWENSTEIN

VOLUME XXXVI. Hormone Action (Part A: Steroid Hormones)
Edited by BERT W. O'MALLEY AND JOEL G. HARDMAN

VOLUME XXXVII. Hormone Action (Part B: Peptide Hormones)
Edited by BERT W. O'MALLEY AND JOEL G. HARDMAN

VOLUME XXXVIII. Hormone Action (Part C: Cyclic Nucleotides)
Edited by JOEL G. HARDMAN AND BERT W. O'MALLEY

VOLUME XXXIX. Hormone Action (Part D: Isolated Cells, Tissues, and Organ Systems)
Edited by JOEL G. HARDMAN AND BERT W. O'MALLEY

VOLUME XL. Hormone Action (Part E: Nuclear Structure and Function)
Edited by BERT W. O'MALLEY AND JOEL G. HARDMAN

VOLUME XLI. Carbohydrate Metabolism (Part B)
Edited by W. A. WOOD

VOLUME XLII. Carbohydrate Metabolism (Part C)
Edited by W. A. WOOD

VOLUME XLIII. Antibiotics
Edited by JOHN H. HASH

VOLUME XLIV. Immobilized Enzymes
Edited by KLAUS MOSBACH

VOLUME XLV. Proteolytic Enzymes (Part B)
Edited by LASZLO LORAND

VOLUME XLVI. Affinity Labeling
Edited by WILLIAM B. JAKOBY AND MEIR WILCHEK

VOLUME XLVII. Enzyme Structure (Part E)
Edited by C. H. W. HIRS AND SERGE N. TIMASHEFF

VOLUME XLVIII. Enzyme Structure (Part F)
Edited by C. H. W. HIRS AND SERGE N. TIMASHEFF

VOLUME XLIX. Enzyme Structure (Part G)
Edited by C. H. W. HIRS AND SERGE N. TIMASHEFF

VOLUME L. Complex Carbohydrates (Part C)
Edited by VICTOR GINSBURG

VOLUME LI. Purine and Pyrimidine Nucleotide Metabolism
Edited by PATRICIA A. HOFFEE AND MARY ELLEN JONES

VOLUME LII. Biomembranes (Part C: Biological Oxidations)
Edited by SIDNEY FLEISCHER AND LESTER PACKER

VOLUME LIII. Biomembranes (Part D: Biological Oxidations)
Edited by SIDNEY FLEISCHER AND LESTER PACKER

VOLUME LIV. Biomembranes (Part E: Biological Oxidations)
Edited by SIDNEY FLEISCHER AND LESTER PACKER

VOLUME LV. Biomembranes (Part F: Bioenergetics)
Edited by SIDNEY FLEISCHER AND LESTER PACKER

VOLUME LVI. Biomembranes (Part G: Bioenergetics)
Edited by SIDNEY FLEISCHER AND LESTER PACKER

VOLUME LVII. Bioluminescence and Chemiluminescence
Edited by MARLENE A. DELUCA

VOLUME LVIII. Cell Culture
Edited by WILLIAM B. JAKOBY AND IRA PASTAN

VOLUME LIX. Nucleic Acids and Protein Synthesis (Part G)
Edited by KIVIE MOLDAVE AND LAWRENCE GROSSMAN

VOLUME LX. Nucleic Acids and Protein Synthesis (Part H)
Edited by KIVIE MOLDAVE AND LAWRENCE GROSSMAN

VOLUME 61. Enzyme Structure (Part H)
Edited by C. H. W. HIRS AND SERGE N. TIMASHEFF

VOLUME 62. Vitamins and Coenzymes (Part D)
Edited by DONALD B. MCCORMICK AND LEMUEL D. WRIGHT

VOLUME 63. Enzyme Kinetics and Mechanism (Part A: Initial Rate and Inhibitor
Methods)
Edited by DANIEL L. PURICH

VOLUME 64. Enzyme Kinetics and Mechanism (Part B: Isotopic Probes and Complex Enzyme Systems)
Edited by DANIEL L. PURICH

VOLUME 65. Nucleic Acids (Part I)
Edited by LAWRENCE GROSSMAN AND KIVIE MOLDAVE

VOLUME 66. Vitamins and Coenzymes (Part E)
Edited by DONALD B. MCCORMICK AND LEMUEL D. WRIGHT

VOLUME 67. Vitamins and Coenzymes (Part F)
Edited by DONALD B. MCCORMICK AND LEMUEL D. WRIGHT

VOLUME 68. Recombinant DNA
Edited by RAY WU

VOLUME 69. Photosynthesis and Nitrogen Fixation (Part C)
Edited by ANTHONY SAN PIETRO

VOLUME 70. Immunochemical Techniques (Part A)
Edited by HELEN VAN VUNAKIS AND JOHN J. LANGONE

VOLUME 71. Lipids (Part C)
Edited by JOHN M. LOWENSTEIN

VOLUME 72. Lipids (Part D)
Edited by JOHN M. LOWENSTEIN

VOLUME 73. Immunochemical Techniques (Part B)
Edited by JOHN J. LANGONE AND HELEN VAN VUNAKIS

VOLUME 74. Immunochemical Techniques (Part C)
Edited by JOHN J. LANGONE AND HELEN VAN VUNAKIS

VOLUME 75. Cumulative Subject Index Volumes XXXI, XXXII, XXXIV–LX
Edited by EDWARD A. DENNIS AND MARTHA G. DENNIS

VOLUME 76. Hemoglobins
Edited by ERALDO ANTONINI, LUIGI ROSSI-BERNARDI, AND EMILIA CHIANCONE

VOLUME 77. Detoxication and Drug Metabolism
Edited by WILLIAM B. JAKOBY

VOLUME 78. Interferons (Part A)
Edited by SIDNEY PESTKA

VOLUME 79. Interferons (Part B)
Edited by SIDNEY PESTKA

VOLUME 80. Proteolytic Enzymes (Part C)
Edited by LASZLO LORAND

VOLUME 81. Biomembranes (Part H: Visual Pigments and Purple Membranes, I)
Edited by LESTER PACKER

VOLUME 82. Structural and Contractile Proteins (Part A: Extracellular Matrix)
Edited by LEON W. CUNNINGHAM AND DIXIE W. FREDERIKSEN

VOLUME 83. Complex Carbohydrates (Part D)
Edited by VICTOR GINSBURG

VOLUME 84. Immunochemical Techniques (Part D: Selected Immunoassays)
Edited by JOHN J. LANGONE AND HELEN VAN VUNAKIS

VOLUME 85. Structural and Contractile Proteins (Part B: The Contractile Apparatus and the Cytoskeleton)
Edited by DIXIE W. FREDERIKSEN AND LEON W. CUNNINGHAM

VOLUME 86. Prostaglandins and Arachidonate Metabolites
Edited by WILLIAM E. M. LANDS AND WILLIAM L. SMITH

VOLUME 87. Enzyme Kinetics and Mechanism (Part C: Intermediates, Stereochemistry, and Rate Studies)
Edited by DANIEL L. PURICH

VOLUME 88. Biomembranes (Part I: Visual Pigments and Purple Membranes, II)
Edited by LESTER PACKER

VOLUME 89. Carbohydrate Metabolism (Part D)
Edited by WILLIS A. WOOD

VOLUME 90. Carbohydrate Metabolism (Part E)
Edited by WILLIS A. WOOD

VOLUME 91. Enzyme Structure (Part I)
Edited by C. H. W. HIRS AND SERGE N. TIMASHEFF

VOLUME 92. Immunochemical Techniques (Part E: Monoclonal Antibodies and General Immunoassay Methods)
Edited by JOHN J. LANGONE AND HELEN VAN VUNAKIS

VOLUME 93. Immunochemical Techniques (Part F: Conventional Antibodies, Fc Receptors, and Cytotoxicity)
Edited by JOHN J. LANGONE AND HELEN VAN VUNAKIS

VOLUME 94. Polyamines
Edited by HERBERT TABOR AND CELIA WHITE TABOR

VOLUME 95. Cumulative Subject Index Volumes 61–74, 76–80
Edited by EDWARD A. DENNIS AND MARTHA G. DENNIS

VOLUME 96. Biomembranes [Part J: Membrane Biogenesis: Assembly and Targeting (General Methods; Eukaryotes)]
Edited by SIDNEY FLEISCHER AND BECCA FLEISCHER

VOLUME 97. Biomembranes [Part K: Membrane Biogenesis: Assembly and Targeting (Prokaryotes, Mitochondria, and Chloroplasts)]
Edited by SIDNEY FLEISCHER AND BECCA FLEISCHER

VOLUME 98. Biomembranes (Part L: Membrane Biogenesis: Processing and Recycling)
Edited by SIDNEY FLEISCHER AND BECCA FLEISCHER

VOLUME 99. Hormone Action (Part F: Protein Kinases)
Edited by JACKIE D. CORBIN AND JOEL G. HARDMAN

VOLUME 100. Recombinant DNA (Part B)
Edited by RAY WU, LAWRENCE GROSSMAN, AND KIVIE MOLDAVE

VOLUME 101. Recombinant DNA (Part C)
Edited by RAY WU, LAWRENCE GROSSMAN, AND KIVIE MOLDAVE

VOLUME 102. Hormone Action (Part G: Calmodulin and Calcium-Binding Proteins)
Edited by ANTHONY R. MEANS AND BERT W. O'MALLEY

VOLUME 103. Hormone Action (Part H: Neuroendocrine Peptides)
Edited by P. MICHAEL CONN

VOLUME 104. Enzyme Purification and Related Techniques (Part C)
Edited by WILLIAM B. JAKOBY

VOLUME 105. Oxygen Radicals in Biological Systems
Edited by LESTER PACKER

VOLUME 106. Posttranslational Modifications (Part A)
Edited by FINN WOLD AND KIVIE MOLDAVE

VOLUME 107. Posttranslational Modifications (Part B)
Edited by FINN WOLD AND KIVIE MOLDAVE

VOLUME 108. Immunochemical Techniques (Part G: Separation and Characterization of Lymphoid Cells)
Edited by GIOVANNI DI SABATO, JOHN J. LANGONE, AND HELEN VAN VUNAKIS

VOLUME 109. Hormone Action (Part I: Peptide Hormones)
Edited by LUTZ BIRNBAUMER AND BERT W. O'MALLEY

VOLUME 110. Steroids and Isoprenoids (Part A)
Edited by JOHN H. LAW AND HANS C. RILLING

VOLUME 111. Steroids and Isoprenoids (Part B)
Edited by JOHN H. LAW AND HANS C. RILLING

VOLUME 112. Drug and Enzyme Targeting (Part A)
Edited by KENNETH J. WIDDER AND RALPH GREEN

VOLUME 113. Glutamate, Glutamine, Glutathione, and Related Compounds
Edited by ALTON MEISTER

VOLUME 114. Diffraction Methods for Biological Macromolecules (Part A)
Edited by HAROLD W. WYCKOFF, C. H. W. HIRS, AND SERGE N. TIMASHEFF

VOLUME 115. Diffraction Methods for Biological Macromolecules (Part B)
Edited by HAROLD W. WYCKOFF, C. H. W. HIRS, AND SERGE N. TIMASHEFF

VOLUME 116. Immunochemical Techniques (Part H: Effectors and Mediators of Lymphoid Cell Functions)
Edited by GIOVANNI DI SABATO, JOHN J. LANGONE, AND HELEN VAN VUNAKIS

VOLUME 117. Enzyme Structure (Part J)
Edited by C. H. W. HIRS AND SERGE N. TIMASHEFF

VOLUME 118. Plant Molecular Biology
Edited by ARTHUR WEISSBACH AND HERBERT WEISSBACH

VOLUME 119. Interferons (Part C)
Edited by SIDNEY PESTKA

VOLUME 120. Cumulative Subject Index Volumes 81–94, 96–101

VOLUME 121. Immunochemical Techniques (Part I: Hybridoma Technology and Monoclonal Antibodies)
Edited by JOHN J. LANGONE AND HELEN VAN VUNAKIS

VOLUME 122. Vitamins and Coenzymes (Part G)
Edited by FRANK CHYTIL AND DONALD B. MCCORMICK

VOLUME 123. Vitamins and Coenzymes (Part H)
Edited by FRANK CHYTIL AND DONALD B. MCCORMICK

VOLUME 124. Hormone Action (Part J: Neuroendocrine Peptides)
Edited by P. MICHAEL CONN

VOLUME 125. Biomembranes (Part M: Transport in Bacteria, Mitochondria, and Chloroplasts: General Approaches and Transport Systems)
Edited by SIDNEY FLEISCHER AND BECCA FLEISCHER

VOLUME 126. Biomembranes (Part N: Transport in Bacteria, Mitochondria, and Chloroplasts: Protonmotive Force)
Edited by SIDNEY FLEISCHER AND BECCA FLEISCHER

VOLUME 127. Biomembranes (Part O: Protons and Water: Structure and Translocation)
Edited by LESTER PACKER

VOLUME 128. Plasma Lipoproteins (Part A: Preparation, Structure, and Molecular Biology)
Edited by JERE P. SEGREST AND JOHN J. ALBERS

VOLUME 129. Plasma Lipoproteins (Part B: Characterization, Cell Biology, and Metabolism)
Edited by JOHN J. ALBERS AND JERE P. SEGREST

VOLUME 130. Enzyme Structure (Part K)
Edited by C. H. W. HIRS AND SERGE N. TIMASHEFF

VOLUME 131. Enzyme Structure (Part L)
Edited by C. H. W. HIRS AND SERGE N. TIMASHEFF

VOLUME 132. Immunochemical Techniques (Part J: Phagocytosis and Cell-Mediated Cytotoxicity)
Edited by GIOVANNI DI SABATO AND JOHANNES EVERSE

VOLUME 133. Bioluminescence and Chemiluminescence (Part B)
Edited by MARLENE DELUCA AND WILLIAM D. MCELROY

VOLUME 134. Structural and Contractile Proteins (Part C: The Contractile Apparatus and the Cytoskeleton)
Edited by RICHARD B. VALLEE

VOLUME 135. Immobilized Enzymes and Cells (Part B)
Edited by KLAUS MOSBACH

VOLUME 136. Immobilized Enzymes and Cells (Part C)
Edited by KLAUS MOSBACH

VOLUME 137. Immobilized Enzymes and Cells (Part D)
Edited by KLAUS MOSBACH

VOLUME 138. Complex Carbohydrates (Part E)
Edited by VICTOR GINSBURG

VOLUME 139. Cellular Regulators (Part A: Calcium- and Calmodulin-Binding Proteins)
Edited by ANTHONY R. MEANS AND P. MICHAEL CONN

VOLUME 140. Cumulative Subject Index Volumes 102–119, 121–134

VOLUME 141. Cellular Regulators (Part B: Calcium and Lipids)
Edited by P. MICHAEL CONN AND ANTHONY R. MEANS

VOLUME 142. Metabolism of Aromatic Amino Acids and Amines
Edited by SEYMOUR KAUFMAN

VOLUME 143. Sulfur and Sulfur Amino Acids
Edited by WILLIAM B. JAKOBY AND OWEN GRIFFITH

VOLUME 144. Structural and Contractile Proteins (Part D: Extracellular Matrix)
Edited by LEON W. CUNNINGHAM

VOLUME 145. Structural and Contractile Proteins (Part E: Extracellular Matrix)
Edited by LEON W. CUNNINGHAM

VOLUME 146. Peptide Growth Factors (Part A)
Edited by DAVID BARNES AND DAVID A. SIRBASKU

VOLUME 147. Peptide Growth Factors (Part B)
Edited by DAVID BARNES AND DAVID A. SIRBASKU

VOLUME 148. Plant Cell Membranes
Edited by LESTER PACKER AND ROLAND DOUCE

VOLUME 149. Drug and Enzyme Targeting (Part B)
Edited by RALPH GREEN AND KENNETH J. WIDDER

VOLUME 150. Immunochemical Techniques (Part K: *In Vitro* Models of B and T Cell Functions and Lymphoid Cell Receptors)
Edited by GIOVANNI DI SABATO

VOLUME 151. Molecular Genetics of Mammalian Cells
Edited by MICHAEL M. GOTTESMAN

VOLUME 152. Guide to Molecular Cloning Techniques
Edited by SHELBY L. BERGER AND ALAN R. KIMMEL

VOLUME 153. Recombinant DNA (Part D)
Edited by RAY WU AND LAWRENCE GROSSMAN

VOLUME 154. Recombinant DNA (Part E)
Edited by RAY WU AND LAWRENCE GROSSMAN

VOLUME 155. Recombinant DNA (Part F)
Edited by RAY WU

VOLUME 156. Biomembranes (Part P: ATP-Driven Pumps and Related Transport: The Na,K-Pump)
Edited by SIDNEY FLEISCHER AND BECCA FLEISCHER

VOLUME 157. Biomembranes (Part Q: ATP-Driven Pumps and Related Transport: Calcium, Proton, and Potassium Pumps)
Edited by SIDNEY FLEISCHER AND BECCA FLEISCHER

VOLUME 158. Metalloproteins (Part A)
Edited by JAMES F. RIORDAN AND BERT L. VALLEE

VOLUME 159. Initiation and Termination of Cyclic Nucleotide Action
Edited by JACKIE D. CORBIN AND ROGER A. JOHNSON

VOLUME 160. Biomass (Part A: Cellulose and Hemicellulose)
Edited by WILLIS A. WOOD AND SCOTT T. KELLOGG

VOLUME 161. Biomass (Part B: Lignin, Pectin, and Chitin)
Edited by WILLIS A. WOOD AND SCOTT T. KELLOGG

VOLUME 162. Immunochemical Techniques (Part L: Chemotaxis and Inflammation)
Edited by GIOVANNI DI SABATO

VOLUME 163. Immunochemical Techniques (Part M: Chemotaxis and Inflammation)
Edited by GIOVANNI DI SABATO

VOLUME 164. Ribosomes
Edited by HARRY F. NOLLER, JR., AND KIVIE MOLDAVE

VOLUME 165. Microbial Toxins: Tools for Enzymology
Edited by SIDNEY HARSHMAN

VOLUME 166. Branched-Chain Amino Acids
Edited by ROBERT HARRIS AND JOHN R. SOKATCH

VOLUME 167. Cyanobacteria
Edited by LESTER PACKER AND ALEXANDER N. GLAZER

VOLUME 168. Hormone Action (Part K: Neuroendocrine Peptides)
Edited by P. MICHAEL CONN

VOLUME 169. Platelets: Receptors, Adhesion, Secretion (Part A)
Edited by JACEK HAWIGER

VOLUME 170. Nucleosomes
Edited by PAUL M. WASSARMAN AND ROGER D. KORNBERG

VOLUME 171. Biomembranes (Part R: Transport Theory: Cells and Model Membranes)
Edited by SIDNEY FLEISCHER AND BECCA FLEISCHER

VOLUME 172. Biomembranes (Part S: Transport: Membrane Isolation and Characterization)
Edited by SIDNEY FLEISCHER AND BECCA FLEISCHER

VOLUME 173. Biomembranes [Part T: Cellular and Subcellular Transport: Eukaryotic (Nonepithelial) Cells]
Edited by SIDNEY FLEISCHER AND BECCA FLEISCHER

VOLUME 174. Biomembranes [Part U: Cellular and Subcellular Transport: Eukaryotic (Nonepithelial) Cells]
Edited by SIDNEY FLEISCHER AND BECCA FLEISCHER

VOLUME 175. Cumulative Subject Index Volumes 135–139, 141–167

VOLUME 176. Nuclear Magnetic Resonance (Part A: Spectral Techniques and Dynamics)
Edited by NORMAN J. OPPENHEIMER AND THOMAS L. JAMES

VOLUME 177. Nuclear Magnetic Resonance (Part B: Structure and Mechanism)
Edited by NORMAN J. OPPENHEIMER AND THOMAS L. JAMES

VOLUME 178. Antibodies, Antigens, and Molecular Mimicry
Edited by JOHN J. LANGONE

VOLUME 179. Complex Carbohydrates (Part F)
Edited by VICTOR GINSBURG

VOLUME 180. RNA Processing (Part A: General Methods)
Edited by JAMES E. DAHLBERG AND JOHN N. ABELSON

VOLUME 181. RNA Processing (Part B: Specific Methods)
Edited by JAMES E. DAHLBERG AND JOHN N. ABELSON

VOLUME 182. Guide to Protein Purification
Edited by MURRAY P. DEUTSCHER

VOLUME 183. Molecular Evolution: Computer Analysis of Protein and Nucleic Acid Sequences
Edited by RUSSELL F. DOOLITTLE

VOLUME 184. Avidin–Biotin Technology
Edited by MEIR WILCHEK AND EDWARD A. BAYER

VOLUME 185. Gene Expression Technology
Edited by DAVID V. GOEDDEL

VOLUME 186. Oxygen Radicals in Biological Systems (Part B: Oxygen Radicals and Antioxidants)
Edited by LESTER PACKER AND ALEXANDER N. GLAZER

VOLUME 187. Arachidonate Related Lipid Mediators
Edited by ROBERT C. MURPHY AND FRANK A. FITZPATRICK

VOLUME 188. Hydrocarbons and Methylotrophy
Edited by MARY E. LIDSTROM

VOLUME 189. Retinoids (Part A: Molecular and Metabolic Aspects)
Edited by LESTER PACKER

VOLUME 190. Retinoids (Part B: Cell Differentiation and Clinical Applications)
Edited by LESTER PACKER

VOLUME 191. Biomembranes (Part V: Cellular and Subcellular Transport: Epithelial Cells)
Edited by SIDNEY FLEISCHER AND BECCA FLEISCHER

VOLUME 192. Biomembranes (Part W: Cellular and Subcellular Transport: Epithelial Cells)
Edited by SIDNEY FLEISCHER AND BECCA FLEISCHER

VOLUME 193. Mass Spectrometry
Edited by JAMES A. McCLOSKEY

VOLUME 194. Guide to Yeast Genetics and Molecular Biology
Edited by CHRISTINE GUTHRIE AND GERALD R. FINK

VOLUME 195. Adenylyl Cyclase, G Proteins, and Guanylyl Cyclase
Edited by ROGER A. JOHNSON AND JACKIE D. CORBIN

VOLUME 196. Molecular Motors and the Cytoskeleton
Edited by RICHARD B. VALLEE

VOLUME 197. Phospholipases
Edited by EDWARD A. DENNIS

VOLUME 198. Peptide Growth Factors (Part C)
Edited by DAVID BARNES, J. P. MATHER, AND GORDON H. SATO

VOLUME 199. Cumulative Subject Index Volumes 168–174, 176–194

VOLUME 200. Protein Phosphorylation (Part A: Protein Kinases: Assays, Purification, Antibodies, Functional Analysis, Cloning, and Expression)
Edited by TONY HUNTER AND BARTHOLOMEW M. SEFTON

VOLUME 201. Protein Phosphorylation (Part B: Analysis of Protein Phosphorylation, Protein Kinase Inhibitors, and Protein Phosphatases)
Edited by TONY HUNTER AND BARTHOLOMEW M. SEFTON

VOLUME 202. Molecular Design and Modeling: Concepts and Applications (Part A: Proteins, Peptides, and Enzymes)
Edited by JOHN J. LANGONE

VOLUME 203. Molecular Design and Modeling: Concepts and Applications (Part B: Antibodies and Antigens, Nucleic Acids, Polysaccharides, and Drugs)
Edited by JOHN J. LANGONE

VOLUME 204. Bacterial Genetic Systems
Edited by JEFFREY H. MILLER

VOLUME 205. Metallobiochemistry (Part B: Metallothionein and Related Molecules)
Edited by JAMES F. RIORDAN AND BERT L. VALLEE

VOLUME 206. Cytochrome P450
Edited by MICHAEL R. WATERMAN AND ERIC F. JOHNSON

VOLUME 207. Ion Channels
Edited by BERNARDO RUDY AND LINDA E. IVERSON

VOLUME 208. Protein–DNA Interactions
Edited by ROBERT T. SAUER

VOLUME 209. Phospholipid Biosynthesis
Edited by EDWARD A. DENNIS AND DENNIS E. VANCE

VOLUME 210. Numerical Computer Methods
Edited by LUDWIG BRAND AND MICHAEL L. JOHNSON

VOLUME 211. DNA Structures (Part A: Synthesis and Physical Analysis of DNA)
Edited by DAVID M. J. LILLEY AND JAMES E. DAHLBERG

VOLUME 212. DNA Structures (Part B: Chemical and Electrophoretic Analysis of DNA)
Edited by DAVID M. J. LILLEY AND JAMES E. DAHLBERG

VOLUME 213. Carotenoids (Part A: Chemistry, Separation, Quantitation, and Antioxidation)
Edited by LESTER PACKER

VOLUME 214. Carotenoids (Part B: Metabolism, Genetics, and Biosynthesis)
Edited by LESTER PACKER

VOLUME 215. Platelets: Receptors, Adhesion, Secretion (Part B)
Edited by JACEK J. HAWIGER

VOLUME 216. Recombinant DNA (Part G)
Edited by RAY WU

VOLUME 217. Recombinant DNA (Part H)
Edited by RAY WU

VOLUME 218. Recombinant DNA (Part I)
Edited by RAY WU

VOLUME 219. Reconstitution of Intracellular Transport
Edited by JAMES E. ROTHMAN

VOLUME 220. Membrane Fusion Techniques (Part A)
Edited by NEJAT DÜZGÜNEŞ

VOLUME 221. Membrane Fusion Techniques (Part B)
Edited by NEJAT DÜZGÜNEŞ

VOLUME 222. Proteolytic Enzymes in Coagulation, Fibrinolysis, and Complement Activation (Part A: Mammalian Blood Coagulation Factors and Inhibitors)
Edited by LASZLO LORAND AND KENNETH G. MANN

VOLUME 223. Proteolytic Enzymes in Coagulation, Fibrinolysis, and Complement Activation (Part B: Complement Activation, Fibrinolysis, and Nonmammalian Blood Coagulation Factors)
Edited by LASZLO LORAND AND KENNETH G. MANN

VOLUME 224. Molecular Evolution: Producing the Biochemical Data
Edited by ELIZABETH ANNE ZIMMER, THOMAS J. WHITE, REBECCA L. CANN, AND ALLAN C. WILSON

VOLUME 225. Guide to Techniques in Mouse Development
Edited by PAUL M. WASSARMAN AND MELVIN L. DEPAMPHILIS

VOLUME 226. Metallobiochemistry (Part C: Spectroscopic and Physical Methods for Probing Metal Ion Environments in Metalloenzymes and Metalloproteins)
Edited by JAMES F. RIORDAN AND BERT L. VALLEE

VOLUME 227. Metallobiochemistry (Part D: Physical and Spectroscopic Methods for Probing Metal Ion Environments in Metalloproteins)
Edited by JAMES F. RIORDAN AND BERT L. VALLEE

VOLUME 228. Aqueous Two-Phase Systems
Edited by HARRY WALTER AND GÖTE JOHANSSON

VOLUME 229. Cumulative Subject Index Volumes 195–198, 200–227

VOLUME 230. Guide to Techniques in Glycobiology
Edited by WILLIAM J. LENNARZ AND GERALD W. HART

VOLUME 231. Hemoglobins (Part B: Biochemical and Analytical Methods)
Edited by JOHANNES EVERSE, KIM D. VANDEGRIFF, AND ROBERT M. WINSLOW

VOLUME 232. Hemoglobins (Part C: Biophysical Methods)
Edited by JOHANNES EVERSE, KIM D. VANDEGRIFF, AND ROBERT M. WINSLOW

VOLUME 233. Oxygen Radicals in Biological Systems (Part C)
Edited by LESTER PACKER

VOLUME 234. Oxygen Radicals in Biological Systems (Part D)
Edited by LESTER PACKER

VOLUME 235. Bacterial Pathogenesis (Part A: Identification and Regulation of Virulence Factors)
Edited by VIRGINIA L. CLARK AND PATRIK M. BAVOIL

VOLUME 236. Bacterial Pathogenesis (Part B: Integration of Pathogenic Bacteria with Host Cells)
Edited by VIRGINIA L. CLARK AND PATRIK M. BAVOIL

VOLUME 237. Heterotrimeric G Proteins
Edited by RAVI IYENGAR

VOLUME 238. Heterotrimeric G-Protein Effectors
Edited by RAVI IYENGAR

VOLUME 239. Nuclear Magnetic Resonance (Part C)
Edited by THOMAS L. JAMES AND NORMAN J. OPPENHEIMER

VOLUME 240. Numerical Computer Methods (Part B)
Edited by MICHAEL L. JOHNSON AND LUDWIG BRAND

VOLUME 241. Retroviral Proteases
Edited by LAWRENCE C. KUO AND JULES A. SHAFER

VOLUME 242. Neoglycoconjugates (Part A)
Edited by Y. C. LEE AND REIKO T. LEE

VOLUME 243. Inorganic Microbial Sulfur Metabolism
Edited by HARRY D. PECK, JR., AND JEAN LEGALL

VOLUME 244. Proteolytic Enzymes: Serine and Cysteine Peptidases
Edited by ALAN J. BARRETT

VOLUME 245. Extracellular Matrix Components
Edited by E. RUOSLAHTI AND E. ENGVALL

VOLUME 246. Biochemical Spectroscopy
Edited by KENNETH SAUER

VOLUME 247. Neoglycoconjugates (Part B: Biomedical Applications)
Edited by Y. C. LEE AND REIKO T. LEE

VOLUME 248. Proteolytic Enzymes: Aspartic and Metallo Peptidases
Edited by ALAN J. BARRETT

VOLUME 249. Enzyme Kinetics and Mechanism (Part D: Developments in Enzyme Dynamics)
Edited by DANIEL L. PURICH

VOLUME 250. Lipid Modifications of Proteins
Edited by PATRICK J. CASEY AND JANICE E. BUSS

VOLUME 251. Biothiols (Part A: Monothiols and Dithiols, Protein Thiols, and Thiyl Radicals)
Edited by LESTER PACKER

VOLUME 252. Biothiols (Part B: Glutathione and Thioredoxin; Thiols in Signal Transduction and Gene Regulation)
Edited by LESTER PACKER

VOLUME 253. Adhesion of Microbial Pathogens
Edited by RON J. DOYLE AND ITZHAK OFEK

VOLUME 254. Oncogene Techniques
Edited by PETER K. VOGT AND INDER M. VERMA

VOLUME 255. Small GTPases and Their Regulators (Part A: Ras Family)
Edited by W. E. BALCH, CHANNING J. DER, AND ALAN HALL

VOLUME 256. Small GTPases and Their Regulators (Part B: Rho Family)
Edited by W. E. BALCH, CHANNING J. DER, AND ALAN HALL

VOLUME 257. Small GTPases and Their Regulators (Part C: Proteins Involved in Transport)
Edited by W. E. BALCH, CHANNING J. DER, AND ALAN HALL

VOLUME 258. Redox-Active Amino Acids in Biology
Edited by JUDITH P. KLINMAN

VOLUME 259. Energetics of Biological Macromolecules
Edited by MICHAEL L. JOHNSON AND GARY K. ACKERS

VOLUME 260. Mitochondrial Biogenesis and Genetics (Part A)
Edited by GIUSEPPE M. ATTARDI AND ANNE CHOMYN

VOLUME 261. Nuclear Magnetic Resonance and Nucleic Acids
Edited by THOMAS L. JAMES

VOLUME 262. DNA Replication
Edited by JUDITH L. CAMPBELL

VOLUME 263. Plasma Lipoproteins (Part C: Quantitation)
Edited by WILLIAM A. BRADLEY, SANDRA H. GIANTURCO, AND JERE P. SEGREST

VOLUME 264. Mitochondrial Biogenesis and Genetics (Part B)
Edited by GIUSEPPE M. ATTARDI AND ANNE CHOMYN

VOLUME 265. Cumulative Subject Index Volumes 228, 230–262

VOLUME 266. Computer Methods for Macromolecular Sequence Analysis
Edited by RUSSELL F. DOOLITTLE

VOLUME 267. Combinatorial Chemistry
Edited by JOHN N. ABELSON

VOLUME 268. Nitric Oxide (Part A: Sources and Detection of NO; NO Synthase)
Edited by LESTER PACKER

VOLUME 269. Nitric Oxide (Part B: Physiological and Pathological Processes)
Edited by LESTER PACKER

VOLUME 270. High Resolution Separation and Analysis of Biological Macromolecules (Part A: Fundamentals)
Edited by BARRY L. KARGER AND WILLIAM S. HANCOCK

VOLUME 271. High Resolution Separation and Analysis of Biological Macromolecules (Part B: Applications)
Edited by BARRY L. KARGER AND WILLIAM S. HANCOCK

VOLUME 272. Cytochrome P450 (Part B)
Edited by ERIC F. JOHNSON AND MICHAEL R. WATERMAN

VOLUME 273. RNA Polymerase and Associated Factors (Part A)
Edited by SANKAR ADHYA

VOLUME 274. RNA Polymerase and Associated Factors (Part B)
Edited by SANKAR ADHYA

VOLUME 275. Viral Polymerases and Related Proteins
Edited by LAWRENCE C. KUO, DAVID B. OLSEN, AND STEVEN S. CARROLL

VOLUME 276. Macromolecular Crystallography (Part A)
Edited by CHARLES W. CARTER, JR., AND ROBERT M. SWEET

VOLUME 277. Macromolecular Crystallography (Part B)
Edited by CHARLES W. CARTER, JR., AND ROBERT M. SWEET

VOLUME 278. Fluorescence Spectroscopy
Edited by LUDWIG BRAND AND MICHAEL L. JOHNSON

VOLUME 279. Vitamins and Coenzymes (Part I)
Edited by DONALD B. MCCORMICK, JOHN W. SUTTIE, AND CONRAD WAGNER

VOLUME 280. Vitamins and Coenzymes (Part J)
Edited by DONALD B. MCCORMICK, JOHN W. SUTTIE, AND CONRAD WAGNER

VOLUME 281. Vitamins and Coenzymes (Part K)
Edited by DONALD B. MCCORMICK, JOHN W. SUTTIE, AND CONRAD WAGNER

VOLUME 282. Vitamins and Coenzymes (Part L)
Edited by DONALD B. MCCORMICK, JOHN W. SUTTIE, AND CONRAD WAGNER

VOLUME 283. Cell Cycle Control
Edited by WILLIAM G. DUNPHY

VOLUME 284. Lipases (Part A: Biotechnology)
Edited by BYRON RUBIN AND EDWARD A. DENNIS

VOLUME 285. Cumulative Subject Index Volumes 263, 264, 266–284, 286–289

VOLUME 286. Lipases (Part B: Enzyme Characterization and Utilization)
Edited by BYRON RUBIN AND EDWARD A. DENNIS

VOLUME 287. Chemokines
Edited by RICHARD HORUK

VOLUME 288. Chemokine Receptors
Edited by RICHARD HORUK

VOLUME 289. Solid Phase Peptide Synthesis
Edited by GREGG B. FIELDS

VOLUME 290. Molecular Chaperones
Edited by GEORGE H. LORIMER AND THOMAS BALDWIN

VOLUME 291. Caged Compounds
Edited by GERARD MARRIOTT

VOLUME 292. ABC Transporters: Biochemical, Cellular, and Molecular Aspects
Edited by SURESH V. AMBUDKAR AND MICHAEL M. GOTTESMAN

VOLUME 293. Ion Channels (Part B)
Edited by P. MICHAEL CONN

VOLUME 294. Ion Channels (Part C)
Edited by P. MICHAEL CONN

VOLUME 295. Energetics of Biological Macromolecules (Part B)
Edited by GARY K. ACKERS AND MICHAEL L. JOHNSON

VOLUME 296. Neurotransmitter Transporters
Edited by SUSAN G. AMARA

VOLUME 297. Photosynthesis: Molecular Biology of Energy Capture
Edited by LEE MCINTOSH

VOLUME 298. Molecular Motors and the Cytoskeleton (Part B)
Edited by RICHARD B. VALLEE

VOLUME 299. Oxidants and Antioxidants (Part A)
Edited by LESTER PACKER

VOLUME 300. Oxidants and Antioxidants (Part B)
Edited by LESTER PACKER

VOLUME 301. Nitric Oxide: Biological and Antioxidant Activities (Part C)
Edited by LESTER PACKER

VOLUME 302. Green Fluorescent Protein
Edited by P. MICHAEL CONN

VOLUME 303. cDNA Preparation and Display
Edited by SHERMAN M. WEISSMAN

VOLUME 304. Chromatin
Edited by PAUL M. WASSARMAN AND ALAN P. WOLFFE

VOLUME 305. Bioluminescence and Chemiluminescence (Part C)
Edited by THOMAS O. BALDWIN AND MIRIAM M. ZIEGLER

VOLUME 306. Expression of Recombinant Genes in Eukaryotic Systems
Edited by JOSEPH C. GLORIOSO AND MARTIN C. SCHMIDT

VOLUME 307. Confocal Microscopy
Edited by P. MICHAEL CONN

VOLUME 308. Enzyme Kinetics and Mechanism (Part E: Energetics of Enzyme
Catalysis)
Edited by DANIEL L. PURICH AND VERN L. SCHRAMM

VOLUME 309. Amyloid, Prions, and Other Protein Aggregates
Edited by RONALD WETZEL

VOLUME 310. Biofilms
Edited by RONALD J. DOYLE

VOLUME 311. Sphingolipid Metabolism and Cell Signaling (Part A)
Edited by ALFRED H. MERRILL, JR., AND YUSUF A. HANNUN

VOLUME 312. Sphingolipid Metabolism and Cell Signaling (Part B) (in preparation)
Edited by ALFRED H. MERRILL, JR., AND YUSUF A. HANNUN

VOLUME 313. Antisense Technology (Part A: General Methods, Methods of Delivery, and RNA Studies)
Edited by M. IAN PHILLIPS

VOLUME 314. Antisense Technology (Part B: Applications)
Edited by M. IAN PHILLIPS

VOLUME 315. Vertebrate Phototransduction and the Visual Cycle (Part A)
Edited by KRZYSZTOF PALCZEWSKI

VOLUME 316. Vertebrate Phototransduction and the Visual Cycle (Part B)
Edited by KRZYSZTOF PALCZEWSKI

VOLUME 317. RNA–Ligand Interactions (Part A: Structural Biology Methods)
Edited by DANIEL W. CELANDER AND JOHN N. ABELSON

VOLUME 318. RNA–Ligand Interactions (Part B: Molecular Biology Methods)
Edited by DANIEL W. CELANDER AND JOHN N. ABELSON

VOLUME 319. Singlet Oxygen, UV-A, and Ozone
Edited by LESTER PACKER AND HELMUT SIES

VOLUME 320. Cumulative Subject Index Volumes 290–319 (in preparation)

VOLUME 321. Numerical Computer Methods (Part C) (in preparation)
Edited by MICHAEL L. JOHNSON AND LUDWIG BRAND

VOLUME 322. Apoptosis (in preparation)
Edited by JOHN C. REED

VOLUME 323. Energetics of Biological Macromolecules (Part C) (in preparation)
Edited by MICHAEL L. JOHNSON AND GARY K. ACKERS

VOLUME 324. Branched-Chain Amino Acids (Part B) (in preparation)
Edited by ROBERT A. HARRIS AND JOHN R. SOKATCH

VOLUME 325. Regulators and Effectors of Small GTPases (Part D: Rho Family) (in preparation)
Edited by W. E. BALCH, CHANNING J. DER, AND ALAN HALL

VOLUME 326. Applications of Chimeric Genes and Hybrid Proteins (Part A: Gene Expression and Protein Purification) (in preparation)
Edited by JEREMY THORNER, SCOTT D. EMR, AND JOHN N. ABELSON

VOLUME 327. Applications of Chimeric Genes and Hybrid Proteins (Part B: Cell Biology and Physiology) (in preparation)
Edited by JEREMY THORNER, SCOTT D. EMR, AND JOHN N. ABELSON

VOLUME 328. Applications of Chimeric Genes and Hybrid Proteins (Part C: Protein–Protein Interactions and Genomics) (in preparation)
Edited by JEREMY THORNER, SCOTT D. EMR, AND JOHN N. ABELSON

VOLUME 329. Regulators and Effectors of Small GTPases (Part E: GTPases Involved in Vesicular Traffic) (in preparation)
Edited by W. E. BALCH, CHANNING J. DER, AND ALAN HALL

VOLUME 330. Hyperthermophilic Enzymes (Part A) (in preparation)
Edited by MICHAEL W. W. ADAMS AND ROBERT M. KELLY

VOLUME 331. Hyperthermophilic Enzymes (Part B) (in preparation)
Edited by MICHAEL W. W. ADAMS AND ROBERT M. KELLY

VOLUME 332. Regulators and Effectors of Small GTPases (Part F: Ras Family I) (in preparation)
Edited by W. E. BALCH, CHANNING J. DER, AND ALAN HALL

VOLUME 333. Regulators and Effectors of Small GTPases (Part G: Ras Family II) (in preparation)
Edited by W. E. BALCH, CHANNING J. DER, AND ALAN HALL

VOLUME 334. Hyperthermophylic Enzymes (Part C) (in preparation)
Edited by MICHAEL W. W. ADAMS AND ROBERT M. KELLY

Section I

Singlet Oxygen

A. Generation and Detection: Chemical Systems
Articles 1 through 5

B. Generation and Detection: Biological Systems
Articles 6 through 11

C. Biological Effects: Role in Signaling
Articles 12 and 13

D. Toxic Effects
Articles 14 through 20

E. Protection: Singlet Oxygen Quenchers
Articles 21 and 22

[1] Naphthalene Endoperoxides as Generators of Singlet Oxygen in Biological Media

By CHRISTEL PIERLOT, JEAN-MARIE AUBRY, KARLIS BRIVIBA, HELMUT SIES, and PAOLO DI MASCIO

Introduction

Molecular oxygen exhibits a remarkable electronic structure, as its higher occupied electronic level is constituted of two π^* orbitals of the same energy, so-called degenerated, filled with only two electrons. In the ground state, each of these electrons lies in one π^* orbital and their spins are parallel, hence it is a triplet state denoted $^3O_2(^3\Sigma_g^-)$. It behaves chemically as a poorly reactive oxidizing diradical despite its high oxidation potential. The first excited state has both electrons in the same orbital with opposite spins. It is a singlet state, denoted $^1O_2(^1\Delta_g)$, with a relatively long lifetime (45 min *in vacuo* and 4 μsec in water) and a substantial reactivity toward electron-rich organic molecules such as olefins, dienes, polycyclic aromatic compounds, sulfides, or phenols.[1]

Singlet oxygen has been shown to be generated in biological systems and can be implicated in defense mechanisms against viruses and bacteria by phagocytic cells.[2] Dark reactions (chemiexcitation), such as reactions catalyzed by peroxidases (myeloperoxidase) or oxygenases (lipoxygenase or cyclooxygenase), or the reaction of hydrogen peroxide with hypochlorite or peroxynitrite[3] or thermodecomposition of dioxetanes[4,5] can be responsible for $^1O_2$ generation in biological systems. Singlet oxygen is one of the major species mediating cytotoxic effects of photodynamic treatment. In addition to these cytotoxic effects, $^1O_2$ can be responsible for ultraviolet A-induced activation of gene expression (for review, see Ref. 6).

Biological targets for $^1O_2$ include unsaturated fatty acids, proteins, and

---

[1] A. A. Frimer, ed., "Singlet Oxygen," Vols. 1–3. CRC Press, Boca Raton, FL, 1985.

[2] M. J. Steinbeck, A. U. Khan, and M. J. Karnovsky, *J. Biol. Chem.* **268**, 15649 (1993).

[3] P. Di Mascio, E. J. H. Bechara, M. H. G. Medeiros, K. Briviba, and H. Sies, *FEBS Lett.* **355**, 287 (1994).

[4] K. Briviba, C. R. Saha-Möller, W. Adam, and H. Sies, *Biochem. Mol. Biol. Int.* **38**, 647 (1996).

[5] G. Cilento, *in* "Chemical and Biological Generation of Excited States" (W. Adam and G. Cilento, eds.), p. 279. Academic Press, New York, 1982.

[6] K. Briviba, L. O. Klotz, and H. Sies, *Biol. Chem.* **378**, 1259 (1997).

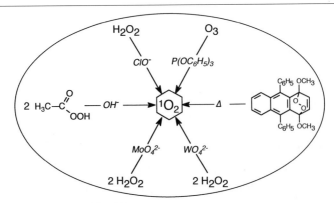

FIG. 1. Most efficient chemical sources of singlet oxygen $^1O_2$.

DNA.[7-10] Considering the complexity of biological systems and the great variety of reactive species generated by photochemistry, it is difficult to assess unambiguously the role of $^1O_2$ in the resulting biological effects. Therefore, alternative methods to generate $^1O_2$ without concomitant oxidants or toxic compounds have been sought. Several chemical sources of $^1O_2$ are able to convert in the dark an oxygen precursor into $^1O_2$ with almost quantitative yields[11] (Fig. 1). The conversion to $^1O_2$ by these compounds involves the oxidation (ClO^-), the disproportionation of hydrogen peroxide catalyzed by MoO_4^{2-}, WO_4^{2-}, or the reduction of ozone by triphenylphosphite, the base-catalyzed disproportionation of peracids, or the thermolysis of polycyclic aromatic endoperoxides. Unfortunately, conditions required by biological systems (aqueous environment, neutral pH, moderate temperature) are not compatible with these chemical sources of $^1O_2$. Further, these compounds are also toxic and are strong oxidants. Therefore, efforts have been devoted to develop suitable $^1O_2$ generators based on the thermolysis of endoperoxides. These compounds are chemically inert and have been employed as versatile sources of $^1O_2$.

The reversible binding of oxygen to polycyclic aromatic compounds was

[7] H. Sies and C. F. M. Menck, *Mutat. Res.* **275,** 367 (1992).

[8] H. Sies, *Mutat. Res.* **299,** 183 (1993).

[9] C. Menck, P. Di Mascio, L. F. Agnez, D. T. Ribeiro, and R. C. de Oliveira, *Quím. Nova* **16,** 328 (1993).

[10] P. Di Mascio, M. H. G. Medeiros, E. J. H. Bechara, and L. H. Catalani, *Ciência Cultura* **47,** 297 (1995).

[11] J. M. Aubry, in "Membrane Lipid Oxidation" (C. Vigo-Pelfrey, ed.), Vol. 2, p. 65. CRC Press, Boca Raton, FL, 1991.

discovered in 1926 by Dufraisse.[12,13] He also demonstrated that the oxygen released was in an "activated" state, but the precise nature of the species involved, i.e., $^1O_2(^1\Delta_g)$, was established in 1972 by Wasserman and Larsen[14] using the endoperoxide of 9,10-diphenylanthracene. When this endoperoxide is warmed to 80°, it decomposes into the parent compound and oxygen, 32% of which is in the singlet state.[11] The endoperoxide of 1,4-dimethylnaphthalene (DMN) evolves $^1O_2$ at lower temperatures and with a higher yield (76%).

Thus, water-soluble and nontoxic derivatives of DMN may act as carriers of $^1O_2$, as they trap this species at low temperature (0–5°). These endoperoxides, which may be stored for months at −80°, release a definite amount of $^1O_2$ on warming at 37° [reaction (1)], once dissolved in biological systems.

$$\text{Hyd, Hyd' = hydrophilic groups}$$

REACTION 1

The first water-soluble $^1O_2$ carriers reported by Nieuwint et al.[15] and Saito et al.[16] compounds **4** and **7** in Fig. 3, were designed to possess the properties described previously. These compounds bear one or two sodium propanoate substituents, respectively, grafted onto the 1,4 positions of the naphthalene core.[15,16] The corresponding endoperoxides have been used as chemical sources of $^1O_2$ to assess the activity of $^1O_2$ toward chemical, biochemical, or biological targets. However, these anionic compounds release $^1O_2$ in the aqueous phase, potentially far from the target. Subsequently, a second generation of carriers has been synthesized. They bear specific groups such as acridine moiety,[17] quaternary ammonium group **5**,[17]

[12] C. Moureu, C. Dufraisse, and P. M. Dean, *C.R. Acad. Sci.* **182**, 1140 (1926).

[13] C. Dufraisse and L. Velluz, *C.R. Acad. Sci.* **208**, 1822 (1939).

[14] B. H. Wasserman and D. L. Larsen, *J. C. S. Chem. Comm.* **253** (1972).

[15] A. W. M. Nieuwint, J. M. Aubry, F. Arwert, H. Kortbeek, S. Herzberg, and H. Joenje, *Free Radic. Res. Commun.* **1**, 1 (1985).

[16] I. Saito, T. Matsuura, and K. Inoue, *J. Am. Chem. Soc.* **103**, 188 (1981).

[17] T. Matsuura and K. Inoue, *Free Radic. Res. Commun.* **2**, 327 (1987).

TABLE I
HYDROPHILIC FUNCTIONS INSENSITIVE
TO SINGLET OXYGEN

Function	Formula
Carboxylate	$-COO^-$, Na^+
Amide	$-CONHR_1R_2$
Phosphate	$-PO_4^{2-}$, $2\ Na^+$
Sulfonate	$-SO_3^-$, Na^+
Sulfate	$-SO_4^-$, Na^+
Sulfonamide	$-SO_2NH_2$
Quaternary ammonium	$-N(CH_3)_3^+$, Cl^-
Alcohol	$-OH$

or nonionic hydrophilic groups (**6** and **8**) (Fig. 3)[18,19] in order to confer a particular affinity for polynucleotides, negatively charged sites, or intracellular targets to these molecules.

Strategy for Synthesis of Water-Soluble Carriers

Hydrophilic substituents grafted onto the naphthalene backbone must be insensitive to $^1O_2$ and to the photosensitizer or the chemical sources of $^1O_2$ that are required to prepare the endoperoxide. Moreover, the hydrophilic function must not act as a $^1O_2$ quencher as amino or phenol groups do. Table I presents the main functions meeting these requirements.

The naphthalene itself does not react with $^1O_2$; the direct binding of one of these electron-attractive groups to the aromatic core decreases its reactivity further. Therefore, at least one, and preferably two, electron-donating groups must be present on the positions 1,4 to allow the [4 + 2] cycloaddition of $^1O_2$ and to stabilize the endoperoxide. Thus, the relevant starting structure is DMN to which water-solubilizing groups must be added.

Figure 2 explains why the best position for anchoring lies on the 1,4-methyl group of DMN, provided that a sufficiently long alkyl spacer separates the hydrophilic group from the 1,4 carbons of the naphthalene moiety.

All the naphthalenic carriers of $^1O_2$ used in biological media can be prepared from the (inexpensive) 1-methylnaphthalene (compound **1**) using

[18] C. Pierlot, S. Hajjam, C. Barthélémy, and J. M. Aubry, *J. Photochem. Photobiol. B* **36,** 31 (1996).

[19] C. Pellieux, A. Dewilde, C. Pierlot, and J. M. Aubry, *Methods Enzymol.* **319,** [18] (2000) (this volume).

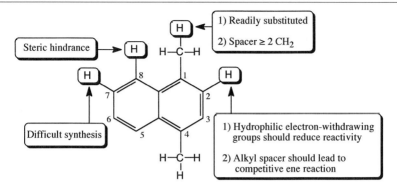

FIG. 2. Advantages and disadvantages of positions on DMN regarding the grafting of hydrophilic groups.

the general scheme presented in Fig. 3.[20–23] In the first step, compound **1** is chloromethylated, giving 4-methyl-1-chloromethylnaphthalene **2**, which leads to sodium 4-methyl-1-naphthalenepropanoate (MNP, **4**)[16] and 4-methyl-*N,N,N*-trimethyl-1-naphthaleneethanaminium chloride (MNEA, **5**),[18] anionic or cationic compounds, respectively, which bear only one water-solubilizing group. The conventional route to bifunctional naphthalene derivatives **6–8** was started by a double bromination of the (costly) DMN,[20,21] which, however, is no longer available commercially. An alternative access to 1,4-dihalogenomethylnaphthalene **3** is by monobromination of **2**, which then is converted into nonionic 1,4-naphthalenedimethanol (NDMOL, **6**) or anionic disodium 1,4-naphthalenedipropanoate (NDP, **7**) according to reported methods.[18,24] Synthesis of the new nonionic carrier *N,N′*-di(2,3-dihydroxypropyl)-1,4-naphthalenedipropanamide (DHPN, **8**), involving amidation of the diethyl ester of compound **7**, is described in detail later.

All five $^1O_2$ carriers **4–8** mentioned previously exhibit sufficient water solubility (10^{-2} *M*) for most biological applications (Table II). However, the solubility of MNP, **4**, and NDMOL, **6**, is sometimes insufficient and they may be advantageously replaced by NDP, **7**, and DHPN, **8**.

[20] C. S. Marvel and B. D. Wilson, *J. Org. Chem.* **23**, 1483 (1958).

[21] G. Lock and E. Walter, *Chem. Ber.* **75B**, 1158 (1942).

[22] G. Lock and R. Schneider, *Chem. Ber.* **91**, 1770 (1958).

[23] Kessar, P. Jit, K. Mundra, and A. Lumb, *J. Chem. Soc.* **2**, 266 (1971).

[24] P. Di Mascio and H. Sies, *J. Am. Chem. Soc.* **111**, 2909 (1989).

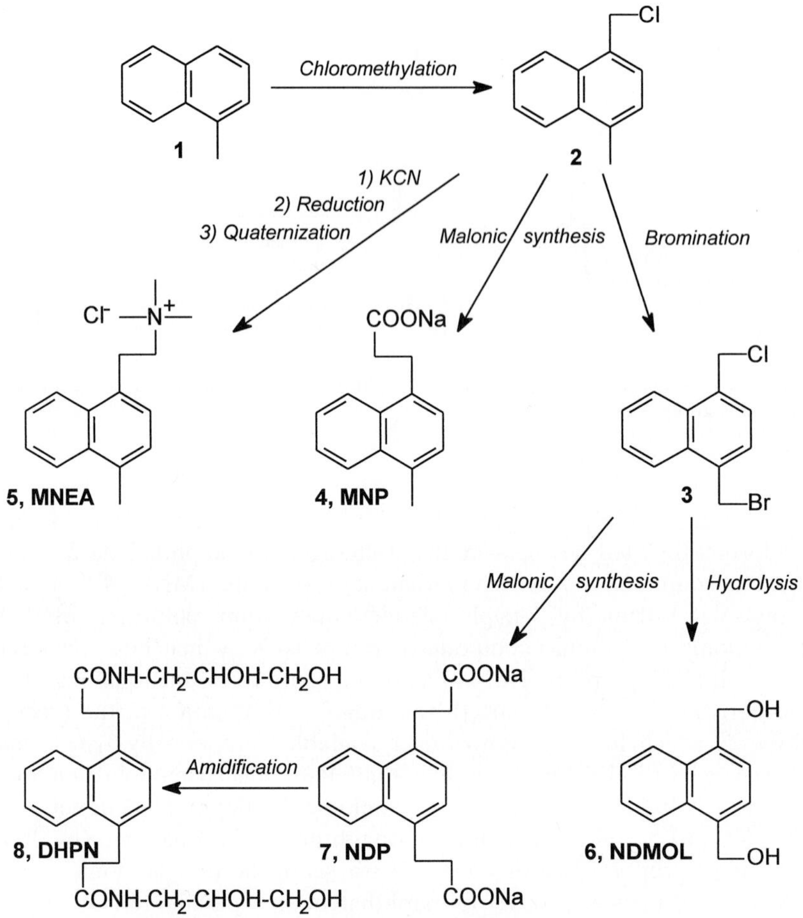

Fig. 3. Synthesis of water-soluble naphthalenic carriers of $^1O_2$.

TABLE II
MAIN PHYSICOCHEMICAL PROPERTIES OF $^1O_2$ CARRIERS

	MNP (4)	MNEA (5)	NDMOL (6)	NDP (7)	DHPN (8)
Water solubility $(M)^a$	10^{-2}	>1	0.95×10^{-2}	>1	3.9×10^{-2}
$10^{-5} k_t$ $(M^{-1} sec^{-1})^b$	70	14	4	28	10

^{a} In H_2O at 20°.
^{b} Overall quenching rate constants of $^1O_2$ measured by flash photolysis in D_2O.[18,65]

Peroxidation of Naphthalenic Carriers

Two primary sources of $^1O_2$ can be used to prepare endoperoxides on a preparative scale: the regular photochemical method[25] [reaction (2)] and the molybdate-catalyzed disproportionation of hydrogen peroxide[26] [reaction (3)]. In both cases, the oxidation proceeds more rapidly in deuterated

$$^3O_2 + h\nu \xrightarrow{\text{Photosensitizer}} {}^1O_2$$

REACTION 2

$$2 H_2O_2 \xrightarrow[\text{Water}]{MoO_4^{2-}} 2 H_2O + {}^1O_2 \quad (100\%)$$

REACTION 3

solvents (D_2O and CD_3OD), as the lifetime of $^1O_2$ is much longer in these solvents than in protonated ones.

The choice of the oxidizing method depends on the physicochemical properties of the naphthalenic compound (water solubility, nature of the hydrophilic functions, and reactivity toward $^1O_2$). For instance, this chemical process is ideal for peroxidizing salts of carboxylated naphthalene derivatives (MNP **4** and NDP **7**) because the endoperoxides can be recovered readily by precipitating the acidic forms. For other $^1O_2$ carriers, photooxidation should be used on the condition that the photosensitizer can be eliminated at the end of the reaction.

The interaction of $^1O_2$ with naphthalenic compound **N** can be described by reactions (4) and (5). Singlet oxygen produced by a chemical or a

$$^1O_2 + N \xrightarrow{k_q} {}^3O_2 + N$$

REACTION 4

$$^1O_2 + N \xrightarrow{k_r} NO_2$$

REACTION 5

photochemical source is either quenched by N with rate constant k_q [reaction (4)] or reacted with N with rate constant k_r [reaction (5)]. Thus, the

[25] B. H. Wasserman and R. W. Murray, eds., "Singlet Oxygen." Academic Press, New York, 1979.

[26] J. Aubry, B. Cazin, and F. Duprat, *J. Org. Chem.* **54,** 726 (1989).

TABLE III
THERMOLYSIS OF NAPHTHALENE ENDOPEROXIDES IN WATER AT 37°

	MNPO$_2$	NDPO$_2$	MNEAO$_2$	NDMOLO$_2$	DHPNO$_2$
$t_{50\%}$ (min)[a]	23	23	22	70	23
$t_{95\%}$ (min)[b]	99	99	95	300	99
$^1O_2$ yield (%)[c]	45	50	65	51	59

[a] Half-time of decomposition.
[b] Time necessary to decompose 95% of naphthalenic endoperoxides.
[c] Cumulative yields of $^1O_2$ produced on thermolysis.

overall reactivity of N toward $^1O_2$ can be expressed by the sum $k_t = k_r + k_q$ (Table II).

To explain the difference in reactivity of 1,4-substituted carriers **4–8** toward $^1O_2$, two phenomena have to be considered: the electron density of the naphthalene core and the steric hindrance induced by the 1,4 substituents themselves or by additional groups located on the 2,3,5,8 positions. Electronic effects are of primary importance when a short spacer separates the hydrophilic groups from the naphthalene core. Thus, the electron-attractive effect caused by the quaternary ammonium group of MNEA **5** or by the OH functions of NDMOL **6** leads to molecules 5 and 17 times less reactive than MNP **4**, respectively. Longer alkyl spacers increase the electron density of the naphthalene core, but their steric hindrance lowers the rate of reaction with $^1O_2$ significantly. Thus, the weak overall rate constants ($k_r + k_q$) of NDP **7** ($2.8 \times 10^6\ M^{-1}\ \text{sec}^{-1}$) and of the more crowded DHPN **8** ($1.0 \times 10^6\ M^{-1}\ \text{sec}^{-1}$) are much lower than the value obtained with MNP **4** ($7.0 \times 10^6\ M^{-1}\ \text{sec}^{-1}$).

Thermolysis of the endoperoxide NO$_2$ [reaction (6)] follows a first-order

$$NO_2 \xrightarrow{\ \ k\ \ } N \ + \ \alpha\,^1O_2 \ + \ (1-\alpha)\,^3O_2$$

REACTION 6

kinetics with a rate constant k. More telling data, such as the half-time of decomposition ($t_{50\%} = \ln 2/k$) or the time necessary to decompose 95% of the starting endoperoxide ($t_{95\%} = \ln 20/k$), can be calculated. Table III indicates that most of the endoperoxides release 95% of their oxygen within 2 hr at 37°. This value is convenient for carrying out biological tests.[19]

A part of oxygen formed during the thermolysis is in the singlet excited state [reaction (6)]. It was measured by complete trapping with tetrapotassium rubrene-2,3,8,9-tetracarboxylate (RTC).[18,27] Finally, it can be consid-

[27] J. Aubry, J. Rigaudy, and K. C. Nguyen, *Photochem. Photobiol.* **33**, 149 (1981).

ered roughly that all the known naphthalenic carriers release $^1O_2$ in water with an approximate 50% yield (Table III).

Procedures for Synthesis and Oxidation of $^1O_2$ Carriers

Synthesis of naphthalenic carriers **4–7** and their corresponding endoperoxides has been reported previously,[18] and more detailed procedures for the preparation of endoperoxides of disodium 1,4-naphthalenedipropanoate (NDPO$_2$) and of sodium 4-methyl-1-naphthalenepropanoate (MNPO$_2$) are presented in Refs. 15 and 24 and 15 and 26, respectively. This article describes the synthesis of both reduced and oxidized forms of the new nonionic compound DHPN, **8**. It is noteworthy that both photochemical and chemical methods yield the wanted endoperoxide, but only photooxygenation leads to a pure sample suitable for biological assay.[19] We also describe the synthesis of 1-bromomethyl-4-chloromethylnaphthalene, **3**, which opens a new access to 1,4-disubstituted compounds **6–8**.

1-Bromomethyl-4-chloromethylnaphthalene 3

A mixture of 1-chloromethylnaphthalene **2** (292 g, 0.48 mol) and N-bromosuccinimide (86.6 g, 0.48 mol) in 1 liter of CCl$_4$ is stirred for 15 min. Then 1.5 g of benzoyl peroxide and 1.5 g of azobisisobutyronitrile are added to the solution, which is refluxed for 4 hr under magnetic stirring. The hot solution is filtered, and the cake is extracted with boiling CCl$_4$. The combined filtrates are evaporated, and the white solid is recrystallized from MeOH, affording 65 g (50%) of **3**. TLC: cyclohexane (R_f = 0.3); ^1H NMR (CDCl$_3$): d (ppm) = 8.1–8.3 (m; 2H, Ar), 7.5–7.9 (m; 2H, Ar), 5.05 (s; 2H, CH$_2$-Br), 4.95 (s; 2H, CH$_2$-Cl).

N,N'-Di(2,3-dihydroxypropyl)-1,4-naphthalenedipropanamide DHPN 8

A solution of diethyl 1,4-naphthalenedipropanoate (1 g, 2.6 mmol) and 3-amino-1,2-propanediol (3.78 g, 42 mmol) in 10 ml MeOH is stirred for 24 hr under reflux. After evaporation of the solvent, the residue is triturated with 50 ml of acetone. The colorless precipitate is filtered by suction and rinsed with acetone. Recrystallization from MeOH affords 685 mg (63%) of **8**. mp: (165°); ^1H NMR (300 MHz, DMSO, TMS): d = 8.10–8.15 (m, 2H, H$_5$), 7.93 (m, 2H, CONH), 7.55–7.60 (m, 2H, H$_6$), 7.28 (s, 2H, H$_2$), 4.75 (d, 2H, CHOH), 4.55 (t, 2H, CH$_2$OH), 3.45 (m, 2H, CHOH), 3.2 (m, 8H, NHCH$_2$-CH$_2$OH), 3.0 (m, 4H, Ar-CH$_2$), 2.5 (m, 2H, CH$_2$CO); ^{13}C NMR (300 MHz, DMSO, TMS): d = 172.0 (s, CO), 135.9 and 131.6 (s, C1, C9), 125.7, 125.5 and 124.4 (s, C2, C5, C6), 70.6 (CH(OH)), 63.7 (CH$_2$(OH)),

42.2 (CH_2-N), 36.5 (s, Ar-CH_2), 28.4 (CHCO); MS: m/z 441 (MNa^+), 419 (MH^+).

Endoperoxide of N,N′-Di(2,3-dihydroxypropyl)-1,4-naphthalenedipropanamide $DHPNO_2$

DHPN, **8** (1 g, 2.3 mmol), 160 μl of an aqueous (D_2O) solution of 10^{-3} M methylene blue and 10 ml of deuterated water are introduced into a cylindrical glass cell (F = 4 cm; thickness = 1 cm). After gentle warming to dissolve **8**, the limpid solution is cooled at 5° and irradiated with a sodium lamp (150 W) under continuous bubbling of oxygen. Some methylene blue is added periodically during photooxygenation to compensate for its fading. After 4 hr, HPLC analysis shows 80% transformation. One hundred milligrams of cationic resin (Touzart-Matignon) is then added to the mixture, and the blue solution is stirred for 15 min at 0° until complete fixation of the sensitizer onto the insoluble polymer. The solution is filtered onto a polymeric membrane (porosity, 0.45 μm) and stored at −80°.

Analysis

The rate of formation and purity of the endoperoxide may be determined either by UV spectroscopy or by HPLC. Ultraviolet spectra of diluted aqueous solutions (2 × 10^{-4} M) are recorded between 250 and 350 nm. The naphthalenic compound shows a strong maximum at 288 nm (ε = 7780 M^{-1} cm^{-1}), whereas the absorbance of the endoperoxide is negligible at this wavelength. Hence, the purity of the sample may be calculated by comparing the absorbance at 288 nm of fresh and preheated (3 hr at 37°) solutions. HPLC analysis of aqueous solutions (2 × 10^{-4} M) is carried out with a Waters system and a reversed-phase column (Nova-Pack C-18, 25 cm) using a mixture of H_2O 59.9/MeOH 40/H_3PO_4 0.1 as eluent and UV detection at 210 nm. The purity of the endoperoxides is determined by comparison of the chromatograms of the fresh aqueous solution before and after warming at 37° for 3 hr.

Detection and Identification of $^1O_2$

An important method for the detection and characterization of $^1O_2$ is to measure chemiluminescence arising from the radioactive transition of $^1O_2$ to the ground state. There are two types of chemiluminescence derived from $^1O_2$.[28]

---

[28] M. Kasha and A. U. Khan, *Ann. N.Y. Acad. Sci.* **171**, 5 (1970).

Dimol Emission

The bimolecular transition [$^1O_2$ + $^1O_2$ → 2 $^3O_2$ + hν (λ = 634 and 703 nm)] can be monitored by means of a red-sensitive, thermoelectrically cooled photomultiplier tube connected to a discriminator, amplifier, and recording system as developed by Boveris et al.[29] Dimol emission has often been used in complex systems such as enzymatic model reactions, suspensions of subcellular fractions and cells, perfused organ, or in situ for an exposed organ.[30]

Monomol Emission

In 1979, Khan and Kasha[31] developed spectroscopic instrumentation capable of direct solution spectral studies of $^1O_2$ emission [$^1O_2$ → $^3O_2$ + hν (λ = 1270 nm)] using a thermoelectrically cooled lead sulfite detector. The further development of a more sensitive spectrometer by Khan,[32] based on a germanium diode photodetector, augmented the capability for the examination of many reactions generating $^1O_2$. The intensity of this emission is directly proportional to the concentration of $^1O_2$, e.g., using the endoperoxide NDPO$_2$ [reaction (1)] as a source of $^1O_2$,[24] and provides a direct measure of the amount produced.

Chemical Traps

Trapping techniques are based on detection of the chemical product resulting from $^1O_2$ added to an appropriate substrate. The reactions of various types of substrates with $^1O_2$ are quite well established. They include the Diels–Alder reaction of dienes to form endoperoxides ([2 + 4] cycloaddition)[33] and the "ene" reaction of alkenes to give allylic hydroperoxides.[34] In the $^1O_2$ "ene" reaction, olefins containing allylic hydrogens are oxidized to the corresponding allylic hydroperoxides in which the double bond is shifted to the adjacent position. In addition, $^1O_2$ reacts with electron-rich alkenes without allylic hydrogens or sterically hindered to form 1,2-dioxetanes ([2 + 2] cycloaddition).[35]

[29] A. Boveris, E. Cadenas, and B. Chance, Fed. Proc. **40**, 195 (1981).
[30] E. Cadenas and H. Sies, Methods Enzymol. **105**, 221 (1984).
[31] A. U. Khan and M. Kasha, Proc. Natl. Acad. Sci. U.S.A. **76**, 6046 (1979).
[32] A. U. Khan, J. Am. Chem. Soc. **103**, 6516 (1981).
[33] A. J. Bloodworth and H. J. Eggelte, in "Singlet O$_2$" (A. A. Frimer, ed.), Vol. II. CRC Press, Boca Raton, FL, 1985.
[34] K. Alder, F. Pascher, and A. Schmitz, Ber. Dtch. Chem. Ges. **76**, 27 (1943).
[35] A. L. Baumstark, in "Singlet O$_2$" (A. A. Frimer, ed.), Vol. II. CRC Press, Boca Raton, FL, 1985.

Use of Deuterated Solvent

The use of deuterated solvent as a tool to characterize the presence of $^1O_2$ has become universal. It is based on the fact that the lifetime of $^1O_2$ in D_2O is approximately 15–18 times longer than in water and in deuterated organic solvents.[36] In cases where either $^1O_2$ and O_2^- might be involved and the reaction is accompanied by product formation, the technique based on deuterated solvents cannot be used because both $^1O_2$ and O_2^- lifetimes are longer in those solvents.[37]

Exposure of Nucleic Acids or Cells to Naphthalene Endoperoxides

In order to characterize the molecular nature of $^1O_2$-induced DNA damage and mutations in mammalian cells, plasmid DNA or a SV40-based shuttle vector were reacted with $^1O_2$ arising from the thermal decomposition of $NDPO_2$.[24] The reactivity of $^1O_2$ toward nucleic acids has been reported[6–8,38,39] and is discussed elsewhere in this volume.[40] Strand breaks and alkali-labile sites were identified on DNA molecules exposed to different sources of $^1O_2$, including $NDPO_2$.[24,41] Guanine oxidation sites were found to be adjacent phosphodiester breaks.[39] Using *in vitro* models treated with $NDPO_2$, it has been shown that DNA polymerases may be blocked when replicating single-stranded DNA templates containing lesions induced by $^1O_2$.[42,43] These base lesions in DNA are premutagenic candidates *in vivo,* and $^1O_2$ mutagenicity was reported in bacteria[44] or in mammalian cells.[45] Similar to $NDPO_2$, $^1O_2$ can be generated by thermodecomposition of $DHPNO_2$. As with $NDPO_2$, $DHPNO_2$ releases $^1O_2$ on incubation at 37° with a similar time course.[46]

[36] T. Kajiwara and D. R. Kearns, *J. Am. Chem. Soc.* **95,** 5886 (1973).

[37] B. H. J. Bielski and E. Saito, *J. Phys. Chem.* **75,** 2263 (1971).

[38] P. Di Mascio, H. Wefers, H.-P. Do-Thi, M. V. M. Lafleur, and H. Sies, *Biochem. Biophys. Acta* **1007,** 151 (1989).

[39] J. Cadet, M. Berger, T. Douki, B. Morin, S. Raoul, J.-L. Ravanat, and S. Spinelli, *Biol. Chem.* **378,** 1275 (1997).

[40] J. Cadet, T. Douki, J.-P. Pouget, and J.-L. Ravanat, *Methods Enzymol.* **319,** [14] (2000) (this volume).

[41] E. R. Blazek, J. G. Peak, and M. J. Peak, *Photochem. Photobiol.* **49,** 607 (1989).

[42] J. Piette, C. M. Calberg-Bacq, M. Lopez, and A. Van de Vorst, *Biochem. Biophys. Acta* **781,** 257 (1984).

[43] D. T. Ribeiro, F. Bourre, A. Sarasin, P. Di Mascio, and C. F. M. Menck, *Nucleic Acids Res.* **20,** 2465 (1992).

[44] P. Di Mascio, C. F. M. Menck, R. G. Nigro, A. Sarasin, and H. Sies, *Photochem. Photobiol.* **51,** 293 (1990).

[45] D. T. Ribeiro, C. Madzak, A. Sarasin, P. Di Mascio, and H. Sies, *Photochem. Photobiol.* **55,** 39 (1992).

[46] L. O. Klotz, C. Pellieux, K. Briviba, C. Pierlot, J. M. Aubry, and H. Sies, *Eur. J. Biochem.* **260,** 917 (1999).

Reaction of Deoxyguanosine with $^1O_2$

In DNA, $^1O_2$ reacts preferentially with deoxyguanosine residues, leading to the formation of at least four different reaction products: two 4R* and 4S* diastereomers of 4-hydroxy-8-oxo-7,8-dihydro-2'-deoxyguanosine, the main product, and 8-oxo-7,8-dihydro-2'-deoxyguanosine (8-oxodGuo).[39,40] The amount of 8-oxodGuo present in the solution is analyzed by HPLC using a Shimadzu (Kyoto, Japan) system, connected to a UV detector, and set at 285 nm and an electrochemical detector at a potential of 650 mV. A reversed-phase column C_{18} (Spherex, 250 × 4.6 mm, 5 μm) is used, and the mobile phase is KH_2PO_4, 50 m*M,* pH 5.5, with 10% methanol and 2.5 m*M* EDTA.[39] The 4-OH-8-oxodGuo is measured by HPLC with the UV detector using a normal phase amino-substituted silica gel Hypersil NH_2 column (250 × 4.6 mm, 5 μm) and a mobile phase of 25 m*M* ammonium formate and acetonitrile (40 : 60). Electrospray ionization mass spectrometry is also used to identify the oxidation products of dGuo after reaction with $^1O_2$. Samples are analyzed with a Quattro II (Micromass, Manchester, UK) mass spectrometer with an electrospray ion source. A 10-μl sample of the mixture of dGuo and **1** optical density of methylene blue or Rose Bengal in H_2O, pH 7.0, irradiated for 30 min or treated with $NDPO_2$ is injected. Positive-ion electrospray spectra are recorded at a capillary voltage of 3.5 kV, a cone voltage of 50 V, and a source temperature of 80°. Data are processed with MassLynx software. The mass spectrum obtained exhibits peaks at m/z = 268.11, 284.12, and 300.14 attributed to $[M + H]^+$ of dGuo, 8-oxodGuo,[47] and 4-OH-8-oxodGuo, respectively. The peak at m/z = 274.09 is attributed to 2'-deoxyguanidinohydantoin.[48] The loss of the sugar ring produces peaks at m/z = 151.90, 157.94, and 183.93, corresponding to guanine, guanidinohydantoin, and 4-OH-8-oxoGuo, respectively.

Exposure of Plasmid DNA to $^1O_2$ Generated by $NDPO_2$

DNA samples (1–2 μg/200 ml) are incubated at 37° for 90 min in 50 m*M* sodium phosphate buffer in H_2O or D_2O, pH/pD 7.4 in the presence of 10 m*M* $NDPO_2$. During the first 30 min of incubation, the samples are vortexed every 2 min in order to distribute evenly the $^1O_2$ produced in the solution. After treatment, the DNA samples may be used directly for analysis or sterilized and purified with chloroform before transfection into mammalian cells and/or shutting into *Escherichia coli.* Separation of the different pBR322 DNA conformations is performed

[47] J.-L. Ravanat, A. Duretz, T. Guiller, T. Douki, and J. Cadet, *J. Chromatogr. B* **715,** 349 (1998).
[48] C. J. Burrows and J. G. Muller, *Chem. Rev.* **98,** 1109 (1998).

by gel electrophoresis using 0.7% agarose gels in a horizontal gel electrophoresis chamber at 30 mA for 120 min in 89 mM Tris borate/2 mM EDTA buffer. DNA (10 μl, 100 ng) is mixed with 2 μl of gel-loading buffer (15% Ficoll type 400/0.25% bromphenol blue) and applied to the gel. After gel electrophoresis, the DNA bands are stained with ethidium bromide and visualized by fluorescence in a UV DNA transilluminator. The DNA bands are quantified by scanning the negatives of the gel pictures using a densitometer. The number of single-stranded breaks is calculated based on the Poisson distribution. For supercoiled double-stranded DNA (FI), a correction factor of 1.4 is applied to account for its relatively lower fluorescence compared to the open circular form (FII)[49]: $e^{-x} = 1.4 \times FI/(1.4 \times FI + FII)$. For single-stranded DNA, the number of breaks is calculated, taking the ratio of molecules with zero or one break from treated DNA (T-DNA) to untreated DNA (C-DNA): $e^{-x} (1 + x) = T\text{-DNA}/C\text{-DNA}$. A role of $^1O_2$ in DNA damage, such as breaking the DNA backbone, was studied by replacing H_2O by D_2O in the buffer, leading to an increased percentage of breaks generated in keeping with the longer lifetime of $^1O_2$. To study the effect of inhibitors, some of the solutions contained sodium azide (1 mM) or freshly dissolved spermine or spermine hydrochloride,[50] in addition to 10 mM NDPO$_2$.

Virucidal Activity of Singlet Oxygen Generated by NDPO$_2$

A broad range of enveloped viruses (e.g., human herpes simplex virus type 1 or human immunodeficiency virus type 1) are inactivated rapidly when irradiated in the presence of photoactive dyes.[51–53] Singlet oxygen is suggested to have antiviral activity in these photodynamic procedures; however, apart from the direct generation of $^1O_2$ by the type II reaction, the excited sensitizer may also react in type I (free radical) reactions, interacting directly with the substrate to yield reactive-free radicals. A direct antiviral effect of $^1O_2$ was demonstrated using NDPO$_2$ to exclude the participation of type I reactions, which may occur using photoactive dyes. Singlet oxygen generated by NDPO$_2$ is capable of killing enveloped viruses, herpes simplex virus type 1 or suid herpes virus type 1,[54] human

[49] R. S. Lloyd, C. W. Haidle, and D. L. Robberson, *Biochemistry* **17**, 1830 (1978).

[50] A. U. Khan, P. Di Mascio, M. H. G. Medeiros, and T. Wilson, *Proc. Natl. Sci. U.S.A.* **89**, 11428 (1992).

[51] J. R. Perdrau and C. Todd, *Proc. Roy. Soc. Ser. B* **112**, 288 (1993).

[52] J. Lenard, A. Rabson, and R. Vanderoef, *Proc. Natl. Acad. Sci. U.S.A.* **90**, 158 (1993).

[53] H. Margolis-Nunno, E. Ben-Hur, P. Gottlieb, R. Robinson, J. Oetjen, and B. Horowitz, *Transfusion* **36**, 743 (1996).

[54] K. Müller-Breitkreutz, H. Mohr, K. Briviba, and H. Sies, *J. Photochem. Photobiol. B Biol.* **30**, 63 (1995).

immunodeficiency virus type 1, and cytomegalovirus,[55,56] but has little effect on nonenveloped virus such as adenovirus and poliovirus 1. In the study using herpes simplex virus type 1, 90% inactivation was observed after treatment with 3 mM NDPO$_2$, which generates a steady-state concentration of $^1O_2$ of about 10^{-12} M in H$_2$O-based phosphate-buffered saline (PBS). In the presence of human blood plasma (80%, v/v), which quenches $^1O_2$, virus killing was diminished substantially, e.g., 10 mM NDPO$_2$ was necessary to induce a loss of viral infectivity by 90%. In contrast to viruses, mammalian cells in culture are not as sensitive to NDPO$_2$; concentrations up to 20 mM were not toxic in various cell culture systems.

Regarding the mechanism of virus inactivation, it appears that $^1O_2$ causes inactivation by damaging viral envelopes, as can be deduced from the enhanced susceptibility of enveloped viruses in comparison to nonenveloped viruses to $^1O_2$.[55] However, $^1O_2$ may also modify DNA, particularly guanine residues, leading to the formation of 8-oxodG and other products. The exact mechanism by which $^1O_2$ inactivates viruses remains to be established.

Exposure of Cultured Cells to Singlet Oxygen Generated by Endoperoxides

Cells are grown to approximately 80% confluence in petri dishes, washed with PBS, and covered with PBS containing Ca^{2+} and Mg^{2+} or with serum-free medium at a temperature of 37°. Chemical generation of $^1O_2$ is achieved by adding NDPO$_2$ or DHPNO$_2$ over a concentration range of 0.3–10 mM to cells. The cells are incubated with NDPO$_2$ or DHPNO$_2$ for up to 1 hr at 37°. Control experiments are performed with solutions of preheated NDPO$_2$ and DHPNO$_2$ for 24 hr at 37°. These solutions contain the decomposition products NDP and DHPN, respectively. After treatment, the cells are covered with medium and are incubated at 37° for a desired time, washed with PBS at room temperature, and used for analysis as described in Refs. 57–61.

[55] A. Dewilde, C. Pellieux, S. Hajjam, P. Wattre, C. Pierlot, D. Hober, and J. M. Aubry, *J. Photochem. Photobiol. B. Biol.* **36,** 23 (1996).
[56] A. Dewilde, C. Pellieux, C. Pierlot, P. Wattre, and J. M. Aubry, *Biol. Chem.* **379,** 1377 (1998).
[57] K. Scharffetter-Kochanek, M. Wlaschek, K. Briviba, and H. Sies, *FEBS Lett.* **331,** 304 (1993).
[58] M. Wlaschek, K. Briviba, G. P. Stricklin, H. Sies, and K. Scharffetter-Kochanek, *J. Invest. Dermatol.* **104,** 194 (1995).
[59] M. Wlaschek, J. Wenk, P. Brenneisen, K. Briviba, A. Schwarz, H. Sies, and K. Scharffetter-Kochanek, *FEBS Lett.* **413,** 239 (1997).
[60] S. Grether-Beck, S. Olaizola-Horn, H. Schmitt, M. Grewe, A. Jahnke, J. P. Johnson, K. Briviba, H. Sies, and J. Krutmann, *Proc. Natl. Acad. Sci. U.S.A.* **93,** 14586 (1996).

The treatments of NDPO$_2$ or DHPNO$_2$ are nontoxic to human skin fibroblasts or keratinocytes tested up to 10 mM, as determined by means of the reduction of [3-(4,5-dimethylthiazol-2-yl)-2,5-diphenyl-2H-tetrazolium bromide (MTT)]. The cellular accumulation of $^1O_2$ sources (NDPO$_2$ or DHPNO$_2$) is discussed elsewhere in this volume.[62]

Singlet oxygen can also be produced by irradiating PBS containing 0.3 μM of Rose Bengal (RB; Sigma) or RB-agarose (Molecular Probes, Eugene, OR) for 10 min with a commercially available 500-W halogen lamp from a fixed distance of 66 cm. Approximately 130 μM (cumulative concentration) of $^1O_2$ is generated during 10 min of irradiation of 0.3 μM RB as determined by the bleaching of p-nitrosodimethylaniline.[62]

Activation of Gene Expression by NDPO$_2$

Singlet oxygen generated by thermodecomposition of NDPO$_2$ has been shown to induce the gene expression of interstitial collagenase (matrix metalloproteinase-1, MMP-1) in skin fibroblasts,[57,58] IL-1α/β, IL-6,[59] intercellular adhesion molecule-1 (ICAM-1) in keratinocytes,[60] and FAS ligand in skin-infiltrating T-helper cells Fas-ligand.[61] For example, using 3 mM NDPO$_2$ in H$_2$O-based PBS at 37°, an increase in mRNA of interstitial collagenase (MMP-1) in cultured human fibroblasts was observed.[57] Further, the expression of heme oxygenase-1 is enhanced by $^1O_2$ generated photochemically.[63]

Collagenase (MMP-1) induction by $^1O_2$ and UVA seems to be an indirect effect mediated by interleukins IL-1α/β and IL-6. Singlet oxygen generated by the thermodecomposition of NDPO$_2$ induced expression of IL-1α/β and IL-6. Singlet oxygen is made responsible for a leakage of preformed cytosolic IL-1, which then binds to its receptor, thereby stimulating the gene transcription of collagenase, its own gene, and IL-6.[59] IL-6, in turn, once synthesized and secreted, binds to the IL-6 receptor and stimulates collagenase synthesis.[64]

Transcription factor AP-2 but neither AP-1 nor NF-κB has been proven

[61] A. Morita, T. Werfel, H. Stege, C. Ahrens, K. Karmann, M. Grewe, S. Grether-Beck, T. Ruzicka, A. Kapp, L. O. Klotz, H. Sies, and J. Krutmann, *J. Exp. Med.* **186,** 1763 (1997).

[62] L. O. Klotz, K. Briviba, and H. Sies, *Methods Enzymol.* **319,** [13] (2000) (this volume).

[63] S. Basu-Modak and R. M. Tyrrell, *Cancer Res.* **53,** 4505 (1993).

[64] M. Wlaschek, G. Heinen, A. Poswig, A. Schwarz, T. Krieg, and K. Scharffetter-Kochanek, *Photochem. Photobiol.* **59,** 550 (1994).

[65] J. M. Aubry, B. Mandard-Cazin, M. Rougee, and R. V. Bensasson, *J. Am. Chem. Soc.* **117,** 9159 (1995).

to be responsible for the induction of ICAM-1 by singlet oxygen generated by NDPO$_2$.[60]

Some of the intermediate steps in the signaling network, such as the activation of mitogen-activated protein kinases, which can be involved in induction processes leading from $^1O_2$ generated by NDPO$_2$, DMNO$_2$, Rose Bengal plus light (RB/hν), methylene blue plus light (MB/hν), or UVA (320–400 nm) to induced gene expression, are shown in Fig. 4 and are also reviewed in Klotz *et al.*[62]

Conclusion

Due to the difficulties involved in obtaining $^1O_2$ free from other reactive contaminants, there is a paucity of detailed studies on the aspects of $^1O_2$ biochemistry mentioned earlier. The aim of this article is to present a useful tool to generate $^1O_2$ using endoperoxides of water-soluble naphthalene

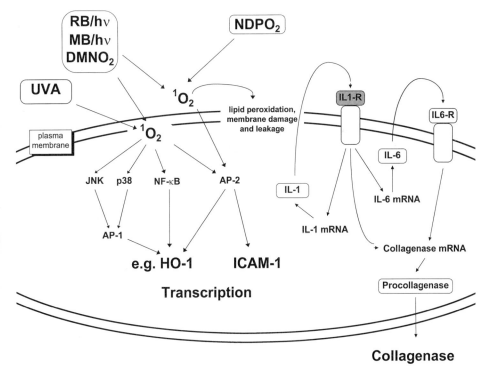

FIG. 4. Scheme outlining effects on gene expression of singlet oxygen generated by NDPO$_2$, DMNO$_2$, Rose Bengal plus light (RB/hν), methylene blue plus light (MB/hν), or UVA (320–400 nm). Modified from Briviba *et al.*[6]

derivatives. The physicochemical properties are suitable as chemical sources of $^1O_2$ in biological media.

Acknowledgments

Supported by the Fundação de Amparo à Pesquisa do Estado de São Paulo, FAPESP (Brazil), the Conselho Nacional para o Desenvolvimento Cientifico e Tecnológico, CNPq (Brazil), and the Programa de Apoio aos Núcleos de Excelência, PRONEX/FINEP (Brazil), Universidade de São Paulo/Comité Français d'Evaluation de la Coopération Universitaire avec le Brésil (USPCOFECUB), by the Deutsche Forschungsgemeinschaft, SFB 503, Project B1. Flash photolysis experiments were performed at the Paterson Institute for Cancer Research Free Radical Research Facility (Manchester, UK) with the support of the European Commission through access to large-scale facilities activity of the TMR program.

[2] Photosensitized Production of Singlet Oxygen

By Irene E. Kochevar and Robert W. Redmond

Introduction

Photosensitization provides a simple, clean, and controllable method for producing singlet oxygen in solution or in cultured cells. In this process, the photosensitizer molecule, PS, absorbs the energy of a photon ($h\nu$) of ultraviolet or visible radiation to become an excited singlet state, $^1PS^*$, which rapidly converts into an excited triplet state, $^3PS^*$. The lifetime of $^3PS^*$ is longer (typically microseconds) than that of $^1PS^*$ (typically nanoseconds), so that energy transfer from ^3PS to dissolved oxygen molecule to form singlet oxygen, $O_2(^1\Delta_g)$, is possible [Eq. (1)]. In this

$$PS \xrightarrow{h\nu} {}^1PS^* \to {}^3PS^* \xrightarrow{O_2} PS + O_2(^1\Delta_g) \qquad (1)$$

process, $^3PS^*$ is converted back to PS, the initial (ground) state, which can subsequently absorb another photon to begin the cycle again. Thus, each PS molecule can generate many (typically 10^3–10^5) singlet oxygen molecules before being bleached by reacting itself with singlet oxygen or by another reaction.

The amount of singlet oxygen generated by a PS is determined by the rate of absorption of photons, the triplet quantum yield, and the efficiency of the energy transfer process. The rate of absorption of photons is dependent on the concentration and molar absorption coefficient of the PS, the overlap between the emission spectrum of the light source and the absorption spectrum of the PS, and the intensity of the light. The efficiency for

TABLE I

RECOMMENDED PHOTOSENSITIZERS FOR GENERATION OF SINGLET OXYGEN

Photosensitizer	Absorption range (nm)	Singlet oxygen quantum yield (Φ_Δ)	Solvents
Rose bengal	450–580	0.76	Water, alcohols
Methylene blue	550–700	0.52	Water, alcohols
Phenalenone	<400	0.95	Organic solvents
meso-Tetraphenyl porphyrin	300–650	0.62	Organic solvents

energy transfer can be close to unity by choosing an appropriate PS. Typically, sufficient oxygen is present in air-saturated solution to intercept all ^{3}PS*. Thus, factors influencing the amount of singlet oxygen generated by photosensitization can be regulated easily.

After singlet oxygen is generated it either reacts with a substrate or loses its excitation energy as heat; in comparison to these processes, light emission (phosphorescence) is a very inefficient process. When singlet oxygen is generated by photosensitization in cells, it reacts rapidly with nearby biomolecules and is probably entirely consumed during this process rather than returning to the ground state.[1]

Photosensitizers

A variety of ultraviolet (UV) and visible-absorbing compounds are available for singlet oxygen photosensitization. Some of the most commonly used are listed in Table I, along with relevant photophysical parameters. In general, it is best to use a compound with a high absorption coefficient in the spectral region of the excitation light, a high quantum yield for singlet oxygen formation (Φ_Δ), and that is photostable.

Light Sources

A large variety of light sources are applicable to the study of photosensitized singlet oxygen generation. Both polychromatic (lamp) and monochromatic (laser) sources may be used as long as the source is sufficiently intense to produce enough singlet oxygen in a practical time duration. These include very inexpensive, broad-spectrum sources such as slide-projector lamps, moderately expensive filtered xenon arc lamps, and expensive monochromatic laser sources, both continuous wave (CW) or pulsed. In many experiments it is important to quantitatively analyze the effects of singlet oxygen,

[1] A. Baker and J. R. Kanofsky, *Photochem. Photobiol.* **55,** 523 (1992).

thus, actinometry (the measurement of number of photons absorbed in the system) is required. For monochromatic sources this is easily done by simple measurement of the power of the laser using a relatively inexpensive, calibrated detector. The use of bandpass filters with polychromatic sources is recommended to simplify actinometry. Filters are also recommended to remove short wavelength UV light, which can cause significant effects on direct absorption in biological systems, and infrared light, which can cause a change in temperature of the sample.

The light dose incident on the sample (D) in Joules is calculated from the product of the average power of the source, in watts, and the duration of the irradiation in seconds. The *absorbed* light dose (D_{abs}) is a function of the incident light dose (D) and the absorbance (A) of the sample at the excitation wavelength [Eq. (2)].

$$D_{abs} = D \times (1-10^{-A}) \qquad (2)$$

Methods

Photosensitization in Solution

Singlet oxygen reactions with biological molecules are often characterized in simple homogeneous solutions containing only the PS, substrates, oxygen, and solvent. Two common experiments involving singlet oxygen photosensitization are (1) the determination of the yield of singlet oxygen produced and (2) the measurement of the rate of reaction between singlet oxygen and a given substrate. Various techniques are available for these purposes, but this article describes general methods that utilize commonly available laboratory equipment. Both methods described for the measurement of singlet oxygen production are based on the change in absorbance of a probe compound on reaction with singlet oxygen, with one being appropriate for aqueous systems and the other for organic solvents.

1,3-Diphenylisobenzofuran (DPBF).[2] In this case the probe molecule is soluble in organic solvents and undergoes direct reaction with singlet oxygen, resulting in a loss of absorbance at 410 nm that can be followed with a spectrophotometer. Prepare a solution containing a PS (see Table I for recommended PS) and 15 μM DPBF in the solvent of interest. Irradiate using a wavelength where the PS absorbs and take readings of the DPBF absorbance at various time intervals. Continue irradiation until 10–15% of the initial absorbance has been lost. The slope (S) of the plot of bleached absorption ($-\Delta A$) vs irradiation time (equivalent to light dose) is propor-

[2] R. H. Young, K. Wehrly, and R. Martin, *J. Am. Chem. Soc.* **93,** 5774 (1971).

tional to the rate of production of singlet oxygen. Controls should be carried out to ensure that no direct interaction between the excited state PS and DPBF results in loss of DPBF absorbance. This can be verified by saturating the solution with nitrogen to remove oxygen from the sample. Under these conditions, loss of DPBF should be negligible.

p-Nitrosodimethylaniline (RNO) Bleaching.[3] This method is applicable to aqueous systems. The probe molecule in this case does not react directly with $O_2(^1\Delta_g)$ but with the product formed by the reaction of $O_2(^1\Delta_g)$ with imidazoles such as histidine. The reaction of RNO with this product results in loss of the RNO absorption measured at 440 nm. Prepare a sample of the PS that absorbs at the wavelength of the excitation light in neutral aqueous buffer containing 0.01 M histidine and 50 μM RNO. Irradiate samples in 10-mm path length cuvettes with periodic measurement of the sample absorbance at 440 nm in a spectrophotometer. Continue until the extent of bleaching reaches 10–15% of the initial RNO absorbance. The slope (S) of the plot of bleached absorption $(-\Delta A)$ vs irradiation time (equivalent to light dose) is proportional to the rate of production of singlet oxygen. An important control for this experiment is the verification that no RNO bleaching takes place in the absence of histidine (which would indicate another process other than involving singlet oxygen such as direct interaction between the excited PS and RNO or reaction of photogenerated radical intermediates with RNO). A control should also be carried out in the absence of the PS when the irradiation is carried out at a wavelength where RNO itself absorbs.

In these experiments it is preferable to excite the PS at a wavelength where the probe molecule does not absorb. Similarly, analysis is simplified when the PS does not absorb significantly at the wavelength used to monitor photobleaching of the probe.

Singlet Oxygen Quantum Yield Measurement. To determine quantum yields of singlet oxygen produced (Φ_Δ) by a PS of interest it is necessary to determine the number of molecules of singlet oxygen produced as a function of the number of photons absorbed. To measure these parameters in absolute terms is complex. In practice, comparative actinometry is usually carried out where a standard PS of known Φ_Δ is irradiated under the same conditions and the samples are optically matched, i.e., the same number of photons is absorbed in each case. This is simple when a monochromatic light source is used because the absorbance of the two solutions at that wavelength can be matched exactly. For broad band sources, the use of narrow bandpass filters and a standard that has a similar absorption spectrum to the unknown is recommended for improved accuracy. The solvent-

[3] I. Kraljic and S. E. Mohsni, *Photochem. Photobiol.* **28,** 577 (1978).

dependent variation in singlet oxygen lifetime requires that both systems should be studied in the same solvent. Phenalenone is a widely used standard and has a Φ_Δ of 0.95 in a large variety of organic solvents.[4] The quantum yield of singlet oxygen formation, Φ_Δ, for the compound of interest can be calculated from Eq. (3):

$$\Phi_\Delta(U) = \Phi_\Delta(St) \times \frac{S(U)}{S(St)} \tag{3}$$

where U and St denote unknown and standard, respectively, and S is the slope of the bleaching of the probe absorbance with irradiation time, as described earlier.

Reactions of Singlet Oxygen with Substrate. The rate constant of reaction between singlet oxygen and a substrate (A) can be determined using the approach described previously with solutions containing PS and probe molecule and carrying out the experiment as a function of substrate concentration. With DPBF as the probe molecule P, the overall reduction in DPBF concentration is proportional to the product of the yield of singlet oxygen and the parameter $k_Q[P]/(k_d + k_Q[P])$, where k_d and k_Q are defined by Eqs. (4) and (5), respectively.

$$O_2(^1\Delta_g) \xrightarrow{k_d} O_2 \tag{4}$$

$$O_2(^1\Delta_g) + P \xrightarrow{k_Q} PO_2 \tag{5}$$

In the presence of a substrate (A) that also reacts with singlet oxygen, a further reaction [Eq. (6)] is added, which results in a reduction in the rate of bleaching of the probe, P.

$$O_2(^1\Delta_g) + A \xrightarrow{k_R} AO_2 \tag{6}$$

For each substrate concentration the slope (S) of the plot of bleached absorbance vs irradiation time is a measure of singlet oxygen trapping by the probe in that system. A minimum of five different substrate concentrations is recommended for accurate determination of k_R. The slope (S_0) of the same plot in the absence of substrate is also required. Considering Eqs. (4–6) the following relationship (Eq. 7) applies, and a plot of S/S_0 vs $[A]$ yields a straight line plot with a gradient of $k_R/(k_d + k_Q[P])$.

$$\frac{S_0}{S} = 1 + \frac{k_R[A]}{k_d + k_Q[P]} \tag{7}$$

[4] R. Schmidt, C. Tanielian, R. Dunsbach, and C. Wolff, *J. Photochem. Photobiol. A Chem.* **79**, 11 (1994).

TABLE II
RATE OF DECAY OF SINGLET OXYGEN, k_d, IN
VARIOUS SOLVENTS

Solvent	k_d (sec^{-1})
Water	3.0×10^5
Methanol	9.6×10^4
Ethanol	6.7×10^4
Benzene	3.1×10^4
Acetonitrile	1.5×10^4
Chloroform	4.0×10^3

For the DPBF method the probe (P) is DPBF itself. For the RNO method, RNO bleaching is due to its reaction with an intermediate product of the reaction of singlet oxygen with histidine, thus the probe (P) in the just-described scheme is actually histidine, and k_Q refers to the reaction of singlet oxygen with histidine. The rate constant, k_R, can be evaluated using k_d (the reciprocal singlet oxygen lifetime) and k_Q for the probe used (8 × 10^8 and 7 × $10^7 M^{-1}$ sec^{-1} for DPBF and histidine, respectively). It should be noted that k_d varies significantly with solvent and that k_R reflects the sum of chemical and physical reaction of singlet oxygen with the substrate A. Values of k_d for many solvents are available in the literature,[5] and Table II gives k_d in some of the more commonly used solvents.

Photosensitization of Cells

Singlet oxygen can be generated in eukaryotic and prokaryotic cells by extending and adapting the methods described earlier for studies in solution. Special procedures must be followed to maintain cell viability and to eliminate effects due to the PS alone and light treatment alone.

Photosensitizers can generate singlet oxygen both in the cell culture medium and within the cells, depending on the location of the PS. In the former case, very few of the singlet oxygen molecules generated actually reach the cells, as singlet oxygen molecules exist ~4 μsec in aqueous solution; during this time they diffuse only about 100 nm. The singlet oxygen-induced alterations are limited mainly to plasma membrane-associated molecules, as most of the singlet oxygen molecules reaching the cells encounter the membrane first and react with membrane components.[6]

Photosensitizers that enter cells tend to concentrate in one or more subcellular locations, e.g., inner mitochondrial membrane, lysosomes, or

[5] F. Wilkinson, W. P. Helman, and A. B. Ross, *J. Phys. Chem. Ref. Data* **24**, 663 (1995).
[6] J. R. Kanofsky, *Photochem. Photobiol.* **53**, 93 (1991).

nucleus, depending on physical properties such as solubility, polarity, and specific binding characteristics. Reactions of singlet oxygen with cell components are generally limited to the subcellular localization sites of the PS because of the high reactivity of singlet oxygen with certain biomolecules. Rose bengal (RB) and methylene blue are two dyes that are taken up by cells, produce singlet oxygen efficiently, and are relatively nontoxic to cells in the absence of light.

Experiments are frequently conducted to determine whether singlet oxygen affects a particular cellular molecule or cell function or to compare the effectiveness of PS that are believed to generate singlet oxygen. For appropriate interpretation of the results of these experiments, at least two conditions must be met: the response being measured should be a linear function of the dose of light delivered and all the cells in the sample should receive the same amount of radiation.

Linear Dose–Response. The effect of singlet oxygen on cells, in most cases, should be proportional to the number of photons absorbed by the PS. Irradiating samples for different time periods while holding the PS concentration and light intensity constant is the best way to vary the absorbed dose. This is because changing the PS concentration may alter its subcellular distribution as the sites with the highest binding constants are occupied at the lowest concentrations and because varying the light intensity, especially laser intensity, may alter the photochemical reactions. In addition, the intensity of the light source, in general, is less readily varied and accurately measured than the irradiation time. A nonlinear dose–response relationship indicates a complex process such as change in the location of the PS as photodamage occurs, bleaching of the PS, or a threshold phenomenon. It is especially important to establish the dose–response relationship when the efficiencies of different PS are being compared.

Equal Illumination of Cells in a Sample. Most light sources do not produce a uniform illumination area and some, especially lasers, produce a beam with a cross section that is often smaller than the area of the sample. Measures must be taken to ensure that all cells in a sample are exposed to the same light dose or, if multiple samples are being irradiated concurrently, that all samples receive the same dose. For cells grown in suspension, these problems can be minimized by stirring the cell suspension gently with a Teflon-coated stirring bar. For attached cells, achieving uniform illumination is more difficult; rotating the sample only approximates a uniform illumination field.

Generation of Singlet Oxygen in Cells Grown in Suspension Using Rose Bengal as the Photosensitizer. Rose bengal is frequently used to generate singlet oxygen in cells and is used in this example. The effect of RB in the dark is first assessed on the cell response of interest (e.g., enzyme activity, apoptosis, gene expression) to determine the appropriate concentration for irradiation experiments. A 0.5 mM stock solution of RB in PBS is prepared; the absorption at 548 nm on dilution of this stock solution is used to determine the actual concentration using $5.4 \times 10^4 \, M^{-1} \, cm^{-1}$ as the absorption coefficient. Exponentially growing cells are centrifuged, washed with a physiological buffer appropriate to the cell line being used, and resuspended at $5 \times 10^5 - 1 \times 10^7$ cells/ml in buffer containing 1–10 μM RB. The number of cells to be used will vary with the needs of the assay employed. After incubation for a time equivalent to an initial uptake period (typically 15–30 min for RB) plus the expected irradiation time and postirradiation incubation time, the cells are assayed for the response of interest.

For irradiations, a cell suspension is prepared containing $5 \times 10^5 - 1 \times 10^7$ cells/ml and a concentration of RB that does not alter the cell response in the dark. This suspension is incubated for 15–30 min in the dark at room temperature. Aliquots (1–2 ml) of the cell suspension are transferred in dim light to 1-cm^2 glass spectrophotometer cuvettes and a microstirring bar is added. The suspensions are stirred slowly and exposed in duplicate to light, e.g., 514-nm light from an argon ion laser with a 6-cm beam diameter. At least five different light doses are delivered. Controls include unirradiated samples containing RB and samples irradiated with the highest dose without RB. The irradiation times are calculated by dividing the dose to be delivered by the laser power. The doses required will vary depending on the sensitivity of the cell response of interest to singlet oxygen.[7] The power of the light source is measured before and after the experiment and more frequently if necessary. If the irradiation time is greater than about 10 min, care should be taken against heating of the samples as some light sources (e.g., xenon lamps and halogen lamps) emit a high percentage of their radiation in the infrared; use of filters, as described earlier, reduces this potential problem.

Comparison of the Effectiveness of Singlet Oxygen Generated by Different Photosensitizers. Even PS with the same efficiency for the generation of singlet oxygen may vary in their ability to alter a cell function because of differences in uptake by the cells or by differences in subcellular location. Relative efficiencies of photosensitization (i.e., effect per photon absorbed) can be best estimated using relatively narrow band light sources such as a xenon lamp with a monochromator or narrow band filter; laser radiation is ideal but not always available. The irradiation wavelength(s) used for each PS should center on the absorption maximum and can differ between PS. The concentration of each PS that is nontoxic in the dark is first

[7] I. E. Kochevar, J. Bouvier, M. Lynch, and C.-W. Lin, *Biochim. Biophys. Acta* **1196,** 172 (1994).

established as described earlier. Next, the absorption of incident light by each PS within the cells at the irradiation wavelength(s) is estimated. To do this, cell suspensions containing each PS are centrifuged and the dye is extracted from the cell pellet using an appropriate solvent; the solvent will vary with the PS because of differences in polarity and solubility. The absorption spectrum of the PS in this extract is measured, and the dose of light absorbed at each incident dose to be used is calculated from Eq. (2). If different wavelengths are used for different PS, correction for differences in the energy per photon is made using Eq. (8):

$$\text{energy (joules)} = 5 \times 10^{15} \times \text{wavelength (nm)} \qquad (8)$$

To compare the relative efficiencies of different PS, dose–response curves are established using doses equivalent to the same number of absorbed photons for each PS.[8]

Summary

Photosensitization is a simple and controllable method for the generation of singlet oxygen in solution and in cells. Methods are described for determining the yield of singlet oxygen in solution, for measurement of the rate of reaction between singlet oxygen and a substrate, and for comparing the effectiveness of singlet oxygen generated by different photosensitizers in cells. These quantitative measurements can lead to better understanding of the interaction of singlet oxygen with biomolecules.

Acknowledgment

This work was supported by Grants NIH GM30755 (I.E.K.) and NIH CA68524 (R.W.R.).

[8] I. E. Kochevar, C. Lambert, M. Lynch, and A. C. Tedesco, *Biochim. Biophys. Acta* **1280,** 223 (1996).

[3] Singlet Oxygen from Irradiated Titanium Dioxide and Zinc Oxide

By YORIHIRO YAMAMOTO, NORITAKA IMAI, RYUICHI MASHIMA, RYUSEI KONAKA, MASAYASU INOUE, and WALTER C. DUNLAP

Introduction

Titanium dioxide (TiO$_2$) and zinc oxide (ZnO) are photo-semiconducting catalysts used in the oxidation of various organic compounds. Spin-trapping studies using electron paramagnetic resonance (EPR) spectroscopy have demonstrated the formation of active oxygen species on ultraviolet (UV) irradiation: hydroxyl radicals from TiO$_2$[1-3] and superoxide anion radicals or protonated perhydroxyl radicals from TiO$_2$[1,2,4] and ZnO.[5] The photochemical generation of hydrogen peroxide[6-8] and singlet oxygen[6] by irradiation of TiO$_2$ and ZnO has also been reported. The cytotoxicity of irradiated TiO$_2$ to prokaryotic and eukaryotic cells[9-14] has been ascribed to the production of these active oxygen species, especially to that of highly reactive hydroxyl radicals.

Despite their known photoreactivity, TiO$_2$ and ZnO have been widely used in sunscreening and cosmetic products without complication. Given

[1] C. D. Jaeger and J. A. Bard, *J. Phys. Chem.* **83**, 3146 (1979).

[2] H. Noda, K. Oikawa, and H. Kamada, *Bull. Chem. Soc. Jpn.* **65**, 2505 (1992).

[3] J. R. Harbour, J. Tromp, and M. L. Hair, *Can. J. Chem.* **63**, 204 (1985).

[4] H. Noda, K. Oikawa, H. Ohya-Nishiguchi, and H. Kamada, *Bull. Chem. Soc. Jpn.* **66**, 3542 (1993).

[5] C. Chen, R. P. Veregin, J. R. Harbour, M. L. Hair, S. L. Issler, and J. Tromp, *J. Phys. Chem.* **93**, 2607 (1989).

[6] S. P. Pappas and R. M. Fisher, *J. Paint Technol.* **46**, 65 (1974).

[7] J. R. Harbour and M. L. Hair, *J. Phys. Chem.* **83**, 652 (1979).

[8] M. V. Rao, K. Rajeshwar, V. R. P. Verneker, and J. DuBow, *J. Phys. Chem.* **84**, 1987 (1980).

[9] R. Cai, K. Hashimoto, K. Itoh, Y. Kubota, and A. Fujishima, *Bull. Chem. Soc. Jpn.* **64**, 1268 (1991).

[10] R. Cai, Y. Kubota, T. Shuin, H. Sakaki, K. Hashimoto, and A. Fujishima, *Cancer Res.* **52**, 2346 (1992).

[11] J. C. Ireland, P. Klostermann, E. Rice, and R. M. Clark, *Appl. Environ. Microbiol.* **59**, 1668 (1993).

[12] Y. Kubota, T. Shuin, C. Kawasaki, M. Kosaka, H. Kitamura, R. Cai, H. Sakai, K. Hashimoto, and A. Fujishima, *Br. J. Cancer* **70**, 1107 (1994).

[13] H. Sakai, E. Ito, R. Cai, T. Yoshioka, Y. Kubota, K. Hashimoto, and A. Fujishima, *Biochim. Biophys. Acta* **1201**, 259 (1994).

[14] W. G. Wamer, J. Yin, and R. R. Wei, *Free Radic Biol. Med.* **23**, 851 (1997).

that studies have shown that singlet oxygen causes the oxidation of nitrone spin-trapping reagents to yield apparent hydroxyl radical adducts,[15–17] we have reexamined the photochemical production of active oxygen species from irradiated TiO_2 and ZnO. We demonstrated the formation of singlet oxygen by examination of the oxidation products of methyl oleate and 2,2,6,6-tetramethyl-4-piperidone (4-oxo-TMP) and compared the rate of oxidation of uric acid and 2,6-di-*tert*-butyl-4-methylphenol (butylated hydroxytoluene, BHT).[18]

Methods

Materials

Commercial grades of TiO_2 (anatase) and ZnO are from Miyoshi Kasei (Urawa, Saitama). All are submicron white powders and their average diameters are 30 nm for TiO_2 and 40 nm for ZnO. Uric acid, BHT, palladium (10%) on activated carbon, and 2,2,6,6-tetramethyl-4-piperidone-N-oxyl (4-oxo-TEMPO) are from Wako (Osaka). Methyl oleate is from Sigma (Tokyo) and purified by a reversed-phase, high-performance liquid chromatography (HPLC) prior to use.[19] Bis(trimethylsilyl)trifluoroacetamide (BSTFA) is from Supelco (Tokyo) and used as received. 4-Oxo-TMP hydrochloride is from Sigma and purified as described previously.[20] All other reagents and solvents are analytical grade, and water is double-distilled or Super-Q (Millipore) grade.

Light Source

A UVA fluorescent tube (BLE-270W, Spectronics, New York) is used as a light source in all experimental procedures with the exception of EPR spectroscopy (see later). The lamp emission spectrum produces maximum output at 350 nm and no emission occurs below 300 nm: total light intensity is 1.1 mW/cm^2.

[15] J. R. Harbour, S. L. Issler, and M. L. Hair, *J. Am. Chem. Soc.* **102,** 7778 (1980).

[16] J. B. Feix and B. Kalyanaraman, *Arch. Biochem. Biophys.* **291,** 43 (1991).

[17] P. Bilski, K. Reszka, M. Bilska, and C. F. Chignell, *J. Am. Chem. Soc.* **118,** 1330 (1996).

[18] Y. Yamamoto, W. C. Dunlap, N. Imai, R. Mashima, R. Konaka, M. Inoue, Y. Hasegawa, and T. Miyoshi, *in* "20th IFSCC International Congress, Cannes, 1998. Platform Presentation Preprints," Vol. 1, p. 53, 1998.

[19] Y. Yamamoto, M. H. Brodsky, J. C. Baker, and B. N. Ames, *Anal. Biochem.* **160,** 7 (1987).

[20] W. C. Dunlap, Y. Yamamoto, M. Inoue, M. Kashiba-Iwatsuki, M. Yamaguchi, and K. Tomita, *Int. J. Cosmetic Sci.* **20,** 1 (1998).

Oxidation of Methyl Oleate and Determination of Its Hydroperoxides

Photooxidation of methyl oleate (100 mM) is carried out in ethanol (or acetonitrile) in the presence or absence of 10 mg/ml TiO$_2$ or 0.5 mg/ml ZnO under aerobic conditions at room temperature. Methyl oleate hydroperoxides are quantified by HPLC separation on an octadecylsilyl column (5 μm, 4.6 × 250 mm, Supelco) with methanol as the mobile phase (1 ml/min) and detection using a hydroperoxide-specific chemilumines-cence assay.[19]

GC/MS Analysis of Methyl Oleate Hydroperoxides

Gas chromatographic and mass spectrometric analysis (GC/MS) are obtained on a Hitachi (Tokyo) system consisting of a G-5000M GC and M-7200 MS. Methyl oleate hydroperoxides are first reduced to hydroxides with sodium borohydride and then purified on a semipreparative octadecyl-silyl column (10 μm, 20 × 250 mm, Shiseido, Tokyo) using methanol as the mobile phase (8 ml/min). The methyl oleate hydroxides are hydrogenated in the presence of palladium (10%) on activated carbon, and the saturated aliphatic alcohols are derivatized to trimethylsilyloxyl ethers by BSTFA prior to GC/MS analysis. A capillary column (CP-Si 24 CB, 0.25 mm × 30 m, Chrompack, The Netherlands) is used. Operating temperatures of the injector and detector are 250 and 200°, respectively. Column tempera-ture is maintained at 200° for 4 min after injection and is then raised to 280° at a rate of 10°/min.

Electron Paramagnetic Resonance Measurements

Electron paramagnetic resonance measurements are recorded at 25° with a JEOL JES-TE200 spectrometer operating at X-band with a micro-wave power of 8.0 mW. A xenon arc lamp (Ushio, ES-UXL10, operating at 500 W) is focused directly onto a flat quartz cell within the microwave cavity of the spectrometer. Signal intensities are normalized to a manganese internal standard, and concentrations of reagent adducts are determined by external calibration.

Photooxidation of Urate and BHT

Photooxidation of 10 μM urate (or BHT) is conducted in aqueous 80% methanol in the presence or absence of TiO$_2$ (or ZnO) at room temperature under aerobic conditions.[20] Urate is quantified by HPLC separation[19] on an aminopropylsilyl column (5 μm, 4.6 × 250 mm, Supelco) using methanol/ 40 mM aqueous monobasic sodium phosphate (90/10) as the mobile phase (1 ml/min) with detection at 252 nm. BHT is measured similarly by UV

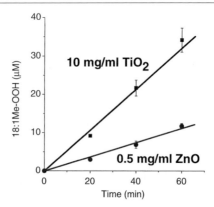

FIG. 1. Formation of methyl oleate hydroperoxides (18:1 Me-OOH) during the photooxi-dation of methyl oleate (100 mM) in ethanol in the presence of 10 mg/ml TiO$_2$ or 0.5 mg/ml ZnO under aerobic conditions at room temperature. Values are the mean of three independent analyses. Data points are mean values ($n = 3$) with standard deviation bars.

detection at 277 nm after separation on an octadecylsilyl column (5 μm, 4.6 $\times$ 250 mm, Supelco) using methanol/water (90/10) as the mobile phase (1 ml/min).

Evidence for Singlet Oxygen Production

The oxidation of methyl oleate by singlet oxygen is known to give 9- and 10-hydroperoxyoctadecenoates,[21] whereas autoxidation provides four hydroperoxides (8-, 9-, 10-, and 11-hydroperoxyoctadecenoates) in equal quantities.[22] Therefore, methyl oleate can serve as a good probe to differen-tiate between singlet oxygen-induced oxidation and free radical-mediated autoxidation. Figure 1 shows the production of methyl oleate hydroperox-ides during irradiation of TiO$_2$ (10 mg/ml) or ZnO (0.5 mg/ml) in ethanolic suspension with 100 mM methyl oleate under aerobic conditions at room temperature. No formation of methyl oleate hydroperoxides was observed in the absence of TiO$_2$ and ZnO. The rate of methyl oleate hydroperoxide formation was time dependent (Fig. 1) and concentration dependent with TiO$_2$ and ZnO (data not shown). In this study, ZnO was found to have greater efficiency for the production of methyl oleate hydroperoxides.

The reduced trimethylsilyl derivatives of methyl oleate hydroperoxides were prepared after photooxidation of methyl oleate in the presence of

[21] E. N. Frankel, W. E. Neff, and E. Selke, *Lipids* **16,** 279 (1981).
[22] E. N. Frankel, W. E. Neff, and W. K. Rowedder, *Lipids* **12,** 901 (1977).

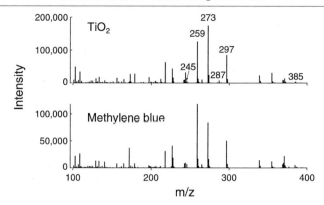

FIG. 2. GC/MS spectra of methyl oleate hydroperoxides (derivatized as reduced TMS ethers) produced in the UVA photooxidation of methyl oleate in acetonitrile in the presence of TiO₂ or methylene blue.

either TiO₂ or methylene blue, an efficient photosensitizing dye for the production of singlet oxygen. The isomeric hydroperoxide derivatives eluted together at 12.8 min under our GC conditions. The near identical spectra of hydroperoxide derivatives (Fig. 2) suggest the formation of singlet oxygen from irradiated TiO₂.

The positions where the hydroperoxyl groups were introduced at the oleate unsaturation were identified by mass fragmentation of reduced and derivatized hydroperoxides. As shown in Fig. 3, the 8-hydroperoxide pro-

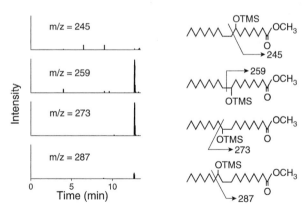

FIG. 3. Expected mass fragments from methyl oleate hydroperoxide (derivatized as reduced TMS ethers) and GC/MS selective ion chromatograms of methyl oleate hydroperoxide derivatives obtained from methyl oleate on coirradiation with TiO₂ in acetonitrile; isomeric derivatives eluted at 12.8 min.

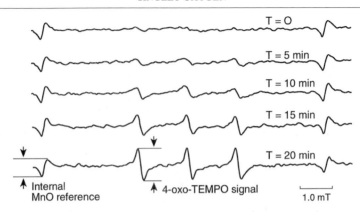

FIG. 4. Sequential EPR spectra for the production of 4-oxo-TEMPO by the reaction of 4-oxo-TMP with singlet oxygen on coirradiation with 1.6 mg/ml TiO$_2$.

duces a major fragment at m/z 245, the 9-hydroperoxide at m/z 259, the 10-hydroperoxide at m/z 273, and the 11-hydroperoxide at m/z 287. Corresponding selective ion chromatograms monitored at $m/z = 245, 259, 273$, and 287 are also provided in Fig. 3. At elution of the hydroperoxide derivatives ($t = 12.8$ min), GC peaks obtained with mass detection at $m/z = 259$ and 273 were large in comparison to those peaks obtained at $m/z = 245$ and 287, indicating a predominance of the formation of the 9- and 10-hydroperoxides. Photooxidation of methyl oleate by photoexcited ZnO also showed a predominance of the formation of 9- and 10-hydroperoxides

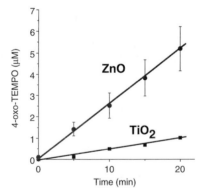

FIG. 5. Formation of 4-oxo-TEMPO on coirradiation of 4-oxo-TMP with 1.6 mg/ml TiO$_2$ or ZnO in ethanol at room temperature under aerobic conditions. Data points are mean values ($n = 4$) with standard error bars. Control experiment (4-oxo-TMP without microfine oxides) did not produce EPR-detectable products (data not shown).

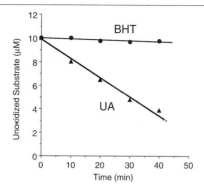

FIG. 6. Oxidation of uric acid (UA; 10 μM) or 2,6-di-*tert*-butyl-4-methylphenol (BHT; 10 μM) on coirradiation with 4 mg/ml TiO$_2$ in aqueous 80% methanol at room temperature under aerobic conditions. Values are the mean of two independent analyses.

(data not shown). These results provide strong evidence for the photogeneration of singlet oxygen by TiO$_2$ and ZnO.

The generation of singlet oxygen by photoexcited TiO$_2$ and ZnO was validated using the singlet oxygen-specific reagent 4-oxo-TMP. Singlet oxygen causes oxidation of this reagent to give the paramagnetic N-oxide (4-oxo-TEMPO), which is readily detected by its characteristic EPR spectra (Fig. 4). The rate of 4-oxo-TEMPO formation from irradiated TiO$_2$ or ZnO in ethanolic suspension is given in Fig. 5. As in the photooxidation of methyl oleate by TiO$_2$ and ZnO (Fig. 1), ZnO showed the greater quantum efficiency for generating singlet oxygen as measured by the formation of 4-oxo-TEMPO.

If hydroxyl radicals were formed on irradiation of TiO$_2$ as previously reported,[1–3] they would react with alcoholic solvents to produce peroxyl radicals; in the case of methanol, HOCH$_2\cdot$ is first produced and then converted to HOCH$_2$OO$\cdot$ by rapid combination with oxygen. The resultant formation of peroxyl radicals should cause the oxidative degradation of both uric acid and BHT.[23,24] However, irradiation of TiO$_2$ (4 mg/ml) in the presence of 10 μM uric acid or BHT in aqueous 80% methanol only affected urate oxidation, whereas BHT remained unchanged (Fig. 6). These data substantiate the production of singlet oxygen and provide little evidence for the generation of hydroxyl radicals by photoexcited TiO$_2$, as BHT is

[23] E. Niki, M. Saito, Y. Yoshikawa, Y. Yamamoto, and Y. Kamiya, *Bull. Chem. Soc. Jpn.* **59,** 471 (1996).

[24] G. W. Burton, T. Doba, E. J. Gabe, L. Hughes, F. L. Lee, L. Prasad, and K. U. Ingold, *J. Am. Chem. Soc.* **107,** 7053 (1985).

reactive with peroxyl radicals but not with singlet oxygen.[25] Uric acid must be decomposed as it is very reactive toward singlet oxygen.[26] Similarly, BHT was stable on coirradiation with ZnO (data not shown).

Discussion

In contrast to oxygen-centered radicals, the production of singlet oxygen from photoexcited TiO_2 or ZnO has received little attention. In an early study on the chalking of paints, Pappas and Fisher[6] observed the single oxygen-specific conversion of furan to 2-methoxy-5-oxo-2,5-dihydrofuran on irradiation of TiO_2 or ZnO and its inhibition by the addition of triethylamine. These results suggest the formation of singlet oxygen from photoexcited TiO_2 and ZnO, but their report has generally been overlooked.

The cytotoxicity of irradiated TiO_2 to various cells in culture[9–14] has been attributed to the generation of hydroxyl radicals, although direct evidence is lacking in these studies. Our evidence based on the stability of BHT on irradiation of TiO_2 or ZnO does not support significant production of hydroxyl radicals. The interpretation of results from previous EPR studies[1–3] identifying the photoproduction of hydroxyl radicals using nitrone spin-trapping reagents is therefore likely to be incorrect given that singlet oxygen-induced oxidation of these reagents affords a product resembling the hydroxyl radical spin adduct.[15–17]

Photoexcited TiO_2 or ZnO may be used as a convenient source of singlet oxygen. However, it should be kept in mind that they may also produce superoxide and hydrogen peroxide. In fact, removal of superoxide with a spin-trapping reagent was found to enhance the rate of formation of singlet oxygen,[27] as superoxide effectively quenches singlet oxygen.[28]

Finally, it is noteworthy that singlet oxygen is very toxic to prokaryotic cells but is almost nontoxic to eukaryotic cells.[29] This may provide a reason why TiO_2 and ZnO have been successful in sunscreening and cosmetic use without causing apparent complications.

[25] Y. Kohno, Y. Egawa, S. Itoh, S. Nagaoka, M. Takahashi, and K. Mukai, *Biochim. Biophys. Acta* **1256,** 52 (1995).

[26] J. R. Wagner, P. A. Motchnik, R. Stocker, H. Sies, and B. N. Ames, *J. Biol. Chem.* **268,** 18502 (1993).

[27] R. Konaka, E. Kasahara, W. C. Dunlap, Y. Yamamoto, K. C. Chien, and M. Inoue, unpublished results.

[28] H. L. Guiraud and C. S. Foote, *J. Am. Chem. Soc.* **98,** 1984 (1976).

[29] M. Nakano, Y. Kambayashi, H. Tatsuzawa, T. Komiyama, and K. Fujimori, *FEBS Lett.* **432,** 9 (1998).

Conclusion

The generation of singlet oxygen by irradiation of TiO_2 and ZnO dispersions was confirmed by evaluating the singlet oxygen-specific products arising from the oxidation of methyl oleate and 4-oxo-TMP. Furthermore, coirradiation of TiO_2 or ZnO with BHT did not cause oxidative decomposition of BHT, thus providing little evidence for the photoproduction of hydroxyl radicals by these microfine oxides.

Acknowledgments

We thank Mr. Tim Simmons for his help in preparing several figures. This work was presented[18] in part at the 20th International Federation Societies of Cosmetic Chemists (IFSCC) International Congress, Cannes, on September 15, 1998.

[4] Time-Resolved Singlet Oxygen Detection

By SANTI NONELL and SILVIA E. BRASLAVSKY

Aims and Scope

Techniques for the detection of singlet oxygen [hereafter referred to as $O_2(^1\Delta_g)$, which in condensed phase is the most abundant of the two lowest excited oxygen especies[1]] are increasingly implemented in many laboratories, reflecting the growing awareness that $O_2(^1\Delta_g)$ plays crucial roles in several photoexcited systems. Time-resolved detection techniques have been used for (1) the identification of $O_2(^1\Delta_g)$, (2) the measurement of quantum yields of $O_2(^1\Delta_g)$ production in photosensitized processes, Φ_Δ, and (3) the determination of rate constant for the interaction of $O_2(^1\Delta_g)$ with substrates.

This article describes the methods used most commonly to attain the just-described goals associated with the technique of time-resolved near-infrared detection of $O_2(^1\Delta_g)$ phosphorescence at 1.27 μm (TRNIR). Focus is placed on laboratory procedures rather than on techniques. The presentation of background theory has been kept to a minimum as a number of excellent reviews already exist that fulfill that purpose. Only methods for homogeneous solutions are discussed. Clues are given for extending them to more complex systems, but the number of possible situations precludes general statements. We have restricted ourselves to systems in which

[1] R. Schmidt and M. Bodesheim, *J. Phys. Chem.* **98**, 2874 (1994).

$O_2(^1\Delta_g)$ is produced by photosensitization, thus excluding chemical or enzymatic production.

Background

Photoexcitation of chromophores in the presence of oxygen often leads to the production of $O_2(^1\Delta_g)$ by a bimolecular energy-transfer process involving the excited chromophore and ground-state triplet oxygen, $O_2(^3\Sigma_g^-)$ or $^3O_2$. This process is referred to as photosensitization and the excited chromophore is then the sensitizer. With most organic sensitizers the relevant excited state is the lowest triplet (hereafter ^3M) and we will assume that no sensitization occurs from other excited states. Production of additional $O_2(^1\Delta_g)$ by quenching of the sensitizer first excited singlet state may occur in some cases.[2]

The quantum yield for $O_2(^1\Delta_g)$ production, Φ_Δ, is defined as the number of $O_2(^1\Delta_g)$ molecules formed per absorbed photon. Its value is determined by three factors, as shown in Eq. (1):

$$\Phi_\Delta = \Phi_T \cdot f_{T\Sigma} \cdot S_\Delta \tag{1}$$

The triplet quantum yield, Φ_T, is the amount of triplet chromophores produced relative to the amount of originally photoexcited chromophores. Intersystem crossing to the triplet state occurs in competition with the fluorescence and internal conversion, which return the sensitizer back to its ground state. As a rule of thumb, the smaller the singlet-to-triplet energy gap the faster the intersystem-crossing process and hence the larger Φ_T.[3] Triplet formation occurs in most cases in 0.1–10 ns. Quenching of the sensitizer singlet state by $O_2(^3\Sigma_g^-)$ is generally negligible, but must be considered when the singlet lifetime exceeds 10 ns as it might affect the Φ_T value, generally increasing it, and could also result in the additional production of $O_2(^1\Delta_g)$.[2]

The second factor, $f_{T\Sigma}$, is the fraction of triplets trapped by $O_2(^3\Sigma_g^-)$ and reflects the fact that energy transfer from ^3M to $O_2(^3\Sigma_g^-)$ is a collisional process that requires the formation of encounter complexes between both entities within the lifetime of ^3M, as shown by Eq. (2):

$$f_{T\Sigma} = \frac{k_q^{O_2}[O_2(^3\Sigma_g^-)]}{k_T + k_q^{O_2}[O_2(^3\Sigma_g^-)]} = \tau_T k_q^{O_2}[O_2(^3\Sigma_g^-)] \tag{2}$$

[2] F. Wilkinson, W. P. Helman, and A. B. Ross, *J. Phys. Chem. Ref. Data* **22,** 113 (1993).
[3] N. J. Turro, "Modern Molecular Photochemistry." University Science Books, Mill Valley, CA, 1991.

where $k_q^{O_2}$ is the rate constant for triplet quenching by oxygen, k_T is the rate constant for triplet decay by all other deactivation processes, and τ_T is the triplet lifetime in the actual system.

Under conditions of low oxygen concentration, high viscosity (low $k_q^{O_2}$), or short intrinsic triplet lifetime (large k_T), $f_{T\Sigma}$ is expected to be low and so is Φ_Δ. The third factor, S_Δ, is the fraction of encounter complexes that yield $O_2(^1\Delta_g)$ on dissociation. It is now well established that these complexes have substantial charge-transfer character and stability, i.e., they are real exciplexes, and that the value of S_Δ depends essentially on the efficiency of energy transfer within the singlet exciplex relative to the efficiency of quenching to the ground state via the triplet exciplex.[4–6] Charge-transfer interactions are more important for molecules with high triplet energies, hence correlations between S_Δ and the sensitizer triplet energy are observed for a series of related compounds, e.g., ketones.[7,8] A list of S_Δ values can be found in Wilkinson et al.[2]

Thus, Φ_Δ is an oxygen-dependent parameter rather than a unique property of a sensitizer. This must be taken into account when measuring and reporting Φ_Δ data. The kinetics of $O_2(^1\Delta_g)$ in a laser flash photosensitization experiment in homogeneous medium is used to obtain information on $O_2(^1\Delta_g)$ formation (τ_T, Φ_Δ) and decay (τ_Δ) [Eq. (3)]:

$$[O_2(^1\Delta_g)](t) = \Phi_\Delta \frac{E_1(1 - 10^{-A})}{Lh\nu V} \cdot \frac{\tau_\Delta}{\tau_T - \tau_\Delta} (e^{-t/\tau_T} - e^{-t/\tau_\Delta}) \qquad (3)$$

where E_1 is the energy of the laser flash ($= nLh\nu$, with n the amount of einsteins of the laser pulse, L the Avogadro's constant, h the Planck's constant, and ν the laser frequency), A is the solution absorbance at the excitation wavelength, V is the irradiated volume, and τ_Δ is the $O_2(^1\Delta_g)$ lifetime.

More complex rate laws are needed to describe $O_2(^1\Delta_g)$ kinetics in microheterogeneous systems, such as biological media, due to diffusion and partition between the different regions.[9] A detailed analysis of such cases is beyond the scope of this article.

The time-resolved detection of $O_2(^1\Delta_g)$ is usually accomplished by taking advantage of its weak phosphorescence at 1.27 μm. The intensity of this emission, and therefore the electrical signal produced by a detector, is

[4] F. Wilkinson, D. J. McGarvey, and A. F. Olea, *J. Phys. Chem.* **98,** 3762 (1994).
[5] A. J. McLean and M. A. J. Rodgers, *J. Am. Chem. Soc.* **115,** 4786 (1993).
[6] C. Grewer and H. D. Brauer, *J. Phys. Chem.* **98,** 4230 (1994).
[7] O. L. J. Gijzeman, F. Kaufman, and G. Porter, *J. Chem. Soc Faraday Trans.* **69,** 708 (1973).
[8] M. N. Nau and J. C. Scaiano, *J. Phys. Chem.* **100,** 11360 (1996).
[9] Y. Fu and J. R. Kanofsky, *Photochem. Photobiol.* **62,** 692 (1995).

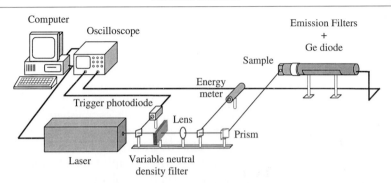

FIG. 1. Layout of a $O_2(^1\Delta_g)$ laser flash kinetic spectrophotomer. The output of the pulsed laser, conveniently attenuated with neutral density filters, excites the sample in a standard fluorescence cuvette. Two beam splitters sample the beam for the purpose of triggering the oscilloscope with a fast silicon photodiode and for measuring the pulse energy. The concomitant sample emission is passed through silicon and interference filters (1270 nm) and is detected by a liquid N_2-cooled germanium diode (North Coast EO-817P).

proportional to the $O_2(^1\Delta_g)$ concentration [Eq. (4) for homogeneous systems],

$$S(t) = \frac{\kappa}{n_r^2} k_R [O_2(^1\Delta_g)](t) = \frac{\kappa}{n_r^2} k_R \Phi_\Delta \frac{E_l(1 - 10^{-A})}{Lh\nu V} \cdot \frac{\tau_\Delta}{\tau_T - \tau_\Delta} (e^{-t/\tau_T} - e^{-t/\tau_\Delta})$$

$$(4)$$

where κ is a proportionality constant that contains geometrical and electronic factors of the detection system, n_r is the solvent refractive index, and k_R is the solvent-dependent $O_2(^1\Delta_g)$ radiative decay rate constant.[10]

The following sections describe how to use Eq. (4) with TRNIR for the determination of Φ_Δ and τ_Δ data.

Time-Resolved Near-Infrared Technique

The TRNIR setup is implemented readily in a laboratory having pulsed lasers (Fig. 1).

All that is required is a wavelength within the absorption spectrum of the photosensitizer, a pulse duration in the nanosecond time scale or lower, and approximately 1 mJ energy per pulse. It is convenient to use wavelengths in relatively flat absorbance regions in order to minimize errors due to different bandwidths of spectrometer and excitation laser and in the absorbance of different sensitizers. Nowadays the most commonly used

[10] R. D. Scurlock, S. Nonell, S. E. Braslavsky, and P. R. Ogilby, *J. Phys. Chem.* **99**, 3521 (1995).

is the Nd:YAG laser, either frequency doubled or tripled, or pumping a dye laser or an optical parametric oscillator (OPO) for tunability.

The most common arrangement between excitation and observation ports is the right-angle geometry. A standard fluorescence cell is used for the sample. Placing the cell in a mirrored holder with just two holes drilled for entry and exit of the laser beam is advantageous in order to (1) increase the signals and (2) shield the detector from spurious light. Some authors prefer to use more sophisticated combinations of mirrors and lenses to collect as large a fraction as possible of the weak emission.[11]

Emission filters are used to prevent both scattered laser light and sensitizer fluorescence or phosphorescence from impinging onto the detector. Observation of large spikes in the early stages of the signal is one of the most common problems as they often produce saturation of the detector electronics and obscure the early stages of the signal. It cannot be overstressed that shielding the detector from stray light is an essential requirement for obtaining good kinetic traces. The simplest approach is to use a cutoff silicon filter, which blocks light below 1050 nm, in combination with an interference filter for selection of a narrow range around 1.27 μm. It is, however, known that silicon filters do emit light when illuminated with UV-VIS light.[12] A simple trick that has proven highly successful is to use an array of long-pass filters with increasingly higher cutoff wavelengths.[13] It is also essential to eliminate NIR background radiation, a problem often encountered when using a Nd:YAG laser. It is advisable to tape an IR-blocking filter (e.g., Schott KG5) at every port of the laser.

The detector of choice for $O_2(^1\Delta_g)$–NIR emission is still the germanium photodiode. A number of manufacturers supply the detector "ready to use," i.e., wired properly for time-resolved measurements, coupled to a preamplifier, and even encased in a cooling unit such as a Dewar flask for noise reduction. The size of the diode determines its sensitivity but also its response time. A convenient trade-off is a 3-mm-diameter diode with a time response of approximately 250 nsec. Figure 2 shows a circuit diagram of a home-made germanium detector and its preamplifier, an inexpensive alternative to commercial mounted detectors providing approximately 1-μsec time resolution.

Alternative detectors replacing germanium photodiodes are InGaAs diodes.[14] Hamamatsu has released a photomultiplier tube that will probably

[11] P. Bilski and C. F. Chignell, *J. Biochem. Biophys. Methods* **33**, 73 (1996).

[12] R. D. Scurlock, K.-K. Iu, and P. R. Ogilby, *J. Photochem.* **37**, 247 (1987).

[13] J. L. Bourdelande, J. Font, R. González-Moreno, and S. Nonell, *J. Photochem. Photobiol. A Chem.* **115**, 69 (1998).

[14] A. Michaeli and J. Feitelson, *Photochem. Photobiol.* **61**, 255 (1995).

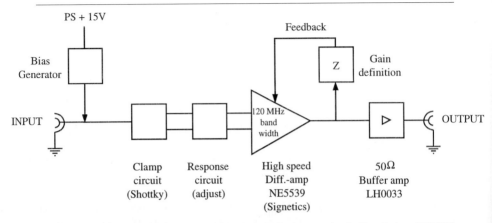

Fig. 2. Circuit diagram for a germanium photodiode detector (typically a Judson J16/8SP) amplifier. Clamp circuit: made with Schottky diodes arranged in antiparallel fashion. Response circuit: the response is defined in low-pass and in high-pass frequencies in order to inhibit negative overshooting, 200 Hz < transit frequency < 8 MHz. High-speed amplifier: NE5539 (Signetics) for video and high frequencies, an integrated circuit with defined gain (10-400). Buffer amplifier: LH0033 (various brands): impedance transformer to 50 Ω. Clamp circuit and response circuit are needed in view of the high dynamics of the signal. Total gain: 120.

replace photodiodes due to its higher gain and faster response time.[15] The availability of such tubes opens new possibilities to the field as they allow for time-correlated single-photon counting detection of $O_2(^1\Delta_g)$ phosphorescence, an approach that has already been applied with custom-made tubes.[16]

The analog output of the diode is fed to a transient recorder. A 100-MHz sampling rate capability suffices for most applications. It is customary to use a computer for data storage and analysis.

It is recommended to add a laser energy meter. This is especially important for determining Φ_Δ values (vide infra) and also to rule out undesired phenomena, such as two-photon absorption or ground-state depletion.

Other Techniques

Time-resolved detection of $O_2(^1\Delta_g)$ can also be accomplished using its absorption of light or its emission of heat. Time-resolved thermal lensing, which monitors the release of heat accompanying $O_2(^1\Delta_g)$ formation and

[15] O. Shimizu, J. Watanabe, and K. Imakubo, *J. Phys. Soc. Jpn.* **66**, 268 (1997).

[16] S. Y. Egorov, V. F. Kamalov, N. I. Koroteev, A. A. Krasnovsky, Jr., B. N. Toleutaev, and S. V. Zinukov, *Chem. Phys. Lett.* **163**, 421 (1989).

decay, has been used for the determination of Φ_Δ and τ_Δ values in a single experiment, without the need for an external reference.[17,18] This technique, however, requires transparency of the solution. Laser-induced optoacoustic spectroscopy is also used for the determination of Φ_Δ values.[19,20] More recently, time-resolved $O_2(^1\Delta_g)$ absorption in the IR has been demonstrated.[21]

General Remarks on Sample Preparation and Handling

Solvents should be purified following standard procedures.[22] The value of τ_Δ is especially sensitive to impurities containing OH groups.[23] Hence, every effort must be made to eliminate residual water from the solvent of choice. Solutions should always be prepared immediately before use and handled under dim light. Storing solutions over long periods of time is discouraged.

It is often desirable to adjust the O_2 content of the solutions by bubbling with mixtures containing an inert gas such as nitrogen, helium, or argon. A convenient way of achieving this is with the use of flow meters to adjust the proportions of the two gases. The gases should always be blown through at least two wash bottles containing the desired solvent for minimizing solvent evaporation in the sample.

Recording $O_2(^1\Delta_g)$ Kinetic Traces

The diode signal should be coupled to the digitizer through a 50-Ω entrance resistor for better time resolution, always paying attention to achieving good impedance matching. Impedance mismatch often results in signals "going below zero." The time base in the digitizer should be chosen such that four to five lifetimes are recorded. The signal should be authenticated to rule out artifacts or spurious emissions: it should disappear when oxygen is excluded from the system (e.g., upon inert gas saturation) and it should be possible to reduce its lifetime (not its amplitude) adding well-

[17] G. Rossbroich, N. A. García, and S. E. Braslavsky, *J. Photochem.* **31**, 35 (1985).

[18] R. W. Redmond and S. E. Braslavsky, *Chem. Phys. Lett.* **148**, 523 (1988).

[19] S. E. Braslavsky and G. E. Heibel, *Chem. Rev.* **92**, 1381 (1992).

[20] C. Martí, O. Jürgens, O. Cuenca, M. Casals, and S. Nonell, *J. Photochem. Photobiol. A Chem.* **97**, 11 (1996).

[21] D. Weldon and P. R. Ogilby, *J. Am. Chem. Soc.* **120**, 12978 (1998).

[22] S. E. Braslavsky and W. Frohn, *in* "CRC Handbook of Organic Photochemistry" (J. C. Scaiano, ed.), Vol. 2, p. 347. CRC Press, Boca Raton, FL, 1989.

[23] R. Schmidt and H. D. Brauer, *J. Am. Chem. Soc.* **109**, 6976 (1987).

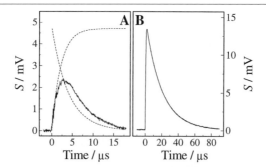

FIG. 3. Typical $O_2(^1\Delta_g)$ phosphorescence signals. (A) Rise and decay occur with similar rate constants. The signal is for 10 μM 1H-phenalen-1-one-2-sulfonic acid in air-saturated water excited with a nitrogen laser at 337 nm. Dashed lines are the pure rise and decay components obtained by fitting Eq. (5) to the signal. (B) Rise is much faster than decay, hence the decay part can be analyzed with Eq. (6). The signal is for 1H-phenalen-1-one in air-saturated benzene, all other conditions equal.

established $O_2(^1\Delta_g)$ quenchers such as azide, 1,4-diazabicyclo[2.2.2]octane (DABCO), or β-carotene.[24]

Measuring Φ_Δ in Homogeneous Systems

For the purpose of obtaining Φ_Δ, Eq. (4) is usually written in the more compact form of Eq. (5):

$$S(t) = S(0) \frac{\tau_\Delta}{\tau_T - \tau_\Delta} (e^{-t/\tau_T} - e^{-t/\tau_\Delta}) \tag{5}$$

For $\tau_T \ll \tau_\Delta$, the decay portion of the signal ($t > 5\tau_T$) can be described by Eq. (6):

$$S(t) = S(0)e^{-t/\tau_\Delta} \tag{6}$$

Examples of signals fitted with both equations are shown in Fig. 3.

Using Eq. (6) to analyze the decay portion of the curve when the condition $\tau_T \ll \tau_\Delta$ is not fulfilled can lead to serious errors in the estimation of $S(0)$. In Eqs. (5) and (6), $S(0)$ refers to the amplitude of the signal at "zero" time, which is extrapolated from the fitting of data with the decay function. $S(0)$ contains information on Φ_Δ, Eq. (7):

$$S(0) = \frac{\kappa}{n_r^2} k_R \Phi_\Delta \frac{E_1(1 - 10^{-A})}{Lh\nu V} \tag{7}$$

[24] F. Wilkinson, W. P. Helman, and A. B. Ross, *J. Phys. Chem. Ref. Data* **24,** 663 (1995).

Thus, the ratio of the $S(0)$ values for two sensitizers in the same solvent, excited under the same geometrical conditions, at the same wavelength, with the same laser energy, and having the same absorbance should afford their relative Φ_Δ values. This is the rationale for the comparative method for determination of Φ_Δ. In practice, one prefers to measure $S(0)$ as a function of E_1 and A, and determine Φ_Δ from the analysis of these dependencies. An extensive list of Φ_Δ values has been compiled by Wilkinson et al.[2]

Procedure

1. Choose a reference sensitizer that meets the following criteria: good solubility in the desired solvent, high and precisely known Φ_Δ, and chemically and photochemically stable under experimental conditions. For UV excitation, 1 *H*-phenalen-1-one, also known as perinaphthenone, is widely accepted as the reference compound of choice with $\Phi_\Delta = 0.95 \pm 0.05$ in most solvents studied so far, ranging from water to benzene.[20,25–27] Its 2-sulfonic acid derivative has an even higher solubility in aqueous solutions.[28] For excitation in the visible there is no universal reference comparable to perinaphthenone, although Wilkinson et al.[2] provide a list of sensitizers with self-consistent Φ_Δ values, e.g., rose bengal for water ($\Phi_\Delta = 0.76$) and methanol ($\Phi_\Delta = 0.80$) and *meso*-tetraphenylporphine for benzene ($\Phi_\Delta = 0.66$).

2. Prepare stock solutions of both sensitizers in the same solvent, with A approximately 1 in a 1-cm cuvette.

3. Prepare at least five dilutions of each stock solution, covering the linear absorption range of the TRNIR spectrometer. A good set of A values is often 0.05, 0.10, 0.15, 0.22, and 0.30.

4. Register the absorption spectrum of each solution and read the absorbance at the laser excitation wavelength. Use slits as narrow as possible and make sure the spectrophotometer wavelength matches that of the laser. Watch for "hidden" absorptions, e.g., due to the solvent-oxygen charge-transfer band, which may contribute to the formation of $O_2(^1\Delta_g)$.[29]

[25] E. Oliveros, P. S. Murasecco, T. A. Saghafi, A. M. Braun, and H. J. Hansen, *Helv. Chim. Acta* **74**, 79 (1991).

[26] R. Schmidt, C. Tanielian, R. Dunsbach, and C. Wolff, *J. Photochem. Photobiol. A Chem.* **79**, 11 (1994).

[27] E. Oliveros, S. H. Bossmann, S. Nonell, C. Martí, G. Heit, G. Tröscher, A. Neuner, C. Martínez, and A. M. Braun, *New J. Chem.* 85 (1999).

[28] S. Nonell, M. González, and F. R. Trull, *Afinidad* **50**, 445 (1993).

[29] R. D. Scurlock and P. R. Ogilby, *J. Phys. Chem.* **93**, 5493 (1989).

For each of the just-described solutions:

5. Record the singlet oxygen transient signals at several values of E_1, up to a laser fluence of approximately 1 mJ cm^{-2} per pulse. Average multiple signals until the signal-to-noise ratio >5:1. Register the absorption spectrum after the experiments to rule out sensitizer photodegradation. Should bleaching occur, change the solution after each laser shot (e.g., by flowing the solution through a flow-through cell).

6. Reduce data by fitting an appropriate kinetic function such as Eq. (5) or Eq. (6) and determine $S(0)$. Plot $S(0)$ as a function of E_1. At low E_1 a straight line with zero intercept should be observed (Fig. 4). Positive, nonzero intercepts may indicate the onset of negative deviations from linearity at the highest laser fluences, often due to ground-state depletion and/or triplet–triplet annihilation. The former situation is especially favored when highly diluted solutions are used, i.e., when the absorption coefficient of the sensitizer is large. Measure with lower laser fluences.

7. After repeating steps 5 and 6 for each solution, plot the slopes of the energy curves obtained, $S(0)_E$, vs the absorption factor $(1-10^{-A})$. A straight line with a zero intercept for both the sample and reference sensitizers should be obtained. Nonzero intercepts may indicate the presence of an absorbing impurity in the solvent or the mismatch between laser and spectrophotometer wavelengths. Always run a "blank" of the solvent and calibrate the spectrophotometer. A second and frequent problem is nonlinearity of the plots, which may indicate concentration effects on Φ_Δ, such as dye aggregation or triplet (or

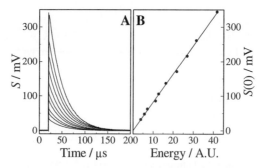

FIG. 4. (A) Laser-fluence dependence of the emission signal. (B) Zero-time amplitude $S(0)$, as a function of the laser energy. Observation of linearity with zero intercept rules out unwanted phenomena such as ground-state depletion, excited-state self-annihilation, or multiphotonic absorption. The signal is for 10 μM 1H-phenalen-1-one in air-saturated benzene, excited with a nitrogen laser at 337 nm.

singlet) quenching by the ground-state sensitizer. Take the slope at zero concentration as the measure of Φ_Δ (Fig. 5).

8. Calculate Φ_Δ from the ratio of the slopes $S(0)_{EA}$ of the linear plots obtained in step 7 using Eq. (8):

$$\Phi_{\Delta,\text{sample}} = \Phi_{\Delta,\text{ref}} \frac{S(0)_{EA,\text{sample}}}{S(0)_{EA,\text{ref}}} \tag{8}$$

A problem frequently arises when a solvent is used for which there is no well-established reference. In such cases it is still possible to determine Φ_Δ by the comparative method using a reference in a different solvent, but solvent differences in both refractive index and radiative rate constant must be taken into account by using Eq. (7).[10]

Measuring τ_Δ and $O_2(^1\Delta_g)$ Quenching Rate Constants

The method is based on the Stern–Volmer relationship, which in turn describes the increase in the pseudo-first-order $O_2(^1\Delta_g)$ decay rate constant k_Δ on addition of a quencher in a homogeneous system, as shown by Eq. (9). An extensive compilation of quenching rate constants is available.[24]

$$k_\Delta = \tau_\Delta^{-1} = k_d + k_q[Q] \tag{9}$$

k_d is the $O_2(^1\Delta_g)$ intrinsic decay rate constant and k_q is the bimolecular rate constant for $O_2(^1\Delta_g)$ quenching by Q (the sum of physical and chemical quenching).

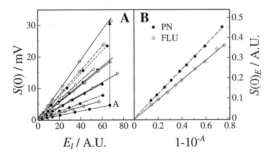

FIG. 5. (A) Zero-time amplitude, $S(0)$, as a function of the laser fluence for a set of sample and reference solutions with different absorbance values. PN, $1H$-phenalen-1-one; FLU, fluorenone. The solvent is air-saturated N,N'-dimethylacetamide. Excitation was with a nitrogen laser at 337 nm. (B) Slopes of fluence-dependent plots (from (A)) as a function of the absorption factor $(1-10^A)$. Observation of linearity with zero intercept rules out unwanted phenomena such as quenching of the sensitizer excited states by the sensitizer itself, equilibria among several forms of the sensitizer (e.g., acid–base or monomer–dimer), or production of $O_2(^1\Delta_g)$ by a solvent impurity.

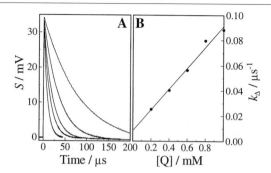

FIG. 6. (A) $O_2(^1\Delta_g)$ kinetics in air-saturated acetonitrile as a function of added sodium azide. The sensitizer is 10 μM 1H-phenalen-1-one. (B). Stern–Volmer-like plot for the increase in the decay rate constant on addition of quencher.

Procedure

1. Choose a sensitizer that meets the following criteria: good solubility in the desired solvent; Φ_Δ close to 1; high molar absorption coefficient at the excitation wavelength, aiming at minimizing photoexcitation of the quencher; poor $O_2(^1\Delta_g)$ quencher; and no interaction with the actual quencher, either in the ground or in the excited state. 1H-phenalen-1-one or its 2-sulfonic acid derivative are good sensitizers for UV excitation. The extremely low rate constant for buckminster-fullerene C_{60} quenching of $O_2(^1\Delta_g)$ makes it an ideal candidate for both UV and VIS excitation.[30]

2. Prepare a stock solution of the sensitizer, with A approximately 1 (in a 1-cm pathlength cuvette) when diluted to its final concentration. It works well to prepare the solution such that it has an absorbance of 1 in a 1-mm cuvette.

3. Prepare a stock solution of the quencher. If the quenching rate constant is roughly known, a simple rule of thumb for calculating the concentration of this solution is ([Q]/mol dm^{-3}) $\approx$ $10^6/(k_q/dm^3$ mol^{-1} sec^{-1}), i.e., at this concentration τ_Δ is reduced to approximately 1 μsec. If solubility precludes reaching this value, try using a deuterated counterpart of the solvent in order to increase the intrinsic τ_Δ.[23]

4. Prepare the final solutions by mixing the same volume of the sensitizer solution, variable amounts of the quencher solution, and the necessary solvent to reach the same final volume.

[30] F. Prat, C. Hou, and C. S. Foote, *J. Am. Chem. Soc.* **119**, 5051 (1997).

5. Record the singlet oxygen decay profile. If signal averaging is needed to improve the signal-to-noise ratio, use fresh solution for each laser shot (e.g., by flowing the solution through a flow-through cell).
6. Repeat step 5 for each solution. Pick them at random to avoid following a trend.
7. Reduce data by fitting an appropriate kinetic model such as Eq. (5) or Eq. (6). Plot the reciprocal of the measured τ_Δ vs the quencher concentration as in Fig. 6. Observation of a straight line validates the use of the Stern–Volmer Eq. (9) for determining k_q.

Concluding Remarks

Time-resolved $O_2(^1\Delta_g)$ detection methods are now well established for homogeneous systems. Their use has provided a wealth of data that conforms most of the body of knowledge about $O_2(^1\Delta_g)$. There is still much to do, especially in the areas of heterogeneous systems and microheterogeneous such as biological media, where technical improvements as well as new methodologies are still needed. On the technical side, new detectors with better time resolution and sensitivity are required, a goal that may be fulfilled with the new photomultiplier tubes. In addition, new methods providing space resolution as well as time resolution, the so-called singlet oxygen microscope, will find their application in heterogeneous media.[31]

Acknowledgments

S.N. acknowledges grants from the Spanish funding agencies CICYT-CIRIT and MEC, refs. QFN94-4613-C02-01 and PM98-0017-C02-02. S.E.B. is deeply indebted to Kurt Schaffner for his support. We thank Uli Paul for the amplifier development and Willi Schlamann for excellent technical assistance.

[31] T. Keszthelyi, D. Weldon, T. N. Andersen, T. D. Poulsen, K. V. Mikkelsen, and P. R. Ogilby, *Photochem. Photobiol.* **70,** 531 (1999).

[5] Measurement of Photogenerated Singlet Oxygen in Aqueous Media

By Véronique Nardello and Jean-Marie Aubry

Introduction

For many years, much work has been devoted to the involvement of singlet molecular oxygen, $^1O_2(^1\Delta_g)$, in many processes occurring in aqueous environments (biological systems, natural waters, reaction media).[1–10] Chemical trapping is the most widely used and sensitive method to detect and quantify this excited species. Accordingly, a large number of water-soluble traps have been developed to measure the cumulative amount of $^1O_2$ in an aqueous environment.[11–18] The most selective ones belong to the series of polycyclic aromatic compounds, which react with $^1O_2$ according to a [4 + 2] cycloaddition, affording a stable endoperoxide highly specific for $^1O_2$. Popular examples of this type of trap are tetrapotassium rubrene-2,3,8,9-tetracarboxylate (RTC)[11] and disodium 9,10-anthracenedipro-

[1] N. A. Garcia, J. Photochem. Photobiol. B Biol. 22, 185 (1994).

[2] J. M. Aubry, in "Membrane Lipid Oxidation" (C. Vigo-Pelfrey, ed.), Vol. II, p. 65. CRC Press, Boca Raton, FL, 1991.

[3] J. M. Aubry and B. Cazin, Inorg. Chem. 27, 2013 (1988).

[4] R. C. Straight and J. D. Spikes, "Singlet Oxygen" (A. A. Frimer, ed.), Vol. 4, p. 91. CRC Press, Boca Raton, FL, 1985.

[5] J. M. Aubry, J. Am. Chem. Soc. 107, 5844 (1985).

[6] H. Sugimoto and D. T. Sawyer, J. Am. Chem. Soc. 106, 4283 (1984).

[7] J. R. Kanofsky, J. Photochem. 25, 105 (1984).

[8] A. U. Khan, J. Am. Chem. Soc. 105, 7195 (1983).

[9] R. W. Murray, "Singlet Oxygen" (H. H. Wasserman and R. W. Murray, eds.), p. 59. Academic Press, New York, 1979.

[10] R. G. Zepp, N. L. Wolfe, G. L. Baughman, and R. C. Hollis, Nature 267, 421 (1977).

[11] J. M. Aubry, J. Rigaudy, and N. K. Cuong, Photochem. Photobiol. 33, 149 (1981).

[12] B. A. Linding, M. A. J. Rodgers, and A. P. Schaap, J. Am. Chem. Soc. 102, 5590 (1980).

[13] A. P. Schaap, A. L. Thayer, K. A. Zaklika, and P. C. Valenti, J. Am. Chem. Soc. 101, 4016 (1979).

[14] K. Müller and K. Ziereis, Arch. Pharm. 326, 369 (1993).

[15] A. W. M. Nieuwint, J. M. Aubry, F. Arwert, U. Kortbeek, S. Herzberg, and H. Joenje, Free Radic Res. Commun. 1, 1 (1985).

[16] J. M. Aubry, B. Cazin, and F. Duprat, J. Org. Chem. 54, 726 (1989).

[17] V. Nardello, N. Azaroual, I. Cervoise, G. Vermeersch, and J. M. Aubry, Tetrahedr. 52, 2031 (1996).

[18] V. Nardello, D. Brault, P. Chavalle, and J. M. Aubry, J. Photochem. Photobiol. B Biol. 39, 146 (1997).

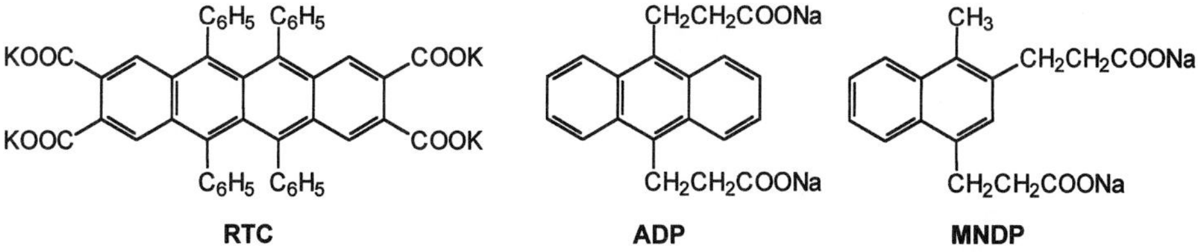

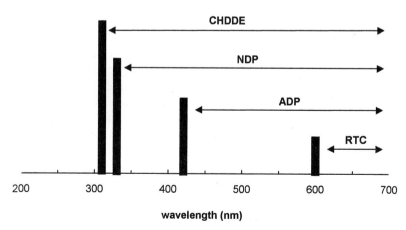

SCHEME 1

panoate (ADP) (Scheme 1).[12,13] However, these compounds absorb light in the visible range and, accordingly, may themselves act as photosensitizers (Fig. 1). Hence, they are unsuitable for the measurement of photogenerated $^1O_2$. Naphthalene derivatives such as disodium 4-methyl-1,3-naphthalenedipropanoate (MNDP)[14] (Scheme 1) or disodium 1,4-naphthalenedipropanoate (NDP)[15,16] overcome this problem, but these compounds react slowly with $^1O_2$ and the endoperoxides thus obtained are unstable.

To overcome these problems, a new colorless trap, the disodium 1,3-cyclohexadiene-1,4-diethanoate (CHDDE) **4**,[17,18] has been designed on the basis of a 1,3-cyclohexadienic core. This trap has several advantages: it is readily water soluble, it does not absorb visible light (Fig. 1), and it is highly reactive toward $^1O_2$ by giving mainly a specific and stable endoperoxide (Scheme 2).

FIG. 1. Transparency areas of the water-soluble traps of $^1O_2$: RTC, ADP, NDP, and CHDDE **4**.

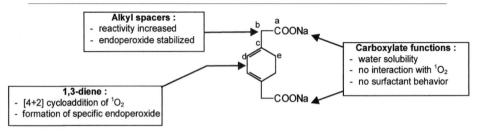

SCHEME 2

Synthesis of CHDDE 4

CHDDE is prepared in three steps from 1,4-cyclohexanedione **1** as described in Scheme 3. First, according to the method described by Engel *et al.*,[19] *E* and *Z* isomers of 1,4-bis(carbethoxymethylene) cyclohexane **2** are prepared by a double Wittig reaction of triethyl phosphonoacetate on 1,4-cyclohexanedione **1**. The diadducts **2** (*E* + *Z*) are simultaneously isomerized and saponified with KOH/MeOH into the more stable conjugated diene **3**. The diacid **3** is then neutralized with two equivalents of sodium methanoate, leading to anhydrous sodium 1,3-cyclohexadiene-1,4-diethanoate (CHDDE) **4**.

During this preparation, a small amount (5%) of sodium 1,4-phenylene-diethanoate (PDE) **5** is formed. It is not necessary to remove it because it may be used as an internal standard in high-performance liquid chromatography (HPLC) analysis.

Physicochemical Properties of CHDDE 4

Thanks to the two sodium carboxylate groups, CHDDE is highly soluble (>1 mol $\cdot$ liter^{-1}) in basic and neutral (pH ≥ 6) aqueous medium and in methanol. However, its solubility is much lower in acidic water (2×10^{-2} mol $\cdot$ liter^{-1} at pH 2). As indicated in Fig. 1, CHDDE does not absorb above 310 nm but shows a well-defined absorption maximum at 270 nm characteristic of its 1,3-cyclohexadienic structure. In addition, ^{1}H and ^{13}C nuclear magnetic resonance (NMR) spectra are very simple according to the symmetry of this molecule (Fig. 2).

The disappearance of photogenerated $^1O_2$ in aqueous solutions containing CHDDE **4** may occur through three main pathways shown in Eqs. (1)–(3)[20]:

[19] P. S. Engel, R. L. Allgren, W.-K. Chae, R. A. Leckonby, and N. A. Marron, *J. Org. Chem.* **44**, 4233 (1979).
[20] F. Wilkinson and J. G. Brummer, *J. Phys. Chem. Ref. Data* **10**, 809 (1981).

1 2 (*E* + *Z*) 3 4 5

95 % 5 %

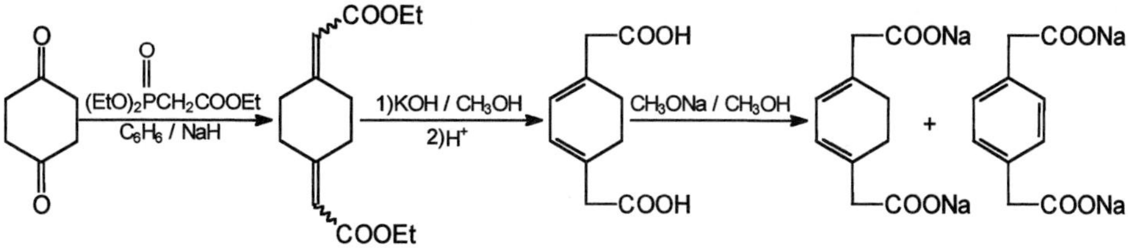

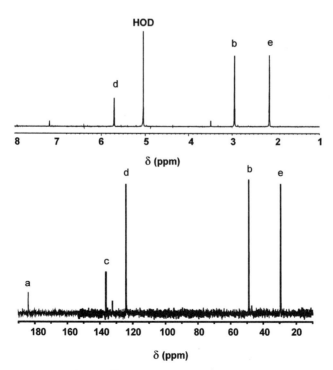

SCHEME 3

$$^1O_2 + \text{solvent} \xrightarrow{k_d} {}^3O_2 \tag{1}$$

$$^1O_2 + \text{CHDDE} \xrightarrow{k_q} {}^3O_2 + \text{CHDDE} \tag{2}$$

$$^1O_2 + \text{CHDDE} \xrightarrow{k_r} \text{CHDDEO}_2 \tag{3}$$

FIG. 2. ^1H and ^{13}C NMR spectra of CHDDE **4** in D_2O.

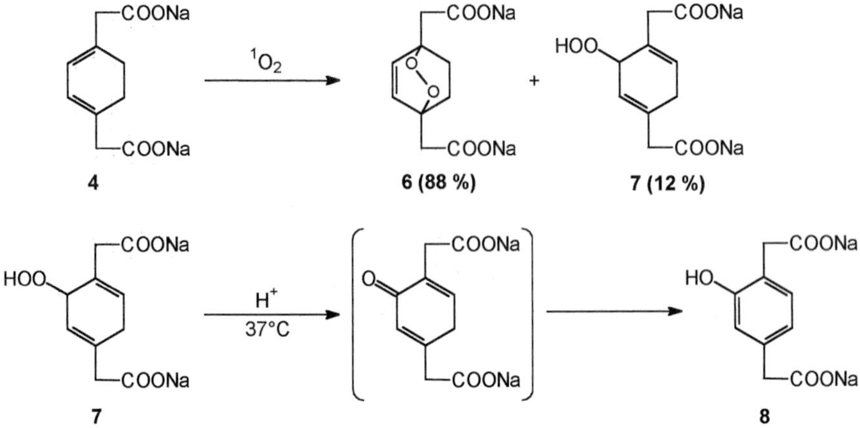

4 $^1O_2$ **6 (88 %)**   +   **7 (12 %)**

7 $\dfrac{H^+}{37°C}$ ⟶   **8**

SCHEME 4

The overall rate constant, $k_r + k_q$, of CHDDE was measured by flash photolysis, and the chemical rate constant k_r was determined by comparison with RTC. Both constants are equal to $(2.6 \pm 0.3) \times 10^7$ mol · liter^{-1} · sec^{-1}.[17,21] This result strongly suggests that the interaction of CHDDE with $^1O_2$ mainly occurs through a pure chemical process ($k_r \gg k_q$).

Peroxidation of CHDDE 4

The peroxidation of CHDDE **4** by $^1O_2$ occurs via two competitive pathways. The main one is a [4 + 2] cycloaddition of $^1O_2$ on the diene moiety leading to the endoperoxide **6** (88%), whereas the side reaction is an ene reaction of $^1O_2$ on a single double bond giving the hydroperoxide **7** (12%) with concomitant migration of the double bond (Scheme 4). Under biological conditions ($\theta = 37°$), hydroperoxide **7** is gradually ($t_{1/2} = 5$ hr) converted into **8**, whereas CHDDEO$_2$ **6** and PDE **5** remain unchanged.

The reaction of CHDDE **4** with photogenerated $^1O_2$ may be monitored either by ultraviolet (UV) spectroscopy or by HPLC (Fig. 3). The disappearance of CHDDE **4** is detected easily by UV spectroscopy at 270 nm ($\varepsilon = 7340$ mol^{-1} · liter · cm^{-1}). However, this simple method is not recommended because it does not distinguish reactions due to $^1O_2$ from those involving other species. However, HPLC allows the quantification not only of the disappearance of the trap, but also of the appearance of both hydroperoxide **7** and endoperoxide **6**, specific of the reaction with $^1O_2$. In addition, PDE

[21] J. M. Aubry, B. Mandard-Cazin, M. Rougee, and R. Bensasson, *J. Am. Chem. Soc.* **117**, 9159 (1995).

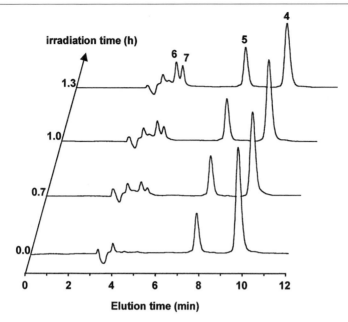

FIG. 3. Evolution of HPLC chromatograms during the photooxidation of 10^{-2} mol $\cdot$ liter^{-1} CHDDE **4** in D$_2$O in the presence of 10^{-4} mol $\cdot$ liter^{-1} TPPS.

5, which does not interact with $^1O_2$, can be used as an internal standard to quantify precisely ($\pm2\%$) CHDDE **4** and its oxidation products.

Measurement of the Amount of $^1O_2$

Measurement of the cumulative amount of photogenerated $^1O_2$ in aqueous solution may be achieved by chemical trapping with CHDDE **4** according to two procedures, depending on the experimental conditions. From Eqs. (1) and (3) and assuming pseudo-stationary conditions ($d[^1O_2]/dt = 0$), the cumulative amount of photogenerated $^1O_2$, $[^1O_2]_t$ may be expressed as shown in Eq. (4):

$$[^1O_2]_t = ([CHDDE]_0 - [CHDDE]_t) + \frac{k_d}{k_r} \ln \left(\frac{[CHDDE]_0}{[CHDDE]_t} \right) \qquad (4)$$

Both methods are based on Eq. (4), which relates the disappearance of the trap with the total amount of $^1O_2$ formed during a period of irradiation. When $[CHDDE]_0 \gg k_d/k_r$, all the singlet oxygen formed is trapped by

CHDDE. In that case, the disappearance of the trap varies linearly as a function of the irradiation time (Fig. 4) [Eq. (5)]:

$$\frac{[CHDDE]_t}{[CHDDE]_0} = 1 - \frac{v \times t}{[CHDDE]_0} \tag{5}$$

where v is the rate of formation of $^1O_2$ (mol $\cdot$ liter^{-1} $\cdot$ sec^{-1})

Such a situation occurs when the reaction takes place in D_2O in which the lifetime of $^1O_2$ is relatively long ($\tau_\Delta = 1/k_d^{D_2O} = 65$ μsec). In that case, $k_d^{D_2O}/k_r = 7 \times 10^{-4}$ mol $\cdot$ liter^{-1} and an initial concentration of CHDDE equal to 10^{-2} mol $\cdot$ liter^{-1} is sufficient to trap all the available singlet oxygen. Measurement of the amount of $^1O_2$ is then very accurate ($\pm 5\%$) and is only limited by the precision of the analytical method used to monitor the disappearance of CHDDE (UV or preferably HPLC). However, it is necessary to split the irradiation time in order to avoid oxygen depletion, which would decrease the rate of $^1O_2$ generated by photosensitization and which could induce side reactions of type I (electron transfer). Practically, the irradiation times must be chosen so that less than half the oxygen present in solution ($\approx 10^{-3}$ mol $\cdot$ liter^{-1}) is consumed.

Sometimes, the photosensitizer cannot be obtained in deuterated water and the measurement of $^1O_2$ photogenerated has to be realized in ordinary water. In that solvent, the lifetime of $^1O_2$ is rather short ($\tau_\Delta = 1/k_d^{H_2O} =$

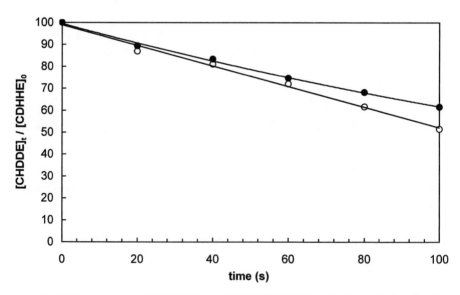

FIG. 4. Disappearance of CHDDE **4** from H_2O (●) ([CHDDE] = 5×10^{-4} mol $\cdot$ liter^{-1}) and D_2O (○) ([CHDDE] = 10^{-2} mol $\cdot$ liter^{-1}) solutions, during irradiation, monitored by HPLC.

4.4 μsec and $k_d^{H_2O}/k_r = 10^{-2}$ mol $\cdot$ liter^{-1}. As it is not recommended to introduce CHDDE at a concentration much higher than 10^{-2} mol $\cdot$ liter^{-1}, it is better to use a much more dilute solution (5×10^{-4} mol $\cdot$ liter^{-1}).[18] Under such conditions, the first term of Eq. (4) is negligible compared to the logarithmic term and the disappearance of CHDDE follows an exponential law [Eq. (6)]:

$$\frac{[\text{CHDDE}]_t}{[\text{CHDDE}]_0} = \exp\left(-\frac{v \times k_r}{k_d^{H_2O}} \times t\right) \tag{6}$$

Values of $\log[\text{CHDDE}]_t/[\text{CHDDE}]_0$ reported as a function of time provide a straight line, the slope of which allows the calculation of the rate of $^1O_2$ formation (v), knowing $k_d^{H_2O}/k_r$. This method is less accurate ($\pm$20%) than the previous one, as it is limited by the precision of k_r and k_q, but it is very reliable when several photosensitizers have to be compared.

Finally, this method has the advantage of using ordinary water to consume a small amount of trap and avoiding the problems of oxygen depletion during irradiation.

Materials and Methods

Instrumentation

Ultraviolet-visible spectra are recorded with a Milton Roy Spectronic 3000 spectrophotometer equipped with a diode array photodetector. High-performance liquid chromatography analyses are performed with a reversed-phase column (Nova-Pak C18, 4 μm, 4.6 $\times$ 250 mm) using a 600 controller pump from Waters, a mixture of $CH_3OH/H_2O/H_3PO_4$ (40:60:0.2) as the eluent, and UV detection at 200 and 270 nm with a Waters 490E programmable multiwavelength detector.

Reactants

1,4-Cyclohexanedione (98%), triethyl phosphonoacetate (99%), benzene (99%), sodium hydride (60%), and sodium methoxide (97.5%) are from Aldrich-Chemie. Potassium hydroxide (rectapur) and methanol (R. P. Normapur) are from Prolabo.

1,4-Bis(carbethoxymethylene)cyclohexane 2 (E + Z) (adapted from reference 19)

A dry, 500-ml, three-necked, round-bottom flask equipped with a mechanical stirrer and an addition funnel is charged with 7.3 g (0.18 mol) of

60% NaH in mineral oil and 30 ml of dry benzene under a N_2 stream. The liquid phase is discarded and is replaced by 50 ml of dry benzene. To this stirred mixture is added 40.8 g (0.18 mol) of triethyl phosphonoacetate dropwise for 30 min with ice bath cooling. Vigorous evolution of hydrogen occurs. At the end of the addition, a brown solution is obtained. The mixture is then stirred for 1 hr at room temperature until hydrogen evolution ceases. To this solution is added 10 g (0.09 mol) of 1,4-cyclohexanedione **1** in 90 ml of benzene dropwise for 1 hr with ice bath cooling (T = 10–15°). The reaction medium becomes fluorescent orange and russet when approximately 50% of diketone **1** has been added, and a clear brown precipitate is formed. When addition of dione **1** is complete, the mixture is stirred for 15 min at 60°. The supernatant is then decanted, and the precipitate is washed with 2 × 30 ml of benzene and heated again at 60° for 15 min. The three combined benzene solutions are concentrated with a rotary evaporator yielding quantitatively 20.5 g of **2** (*E* + *Z*) (yield = 91%). The resulting oil is allowed to stand in the refrigerator where the *E* isomer could crystallize out.

1,3-Cyclohexadiene-1,4-diethanoic acid 3

A dry, 100-ml, two-necked, round-bottom flask equipped with a refrigerant is charged with 1 g of **2** (*E* + *Z*) (4 mmol). A solution of 1 mol · liter^{-1} of potassium hydroxide (50 mmol) in methanol (50 ml) is added and refluxed under a N_2 stream. Evolution of the reaction medium is followed by HPLC (eluent: $CH_3OH/H_2O/H_3PO_4$ = 60:40:0.2, a detection wavelength of 270 nm). After 1 hr, the peak corresponding to the formed cyclohexadiene **3** remains constant, and the reaction is stopped. Methanol is then removed with a rotary evaporator and a clear brown residue is obtained. It is then solubilized in 100 ml of H_2O and acidified with 3.4 ml of concentrated H_3PO_4 up to pH ≈2.0. The acid solution is extracted with 3 × 50 ml of ether, which is dried with $MgSO_4$. After removal of the ether with a rotary evaporator, a white residue is obtained. The powder is ground and triturated with ether. The diacid **3** is recovered as a white powder by centrifugation (yield = 68%).

Disodium 1,3-cyclohexadiene-1,4-diethanoate (CHDDE) 4

A flask containing 3.77 g (20 mmol) of acid **3** dissolved into 50 ml of THF is cooled with an ice bath. To this solution 1.08 g of sodium methanoate (40 mmol) in 10 ml of methanol is added dropwise. Solvents are then removed with a rotary evaporator, and the residue is triturated three times with ether, yielding 2.5 g of CHDDE **4** as a white powder (yield = 54%).

[6] Assay for Singlet-Oxygen Generation by Peroxidases Using 1270-nm Chemiluminescence

By Jeffrey R. Kanofsky

Introduction

Singlet oxygen ($^1\Delta_g$) represents one possible product of systems composed of a peroxidase, hydrogen peroxide, and a halide,

$$H_2O_2 + H^+ + X^- \xrightarrow{\text{peroxidase}} H_2O + HOX \tag{1}$$

$$H_2O_2 + HOX \longrightarrow O_2(^1\Delta_g) + H_2O + H^+ + X^- \tag{2}$$

where X^- represents a chloride or bromide ion.[1-7] Both heme-containing peroxidases and nonheme vanadium bromoperoxidases can produce singlet oxygen.[1-7] Peroxidases can also oxidize iodide anion. However, with hypoiodous acid, reaction (2) is endothermic and singlet oxygen is not produced.[8] In order to distinguish singlet oxygen from a variety of other reactive products generated by peroxidases, including radicals and various oxidized halide species, a technique with extreme selectivity is required. The measurement of singlet oxygen phosphorescence at 1270 nm is one such technique. Even in complex biochemical systems, chemiluminescence near 1270 nm from other excited species is extremely uncommon.

Singlet oxygen ($^1\Delta_g$) is an electronically excited oxygen molecule with 23 kcal per mole more energy than ground-state oxygen. This energy can be released as a 1270-nm photon,

$$O_2(^1\Delta_g) + H_2O \rightarrow O_2(^3\Sigma_g^-) + H_2O + h\nu(1270\,\text{nm}) \tag{3}$$

but the transition, which requires an electron spin change, is not allowed by the rules of quantum mechanics. Thus, the radiative rate is low. Most deactivating collisions of singlet oxygen with water molecules generate heat via a nonradiative transition.

[1] J. R. Kanofsky, *J. Biol. Chem.* **258,** 5991 (1983).
[2] A. U. Khan, P. Gebauer, and L. P. Hager, *Proc. Natl. Acad. Sci. U.S.A.* **80,** 5195 (1983).
[3] J. R. Kanofsky, *J. Biol. Chem.* **259,** 5596 (1984).
[4] J. R. Kanofsky, J. Wright, G. E. Miles-Richardson, and A. I. Tauber, *J. Clin. Invest.* **74,** 1489 (1984).
[5] A. U. Khan, *Biochem. Biophys. Res. Commun.* **122,** 668 (1984).
[6] J. R. Kanofsky, H. Hoogland, R. Wever, and S. J. Weiss, *J. Biol. Chem.* **263,** 9692 (1988).
[7] R. R. Everett, J. R. Kanofsky, and A. Butler, *J. Biol. Chem.* **265,** 4908 (1990).
[8] W. H. Koppenol and J. Butler, *Adv. Free Radic. Biol. Med.* **1,** 91 (1985).

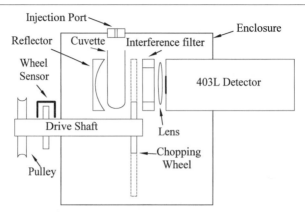

FIG. 1. Near-infrared chemiluminescence spectrometer.

$$O_2(^1\Delta_g) + H_2O \rightarrow O_2(^3\Sigma_g^-) + H_2O + heat \qquad (4)$$

Reaction (4) is efficient, and the lifetime of singlet oxygen in water is only 3.1 μsec.[9] The modest singlet oxygen production rates found in peroxidase systems and the short singlet oxygen lifetime result in low steady-state singlet oxygen concentrations. The combination of a low steady-state concentration and of a low radiative rate causes the 1270-nm emission from peroxidase systems to be very weak. Thus, measuring the weak 1270-nm emission from singlet oxygen generated by peroxidases would appear to represent a significant technical challenge, but modern solid-state near-infrared detectors have made the measurement of singlet oxygen production by various biochemical systems straightforward.

Method

Near-Infrared Chemiluminescence Spectrometer

A spectrometer designed for detecting the singlet oxygen emission at 1270 nm from biological systems is shown in Fig. 1. The reflector behind the sample cuvette and the lens in front of the infrared detector increase the efficiency of the light collection. Wavelength selection is made with a series of interference filters having 50-nm bandwidths. Interference filters, with a relatively large light-acceptance angle, contribute to the high light collection. Five interference filters are mounted in a plate that can be moved

[9] S. Y. Egorov, V. F. Kamalov, N. I. Koroteev, A. A. Krasnovsky, Jr., B. N. Toleutaev, and S. U. Zinukou, *Chem. Phys. Lett.* **163,** 421 (1989).

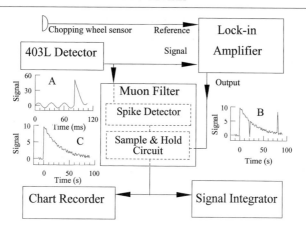

Fig. 2. Block diagram of electronics for near-infrared chemiluminescence spectrometer. (A) Simulated output of 403L detector, (B) simulated output of lock-in amplifier, and (C) simulated output of muon filter.

from the outside of the instrument to select a given filter without opening the spectrometer case. Other spectrometers described in the literature have used diffraction gratings or interferometers for wavelength selection.[2,5,10] In the spectrometer shown in Fig. 1, a chopping wheel modulates the light emission from the cuvette at 37 Hz. Light modulation combined with a lock-in amplifier improves the signal-to-noise ratio of the spectrometer.

The spectrometer contains an injection port with a black rubber septum. Generally, the peroxidase reaction is initiated by the rapid injection of hydrogen peroxide into a cuvette already containing a peroxidase and a halide ion.

By far the most critical component of the spectrometer is the near-infrared detector. The spectrometer shown in Fig. 1 uses a 403L detector (Applied Detector Corp., Fresno, CA). This hyperpure germanium photodiode has an integrated Dewar flask and transimpedance amplifier. The photodiode, input stage of the detector's transimpedance amplifier, and the amplifier's feedback resistor are held at 77°K in order to reduce thermal noise. The transimpedance amplifier of the 403L detector can be set for one of two time constants. For the detection of near-infrared chemiluminescence, the detector works best with the shorter time constant selected. The noise equivalent power of the detector is 1×10^{-15} WHz$^{-1/2}$. The detector area is 25 mm^2 and the detector response time is 1 to 2 msec.

Figure 2 shows a block diagram of the spectrometer electronics. The

[10] J. M. Wessels and M. A. J. Rodgers, *J. Phys. Chem.* **99**, 15725 (1995).

output of the 403L detector is connected to a lock-in amplifier. The lock-in amplifier selectively amplifies the electrical output of the detector that has the same frequency and phase as the reference signal of the chopping wheel.

The output of the 403L detector is also connected to a muon filter (Applied Detector Corp). This electronic device reduces artifacts caused by ionizing radiation. Biased photodiodes, such as the 403L, respond to various forms of ionizing radiation with the production of large electrical transients that decay with the time constant of the detector. Insert A in Fig. 2 shows a simulated output from the 403L detector. Note that the output is modulated at the frequency of the chopping wheel. A radiation-induced transient is also shown. Insert B in Fig. 2 shows a simulated output from the lock-in amplifier. The output shows chemiluminescence from a reaction initiated at zero time. Two radiation-induced transients are also shown. Transients in the lock-in amplifier output may be either positive or negative depending on the position of the chopping wheel at the time the ionizing radiation strikes the detector.

The muon filter consists of two functional parts, a spike detector and a sample-and-hold circuit. The spike detector of the muon filter can be adjusted to detect all electrical transients above a selected voltage. The output from the lock-in amplifier is connected to the sample-and-hold circuit of the muon filter. During the time period of these electrical transients, the electrical output of the sample-and-hold circuit is held at the value immediately preceding the transient. The net result is a nearly complete elimination of radiation-induced transients in the electrical signal sent to the recorder and electronic integrator. Compare inserts B and C in Fig. 2. It is important to recognize, however, that during the time periods when ionizing radiation strikes the detector, the spectrometer is not actually measuring the near-infrared chemiluminescence.

The muon filter is most useful for measuring very low levels of chemiluminescence and decreases the detection limit for singlet oxygen by about a factor of 10. For more intense chemiluminescence, it is sometimes better not to use the muon filter. The muon filter cannot distinguish relatively intense chemiluminescence from radiation-induced transients. When attempting to measure relatively intense emission, the muon filter output may lock at a constant value and no signal measurements will occur until the chemiluminescence intensity falls below the threshold of the spike detector.

The spectrometer shown in Fig. 1 can detect singlet oxygen production rates on the order of 4 nmol min^{-1} in a 2-ml reaction volume when using water as a solvent. This corresponds to a steady-state concentration for singlet oxygen on the order of 1×10^{-13} M. The use of deuterium oxide as a solvent instead of water increases the intensity of singlet oxygen emission. In deuterium oxide, the singlet oxygen detection limit is on the order of

0.1 nmol min^{-1}. This corresponds to a steady-state concentration of 3 × 10^{-15} M.

Calibration of Spectrometer

Quantitative measurements of singlet oxygen generation require calibration of the spectrometer with a singlet oxygen generating standard. The reaction of hydrogen peroxide with hypochlorous acid is known to generate stoichiometric amounts of singlet oxygen via reaction (2).[11] Hypochlorous acid can be distilled under reduced pressure from a 5.25% commercial bleach solution (e.g., Clorox, Clorox Company, Oakland, CA) that has been acidified to pH 8.0 with phosphoric acid. The distillate, collected at 0°, is roughly 300 mM. Then, the pH of the distillate is adjusted to 11 with sodium hydroxide. The sodium hypochlorite solution is stable in polyethylene bottles for a few weeks at 4°. The sodium hypochlorite solution should be assayed on the day of use, using an extinction coefficient of 391 M^{-1} at 292 nm.[12] Hydrogen peroxide should also be assayed prior to its use as a calibrating standard. This can be done iodometrically.[13]

The intensity of singlet oxygen emission from a biological system is given by Eq. (5):

$$I = Ark_r\tau \tag{5}$$

where I is the intensity of the singlet oxygen emission, A is a calibration constant that depends on the geometry of the spectrometer cuvette and the light-collection optics, r is the rate of singlet oxygen generation per unit volume, k_r is the radiative rate for singlet oxygen, and τ is the lifetime of singlet oxygen in the system being studied.[14] Compared to many organic solvents, the product of k_r and τ for water is particularly small, and consequently the singlet oxygen emission is of low intensity.[15]

Singlet oxygen yields may need to be corrected for the quenching of singlet oxygen by the peroxidase or other components in the system being studied when the lifetime of singlet oxygen in the peroxidase system differs significantly from the lifetime of singlet oxygen in the calibration standard.[16] The corrected singlet oxygen yield, Y, for a peroxidase reaction started at zero time is given by Eq. (6):

[11] A. M. Held, D. J. Halko, and J. K. Hurst, *J. Am. Chem. Soc.* **100,** 5732 (1978).
[12] T. Chen, *Anal. Chem.* **39,** 804 (1967).
[13] M. L. Cotton and H. B. Dunford, *Can. J. Chem.* **51,** 582 (1973).
[14] A. A. Gorman and M. A. J. Rodgers, *J. Photochem. Photobiol. B Biol.* **14,** 159 (1992).
[15] A. A. Gorman, I. Hamblett, C. Lambert, A. L. Prescott, M. A. J. Rodgers, and H. M. Spence, *J. Am. Chem. Soc.* **109,** 3091 (1987).
[16] J. R. Kanofsky, *J. Biol. Chem.* **263,** 14171 (1988).

$$Y = B \left(\frac{k_c \tau_c}{k_p \tau_p} \right) \int_0^{t_{max}} I dt \qquad (6)$$

where B is the calibration constant obtained using the hydrogen peroxide–hypochlorous acid reaction, I is the intensity of the 1270-nm emission, t is time, t_{max} is the time when the light integration is stopped, k_c is the radiative rate for singlet oxygen in the calibrating system, τ_c is the lifetime of singlet oxygen in the calibrating system, k_p is the radiative rate for singlet oxygen in the peroxidase system studied, and τ_p is the lifetime of singlet oxygen in the peroxidase system. It is generally assumed that k_c is equal to k_p.

The lifetime of singlet oxygen in the calibrating system and in the peroxidase system can be measured directly using time-resolved methods.[16] The lifetimes can also be calculated from values of single oxygen-quenching constants in the literature using formula (7),

$$\frac{1}{\tau_p} = \frac{1}{\tau_s} + \sum_{i=1}^{n} k_i c_i \qquad (7)$$

where τ_s is the lifetime of singlet oxygen in the solvent, k_i is the singlet oxygen-quenching constant for system component i, and c_i is the concentration of component i.

Results and Discussion

Singlet Oxygen Yields and Spectral Analysis

Under appropriate conditions, chloroperoxidase, eosinophil peroxidase, horseradish peroxidase, lactoperoxidase, myeloperoxidase, and vanadium bromoperoxidases have been shown to produce 1270-nm chemiluminescence characteristic of singlet oxygen.[1–7,17] Figure 3 shows the time course of 1270-nm emission from a system composed of eosinophil peroxidase, hydrogen peroxide, and bromide ion. Under the conditions shown in Fig. 3, the yield of singlet oxygen was 20% of that predicted by Eqs. (1) and (2); however, under other conditions and with other peroxidases, near-stoichiometric yields of singlet oxygen have been reported.[3,4,7]

Figure 4 shows low-resolution spectra of near-infrared chemiluminescence from the reaction of hydrogen peroxide with hypochlorous acid and from a system containing eosinophil peroxidase, hydrogen peroxide, and bromide ion. Note the expected peak emission near 1270 nm in each case. Also note that the emission peaks are asymmetric with a relatively large

[17] J. R. Kanofsky, in "Peroxidases in Chemistry and Biology" (J. Everse, K. E. Everse, and M. B. Grisham, eds.), Vol. II, Chap. 9. CRC Press, Boca Raton, FL, 1991.

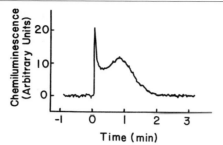

Fig. 3. Time course of 1268-nm chemiluminescence from the eosinophil peroxidase system. The peroxidase system contained 60 n*M* eosinophil peroxidase, 100 μ*M* hydrogen peroxide, 100 μ*M* sodium bromide, 100 m*M* sodium chloride, and 50 m*M* sodium phosphate, pH 7.0, in deuterium oxide solvent. Reprinted with permission from J. R. Kanofsky *et al., J. Biol. Chem.* **276,** 9692 (1988).

trailing edge on the long wavelength side of the peak. This is an artifact due to the transmission of off-axis light by interference filters. Collimation of the light from the cuvette will improve the symmetry of the peaks, but at a cost of decreased light collection and consequently lower spectrometer sensitivity.

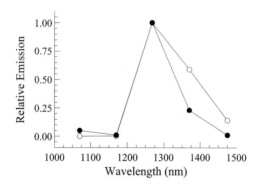

Fig. 4. Spectral analysis of near-infrared chemiluminescence from the reaction of hydrogen peroxide with hypochlorous acid and from a peroxidase system. ○, reaction of hydrogen peroxide with hypochlorous acid, 500 μ*M* hydrogen peroxide, 500 μ*M* hypochlorous acid, pH 5.0, 100 m*M* sodium acetate, 100 m*M* sodium chloride, deuterium oxide solvent. ●, peroxidase system, 30 n*M* eosinophil peroxidase, 100 μ*M* hydrogen peroxide, 100 m*M* sodium bromide, pH 7.0, 50 m*M* sodium phosphate, deuterium oxide solvent. Each point was obtained using an interference filter with a 50-nm bandwidth. The size of the symbol approximates the standard error for each data point.

*Experimental Manipulations Useful for Confirming the Assignment of
1270-nm Emission to Singlet Oxygen*

The assignment of 1270-nm chemiluminescence to singlet oxygen can
be strengthened by experimental manipulations that alter the lifetime of
singlet oxygen in the peroxidase system being studied. Changes in singlet
oxygen lifetime will alter the intensity of the singlet oxygen emission as
shown by Eq. (5).

One experimental manipulation is to change the solvent from water to
deuterium oxide. The lifetime of singlet oxygen is 3.1 μsec in water, but
68 μsec in deuterium oxide.[9,18,19] Considering only the singlet oxygen lifetime, the emission intensity should increase by a factor of 21. However,
the use of deuterium oxide as a solvent may affect the intensity of singlet
oxygen emission by three other mechanisms.

First, the transmission of light at 1270 nm through deuterium oxide is
greater than through water.[20] Typically, this difference causes an additional
factor of 2 increase in the intensity of the 1270-nm emission, but the actual
increase depends on the cuvette geometry and on the light collection optics.[21] Second, the radiative rate for singlet oxygen in deuterium oxide may
be larger than the radiative rate in water.[22] An increase in the radiative
rate will also increase in the intensity of the 1270-nm emission. However,
work by Losev *et al.*[23] suggests that the radiative rate in the two solvents
is the same. Studies of 1270-nm chemiluminescence from the reaction of
hydrogen peroxide with hypochlorous acid, as well as studies using various
peroxidases, are most consistent with the radiative rates being the same in
water and deuterium oxide.[1,21] For example, changing the solvent from
water to deuterium oxide increases the 1270-nm emission from a lactoperoxidase system by a factor of 42 ± 5, a value roughly equal to the product
of the lifetime ratio and the increased optical transmission at 1270 nm.[1]

There is a third additional mechanism by which the use of deuterium
oxide as a solvent may affect the 1270-nm emission. Replacement of water
solvent with deuterium oxide can affect the kinetics of peroxidase reactions.[1] In comparing the 1270-nm emission from a peroxidase system using
water solvent with the emission from a system using deuterium oxide solvent, it is important to make sure that the integration of light emission

[18] P. R. Ogilby and C. S. Foote, *J. Am. Chem. Soc.* **104**, 2069 (1982).
[19] J. G. Parker and W. D. Stanbro, *J. Photochem.* **25**, 545 (1984).
[20] J. G. Bayly, V. B. Kartha, and W. H. Stevens, *Infrared Phys.* **3**, 211 (1963).
[21] J. R. Kanofsky and B. Axelrod, *J. Biol. Chem.* **261**, 1099 (1986).
[22] R. Schmidt and E. Afshari, *J. Phys. Chem.* **94**, 4377 (1990).
[23] A. P. Losev, I. M. Byteva, and G. P. Gurinovich, *Chem. Phys. Lett.* **143**, 127 (1988).

continues until the reaction is complete both in deuterium oxide and in water.[1]

A second type of experimental manipulation that alters the singlet oxygen lifetime is the addition of a singlet oxygen quencher. As shown by Eq. (5), there should be quantitative agreement between the fractional reduction in emission intensity and the fractional reduction in singlet oxygen lifetime of the peroxidase system being studied. However, caution is needed in interpreting the effect of singlet oxygen quenchers. Some singlet oxygen quenchers are oxidized readily and thus may be substrates for peroxidase compounds I or II or may react with the oxidized halogen species generated by the peroxidase. Other singlet oxygen quenchers, such as azide ion, may inhibit peroxidases.

In summary, the combination of a spectral analysis showing an emission maximum near 1270 nm and quantitative agreement between theory and experiment for the emission intensity changes caused by deuterium oxide solvent and by the addition of singlet oxygen quenchers provides an exceptionally specific method for documenting singlet oxygen generation by peroxidases.

Acknowledgments

This material is based on work supported by the office of Research and Development, Medical Research Service, Department of Veterans Affairs. The opinions expressed are those of the author and not necessarily the opinions of the Department of Veterans Affairs or of the United States government.

[7] Formation of Electronically Excited States during the Oxidation of Arachidonic Acid by Prostaglandin Endoperoxide Synthase

By ENRIQUE CADENAS and HELMUT SIES

Prostaglandin endoperoxide H synthase (PGHS) is a bifunctional enzyme that catalyzes the oxidation of arachidonic acid to prostaglandin H_2 (PGH_2), the committed step in prostanoid biosynthesis. The enzyme contains in a single protein,[1,2] a cyclooxygenase activity, which catalyzes the

[1] G. J. Roth, C. J. Siok, and J. Ozols, J. Biol. Chem. **255**, 1301 (1980).
[2] F. J. Van der Ouderaa, M. Buytenhek, D. H. Nugteren, and D. A. van Dorp, Eur. J. Biochem. **109**, 1 (1980).

incorporation of O_2 into arachidonic acid to form prostaglandin G_2 (PGG_2) [reaction (1)], and a peroxidase activity that catalyzes the reduction of PGG_2 to PGH_2 [reaction (2)].

The cyclooxygenase and peroxidase sites are physically and functionally separate.[3,4] The mechanistic interplay between cyclooxygenase and peroxidase activities is explained by a branched chain model.[5-7]

A feature of prostaglandin endoperoxide synthase activity is suicide inactivation,[8] probably due to the formation of a potent oxidant[9] produced during the peroxidase-catalyzed conversion of PGG_2 to PGH_2. Early reports attempted to identify this oxidant as an oxygen-centered radical[10,11] or as singlet molecular oxygen.[9,12] Electron paramagnetic resonance (EPR) studies[11] suggested the formation of hydroxyl radical, although this view was difficult to sustain due to the lack of a discernible hyperfine structure of the EPR spectrum. Likewise, studies with arachidonic acid-supplemented sheep vesicular gland microsomes (containing a high PGHS activity) failed to provide unequivocal evidence for the generation of singlet oxygen ($^1O_2$) in terms of photoemission[9,12] or of formation of the 5α-hydroperoxide adduct of cholesterol, a specific probe for singlet oxygen.[12] Enzyme inactivation was later considered to occur via destructive side reactions of enzyme intermediates formed in the cyclooxygenase catalytic cycle.[13] More recently,

[3] D. Picot, P. J. Loll, and R. M. Garavito, *Nature* **367,** 243 (1994).

[4] W. L. Smith, R. M. Garavito, and D. L. DeWitt, *J. Biol. Chem.* **271,** 33157 (1996).

[5] R. Dietz, W. Nastainczyk, and H. H. Ruf, *Eur. J. Biochem.* **171,** 321 (1988).

[6] A. Guo, L. H. Wang, K. H. Ruan, and R. J. Kulmacz, *J. Biol. Chem.* **271,** 19134 (1996).

[7] J. A. Mancini, D. Riendeau, J. P. Falgueyret, P. J. Vickers, and G. O'Neill, *J. Biol. Chem.* **270,** 29372 (1995).

[8] M. E. Helmer and W. E. M. Lands, *J. Biol. Chem.* **255,** 6253 (1980).

[9] L. J. Marnett, P. Wlodawer, and B. Samuelsson, *Biochem. Biophys. Res. Commun.* **60,** 1286 (1974).

[10] L. J. Marnett, M. J. Bienkowski, M. Leithauser, N. R. Pagels, A. Panthananickal, and G. A. Reed, *in* "Prostaglandins and Related Lipids" (T. J. Powles, R. S. Bockman, K. V. Honn, and P. Pamwell, eds.), Vol. 2, p. 97. A. R. Liss, New York, 1982.

[11] R. W. Egan, J. Paxton, and F. A. Kuehl, *J. Biol. Chem.* **251,** 7329 (1976).

[12] L. J. Marnett, P. Wlodawer, and B. Samuelsson, *J. Biol. Chem.* **250,** 8510 (1975).

[13] R. J. Kulmacz, *Prostaglandins* **34,** 225 (1987).

attention was focused on the formation of a tyrosyl radical during enzyme catalysis, although no clear relationship has been established between, on the one hand, the production of this radical and, on the other hand, its role in either enzyme inactivation or enzyme catalysis.[4,14]

This article addresses the formation of electronically excited states whose relaxation yields low-level chemiluminescence during the oxidation of arachidonic acid by PGHS. This is preceded by a brief description of the photon-counting apparatus used to measure low-level or ultraweak chemiluminescence and is followed by a discussion of the possible molecular mechanism accounting for the generation of electronically excited states by the enzymatic reaction described previously.

Single Photon-Counting Apparatus

The photon-counting apparatus used for these studies has been described previously in this series.[15,16] It consists of a red-sensitive photomultiplier cooled to −40° by a thermoelectric cooler (FACT 50 MKIII from EMI Gencom, Plainview, NY) in order to reduce the dark current. Suitable photomultipliers with a S-20 response (usually 300–800 nm) that have been utilized in similar studies are EMI 9658AM[15,16] and 9814,[17] RCA4832 and 8852 (RCA, Lancaster, PA),[18] and Hamamatsu HTV R374, R375, R550, and R878 (Hamamatsu Photonics Co., Hamamatsu City, Japan).[19] The applied potential to the photomultiplier varies between −0.9 and −1.2 kV (dark current 20–80 counts sec^{-1}). The phototube output is connected to an amplifier discriminator (EG&G Princeton Applied Research, Princeton, NJ) adjusted for single photon counting and connected to both a frequency counter[15,16] and a recorder. Light gathering and thermal isolation from the sample are established by using a Lucite rod[15,16] as an optical coupler or an ellipsoidal light reflector.[19] Application of a multichannel plate and a position-sensitive detector to studies of ultraweak light emission of biological materials permits the tissue-specific localization of photon emission as well as obtaining well-resolved spectra of some ultraweak emission processes.[20] Two-dimensional photon counting, a variation of the photon-

[14] D. C. Goodwin, M. R. Gunther, L. C. Hsi, B. C. Crews, T. E. Eling, R. P. Mason, and L. J. Marnett, *J. Biol. Chem.* **273,** 8903 (1998).

[15] E. Cadenas and H. Sies, *Methods Enzymol.* **105,** 221 (1983).

[16] M. E. Murphy and H. Sies, *Methods Enzymol.* **186,** 595 (1990).

[17] R. Barsacchi, P. Camici, U. Bottigli, P. A. Salvatori, G. Pelosi, M. Maiorino, and F. Ursini, *Biochim. Biophys. Acta* **762,** 241 (1983).

[18] C. F. Deneke and N. I. Krinsky, *Photochem. Photobiol.* **25,** 299 (1977).

[19] H. Inaba, Y. Shimizu, Y. Tsuji, and A. Zamagishi, *Photochem. Photobiol.* **30,** 169 (1979).

[20] R. Q. Scott and H. Inaba, *J. Biolum. Chemilum.* **4,** 507 (1989).

counting device described previously,[15,16] may amplify the potential of single photon-counting imaging considerably in chemiluminescence research inasmuch as it provides a means to obtain information about the spatial distribution of ultraweak emission intensity in tissues.[20]

Low-Level Chemiluminescence of Biological Systems

Photoemission from biological systems indicates that the formation of electronically excited states is a functional manifestation of oxidative conditions. In order to understand the molecular mechanisms underlying the formation of excited species, three approaches—whose applicability depends on the experimental models—are desirable: (1) to establish a correlation between electronically excited state formation and accumulation of stable molecular products,[21] (2) to assess the feasibility of energy transfer to different acceptors (sensitized emission) in order to gather knowledge on the nature and mechanisms of formation of excited carbonyls,[22,23] and (3) to identify the wavelength(s) of maximal chemiluminescence emission, which can be attributed to a particular excited state.

For spectral analysis of emitted light, optical filters are placed into the light path and, according to the type of filter used, a broad or rather resolved spectral distribution of photoemission may be obtained. Thus, interference filters, having a narrow bandwidth (8–22 nm), are expected to allow for a more resolved spectral analysis of chemiluminescence. However, because of their low transmission value (usually 45–55%), their use would be restricted to those systems with high chemiluminescence yield. Alternatively, cutoff filters, consisting of colored glass (Toshiba Electric Co., Minato ku, Japan, or Jenaer Glaswerke Schott, Mainz, FRG) or gelatin filters (Wratten gelatin filters, Eastman-Kodak Co., Rochester, NY), provide a 1–99% transmission range. The use of cutoff filters does not provide a satisfactory spectral resolution, but it offers a rather valuable clue as to the maximal wavelength of photoemission. The difference between spectral characteristics of two successive cutoff filters [i.e., $\Delta F1,2(\lambda) = F1(\lambda) - F2(\lambda)$] is corrected for the transmittance characteristics of the filters and for the photomultiplier tube quantum efficiency at that particular wavelength.[24]

[21] G. Cilento, N. Durán, K. Zinner, C. C. C. Vidigal, O. M. M. Faria Oliveira, M. Haun, A. Faljoni, O. Augusto, R. Casadei de Baptista, and E. H. J. Bechara, *Photochem. Photobiol.* **28**, 445 (1978).

[22] G. Cilento, *in* "Chemical and Biochemical Generation of Excited States" (W. Adam and G. Cilento, eds.), p. 278. Academic Press, New York, 1982.

[23] G. Cilento, *Pure Appl. Chem.* **56**, 1179 (1984).

[24] B. R. Anderson, T. F. Lint, and A. M. Brendzel, *Biochim. Biophys. Acta* **542**, 527 (1978).

Although low-level chemiluminescence, if aided by spectral analysis, can point to the involvement of a particular excited species, the task becomes problematical with increasing complexity of the biological system. Photoemission from living systems exhibits an intricate spectral distribution, thus revealing that light emission might be more complex than that observed in the relaxation of a sole excited state to the ground state. Therefore, the unambiguous identification of the main excited species, whose relaxation yields chemiluminescence, is likely to require the assistance of the technical approaches described earlier, along with those pertaining to the particular chemistry with which the excited state is endowed, such as its chemical reactivity and specific quenching.

The formation of $^1O_2$ in biosystems has been attributed, among others, to phagocytosis, photosensitization reactions, recombination of peroxyl radicals (e.g., arising from lipid peroxidation), and peroxidase-catalyzed reactions. The detection of $^1O_2$ in biological systems[25] has been aided by the use of specific $^1O_2$ scavengers and quenchers, solvent (D_2O) effects, the detection of specific products of cholesterol oxidation (5α-hydroperoxide adduct, distinct from those formed on reaction with hydroxyl radical) or chemical trapping by the water-soluble 9,10-bis(ethylene) anthracene disulfate (which yields specifically an endoperoxide as oxidation product),[26] and the observation of low-level chemiluminescence. The latter approach follows the decay of $^1O_2$ via its luminescence and is expressed as $^1O_2$ monomol emission (due to $^1\Delta O_2$ itself; 1268 nm) and dimol emission (resulting from the dimer $^1\Delta O_2 \cdots {}^1\Delta O_2$; 634 nm) [reactions (3) and (4)].

$$^1O_2 \to {}^3O_2 + h\nu_{1268\,nm} \tag{3}$$
$$^1O_2 + {}^1O_2 \to 2\,{}^3O_2 + h\nu_{634\,nm} \tag{4}$$

Although the intensity in the visible region is higher than that in the infrared region, dimol emission requires collision involving orbital overlap, a process that may be restricted by the lifetime of $^1O_2$ (~1 μsec). The observation of $^1O_2$ monomol emission in biosystems is possible because of the development of sensitive infrared spectroscopic techniques.[26-28] The 1268-nm emission, along with isotope effects, lends strong support to the identification of $^1O_2$ in biosystems because this is the only excited state capable of emitting in this region.

[25] C. S. Foote, F. C. Shook, and R. B. Akaberli, *Methods Enzymol.* **105**, 36 (1984).
[26] P. Di Mascio and H. Sies, *J. Am. Chem. Soc.* **111**, 2909 (1989).
[27] J. R. Kanofsky, *J. Photochem.* **25**, 105 (1984).
[28] A. U. Khan, *J. Photochem.* **25**, 327 (1984).

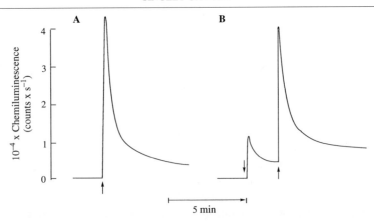

FIG. 1. Low-level chemilumiescence associated with arachidonic acid oxidation by purified PGHS or ram vasicular gland microsomes. Assay conditions: (A) isolated PGHS (58 μg/ml) or (B) ram vesicular gland microsomes (0.5 mg protein/ml) (downward arrow) in 0.1 M Tris–HCl buffer, pH 8.1, were supplemented with 54 μM arachidonic acid (indicated by the upward arrow); 1 μM hematin was present in the solution containing the isolated PGHS.

Low-Level Chemiluminescence Associated with Arachidonic Acid Oxidation by Prostaglandin Endoperoxide H Synthase

Figure 1A shows the time course of light emission associated with arachidonic acid oxidation by purified PGHS: maximal light emission intensity is obtained within 15–30 sec after the addition of arachidonic acid and is followed by a rapid decay to the background level. The lack of persistence of maximal chemiluminescence intensity seems to result from the rapid inactivation of the enzyme.[29] Accordingly, the subsequent addition of arachidonic acid after the maximal emission intensity has been achieved is not followed by a new photoemission burst. This time course agrees with the rapid formation of product described, where the maximal activity is reached within 30 sec.[30] Figure 1B shows a similar time course of low-level chemiluminescence arising from the supplementation of ram vesicular gland microsomes with arachidonic acid.

Chemiluminescence intensity is dependent on both the amount of isolated PGHS or ram vesicular gland microsomes and arachidonic acid concentration. Kinetic analysis of photoemission performed on the isolated

[29] R. W. Egan, P. H. Gale, and F. A. Kuehl, Jr., *J. Biol. Chem.* **254,** 3295 (1979).
[30] S. Ohki, N. Ogino, S. Yamamoto, and O. Hayishi, *J. Biol. Chem.* **254,** 829 (1979).

enzyme showed a K_M value for arachidonic acid of about 6 μM, in agreement with the K_M reported for enzymatic activity assessed by O_2 consumption.[31,32]

Several lines of evidence permit ascribing the low-level chemiluminescence of arachidonic acid-supplemented ram seminal gland microsomes to an intrinsic PGHS activity. (a) The time course of photoemission is similar to that observed with purified PGHS measured by oxygen consumption and product formation (both reach a maximum within 15 30 sec).[29,33] (b) The pH dependence of chemiluminescence agrees with the reported pH dependence of the cyclooxygenase reaction.[30] The 703-nm emission (see later) and total photoemission show similar pH dependence patterns. (c) Further support for an involvement of cyclooxygenase activity in the chemiluminescence of ram seminal gland microsomes is provided by the inhibitory effect of indomethacin, a compound that elicited a half-maximal inhibitory effect at 0.8–1.2 μM, a range similar to that required for the inhibition of prostaglandin biosynthesis.[34] More recently, indomethacin was found to block nitration of the tyrosine residues in PGHS.[35]

Molecular O_2 is a requisite condition to observe light emission with arachidonic acid as a substrate for both ram vesicular gland microsomes and purified PGHS. Photoemission intensity decreases by ~90% in anerobic conditions, and reoxygenation of anaerobic mixtures is accompanied by a burst of chemiluminescence. This implies substrate availability for light emission and, possibly, preservation of the enzyme activity in anaerobiosis. The requirement of O_2 for chemiluminescence may be overcome when PGG_2 substitutes for arachidonic acid, thus suggesting that PGG_2 is critical for the formation of an oxoferryl complex and that the formation of the oxidant generated from the reaction does not require atmospheric O_2.

Spectral analysis of photoemission arising from arachidonic acid/isolated PGHS mixtures in the 600- to 750-nm range shows two distinctive peaks at 634 and 703 nm and lower intensities at 640–690 nm. A small peak is also observed at about 654 nm (Fig. 2A). The ratios of intensities at 634/668 and 703/668 nm are 3.29 and 5.17, respectively (703/634 nm is about 1.57). The similar relative intensities of these peaks, along with the lower intensity at 668–670 nm, seem indicative of $^1O_2$ dimol emission.

Spectral analysis of photoemission arising from arachidonic acid-supplemented ram vesicular gland microsomes shows a prominent peak at 703

[31] F. J. van der Ouderaa and M. Buytenhek, *Methods Enzymol.* **86,** 60 (1982).
[32] O. Laneuville, D. K. Breuer, D. L. DeWitt, T. Hla, C. D. Funk, and W. L. Smith, *J. Pharmacol. Exp. Ther.* **281,** 927 (1994).
[33] M. Hamberg and B. Samuelsson, *Proc. Natl. Acad. Sci. U.S.A.* **70,** 899 (1973).
[34] R. J. Flower and J. R. Vane, *Biochem. Pharmacol.* **23,** 1439 (1974).
[35] T. Shimokawa, R. J. Kulmacz, D. L. Dewitt, and W. L. Smith, *J. Biol. Chem.* **265,** 20073 (1990).

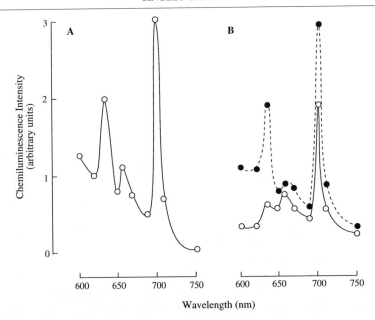

F$_{\text{IG}}$. 2. Spectral analysis of red photoemission. Assay conditions as in Fig. 1. (A) Spectral distribution of chemiluminescence obtained with (A) isolated PGHS and (B) ram vesicular gland microsomes (●) plus 60 mM DABCO (○). Spectral analysis was obtained with interference filters (see text).

nm; the intensity of the 634-nm red emission, however, was as low as that of the 668-nm trough (Fig. 2B). Diazabicyclo[2.2.2]octane (DABCO), a $^1O_2$ dimol emission enhancer in aqueous solutions,[18] altered this spectral distribution, rendering it similar to that obtained with the purified enzyme (Fig. 2A). The typical profile of $^1O_2$ dimol emission can be altered as the system gains in complexity, as illustrated by the effect (shift to shorter wavelengths) of several amino acids and arachidonic acid on the $^1O_2$ dimol emission originating from HOCl/H_2O_2 mixtures.[24] In addition, NaN$_3$ and β-carotene decreased the photoemission intensity from arachidonic acid/ ram seminal gland microsome mixtures, thus strengthening the contribution of $^1O_2$ to this chemiluminescence. These compounds are not expected to modify the spectral profile characteristic of dimol emission but affect the intensities of maximal emission.

Cooxidation of various compounds is related to the peroxidase activity of PGHS; this may be considered a free radical pathway for the activation of xenobiotics[36] that results in enhanced prostanoid biosynthesis (possibly

[36] L. J. Marnett and T. E. Eling, *Rev. Biochem. Toxicol.* **5,** 135 (1983).

by protecting cyclooxygenase from the oxidant generated). This effect is usually associated with an inhibition of chemiluminescence[37] or a decrease of an EPR signal.[29] For example, phenol, p-benzohydroquinone, and GSH inhibit the chemiluminescence originating from the oxidation of arachidonic acid by ram seminal vesicle microsomes[37] and, conversely, these compounds are known to stimulate prostaglandin biosynthesis.[30] These effects do not distinguish a reaction of these compounds with the oxoferryl moiety inherent in peroxidase-catalyzed reactions from a quenching of the oxidant released from the reaction. An interpretation of these effects remains elusive, although it may be argued that phenol oxidation products are not found during the conversion of PGG_2 to PGH_2 and that the peroxidase in PGHS has none[30] or very little[38] GSH peroxidase activity. One-electron oxidation of GSH to its thiyl-free radical occurs during the peroxidase reaction,[38] likely involving the interaction of compound I (see later) with the thiol.

Mechanistic Aspects

It may be concluded that $^1O_2$ dimol emission contributes to the low-level chemiluminescence associated with the oxidation of arachidonic acid by PGHS. Because the product of the cyclooxygenase reaction, PGG_2, gives rise to electronically excited state formation in the presence of purified PGHS under anaerobic conditions, it may be surmised that the peroxidase activity of PGHS is primarily responsible for the formation of the excited species. In accordance with the commonly accepted peroxidase reaction, the first step in the reaction of PGG_2 and the peroxidase is the formation of compound I [reaction (5), where P stands for porphyrin] involving the transfer of two oxidizing equivalents onto the iron porphyrin ring to form the oxoferryl complex and a porphyrin cation radical.

$$[P]\text{-}Fe^{III} + PGG_2 \rightarrow [P^{\cdot+}]\text{-}Fe^{IV} = O + PGH_2 \qquad (5)$$

(Subsequent transfer of the radical character from the porphyrin radical to a tyrosine to form a tyrosyl radical, whose involvement in enzyme catalysis or enzyme inactivation is controversial, is not considered in these equations.)

The oxoferryl moiety in compound I of the peroxidase may react with a second molecule of the hydroperoxy endoperoxide (PGG_2) [reaction (6)] to yield, partly, oxygen in a singlet state.

[37] E. Cadenas, H. Sies, W. Nastainczyk, and V. Ullrich, *Hoppe-Seyler's Z. Physiol. Chem.* **364**, 519 (1983).

[38] T. E. Eling, J. F. Curtis, L. S. Harman, and R. P. Mason, *J. Biol. Chem.* **261**, 5023 (1986).

$$[P]\text{-}Fe^{IV} = O + PGG_2 \rightarrow [P]\text{-}Fe^{III} + PGH_2 + {}^1O_2 \qquad (6)$$

These reactions are analogous to the reaction of cytochrome P450 with organic hydroperoxides, in which an oxoferryl or oxenoid complex of cytochrome P450 is the active oxygenating intermediate.[39,40] Furthermore, the reaction of metmyoglobin with H_2O_2 yields an oxoferryl complex (ferrylmyoglobin),[41] which, in the absence of suitable electron donors, may react with another molecule of H_2O_2 to form electronically excited states.[42] (At variance with PGHS, the second oxidizing equivalent in ferrylmyoglobin is not located on the porphyrin ring, but is delocalized onto an aromatic amino acid.) This reaction promotes formation of the 5α-hydroperoxide adduct of cholesterol, thereby furnishing evidence for ${}^1O_2$ release in the overall hydroperoxide disproportionation.[42]

Reactions (5) and (6) and the metabolism of hydroperoxides by several hemoproteins and halide-dependent enzymic reactions can proceed, at least formally, as a hydroperoxide disproportionation with the release of oxygen in an excited state, ${}^1O_2$. Examples are given by the halide-mediated H_2O_2 decomposition by lacto-, myelo-, and chloroperoxidase,[27,28] the metabolism of hydroperoxides by cytochrome P450 (in the absence of electron donors and hydroxylatable substrates),[39] and the PGHS reaction.[37] As mentioned previously, the H_2O_2-mediated oxidation of metmyoglobin to ferrylmyoglobin or its transient-free radical form resembles these reactions and proceeds with the formation of electronically excited states and the promotion of cholesterol oxidation to the 5α-hydroperoxide adduct, the latter product being a specific probe for singlet oxygen.[42] Also, the interaction of methemoglobin with 15-hydroperoxyarachidonate was shown to be a source of ${}^1O_2$ measured as monomol emission.[43]

In summary, the enzymic conversion of PGG_2 to PGH_2 is associated with photoemission in the visible region with a spectral distribution resembling the pattern of ${}^1O_2$ dimol emission.[37] However, no photoemission was observed in the infrared region (1268 nm) that can be ascribed to ${}^1O_2$ monomol emission.[44] ${}^1O_2$ dimol emission [reaction (4)] may prevail over monomol emission [reaction (3)] if a relatively high level of ${}^1O_2$ builds up in the vicinity of its generation site; in this situation, a bimolecular collision may be favored despite the short lifetime of ${}^1O_2$ in solution. Clearly, the

[39] E. Cadenas, H. Sies, H. Graf, and V. Ullrich, *Eur. J. Biochem.* **130,** 117 (1983).

[40] E. Cadenas and H. Sies, *Eur. J. Biochem.* **124,** 349 (1982).

[41] C. Giulivi and E. Cadenas, *Methods Enzymol.* **233,** 189 (1994).

[42] D. Galaris, D. Mira, A. Sevanian, E. Cadenas, and P. Hochstein, *Arch. Biochem. Biophys.* **262,** 221 (1988).

[43] J. R. Kanofsky, *J. Biol. Chem.* **261,** 13546 (1986).

[44] J. R. Kanofsky, *Photochem. Photobiol.* **47,** 605 (1988).

visible photoemission cannot be ascribed uniquely to $^1O_2$, for a large fraction of it occurs below 600 nm with a spectral pattern difficult to attribute to any particular excited state. To assess the biological significance of $^1O_2$ or any other electronically excited state in a biological setting requires identification of the enzymic reaction leading to its generation, quantification of the excited state yield in relation to stable molecular products, and consideration of the decay pathways of the excited state, including its chemical reactivity.

[8] Singlet Oxygen Detection with Sterically Hindered Amine Derivatives in Plants under Light Stress*

By ÉVA HIDEG, IMRE VASS, TAMÁS KÁLAI, and KÁLMÁN HIDEG

The production of reactive oxygen species (ROS) is often presumed in plants to be a consequence of a variety of both natural and man-made environmental stress conditions.[1-3] However, the techniques developed for trailing oxidative stress in mammalian tissues usually do not apply directly. In plant membranes, especially in chloroplasts, strong reducing or oxidative conditions, as well as extreme pH, may modify ROS probes, canceling their function. Using the example of singlet oxygen, this article shows a series of approaches to overcome such problems.

Singlet oxygen is produced in the photosynthetic membrane of plants exposed to stress by excess photosynthetically active radiation (400–700 nm) (PAR). This was established theoretically, and a type II photodynamic reaction[4] between the reaction center chlorophyll of photosystem (PS) II

---

* Abbreviations used: DanePy, 3-[N-(β-diethylaminoethyl)-N-dansyl]aminomethyl-2,2,5,5-tetramethyl-2,5-dihydro-1H-pyrrole; DanePyo, 3-[N-(β-diethylaminoethyl)-N-dansyl]amino-methyl-2,2,5,5-tetramethyl-2,5-dihydro-1H-pyrrole-1-oxyl; Tepy, 2,2,5,5-tetramethylpyrroli-dine; Tepyo, 2,2,5,5-tetramethylpyrrolidine-1-oxyl; Temp, 2,2,6,6-tetramethylpiperidine; Tempo, 2,2,6,6-tetramethylpiperidine-1-oxyl.

[1] K. Asada, in "Causes of Photooxidative Stress in Plants and Amelioration of Defence Systems" (Ch. H. Foyer and P. M. Mullineaux, eds.), p. 77. CRC Press, Boca Raton, FL, 1994.

[2] G. A. F. Hendry, Proc. Roy. Soc. Edinburgh 102B, 155 (1994).

[3] É. Hideg, in "Handbook of Photosynthesis" (M. Pessarakli, ed.), p. 911. Dekker, New York, 1996.

[4] E. F. Elstner, in "The Biochemistry of Plants" (D. D. Davies, ed.), Vol. II, p. 253. Academic Press, San Diego, 1987.

and molecular oxygen was hypothesized as the source.[5,6] Singlet oxygen was proposed to be involved in membrane structural damage by initiating the selective degradation of one substantial reaction center protein—D1—of PS II.[5,7] Subsequent general protein and membrane lipid damage is likely to be prompted by singlet oxygen and other ROS as well. In this way, the identification of ROS is a central issue in understanding the mechanism of light stress. Both triplet chlorophyll[8,9] and singlet oxygen production[10–13] were confirmed experimentally in photosynthetically active membrane preparations.

Singlet oxygen can be detected in basically two different ways: by monitoring far-red luminescence arising from the monomolar conversion of oxygen from $^1\Delta_g$ to $^3\Sigma_g$ state[4] or by applying EPR spectroscopy. The first method has the advantage of making fast kinetic studies possible.[10] However, its application to photosynthetic systems is seriously limited by the substantial fluorescence background from bulk chlorophyll. As a consequence, this method can be used only in highly purified PS II reaction center or core preparations, which have limited electron transport activity, and thus is not suitable for photoinhibitory studies in more intact systems.

A successful approach to overcome these difficulties is the application of EPR spectroscopy, namely trapping singlet oxygen with *Temp,* a sterically hindered amine, to detect its production in thylakoid membranes on photoinhibition.[12,13] This technique is based on the conversion of *Temp* into the nitroxide radical *Tempo* when reacting with singlet oxygen[14,15] (Scheme 1), which has a widespread application in mammalian systems.[16]

In our experiments on photoinhibition, spinach thylakoids or PS II-enriched subthylakoid membranes are used in a potassium phosphate buffer

[5] I. Vass, S. Styring, T. Hundall, A. Koivuniemi, E.-M. Aro, and B. Andersson, *Proc. Natl. Acad. Sci. U.S.A* **89,** 1408 (1992).

[6] J. R. Durrant, L. B. Giorgi, J. Barber, D. R. Klug, and G. Porter, *Biochim. Biophys. Acta* **1017,** 167 (1990).

[7] E.-M. Aro, I. Virgin, and B. Andersson, *Biochim. Biophys. Acta* **1143,** 113 (1993).

[8] I. Vass and S. Styring, *Biochemistry* **31,** 5957 (1992).

[9] I. Vass and S. Styring, *Biochemistry* **32,** 3334 (1993).

[10] A. N. Macpherson, A. Telfer, J. Barber, and T. G. Truscott, *Biochim. Biophys. Acta* **1143,** 301 (1993).

[11] A. Telfer, S. Dhami, S. M. Bishop, D. Philips, and J. Barber, *Biochemistry* **33,** 14469 (1994).

[12] É. Hideg, C. Spetea, and I. Vass, *Photosynth. Res.* **39,** 191 (1994).

[13] É. Hideg, C. Spetea, and I. Vass, *Biochim. Biophys. Acta* **1186,** 143 (1994).

[14] Y. Lion, M. Delmelle, and A. van de Vorst, *Nature* **263,** 442 (1976).

[15] Y. Lion, E. Gandin, and A. van de Vorst, *Photochem. Photobiol.* **31,** 305 (1980).

[16] C. A. Rice-Evans, A. T. Diplock, and M. C. R. Symons, *in* "Laboratory Techniques in Biochemistry and Molecular Biology" (R. H. Burdon and P. H. van Knippenberg, eds.), Vol. 22. Elsevier, Amsterdam, 1991.

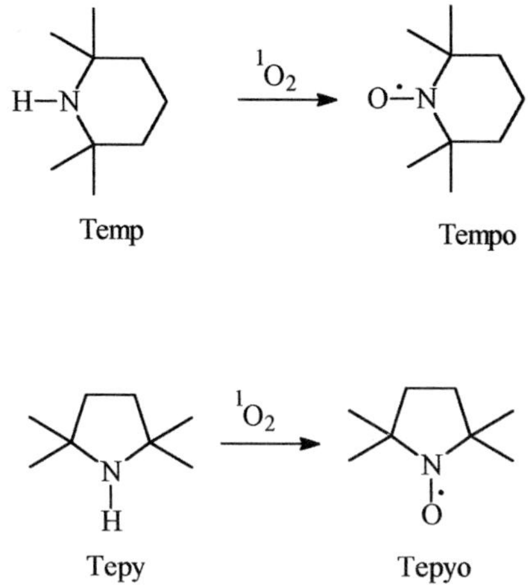

SCHEME 1. Chemical structure of the singlet oxygen traps *Temp* and *Tepy* and their corresponding nitroxide radicals, *Tempo* and *Tepyo*, respectively.

(pH 7.2, 0.2 M sucrose, 5 mM MgCl$_2$, and 10 mM NaCl). HEPES or Mes buffers can also be applied, but these may interact with singlet oxygen and lessen *Tempo* production. Photoinhibitory treatment is usually carried out in samples containing 100–200 μg ml^{-1} chlorophyll and 5–20 mM *Temp*, illuminated by 1000 μM m^{-2} sec^{-1} PAR while stirred in a temperature-controlled glass cuvette at 20–22°. EPR spectra of nitroxide radicals are measured with a Brucker ECS-106 spectrometer (Bruker, Germany) utilizing the ECS-106 data acquisition program. X-band spectra are recorded at room temperature with microwave frequency of 9.43 GHz, microwave power of 16 mW, and modulation frequency of 100 kHz. *Tempo* can be identified from its three-line EPR spectrum with hyperfine line-splitting parameter $a_N = 1.59$ mT (Fig. 1). The amount of trapped singlet oxygen can be estimated from comparing spectra (using the area under the measured EPR absorption spectrum, i.e., the double integral of the detected signal for quantification) with that from known amounts of *Tempo*.

A critical experimental problem of using *Temp* in plant systems is the conversion of the produced *Tempo* into EPR silent hydroxylamine (Scheme 2) by the reducing compounds formed in illuminated thylakoid membrane, which leads to the underestimation of singlet oxygen production. This problem can be resolved by extracting samples after photoinhibition into an organic solvent, e.g., ethyl acetate, in order to reoxidize *N*-hydroxilamines to *Tempo* with catalytic amounts of PbO$_2$ before EPR spectroscopy.[12] This also allows taking EPR spectra of larger volumes, which may be advanta-

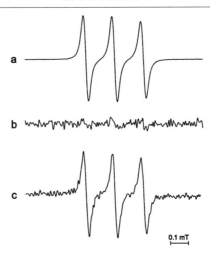

a

b

c

0.1 mT

Magnetic field (mT)

FIG. 1. EPR spectra of Tempo (a) and isolated PS II exposed to 1200 μM m^{-2} sec^{-1} PAR 0 (b) and 22 (c) min in the presence of 12 mM Temp.

geous using less sensitive EPR spectrometers, especially ones with cavities hard to tune when measuring samples in water-based buffers.

Studies with *Temp* revealed that singlet oxygen production in isolated systems appears as a unique characteristic of a pathway known as the acceptor side-induced photoinhibition of PS II, which involves the formation of triplet reaction center chlorophyll.[12,13,17] Singlet oxygen was not detected in samples incapable of photosynthetic oxygen evolution[13,17] nor elsewhere except in PS II.[18] However, the occurrence of singlet oxygen production under photoinhibitory conditions *in vivo* has remained an open question for a long time. Leaves may certainly experience high PAR fluxes, sometimes close to 2000 μM m^{-2} sec^{-1}, but they are also equipped with means of energy dissipation and both enzymatic and nonenzymatic defenses against ROS.[19,20] However, the combination of light and other stress conditions, such as high or low temperature and UV-B irradiation, may lessen

[17] A. Krieger, A. W. Rutherford, I. Vass, and É. Hideg, *Biochemistry* **37,** 16262 (1998).

[18] É. Hideg and I. Vass, *Photochem. Photobiol.* **62,** 949 (1995).

[19] K. Asada, *in* "Causes of Photooxidative Stress in Plants and Amelioration of Defence Systems" (Ch. H. Foyer and P. M. Mullineaux, eds.), p. 77. CRC Press, Boca Raton, FL, 1994.

[20] G. H. Krause, *in* "Causes of Photooxidative Stress in Plants and Amelioration of Defence Systems" (Ch. H. Foyer and P. M. Mullineaux, eds.), p. 43. CRC Press, Boca Raton, FL, 1994.

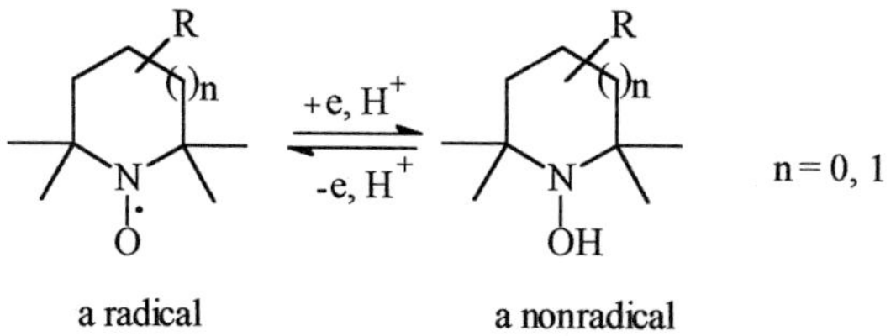

a radical   a nonradical

SCHEME 2. Conversion of nitroxide radicals into nonradical N-hydroxy compounds by one electron reduction.

the capacity of antioxidants and/or may create conditions when energy dissipation fails and PAR becomes excess, thus making ROS-mediated photoinhibitory damage feasible. Also, at early stages of plant development, young leaves with an immature antioxidant system or with an incomplete greening process may be prone to singlet oxygen-mediated oxidative damage.

An obvious attempt to utilize singlet oxygen trapping *in vivo* is to infiltrate the traps into leaves. In these experiments, segments were cut from *Temp*-infiltrated leaves after preset times of photoinhibition, pasted on an open tissue cell holder with the adaxial side up, and placed in the EPR cavity. Spectra were taken under conditions described earlier to find any *Tempo* signal as marker of singlet oxygen production. Unfortunately, rapid conversion of any *Tempo* into *TempOH* when infiltrated into leafs obstructed such attempts. Moreover, we found that even leaves performing photosynthesis under physiological light conditions (80–200 μM m^{-2} sec^{-1} PAR) were able to reduce 50–100 mM infiltrated *Tempo* within a few minutes (unpublished). This makes detecting orders of magnitude smaller concentrations of *Tempo*—yielded in a presumed reaction between infiltrated *Temp* and internally produced singlet oxygen—unlikely.

Using *Tepy*, which has a five-membered ring instead of the six-membered ring of *Temp* (Scheme 1), helps the situation somewhat but it does not solve the problem. *Tepyo*, the singlet oxygen adduct of *Tepy*, appears less prone to oxidation or decomposition by products of plant metabolism: three-line EPR spectra of *Tepyo* can be observed temporarily in segments of a *Tepy*-infiltrated leaf on exposure to photoinhibitory conditions (Fig. 2). This may be regarded as a strong indication for singlet oxygen production in photoinhibited leaves. However, we found that this experiment was hard to reproduce quantitatively, although to a lesser extent than *Tempo*; *Tepyo* was converted into EPR silent forms too fast for reliable analysis. Also, in general, leaf samples are not easy to handle in EPR measurements due to their high water content.

A potentially powerful approach used to overcome the technical diffi-

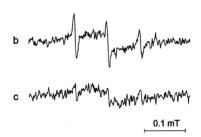

FIG. 2. EPR spectra of a broad bean leaf segment infiltrated with *Tepy* and exposed to photoinhibition by 2000 μM m^{-2} sec^{-1} PAR for (a) 0, (b) 10, and (c) 25 min.

culties of EPR spin trapping in intact leaves is the application of fluorescent probes. Singlet oxygen quenches the fluorescence of many aromatic compounds, which could all, in principle, be applied as probes.[21] The choice of the probe depends on the macromolecule environment in which it is designed to function. It should also be kept in mind that, besides reacting with singlet oxygen, several other factors may decrease fluorescence. In order to avoid such artifacts, a special probe, a double (spin and fluorescence) sensor, was designed.[22]

A double sensor is fluorescent and EPR inactive. On reacting with singlet oxygen, it is converted into a nitroxide radical, which has a lower fluorescence yield than the nonactive nitroxide and at same time is EPR active (Scheme 3). This makes it possible to compare the two reporter functions (i.e., the decrease of fluorescence yield and the induction of EPR signal) in the same experiment to check to what extent fluorescence quenching corresponds to nitroxide activation. Once this is established, double sensors offer the possibility of choosing the function more convenient for the application. Using *DanePy,* a probe containing the fluorescent dye dansyl and pyrroline nitroxide (Scheme 3), it is possible to compare the two sensor functions in response to singlet oxygen generated in PS II preparations by photoinhibitory conditions and chemically.[22] These measurements demonstrated that the two functions were equal and also showed that the active nitroxide radical form of the probe (*DanePyo*) was more

[21] E. Gandin, Y. Lion, and A. van de Vorst, *Photochem. Photobiol.* **37,** 271 (1983).
[22] T. Kálai, É. Hideg, I. Vass, and K. Hideg, *Free Radic. Biol. Med.* **24,** 649 (1998).

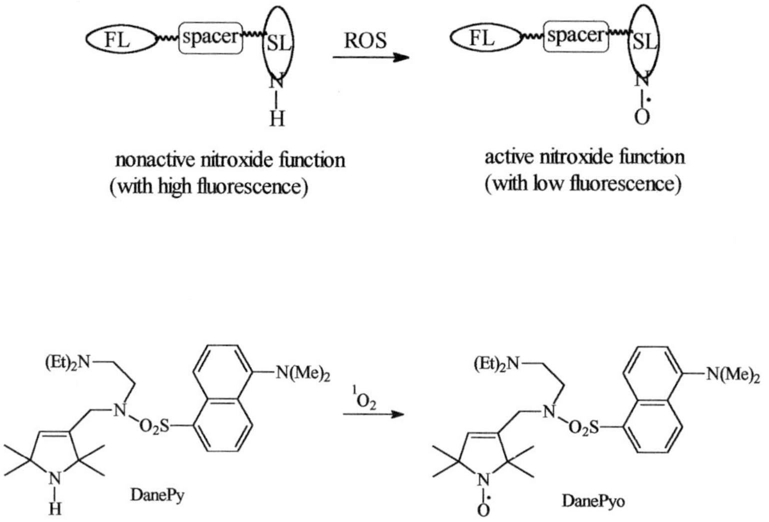

nonactive nitroxide function
(with high fluorescence)

active nitroxide function
(with low fluorescence)

SCHEME 3. Principle of double sensor functioning (top) and chemical structure of the double singlet oxygen sensor *DanePy* and its nitroxide derivative *DanePyo* (bottom).

stable in photosynthetic membranes than either *Tempo* or *Tepyo,* offering the possibility of using this double sensor to detect singlet oxygen in leaves.[23]

In these studies, detached whole leaves of broad bean (*Vicia faba* L.) plants or small segments of tobacco (*Nicotiana tabacum* L.) leaves are infiltrated using a large (50 ml) plastic syringe filled with 10–15 ml 50 mM solution of the trap (ethanol concentration <1%). The leaf is placed inside from the back and the syringe is closed. After removing the air, the plunger is pulled rapidly while the nozzle is kept closed. This infiltration process does not harm the leaf; it reduces the oxygen-evolving ability of PS II by only 10–15% as compared to untreated ones. For photoinhibition, infiltrated leaves are placed on wet tissue paper, and their adaxial sides are exposed to high-intensity (1500–2000 μM m^{-2} sec^{-1}) cool (or heat-filtered) light through an optical fiber guide from a KL-1500 (DMP, Switzerland) lamp. This method assures that leaves are neither dried out nor heated during photoinhibition.

In leaves infiltrated with *DanePy,* fluorescence emission spectra are recorded with Quanta Master QM-1 (Photon Technology Int. Inc.) using 345 nm excitation wavelength, 3 nm excitation, and emission slits. Figure

[23] É. Hideg, T. Kálai, K. Hideg, and I. Vass, *Biochemistry* **37,** 11405 (1998).

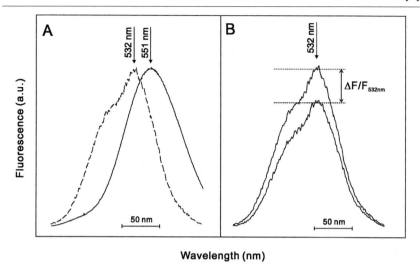

FIG. 3. (A) Fluorescence emission spectra of *DanePy* in water solution (solid line) and after infiltration into a broad bean leaf (dashed line). (B) Fluorescence emission spectra of *DanePy* in a broad bean leaf before (0 min) and after exposure to photoinhibition by 1500 μM m^{-2} sec^{-1} PAR for 60 min.

3A shows that the original fluorescence maximum of *DanePy* was shifted from 551 to 532 nm when infiltrated into a leaf. Also, a shoulder appeared around 490 nm in the spectrum of *DanePy* in the leaf, which was not present in that of the sensor in solution.[23] Because fluorescence emission maxima of *DanePy* in leaf were broad, the signal-noise ratio of fluorescence-quenching data was improved by using the average of six data points around the emission maximum instead of reading a single value. When the infiltrated leaves were photoinhibited, *DanePy* fluorescence was quenched (Fig. 3B). This fluorescence quenching is the first direct experimental evidence that photoinhibition of photosynthesis *in vivo* is accompanied by ^{1}O$_2$ production and is, at least partly, governed by the same process as photoinhibition *in vitro* in isolated membranes.[23,24]

In summary, the oxidation of five- and six-membered sterically hindered amines into stable nitroxide-free radicals provides the possibility of detecting the singlet oxygen produced in PS II on photoinhibition by excess PAR. Five-membered nitroxides are generally more stable than six-

[24] É. Hideg, T. Kálai, I. Vass, and K. Hideg, *in* "Photosynthesis: Mechanisms and Effects" (Gy. Garab, ed.), Vol. III, p. 2139. Kluwer Academic, Dordrecht/Norwell, MA, 1998.

membered ones,[25,26] which is of special importance in studies on complex biological systems when ascorbate or other products convert nitroxides into EPR silent *N*-hydroxy compounds in a one-electron reduction. For *in vivo* plant studies, a five-membered secunder amine was combined with dansyl, the fluorescent probe frequently applied in peptide chemistry. This double (fluorescencent and spin) probe was successfully applied in identifying singlet oxygen production in photoinhibited leaves. For future studies, the spacer connecting the two sensors can be altered in order to suit the solubility, polarity, and thus the sterical specificity of the probe to the biological problem.

Acknowledgment

This work was supported by grants from the Hungarian Scientific Research Fund (OTKA T-17455), the Hungarian Academy of Sciences (AKP 97-13 2,4/23) and ETT (80-4/1998).

[25] R. C. Brasch, M. T. McNamara, R. L. Ehman, W. R. Couet, T. N. Tozer, G. Sosnovsky, N. U. Maheswara Rao, and I. Prakash, *Eur. J. Med. Chem.* **24,** 335 (1989).
[26] M. C. Krishna, W. Degraff, H. O. Hankovszky, P. C. Sár, T. Kálai, J. Jekô, A. Russo, J. B. Mitchell, and K. Hideg, *J. Med. Chem.* **41,** 3477 (1998).

[9] Cholesterol as a Singlet Oxygen Detector in Biological Systems

By Albert W. Girotti and Witold Korytowski

Introduction

Delta-state singlet molecular oxygen ($^1\Delta_g O_2$, or simply $^1O_2$) is a nonradical activated oxygen species that is generated by 22.5 kcal of energy transfer to ground state triplet oxygen, $^3O_2$.[1] Like other activated species, $^1O_2$ can react with vital targets such as lipids, proteins, and nucleic acids, thereby producing cytopathological effects.[2] The most prominent source of $^1O_2$ with biological relevance is photodynamic action,[1,2] although significant amounts may also be produced by stimulated phagocytes, particularly eosinophils.[3]

[1] C. S. Foote, *Science* **162,** 963 (1968).
[2] J. D. Spikes, *in* "The Science of Photobiology" (K. C. Smith, ed), p. 79. Plenum Press, New York, 1989.
[3] J. R. Kanofsky, *in* "Oxygen Free Radicals in Tissue Damage" (M. Tarr and F. Sampson, eds.), p. 77. Birkhauser, Boston, 1993.

Photodynamic generation of $^1O_2$ occurs when an appropriate sensitizing agent is photoexcited in the presence of $^3O_2$.[1-3] $^1O_2$ produced by endogenous sensitizers and 320–400 nm UVA radiation, which passes through the stratospheric ozone layer, is of particular biomedical interest because of its association with skin aging and cancer.[4,5] However, when generated under carefully controlled conditions using exogenous sensitizers and light in the visible range (400–700 nm), $^1O_2$ can be exploited for therapeutic purposes, as, for example, in antineoplastic photodynamic therapy (PDT).[6,7]

The lifetime of $^1O_2$ in pure water is 4–5 μsec, which corresponds to a mean diffusion distance of 90–100 nm.[8] In a cell environment where physical and chemical quenchers abound, the lifetime has been estimated to be <50 nsec.[9] Under these conditions, $^1O_2$ would survive an excursion of <10 nm from its point of origin, which is <0.1% of the radius of an average eukaryotic cell. Clearly, therefore, any $^1O_2$ generated within a cell would have little chance of escaping and being detected in the extracellular compartment. Under these circumstances, it would make little sense to test for $^1O_2$ involvement in some effect, say cell killing, by introducing scavengers such as azide or imidazole.[10] These agents not only lack absolute specificity for $^1O_2$, but most likely could not be used in high enough concentrations to compete with natural acceptors. Likewise, D_2O enhances $^1O_2$ lifetime ~15-fold relative to H_2O,[8] but it would be ill-advised to use D_2O to probe for $^1O_2$ intermediacy in cells because, as already intimated, its lifetime in a cellular milieu would not be determined by aqueous decay.[9] Similar reasoning would apply for physical detection methods, e.g., steady state or time-resolved measurement of 1270-nm luminescence.[8,11] Using this highly sensitive approach with D_2O buffers and lipophilic sensitizers, workers have detected photodynamic $^1O_2$ in isolated membrane suspensions,[12] but not in suspensions of intact cells.[13] A small 1270-nm emission has been detected,[14] but only under stringently specified conditions, i.e., with sensitizer tethered to the extracellular face of the plasma membrane.

[4] R. M. Tyrrell and S. M. Keyse, *J. Photochem. Photobiol. B Biol.* **4,** 349 (1990).

[5] R. M. Tyrrell, *Mol. Aspects Med.* **15,** 1 (1994).

[6] B. W. Henderson and T. J. Dougherty, *Photochem. Photobiol.* **55,** 145 (1992).

[7] F. Sieber, *in* "Bone Marrow Processing and Purging: A Practical Guide" (J. Gee, ed.), p. 263. CRC Press, Boca Raton, FL, 1991.

[8] M. A. J. Rodgers and P. T. Snowden, *J. Am. Chem. Soc.* **104,** 5541 (1982).

[9] J. Moan and K. Berg, *Photochem. Photobiol.* **53,** 549 (1991).

[10] C. S. Foote, F. C. Shook, and R. B. Abakerli, *Methods Enzymol.* **105,** 36 (1984).

[11] R. D. Hall and C. F. Chignell, *Photochem. Photobiol.* **45,** 459 (1987).

[12] J. P. Thomas, R. D. Hall, and A. W. Girotti, *Cancer Lett.* **35,** 295 (1987).

[13] P. A. Firey, T. W. Jones, G. Jori, and M. A. J. Rodgers, *Photochem. Photobiol.* **48,** 357 (1988).

[14] A. Baker and J. R. Kanofsky, *Photochem. Photobiol.* **57,** 720 (1993).

A logical alternative to physical detection with its many difficulties would be a chemical approach involving selection of a naturally occurring target molecule and identification of signature products that arise only when this molecule reacts with $^1O_2$. Several different types of oxidizable target could qualify for this purpose.[15] However, cholesterol (Ch), a monounsaturated lipid found in all eukaryotic cell membranes, is especially well suited for the following reasons: (i) unlike other membrane lipids, it exists as a single molecular species, making product determination much less complicated; (ii) its oxidation products are ready for analysis without the need for potentially artifactual hydrolysis steps; and (iii) unlike phospholipids, it can be transfer radiolabeled spontaneously in cells, i.e., without a transfer protein requirement.[16] In $^1O_2$-mediated reactions, e.g., type II photodynamic reactions, only three primary oxidation products of Ch are possible: 3β-hydroxy-5α-cholest-6-ene-5-hydroperoxide (5α-OOH), 3β-hydroxycholest-4-ene-6α-hydroperoxide(6α-OOH), and 3β-hydroxycholest-4-ene-6β-hydroperoxide(6β-OOH), each of which arises via the ene addition of $^1O_2$.[17] This contrasts with free radical-mediated reactions, including type I photoreactions,[18,19] which give rise to a variety of other primary species, two of the most prominent of which are 3β-hydroxycholest-5-ene-7α-hydroperoxide(7α-OOH) and 3β-hydroxycholest-5-ene-7β-hydroperoxide (7β-OOH). Thus, identification of 5α-OOH, 6α-OOH, or 6β-OOH in a reaction system unambiguously signifies $^1O_2$ intermediacy, whereas 7α-OOH and 7β-OOH signify the involvement of free radicals such as hydroxyl (HO·), lipid oxyl (LO·), or lipid peroxyl (LOO·) radical. One should be aware that 5α-OOH generated in a purely $^1O_2$ reaction might partially rearrange to 7α-OOH,[20] thereby opening the possibility for misinterpretation. However, this occurs mainly during lipid extraction rather than *in situ* and can be minimized with certain precautions (to be described).

This article outlines procedures for using Ch as a mechanistic reporter, using photosensitization as a means of oxygen activation in two different test systems: isolated erythrocyte membranes and L1210 leukemia cells. Two techniques for analyzing $^1O_2$-derived cholesterol hydroperoxides

[15] B. Halliwell and J. M. C. Gutteridge, "Free Radical Biology and Medicine." Clarendon Press, New York, 1989.

[16] G. J. Bachowski, W. Korytowski, and A. W. Girotti, *Lipids* **29**, 449 (1994).

[17] M. J. Kulig and L. L. Smith, *J. Org. Chem.* **38**, 3639 (1973).

[18] L. L. Smith, J. I. Teng, M. J. Kulig, and F. L. Hill, *J. Org. Chem.* **38**, 1763 (1973).

[19] J. E. van Lier, *in* "Photobiological Techniques" (D. P. Valenzeno, R. H. Pottier, P. Mathis, and R. H. Douglas, eds.), p. 85. Plenum Press, New York, 1991.

[20] A. L. J. Beckwith, A. G. Davies, I. G. E. Davison, A. Maccoll, and M. H. Mruzak, *J. Chem. Soc. Perkin Trans.* **II,** 815 (1985).

(ChOOHs) and distinguishing them from free radical-derived counterparts are described: (i) a relatively simple thin-layer chromatographic (TLC) method using colorimetric detection and (ii) a state-of-the-art method involving high-performance liquid chromatography with mercury cathode electrochemical detection [HPLC-EC(Hg)].

Materials and General Methods

Chemicals and Reagents

Cholesterol, 7-ketocholesterol (7-one), 8-hydroxyquinoline, sodium ascorbate, N,N,N',N'-tetramethyl-p-phenylenediamine (TMPD), Chelex-100, and RPMI 1640 medium are obtained from Sigma (St. Louis, MO). Desferrioxamine (DFO) is from Ciba-Geigy (Suffern, NY), and fetal calf serum is from Hyclone (Logan, UT). The photosensitizing dyes merocyanine 540 (MC540) and chloroaluminum phthalocyanine tetrasulfonate (AlPcS$_4$) are supplied, respectively, by Eastman Kodak (Rochester, NY) and Porphyrin Products (Logan, UT). HPLC grade solvents, including methanol, acetonitrile, isopropanol, benzene, heptane, and ethyl acetate, are obtained from Burdick and Jackson (Muskegon, MI). Phosphate-buffered saline (PBS) solutions are treated with Chelex-100 in order to remove trace metal ions that might otherwise catalyze peroxide decomposition.[21] A/B-ring ChOOH standards (5α-OOH, 6α-OOH, 6β-OOH, and 7$\alpha\beta$-OOH) are prepared by dye-sensitized photooxidation of Ch-containing liposomes, the different isomers being separated by means of semipreparative HPLC[22,23] and identified by nuclear magnetic resonance.[22] 3β-Hydroxy-cholest-5-ene-25-hydroperoxide (25-OOH) is provided gratis by Dr. J. van Lier (Sherbrooke). Stock 1.0 mM ferric-8-hydroxyquinoline [Fe(HQ)$_2$] is prepared by mixing 1.0 mM FeCl$_3$ in 4 mM HCl with 2.0 mM 8-hydroxyquinoline in 50% ethanol. Stock solutions of 10 mM sodium ascorbate in PBS are prepared immediately before use.

Membrane Preparation

Plasma membranes of human erythrocytes (white ghosts) are isolated by conventional hypotonic lysis, followed by extensive washing to deplete hemoglobin.[24] The membranes are incubated in the presence of 0.1 mM DFO for 1 hr to remove nonheme Fe^{3+} and then washed with and resus-

[21] G. R. Buettner, *J. Biochem. Biophys. Methods* **16,** 27 (1988).
[22] W. Korytowski, G. J. Bachowski, and A. W. Girotti, *Anal. Biochem.* **197,** 149 (1991).
[23] W. Korytowski, P. E. Geiger, and A. W. Girotti, *Methods Enzymol.* **300,** 23 (1998).
[24] G. Fairbanks, T. L. Steck, and D. F. H. Wallach, *Biochemistry* **10,** 2606 (1971).

pended in PBS. Total protein is determined by the method of Lowry et al.[25] Stored under argon at 4°, the membranes are typically used for experiments within 2 weeks.

Cultured Cells

Murine leukemia L1210 cells (ATTC No. CCL219) are grown in suspension culture at 37° using RPMI 1640 medium supplemented with 1% serum, 10 ng/ml sodium selenite, and various other factors.[26] Cells are reseeded into fresh supplemented medium every 2 days, typically exhibiting a doubling time of 1.2 days. All experiments are carried out on logarithmically growing cells.

Photooxidation Procedures

A typical protocol for photoperoxidizing erythrocyte ghosts is as follows. A stock membrane suspension in PBS (1.0 mg protein/ml corresponding to ~0.9 mg total lipid/ml or ~0.23 mg Ch/ml) is preincubated in the dark (15–30 min) with a sensitizing agent, exemplified here by 10–25 μM MC540 or AlPcS$_4$ taken from a 5 mM stock solution in 50% ethanol or PBS, respectively. Aerobic reactions at constant temperature are conveniently carried out in thermostatted Stirrer Bath vials (YSI Instruments, Yellow Springs, OH), using one or more four-place units.[27] Reaction mixtures (1.5–3.0 ml) are stirred magnetically at a low rate, permitting ready mixing of additives [e.g., Fe(HQ)$_2$ and ascorbate] and facilitating oxygen exchange. The mixtures are irradiated as described earlier, using a fan-cooled 90-W quartz-halogen lamp and a window glass filter to exclude wavelengths below ~300 nm. Light intensity (fluence rate) near suspension surfaces, typically maintained at ~150 mW/cm^2 for AlPcS$_4$-sensitized reactions, is measured with a YSI Model 65A radiometer.[27] Samples are removed periodically during irradiation, treated with EDTA, and extracted for TLC or HPLC-EC(Hg) separations (see Materials and General Methods).

Cell photooxidation is typically carried out as follows, with the MC540-sensitized process being used as a well-studied example.[26,27] Exponentially growing cells are centrifuged (1000g, 5 min), resuspended to a titer of ~1.0 × 10^7/ml in serum-free PBS or RPMI medium, and dark incubated for 30 min with MC540 (10 or 25 μM). Aliquots (15–20 ml) are then

[25] O. H. Lowry, N. J. Rosebrough, A. L. Farr, and R. J. Randall, J. Biol. Chem. 193, 265 (1951).
[26] F. Lin, P. G. Geiger, and A. W. Girotti, Cancer Res. 52, 5282 (1992).
[27] G. J. Bachowski, T. J. Pintar, and A. W. Girotti, Photochem. Photobiol. 53, 481 (1991).

transferred to 75-cm^2 culture flasks and irradiated at room temperature on a translucent plastic platform over a twin bank of 40-W cool white fluorescent tubes. The fluence rate at the platform surface is set at 0.6–0.7 mW/cm^2. Samples (0.5 ml) for HPLC-EC(Hg) are either removed immediately after irradiation or periodically during a period of subsequent dark incubation at 37°. The samples are mixed with EDTA in 1.5-ml microfuge tubes and extracted as described (see Materials and General Methods). Recovered lipid fractions are dried, redissolved in 25 μl of isopropanol, and stored at −20° for HPLC-EC(Hg) analysis. For examining the effects of iron depletion or supplementation on ChOOH status, starting cells are incubated with DFO or Fe(HQ)$_2$ as specified before dye treatment and irradiation.

Lipid Extraction

Extraction of ghost membrane and cellular lipids in preparation for TLC or HPLC-EC(Hg) analysis is carried out as follows.[23] Experimental samples in 1.5-ml polypropylene microfuge tubes are brought to a volume of 0.5 ml with PBS. EDTA (0.1 mM) is included in order to chelate and retain any redox metal ions in the aqueous phase. Samples are then mixed with 0.8 ml of cold chloroform/methanol (2:1, v/v) and extracted by vortexing for ~1 min. After centrifugation, each upper layer is discarded and 0.4 ml of the lower (organic) layer is recovered into a microfuge tube and dried under a stream of nitrogen at room temperature. Simultaneous treatment of several samples can be accomplished with a nine-port Reacti-Vap evaporator from Pierce (Rockford, IL). Because chloroform can facilitate 5α-OOH rearrangement to 7α-OOH,[20] it is important to carry out this step as quickly as possible. Using pure 5α-OOH in a 0.4-ml test extract, we have found that no significant rearrangement occurs when the dry-down period is kept under 15 min. Scaling down the sample and chloroform/methanol volumes can reduce dry-down time further, as the extract volume is proportionately smaller. Membrane ChOOHs are quantitatively extracted using this procedure and are stable for at least 2 weeks when stored at −20° immediately after drying.

Iodometric Assay

This is used for establishing the peroxide content of ChOOH standards and sufficiently oxidized membrane or cell samples.[16,27] Hydroperoxides are reduced anaerobically in the presence of excess potassium iodide, with stoichiometric formation of triiodide, which is measured spectrophotometrically at 353 nm. Quantitation is based on an extinction coefficient of 22.5 mM^{-1} cm^{-1}, with the detection limit being ~1 μM peroxide.

Thin-Layer Chromatographic Analysis of Cholesterol
  Oxidation Products

Photooxidized Erythrocyte Membranes

Thin-layer chromatography is a relatively rapid and inexpensive means of screening for diagnostic ChOOHs.[28,29] This section describes how the photodynamic effects of a lipophilic test dye, MC540, on an easily prepared, well-characterized natural membrane, the erythrocyte ghost, can be examined by TLC. Chromatography is carried out at room temperature using glass-backed silica gel-60 TLC plates (5 × 20 cm or 20 × 20 cm; 0.25 mm thickness) from EM Science (Gibbstown, NJ). Before use, plates are desiccated by warming to 80° (15–30 min). Samples in hexane/isopropanol (1 : 1, v/v), 10–20 μl per lane (<100 μg of total lipid), are spotted onto the plate using a micropipettor. Two mobile phase systems can be used, heptane/ethyl acetate (1 : 1, v/v) or benzene/ethyl acetate (1 : 1, v/v), the latter affording slightly better analyte resolution under otherwise identical conditions.[18] Approximately 50 ml of mobile phase is contained in a conventional all-glass TLC chamber (9 × 23 × 23 cm). The plate is first equilibrated with mobile phase by suspending on a small glass beaker in the closed chamber (~30 min). Irrigation then starts and continues until the solvent front is 2–3 cm from the plate top. The entire chamber is covered during a run to buffer temperature fluctuations and prevent any adventitious photochemistry. After a single irrigation, the plate is dried and sprayed immediately with 1% (w/v) TMPD in methanol/water/acetic acid (50 : 50 : 1, v/v) to visualize peroxides, which oxidize TMPD to a vivid purple product (Wurster dye).[28] Immediately after spraying, plates are clamped under glass to retard background autoxidation and recorded (photographed) as soon as possible. Subsequent to TMPD treatment, plates are sprayed with 50% H_2SO_4 and warmed at 80° in order to visualize parent Ch and ChOOH-derived diols, which appear, respectively, as magenta and blue spots. In addition to direct analysis, most extracted samples are treated with borohydride in order to reduce partially resolvable ChOOHs to well-resolved ChOHs (diols). This typically involves a 5- to 10-min incubation with 1–2 mM borohydride (added as freshly prepared $NaBH_4$ in methanol/10 mM NaOH), followed by solvent evaporation.

A representative TLC profile for MC540/light-treated membranes is shown in Fig. 1. In Fig. 1A, a time-dependent accumulation of TMPD-detectable ChOOH(s) over a 90-min photooxidation period is shown. That

[28] L. L. Smith and F. L. Hill, *J. Chromatogr.* **66,** 101 (1972).
[29] K. Suwa, T. Kimura, and A. P. Schaap, *Biochem. Biophys. Res. Commun.* **75,** 785 (1977).

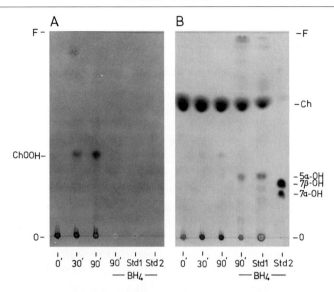

FIG. 1. Silica gel TLC of cholesterol photooxidation products. Ghost membranes (1 mg protein/ml in PBS) were sensitized with 25 μM MC540 and extracted before (0 min) and after being irradiated for 30 and 90 min at 5° (fluence rate ~20 mW/cm²). Recovered lipids were chromatographed using heptane/ethyl acetate (1:1, v/v). Reference standards were as follows: Std 1, ghosts photooxidized with rose bengal; Std 2, reduced 7-one. Where indicated, samples were treated with borohydride (BH₄) before TLC. The plate was sprayed with TMPD (A) to visualize ChOOHs, followed by H_2SO_4 (B) to visualize Ch and diols. Sample load as Ch: 80 μg (lanes 1–4). From G. J. Bachowski, T. J. Pintar, and A. W. Girotti, *Photochem. Photobiol.* **53,** 481 (1991), with permission.

this is ChOOH material has been confirmed by showing identical mobility when [¹⁴C]Ch-labeled ghosts are used.[27] Spots at the origin represent phospholipid-derived hydroperoxides, which are immobile under these TLC conditions. The ChOOH spot disappears after borohydride treatment (Fig. 1A) and is replaced by an H_2SO_4-detectable spot of lower mobility (Fig. 1B), which comigrates with one generated by rose bengal, a well-known ¹O₂ sensitizer.[27] This, along with evidence that 5α-cholest-6-ene-3β,5-diol (5α-OH), obtained by borohydride reduction of authentic 5α-OOH,[22] migrates to this position establishes that MC540-sensitized photooxidation is mediated at least in part by ¹O₂. Little, if any, cholest-5-ene-3β,7α-diol (7α-OH) or cholest-5-ene-3β,7β-diol (7β-OH) can be detected (Fig. 1B), indicating that free radical chemistry is relatively unimportant under the conditions described. Generated in lower yields than 5α-OOH (see later), 6α-OOH and 6β-OOH, which, as diol derivatives, are less mobile than 7α-OH,[22] cannot be seen in Fig. 1B.

Advantages of TLC as illustrated include low cost and relative simplicity for rapid preliminary screening of Ch oxidation products. The Disadvantages include (i) poor separation of A/B-ring hydroperoxides, necessitating reduction to the diols for discrimination, although separation can be improved by using high-performance TLC plates; and (ii) difficulties in quantitation, limiting this mainly to a qualitative technique. Quantitation via densitometric scanning is cumbersome, particularly at the TMPD stage because background (baseline) color is not stable and grows steadily more intense after spraying. One answer to this, but at greater effort and expense, is to use [^{14}C]Ch-labeled membranes and monitor product formation by radioscanning.[27]

HPLC-EC(Hg) Analysis of Cholesterol Hydroperoxides

Like other relatively recent high-performance techniques for lipid hydroperoxide analysis,[30,31] HPLC-EC(Hg) represents a considerable advance over TLC in terms of analyte resolution, detection sensitivity, and accurate quantitation. The idea of using a renewable mercury drop electrode (polarographic approach) for peroxide measurement was introduced in the late 1960s[32] and was later adapted to HPLC separations.[33] This approach has been further developed and refined by our group, with biological applications in mind.[34,35] The following sections outline general operating conditions for HPLC-EC(Hg) and describe its application to diagnostic ChOOH determination, using photooxidized erythrocyte ghosts and leukemia cells as biological examples.

Operating Conditions

Chromatography is accomplished using an Ultrasphere XL-ODS column (4.6 × 150 mm or 250 mm; 5-μm particles) from Beckman Instruments (San Ramon, CA), a Model 2350 pump with injection valve from Isco (Lincoln, NE), and a Model 420 electrochemical detector from EG&G Instruments (Oak Ridge, TN). The detector is equipped with a hanging mercury drop electrode, which is dispensed from a glass capillary and

[30] T. Miyazawa, *Free Radic. Biol. Med.* **7**, 209 (1989).
[31] Y. Yamamoto, B. Frei, and B. N. Ames, *Methods Enzymol.* **186**, 371 (1990).
[32] R. D. Mair and R. T. Hall, *in* "Organic Peroxides" (D. Swern, ed.), p. 578. Wiley-Interscience, New York, 1971.
[33] M. O. Funk, P. Walker, and J. C. Andre, *Bioelectrochem. Bioenerget.* **18**, 127 (1987).
[34] W. Korytowski, G. J. Bachowski, and A. W. Girotti, *Anal. Biochem.* **213**, 111 (1993).
[35] W. Korytowski, P. G. Geiger, and A. W. Girotti, *J. Chromatogr. B* **670**, 189 (1995).

renewed on demand. The operating potential is set at -150 or -300 mV vs Ag/AgCl, as specified. The mobile phase consists of (by volume): (I) 81% methanol, 10% acetonitrile, and 9% 0.25 mM sodium perchlorate/10 mM ammonium acetate in water or (II) 72% methanol, 11% acetonitrile, 8% isopropanol, and 9% 0.25 mM sodium perchlorate/10 mM ammonium acetate in water. The mobile phase is sparged overnight before a run and then continuously throughout the run, using high-purity argon that passes first through an OMI-1 oxygen scrubber (Supelco, Bellefonte, PA) and then through a presaturating mobile phase scrubber.[34,35] This is typically sufficient for a background current of <0.5 nA, permitting the highest detector sensitivity setting (0.1 nA full scale) to be used, if necessary. The elution rate is set at 1.7–1.8 ml/min. Samples are applied manually, using a 10-μl injection loop. A fresh mercury drop is dispensed and equilibrated for 2–4 min before each injection. Under the conditions described, 5α-OOH and 7$\alpha\beta$-OOH have the same EC(Hg) responsiveness, which is ~30% greater than the 6α-OOH or 6β-OOH responsiveness[34]; detection limits for these species are in the 0.1- to 0.3-pmol range.[35] Any sample-to-sample variation in amount of extracted material is corrected for by using A_{212}-detected Ch as an internal standard.[36]

Photooxidized Erythrocyte Membranes: $^1O_2$ vs Free Radical Dominance

Figure 2 shows a reversed-phase HPLC-EC(Hg) pattern of a mixture of four different positional ChOOH standards, including 5α-OOH, 6β-OOH, 7$\alpha\beta$-OOH, and 25-OOH. Baseline resolution of these analytes is accomplished under the conditions used; although 6α-OOH (9.3 min, not shown) is separable from 6β-OOH (10.7 min), 7α-OOH and 7β-OOH are not separable. Using the same HPLC conditions to analyze lipid extracts from AlPcS$_4$/light-treated erythrocyte ghosts, one sees patterns such as shown in Fig. 3. Ghosts photooxidized in the absence of added iron and an iron reductant (ascorbate) exhibit several distinguishable EC(Hg) peaks (Fig. 3b). These disappear[23] when samples are pretreated with triphenylphosphine or with GSH and PHGPX (a selenoperoxidase), consistent with lipid hydroperoxide identity. From retention times (cf. Fig. 2), peaks 1, 2, 3, and 4 are assigned as 7$\alpha\beta$-OOH, 5α-OOH, 6α-OOH, and 6β-OOH. No significant 25-OOH (a free radical indicator) is evident. The broad peaks at >11 min represent phospholipid hydroperoxides. Quantitation based on the responsiveness of standards reveals that the order of ChOOH abundance in Fig. 3b is 5α-OOH > 6α-OOH ≈ 6β-OOH > 7$\alpha\beta$-OOH, indicating $^1O_2$ predominance under these reaction conditions. No peroxides are ob-

[36] P. G. Geiger, W. Korytowski, F. Lin, and A. W. Girotti, *Free Radic. Biol. Med.* **23,** 57 (1997).

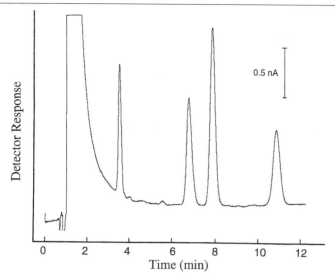

FIG. 2. HPLC-EC(Hg) profiles of ChOOH standards. Identity, injected amounts, and retention times of the various hydroperoxides are as follows: 25-OOH, 50 pmol (3.6 min); $7\alpha\beta$-OOH, 35 pmol (6.7 min); 5α-OOH, 65 pmol (7.8 min); 6β-OOH, 45 pmol (10.7 min). A C-18 column (4.6 × 150 mm; 5-μm particles) with mobile phase I was used. The operating potential and full-scale detector sensitivity were −300 mV and 2 nA, respectively.

served when ghosts are irradiated in the absence of AlPcS₄ (Fig. 3a). Figure 4A shows the time course of ChOOH accumulation for the system represented in Fig. 3b. The observed linearity out to 30 min indicates that there is no significant depletion of membrane Ch in this reaction (~0.3% as 5α-OOH at 30 min). Rate ratios are as follows: 11.8 (5α-OOH/$7\alpha\beta$-OOH) and 2.8 (5α-OOH/6α-OOH). Because of its higher rate of accumulation, 5α-OOH would make a better $^1O_2$ indicator than 6α- or 6β-OOH; however, the advantage of the latter two is that they are not susceptible to re-arrangement.[37]

A shift in ChOOH distribution occurs when membranes are photooxidized in the presence of ascorbate and Fe(HQ)₂, a lipophilic iron complex (Fig. 3c). Most significant is the large increase in $7\alpha\beta$-OOH content, with concomitant decreases in 5α-OOH, 6α-OOH, and 6β-OOH. For carefully prepared and stored membranes (represented here), peroxide formation is negligible during treatment with Fe(HQ)₂/ascorbate alone, i.e., without dye/light (cf. Fig. 3a). Kinetic data (Fig. 4B) reveal that the rate of $7\alpha\beta$-OOH accumulation is increased approximately fivefold by iron/ascorbate,

[37] W. Korytowski, G. J. Bachowski, and A. W. Girotti, *Photochem. Photobiol.* **56**, 1 (1992).

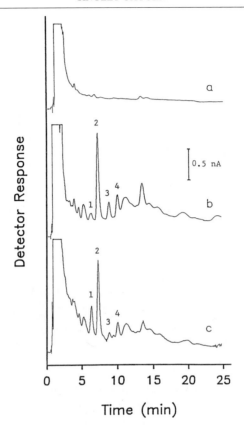

FIG. 3. HPLC-EC(Hg)-detectable hydroperoxides in photooxidized ghost membranes. Membranes (1.0 mg protein/ml in PBS) were sensitized with 10 μM AlPcS$_4$ and exposed at 25° to a 180-J/cm^2 light fluence in the absence (b) or presence (c) of 5 μM Fe(HQ)$_2$ and 100 μM ascorbate. Scan (a) represents a control irradiated without AlPcS$_4$. Extracted lipid fractions were subjected to HPLC-EC(Hg) using a 4.6 × 150-mm column, mobile phase I, and an operating potential of -300 mV. ChOOH assignments are as follows: (1) $7\alpha\beta$-OOH (6.7 min); (2) 5α-OOH (7.5 min); (3) 6α-OOH (9.1 min); and (4) 6β-OOH (10.3 min). Peaks in the 11- to 20-min range represent phospholipid hydroperoxides. The operating potential and full-scale detector sensitivity were -300 mV and 2 nA, respectively. Total lipid per injection was 55 μg. From P. G. Geiger, W. Korytowski, and A. W. Girotti, *Photochem. Photobiol.* **62,** 580 (1995), with permission.

whereas the rates of 5α-OOH and 6α- or 6β-OOH accumulation are decreased by 20 and 45%, respectively. In agreement with other findings,[27] these results are attributed to mechanistic switching from a $^1O_2$-dominated reaction (Fig. 3b) to one in which light-independent, one-electron reduction

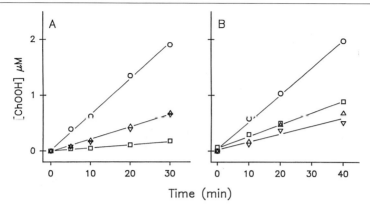

FIG. 4. Time course of ChOOH accumulation during photooxidation of erythrocyte ghosts. Membranes (1.0 mg protein/ml) were irradiated at 25° in the presence of 10 μM AlPcS$_4$ alone (A) or 10 μM AlPcS$_4$ plus 10 μM Fe(HQ)$_2$ and 100 μM ascorbate (B). The fluence rate was ~150 mW/cm^2. Lipid fractions from samples extracted at the indicated time points were analyzed for ChOOHs by HPLC-EC(Hg). Bulk phase ChOOH concentrations in reaction mixtures are indicated. Analytes are denoted as follows: (◯) 5α-OOH, (△) 6α-OOH, (▽) 6β-OOH, and (□) 7$\alpha\beta$-OOH. From P. G. Geiger, W. Korytowski, and A. W. Girotti, *Photochem. Photobiol.* **62,** 580 (1995), with permission.

of $^1O_2$-derived peroxides triggers free radical reactions, which amplify the production of 7$\alpha\beta$-OOH (Fig. 3c). This is illustrated by Eqs. (1)–(5),

$$LH + {}^1O_2 \rightarrow LOOH \tag{1}$$
$$RH + Fe^{3+} \rightarrow R\cdot + H^+ + Fe^{2+} \tag{2}$$
$$Fe^{2+} + LOOH + O_2 \rightarrow Fe^{3+} + OLOO\cdot + OH^- \tag{3}$$
$$OLOO\cdot + LH \rightarrow OLOOH + L\cdot \tag{4}$$
$$L\cdot + O_2 \rightarrow LOO\cdot \tag{5}$$

where RH is an electron donor and LH, LOOH, LOO, and OLOO represent an unsaturated lipid, a hydroperoxide, a peroxyl radical, and an epoxy-allylic peroxyl radical, respectively.[38] It follows that if redox iron is strategically located in a photodynamic system, one could conceivably misinterpret diagnostic data, deducing that the primary chemistry is substantially free radical (type I) in nature, whereas in reality it might be purely $^1O_2$ (type II). Consequently, it would be advisable to examine the effects of iron depletion/inactivation (e.g., with DFO) before drawing conclusions about primary mechanisms.

[38] A. W. Girotti, *J. Lipid Res.* **39,** 1529 (1998).

Photooxidized Cells: Mechanistic Monitoring via Peroxide Ratio

Results of an experiment extending the HPLC-EC(Hg)-based determination of ChOOHs to photodynamically stressed mammalian cells are shown in Fig. 5. In this example, peroxide profiles are compared for MC540-sensitized L1210 cells following irradiation in the absence or presence of added iron in the form of $Fe(HQ)_2$. The traces in Fig. 5 show several well-resolved peaks, including those of $7\alpha\beta$-OOH, 5α-OOH, and 6β-OOH, the identities of which are established by spiking samples with standards. For non-$Fe(HQ)_2$-treated cells (top trace), the 5α-OOH peak is considerably larger than the $7\alpha\beta$-OOH, suggesting major $^1O_2$ intermediacy as in the ghost membrane model (Fig. 3b). Other conditions remaining equal, irradiation in the presence of $Fe(HQ)_2$, which enhances dye/light-induced lethality greatly,[36] results in a large increase in $7\alpha\beta$-OOH relative to 5α-OOH or 6β-OOH, signifying the preponderance of radical chemistry. No significant

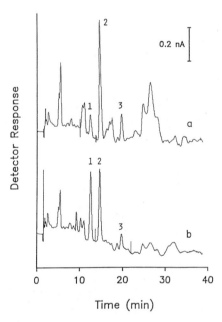

FIG. 5. Lipid hydroperoxide profile of L1210 cells after photooxidation in the absence (a) or presence (b) of $Fe(HQ)_2$. Cells (1.0×10^7/ml in PBS) were preincubated for 30 min with 25 μM MC540 without (a) or with (b) 5 μM $Fe(HQ)_2$ and then exposed to a 0.9-J/cm² light fluence. Extracted lipids were analyzed by HPLC-EC(Hg) using a 4.6 × 250-mm column, mobile phase II, and an operating potential of −150 mV. Total lipid per injection was 110 μg. Peak assignments: (1) $7\alpha\beta$-OOH, (2) 5α-OOH, and (3) 6β-OOH. The ratio of $7\alpha\beta$-OOH peak area to 5α-OOH peak area is 0.2 for scan (a) and 1.1 for scan (b).

$7\alpha\beta$-OOH is generated by Fe(HQ)$_2$ alone, i.e., without dye/light treatment (not shown). A mechanistic shift such as this can be conveniently monitored in terms of a peroxide ratio, e.g., $7\alpha\beta$-OOH/5α-OOH or $7\alpha\beta$-OOH/6β-OOH, which is independent of any sampling discrepancies. For the experiment shown in Fig. 5, the $7\alpha\beta$-OOH/5α-OOH ratio is increased more than fivefold by added iron.

In addition to assessing ChOOH status during light exposure, one can monitor it afterward to acquire postirradiation mechanistic information. Results of such an experiment for MC540/light-treated L1210 cells are represented in Fig. 6, where peroxide ratios are plotted as a function of postirradiation (dark) incubation time at 37°. As shown in Fig. 6A, the $7\alpha\beta$-OOH/5α-OOH ratio for control cells increases slowly and linearly over time, reaching 1.5 times its starting value after 2 hr. The rate of increase is ~6.5 times greater for Fe(HQ)$_2$-treated cells, but is reduced to practically zero by preincubation with DFO. Added immediately after irradiation, the chain-breaking antioxidant butylated hydroxytoluene (e.g., at 10 μM) also reduces the rate substantially (not shown). Qualitatively similar trends

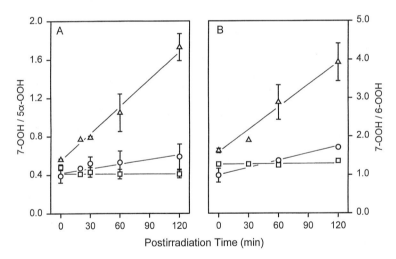

FIG. 6. Postirradiation changes in relative ChOOH contents of L1210 cells. Cells (1.0 × 10^6/ml in 5% serum/RPMI medium) were preincubated with 100 μM DFO (20 hr) or 0.5 μM Fe(HQ)$_2$ (30 min), pelleted, and resuspended in serum-free RPMI to 1.0 × 10^7 cells/ml. Controls without DFO or Fe(HQ)$_2$ were prepared alongside. Following sensitization with 10 μM MC540 and exposure to a 0.6-J/cm^2 light fluence, lipids were extracted and analyzed by HPLC-EC(Hg) immediately or after the indicated periods of dark incubation at 37°. 7-OOH/5α-OOH and 7-OOH/6-OOH ratios are plotted for each condition: (○) control, (△) Fe(HQ)$_2$, and (□) DFO. (6-OOH and 7-OOH denote the respective α and β epimers.) Data points are means ±SD of values from five experiments.

are observed when the $7\alpha\beta$-OOH/6β-OOH ratio is tracked (Fig. 6B). In accordance with Eqs. (1)–(5) and other supporting evidence,[36] these findings suggest that primary photoperoxides generated by $^1O_2$ attack undergo iron-dependent turnover *in situ,* giving rise to light-independent peroxidative chains propagated by free radicals. Such secondary reactions appear to play a significant role in photodynamically initiated cell killing.[36] It is apparent from these results that the peroxide ratio, e.g., $7\alpha\beta$-OOH/5α-OOH, can serve as a highly sensitive index for monitoring $^1O_2$ vs free radical dominance in cells, both during and after a photooxidative insult.

Summary

In cells under oxidative attack, membrane Ch, through the formation of its signature hydroperoxide and diol products, can serve as a unique detector *in situ,* allowing discrimination between $^1O_2$ and free radical intermediacy. Of the two techniques described for analyzing Ch oxidation products, TLC with color development suffices for preliminary, mainly qualitative product screening, whereas a high-performance approach such as HPLC-EC(Hg) is advised when maximum resolution and sensitivity of quantitation are necessary. By using these strategies, one can monitor the formation of $^1O_2$, for example, in a biologically relevant milieu (membrane), thus avoiding the difficulties associated with external detection, e.g., by physical means. These approaches would be valuable for assessing reaction mechanisms for various oxidative agents of biomedical importance, including environmental phototoxins and the rapidly emerging family of phototherapeutic drugs.[6] Although photodynamic stress has been emphasized, the methods described should have broad applicability in the elucidation of oxidative mechanisms.

Acknowledgments

We appreciate Pete Geiger's expert technical assistance. This work was supported by USPHS Grants CA49089, CA70823, and CA72630 from the National Cancer Institute and by FIRCA Grant TW00424 from the Fogarty International Center.

[10] Singlet Oxygen Scavenging in Phospholipid Membranes

By KENJI FUKUZAWA

Introduction

Singlet oxygen ($^1O_2$) has aroused much interest as a biological oxidant. It is generated in a variety of biological systems and by photosensitization on absorption of light. There are many reports that $^1O_2$ is scavenged by lipid-soluble antioxidants such as α-tocopherol (α-Toc) and carotenoids, but in most previous studies a homogeneous solution such as ethanol, not membranes, has been used, although these compounds are mainly located in membranes. This article describes a method for measuring the $^1O_2$ scavenging rate constants (k_s values) of α-Toc and carotenoids in model membranes using phosphatidylcholine liposomes. The dynamics and kinetics of the compounds for $^1O_2$ scavenging in membranes are compared to those in ethanol (EtOH) solution and are interpreted with respect to the generation sites of $^1O_2$ and the concentrations of antioxidants, especially the local concentrations and mobilities of their active moieties in the membranes.

Photosensitized $^1O_2$ Generation in Phospholipid Membranes

The thermally induced decomposition of endoperoxides has been adopted as a method for measuring only $^1O_2$ extrusion in organic solvents.[1,2] However, this method cannot be applied to liposome systems because insufficient amounts of endoperoxides for k_s measurements can be incorporated into the membranes. Therefore, the author used a photosensitizing method by which $^1O_2$ is mainly generated, but some radical compounds, such as superoxide and hydroxyl radical, are produced in small amounts. $^1O_2$ is site-specifically generated in liposomes depending on the photosensitizer[3–5]: water-soluble methylene blue (MB) and Rose bengal (RB) generate $^1O_2$ at the membrane surface, whereas lipid-soluble pyrene compounds, such as 12-(1-pyrene)dodecanoic acid (PDA), 4-(1-pyrene)butyric acid, 2-

[1] P. Mascio, S. Kaiser, and H. Sies, *Arch. Biochem. Biophys.* **274,** 532 (1989).
[2] K. Mukai, K. Daifuku, K. Okabe, T. Tanigaki, and K. Inoue, *J. Org. Chem.* **56,** 4188 (1991).
[3] K. Fukuzawa, K. Matsuura, A. Tokumura, A. Suzuki, and J. Terao, *Free Radic. Biol. Med.* **22,** 923 (1997).
[4] K. Fukuzawa, Y. Inokami, A. Tokumura, J. Terao, and A. Suzuki, *BioFactors* **7,** 31 (1998).
[5] K. Fukuzawa, Y. Inokami, A. Tokumura, J. Terao, and A. Suzuki, *Lipids* **33,** 751 (1998).

METHODS IN ENZYMOLOGY, VOL. 319

(1-pyrene)acetic acid, and β-(pyrene-1-yl)-dodecanoyl-γ-palmitoyl phosphatidylcholine, generate it at different depths in the membrane.

Calculation of Rate Constant for Scavenging $^1O_2$

The method of Young et al.[6] is usually used to measure the $^1O_2$ scavenging rate constant (k_s) by determining the competitive inhibition by a lipid-souble antioxidant (ANT) of photooxidation of 1,3-diphenylisobenzofuran (DPBF), a specific $^1O_2$ trap [Eqs. (1)–(4)]:

$$^1O_2 + ANT \xrightarrow{k_{q(ANT)}} ANT + {}^3O_2 \qquad (1)$$

$$^1O_2 + ANT \xrightarrow{k_{r(ANT)}} ANT\text{-}O_2 \qquad (2)$$

$$^1O_2 + DPBF \xrightarrow{k_{r(DPBF)}} DPBF\text{-}O_2 \qquad (3)$$

$$^1O_2 \xrightarrow{k_d} {}^3O_2 \qquad (4)$$

where $k_{r(ANT)}$ and $k_{r(ANT)}$ are the $^1O_2$ scavenging rate constants of ANT by quenching and an irreversible reaction, respectively. The $k_{r(DPBF)}$ is the rate constant of reaction of $^1O_2$ with DPBF, and k_d is the rate constant for the natural decay of $^1O_2$ to ground state oxygen ($^3O_2$). The sum of $k_{q(ANT)}$ and $k_{r(ANT)}$ is equal to $k_{s(ANT)}$ [$k_{s(ANT)} = k_{q(ANT)} + k_{r(ANT)}$]. $k_{q(ANT)}$ values are several orders of magnitude higher than those of $k_{r(ANT)}$ for α-Toc and carotenoids: for example, α-Toc is supposed to deactivate approximately 40 molecules of $^1O_2$ before being destroyed in lipid bilayers.[7] There are no significant differences in $k_{r(DPBF)}$ values in photosensitizing systems.[3]

Equation (5) is used to determine the k_s value of ANT:

$$S_0/S_{ANT} = 1 + (k_s/k_d)[ANT] \qquad (5)$$

where S_0 and S_{ANT} denote the slopes of first-order plots of disappearance of DPBF in the absence and presence of ANT in experiments. The following factors are modulated to obtain linear slopes of the plots of S_0/S_{ANT}: photoirradiation time, illumination power, distance from the light source, and concentrations of DPBF and ANT. The k_d value in liposomes has not yet been determined. The author used the k_d value in EtOH solution[8] (8.3×10^4 sec^{-1}) as the k_d value in liposomes in an MB or RB system and the k_d value in tert-BuOH[9] (3.0×10^4 sec^{-1}) as that in a PDA system because

[6] R. H. Young, K. Wehrly, and R. L. Martin, J. Am. Chem. Soc. 93, 5774 (1971).
[7] M. Fragata and F. Bellemare, Chem. Phys. Lipids 27, 93 (1980).
[8] P. B. Merkel and D. R. Kearns, J. Am. Chem. Soc. 94, 7244 (1972).
[9] R. H. Young, D. Brewer, and R. A. Keller, J. Am. Chem. Soc. 95, 375 (1973).

$^1O_2$ is generated at the polar membrane surface in the MB or RB system and in the hydrophobic inner region of the membrane in the PDA system. Micropolarity close to the membrane surface is thought to be equal to that in EtOH,[10] in which the dielectric constant is higher than that in *tert*-BuOH.

Preparation of Liposomes[3–5]

A solution (700–800 μl) of dimyristoylphosphatidylcholine (DMPC) with or without stearylamine (SA) or dicetyl phosphate (DCP) in chloroform is mixed with a lipid-soluble antioxidant, such as α-Toc or carotenoids, and pyrene compounds (for studies on pyrene-sensitized photooxidation) in a light-proof brown test tube and dried under a stream of nitrogen at reduced pressure. The resulting thin film is treated with 250 μl of a solution of DPBF in EtOH in a voltex mixer and evaporated under a stream of nitrogen at reduced pressure. The resulting thin film is dispersed in 2 ml of 10 mM N-2-hydroxyethylpiperazine-N-2-ethanesulfonic acid (HEPES) buffer, pH 7.0, containing 0.5 mM diethylenetriaminepentaacetic acid (DTPA) in a vortex mixer and ultrasonicated in a Bransonic bath (Yamato, Tokyo, Japan) at 40° for 10 min (neutrally charged DMPC liposomes). For studies on MB- or RB-sensitized photooxidation, a solution (10–20 μl) of MB or RB in HEPES buffer, pH 7.0, is added to the DCP- or SA-DMPC liposome solution (2 ml) prepared by ultrasonication for 2 min at 40° and further ultrasonicated for 15 sec to distribute MB or RB at the surface of the outer and inner layers of the membrane. During liposome preparation, care is taken to exclude light. Liposomes are prepared from DMPC, which are insensitive to lipid chain peroxidation and are in a liquid crystalline state at 37°. Positively charged MB generates $^1O_2$ at the negatively charged membrane surface of DCP-DMPC liposomes and negatively charged RB generates $^1O_2$ at the positively charged membrane surface of SA-DMPC liposomes because these compound interact electrostatically at the membrane surface of oppositely charged liposomes. PDA generates $^1O_2$ in the inner region of the membrane irrespective of the membrane charge. The concentrations of reactants concentrated in membranes were 241 (DCP-DMPC liposomes), 254 (SA-DMPC liposomes), and 266 (DMPC liposomes) times those in the bulk phase, as calculated from the volume of the membrane layer of liposomes,[11] assuming that the volume of one SA molecule is 557 Å^3 and that of DCP and DMPC molecules is 1253 Å.^{3}

[10] F. Bellemare and M. J. Fragata, *Colloid Interface Sci.* **77,** 243 (1980).
[11] C. Huang and J. T. Mason, *Proc. Natl. Acad. Sci. U.S.A.* **75,** 308 (1978).

Practical Instances for k_s Measurements in Liposomes

Figure 1 shows measurements of the k_s value of β-carotene in liposomes.[5] Samples of 2 ml of liposome suspension are placed 4 cm from an 85-W halogen-tungsten lamp (Toshiba JD 100W 110V 85WN-E) and photoirradiated at 37° at 1564 lumen, as monitored with an Optical Multi Power Meter (ADVANTEST Q8221-82214). DPBF is quantitated as follows. Samples of 20 μl after photooxidation for given times are subjected to HPLC on an ODS column (YMC-Pack A-312, Yamamura Chemical Laboratory, Kyoto, Japan; 6 × 150 mm) with a UV detector at 411 nm and eluted with MeOH/H$_2$O/triethylamine (95:5:0.01, v/v) at a flow rate of 1 ml/min.[3–5] The k_s value is calculated as $5.2 \times 10^9\ M^{-1}\ sec^{-1}$ using a k_d value in tert-BuOH of $3.0 \times 10^4\ sec^{-1}$. The k_s value of β-carotene in DMPC liposomes is revised as $2.05 \times 10^7\ M^{-1}\ sec^{-1}$, taking into account the concentrations of reactants, all of which were assumed to be present in the membranes at 265 times those in the bulk phase.

Table I shows k_s values for carotenoids such as β-Car, astaxanthin, and canthaxanthin in EtOH solution and liposomes in RB and PDA systems. The experimental conditions are as follows. The concentrations of carotenoids are 2.5–15 μM; the concentrations of DPBF, RB, and PDA are 250,

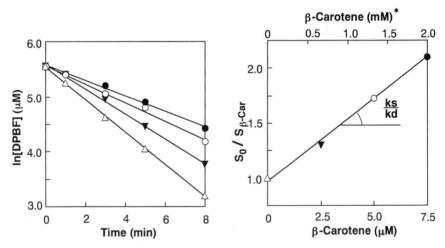

Fig. 1. Time course of photosensitized oxidation of DPBF in DMPC liposomes in the presence of various concentrations of β-Car, and plots of $S_0/S_{\beta\text{-Car}}$ vs β-Car concentration. (●) 0 μM β-Car, (○) 2.5 μM β-Car, (▼) 5 μM β-Car, and (△) 7.5 μM β-Car. The concentrations of other reactants were 5 mM DMPC, 0.5 mM DTPA, and 250 μM PDA in 10 mM HEPES buffer (pH 7.0). Photoirradiation was carried out at 37° for 8 min. *Concentration of β-Car in liposome membranes (mol of β-Car/liter of DMPC liposomes), which was 265 times that in the bulk water phase.

TABLE I
RATE CONSTANTS (k_s) OF CAROTENOIDS FOR SCAVENGING $^1O_2$ GENERATED
WITH RB, PDA, AND EP IN EtOH SOLUTION AND LIPOSOMES

	k_s (M^{-1} sec^{-1})		
	β-Carotene	Canthaxanthin	Astaxanthin
EtOH			
RB	1.3×10^{10}	1.3×10^{10}	2.4×10^{10}
PDA	1.0×10^{10}	1.4×10^{10}	2.1×10^{10}
Average	1.2×10^{10}	1.4×10^{10}	2.3×10^{10}
EP (reported[a])	1.4×10^{10}	2.1×10^{10}	2.4×10^{10}
Liposomes			
RB[b]	14.4×10^9	12.5×10^9	12.7×10^9
RB[c]	5.2×10^9	4.5×10^9	4.6×10^9
PDA[c]	4.5×10^9	4.5×10^9	5.1×10^9
RB[c] (revised[d])	2.05×10^7	1.77×10^7	1.81×10^7
PDA[c] (revised[d])	1.70×10^7	1.70×10^7	1.92×10^7
Average[c] (revised[d])	1.88×10^7	1.74×10^7	1.87×10^7

[a] From P. Mascio et al., Arch. Biochem. Biophys. **274**, 532 (1989).
[b] Values calculated using the k_d value in EtOH[8] (8.3×10^4 sec^{-1})
[c] Values calculated using the k_d value in tert-BuOH[9] (3.0×10^4 sec^{-1})
[d] Values in liposomes revised considering that the concentrations of reactants
in membranes were 254 and 265 times higher in the RB system (SA-DMPC
liposomes) and PDA system (DMPC liposomes), respectively, than in the
bulk phase.

5, and 250 μM, respectively; the concentrations of other reactants in the
liposome systems are 5 mM DMPC with or without 0.5 mM SA, and 10
mM HEPES buffer (pH 7.0) containing 0.5 mM DTPA; the EtOH solution
and liposomes (2 ml) are photoirradiated for ~30 sec and ~8 min in the
RB and PDA systems, respectively, at 37°. There is no difference in k_s
values in the EtOH solution in RB and PDA systems, which are almost
the same as the values in systems using the endoperoxide (EP) 3,3'-(1,4-
naphthylidene) dipropionate,[1] which thermally induces the generation of
only $^1O_2$, indicating that the active oxygen species generated in these photosensitized systems is almost all $^1O_2$. However, k_s values in liposomes calculated using k_d values in EtOH (8.3×10^4 sec^{-1}) and tert-BuOH (3.0×10^4
sec^{-1}) in the RB system[8] and PDA system,[9] respectively, are different in
the two photoirradiation systems. k_s values in liposomes in the RB system
are recalculated using the k_d value as 3.0×10^4 sec^{-1}, which is almost the
same as those in the PDA system. The active site of carotenoids, the
center of the conjugated polyene chain, is reported to be localized in the

TABLE II

RATE CONSTANTS (k_s) OF α-Toc FOR SCAVENGING $^1O_2$ GENERATED WITH MB,
RB, PDA, AND EP IN EtOH SOLUTION AND LIPOSOMES

		k_s (M^{-1} sec^{-1})		
		Liposomes		
	EtOH[b]	Apparent	Revised 1[d]	Revised 2[e]
MB	3.1×10^8	$3.5 \times 10^{9\,b}$	1.45×10^7	3.10×10^6
RB	3.6×10^8	$3.4 \times 10^{9\,b}$	1.38×10^7	2.95×10^6
PDA	2.1×10^8	$0.66 \times 10^{9\,c}$	0.25×10^7	3.25×10^6
Average	2.9×10^8			
EP[a]	3.0×10^8			

[a] Reported value (Ref. 1).
[b] Values calculated using the k_d value in EtOH[8] (8.3×10^4 sec^{-1}).
[c] Values calculated using the k_d value in *tert*-BuOH[9] (3.0×10^4 sec^{-1}).
[d] Values in liposomes revised considering that the concentrations of reactants in membranes were 241, 254, and 265 times higher in the MB system (DCP-DMPC liposomes), RB system (SA-DMPC liposomes), and PDA system (DMPC liposomes), respectively, than in the bulk phase.
[e] Values in liposomes revised considering that the local concentrations of the OH groups of α-Toc in membranes were 55 and 45% in HB and HC of DMPC liposomes, respectively (Fig. 2). k_s (revised 2) = k_s (revised 1) $\times$ 4/34 $\times$ 100/55 (MB or RB system) and = k_s (revised 1) $\times$ 20/34 $\times$ 100/45 (PDA system).

hydrophobic region,[12,13] where $^1O_2$ is also highly localized irrespective of its generation site because of its higher solubility in the hydrophobic region[14] than in the polar region, resulting in similar k_s values for carotenoids in both photosensitizing systems calculated using the k_d value in *tert*-BuOH.

Table II shows k_s values for α-Toc in EtOH solution and liposomes in MB, RB, and PDA systems. The experimental conditions are as follows: the concentration of α-Toc is ~100 μM and the concentration of MB is 50 μM. The MB system is photoirradiated for ~20 sec in EtOH solution and ~90 sec in liposomes. Other experimental conditions are as for Table I. k_d values in liposomes are taken as 8.3×10^4 sec^{-1} and 3.0×10^4 sec^{-1} in the MB or RB system[8] and in the PDA system,[9] respectively. k_s values in liposome membranes are revised, taking into account the concentrations of reactants in membranes (revised 1 in Table II).

[12] G. Britton, *FASEB J.* **9**, 1551 (1995).
[13] J. Gabrielska and W. I. Gruszecki, *Biochim. Biophys. Acta* **1285**, 167 (1996).
[14] R. Battino (ed.), "Oxygen and Ozone: Solubility Data Series," Vol. 7. Pergamon Press, Oxford, 1981.

In EtOH solution, k_s values for α-Toc are almost the same in the three photoirradiation systems. They are also similar to the reported values in the EP system. In liposomes, the revised k_s value (revised 1) in the MB or RB system is about six times that in the PDA system. Because the OH groups of α-Toc are localized mainly in the region close to the membrane surface,[7,15–18] their concentration in this region, where $^1O_2$ is generated by MB or RB, is higher than that postulated, assuming that they are distributed uniformly in the membrane. On the contrary, PDA generates $^1O_2$ in the hydrophobic region where the concentration of OH groups of α-Toc is lower than that postulated, as described earlier, resulting in relatively lower k_s values in the PDA system. The k_s value of α-Toc in liposomes is revised further in consideration of the site-specific $^1O_2$ scavenging reaction of α-Toc, which depends on the localization of the OH groups of α-Toc and the generation sites of $^1O_2$ in the membranes. Equation (6) was used to revise the k_s value of α-Toc in liposomes, assuming the following two points: (i) the local concentrations of the OH groups of α-Toc in DMPC liposomes are about 0, 50–60, and 40–50% in the polar zone (PZ), hydrogen belt (HB), and hydrophobic core (HC) of the membranes, respectively, and the lengths of these membrane regions of DMPC liposomes are about 10 Å (PZ), 4 Å (HB), and 20 Å (HC) (Fig. 2)[7,15–18]; (ii) $^1O_2$ generated by RB or MB is only scavenged by the OH groups localized in the HB of membranes, whereas that generated by PDA is mainly scavenged by the OH groups in the HC of membranes because of the slight diffusion of $^1O_2$ generated in the HC into the HB due to its higher solubility in solvents with higher hydrophobicities.[14]

$$S_0/S_{\alpha-T} = 1 + (k_s/k_d)[\text{OH groups of } \alpha\text{-Toc}] \qquad (6)$$

where [OH groups of α-Toc] = [α-Toc] $\times$ (10 + 4 + 20)/4 $\times$ 50 $\sim$ 60/100 (MB or RB system) and = [α-Toc] $\times$ (10 + 4 + 20)/20 $\times$ 40 $\sim$ 50/100 (PDA system). The revised k_s values (revised 2 in Table II) for α-Toc in liposomes were recalculated as 2.84–3.41 and 2.71–3.25 $\times$ 10^6 M^{-1} sec^{-1} in the MB and RB systems, respectively, and 2.93–3.66 $\times$ 10^6 M^{-1} sec^{-1} in the PDA system. OH groups of α-Toc in the region close to the membrane surface are supposed to have higher reactivity but lower mobility than those in the inner membrane region because the k_s value of α-Toc is higher in solvents with higher dielectric constants (ε) [ε = 32.6, k_s = 5.3–6.7 in

[15] M. Takahashi, J. Tsuchiya, and E. Niki, *J. Am. Chem. Soc.* **111,** 6350 (1989).
[16] F. J. Aranda, A. Coutinho, M. N. Berberan-Santos, M. J. E. Prieto, and J. C. Gomez-Fernandez, *Biochim. Biophys. Acta* **985,** 26 (1989).
[17] K. Fukuzawa, W. Ikebata, A. Shibata, I. Kumadaki, T. Sakanaka, and S. Urano, *Chem. Phys. Lipids* **63,** 69 (1992).
[18] K. Fukuzawa, W. Ikebata, and K. Sohmi, *J. Nutr. Sci. Vitaminol.* **39,** S9 (1993).

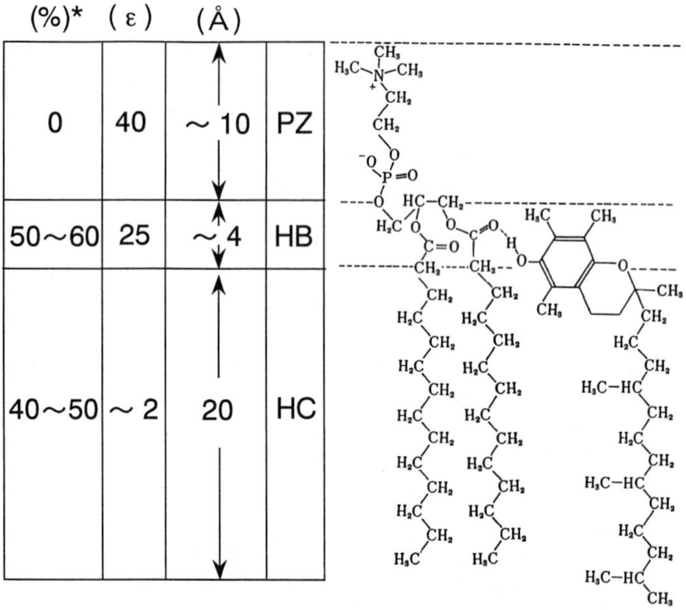

FIG. 2. Membrane location of α-Toc in DMPC liposomes. Lengths (Å) and dielectric constants (ε) in the polar zone (PZ), hydrogen belt (HB), and hydrophobic core (HC) of the membranes were referred to the report by Fragata and Bellemare.[7] *Percentage localization of the OH groups of α-Toc were deduced from the results in Refs. 15–18.

methanol; $\varepsilon = 25.5$, $k_s = 2.1–3.6$ in ethanol; $\varepsilon = 2.02$, $k_s = 0.9$ in isooctane],[7] but the degree of packing of phosphatidylcholine molecules is higher near the membrane surface,[19] and the chromanol moiety of α-Toc is also near the membrane surface and its mobility is restricted, possibly by hydrogen bonding of its OH groups with ester carbonyl moieties of membrane phosphatidylcholine,[7,20] resulting in unexpectedly similar revised k_s values for α-Toc in membranes in the MB or RB system and PDA system.

The revised k_s values in liposomes for α-Toc and carotenoids were 94 and 600–1200 times lower, respectively, than those in EtOH solution (Table III). The lower k_s values in liposomes than in EtOH solution may have been due to the lower diffusion rates of reactants in membranes than in EtOH solution and their lateral diffusion in membranes but their three-dimensional diffusion in EtOH solution.

[19] I. C. P. Smith, A. P. Tulloch, G. W. Stockton, S. Schreier, A. Joyce, K. W. Butler, Y. Boulanger, B. Blackwell, and L. G. Bennett, *Ann. N.Y. Acad. Sci.* **308,** 8 (1978).

[20] S. Urano, M. Matsuo, T. Sakanaka, I. Uemura, M. Koyama, I. Kumadaki, and K. Fukuzawa, *Arch. Biochem. Biophys.* **303,** 10 (1993).

TABLE III

RATIOS OF k_s VALUES IN THE MB OR RB SYSTEM TO THOSE IN THE PDA SYSTEM, THOSE IN
LIPOSOMES TO THOSE IN EtOH SOLUTION, AND THOSE OF CAROTENOIDS TO THOSE OF α-Toc

	Ratio of k_s value				
	RB system/ PDA system		Lipsomes/ EtOH	Carotenoids/α-Toc	
	EtOII	Lipsomes		EtOH[e,g]	Liposomes[f,h]
α-Tocopherol	1.7 (1.5)[a]	5.5,[b] 0.92[c] (5.8,[b] 0.97[c])[a]	1/94[e,f]	1	1.0
β-Carotene	1.3	1.2[d]	1/638[g,h]	41	6.1
Canthaxanthin	0.93	1.0[d]	1/805[g,h]	48	5.6
Astaxanthin	1.1	0.94[d]	1/1230[g,h]	79	6.0

[a] MB system/PDA system.
[b] Ratio of revised k_s value (revised 1 in Table II).
[c] Ratio of revised k_s value (revised 2 in Table II).
[d] Ratio of revised k_s value (Table I).
[e] Averaged k_s values for α-Toc in EtOH solution in MB, RB, and PDA systems (Table II).
[f] Averaged k_s value for α-Toc in liposomes: $3.1 \times 10^6\ M^{-1}\ sec^{-1}$ (revised 2 in Table II).
[g] Averaged k_s values for carotenoids in EtOH solution in RB and PDA systems (Table I).
[h] Averaged k_s values (revised) for carotenoids in liposomes in RB and PDA systems (Table 1).

k_s values of carotenoids were about 40–80 times that of α-Toc in EtOH solution, but only six times that of α-Toc in liposomes (Table III). These results on the k_s values are consistent with findings that the activities of carotenoids for the inhibition of $^1O_2$-dependent lipid peroxidation in egg phosphatidylcholine liposomes were decreased more than that of α-Toc in liposomes than in EtOH.[5] The mobilities of carotenoids would be suppressed more than that of α-Toc in membranes, resulting in greater decreases in their k_s values than that of α-Toc in liposomes than in EtOH solution. This is supported by the report[13] that carotenoid molecules are anchored across the hydrophobic bilayer, bringing their two head groups in contact with opposite polar sides of the membranes, but the α-Toc molecule is located in the monolayer of the membranes.

Suggestions

The following points should be taken into consideration when measuring the k_s value of antioxidants in membranes: (i) the concentrations of antioxidants, which are concentrated in membranes, (ii) the membrane localizations of active groups of antioxidants and their local concentrations in membranes, (iii) the generation site of $^1O_2$ in membranes, depending on the localizations of photosensitizers in the surface or inner region, and (iv) the k_d value in membranes, which is substituted for the value in organic solvents with a similar dielectric constant for that around the membrane region where the active sites of antioxidants are distributed. The following factors have important influences on the k_s values of antioxidants in membranes: (i) the mobility of antioxidants in membranes, which is lower than that in organic solvents, especially the mobility of active sites of antioxidants, and is suppressed more in the region of polar head groups than in the hydrophobic core, (ii) the dielectric constant in membrane regions where $^1O_2$ is scavenged, which is high at the membrane surface and low in the inner hydrophobic region, and (iii) the solubility of $^1O_2$ in membranes, which is high in the inner region with low polarity.

[11] Catalase Modification as a Marker for Singlet Oxygen

By Fernando Lledias and Wilhelm Hansberg

Introduction

Singlet oxygen ($^1O_2$) can originate as a consequence of photosensitization reactions. Visible light and the ultraviolet-A (320–380 nm) component of sunlight can produce $^1O_2$ through interaction with endogenous photosensitizers. The principal cellular photosensitizers are flavines, porphyrines, chlorophylls, quinones, bilirubin, and retinal. When excited by light, these compounds can transfer their excitation energy onto an adjacent dioxygen molecule, converting it to the singlet state while the photosensitizer molecule returns to its ground state. Thus, sunlight may be an important source of $^1O_2$ in biological systems.[1–4] In nutrient-starved bacteria, $^1O_2$ may be an important mutagen.[5] Singlet oxygen can also arise from membrane lipid peroxides.[6] Activated leukocytes in mammals, particularly eosinophils, can

[1] R. M. Tyrrell, *BioEssays* **18,** 139 (1995).
[2] E. Hideg, K. Tamás, K. Hideg, and I. Vass, *Biochemistry* **37,** 11405 (1998).
[3] P. E. Hartman, W. J. Dixon, T. A. Dahl, and M. E. Daub, *Photochem. Photobiol.* **47,** 699 (1988).
[4] P. W. Albro, P. Bilski, J. T. Corbett, J. L. Schroeder, and C. F. Chignell, *Photochem. Photobiol.* **66,** 316 (1997).
[5] B. A. Bridges and A. Timms, *Mutat. Res.* **403,** 21 (1998).
[6] Y. Liu, K. Stolze, A. Dadak, and H. Nohl, *Photochem. Photobiol.* **66,** 443 (1997).

generate $^1O_2$.[7] Some enzymatic reactions have been shown to produce $^1O_2$.[8-10]

Singlet oxygen is a highly reactive form of dioxygen that may harm cells by oxidizing critical organic molecules. Singlet oxygen can react with guanine residues in nucleic acids forming 8-oxoguanine,[11-14] with Trp, Tyr, His, Lys, Met, and Cys in proteins generating different amino acid derivatives,[15-18] and with unsaturated carbon chains in lipids to form hydroperoxides.[6,19-22]

Under $^1O_2$ generating conditions there is increased transcription and translation of distinct stress-related proteins (heat shock proteins, glucose regulated proteins, heme oxygenase-1, collagenase, and adhesion molecules) through the activation of cytokines, receptors, kinases, and transcription factors (NF-κB, AP-1, and AP-2).[23-28]

Singlet oxygen in cells may be quenched to $^3O_2$ in energy transfer

[7] J. R. Kanofsky, H. Hoogland, R. Wever, and S. J. Weiss, *J. Biol. Chem.* **263**, 9692 (1988).

[8] J. R. Kanofsky and B. Axelrod, *J. Biol. Chem.* **261**, 1099 (1986).

[9] R. D. Hall, W. Chamulitrat, N. Takahashi, C. F. Chignell, and R. P. Mason, *J. Biol. Chem.* **264**, 7900 (1989).

[10] J. Durner, V. Gailus, and P. Boger, *FEBS Lett.* **354**, 71 (1994).

[11] M. Pflaum, S. Boitreux, and B. Epe, *Carcinogenesis* **15**, 297 (1994).

[12] J. L. Ravanat and J. Cadet, *Chem. Res. Toxicol.* **8**, 379 (1995).

[13] B. A. Bridges, M. Sekiguchi, and T. Tajiri, *Mol. Gen. Genet.* **251**, 352 (1996).

[14] X. Zhang, B. S. Rosenstein, Y. Wang, M. Lebwohl, and H. Wei, *Free Radic. Biol. Med.* **23**, 980 (1997).

[15] A. Michaeli and J. Feitelson, *Photochem. Photobiol.* **65**, 309 (1997).

[16] H. R. Shen, J. D. Spikes, P. Kopeckova, and J. Kopecek, *Photochem. Photobiol. B* **35**, 213 (1996).

[17] C. Salet, M. Moreno, F. Ricchelli, and P. Bernardi, *J. Biol. Chem.* **272**, 21938 (1997).

[18] J. A. Silvester, G. S. Timmins, and M. J. Davies, *Arch. Biochem. Biophys.* **350**, 249 (1998).

[19] C. Tanielian and R. Mechin, *Photochem. Photobiol.* **59**, 263 (1994).

[20] G. F. Vile and R. M. Tyrrell, *Free Radic. Biol. Med.* **18**, 721 (1995).

[21] S. P. Stratton and D. C. Lieber, *Biochemistry* **36**, 12911 (1997).

[22] A. W. Girotti, *J. Lipid Res.* **39**, 1529 (1998).

[23] S. W. Ryter and R. M. Tyrrell, *Free Radic. Biol. Med.* **24**, 1520 (1998).

[24] M. Walaschek, J. Wenk, P. Brenneisen, K. Briviba, A. Schwarz, H. Sies, and K. Schaffetter-Kochanek, *FEBS Lett.* **413**, 239 (1997).

[25] G. L. Schieven, R. S. Mittler, S. G. Nadler, J. M. Kirihara, J. B. Bolen, S. B. Kanner, and J. A. Ledbetter, *J. Biol. Chem.* **271**, 20718 (1996).

[26] C. Rosette and M. Karin, *Science* **274**, 1194 (1996).

[27] B. Derijard, M. Hibi, I. H. Wu, T. Barrett, B. Su, T. Deng, M. Karin, and R. J. Davis, *Cell* **76**, 1025 (1994).

[28] L.-O. Klotz, K. Briviba, and H. Sies, *FEBS Lett.* **408**, 289 (1997).

reactions with compounds such as carotenoids,[29–32] tocopherols,[33–36] and plasmalogen.[37]

Pure Source of Singlet Oxygen

To identify $^1O_2$ as the reactive intermediate responsible for a cellular effect, a pure source of $^1O_2$ is required as are specific markers to detect it. Illuminated photosensitizers will produce $^1O_2$, as will electron transfer products,[38] such as O_2^-, H_2O_2, $\cdot OH$, $ROO\cdot$, and the sensitizer radical, especially when in the presence of reducing agents. Thus, researchers have used ingenious procedures to assure a pure source of $^1O_2$ in a system. This has been done by separating the sensitizer from the target sample by a gas phase, disabling the direct substrate–sensitizer interaction. Singlet oxygen is the only reactive oxygen species capable of diffusing across the intervening gas phase separation to react with a target sample placed a short distance away. Besides, a dry immobilized photosensitizer will still be active and there is no solvent to mediate electron transfer reactions.

Midden and Wang[39] described a simple method to determine the rate of $^1O_2$ reactions using an illuminated sensitizer immobilized on a glass plate close to the surface of a solution containing the target. This method was further adapted to expose bacteria [40–42] and mammalian cells[43,44] to $^1O_2$.

Although the reaction with a target in such systems is less efficient than

[29] O. Hirayama, K. Nakamura, S. Hamada, and Y. Kobayasi, *Lipids* **29**, 149 (1994).

[30] W. A. Schroeder and E. A. Johnson, *J. Biol. Chem.* **270**, 18374 (1995).

[31] H. Tatsuzawa, T. Mayurama, N. Misawa, K. Fujimori, K. Hori, Y. Sano, Y. Kambayashi, and M. Nakano, *FEBS Lett.* **439**, 329 (1998).

[32] H. Sies and W. Stahl, *Proc. Soc. Exp. Biol. Med.* **218**, 121 (1998).

[33] S. Kaiser, P. DiMascio, M. E. Murphy, and H. Sies, *Arch. Biochem. Biophys.* **277**, 101 (1990).

[34] S. Itoh, S. Nagaoka, K. Mukai, S. Ikesu, and Y. Kaneko, *Lipids* **29**, 799 (1994).

[35] A. Kamal-Eldin and L. A. Appelqvist, *Lipids* **31**, 671 (1996).

[36] K. Fukuzawa, K. Matsuura, A. Tokumura, A. Suzuki, and J. Terao, *Free Radic. Biol. Med.* **22**, 923 (1997).

[37] N. Nagan, A. K. Hajra, L. K. Larkins, P. Lazarow, P. E. Purdue, W. B. Rizzo, and R. A. Zoeller, *Biochem. J.* **332**, 273 (1998).

[38] V. S. Srinivasan, D. Podolski, N. J. Westric, and D. C. Neckers, *J. Am. Chem. Soc.* **100**, 6513 (1978).

[39] W. R. Midden and S. Y. Wang, *J. Am. Chem. Soc.* **105**, 4129 (1983).

[40] T. A. Dahl, W. R. Midden, and P. E. Hartman, *Photochem. Photobiol.* **46**, 346 (1987).

[41] T. A. Dahl, W. R. Midden, and P. E. Hartman, *J. Bacteriol.* **171**, 2188 (1989).

[42] T. A. Dahl, W. R. Midden, and P. E. Hartman, *Mutat. Res.* **201**, 127 (1988).

[43] T. P. Wang, J. Kagan, S. Lee, and T. Keiderling, *Photochem. Photobiol.* **52**, 753 (1990).

[44] T. A. Dahl, *Photochem. Photobiol.* **57**, 248 (1993).

in solution, $^1O_2$ has a much longer lifetime in the gaseous phase (about 10 times) than in a liquid phase.[45] Singlet oxygen in atmospheric air can diffuse some millimeters [diffusive transport $\delta = (D\tau)^{1/2} \approx 1.5$ mm; collisional lifetime τ (approximately 0.1 sec), D diffusion coefficient of oxygen in air[45]]. Thus, it is possible to detect the reaction of $^1O_2$ with a substrate separated by several millimeters (1–3 mm, depending on the system).

Singlet oxygen probably reacts only at the surface of the condensed phase.[46] However, diffusion of target molecules within the volume will be much faster than its rate of reaction with $^1O_2$ at the surface. Thus, with time, reaction of all the substrate molecules can be obtained.

Cellular Markers for Singlet Oxygen

In order to identify the reactive intermediate responsible for the effect of photosensitization, $^1O_2$ quenchers and enhancers of $^1O_2$ lifetime (deuterated solvents) have been employed. Although indicative of $^1O_2$, these agents are not specific enough for $^1O_2$ to allow conclusive identification.[47] Besides, deuterated water by itself causes dramatic effects on the cytoskeleton of cells[48] and arrests cells during interphase.[49]

Specific cellular markers for $^1O_2$ are required. Because the main effect of an external $^1O_2$ source on cells mainly involves the plasma membrane and because O_2 is much more soluble in lipids than in water, membrane lipids are effective targets for $^1O_2$. Cholesterol hydroperoxides[50] and plasmalogen[51] oxidation products have been proposed as markers for $^1O_2$.

By using a photosensitization reaction with separation of the target from the $^1O_2$ source, we have found that catalase, either purified or *in vivo*, is specifically oxidized by $^1O_2$.[52,53] Thus, this reaction can be used as a specific marker for $^1O_2$.

[45] J. G. Parker, *Photochem. Photobiol.* **48,** 225 (1988).

[46] G. Deadwyler, P. D. Sima, Y. Fu, and J. R. Kanofsky, *Photochem. Photobiol.* **65,** 884 (1997).

[47] T. A. Dahl, W. R. Midden, and P. E. Hartman, *Photochem Photobiol.* **47,** 357 (1988).

[48] H. Omori, M. Kuroda, H. Naora, H. Takeda, Y. Nio, H. Otani, and K. Tamura, *Eur. J. Cell Biol.* **74,** 273 (1997).

[49] D. Schroeter, J. Lamprecht, R. Eckhardt, G. Futterman, and N. Paweletz, *Eur. J. Cell Biol.* **58,** 365 (1992).

[50] P. G. Geiger, W. Korytowski, F. Lin, and A. W. Girotti, *Free Radic. Biol. Med.* **23,** 57 (1997).

[51] O. H. Morand, *Methods Enzymol.* **234,** 603 (1994).

[52] F. Lledías, P. Rangel, and W. Hansberg, *J. Biol. Chem.* **273,** 10630 (1998).

[53] F. Lledías, P. Rangel, and W. Hansberg, *Free Radic. Biol. Med.* **26,** 1396 (1999).

Modification of Catalase by Singlet Oxygen

Spontaneous modification of purified Cat-1 from *Neurospora crassa* is dependent on O_2; no modification occurs under Ar. Singlet oxygen generated by photosensitization reactions separated from the enzyme brings about a rapid sequential shift in the electrophoretic mobility of purified Cat-1[52] similar to the ones observed *in vivo*.[53,54] Cat-1 does not change with light alone. The electrophoretic mobility of Cat-1 changes due to a gain in net negative charges, shifting the pI from 5.45 to 5.25, a change of 0.05 pH units per modified monomer in a sequential transformation of all monomers in the tetramer.[52]

The probable modification site in Cat-1 is the heme group. Results indicate that the heme is altered by $^1O_2$ in what seems to be a three-step reaction. Modification increases the asymmetry of the heme as suggested by absorbance spectra.[52]

All catalases tested so far are modified by $^1O_2$. Bacterial, fungal, plant, and animal catalases are susceptible to $^1O_2$ oxidation, giving rise to enzyme conformers with a higher migration rate toward the cathode than the unmodified enzyme.[52] The most remarkable result is that all catalases tested remain active after $^1O_2$ oxidation.

In Vivo Oxidation of Catalase

Using an adaptation of the method described by Dahl *et al.*[40] for exposure to a pure source of $^1O_2$ of *N. crassa* conidia collected on a membrane filter, Cat-1 is oxidized within minutes.[53] Light by itself increases Cat-1 modification in germinating conidia. After 1 hr of illumination of conidia with intense light (1.5 W/cm²), Cat-1a is completely transformed into Cat-1c and Cat-1e conformers (Fig. 1A), whereas under laboratory light (0.036 W/cm²) it takes 3 hr to observe some Cat-1a modification.[53] Not only during germination of conidia is Cat-1 oxidized but also in the aerial hyphae just before forming conidia. Air exposure of hyphae starts conidiation. After 10 hr air exposure, Cat-1a in the aerial hyphae is transformed into Cat-1c and Cat-1e (Fig. 1B).[54]

The carotenoid content increases 10-fold during conidiation. Carotenoid mutant strains in *N. crassa* can be used to enhance *in vivo* catalase modification. A rapid Cat-1 oxidation is observed in germinating conidia from an albino strain (*al-1*) in liquid culture illuminated for 20 min (Fig. 1C). In the wild-type strain, after 1 hr under the same conditions, only some Cat-1a transformation to Cat-1c is observed.[53]

Other stress conditions in which Cat-1 modification in *N. crassa* is ob-

---

[54] W. Hansberg, *Ciência Cult.* **48,** 68 (1996).

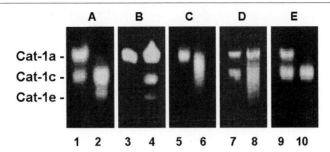

FIG. 1. *In vivo* modification of Cat-1 during cell differentiation and under stress conditions. (A) Cat-1 from germinating conidia from the wild-type strain before (lane 1) and after 1 hr illumination under intense (1.5 W/m^2) light (lane 2). Adapted from Lledías *et al.*,[53] with permission. (B) Cat-1 from aerial mycelium before conidiation (lane 3) and during conidiation (lane 4). Adapted from Hansberg,[54] with permission. (C) Cat-1 from albino (*al-1*) conidia before germinating (lane 5) and after 20 min germination under illumination at normal laboratory light (0.036 W/cm^2) (lane 6). Adapted from Lledías *et al.*,[53] with permission. (D) Cat-1 from hyphae after 24 hr growth at 30° (lane 7) and after 20 hr growth at 30° plus 4 hr at 48° (lane 8). (E) Cat-1 from hyphae after 22 hr growth (lane 9) and after 20 hr growth plus 2 hr with 5 m*M* paraquat (lane 10).

served are heat shock and treatment with paraquat. When a liquid culture after 20 hr growth at 30° is subjected for 4 hr to a heat shock of 48°, Cat-1a is transformed into other conformers with higher electrophoretic mobility as compared with a 24-hr culture without heat shock treatment (Fig. 1D). When a 20-hr culture is incubated with 5 m*M* paraquat for 2 hr, all the Cat-1a enzyme is transformed into other conformers (Fig. 1E). However, not all the activity seen in Figs. 1D and 1E is due to Cat-1. Recent experiments using 2D-electrophoresis indicate that another catalase (Cat-3) is expressed during growth and comigrates with Cat-1c.

The *in vivo* catalase modification by $^1O_2$ in *N. crassa* appears to be indicative of stress conditions. Different catalase activity bands in zymograms have been detected in bacteria under stress or cell differentiation conditions.[55,56] New catalase activities appeared on germination of some plant seeds, [57,58] and some plant catalases are induced by light.[58,59] Erythrocyte catalase from HIV(+) patients has an acidic isoform that is not present in uninfected individuals.[60]

[55] P. C. Loewen and J. Switala, *J. Bacteriol.* **169,** 3601 (1987).

[56] H.-P. Kim, J.-S. Lee, J. Ch. Hah, and J.-H. Roe, *Micobiology* **140,** 3391 (1994).

[57] P. H. Quail and J. G. Scandalios, *Proc. Natl. Acad. Sci. U.S.A.* **68,** 1402 (1971).

[58] R. Eising, R. N. Trelease, and W. Ni, *Arch. Biochem. Biophys.* **278,** 258 (1990).

[59] J. G. Scandalios, L. Guan, and A. N. Polidoros, *in* "Oxidative Stress and the Molecular Biology of Antioxidant Defenses" (J. G. Scandalios, ed.), p. 343. Cold Spring Harbor Laboratory, Cold Spring Harbor, NY, 1997.

[60] S. Yano, M. Colon, and N. Yano, *Mol. Cell. Biochem.* **165,** 77 (1996).

Methods

In Vivo Oxidation of Catalase

Microorganisms or mammalian cells are cultured and then treated under conditions in which $^1O_2$ is expected to be formed, such as intense light, photosensitizing conditions, or in the presence of a chemical source of $^1O_2$. Cell extracts are prepared and analyzed by polyacrylamide gel electrophoresis and zymograms.

Preparation of Cell Extracts. Cells are homogenized five times for 30 sec in a Bead-Beater or Vortex with glass beads (710–1180 μm for fungi and 150–212 μm for bacteria) at a ratio of 1 g of dried weight per 7.5 ml of 20 mM 4-(2-hydroxy)-1-piperazineethanesulfonic acid, pH 7.2, containing 1 mM phenylmethylsulfonyl fluoride, 1 mM dithiothreitol, and 0.1 mM deferriferrioxamine B mesylate (Desferal). Plant seeds are ground in the same buffer in a mortar. Human catalase extracted from blood clots or myeloid leukemia U937 cells are homogenized in a Potter–Elvejem homogenizer in the same buffer. Cell extracts are centrifuged for 20 min at 6000g and 4°. Supernatants usually contain 85% of the initial catalase activity and are used directly for catalase determinations.

Determination of Catalase Activity. Catalase activity is measured by determining the initial rate of O_2 production with a Clark microelectrode.[61] The reaction is started by injecting catalase, usually 5 μl or less, into a sealed chamber filled with 2 ml of 10 mM of H_2O_2 in 10 mM phosphate buffer (PB), pH 7.8 (adjusted by mixing Na_2HPO_4 and KH_2PO_4 solutions). Units are defined as micromoles O_2 produced per minute per milligram protein under these conditions. Activity is measured in samples just before loading them on a gel for electrophoresis.

Catalase Activity in Polyacrylamide Gels. Minigels (8 × 9 cm and 0.75 mm thick) of 8% polyacrylamide and 0.2% bisacrylamide are made according to the Laemmli procedure, but without adding 2-mercaptoethanol or boiling the samples. Forty to 100 U of catalase are loaded in each lane. It is convenient to load different amounts of the sample to obtain good resolution of the different bands in the gel. Gels are run at 150 V for 2.5 hr and, immediately after electrophoresis, are stained for catalase activity.

Catalase activity in polyacrylamide gels is detected by incubating the gel after electrophoresis for 5 min in 5% methanol and then, after rinsing three times with tap water, for 10 min in 10 mM H_2O_2. The gel, rinsed with tap water, is incubated in a 1/1 mixture of freshly prepared 2% potassium ferric cyanide and 2% ferric chloride. Blue color develops in the gel except at zones where H_2O_2 was decomposed by catalase. Staining is stopped by soaking the gel in a 10% acetic acid and 5% methanol solution.

---

[61] M. Rørth and P. K. Jensen, *Biochim. Biophys. Acta* **139**, 171 (1967).

In Vitro Oxidation of Catalase

Purified catalases can be used as an *in vitro* marker for $^1O_2$. We have used a pure source of $^1O_2$, generated either through a photosensitization reaction or chemically, to follow the oxidation of Cat-1 by $^1O_2$.

Catalase Purification. The cell extract is frozen and thawed twice and centrifuged for 20 min at 6000g and 4°. The resulting supernatant can be heated for 5 min at 60° (for heat resistant catalases only). After centrifugation, the supernatant is precipitated with acetone. The precipitate is resuspended in the same cell extract buffer and centrifuged. Catalase usually dissolves without activity loss. However, if catalase is inactivated by acetone, 1 volume of hexane, instead of acetone, is a convenient purification step: catalase remains active in the aqueous phase. The extract is fractionated with ammonium sulfate. The ammonium sulfate-precipitated fraction containing most of the activity (the concentration varies for different catalases) is resuspended in 3 ml of 0.5 M ammonium sulfate and 50 mM PB, and then 1 volume of phenyl-Sepharose CL-4B, equilibrated in the same solution, is added. The resulting sludge is stirred for 30 min at room temperature, washed with 50 ml of the same solution, and then loaded on a small column. The enzyme is usually eluted with 20 ml of 50 mM PB and 20 ml 10 mM PB, but this can vary for different catalases. Fractions having most of the activity are pooled and concentrated with an Amicon ultrafiltration cell, holding a YM30 filter, and applying N_2 at a pressure of 25 psi. The filter is washed with 1 volume of 10 mM PB and added to the concentrated enzyme. This procedure gives a catalase that is usually at least 90% purified, with a yield of about 70%, depending on the source. A second passage through the phenyl-Sepharose column can increase the purity of the catalase.

Catalase Hanging Drop System. A drop of purified catalase (40 units) in 2.5 mM Tris, pH 8.9, 19 mM glycine, or in 10 mM PB is suspended from a 1-ml micropipette tip, outlet closed with Parafilm, in a 2-ml glass vial in close vicinity to a dry filter paper impregnated with a photoactive substance (rose bengal, riboflavin, acridine orange, methylene blue, or toluidine blue) (Fig. 2A). Illumination is done for 3 hr at 40 cm distance. To avoid quenching by water vapor in the system, it is recommended that the vial be cooled. In the experiment shown in Fig. 2, a fan was used to keep the vial at room temperature. It is also convenient to have the catalase in a small volume (15 μl) in order to have a high surface area-to-volume ratio. After exposure, the drop is recovered by piercing the Parafilm from the inside of the tip with a Hamilton syringe.

With an incandescent light bulb of 100 W, rose bengal is the most effective photosensitizer. However, when using a tungsten-halogen lamp of 75 W at the same distance, rose bengal is not effective, but instead

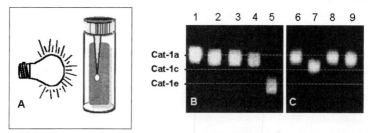

FIG. 2. Exposure of Cat-1 to a pure source of $^1O_2$. (A) Scheme of the system used. A drop of 15 μl of purified Cat-1 was suspended from a 1-ml micropipette tip, with the outlet closed with Parafilm, in a 2-ml glass vial in close vicinity to a dry filter paper impregnated with a photoactive substance. Singlet oxygen is the only reactive oxygen species in this system that can diffuse through air. A fan was used to keep the vial at room temperature. (B) Zymogram of purified Cat-1a after exposure to a pure source of singlet oxygen using an incandescent light bulb of 100 W for 3 hr at 40 cm distance. Lane 1, control of Cat-1a illuminated in the absence of photosensitizing agent; lanes 2–5, in the presence of acridine orange, methylene blue, riboflavin, and rose bengal, respectively. (C) Zymogram of purified Cat-1a after exposure to a pure source of singlet oxygen using a tungsten-halogen lamp of 75 W for 3 hr at 40 cm distance. Lanes 6–9, acridine orange, toluidine blue, riboflavin, and rose bengal, respectively (A and B, adapted from Lledías *et al.,*[52] with permission).

toluidine blue brings about the same modification of Cat-1, although with a low efficiency (Figs. 2B and 2C). To avoid photoinhibition of plant and animal catalases, a yellow cutoff filter at 515 nm (Schott OG515) is placed in front of the light bulb to avoid heme excitation, and rose bengal is used as the photosensitizer.

Concluding Remarks

Most aerial organisms have a typical catalase, along with catalase peroxidases and peroxidases. Typical catalases are a conserved family of proteins, sharing over 35% amino acid identity from bacteria to humans. Catalase is one of the most active enzymes known and is usually stable. Because of this, its activity can be detected readily and the enzyme is purified easily. Catalases from bacteria, fungi, plants, and animals are oxidized by $^1O_2$, giving rise to more acidic, fully active conformers.[52] Total cell extracts can be analyzed by gel electrophoresis to detect the catalase mobility shift caused by $^1O_2$. The oxidation of catalase is specific for $^1O_2$ and is probably due to heme modification.[52] An electrophoretic mobility shift of catalase from cells induced to differentiate or subjected to stress conditions could be indicative of $^1O_2$. Catalase modification can be increased or decreased according to the amount of effectively quenching carotenoids in the cell.

A purified catalase can also serve as a marker to detect *in vitro* generation of $^1O_2$.

Acknowledgments

Part of this work was supported by Grant IN-206097 from DGAPA/UNAM. We thank Dr. Jesús Aguirre for critically reading the manuscript.

[12] Nuclear Factor-κB Activation by Singlet Oxygen Produced during Photosensitization

By JEAN-YVES MATROULE and JACQUES PIETTE

Introduction

NF-κB transcription factors bind DNA as dimers constituted from a family of proteins designated as the Rel/NF-κB family. In mammals, this family contains proteins p50, p52, p65 (RelA), RelB, and c-Rel (Rel).[1,2] These five proteins harbor a related, but nonidentical 300 amino acid long Rel homology domain (RelHD), which is responsible for dimerization, nuclear translocation, and specific DNA binding. In addition, RelA, RelB, and c-Rel, but not p50 or p52, contain one or two transactivating domains. NF-κB complexes are sequestered in the cytoplasm of most resting cells by inhibitory proteins belonging to the IκB family.[3–5] Members of the IκB family are IκBα, IκBβ, IκBε, p100, and p105. Following various stimuli, including the interaction of TNF-α and IL-1β with their receptors, IκBα is first phosphorylated on serines 32 and 36, ubiquitinated at lysines 21 and 22, and degraded rapidly by the proteasome, allowing NF-κB nuclear translocation and gene activation to take place.[6,7] In the case of these two types of cytokines, the signal transduction pathway leading to the phosphorylation and degradation of IκB proteins has been clarified in HeLa

[1] P. A. Baeuerle and T. Henkel, *Annu. Rev. Immunol.* **12,** 141 (1994).

[2] S. Miyamoto and I. M. Verma, *Adv. Cancer Res.* **66,** 255 (1996).

[3] A. A. Beg and A. S. Baldwin, *Genes Dev.* **7,** 2064 (1995).

[4] F. Mercurio, J. A. DiDonato, C. Rosette, and M. Karin, *Genes Dev.* **7,** 705 (1993).

[5] S. Haskill, A. A. Beg, S. M. Tompkins, J. S. Morris, A. D. Yurochko, A. Sampson-Johannes, K. Mondal, P. Ralph, and A. S. Baldwin, *Cell* **65,** 1281 (1991).

[6] K. Brown, S. Park, T. Kanno, G. Franzoso, and U. Siebenlist, *Proc. Natl. Acad. Sci. U.S.A.* **90,** 2532 (1993).

[7] E. B.-M. Traenckner, H. L. Pahl, T. Henkel, K. N. Schmidt, S. Wilk, and P. A. Baeuerle, *EMBO J.* **14,** 2876 (1995).

and L293 cells.[8–10] It is included in a 700- to 900-kDa complex called signalosome whose important partners are proteins associated to the TNF-α or IL-1 receptors, NIK, IKK-α, -β, and -γ.

It has been shown that NF-κB activation in response to UV irradiation of HeLa cells or of primary skin fibroblasts occurs via a slow and sustained kinetics involving dissociation of NF-κB from IκBα and degradation of the latter by the proteasome.[11,12] Early IκBα degradation does not require IKK and does not depend on IκBα phosphorylation at positions 32 and 36. However, induced IκBα degradation requires intact N (1–36)- and C (277–287)-terminal sequences.[11]

Proinflammatory cytokines such as TNF-α or IL-1β or the bacterial outer membrane component (LPS) are potent activators of NF-κB, which mediates several of their biological activities, such as stimulation of the transcription in lymphocytes through the intracellular generation of oxidative stress.[13,14] The reactive oxygen species (ROS) generated intracellularly by proinflammatory cytokines and required for NF-κB activation is suspected of being peroxide in nature, but no definitive demonstration has yet been provided. Schreck et al.[15,16] have proposed peroxides as secondary messengers in NF-κB activation by showing that (i) hydrogen peroxide alone can induce NF-κB and (ii) other ROS such as superoxide anion and singlet oxygen ($^1O_2$) have no effect. The exclusion of $^1O_2$ as a mediator in NF-κB activation is based on the use of thermal decomposition of the endoperoxide of naphthalene-3,4-dipropionate, which does not enter the cells (see also Klotz et al., article [13], this volume). However, using photosensitizers such as methylene blue, proflavin, and rose bengal, which are uptaken by many cell types and are good $^1O_2$ generators, it has been shown

[8] C. H. Régnier, H. Y. Song, X. Gao, D. V. Goeddel, Z. Cao, and M. Rothe, *Cell* **90,** 373 (1997).

[9] H. Y. Song, C. Régnier, C. J. Kirschning, D. V. Goeddel, and M. Rothe, *Proc. Natl. Acad. Sci. U.S.A* **94,** 9792 (1997).

[10] F. Mercurio, H. Zhu, B. W. Murray, A. Shevchenko, B. L. Bennett, J. W. Li, D. B. Young, M. Barbosa, and B. Mann, *Science* **278,** 860 (1997).

[11] K. Bender, M. Göttlicher, S. Whiteside, H. J. Rahmsdorf, and P. Herrlich, *EMBO J.* **17,** 5170 (1998).

[12] N. Li and M. Karin, *Proc. Natl. Acad. Sci. U.S.A* **95,** 13012 (1998).

[13] J. A. Satriano, M. Shuldiner, K. Hora, Y. Xing, Z. Shan, and D. Schlondorff, *J. Clin. Invest.* **92,** 1564 (1995).

[14] L. Feng, Y. Xia, G. E. Garcia, D. Hwang, and C. B. Wilson, *J. Clin. Invest.* **95,** 1669 (1995).

[15] R. Schreck, P. Rieber, and P. A. Baeuerle, *EMBO J.* **10,** 2247 (1991).

[16] R. Schreck, B. Meier, D. N. Mannel, W. Dröge, and P. A. Baeuerle, *J. Exp. Med.* **175,** 1181 (1992).

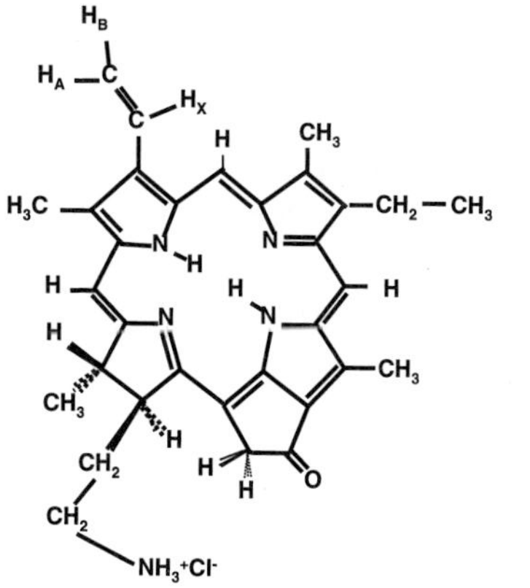

FIG. 1. Chemical structure of aminopyropheophorbide.

that NF-κB can be activated under these experimental conditions together with a concomitant degradation of IκBα.[17,18]

One standard experimental tool used to demonstrate $^1O_2$ involvement in a biological process is the replacement of water by deuterium oxide, which is known to increase its lifetime.[19] Deuterium substitution has been shown on several occasions to increase cell lethality triggered by photosensitization,[20] but it has rarely been used to demontrate the involvement of singlet oxygen in transcription factor activation, except for AP-2 activation after UV-A treatment of keratinocytes.[21]

This article demontrates the involvement of $^1O_2$ in NF-κB activation in a human colon cancer cell line photosensitized by a promising new drug (aminopyropheophorbide, APP; Fig. 1) for photodynamic therapy (PDT).[22] Several lines of evidence suggest that $^1O_2$ is the major damaging species in PDT.[23–25] Biological effects resulting from PDT-induced changes in gene

[17] S. Legrand-Poels, V. Bours, B. Piret, M. Pflaum, B. Epe, B. Rentier, and J. Piette, *J. Biol. Chem.* **270**, 6925 (1995).

[18] B. Piret, S. Legrand-Poels, C. Sappey, and J. Piette, *Eur. J. Biochem.* **228**, 44 (1995).

[19] C. S. Foote and E. L. Clennan, *in* "Active Oxygen Chemistry" (C. S. Foote, J. S. Valentine, A. Greenberg, and J. F. Libman, eds.), p. 105. Blakie Academic and Professional, London, 1995.

[20] J. Piette, C. M. Calberg-Bacq, and A. Van de Vorst, *Photochem. Photobiol.* **26**, 377 (1977).

[21] S. Grether-Beck, S. Olaizola-Horn, H. Schmidt, M. Grewe, A. Jahnke, J. P. Johnson, K. Briviba, H. Sies, and J. Krutmann, *Proc. Natl. Acad. Sci. U.S.A* **93**, 14586 (1996).

[22] A.-S. Fabiano, D. Allouche, Y.-H. Sanejouand, and N. Paillous, *Photochem. Photobiol.* **66**, 336 (1997).

[23] K. R. Weishaupt, C. J. Gomer, and T. J. Dougherty, *Cancer Res.* **36**, 2326 (1976).

[24] T. Ito, *Photochem. Photobiol.* **28**, 493 (1978).

[25] M. Athar, H. Muktar, and D. Bickers, *J. Invest. Dermatol.* **90**, 652 (1988).

expression and signal transduction are still largely unknown [for effect of 5-aminolevulinate-PDT on cellular signaling (MAPK activation), see Klotz *et al.,* article [13], this volume]. Because cytokine release during PDT may have important biological effects for surrounding cells, we will focus our attention on transcription factor NF-κB because it is a redox-activated transcription factor involved in the control of genes encoding several important cytokines such as interleukin (IL)-1, IL-2, IL-6, and tumor necrosis factor (TNF)-α as well as chemokines such as IL-8, RANTES, and MIP-1.[26]

$^1O_2$ Involvement in Cell Killing Mediated by Aminopyropheophorbide (APP) Photosensitization

Procedures

Aminopyropheophorbide (APP). Aminopyropheophorbide chlorhydrate, received from Dr. N. Paillous (University of Toulouse, France), is used without any further purification.[22] APP is obtained by substituting the acid group of pyropheophorbide-*a* by an ammonium group via a Curtius rearrangement in mild conditions. An APP stock solution is made in ethanol and its concentration is determined by absorption spectroscopy (UV/visible Perkin-Elmer spectrometer lambda 40). The extinction coefficient at 666 nm is 18,000 M^{-1} cm^{-1}.

APP-Mediated Photosensitization. The human colon carcinoma cell line HCT-116 (from ATCC, Rockville, MD) is grown in McCoy's 5A medium (Gibco BRL, UK) supplemented with 10% fetal calf serum (FCS, Gibco-BRL). Before photosensitization, HCT-116 cells are incubated with 2 μM APP for the last 20 hr. Prior to irradiation, HCT-116 cells are washed twice with phosphate-buffered saline (PBS) and then irradiated with red light ($\lambda > 600$ nm) at a fluence rate of 160 W/m^2 in petri dishes covered with PBS. After irradiation, HCT-116 cells are returned to culture at 37° in McCoy's 5A medium supplemented with 2% FCS. Cell survival is determined after 24 hr using trypan blue exclusion.

Cell Photosensitization in D_2O. HCT-116 cells are grown as just described, but prior to irradiation are washed twice with PBS solubilized in D_2O (PBS-D_2O) and covered with PBS-D_2O before being irradiated as described earlier. After irradiation, HCT-116 cells are covered with culture medium and cultured as described previoulsy.

Results

APP-Mediated Photosensitization Induces HCT-116 Cell Killing. Incubation of HCT-116 cells with 2 μM APP for 20 hr prior to irradiation with

[26] U. Siebenlist, G. Franzoso, and K. Brown, *Annu. Rev. Biol.* **10,** 405 (1994).

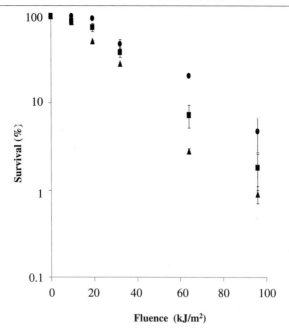

FIG. 2. HCT-116 cell lethality induced by APP-mediated photosensitization. Incubation conditions were as described in the text. ■, irradiation carried out in PBS; ▲, irradiation performed in PBS-D₂O; ●, irradiation done in PBS; HCT-116 cells, however, were preincubated with 100 μM PDTC. Cell survival was determined by trypan blue exclusion.

red light induces lethality, which is light dose dependent (Fig. 2). Irradiation with 100 kJ/m² leads to an approximate two order of magnitude decrease in cell survival. This lethality is dependent on the presence of light and APP as irradiation without either APP or light does not generate any loss of cell survival.

$^1O_2$ Is Involved in HCT-116 Cell Killing by APP Photosensitization. Irradiation of HCT-116 cells incubated with 2 μM APP in PBS-D₂O increases cell lethality significantly (Fig. 2). Because the lifetime of $^1O_2$ is increased by deuterium substitution, it is very likely that $^1O_2$ is the ROS mediating cell killing. It should, however, be kept in mind that D_2O can also enhance the lifetime of triplet excited states of several photosensitizers.[27,28] Using fluorescence microscopy, APP has been shown to localize mainly in the cytoplasmic and internal membranes (lysosome, endoplasmic reticulum, etc.), with slight and diffuse fluorescence also observed in the cytoplasm.

[27] A. Seret, M. Hoebeke, and A. Van de Vorst, *Photochem. Photobiol.* **52,** 601 (1990).
[28] N. J. De Mol, G. M. J. Beijersbergen van Henegouwen, B. Weeda C. N. Knox, and T. G. Truscott, *Photochem. Photobiol.* **44,** 747 (1986)

Because deuterium substitution is more efficient in an aqueous compartment (cytoplasm) than in a rather hydrophobic area such as membranes, one could argue that the deuterium substitution effect shown in Fig. 2 could largely reflect the cytotoxic contribution triggered by APP located in the cytoplasm. However, the effect of solvent isotopic substitution is also influenced by the concentration of the substrate reacting with $^1O_2$, a high substrate concentration yielding a low isotopic effect.[19] Because the ratio of the light dose giving rise to 37% survival in D_2O vs H_2O is between 2 and 3, we can postulate that membranes are likely not primary targets responsible for the cytotoxic effect of APP-mediated photosensitization. Indeed, the substrate concentration for $^1O_2$ is very high in membranes, rendering them unsuitable to detect a isotopic substitution effect.

However, incubation of HCT-116 cells with pyrrolidine 9-dithiocarbamate (PDTC, 100 μM) before and during photosensitization reduces cell killing mediated by APP (Fig. 2). This reduction does not exceed 10%. Because PDTC is a hydrophilic molecule that cannot penetrate into membrane, these data lead us to the idea that, as expected, the cell killing could be due to $^1O_2$ released in the cytoplasm or outside the cells, oxidizing a cytoplasmic or an extramembranous target.

NF-κB Activation by APP-Mediated Photosensitization of HCT-116 Cells

Procedures

APP-Mediated Photosensitization of HCT-116 Cells. Before photosensitization with APP, HCT-116 cells are grown for 1 week in McCoy's 5A medium supplemented with 2% FCS. HCT-116 cells are incubated with APP (2 μM) for 20 hr at 37° in McCoy's 5A medium supplemented with 2% FCS and then washed once in PBS or PBS-D_2O before being covered with either PBS or PBS-D_2O. HCT-116 cells are irradiated with 32 kJ/m^2 with the same light output as described earlier. After irradiation, HCT-116 cells are covered with culture medium supplemented with 2% FCS during various period of times.

Electrophoretic Mobility Shift Assay (EMSA). Nuclear extracts are isolated by a rapid micropreparation technique derived from the large-scale procedure of Dignam *et al.*[29] based on the use of a lysis with detergent followed by high salt extraction of nuclei.[29] In short, HCT-116 cells are washed twice in cold PBS and further washed with 10 mM HEPES, 20 mM KCl, 2 mM MgCl$_2$, 100 μM EDTA, 1 mM dithiothreitol (DTT), and 0.5

---

[29] D. Dignam, R. M. Lebovitz, and R. G. Roeder, *Nucleic Acids Res.* **11,** 1475 (1983).

mM phenylmethylsulfonyl fluoride (PMSF), pH 7.9 (wash buffer). Cells are lyzed in 10 mM HEPES, 10 mM KCl, 2 mM MgCl$_2$, 100 μM EDTA, 1 mM DTT, and 0.5 mM PMSF, pH 7.9, containing Nonidet P-40 (0.2%, v/v) plus a protease inhibitor cocktail (Boehringer Manheim, Germany) and centrifuged at 5000g for 5 min at 4°. Pellets are washed once with the wash buffer and resusupended in the extraction buffer [20 mM HEPES, 1.5 mM MgCl$_2$, 200 μM EDTA, 640 mM NaCl, 1 mM DTT, 0.5 mM PMSF, 25% (v/v) glycerol] and left on ice for 30 min before being centrifuged for 30 min at 15,000g at 4°. The protein content of the nuclear extracts is measured with the Bio-Rad protein assay kit (Bio-Rad). Binding reactions are performed for 25 min at room temperature with 7.5 μg total protein in 20 μl of 20 mM HEPES–KOH (pH 7.9), 75 mM NaCl, 1 mM EDTA, 5% glycerol, 0.5 mM MgCl$_2$, 2 μg acetylated bovine serum albumin, 4 μg poly(dI-dC) (Pharmacia, UK), 1 mM DTT, and 0.2 ng of ^{32}P-labeled oligonucleotides (Eurogentech, Belgium). Oligonucleotides are labeled by end-filling with the Klenow fragment of *Escherichia coli* DNA polymerase I (Boehringer-Mannheim, Germany) with ^{32}P-dATP, ^{32}P-dCTP (New England Nuclear or ICN, UK), and cold dTTP+dGTP. Labeled probes are purified by spin chromatography on G-25 columns (Boehringer-Mannheim). DNA–protein complexes are separated from unbound probe on native 6% polyacrylamide gels at 200 V in 0.25 M Tris, 0.25 M sodium borate, and 0.5 mM EDTA, pH 8.0. Gels are vacuum dried and exposed to Fuji X-ray film at −80° for 16 to 24 hr. The amounts of specific complexes are determined either by counting the radioactivity with a phosphorimager (Molecular Dynamics) or by photodensitometry (LKB, Sweden) of the autoradiography. Supershift experiments are carried out as described,[30] using the same EMSA protocol as described earlier, except for the gel concentration being 4%. The sequences of the probes (Eurogentech, Belgium) used in this work are

Wild-type NF-κB probe: 5′-GGTTACAAGGGACTTTCCGCTG
                                         TGTTCCCTGAAAGGCGACGGTT-5′
Mutated NF-κB probe: 5′-GGTTAACAACTCACTTTCCGCTG
                                         TGTTGAGTGAAAGGCGACGGTT-5′

Results

NF-κB Is Activated by APP Photosensitization. Determining whether photodynamic photosensitizers such as APP, which is mainly localized in cytoplasmic and internal membranes, could activate NF-κB was of interest because it could give important information on possible immunomodulation

---

[30] G. Franzoso, V. Bours, S. Park, M. Tomita-Yamagushi, K. Kelly, and U. Siebenlist, *Nature* **359,** 207 (1992).

A

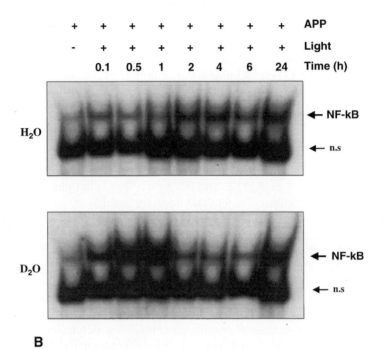

B

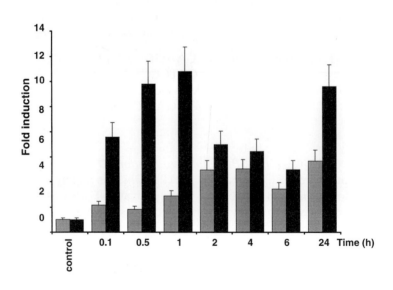

processes triggered by photosensitized tumor cells. To this end, HCT-116 cells were incubated for 20 hr in the dark with 2 μM APP and then irradiated with red light. Nuclear extracts were prepared at various times after photosensitization and analyzed by EMSA. As shown in Fig. 3A, a large retarded band appeared rapidly after irradiation, with its maximal intensity observed between 10 and 30 min. A second wave of NF-κB activation could then be observed 2 hr after irradiation (Fig. 3A). Contrary to the first wave of activation, the second appeared slowly and was sustained up to 24 hr. Competition experiments carried out with a wild-type or a mutated unlabeled NF-κB probe demonstrated that the upper retarded band was specific whereas the lower one was not (data not shown). Using antibodies directed against the various members of the Rel/NF-κB family (p50, RelA, c-Rel, and p52), we also observed that the retarded complex involved the classical p50/RelA heterodimer (data not shown).

In order to evaluate the role of $^1O_2$ in NF-κB activation by APP-mediated photosensitization, HCT-116 cells were irradiated in PBS where H_2O was replaced by D_2O, as the lifetime of $^1O_2$ is increased by this isotopic substitution.[19] EMSA analysis revealed that isotopic substitution increased the intensity of the band corresponding to the p50/RelA complex significantly, demonstrating the involvement of $^1O_2$ in NF-κB activation by APP photosensitization (Fig. 3A). Interestingly, this increase in band intensity was mainly detectable during the first rapid and transient phase (30 and 60 min after irradiation), whereas there was no modification of the band intensity during the second wave (Fig. 3B).

Because APP is mainly located within membranes, we decided to determine whether NF-κB activation would be mediated by $^1O_2$ generated by

---

FIG. 3. Role of $^1O_2$ generated during APP photosensitization on κB-DNA-binding activities in HCT-116 cells. (A) Electromobility shift assay of NF-κB DNA-binding activity. (Top) Induction of a nuclear κB enhancer DNA-binding protein after treatment of HCT-116 cells with 2 μM APP for 17 hr and 32 kJ/m² of irradiation with red light. Nuclear extracts were prepared at various times after irradiation and equal amounts of protein were mixed with a ³²P-labeled κB probe. Samples were loaded on 6% native polyacrylamide gels and electrophoresed at 150 V. The autoradiogram of the gel is shown and the arrow indicates the position of the specific complex; n.s., nonspecific DNA binding. (Bottom) EMSA analysis of NF-κB activity in nuclear extracts of HCT-116 cells treated with 2 μM APP and light in the presence of deuterium oxide. EMSA were carried out as in A and arrows indicate the specific NF-κB complex and a nonspecific band (n.s.). (B) Photodensitometric analysis of the NF-κB specific complex bound to the probe at various times after HCT-116 cell photosensitization mediated by APP. Data are expressed in fold increases of the control (unirradiated HCT-116 cells mixed with APP); gray bar, irradiation done in PBS; black bar, irradiation done in PBS-D_2O. Values represent the mean of three separate experiments with error bars representing the standard error of the mean.

excited APP within membranes or through $^1O_2$ generated in the cytoplasm. If the latter proved to be the case, this effect could be inhibited by an antioxidant molecule that cannot penetrate into the membranes. In order to clarify this point, we chose to investigate the effect of a cytoplasmically located hydrophilic molecule having both antioxidant and metal-chelating properties: PDTC (100 μM, added 90 min before and during irradiation). As shown in Fig. 4, PDTC exhibits no inhibitory effect on NF-κB activation in HCT-116 cells by APP photosensitization (compare top and bottom panels of Fig. 4). These data suggest that $^1O_2$ acts as a second messenger for NF-κB activation through a reaction taking place mainly at the membrane level because a hydrophilic antioxidant, such as PDTC, cannot inhibit its activation.

APP localizes mainly in membranes and, very likely, promotes NF-κB activation through an oxidative stress involving $^1O_2$. The cellular localization of the sensory molecules triggering NF-κB activation still remains a

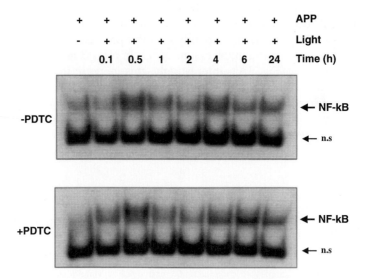

FIG. 4. Effect of PDTC on κB-DNA-binding activities in HCT-116 cells photosensitized by APP. (Top) Induction of a nuclear κB enhancer DNA-binding protein after treatment of HCT-116 cells with 2 μM APP for 17 hr and 32 kJ/m^2 of irradiation with red light. Nuclear extracts were prepared at various times after irradiation and equal amounts of protein were mixed with a ^{32}P-labeled κB probe. Samples were loaded on 6% native polyacrylamide gels and electrophoresed at 150 V. The autoradiogram of the gel is shown and the arrow indicates the position of the specific complex; n.s., nonspecific DNA binding. (Bottom) EMSA analysis of NF-κB activity in nuclear extracts of HCT-116 cells treated with 2 μM APP and light in the presence of 100 μM PDTC. EMSA were carried out as in A and arrows indicate the specific NF-κB complex and a nonspecific band (n.s.).

mystery, but it should be pointed out that extracellular parts of receptors are likely to be involved because a deuterium substitution enhancement effect is observed, ruling out the role of phospholipids or integral membrane proteins as sensors. It has been shown that pyropheophorbide a methyl ester-mediated photosensitization preferentially targets the IL-1 receptor, reinforcing the idea that receptors are probably involved in NF-κB activation mediated by APP photosensitization.[31] $^1O_2$ is known to be the main ROS produced by photosensitization[20] and is proposed by several authors as a second messenger in gene activation in human skin fibroblasts irradiated by UV-A.[21,32,33] Because its lifetime can be increased significantly by deuterium substitution, the observation that NF-κB translocation is greater in a medium where H_2O is substituted by D_2O suggests that $^1O_2$ is involved in the activation mechanism. However, definitive proof of $^1O_2$ involvement in NF-κB activation could issue directly from measuring its emission at 1268 nm. Although this detection is still very difficult to perform with photosensitizers located at the membrane level, data reported with rose bengal irradiated in keratinocyte are encouraging.[34] Coupling molecular biology technologies with direct spectral detection of $^1O_2$ generated inside cells will be a major step toward fully understanding the role of $^1O_2$ in many biological processes, particularly cell signaling.

Acknowledgments

This work was supported by the Belgian National Fund for Scientific Research (NFSR, Brussels, Belgium) and the concerted action program of the University of Liege. J-Y.M. is a Ph.D. student supported by the FRIA (Brussels, Belgium) and J.P. is research director with the NFSR.

[31] J.-Y. Matroule, G. Bonizzi, P. Morliere, N. Paillous, R. Santus, V. Bours, and J. Piette, *J. Biol Chem.* **274,** 2988 (1999).

[32] G. F. Vile and R. M. Tyrrell, *Free Radic. Biol. Med.* **18,** 721 (1995).

[33] K. Scharffetter-Kochanek, M. Wlaschek, P. Brenneissen, M. Schausen, R. Blaudschun, and J. Wenk, *Biol. Chem,* **378,** 1247 (1997).

[34] P. Bilski, B. M. Kukielczak, and C. F. Chignell, *Photochem. Photobiol.* **68,** 675 (1998).

[13] Mitogen-Activated Protein Kinase Activation by Singlet Oxygen and Ultraviolet A

By LARS-OLIVER KLOTZ, KARLIS BRIVIBA, and HELMUT SIES

Introduction

Ultraviolet A (UVA) radiation (320–400 nm) has been connected with processes such as skin aging and photocarcinogenesis.[1] This is due to the fact that UVA penetrates deeper into skin than UVB (280–320 nm) or UVC (<280 nm)[2] and that radiation at wavelengths below 290 nm is completely absorbed by the stratospheric ozone layer[3]; it is also due to its ability to damage DNA and to affect gene expression. For instance, the induction of matrix metalloproteinases by UVA may lead to degradation and disruption of the dermal extracellular matrix,[4] facilitating processes such as wrinkle formation and metastasis of tumor cells. Enhanced expression of a variety of genes, among them heme oxygenase-1,[5,6] interstitial collagenase (MMP-1),[4,7] inflammatory cytokines (IL-1, IL-6),[8] intercellular adhesion molecule-1 (ICAM-1),[9] and Fas-Ligand,[10] can be induced by UVA.

Singlet oxygen ($^1O_2$) is a mediator of UVA-induced expression of these proteins (see Ref. 11 for review). AP-1,[12] AP-2,[9] and NF-κB[13] are transcrip-

[1] K. Scharffetter-Kochanek, M. Wlaschek, P. Brenneisen, M. Schauen, R. Blaudschun, and J. Wenk, *Biol. Chem.* **378,** 1247 (1997).

[2] R. M. Tyrrell, *Bioessays* **18,** 139 (1996).

[3] F. Urbach, *J. Photochem. Photobiol. B Biol.* **40,** 3 (1997).

[4] K. Scharffetter, M. Wlaschek, A. Hogg, K. Bolsen, A. Schothorst, G. Goerz, T. Krieg, and G. Plewig, *Arch. Dermatol. Res.* **283,** 506 (1991).

[5] S. M. Keyse and R. M. Tyrrell, *Proc. Natl. Acad. Sci. U.S.A.* **86,** 99 (1989).

[6] S. Basu-Modak and R. M. Tyrrell, *Cancer Res.* **53,** 4505 (1993).

[7] K. Scharffetter-Kochanek, M. Wlaschek, K. Briviba, and H. Sies, *FEBS Lett.* **331,** 304 (1993).

[8] M. Wlaschek, G. Heinen, A. Poswig, A. Schwarz, T. Krieg, and K. Scharffetter-Kochanek, *Photochem. Photobiol.* **59,** 550 (1994).

[9] S. Grether-Beck, S. Olaizola-Horn, H. Schmitt, M. Grewe, A. Jahnke, J. P. Johnson, K. Briviba, H. Sies, and J. Krutmann, *Proc. Natl. Acad. Sci. U.S.A.* **93,** 14586 (1996).

[10] A. Morita, T. Werfel, H. Stege, C. Ahrens, K. Karmann, M. Grewe, S. Grether-Beck, T. Ruzicka, A. Kapp, L. O. Klotz, H. Sies, and J. Krutmann, *J. Exp. Med.* **186,** 1763 (1997).

[11] K. Briviba, L. O. Klotz, and H. Sies, *Biol. Chem.* **378,** 1259 (1997).

[12] M. Djavaheri-Mergny, J. L. Mergny, F. Bertrand, R. Santus, C. Mazière, L. Dubertret, and J. C. Mazière, *FEBS Lett.* **384,** 92 (1996).

[13] G. F. Vile, A. Tanew-Ilitschew, and R. M. Tyrrell, *Photochem. Photobiol.* **62,** 463 (1995).

tion factors found to be activated by UVA. NF-κB is also activated by $^1O_2$,[14,15] and AP-2 is an integral part of the signaling cascade leading from UVA via $^1O_2$ to an enhanced expression of ICAM-1.[9] In the signaling cascade upstream of AP-1, there are members of the mitogen-activated protein kinase (MAPK) family, the extracellular signal-regulated kinases (ERK), ERK1 and ERK2, the c-Jun-N-terminal kinases (JNKs), and the p38 MAPKs.[16,17] The MAPKs, once activated by dual phosphorylation on a Thr-X-Tyr motif, phosphorylate and thereby activate transcription factors such as c-Jun, ATF-2, and ternary complex factors (such as Elk-1), which, in turn, lead to the induced expression of c-jun and/or c-fos to form the AP-1 proteins, c-Jun and c-Fos.[17,18]

Employing the methods described here, we found that UVA is also capable of activating JNKs[19] and p38-MAPKs[20] in human skin fibroblasts; however, no activation of ERKs was detectable.[19,20] This is a pattern of MAPK activation different from that elicited by UVC, UVB, and hydrogen peroxide. $^1O_2$ was then shown to elicit a similar pattern of MAPK activation[19,20] and to mediate the UVA-induced activation of JNK.[19] Activation of p38 by UVA may also be mediated by $^1O_2$.[20] Furthermore, $^1O_2$ activates JNKs and p38-MAPK only when generated intracellularly.[20] Photodynamic treatment, based on the application of 5-aminolevulinate (ALA) to feed porphyrin biosynthesis and subsequent irradiation with red light, was shown to activate JNKs and p38 but not ERKs in human keratinocytes.[21] ALA-PDT is a treatment proposedly relying on the formation of $^1O_2$, which is in concert with the induced UVA/$^1O_2$ pattern of MAPK activation.

[14] B. Piret, S. Legrand-Poels, C. Sappey, and J. Piette, *Eur. J. Biochem.* **228,** 447 (1995).

[15] S. Legrand-Poels, V. Bours, B. Piret, M. Pflaum, B. Epe, B. Rentier, and J. Piette, *J. Biol. Chem.* **270,** 6925 (1995).

[16] M. Karin, *J. Biol. Chem.* **270,** 16483 (1995).

[17] A. J. Whitmarsh and R. J. Davis, *J. Mol. Med.* **74,** 589 (1996).

[18] A. J. Whitmarsh, S. H. Yang, M. S. Su, A. D. Sharrocks, and R. J. Davis, *Mol. Cell Biol.* **17,** 2360 (1997).

[19] L. O. Klotz, K. Briviba, and H. Sies, *FEBS Lett.* **408,** 289 (1997).

[20] L. O. Klotz, C. Pellieux, K. Briviba, C. Pierlot, J. M. Aubry, and H. Sies, *Eur. J. Biochem.* **260,** 917 (1999).

[21] L. O. Klotz, C. Fritsch, K. Briviba, N. Tsacmacidis, F. Schliess, and H. Sies, *Cancer Res.* **58,** 4297 (1998).

Procedures: Treatment of Cells with UVA and Singlet Oxygen Sources

Ultraviolet A

Cell Culture and UVA Irradiations

Human skin fibroblasts from foreskin biopsies and HaCaT human keratinocytes are cultured in Dulbecco's modified Eagle's medium (DMEM) supplemented with 10% fetal calf serum, L-glutamine (2 mM), streptomycin (0.02 g/liter), and penicillin (20,000 IU/liter).

UVA irradiations are performed at intensities of 40–44 mW/cm^2 for up to approximately 12 min (for a dose of 300 kJ/m^2) using a UVA700 illuminator from Waldmann Lichttechnik (Villingen, Germany). During irradiation, cells are covered with phosphate-buffered saline (PBS). This is necessary because the irradiation of serum-free culture medium [DMEM (without phenol red; Sigma (Deisenhofen, Germany) D5921) + glutamine + penicillin/streptomycin] with UVA (300 kJ/m^2) leads to the generation of hydrogen peroxide (260 μM), which is known to have substantial signaling effects in a variety of cellular systems.[22] For example, we found that the phosphorylation of ERK1 and ERK2 in SkMel-23 melanoma cells is increased approximately 10-fold by irradiation with UVA (300 kJ/m^2) in the presence of DMEM + glutamine + penicillin/streptomycin, whereas the increase in phosphorylation on the irradiation of cells covered with PBS is below twofold (data not shown).

Use of Singlet Oxygen Quenchers and Deuterium Oxide

The intensity of an effect due to singlet oxygen can be modified by the addition of scavengers or enhancers of the lifetime of singlet oxygen. For experiments with MAPK activation by UVA, imidazole and sodium azide are employed as quenchers of $^1O_2$. While the reaction rate constants of imidazole and azide with $^1O_2$ were reported to be $4 \times 10^7 M^{-1} sec^{-1}$ and $5 \times 10^8 M^{-1} sec^{-1}$ respectively,[23] rather high concentrations of these quenchers (20–40 mM) have to be used in biological systems because of competing reactions. Human skin fibroblasts are preincubated with 40 mM imidazole in medium for 30 min and irradiated with UVA in the presence of 40 mM imidazole in PBS. Sodium azide is effective at 10–20 mM when present during irradiation, as it enters the cell rapidly. There are certain caveats regarding the use of sodium azide (e.g., inhibition of mitochondrial respiration), even when the concentrations used are nontoxic under the respective

[22] K. Z. Guyton, Y. Liu, M. Gorospe, Q. Xu, and N. J. Holbrook, *J. Biol. Chem.* **271,** 4138 (1996).
[23] F. Wilkinson, W. P. Helman, and A. B. Ross, *J. Phys. Chem. Ref. Data* **24,** 663 (1995).

conditions. Azide alone must be tested for its effects on the assay system used. As imidazole and azide do not react selectively with $^1O_2$, but also with hydroxyl radicals, an involvement of the latter in the investigated effect has to be tested for. Hydroxyl radical scavengers such as mannitol (at 50–100 mM; preincubation for 30 min and present during irradiation) or dimethyl sulfoxide (DMSO; present at 100–300 mM during irradiation only) can be used.

The lifetime of $^1O_2$ depends on the polarity of the solvent. It is highest (up to >1 msec) in solvents lacking O–H and C–H moieties, such as $CDCl_3$ or CCl_4. A solvent effect exploitable in biological systems is the prolonged lifetime of $^1O_2$ in deuterium oxide (D_2O; approximately 52 μsec[24]) compared to H_2O (approximately 4 μsec[24]). The enhancement of a biological effect in the presence of D_2O may point to the involvement of $^1O_2$. There are, however, reports about the enhancement of the lifetime of triplet excited states of certain photosensitizers by D_2O.[25] Furthermore, a missing D_2O effect does not necessarily exclude the involvement of $^1O_2$; it has been pointed out by Foote and Clennan[24] that solvent isotope effect ratios may vary from as high as about 14 (ratio of the lifetimes of $^1O_2$ in D_2O and H_2O, respectively) to 1.0, depending on the concentration of the substrate reacting with $^1O_2$. The higher the concentration of the target of $^1O_2$, the lower the isotope effect.

For UVA irradiation experiments, we use 90% (v/v) of D_2O in PBS, which is prepared by diluting 10× concentrated PBS in D_2O. It is present during 2 min of preincubation and during irradiation.

Measurement of MAPK Phosphorylation and Activation

Phosphorylation of MAPK. After treatment, cells grown on 30-mm dishes in serum-free medium are incubated at 37° for the desired time, washed with PBS, and lysed by scraping with a pipette tip in 2× sodium dodecyl sulfate–polyacrylamide gel electrophoresis (SDS–PAGE) lysis buffer [125 mM Tris, 4% (w/v) SDS, 20% (v/v) glycerol, 100 mM dithiothreitol (DTT), 0.2% (w/v) bromphenol blue, pH 6.8]. The lysates are heated at 95° for 5 min and the samples are used directly for SDS–PAGE or frozen until use. Samples of 5–20 μl are subjected to gel electrophoresis on 12% SDS–polyacrylamide gels and blotted onto nitrocellulose membranes (ECL nitrocellulose, Amersham). Immunodetection of phosphorylated ERK-1/−2, p38, and JNK are with α-Active-MAPK [Promega, Mannheim,

[24] C. S. Foote and E. L. Clennan, *in* "Active Oxygen in Chemistry" (C. S. Foote, J. S. Valentine, A. Greenberg, and J. F. Liebman, eds.), p. 105. Blackie Academic, London, 1995.
[25] N. J. De Mol, G. M. J. Beijersbergen van Henegouwen, B. Weeda, C. N. Knox, and T. G. Truscott, *Photochem. Photobiol.* **44,** 747 (1986).

Germany; diluted 1:4000 in TBST (10 mM Tris, pH 7.6, 137 mM NaCl, 0.1% (v/v) Tween 20) plus 5% (w/v) low-fat milk powder], α-phospho-p38 (1:1000 in TBST/milk powder; New England Biolabs, Schwalbach, Germany), and α-Active-JNK (1:1000 in TBST/milk powder; Promega) antibodies, respectivley. Incubation with an α-rabbit secondary antibody conjugated to horseradish peroxidase (Cappel/ICN, Eschwege, Germany) is followed by chemiluminescence detection(LumiGlo, New England Biolabs, Beverly, MA). After stripping, the membrane is reprobed with α-p38 (New England Biolabs) or α-JNK2 (Santa Cruz Biotechnology, Santa Cruz, CA) antibodies, which serve as gel loading and protein controls. Stripping is performed by incubating the used nitrocellulose membrane in 100 mM 2-mercaptoethanol, 2% (w/v) SDS, 62.5 mM Tris, pH 6.8, at 55° for 30 min, followed by extensive washing in TBST.

JNK and ERK Activity. Cells are grown to approximately 80% confluency in 60-mm dishes, washed with PBS, and incubated with serum-free medium for an additional 20 to 40 hr. Following the removal of medium and a wash step with PBS, the serum-starved cells are treated as desired. After irradiation, performed with the cells being covered by PBS or PBS plus additives, the cells are covered with serum-free medium, incubated at 37° for the desired time, washed with PBS at room temperature, and lysed with 300 μl of ice-cold RIPA buffer [50 mM Tris, pH 8.0, 150 mM NaCl, 1 mM DTT, 0.5 mM EDTA, 1% (w/v) Nonidet P-40, 0.5% (w/v) sodium desoxycholate, 0.1% (w/v) SDS, 0.2 mM Na$_3$VO$_4$ (Sigma, Deisenhofen, Germany) 0.8 mM phenylmethylsulfonyl fluoride, 1 μg/ml aprotinin (Boehringer, Mannheim, Germany), 2 μg/ml leupeptin (Boehringer)] on ice for 15 min. The lysates are collected and centrifuged (10 min at 4° and 14,000 × g), and the protein content is determined using a commercially available protein assay (Bio-Rad DC-Protein Assay). C-Jun-N-terminal kinase 2 (JNK 2) is immunoprecipitated for 2–3 hr on ice from cell lysates (30–100 μg of protein in a total volume of 100–200 μl RIPA lysis buffer) using 5–10 μl of an anti-SAPK 2 antiserum (rabbit, diluted 1:10 in distilled water), which was a generous gift from Dr. Peter E. Shaw (Department of Biochemistry, University of Nottingham, UK).

Immune complexes are collected at 4° overnight with 30 μl protein A-Sepharose 6MB (Pharmacia, Freiburg, Germany) preequilibrated in RIPA lysis buffer. The pelleted immune complex/protein A conglomerate is washed with both ice-cold RIPA buffer and kinase buffer (10 mM Tris–Cl, pH 7.4, 150 mM NaCl, 10 mM MgCl$_2$, 0.5 mM DTT) three times each.

Thirty-five microliters of kinase buffer is added to the pellets, followed by 5 μl of 150 μM ATP, 6 μl (6 μg) of GST-cJun(1–79) (Alexis, Grünberg, Germany), and 5 μl (5 μCi) of γ-^{32}P-ATP (5 Ci/μmol). The mixtures are incubated at 37° for 25 min and are vortexed repeatedly during this incubation time.

The reaction is stopped by the addition of gel loading buffer and denaturation for 5 min at 95°. The samples are centrifuged for 2 min, and the supernatants are analyzed by electrophoresis on a 12% SDS–polyacrylamide gel. The phosphorylated JNK substrate [GST-cJun(1–79); 37 KDa] is identified by autoradiography and quantitated either by densitometry or by phosphoimaging analysis.

ERK activity is determined by the same procedure but employing an anti-ERK-1/−2 antibody (Upstate Biotechnology, Lake Placid, NY) as the precipitating agent (1 μg/sample) and myelin basic protein (MBP, Sigma, München, Germany) as the substrate (1 mg/ml per assay) for immunoprecipitated ERK.

Some Results: Activation of p38 and JNK but not ERKs by UVA

Employing the methods just described, we demonstrated that UVA is capable of enhancing cellular p38 phosphorylation and JNK activity (Fig. 1) rapidly and transiently. On the contrary, there was neither activation nor increased phosphorylation of ERK-1 or ERK-2 (not shown). Phosphorylation of p38 induced by UVA is diminished in the presence of singlet oxygen quenchers such as sodium azide or imidazole but not by the hydroxyl radical scavengers mannitol (Fig. 1A) or dimethyl sulfoxide (DMSO, not shown), pointing to the involvement of $^1O_2$ in p38 phosphorylation induced by UVA. The same is true for JNK activation induced by UVA[19] (Fig. 1B). JNK activation by UVA is enhanced about twofold in the presence of D_2O,[19] different from p38 whose phosphorylation induced by UVA is not enhanced when D_2O is present.[20]

Singlet Oxygen

Chemical Generation of Singlet Oxygen

Singlet oxygen is generated chemically by thermodecomposition of the 1,4-endoperoxide of N,N'-di(2,3-dihydroxypropyl)-1,4-naphthalene dipropionamide (DHPNO$_2$) or of disodium 3,3'-(1,4-naphthylidene) dipropionate-1,4-endoperoxide (NDPO$_2$). DHPNO$_2$ and NDPO$_2$ release $^1O_2$ on incubation at 37° with similar time courses as determined by measuring $^1O_2$ monomol emission as in Ref. 26. DHPNO$_2$ and NDPO$_2$ are synthesized as described in Ref. 27 and Refs. 28 and 29, respectively.

[26] A. R. Sundquist, K. Briviba, and H. Sies, *Methods Enzymol.* **234**, 384 (1994).
[27] A. Dewilde, C. Pellieux, C. Pierlot, P. Wattre, and J. M. Aubry, *Biol. Chem.* **379,** 1377 (1998).
[28] A. W. M. Nieuwint, J. M. Aubry, F. Arwert, H. Kortbeek, S. Herzberg, and H. Joenje, *Free Radic. Res. Commun.* **1,** 1 (1985).
[29] P. Di Mascio and H. Sies, *J. Am. Chem. Soc.* **111,** 2909 (1989).

A

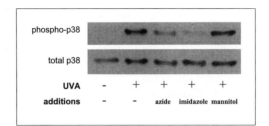

B

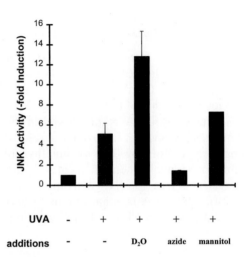

FIG. 1. Phosphorylation of p38 and JNK activation on exposure to UVA. (A) Effects of the singlet oxygen quenchers azide (20 mM) and imidazole (40 mM) and the hydroxyl radical scavenger mannitol (100 mM) on UVA (300 kJ/m^2)-induced p38 phosphorylation. (B) Effect of azide, mannitol, and deuterium oxide (D$_2$O) on UVA (300 kJ/m^2)-induced JNK activation as evaluated by phosphoimaging analysis.

Control experiments are performed with solutions of preheated DHPNO$_2$ and NDPO$_2$ and containing the decomposition products, DHPN and NDP, respectivley. Both DHPNO$_2$ and NDPO$_2$ are nontoxic to skin fibroblasts in concentrations as high as 10 mM within the times investigated.

Photochemical Generation of Singlet Oxygen

Rose Bengal and Rose Bengal-Agarose. Singlet oxygen is produced by irradiating PBS containing 0.3 μM of rose bengal (RB; Sigma, Deisenhofen, Germany) or RB-agarose (Molecular Probes, Eugene, OR) for 10 min with a commercially available 500-W halogen lamp from a fixed distance of 66 cm. No UVA was detectable under these conditions (detection limit of UVA was 0.1 mW/cm^2, or 0.6 kJ/m^2). Approximately 130 μM (cumulative concentration) of $^1O_2$ was generated during 10 min of irradiation of 0.3 μM RB, as determined by the bleaching of p-nitrosodimethylaniline.[30] This estimation of $^1O_2$ amounts generated was done by correlating the bleaching of p-nitrosodimethylaniline induced by RB + light with that induced by $^1O_2$ derived from NDPO$_2$, which decomposes with a half-life of 23.1 min at 37°[31] and forms 0.5 mol of $^1O_2$ per mole of decomposed NDPO$_2$.[29]

5-Aminolevulinate Photodynamic Treatment. Cells are incubated with 1 mM ALA (Merck, Darmstadt, Germany) in culture medium for 24 hr pretreatment to synthesize and accumulate porphyrins, mostly protoporphyrin IX.[32,33] Irradiation with red light is with a PDT 1200 irradiating device from Waldmann (Villingen, Germany) by exposing the culture dishes with cells and medium to doses of 45–75 kJ/m^2 at an intensity of 50 mW/cm^2.

Porphyrins are determined according to Piomelli *et al.*[34] with minor modifications. Directly after treatment, cells are homogenized and extracted in 1 N perchloric acid/methanol (1 : 1, v/v). Following a 10-min centrifugation, porphyrins are measured in the supernatants by fluorescence (excitation: 405 nm, emission 598 nm) using uroporphyrin as a calibration standard. The amount of protein is determined in the pellets according to Lowry *et al.*[35]

Cellular Accumulation of Singlet Oxygen Sources

To assess the delivery of naphthalene derivatives and RB to cells, skin fibroblasts are grown to near confluency in 90-mm culture dishes and incubated with 6 ml of 5 mM DHPN or NDP in PBS for 30 min at 37° or incubated with 6 ml of 0.3 μM RB or RB-agarose in PBS at room temperature in the dark. After incubation, the cells are washed rapidly in PBS and scraped off the plate in 300 μl PBS. The suspension is sonicated for 1 min and centrifuged twice in a microfuge (10 min at 14,000g) to remove particulate fractions. The supernatants are analyzed by reversed-phase high-performance liquid chromatography on an RP C$_{18}$ column for the presence of the tested compounds, which are quantitated by comparison with the

[30] I. Kraljic and S. El Mohsni, *Photochem. Photobiol.* **28,** 577 (1978).

[31] C. Pierlot, S. Hajjam, C. Barthélémy, and J. M. Aubry, *J. Photochem. Photobiol. B Biol.* **36,** 31 (1996).

[32] L. Wyld, J. L. Burn, M. W. Reed, and N. J. Brown, *Br. J. Cancer* **76,** 705 (1997).

[33] C. Fritsch, C. Abels, A. E. Goetz, W. Stahl, K. Bolsen, T. Ruzicka, G. Goerz, and H. Sies, *Biol. Chem.* **378,** 51 (1997).

[34] S. Piomelli, P. Young, and G. Gay, *J. Lab. Clin. Med.* **81,** 932 (1973).

[35] O. H. Lowry, N. J. Rosebrough, A. L. Farr, and R. J. Randall, *J. Biol. Chem.* **193,** 265 (1951).

signals elicited by standard solutions. The detection of eluting compounds is by absorbance at 290 nm. The mobile phase employed is methanol/50 mM ammonium acetate (40/60, pH 7.0) for DHPN, methanol/50 mM ammonium acetate (30/70, pH 7.0) for NDP, and methanol/water (90/10) for RB.

Significant amounts of DHPN and RB were recovered from cells incubated with DHPNO$_2$ and RB; however, neither NDP nor RB could be found associated to cells when incubations were with either NDPO$_2$ or RB-agarose (Table I).

Some Results: Activation of p38 and JNK by Singlet Oxygen

$^1O_2$ generated with RB + light shows a pattern of MAPK activation similar to that observed on UVA irradiation: p38 and JNKs are activated (Fig. 2), but not the ERKs.[19,20] Immobilizing RB on agarose, however, leads to the loss of its p38 and JNK activating capacity on irradiation with light (Fig. 2). Incubation of human skin fibroblasts with DHPNO$_2$ but not its decomposition product, DHPN, leads to an activation of p38 and of JNK, as can be seen from their phosphorylation on treatment (Fig. 2).

Unlike with RB + light and with DHPNO$_2$, there is no increase of p38 or JNK activity with the hydrophilic NDPO$_2$ that was tested at concentrations of 3–10 mM for incubation times of 30 min (Fig. 2). It was shown that $^1O_2$ is generated intracellularly with RB + light and with DHPNO$_2$

TABLE I
DELIVERY TO CELLS OF NAPHTHALENE
DERIVATIVE $^1O_2$ CARRIERS AND RB

Compound	Amount found associated to cells (nmol/10^6 cells)[a]
DHPN	82 ± 3[b]
NDP	<0.1[b]
RB	83 ± 9
RB-agarose	<0.1[c]

[a] The concentrations used were 5 mM for DHPNO$_2$ and NDPO$_2$, 50 μM for RB, and 20 μM for RB-agarose.

[b] Cells were incubated with DHPNO$_2$ or NDPO$_2$, respectively. After incubation, the amount of the decomposition products, DHPN or NDP, associated with the cells was determined.

[c] Cells were incubated with RB-agarose. After incubation, the amount of RB associated with the cells was determined.

A

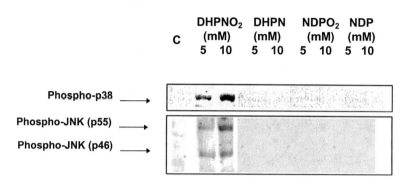

	C	DHPNO$_2$ (mM)		DHPN (mM)		NDPO$_2$ (mM)		NDP (mM)	
		5	10	5	10	5	10	5	10
Phospho-p38 →									
Phospho-JNK (p55) →									
Phospho-JNK (p46) →									

B

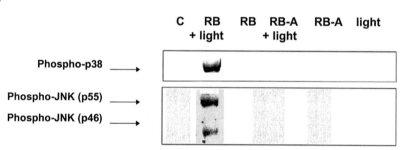

	C	RB + light	RB	RB-A + light	RB-A	light
Phospho-p38 →						
Phospho-JNK (p55) →						
Phospho-JNK (p46) →						

FIG. 2. Activation of p38 and JNK MAP kinases by singlet oxygen. Singlet oxygen was derived from (A) DHPNO$_2$ or NDPO$_2$ and (B) from rose bengal (RB, 0.3 μM) + light or RB-agarose (RB-A; 0.3 μM) + light. Control treatments were with PBS (C), the decomposition products of DHPNO$_2$ or NDPO$_2$, DHPN or NDP, and with RB, RB-agarose, plus light. Activity was assessed as the dual phosphorylation of p38 and JNK in Western blots employing phospho-specific antibodies. Immunodetection of total p38 or JNK served as control (not shown).

but extracellularly with RB-agarose + light or with NDPO$_2$ (Fig. 3): there is significant bleaching of 9,10-diphenylanthracene (DPA) extracted from cells loaded with DPA and treated with RB or DHPNO$_2$[20]; furthermore, RB and DHPN can be extracted from cells incubated with RB or DHPNO$_2$, but only negligible amounts of RB and NDP were extractable from cells treated with RB-agarose or NDPO$_2$ (Table I). Thus, $^1O_2$ has to be generated

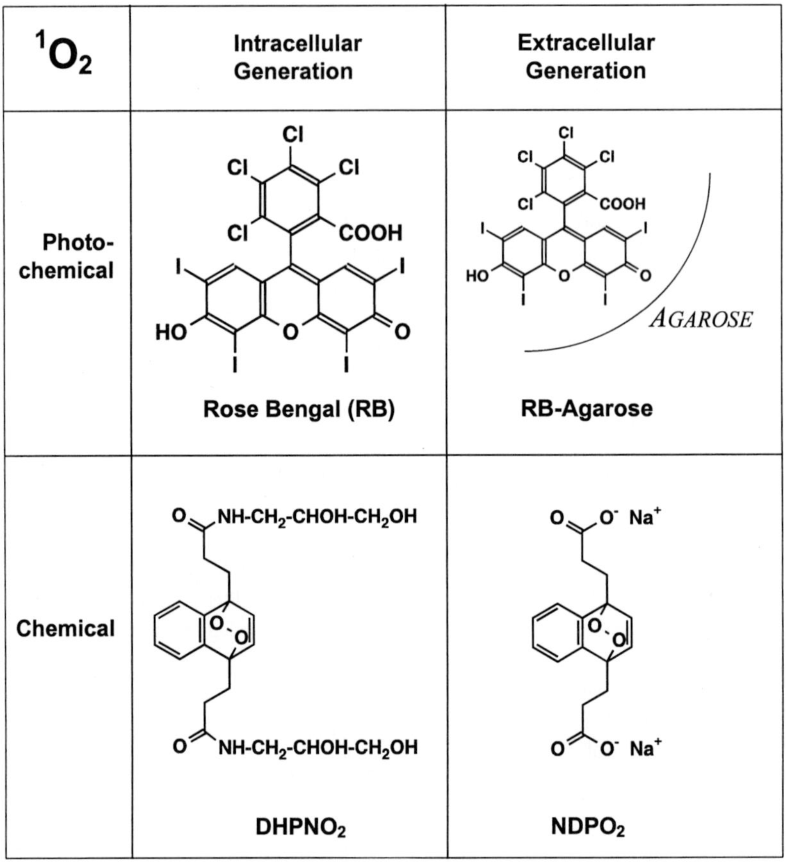

FIG. 3. Compounds used for intracellular and extracellular generation of $^1O_2$.

intracellularly, by either RB + light or DHPNO$_2$, in order to lead to the activation of p38 and JNK.

This pattern of MAP kinase activation (activation of p38 an JNK but not ERKs) may so far be regarded as unique for UVA and $^1O_2$ and as distinct from that observed with reactive oxygen species such as hydroperoxides,[22,36] superoxide generated by redox cycling,[37] nitric oxide and related

[36] K. Z. Guyton, M. Gorospe, T. W. Kensler, and N. J. Holbrook, *Cancer Res.* **56,** 3480 (1996).
[37] A. S. Baas and B. C. Berk, *Circ. Res.* **77,** 29 (1995).

species,[38] or species generated during treatment with UVC[39] or UVB,[40] which also activate ERKs.

Effect of ALA-PDT on MAP Kinases in HaCaT Cells[21]

Photodynamic therapy (PDT) based on the application of the porphyrin precursor 5-aminolevulinic acid (ALA) prior to irradiation with red light circumvents the ALA synthase reaction, the rate-limiting step in mammalian heme synthesis, predominantly leading to protoporphyrin IX.[32,33] The accumulation of photosensitizing porphyrins is more pronounced in cancerous tissue for several reasons, such as the faster uptake of ALA, lowered activity of ferrochelatase catalyzing the formation of heme, and an increased overall metabolic activity (see Peng *et al.*[41] for a review). Neoplastic tissue, therefore, is much more sensitive to ALA-PDT than normal tissue. ALA-PDT has been tested clinically and is efficient in the treatment of solar keratoses and superficial basal cell and squamous cell carcinomas.[42]

HaCaT cells incubated with 1 mM ALA for 24 hr produced a total (cells + incubation medium) of 830 ± 83 pmol fluorescent porphyrin/mg cellular protein. The cells had an average content of 99 ± 30 pmol fluorescent porphyrin/mg protein. The porphyrins undergo photobleaching; after irradiation with 4.5 J/cm^2 or red light only 49 ± 12 pmol/mg protein in the cellular fractions was measured.

Cells are irradiated at 45 and 75 kJ/m^2, with the former resulting in a viability of 65% and the latter leaving only 12% of the cells viable directly after irradiation.

Endogenous c-Jun N-terminal kinase activity is increased substantially on irradiation (45 kJ/m^2) of HaCaT cell preincubated with ALA, whereas control cells show no increase in JNK activity. Phosphoimaging analysis revealed an average induction of about 6-fold. In parallel, the phosphorylation of p38 is enhanced about 5-fold at 75 and at 45 kJ/m^2, with higher variability in the latter case (4.6 ± 3.4-fold induction). However, no activation or increased phosphorylation of the ERK-MAP kinases was observed (Fig. 4).

This UVA/^{1}O$_2$ pattern of MAPK activation (activation of p38 and JNK

[38] H. M. Lander, A. T. Jacovina, R. J. Davis, and J. M. Tauras, *J. Biol. Chem.* **271,** 19705 (1996).
[39] A. Radler-Pohl, C. Sachsenmaier, S. Gebel, H. P. Auer, J. T. Bruder, U. Rapp, P. Angel, H. J. Rahmsdorf, and P. Herrlich, *EMBO J.* **23,** 1005 (1993).
[40] G. J. Fisher, H. S. Talwar, J. Lin, P. Lin, F. McPhillips, Z. Wang, X. Li, Y. Wan, S. Kang, and J. J. Voorhees, *J. Clin. Invest.* **101,** 1432 (1998).
[41] Q. Peng, K. Berg, J. Moan, M. Kongshaug, and J. M. Nesland, *Photochem. Photobiol.* **65,** 235, (1997).
[42] C. Fritsch, G. Goerz, and T. Ruzicka, *Arch. Dermatol.* **134,** 207 (1998).

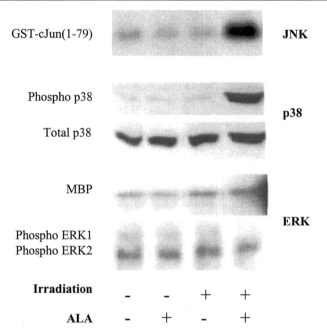

GST-cJun(1-79)   **JNK**

Phospho p38

p38

Total p38

MBP

ERK

Phospho ERK1
Phospho ERK2

Irradiation   −   −   +   +

ALA   −   +   −   +

FIG. 4. Effect of ALA photodynamic treatment on MAP kinase activities and phosphorylation status in HaCaT cells. Irradiation of cells with red light was with a 45-kJ/m² dose. JNK activity was measured in immunocomplex kinase assays using GST-cJun (1–79) as substrate. MBP was the substrate for immunoprecipitated ERK1 and ERK2. The phosphorylation status of p38 and ERK1/2 was tested for using phospho-specific antibodies. Total (phospho- and nonphospho-) p38 served as a loading control.

but not ERKs) is in line with $^1O_2$ being considered as a major mediator of the effects of porphyrin-based PDT.[43] This pattern has also been found in murine keratinocytes treated with benzoporphyrin derivative and light.[44]

Conclusions

Ultraviolet A radiation induces a pattern of MAP kinase activation in human skin fibroblasts consisting of a rapid and transient induction of p38 and JNK MAPK activity but not of ERKs. UVA activation of p38 and JNK is inhibitable by the singlet oxygen quenchers azide and imidazole but not by the hydroxyl radical scavengers mannitol or DMSO, pointing to the involvement of singlet oxygen ($^1O_2$).

[43] R. Bonnett, *Chem. Soc. Rev.* **21**, 19 (1995).
[44] J. Tao, J. S. Sanghera, S. L. Pelech, G. Wong, and J. G. Levy, *J. Biol. Chem.* **271**, 27107 (1996).

Like UVA, $^1O_2$ generated intracellularly on photoexcitation of rose bengal activates p38 and JNK but not ERKs. p38 and JNK activation is also elicited by chemiexcitation for the *intracellular* generation of $^1O_2$ by the lipophilic 1,4-endoperoxide of N,N'-di(2,3-dihydroxypropyl)-1,4-naphthalene dipropionamide. In contrast, *extracellular* generation of $^1O_2$ by irradiation of rose bengal immobilized on agarose beads or by chemiexcitation employing the hydrophilic 1,4-endoperoxide of disodium 3,3' (1,4-naphthylidene) dipropionate is ineffective in activating p38 or JNK. These data suggest that the activation of p38 and JNK by singlet oxygen occurs only when the electronically excited molecule is generated intracellularly.

From these results it can be concluded further that gene expression effects of UVA that were shown previously to be mimicked by NDPO$_2$ or RB-agarose, such as the induction of ICAM-1[9], MMP-1,[45] or the interleukins IL-1 and IL-6,[46] may not be mediated by the MAP kinases studied here. A pathway independent of AP-1 has indeed been identified, involving AP-2 for ICAM-1.[9]

Acknowledgments

We thank Dr. Freimut Schliess and Dr. Clemens Fritsch for fruitful collaborations and Annette Reimann for expert technical assistance. Support by the Deutsche Forschungsgemeinschaft, SFB 503, Project B1, and by the National Foundation for Cancer Research, Bethesda, MD, is gratefully acknowledged.

[45] M. Wlaschek, K. Briviba, G. P. Stricklin, H. Sies, and K. Scharffetter-Kochanek, *J. Invest. Dermatol.* **104,** 194 (1995).
[46] M. Wlaschek, J. Wenk, P. Brenneisen, K. Briviba, A. Schwarz, H. Sies, and K. Scharffetter-Kochanek, *FEBS Lett.* **413,** 239 (1997).

[14] Singlet Oxygen DNA Damage Products: Formation and Measurement

By Jean Cadet, Thierry Douki, Jean-Pierre Pouget, and Jean-Luc Ravanat

Introduction

Evidence is accumulating for the predominant involvement of singlet oxygen ($^1O_2$) in the oxidation reactions of major cellular targets such as

DNA and membranes on exposure to UVA radiation.[1,2] The energy of a photon within the UVA range and the lowest domain of visible light ($\lambda <$ 700 nm) is sufficient to generate $^1O_2$, the lowest excited state of molecular oxygen ($^1\Delta_g$, 94.2 kJ/mol) through a type II photosensitization mechanism.[3,4] Typically, endogenous photosensitizers such as flavins, furocoumarins, quinones, and porphyrins, whose content may vary according to the targeted cells, are excited on absorption of a photon. In a subsequent step, the resulting excited chromophore in a singlet state or more likely in a longer lived triplet state is able, at least, partly to transfer its energy to triplet oxygen (type II mechanism). However, the excited photosensitizers may also promote radical oxidation reactions through the type I mechanism, which mostly involves one-electron or hydrogen abstraction from the cellular targets.[3,4] In addition, $O_2^{-\cdot}$ may be generated as a side product of the type I reaction through oxidation of the radical anion of the photosensitizer by oxygen. Interestingly, comprehensive one-electron and $\cdot$OH-mediated degradation pathways are available for the purine and pyrimidine bases of nucleosides and DNA (for a review, see Cadet *et al.*[5]).

In contrast to $\cdot$OH, which reacts almost indifferently with all the nucleobases and the sugar moiety, $^1O_2$ oxidizes the guanine base of DNA exclusively.[5,6] One electron oxidation is also able to affect all the DNA components with, however, a marked preference for the guanine nucleobase, which exhibits the lowest ionization potential among nucleobases and the sugar moiety. Thus, the ability to generate radical cations within DNA decreases in the following order: guanine $>$ adenine $>$ cytosine $\sim$ thymine $>$ 2-deoxyribose.[5,7] It should be noted that 8-oxo-7,8-dihydroguanine is generated as the main decomposition product of the guanine moiety irrespective of any of the three latter oxidation conditions applied.[5] An increase in the level of 8-oxo-7,8-dihydro-2′-deoxyguanosine (8-oxodGuo **5**) has been observed in cellular DNA on UVA irradiation in the presence[8-10] or absence

[1] C. Kielbassa, L. Roza, and B. Epe, *Carcinogenesis* **18,** 811 (1997).

[2] J. Cadet, M. Berger, T. Douki, B. Morin, S. Raoul, J.-L. Ravanat, and S. Spinelli, *Biol. Chem.* **378,** 1275 (1997).

[3] C. S. Foote, *Photochem. Photobiol.* **54,** 659 (1991)

[4] J. Cadet and P. Vigny, *in* "Bioorganic Photochemistry" (H. Morrison, ed.), Vol. 1, p. 1. Wiley, New York, 1990.

[5] J. Cadet, M. Berger, T. Douki, and J.-L. Ravanat, *Rev. Physiol. Biochem. Pharmacol.* **131,** 1 (1997).

[6] J. Cadet, J.-L. Ravanat, G. W. Buchko, H. C. Yeo, and B. N. Ames, *Methods Enzymol.* **234,** 79 (1994).

[7] L. P. Candeias and S. Steenken, *J. Am. Chem. Soc.* **114,** 699 (1992).

[8] R. J. Mauthe, V. M. Cook, S. L. Coffing, and W. M. Baird, *Carcinogenesis* **16,** 133 (1995).

[9] J. E. Rosen, A. K. Prahalad, G. Schluter, and G. M. Williams, *Photochem. Photobiol.* **65,** 990 (1997).

[10] J. E. Rosen, *Mutat. Res.* **381,** 117 (1997).

of exogenous photosensitizers.[11–15] Support for a predominant involvement of $^1O_2$ in the photosensitized formation of 8-oxodGuo **5** was gained from the measurement of relatively low yields of DNA strand breaks and oxidized pyrimidine bases with respect to those of degraded purine bases in UVA-irradiated cells.[1,16] This achievement, using either the alkaline elution technique or the comet assay associated with DNA repair *N*-glycosylases,[17–19] underlines the low contribution of ·OH radicals to the overall UVA-mediated oxidation of cellular DNA. In this article, emphasis is placed on the measurement of 8-oxodGuo **5** (see Fig. 1), a relevant biomarker of $^1O_2$ oxidation, in cellular DNA using improved chromatographic assays. In addition, a short survey of available information on $^1O_2$ oxidation of the guanine moiety of DNA and model compounds is provided.

Materials and Methods

DNA Extraction and Isolation

The applied protocol, which is derived from the chaotropic method[20] is depicted in Scheme 1. This involves, in the initial step, isolation of nuclei, which is done slightly differently depending on the origin of DNA (cells or tissues).

Typically 1.5 ml of lysis buffer A (320 mM sucrose, 5 mM MgCl$_2$, 10 mM Tris–HCl, 0.1 mM deferroxamine, pH 7.5, 1% Triton X-100) is added to 10^6–10^7 cells, which are collected by centrifugation. The resulting cell suspension is vortexed gently and then nuclei are collected by centrifugation (1500g) for 10 min at 4°. The nuclear pellet is washed with 2 ml of buffer A and nuclei are obtained after centrifugation (1500g) for 10 min at 4°. In the case of tissues, about 100 mg of liver, kidney, etc., are minced with scissors and then homogenized in 2 ml of buffer A. After homogenization, nuclei are collected by centrifugation (1500g) for 10 min at 4° and subsequently washed twice with 2 ml of buffer A.

[11] A. Fischer-Nielsen, S. Loft, and K. G. Jensen, *Carcinogenesis* **14,** 2431 (1993).
[12] J. E. Rosen, A. K. Prahalad, and G. M. Williams, *Photochem. Photobiol.* **64,** 117 (1996).
[13] W. G. Wamer and R. R. Wei, *Photochem. Photobiol.* **65,** 560 (1997).
[14] X. Zhang, B. S. Rosenstein, Y. Wang, M. Lebwohl, D. M. Mitchell, and H. Wei, *Photochem. Photobiol.* **65,** 119 (1997).
[15] E. Kvam and R. M. Tyrrell, *Carcinogenesis* **18,** 2281 (1997).
[16] M. Pflaum, S. Boiteux, and B. Epe, *Carcinogenesis* **15,** 297 (1994).
[17] B. Epe and J. Hegler, *Methods Enzymol.* **234,** 122 (1994).
[18] B. Epe, *Rev. Physiol. Biochem. Pharmacol.* **127,** 223 (1995).
[19] C. Kielbassa and B. Epe, *Methods Enzymol.* **319,** [39], 2000 (this volume).
[20] H. J. Helbock, B. K. Beckman, M. K. Shigenaga, P. B. Walter, A. A. Woodhall, H. C. Yeo, and B. N. Ames, *Proc. Natl. Acad. Sci. U.S.A.* **95,** 288 (1998).

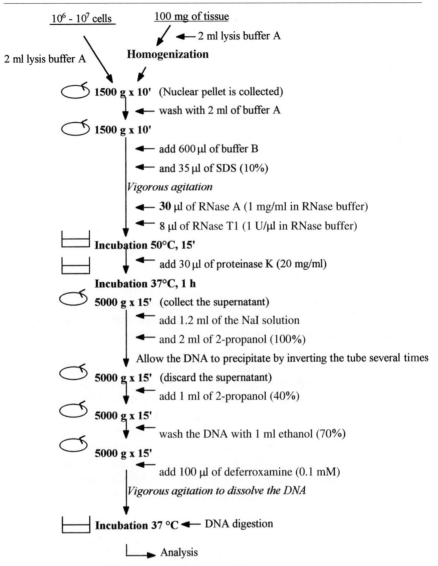

SCHEME 1. General protocol of DNA extraction.

To the nuclear pellet, 600 μl of buffer B (10 mM Tris–HCl, 5 mM EDTA-Na$_2$ 0.15 mM deferroxamine, pH 8.0) and 35 μl of 10% sodium dodecyl sulfate (SDS) are added and the resulting suspension is vortexed to allow lysis of the nuclei. Thereafter, 30 μl of RNase A (1 mg/ml in 10 mM Tris/HCl, 1 mM EDTA, 2.5 mM deferroxamine, pH 7.4) and 8 μl of

RNase A (1 U/μl in the same buffer) are added. The solution is then incubated for 15 min at 50°. Subsequently, 30 μl of Qiagen protease (20 mg/ml H_2O) is added and the resulting mixture is incubated for 1 hr at 37°.

Thereafter, the solution is centrifuged for 15 min at 5000g and the supernatant is collected. Then, 1.2 ml of the NaI solution (7.6 M NaI, 40 mM Tris–HCl, 20 mM EDTA-Na$_2$, 0.3 mM deferroxamine, pH 8.0) and 2 ml of 2-propanol are added. DNA is allowed to precipitate from the solution by inverting the tube gently several times. The DNA is then collected by centrifugation (5000g) for 15 min at 4°. The pellet is washed with 1 ml of 40% 2-propanol and the DNA is collected after centrifugation. The last step is resumed using 70% ethanol instead of 40% 2-propanol.

DNA Hydrolysis

DNA is either digested enzymatically or hydrolyzed chemically, giving rise to nucleosides and nucleobases, respectively. Usually, enzymatic hydrolysis is applied for high-performance liquid chromatography (HPLC) electrochemical detection (ECD) and HPLC with tandem mass spectrometry (HPLC-MS/MS) analyses because nucleosides are better separated than nucleobases on conventional reversed phase columns. It may be added that nucleosides provide a suitable transition as the result of efficient N-glycosidic bond cleavage for tandem MS detection. However, acid hydrolysis is a better alternative for GC-MS analysis as a higher sensitivity is usually observed for the free base compared to the nucleoside.

Enzymatic Digestion. After the last centrifugation, as much of the supernatant as possible is discarded. However, it is important to avoid letting the DNA dry as this induces DNA oxidation. Then, 100 μl of 0.1 mM deferroxamine is added and the resulting solution is vortexed until complete dissolution of DNA. Subsequently, 10 μl of nuclease P1 buffer (300 mM sodium acetate, 1 mM ZnSO$_4$, pH 5.3) that contains 10 U of nuclease P1 is added and the mixture is incubated for 2 hr at 37°. The dephosphorylation of the resulting nucleotides is achieved by adding 12 μl of alkaline phosphatase buffer (500 mM Tris/HCl, 1 mM EDTA-Na$_2$, pH 8) that contains 2 units of alkaline phosphatase. The resulting solution is incubated further for 1 hr at 37°. The mixture of digested nucleosides is then injected directly onto the octadecylsilyl silica gel column for either HPLC-ECD or HPLC-MS/MS analysis. It should be noted that for the HPLC-MS/MS assay, a known amount of [M + 4] isotopically labeled 8-oxodGuo **5** is added as an internal standard prior to enzymatic digestion.

Acidic Hydrolysis. Typically, 500 μl of 88% formic acid and a known amount of [M + 4] 8-oxo-7,8-dihydroguanine (8-oxoGua) are added to the DNA pellet and the resulting solution is heated at 130° for 30 min. Thereafter, the formic acid is cooled and evaporated to dryness under reduced pressure.

HPLC-ECD Measurement

The HPLC-ECD system involves a L-6000 intelligent pump (Hitachi, Tokyo, Japan) connected to a SIL-9A autosampler (Shimadzu, Kyoto, Japan) equipped with an Uptisphere octadecylsilyl silica gel column (250 × 4.6 mm i.d.) from Interchim (Montlucon, France). The column is placed in an oven at 28° during analysis. The isocratic mobile phase consists of 50 mM KH$_2$PO$_4$ that contains 9% of methanol. The coulometric detection is performed using a Coulochem II detector (Esa, Chelmsford, MA). The detection potentials are set at +100 and +400 mV for electrodes 1 and 2, respectively. Unmodified nucleosides are monitored by a UV detector set at 280 nm. Both signals are collected on a D7500 Hitachi (Tokyo, Japan) integrator. The amount of DNA analyzed is determined by measuring the area of the peak of dGuo **1** (see Fig. 1) after appropriate calibration. The amount of 8-oxodGuo **5** is assessed by external calibration.

HPLC-MS/MS Assay

The HPLC-MS/MS apparatus involves a Perkin-Elmer (Foster City, CA) 200 binary pumping system connected to a Perkin-Elmer 200 autosampler and a 785 UV-visible detector set at 280 nm. The separation is performed using an Hypersyl octadecylsilyl silica gel column (150 × 2.0 mm i.d.) from Interchim (Monluçon, France). The mobile phase consists of 10 mM ammonium formate (pH 4.8) and 5% acetonitrile with a flow rate of 0.2 ml/min. Unmodified nucleosides are detected by UV absorption and quantified by external calibration. The quantitative measurement of 8-oxodGuo **5** is achieved using an API365 triple quadrupole mass spectrometer equipped with a turboionspray source (Sciex, Thornill, Canada). The instrument response for 8-oxodGuo **5** is optimized by an infusion experiment of the pure compound dissolved in the mobile phase at a flow rate of 5 μl/min. Electrospray ionization (ESI) is achieved using nitrogen as the nebulizing and curtain gas. An auxiliary gas (air) heated to 450° is also used at a flow rate of 8 liters/min to improve the sensitivity. Two transitions are monitored at m/z 284.0/168.0 for the detection of 8-oxodGuo **5** and at m/z 288.0/172.0 for the measurement of the [M + 4] isotopically labeled internal standard.

GC-MS Measurement

The accurate quantitative determination of 8-oxoGua using the GC-MS assay requires a prepurification of the targeted oxidized guanine base prior to derivatization and analysis. For this purpose, either immunoaffinity chromatography or, as described in the following, HPLC purification can be used.

After formic acid hydrolysis and subsequent evaporation to dryness, the mixture of nucleobases acetate that contains a known amount of [M + 4] 8-oxoGua is dissolved in 100 μl of 25 mM ammonium. Separation of nucleobases is achieved using a Sulpelco LC18-DB 5 μm column (25 cm × 4.6 mm i.d.) under isocratic conditions. The flow rate is set at 1 ml/min with 2% methanol in 25 mM ammonium acetate buffer (pH 5.5) as the mobile phase. Immediately after the elution of guanine, a 1-min fraction that contains 8-oxoGua is collected. Thereafter, the collected HPLC fraction is lyophilized twice and the derivatization is performed for 30 min at 130° using 25 μl of CH$_3$CN and 25 μl of N,O-bis(trimethylsilyl)trifluoroacetamide (Sigma, St. Louis, MO).

GC-MS analyses are performed on a HP 5890 Series II gas chromatograph (Hewlett-Packard, Les Ulis, France) equipped with a capillary column (0.25 mm, 15 m) coated with a 0.1-μm film of methylsiloxane substituted with 5% phenylsiloxane (HP5-trace, Hewlett-Packard). The constant flow of helium is set at 0.8 liter/min. The injection (2 μl) is performed in the splitless mode with the temperature of the injector port set at 260°. The temperature of the GC oven is maintained at 120° for 1 min and is then raised to 300° at a rate of 20°/min and finally left at the latter temperature for 3 min. Detection of positive ions is provided by a HP 5972 mass detector using electron impact ionization (Hewlett-Packard). Detection of 8-oxoGua and its isotopically labeled [M + 4] internal standard is performed in the selected ion monitoring mode, using ions at m/z 440, 455 for 8-oxoGua, on one hand, and ions at m/z 444, 459 for [M + 4] 8-oxoGua, on the other.

Singlet Oxygen Oxidation of the Guanine Moiety of DNA

The main reaction of the dienophile $^1O_2$ with the guanine moiety ($k \approx$ 5.3 × 10^6 M^{-1} sec^{-1} for 5′-dGMP) is accounted for by a Diels-Alder [4 + 2] cycloaddition involving the C-4 and C-8 carbons of the purine ring (Fig. 1).[21] The resulting 4,8-purine endoperoxide, which has been characterized by nuclear magnetic resonance analysis at low temperature on photosensitization of a *tert*-butyldimethylsilyl derivative of 8-methylguanosine in organic solvents, is rather unstable.[21] Thus the decomposition of related endoperoxide **2** derived from 2′-deoxyguanosine (**1**) and short single-stranded oligonucleotides was found to lead to the predominant formation of the 4R and 4S diastereomers of 4-hydroxy-8-oxo-4,8-dihydro-2′-deoxyguanosine (**3**)

[21] C. Sheu and C. S. Foote, *J. Am. Chem. Soc.* **115,** 10446 (1993).

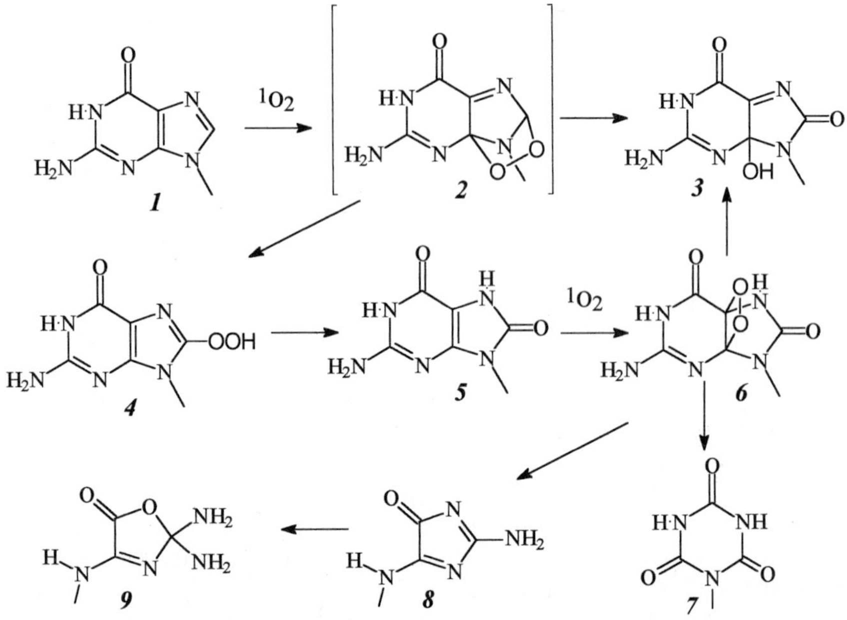

FIG. 1. Main $^1O_2$-mediated oxidation reactions of guanine moiety.

($\approx$85% yield) with a low amount of 8-oxodGuo **5** (yields = 15%).[22–24] The situation is completely different within double-stranded DNA as the level of **5** is enhanced at the expense of **3** whose formation was even not detected.[6,25] Under the latter conditions, endoperoxide **2** is likely to isomerize into 8-hydroperoxy-2'-deoxyguanosine (**4**)[26] prior to undergoing reduction into **5**.

It was shown that 8-oxodGuo **5** is at least 50 times more reactive toward $^1O_2$ than 2'-deoxyguanosine (**1**).[27] The $^1O_2$ oxidation of **5** may be rationalized in terms of the initial formation of an unstable dioxetane **6** through a [2 + 2] cycloaddition across the 4,5-ethylenic bond. The hydrolytic decomposition of **6** was found to involve three main pathways with the subsequent formation of the diastereomeric pair of keto-alcohol **3**, the cyanuric nucleo-

[22] J.-L. Ravanat, M. Berger, F. Benard, R. Langlois, R. Ouellet, and J. E. van Lier, *Photochem. Photobiol.* **55**, 809 (1992).

[23] G. W. Buchko, J. Cadet, M. Berger, and J.-L. Ravanat, *Nucleic Acids Res.* **20**, 4847 (1992).

[24] J.-L. Ravanat and J. Cadet, *Chem. Res. Toxicol.* **8**, 379 (1995).

[25] T. P. A. Devasagayam, S. Steenken, M. S. W. Obendorf, W. A. Schultz, and H. Sies, *Biochemistry* **30**, 6283 (1991).

[26] C. Sheu and C. S. Foote, *J. Am. Chem. Soc.* **117**, 6439 (1995).

[27] C. Sheu and C. S. Foote, *J. Org. Chem.* **60**, 4498 (1995).

side **7**, and the imidazolone derivative **8**.[28,29] The latter compound, which is also the main type I photooxidation product of **1** is then hydrolyzed into the oxazolone derivative **9**.[30,31]

Measurement of 8-OxodGuo 5 in Cellular DNA

Many efforts are currently devoted to the optimization of assays aimed at singling out the formation in cellular DNA of oxidized bases with emphasis on 8-oxoGua, a relevant DNA biomaker of oxidative stress.[32–34] In this respect three chromatographic methods involving HPLC-ECD, HPLC-MS/MS, and HPLC/GC-MS have been compared and critically evaluated.

The measurement of oxidative base damage within cellular DNA is still a challenging analytical problem.[35] It should be remembered that the requested limit of sensitivity of the applied method should be close to one modification per 10^6 to 10^7 DNA bases. In addition, the possibility of artifact formation, due to DNA oxidation during the workup, does not facilitate the achievement of such a goal. This is particularly true for 8-oxodGuo **5**, which is the main oxidation product of 2′-deoxyguanosine (**1**), the most easily oxidizable DNA nucleoside.[5]

As a consequence of such a difficult analytical problem, measurement of the basal level of 8-oxodGuo **5** within cellular DNA over the last decade has led to huge variations, which may reach three orders of magnitude.[32–35] As a general trend, cellular background levels of 8-oxodGuo **5** measured by methods that require DNA extraction were until recently higher than those determined in isolated cells using the comet assay or the alkaline elution technique. This can be explained by the occurrence of an artifactual oxidation of DNA during its extraction and subsequent workup. Interestingly, application of the recently proposed chaotropic method[20] aimed at reducing the artifactual oxidation of normal nucleobases during DNA extraction (vide supra) provides values that are in the same range as those

[28] G. W. Buchko, J. R. Wagner, J. Cadet, S. Raoul, and M. Weinfeld, *Biochim. Biophys. Acta* **1263,** 17 (1995).

[29] S. Raoul and J. Cadet, *J. Am. Chem. Soc.* **118,** 1892 (1996).

[30] J. Cadet, M. Berger, G. W. Buchko, P. C. Joshi, S. Raoul, and J.-L. Ravanat, *J. Am. Chem. Soc.* **116,** 7403 (1994).

[31] S. Raoul, M. Berger, G. W. Buchko, P. C. Joshi, B. Morin, M. Weinfeld, and J. Cadet, *J. Chem. Soc. Perkin Trans.* **2,** 371 (1996).

[32] J. Cadet, T. Douki, and J.-L. Ravanat, *Environ. Health Perspect.* **105,** 1034 (1997).

[33] A. R. Collins, J. Cadet, B. Epe, and C. Gedik, *Carcinogenesis* **18,** 1833 (1997).

[34] J. Cadet, T. Delatour, T. Douki, D. Gasparutto, J.-P. Pouget, J.-L. Ravanat, and S. Sauvaigo, *Mutat. Res.* **424,** 9 (1999).

[35] J. Cadet, C. D'Ham, T. Douki, J.-P. Pouget, J.-L. Ravanat, and S. Sauvaigo, *Free Radic. Res.* **29,** 541 (1998).

reported using the comet assay or the alkaline elution technique. Another major drawback has been identified for the GC-MS assay. It should be remembered that the background levels of 8-oxodGuo 5 assessed by applying the usual GC-MS assay[36] are at least one order of magnitude higher than those determined by HPLC-ECD.[37] In fact the high values from the use of the GC-MS method were shown to result from the oxidation of the large amount of guanine during the derivatization step prior to the chromatographic analysis.[38] Interestingly, GC-MS measurements that involved initial HPLC prepurification of 8-oxoGua (vide supra) in order to remove contaminating guanine were similar[38] to those obtained using the accurate HPLC-EC method.[39–41] Typically the combined application of the chaotropic DNA extraction method with either the HPLC-ECD technique or the recently optimized HPLC/GC-MS assay[38,42] leads to values of basal cellular levels close to one 8-oxodGuo residue per 10^6 base pairs. Interestingly, similar values are obtained by applying the emerging and powerful HPLC-MS/MS method.[43] Among the three methods compared, HPLC-ECD is the easiest one to use. However, it is clear that the use of [M + 4] internal standards in the straightforward HPLC-MS/MS method or the more tedious HPLC/GC-MS assay provides measurements more accurate than those obtained by HPLC-ECD, even if calibration can be done.[44] Another important parameter to be considered is the sensitivity provided by the assays. At the current stage of best optimized conditions and available instruments, the limits of detection of either 8-oxodGuo or 8-oxoGua expressed in femtomoles offered by the three methods are HPLC/GC-MS (300) > HPLC-ECD (50) > HPLC/MS-MS (20).

Further work is required for additional comparison of the three chromatographic techniques with the ^{32}P postlabeling assay, on one hand, and other methods involving isolated cells, on the other. It was also found that the lack of prepurification of 8-oxo-7,8-dihydro-2′-deoxyguanosine 3′-monophosphate (8-oxodGMP) in the ^{32}P postlabeling assay led to arti-

[36] M. Dizdaroglu, *FEBS Lett.* **315,** 1 (1993).

[37] B. Halliwell and M. Dizdaroglu, *Free Radic. Res. Commun.* **16,** 75 (1992).

[38] J.-L. Ravanat, R. J. Tureski, E. Gremaud, L. J. Trudel, and R. H. Stadler, *Chem. Res. Toxicol.* **8,** 1039 (1995).

[39] R. A. Floyd, J. J. Watson, P. T. Wong, D. H. Altmiller, and R. C. Rickard, *Free Radic. Res. Commun.* **1,** 163 (1986).

[40] T. Douki, M. Berger, S. Raoul, J.-L. Ravanat and J. Cadet, *in* "Analysis of Free Radicals in Biological Systems" (A. E. Favier, J. Cadet, B. Kalyanaraman, M. Fontecave, and J.-L. Pierre, eds.), p. 213. Birhauser Verlag, Basel, 1995.

[41] H. Kasai, *Mutat. Res.* **387,** 147 (1997).

[42] T. Douki, T. Delatour, F. Bianchini, and J. Cadet, *Carcinogenesis* **17,** 347 (1996).

[43] J.-L. Ravanat, B. Duretz, A. Guiller, T. Douki, and J. Cadet, *J. Chromatogr. B* **715,** 349 (1998).

[44] J.-L. Ravanat, E. Gremaud, J. Markovic, and R. J. Turesky, *Anal. Biochem.* **260,** 30 (1998).

factual oxidation of contaminating 2'-deoxyguanosine 3'-monophosphate (dGMP), mostly due to the occurrence of self-radiolysis.[45] Calibration of the polynucleotide kinase-mediated phosphorylation of 8-oxodGMP is required as the presence of low amounts of competing dGMP even after the prepurification step may lead to the underestimation of damage. As already discussed not only the elution alkaline technique but also the comet assay associated with DNA repair glycosylases allowed the determination of the number of formamidopyrimidine DNA N-glycosylase sites within cellular DNA under low conditions of oxidative stress exposure.[46–49] However, both qualitative and quantitative calibration of the latter sensitive assay must be completed further using the chromatographic methods discussed in this article that are able to single out specific base damage accurately.

[45] J. Cadet, F. Bianchini, I. Girault, D. Molko, M. Polverelli, J.-L. Ravanat, S. Sauvaigo, N. Signorini, and Z. Tuche, in "DNA and Free Radicals: Techniques, Mechanisms and Applications" (B. Halliwell and O. I. Aruoma, eds.), p. 285. OICA International, Saint Lucia, London, 1998.
[46] M. Pflaum, O. Will, and B. Epe, Carcinogenesis 18, 2225 (1998).
[47] M. Pflaum, C. Kielbassa, M. Garmyn, and B. Epe, Mutat. Res. 408, 137 (1998).
[48] A. R. Collins, M. Dusinska, C. M. Gedik, and R. Stetina, Environ. Health Perspect. 104 (Suppl. 3), 465 (1996).
[49] J.-P. Pouget, J.-L. Ravanat, T. Douki, M.-J. Richard, and J. Cadet, Int. J. Radiat. Biol. 75, 51 (1999).

[15] Ultraviolet A- and Singlet Oxygen-Induced Mutation Spectra

By ANNE STARY and ALAIN SARASIN

Genetic stability is an important prerequisite for the maintenance of the integrity and functions of each cell. Several environmental agents, as well as endogenous sources, are known to play a significant role in the formation of DNA lesions. Most of these lesions are repaired, but a few may escape and can lead to mutagenesis during *trans*-lesion replication or to chromosomal aberrations, and, finally, to tumorigenesis or cell death.

The causal relationship between solar ultraviolet (UV) radiation and carcinogenesis has been well established. UVB radiation (290–320 nm) causes the direct formation of DNA photoproducts and gives rise to mutations in protooncogenes and tumor suppressor genes that eventually lead to skin cancer.[1] Genotoxicity of UVA radiation (320–400 nm), which represents more than 90% of the terrestrial UV solar energy, is most likely induced by indirect mechanisms. After UVA photon absorption by an unidentified endogenous photosensitizer, reactive oxygen species are generated that could react with DNA without further intermediates (type I photosensitized reaction) or via singlet oxygen (type II photosensitized reaction).

The singlet oxygen appears to be an early intermediate of UVA cellular response, judging by its involvement in signal transduction[2,3] and gene induction.[4–7] It belongs to reactive oxygen species (ROS) that are ubiquitous oxidizing agents generated in all aerobic organisms, endogenously as byproducts of respiration and inflammatory response and also exogenously after exposure to a variety of agents.

The goal of this review is to summarize data on DNA lesions, DNA repair, and mutation spectra, which characterize the genotoxic potential of both singlet oxygen and UVA radiation, essentially in mammalian species.

DNA Damage Induced by Singlet Oxygen and Ultraviolet A Radiation

DNA Damage Induced by Singlet Oxygen

In living aerobic organisms, two basically different mechanisms can generate singlet oxygen (for review, see Ref. 8). The major mechanism is photosensitization in which photodynamic compounds, such as flavins or porphyrins, capture photons and transfer energy to ground-state oxygen.[9] Singlet oxygen can also be produced in "dark" reactions that occur during lipid peroxidation and involve several cellular oxidases (peroxidase, cytochromes). The effectiveness of singlet oxygen generation by photosensitization would be expected to depend on the cellular location of the photosensitizer and on the coformation of other ROS as by-products of photosensitization. The specific involvement of singlet oxygen in cellular

[1] A. J. Nataraj, J. C. Trent II, and H. N. Ananthaswamy, *Photochem. Photobiol.* **62,** 218 (1995).

[2] K. M. Hanson and J. D. Simon, *Proc. Natl. Acad. Sci. U.S.A.* **95,** 10576 (1998).

[3] L.-O. Klotz, K. Briviba, and H. Sies, *FEBS Lett.* **408,** 289 (1997).

[4] M. Wlaschek, J. Wenk, P. Brenneisen, K. Briviba, A. Schwarz, H. Sies, and K. Scharffetter-Kochanek, *FEBS Lett.* **413,** 239 (1997).

[5] R. M. Tyrrell, *Photochem. Photobiol.* **63,** 380 (1996).

[6] S. Grether-Beck, S. Olaizola-Horn, H. Schmitt, M. Grewe, A. Jahnke, J. P. Johnson, K. Briviba, H. Sies, and J. Krutmann, *Proc. Natl. Acad. Sci. U.S.A.* **93,** 14586 (1996).

[7] S. Basu-Modak and R. M. Tyrrell, *Cancer Res.* **53,** 4505 (1993).

[8] B. Epe, *Chem.-Biol. Interact.* **80,** 239 (1991).

[9] I. E. Kochevar, *Photochem. Photobiol.* **45,** 891 (1987).

mechanisms, such as cytotoxicity[10] and gene induction,[4,7] has been found by means of the presence of a quencher (sodium azide) or D_2O, an enhancer of singlet oxygen lifetime, but there is no evidence of a direct interaction between singlet oxygen and DNA in a cellular system.

The precise nature of DNA damage induced by singlet oxygen (for review, see Ref. 11) was found by the *in vitro* production of singlet oxygen and direct interaction with either free nucleosides or DNA. Different sources of singlet oxygen have been used such as microwave discharge, thermodissociation of the water-soluble endoperoxide $NDPO_2$,[12] and photosensitized oxidation of dyes such as rose bengal or methylene blue. Specific reaction of singlet oxygen with guanosine on treatment of free nucleosides[13] has been confirmed for DNA.[14] A major type of singlet oxygen-induced DNA damage appeared to be alkali-labile and piperidine-labile sites[15] that were found at guanosine positions by sequencing techniques.[16] The oxidation products included base modifications, 8-hydroxydeoxy-guanine (8-oxodG), and 2,6-diamino-4-hydroxy-5-formamidopyrimidine (FapyGua).[14,17] There is now convincing evidence that 8-oxodG is the major product of base modification caused by singlet oxygen.[8] Singlet oxygen produced by methylene blue and light was shown to lead to single-strand breaks (SSB) in supercoiled plasmid DNA,[12,15] but their number is 17 times lower than the number of 8-oxodG lesions produced in similar experimental conditions.[18] Strand breaks occurred exclusively at guanine residues.[19] Singlet oxygen should also be responsible for lipid peroxidation, leading to formation of 8-oxodG in the presence of 2′-deoxyguanosine.[20]

DNA Damage Induced by UVA

Substantial evidence from *in vitro* studies indicated that oxidative stress plays an important role in biochemical alterations elicited by UVA, includ-

[10] R. M. Tyrrell and M. Pidoux, *Photochem. Photobiol.* **49,** 407 (1989).

[11] H. Sies and C. F. M. Menck, *Mutat. Res.* **275,** 367 (1992).

[12] P. Di Mascio and H. Sies, *J. Am. Chem. Soc.* **111,** 2909 (1989).

[13] J. Cadet and R. Téoule, *Photochem. Photobiol.* **28,** 420 (1978).

[14] R. A. Floyd, M. S. West, K. L. Eneff, and J. E. Schneider, *Arch. Biochem. Biophys.* **273,** 106 (1989).

[15] E. R. Blazek, J. G. Peak, and M. J. Peak, *Photochem. Photobiol.* **49,** 607 (1989).

[16] T. Friedman and D. M. Brown, *Nucleic Acids Res.* **5,** 615 (1978).

[17] S. Boiteux, E. Gajewski, J. Laval, and M. Dizdaroglu, *Biochemistry* **31,** 106 (1992).

[18] J. E. Schneider, S. Price, L. Maidt, J. M. C. Gutteridge, and R. A. Floyd, *Nucleic Acids Res.* **18,** 631 (1990).

[19] T. P. A. Devasagayam, S. Steenken, M. S. W. Obendorf, W. A. Schulz, and H. Sies, *Biochemistry* **25,** 6283 (1991).

[20] N. Sera, H. Tokiwa, and N. Miyata, *Carcinogenesis* **17,** 2163 (1996).

ing membrane lipid damages,[21,22] cell apoptosis,[23,24] and DNA base modifications leading to mutagenesis. An adaptive response of cellular defense against oxidative agents was found to be induced by UVA.[5] Irradiation of plasmid DNA with UVA alone did not result in a genotoxic effect, suggesting that UVA-induced DNA damages were produced indirectly via cellular photosensitizers.[25] The exact action spectrum of UVA remains unclear, reflecting the complex nature of UVA genotoxicity. Some differences found between DNA lesion induction in different experiments might be due to different sensitizing pathways, different skin cell types, and state of growth culture conditions.[26,27]

The induction of DNA single-strand breaks, alkali-labile sites, and DNA–protein cross-links induced by UVA has been investigated extensively in human cells.[28–30] The D_2O environment enhanced the induction of SSB after 365- and 405-nm irradiations, implicating singlet oxygen in the formation of such lesions.[29] Cell cycle kinetics of human fibroblast established that UVA irradiation could inhibit DNA synthesis temporarily, suggesting the presence of a genotoxic damage, with maximal suppression of DNA replication 5 hr after irradiation.[31]

Wavelength dependence of oxidative DNA damage production by UVA and visible light has been quantified in hamster cells by means of several repair endonucleases, as the formamidopyrimidine-DNA glycosylase (Fpg protein), which recognized and excised damaged oxidative base, leading to a single-strand break.[32] Beyond 350 nm, the yield of Fpg-sensitive sites increased linearly with dose and the DNA damage profile resembled that induced by singlet oxygen or type I photosensitizers in cell-free systems.[33] Although the presence of other Fpg-sensitive sites or of base damage insensitive to Fpg endonuclease could not be excluded, 8-oxodG appeared likely to be the major base modification generated by UVA radiation. The

[21] G. F. Vile, A. Tanew-Iliitschew, and R. M. Tyrrell, *Photochem. Photobiol.* **62**, 463 (1995).
[22] F. Gaboriau, P. Morlière, I. Marquis, A. Moysan, M. Gèze, and L. Dubertret, *Photochem. Photobiol.* **58**, 515 (1993).
[23] D. E. Godar and A. D. Lucas, *Photochem. Photobiol.* **62**, 108 (1995).
[24] S. Tada-Oikawa, S. Oikawa, and S. Kawanishi, *Biochem. Biophys. Res. Commun.* **247**, 693 (1998).
[25] T. M. Rünger, B. Epe, and K. Möller, *Rec. Results Cancer Res.* **139**, 31 (1995).
[26] E. Kvam and R. M. Tyrrell, *Carcinogenesis* **18**, 2379 (1997).
[27] T. P. Coohill, M. J. Peak, and J. G. Peak, *Photochem. Photobiol.* **46**, 1043 (1987).
[28] L. Roza, G. P. van der Schans, and P. H. M. Lohman, *Mutat. Res.* **146**, 89 (1985).
[29] M. J. Peak, J. G. Peak, and B. A. Carnes, *Photochem. Photobiol.* **45**, 381 (1987).
[30] J. G. Peak and M. J. Peak, *Mutat. Res.* **246**, 187 (1991).
[31] A. de Laat, M. van Tilburg, J. C. van der Leun, W. A. van Vloten, and F. R. de Gruijl, *Photochem. Photobiol.* **63**, 492 (1996).
[32] B. Epe, M. Pflaum, and S. Boiteux, *Mutat. Res.* **299**, 135 (1993).

yield of 8-oxodG increased up to 2.7-fold in the presence of D_2O during UVA irradiation in human primary skin fibroblasts, demonstrating the role of singlet oxygen in the induction of this oxidative base damage by UVA.[26]

Beyond 350 nm, the induction of pyrimidine dimer, attributed to direct absorption by DNA, has been detected in UVA-exposed hamster cells[33] and human fibroblast cells,[34] but the relative wavelength dependence for cell lethality does not correlate with that of pyrimidine dimer formation. In human skin fibroblasts or lymphocytes, the contribution of pyrimidine dimers versus that of 8-oxodG has been estimated to be 65% at 365 nm and close to or below the detecting limit at wavelengths longer than 365 nm.[26] However, pyrimidine dimer photolesions have also been found to be formed in human keratinocytes and melanocytes in skin exposed to UVA radiation.[35] It appears that there is no clear relationship between the nature of UVA-induced damage formation and the contribution of the pyrimidine dimer as a genotoxic lesion.

The similarity of DNA damage (SSB, alkali-labile sites, and the formation of 8-oxodG) produced by exposure to singlet oxygen and UVA radiation is consistent with the hypothesis that singlet oxygen plays an important role in UVA cytotoxicity.[10]

Repair of DNA Damage Induced by Singlet Oxygen and UVA Radiation

Repair of Singlet Oxygen-Induced DNA Damage

Among cellular defense systems against oxidants are several very efficient components that prevent the interaction of singlet oxygen with guanosines in DNA by quenching, such as α-tocopherol and carotenoids.[36] However, DNA repair systems are still important cellular defense mechanism. The lower level of the transforming capacity of single-strand versus double-strand plasmid into bacteria[37] and simian cells[38] highlights the role of cellular DNA repair machinery to recover damaged DNA exposed to singlet oxygen. The importance of the DNA repair pathway of oxidative lesions is

[33] C. Kielbassa, L. Roza, and B. Epe, *Carcinogenesis* **18,** 811 (1997).

[34] I. C. Enninga, R. T. L. Groenendijk, A. R. Filon, A. A. van Zeeland, and J. W. I. M. Simons, *Carcinogenesis* **7,** 1829 (1986).

[35] A. R. Young, C. S. Potten, O. Nikaido, P. G. Parsons, J. Boenders, J. M. Ramsden, and C. A. Chadwick, *J. Invest. Dermatol.* **111,** 936 (1998).

[36] C. S. Foote and R. W. Denny, *J. Am. Chem. Soc.* **90,** 6233 (1968).

[37] D. Decuyper-Debergh, J. Piette, and A. Van de Vorst, *EMBO J.* **6,** 3155 (1987).

[38] D. T. Ribeiro, C. Madzak, A. Sarasin, P. Di Mascio, H. Sies, and C. F. M. Menck, *Photochem. Photobiol.* **55,** 39 (1992).

supported by the existence of specific DNA repair glycosylases. The repair of oxidized guanosine residues in DNA appears to occur predominantly via a base-excision repair system, functionally conserved in bacteria and mammals (for review, see Ref. 39). The oxidative nucleoside 8-oxodG constitutes the prevalent excised product and is used as an indicator of *in vivo* oxidative DNA damage.[40] Most nucleotide excision repair (NER)-deficient human cell lines resembled normal cells in their ability to repair singlet oxygen-induced DNA damage. Only some xeroderma pigmentosum (XP) group C cells have been reported to show a markedly lower than normal repair of such lesions.[41] However, a role of NER and DNA mismatch repair pathways against oxidative DNA damage processing cannot be excluded.[41–43] In response to a low level of ionizing radiation treatment, mouse ES cells containing defective *Msh2* mismatch genes failed to execute cell death efficiently. DNA from these repair-deficient cells accumulated more 8-oxodG oxidative base damage compared with that of unirradiated cells.[42] Moreover, cells isolated from both Cockayne's syndrome (CS) and patients afflicted with both XP and CS are partially defective in the repair of ionizing radiation-induced lesions as well as 8-oxodG.[44,45]

Repair of UVA-Induced DNA Damage

The repair of SSB in cultured human fibroblasts and keratinocytes after UVA irradiation appears to be a very rapid process, as 90% are removed within 15 min, as observed by alkaline elution[28,30,34] and by alkaline comet assay.[46,47] The induction of UVA-induced SSB can be counteracted in the presence of catalase, an endogeneous antioxidant enzyme, during UVA irradiation[28] or by cell pretreatment with the α-tocopherol antioxidant.[47] No associated decrease of cell survival has been reported, indicating that SSB lesions are not the critical target for UVA-induced cellular lethality.[28,34] Kinetic studies by a comet assay in human fibroblasts[46] and keratinocytes[47] exposed to UVA showed an occurrence of SSB within 10 min and their complete removal within 1 hr. The level of initial DNA damage (SSB and

[39] R. P. Cunningham, *Curr. Biol.* **7,** R576 (1997).
[40] M. K. Shigenaga and B. N. Ames, *Free Radic. Biol. Med.* **10,** 211 (1991).
[41] T. M. Rünger, B. Epe, and K. Möller, *J. Invest. Dermatol.* **104,** 68 (1995).
[42] T. L. DeWeese, J. M. Shipman, N. A. Larrier, N. M. Buckley, L. R. Kidd, J. D. Groopman, R. G. Cutler, H. T. Riele, and W. G. Nelson, *Proc. Natl. Acad. Sci. U.S.A.* **95,** 11915 (1998).
[43] J. T. Reardon, T. Bessho, H. C. Kung, P. H. Bolton, and A. Sancar, *Proc. Natl. Acad. Sci. U.S.A.* **94,** 9463 (1997).
[44] F. Le Page, A. Sarasin, and A. Gentil, *Mutat. Res.* **379,** S167 (1997).
[45] P. K. Cooper, T. Nouspikel, S. G. Clarkson, and S. A. Leadon, *Science* **275,** 990 (1997).
[46] C. Alapetite, T. Wachter, E. Sage, and E. Moustacchi, *Int. J. Radiat. Biol.* **69,** 359 (1996).
[47] J. Lehmann, D. Pollet, S. Peker, V. Steinkraus, and U. Hoppe, *Mutat. Res.* **407,** 97 (1998).

alkali-labile sites) may be attributed to direct UVA-induced lesions as well as transient strand breaks mediated by a repair-associated endonuclease activity.[47] Both very fast repair of SSB and immediate DNA synthesis suppression observed after UVA radiation contrasted with the slower repair of UVB-induced photoproducts and late replication arrest, indicating different mechanisms for DNA damage, DNA repair, and apoptosis. It has been proposed that NER, the major repair system of UVB-induced lesions, does not play an important role in the cellular defense against UVA-induced genotoxic damages.[31,47]

Failure to repair all genomic damages before replication may ultimately lead to cell death and apoptosis. Investigations on the time course of apoptosis in murine lymphoma cells have shown that UVA (340–400 nm) induced an immediate ($<$4 hr) apoptotic response attributed to membrane damage.[23] Fibroblast R6 rat cells exhibited a similar UVA-induced immediate apoptosis that could be abolished by the overexpression of the *bcl-2* anticell death protein.[48] A preventing action of *bcl-2* against UVA-mediated apoptosis via suppression of the formation or of the effects of ROS, such as singlet oxygen, could occur in mitochondrial membranes.[48] A role for UVA-induced oxidative DNA damage in apoptosis has been established in HL-60 human cells by monitoring several markers of apoptotic pathway in a dose-dependent manner, such as an increase of 8-oxodG formation, disruption of the inner mitochondrial membrane potential, and activation of caspase 3.[24]

Mutation Spectra Induced by Singlet Oxygen and UVA Radiation

DNA lesions that escape repair can inhibit replication and transcription and cause mutagenesis. Gene activation or inactivation by mutation may initiate or propagate tumor formation. The determination of mutagenic potential allows the identification and evaluation of the relative contribution toward deleterious biological end points of various sources of DNA damage. The mutational spectrum, which is the result of the number, types, and sites of all mutations observed in a given sequence, reflects not only the types of induced DNA lesions, but also the subsequent processing of damaged DNA following repair, transcription, and replication in the host cells.[49] This can be achieved easily using a particular genomic or vector sequence phenotypically screenable, which allows one to monitor, following replication, the frequency, type, and position of mutations by sequencing.

[48] C. Pourzand, G. Rossier, O. Reelfs, C. Borner, and R. M. Tyrrell, *Cancer Res.* **57,** 1405 (1997).
[49] D. Wang, D. A. Kreutzer, and J. M. Essigmann, *Mutat. Res.* **400,** 99 (1998).

TABLE I
FREQUENCY (%) OF MUTATIONS INDUCED BY *in Vitro* DNA EXPOSURE TO SINGLET
OXYGEN AND UVA CELL IRRADIATION IN MAMMALIAN CELLS

	Singlet oxygen			UVA		
	Escherichia coli JM105 cells[37]	Cos7 monkey cells[50]	293 human cells[51]	293 human cells[64]	CHO (NER⁺) hamster cells[66]	CHO (NER⁻) hamster cells[66]
Base substitution[a]						
Single	73	61	71	48	83	78
Tandem	—	—	12	—	11	14
Multiple	21	21	16	9	4	2
Rearrangement[b]	6	18	2	42	2	6
Transition[a]						
$G:C \rightarrow A:T$	15	15	13	54	35	76
$A:T \rightarrow G:C$	—	2	4	23	8	7
Transversion[a]						
$G:C \rightarrow T:A$	67	51	68	5	3	7
$G:C \rightarrow C:G$	16	33	11	—	10	—
$A:T \rightarrow T:A$	2	—	1	18	3	4
$A:T \rightarrow C:G$	—	—	4	—	41	5
Mutation at $G:Cs$	98	98	91	59	48	83
Mutation compatible with 8-oxodG	67	53	72	5	44	12

[a] Frequencies include the tandem mutations as two single substitutions.
[b] Including frameshift and deletion.

Singlet Oxygen-Induced Mutation Spectrum

To characterize the molecular nature of singlet oxygen-induced mutations, experimental assays have been carried out by the *in vitro* exposure of naked DNA followed by transfection and replication of damaged vectors into host cells. Mutation spectra obtained in bacteria,[37] monkey,[50] and human cells[51] showed that singlet oxygen induced predominantly base substitutions. These point mutations were located at sites that could be oxidized specifically, i.e., $G:C$ base pairs (Table I).

The formation of miscoding guanine on replication in bacteria and mammalian cells is in good agreement with the specific alteration of guanine residues found in singlet oxygen-induced lesion spectrum.[52] Indeed, the

[50] R. C. de Oliveira, D. T. Ribeiro, R. G. Nigro, P. Di Mascio, and C. F. M. Menck, *Nucleic Acids Res.* **20**, 4319 (1992).
[51] J. K. Jeong, M. J. Juedes, and G. N. Wogan, *Chem. Res. Toxicol.* **11**, 550 (1998).
[52] J. Piette, *Photochem. Photobiol. B Biol.* **11**, 241 (1991).

prevalent 8-oxodG lesion has miscoding potential *in vitro*[53] and *in vivo*, as shown by results following the replication of single-stranded DNA vectors carrying a unique 8-oxodG base in both bacteria and monkey cells.[54,55] The presence of 8-oxodG residues in the DNA template during replication before completion of repair induces $G : C \rightarrow T : A$ transversions due to the misincorporation of adenine opposite 8-oxodG.[56,57] Alternatively, as a precursor formed in the nucleotide pool, 8-oxodGTP could be incorporated opposite template adenine leading to $A : T \rightarrow C : G$ transversions.[58,59] As seen in Table I, the predominant mutations are $G : C \rightarrow T : A$ transversions, suggesting that 8-oxodG could be the major premutagenic modification present in DNA damaged by singlet oxygen before any replication. However, the 8-oxodG does not seem to be the only DNA lesion inducing mutations, as other base substitutions at $G : C$ site, $G : C \rightarrow C : G$, and $G : C \rightarrow A : T$ were also recovered. These mutations may be the result of still unknown 8-oxodG processing before or during replication or be due to other lesions induced at low level in singlet oxygen-damaged DNA.[50] The nature of DNA polymerases involved during the *trans*-lesion process also plays an important role in determining the sites and the types of mutations following ROS production.[60]

Intracellular capacities to process oxidative lesions could also play a role in terms of genotoxicity, repair, and *trans*-lesion replication. The influence of the antioxidant defense pathways on mutagenesis has been established by modulation of the cellular redox state. Depletion of the level of the endogeneous glutathione, a major cellular antioxidant, leads to an increased frequency of spontaneous mutations, showing its main role as a protection system against the induction of oxidative premutagenic DNA damage.[61]

UVA-Induced Mutation Spectrum

Although the main UV component of sunlight is UVA, the carcinogenic potential of UVA has been previously overlooked because of its low absorp-

[53] S. Shibutani, M. Takeshita, and A. P. Grollman, *Nature* **349**, 431 (1991).
[54] F. Le Page, A. Margot, A. P. Grollman, A. Sarasin, and A. Gentil, *Carcinogenesis* **16**, 2779 (1995).
[55] M. Moriya, *Proc. Natl. Acad. Sci. U.S.A.* **90**, 1122 (1993).
[56] M. L. Wood, M. Dizdaroglu, E. Gajewski, and J. M. Essigmann, *Biochemistry* **29**, 7024 (1990).
[57] F. Le Page, A. Guy, J. Cadet, A. Sarasin, and A. Gentil, *Nucleic Acids Res.* **26**, 1276 (1998).
[58] Y. I. Pavlov, D. T. Minnick, S. Izuta, and T. A. Kunkel, *Biochemistry* **33**, 4695 (1994).
[59] K. C. Cheng, D. S. Cahill, H. Kasai, S. Nishimura, and L. A. Loeb, *J. Biol. Chem.* **267**, 166 (1992).
[60] D. I. Feig, T. M. Reid, and L. A. Loeb, *Cancer Res.* **54**, 1890 (1994).
[61] L. A. Applegate, D. Lautier, E. Frenk, and R. M. Tyrrell, *Carcinogenesis* **13**, 1557 (1992).

tion by DNA. In the past few years, several studies however, have, demonstrated that broad band UVA sources are carcinogenic in mice[62,63] and mutagenic for mammalian cells.[64–66] Considering the association between carcinogenesis and mutagenesis, action spectra for mutation induction probably are one of the best criteria for the evaluation of carcinogenic risk.

A striking difference between sunlight-induced mutations in human skin tumors[67] and in UVB-induced murine skin is the higher percentage of $G:C \rightarrow T:A$ and the lower percentage of $G:C \rightarrow A:T$ transitions, which are believed to be the UVB fingerprint of dipyrimidine photoproducts.[68,69] The difference may reflect the contribution of ROS, such as 8-oxodG, to UVA in human skin carcinogenesis.[63] Moreover, the precise wavelength dependence of UVA-induced pyrimidine dimer formation, cell killing, and mutation showed that cell lethality and mutation induction at 365 nm are partly due to DNA damage other than pyrimidine dimers.[34] Mutation spectra induced by UVA radiation have been established on the *aprt* gene in hamster cells,[65,66] on the episomal *lacZ'* gene in human cells,[64] and on the *p53* gene in skin tumors of hairless mice.[63]

To determine mutation spectrum induced by UVA in the 293 human epithelial cell line, we developed a shuttle vector system, carrying the *lacZ'* gene as the mutagenesis target.[64] At low cell survival levels ($1 - 0.1\%$), the mutation frequency induced by UVA was higher than that induced by UVB and reached a level 30-fold higher than the background. Thus, for similar cytotoxicity, more mutations were found with UVA than with UVB. Sequence analysis of independent *lacZ'* mutants (36 UVA- and 45 UVB-induced) revealed that frequencies of both point mutations and rearrangements increased with the UV doses. The large number of $G:C \rightarrow A:T$ transitions (Table I, 54%) could be due to the implication of UVA direct absorption–DNA damage, such as pyrimidine dimers, or deamination of methylated cytosine at CpG sites (67% of total $G:C \rightarrow A:T$ transitions). This type of mutation may also be indicative of DNA damage by reactive oxygen species.[60] The relatively high incidence of UVA-induced mutations

[62] A. de Laat, J. C. van der Leun, and F. R. de Gruijl, *Carcinogenesis* **18,** 1013 (1997).

[63] H. J. van Kranen, A. de Laat, J. van de Ven, P. W. Wester, A. de Vries, R. J. W. Berg, C. F. van Kreijl, and F. R. de Gruijl, *Cancer Res.* **57,** 1238 (1997).

[64] C. Robert, B. Muel, A. Benoit, L. Dubertret, A. Sarasin, and A. Stary, *J. Invest. Dermatol.* **106,** 721 (1996).

[65] E. A. Drobetsky, J. Turcotte, and A. Châteauneuf, *Proc. Natl. Acad. Sci. U.S.A.* **92,** 2350 (1995).

[66] E. Sage, B. Lamolet, E. Brulay, E. Moustacchi, A. Châteauneuf, and E. A. Drobetsky, *Proc. Natl. Acad. Sci. U.S.A.* **93,** 176 (1996).

[67] A. Stary, C. Robert, and A. Sarasin, *Mutat. Res.* **383,** 1 (1997).

[68] N. Dumaz, A. Stary, T. Soussi, L. Daya-Grosjean, and A. Sarasin, *Mutat. Res.* **307,** 375 (1994).

[69] E. Sage, *Photochem. Photobiol.* **57,** 163 (1993).

at A : T (Table I, 41%) did not point to one particular UVA-induced DNA lesion and did not favor an important role for pyrimidine dimers. The large contribution of A : T $\rightarrow$ G : C transitions after UVA treatment (23%) was higher compared with that after UVB treatment (6%) and pointed to ROS as a putative intermediate for UVA genotoxicity. Indeed, the A : T $\rightarrow$ G : C mutation has been identified in bacteria as a genetic effect of thymine glycols, one of the principal DNA lesions induced by oxidation.[70] Another ROS-induced DNA damage, the 2-hydroxyadenine (2-OH-Ade), can lead, in bacteria as well as in simian cells, to A : T $\rightarrow$ G : C and G : C $\rightarrow$ A : T mutations due to the mispair of dCMP opposite 2-OH-Ade in DNA and misincorporation of oxidized 2-OH-Ade opposite C residues in DNA, respectively.[71] Although the formation of this DNA damage was less efficient than 8-oxodG in DNA, 2-OH-Ade lesion appeared to be as mutagenic as 8-oxodG, similarly generated in dGTP and incorporated in DNA.[71]

On the *aprt* gene in Chinese hamster ovary cell,[65] the predominant event induced by UVA radiation is the base substitution A : T $\rightarrow$ C : G, constituting 41% of the mutation spectrum in excision repair-proficient rodent cells (Table I). In contrast, the UVA-dependent A : T $\rightarrow$ C : G transversion was absent in the excision repair-deficient strain (except 3 A : T $\rightarrow$ C : G included in tandem mutations).[66] This result suggests that the A : T $\rightarrow$ C : G mutation may be induced by a rare, but highly mutagenic, UVA-specific nondimer lesion, which might be completely hidden by the high level of unrepaired photoproducts in excision repair-deficient rodent cells (Table I). The induction of A : T $\rightarrow$ C : G transversions, typical of singlet oxygen, may be due to the incorporation of 8-oxodGTP, opposite template adenine, from the damaged nucleotide pool generated after UVA radiation (Table I). In the absence of NER, the predominance of G : C $\rightarrow$ A : T transitions (Table I, 76%) can be attributed to the formation of UVA-induced dipyrimidine photoproducts; among the seven tandem mutations, six are located at dipyrimidine sequences.[66]

All mutations detected in the *p53* gene in murine UVA-induced skin tumors have also been found to be G : C $\rightarrow$ A : T transitions, but at a low incidence compared with similar UVB-induced tumors.[63] These mutations are located mainly at the 267 codon, which is also a major hot spot in UVB-induced tumors as well as in internal tumors.

Although the mutagenic effect of UVA radiation is demonstrated clearly, the exact type of premutagenic DNA damage induced by a longer wavelength still remains unknown. Regarding the distinct features from

[70] A. K. Basu, E. L. Loechler, S. A. Leadon, and J. M. Essigmann, *Proc. Natl. Acad. Sci. U.S.A.* **86,** 7677 (1989).
[71] H. Kamiya and H. Kasai, *Biochemistry* **36,** 11125 (1997).

the three different mutation spectra presented here, some questions still remain, such as the nature of the endogenous photosensitizers and that of the DNA lesion involved, the contribution of DNA repair systems, and the effect of *trans*-lesion replication on mutagenesis.

Conclusion

It is clear from various data that both ROS and UVA can lead to deleterious effects on cells, eventually giving rise to cancer induction. In this review, we decided to discuss only their effects on DNA following DNA repair and mutation induction, although it is obvious that some important biological effects should involve proteins, lipides, and membranes. The analysis of mutation spectra is a powerful method for the understanding of the effect of a given genotoxic. However, one needs to know at the chemical level the type of DNA lesions that are produced in order to interpret correctly the types of mutations observed. In the case of both singlet oxygen and UVA-treated cells, it is obvious that not enough is known about the chemistry of their lesions to understand all types of mutations. Very little data are available in mammalian cells concerning UVA-induced mutation spectra, as there is only one study in the adenovirus-transformed 293 human cell line and one set of data in NER$^+$ and NER$^-$ CHO cells. The most important base pair substitution in 293 and NER$^-$ CHO cells is the G:C $\rightarrow$ A:T (Table I), but at a much lower frequency than that observed after UVB treatment.[64] It is clear from these data that UVA produced many more mutations at A:Ts than UVB, indicating that lesions other than (6-4) photoproducts and pyrimidine dimers are mutagenic after UVA irradiation. Interestingly, mutation spectra determined in the *p53* gene of solar-induced skin cancers are identical to the UVB spectrum and different from the UVA one. This analysis permitted us to determine that, during sun-induced skin carcinogenesis in human, the initiating event represented by *p53* mutations is produced mainly by the UVB part and not by the UVA one.[68]

Mutation spectra induced by singlet oxygen in bacteria and in mammalian cells are relatively similar but significantly different from the UVA-induced mutation spectrum, at least in the same human 293 cell line (Table I). The major base pair substitution observed on singlet oxygen-treated template is the G:C $\rightarrow$ T:A transversion, which may correspond to the presence of unrepaired oxidized guanine, such as 8-oxodG. This mutagenic event represents the second most important mutation on the *p53* gene (18.2%) for all classes of human tumors.[72] This mutation is also very impor-

[72] P. Hainaut, T. Hernandez, A. Robinson, P. Rodriguez-Tome, T. Flores, M. Hollstein, C. C. Harris, and R. Montesano, *Nucleic Acids Res.* **26**, 205 (1998).

tant for some specific tumor sites, such as lung cancer (in which the $G:C \rightarrow T:A$ transversion is the first mutagenic event representing 35.6% of all $p53$ base substitutions),[73] where it is supposed to be induced by guanine adducts containing a chemical carcinogen. However, the presence of oxidized guanine, such as 8-oxodG, could still be partly responsible for this transversion.

In conclusion, the molecular biology of DNA repair and mutagenic pathways allows us to establish, in a model system, mutation spectra produced by genotoxic agents, such as UVA or singlet oxygen treatments. These *in vitro* mutation spectra can be compared with *in vivo* mutation spectra determined on "true" human cancers in order to approach the etiologic agent involved in cancer induction. However, a full comprehensive analysis of these mutation spectra necessitates the complete description of DNA lesions produced by these genotoxics. These various DNA lesions should not only be identified, but the chemical and biochemical characteristics determined as well as their mutagenic properties *in vivo*. Complete knowledge of DNA lesions produced by UVA and singlet oxygen exposure is still not adequately known.

Acknowledgments

We thank Professor C. F. M. Menck and Professor J. C. Ehrhart for critical reading of the manuscript. This work has been supported by grants from the Association pour la Recherche sur le Cancer (Villejuif, France) and the Fondation de France (Paris, France).

[73] S. Chevillard, J. P. Radicella, C. Levalois, J. Lebeau, M. F. Poupon, S. Oudard, B. Dutrillaux, and S. Boiteux, *Oncogene* **16,** 3083 (1998).

[16] Damage to DNA by Long-Range Charge Transport

By MEGAN E. NÚÑEZ, SCOTT R. RAJSKI, and JACQUELINE K. BARTON

Introduction

Photochemical reactions on DNA assemblies containing tethered photooxidants, particularly metallointercalating photooxidants, have been critical in establishing that permanent damage to DNA bases can be generated as a result of radical migration from a remote site on the DNA duplex.[1,2]

[1] R. E. Holmlin, P. J. Dandliker, and J. K. Barton, *Angewandte Chem.* **109,** 2830; *Intl. Ed. Eng.* **36,** 2714 (1997).
[2] S. O. Kelley and J. K. Barton, *Metals Biol.* **26,** 211 (1998).

Induction of a 1-electron deficiency in the oxidant attached covalently to the DNA remote from the oxidizable site leads to this "chemistry at a distance," caused by efficient charge transport through the DNA base pair stack. Double helical DNA may be unique as a polymeric assembly in solution because of this interior core of stacked aromatic heterocyclic base pairs. Similarly stacked solid-state materials tend to be conducting along the stacking direction.[3]

Oxidative damage to DNA bases occurs almost exclusively at guanine sites. The preferential damage to guanine is understandable based on its oxidation potential compared to those of the other bases: the oxidation potential of the dG nucleoside as measured by pulse radiolysis is 1.29 V vs NHE, compared to 1.42 V (dA), 1.6 V (dC), and 1.7 V (dT).[4] Furthermore, *ab initio* molecular orbital calculations indicate that guanine stacked 5' to another guanine can be oxidized more easily than guanine stacked next to other bases; for this stacked pair of guanine bases, the bulk of the HOMO is located on the 5' guanine, making the 5' guanine a more easily oxidized target than the 3' guanine.[5,6] This damage at the 5'-G of 5'-GG-3' doublets has become a signature for oxidative damage by electron transfer. Substituted anthraquinones,[7,8] naphthalimide derivatives,[9] and riboflavin[10] have all been shown to generate damage at the 5'-G of 5'-GG-3' doublets, despite the fact that they do not bind these sites on DNA in a sequence-specific manner.

Long-range oxidative damage to DNA by charge transfer along the helix was first demonstrated in an oligonucleotide duplex containing a tethered rhodium intercalator as the photooxidant.[11] In the DNA assembly that was constructed, the metallointercalator was tethered to one end of a 15-mer duplex, separated spatially from two 5'-GG-3' target sites for oxidation. Figure 1 illustrates a general assembly constructed to test for long-range charge transport. Phenanthrenequinone diimine (phi) complexes of rhodium, which intercalate into B-form DNA from the major groove, are effective photooxidants for use in these studies.[12,13] When photoactivated

[3] T. J. Marks, *Science* **227**, 881 (1985).

[4] S. Steenken and S. V. Jovanovic, *J. Am. Chem. Soc.* **119**, 617 (1997).

[5] H. Sugiyama and I. Saito, *J. Am. Chem. Soc.* **118**, 7063 (1996).

[6] P. Ferran, K. N. Houk, and C. S. Foot, *J. Am. Chem. Soc.* **120**, 845 (1998).

[7] D. Breslin and G. B. Schuster, *J. Am. Chem. Soc.* **118**, 2311 (1996).

[8] D. K. Y. Ly, B. Armitage, and G. B. Schuster, *J. Am. Chem. Soc.* **118** 8747 (1996).

[9] I. Saito, M. Takayama, H. Sugiyama, K. Nakatani, A. Tsuchida, and M. Yamamoto, *J. Am. Chem. Soc.* **117**, 6406 (1995).

[10] K. Ito, S. Inoue, K. Yamamoto, and S. Kawanishi, *J. Biol. Chem.* **268**, 13221 (1993).

[11] D. B. Hall, R. E. Holmlin, and J. K. Barton, *Nature* **382**, 731 (1996).

[12] A. Sitlani, E. C. Long, A. M. Pyle, and J. K. Barton, *J. Am. Chem. Soc.* **114**, 2302 (1992).

[13] C. Turro, S. H. Bossmann, S. Niu, J. K. Barton, and N. J. Turro, *Inorg. Chim. Acta* **252**, 333 (1996).

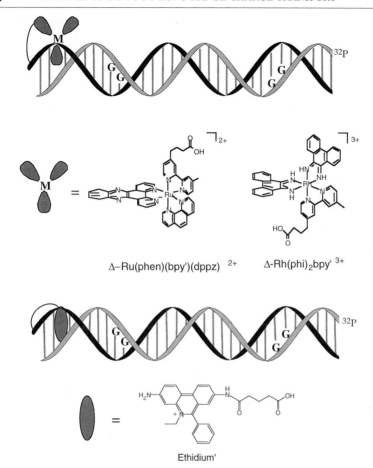

FIG. 1. Schematic of a DNA assembly containing the tethered photooxidant and the chemical structures of the photooxidants employed. The duplex assembly consists of two complementary strands, with a photooxidant tethered covalently to the end of the sugar phosphate backbone of one strand and with the complementary strand radioactively tagged. The labeled strand also contains one or more 5′-GG-3′ sites, which act as sinks for the electronic hole. On photoexcitation, Rh(phi)$_2$bpy′$^{3+}$ (phi = phenanthrenequinone diimine), Ru(phen)(bpy′)(dppz)$^{2+}$ (dppz = dipyridophenazine), and ethidium can all oxidize guanine doublets by long-range charge transfer through DNA.

with high-energy light (313 nm), the metallointercalator promotes direct strand cleavage of DNA, marking the site of intercalation. However, irradiation with lower energy light (365 nm) excites the metallointercalator to oxidize 5′-GG-3′ sites, forming 7,8-dihydro-8-oxoguanine (8-oxo-G) and piperidine-sensitive DNA lesions.[11] These two distinct photochemical reactions of the phi complexes allowed us to show that the metallointercalator could oxidize 5'-GG-3' sites even when the rhodium complex was tethered covalently at a distance of 34 Å (10 bp) from the 5'-GG-3'. More recently, in a series of rhodium-modified DNA assemblies containing two 5'-GG-3' sites, the yield of oxidative damage was relatively insensitive to the distance of charge migration over the range of 41–71 Å.[14] This long-range oxidation by the tethered rhodium intercalator was furthermore seen to yield significant damage over a distance of at least 200 Å.

Importantly, long-range oxidative damage to DNA, like DNA electron transfer chemistry generally, is exquisitely sensitive to the stacking of reactants as well as the stacking of the intervening base pairs. In diastereomeric assemblies containing the tethered metallointercalator, the right-handed (Δ) isomer, which fits more deeply into the major groove of right-handed B DNA, yields more oxidative damage than the left-handed (Λ) isomer.[11] We also find that long-range oxidation is sensitive to disturbances in the DNA π stacking generated by bulges between the oxidant and the 5'-GG-3' site[15]; large bulges that do not stack well into the helix such as the 5'-ATA-3' bulge were shown to diminish oxidation distal to the bulged site compared to that at a proximal 5'-GG-3' doublet. Long-range oxidation also appears to be sensitive to the intervening sequence.[14,16] This effect may reflect the sequence-dependent flexibility in stacking of DNA. Particularly noteworthy in this context is the observation that 5'-TA'-3' steps tend to attenuate long-range oxidative damage, a result that may be attributed to poor base–base overlap. Whether such sequences serve to insulate, or protect, DNA regions within the cell from damage still needs to be determined. Perhaps most intriguing from the standpoint of the *in vivo* consequences of long-range damage to DNA is the observation that DNA-binding proteins can both inhibit and activate long-range oxidative DNA damage.[17] We have found that the *Hha*I methylase, a base-flipping enzyme, can modulate long-range oxidative damage by a tethered rhodium intercalator by binding sequence specifically within a DNA assembly. The native protein, which inserts a nonaromatic glutamine side chain into the DNA stack, inhibits distal oxidation, whereas a mutant protein, which inserts an aromatic tryptophan side chain, stimulates damage at the G, which is 5' to the inserted tryptophan.

By now, a range of oxidants have been shown to initiate long-range oxidative damage mediated by the π-stacked array of heterocyclic base pairs (Table I). A covalently tethered ground-state ruthenium(III) intercalator, generated *in situ,* promotes oxidative damage to guanine doublets in high

[14] M. E. Nunez, D. B. Hall, and J. K. Barton, *Chem. Biol.* **6,** 85 (1999).

[15] D. B. Hall and J. K. Barton, *J. Am. Chem. Soc.* **119,** 5045 (1997).

[16] E. Meggers, M. E. Michel-Beyerle, and B. Giese, *J. Am. Chem. Soc.* **120,** 12950 (1998).

[17] S. R. Rajski, S. Kumar, R. J. Roberts, and J. K. Barton, *J. Am. Chem. Soc.* **121,** 5615 (1999).

TABLE I
DNA TETHERED/INCORPORATED PHOTOOXIDANTS

Structure	Comments	References
	Incorporated at either 5′ or 3′ end of DNA following automated synthesis. Activation requires 365-nm light.	11, 15, 22
	Synthetically incorporated at either 5′ or 3′ of DNA using solid-phase synthesis. Activation requires 436-nm light yielding *in situ* generation of Ru(III) complex. Requires presence of a quencher.	14, 18
	Core structure has been synthetically incorporated into DNA as a phosphoramidite at 5′ end. Also, incorporated into peptide nucleic acids (PNA) for examination of charge migration in PNA/DNA hybrid duplexes. Activation requires 350-nm light.	19
	Synthetically incorporated as phosphoramidite. Activation requires 312-nm light.	21
	Synthetically incorporated at either 5′ or 3′ end of DNA using solid-phase synthesis. Activation requires 313-nm light.	20
	Synthetically incorporated as phosphoramidite. Activation requires light processed through a 320-nm cutoff filter. Formation of a sugar radical proceeds by Norrish I type cleavage of the *t*-butyl ketone to ultimately inject a radical cation into the DNA base stack.	16

yield over a distance of as much as 200Å.[14,18] Covalently bound, photoexcited anthraquinones associated in a capped position on the DNA duplex have been shown to yield piperidine-sensitive damage to guanine and 8-oxoG at long range in analogously prepared DNA assemblies.[19] Long-range guanine damage has been demonstrated by photooxidation of tethered ethidium, the classic organic intercalator,[20] as well as by a uridine base modified with a cyano-benzophenone derivative.[21] It is therefore now clear that oxidative damage to DNA can occur over long distances through DNA-mediated electron transport. Moreover, this long-range chemistry is not a particular characteristic of the oxidant employed. Instead, this chemistry at a distance represents a unique feature of the DNA itself.

This article describes the design and construction of DNA assemblies used to probe long-range oxidative damage in DNA. We also include methodology for the oxidative repair of a thymine dimer lesion in DNA, as this "chemistry at a distance" also depends on long-range charge transfer through the DNA base pair stack.[22]

Experimental Methods

I. Strategy for the Construction of DNA Assemblies with a Tethered Photooxidant

A. Design Considerations. The DNA duplex assembly is composed of two complementary short oligonucleotides (typically about 25 nucleotides in length), one of which is attached covalently to the intercalating photooxidant via a linker to the 5′ or 3′ end of the sugar-phosphate backbone. The complementary oligonucleotide contains 5′-GG-3′ doublets, which serve as targets for oxidative damage by virtue of their low redox potential.[5,6] In addition, the complementary strand bears the radioactive label for detection by denaturing polyacrylamide gel electrophoresis (DPAGE) (Fig. 1).

In an oligonucleotide duplex used to investigate long-range charge transport, design features associated with the sequence are also important to consider. Because the redox potential of a 5′-GG-3′ site varies somewhat depending on its nearest neighbors,[23] all of the 5′-GG-3′ sites in a DNA

[18] M. R. Arkin, E. D. A. Stemp, S. C. Pulver, and J. K. Barton, *Chem. Biol.* **4,** 389 (1997).

[19] S. M. Gasper and G. B. Schuster, *J. Am. Chem. Soc.* **119,** 12762 (1997).

[20] D. B. Hall, S. O. Kelley, and J. K. Barton, *Biochemistry* **37,** 15933 (1998).

[21] K. Nakatani, K. Fujisawa, C. Dohno, T. Nakamura, and I. Saito, *Tetrahedron Lett.* **39,** 5995 (1998).

[22] P. J. Dandliker, R. E. Holmlin, and J. K. Barton, *Science* **275,** 1465 (1997).

[23] I. Saito, T. Nakamura, K. Nakatani, Y. Yoshioka, K. Yamaguchi, and H. Sugiyama, *J. Am. Chem. Soc.* **120,** 12686 (1998).

assembly should be flanked by the same bases if oxidative damage at the two sites is to be best compared directly. We have tended to use C's both to the 3′ and to the 5′ sides of the 5′-GG-3′ doublet. Furthermore, it is critically important that the oxidant be well intercalated into the π stack for long-range charge transfer to occur. We have generally optimized the DNA sequence at the intercalation site empirically through a systematic variation in binding site sequence; once a productive sequence is obtained, it is kept constant. Finally, in order to maintain good signal/noise, the duplex should also not contain more than one 5′-TA-3′ step in the region intervening between the oxidant and the 5′-GG-3′; these 5′-pyrimidine-purine-3′ steps are especially poor conduits for charge transport.[14]

B. Illustrations. Figure 2 shows an example of a polyacrylamide gel autoradiogram used to analyze photooxidative damage to DNA from a distantly bound rhodium intercalator. The tethered assemblies permit intercalation of the photooxidant near the end of the duplex. Photoactivation at low energies (365 nm) of the rhodium complex bound to DNA then promotes oxidation of the 5′ G of 5′-GG-3′ doublets within the duplex, irrespective of whether the doublet sites are proximal or distal to the bound intercalator.[11] Oxidative damage to the guanines is revealed in this assay after base treatment, as piperidine treatment can labilize oxidative base lesions so as to cleave the DNA strand. From the position of bands and their intensity, both the site and the relative extent of oxidative damage can be determined. Using this assay, we have been probing both the distance and the sequence dependence of long-range oxidative damage.[14] Importantly, structural deviations from canonical B-form DNA within the intervening region, which interrupt the π stack, alter the relative yield of distal versus proximal 5′-GG-3′ oxidation, and hence this assay may also be useful in probing local perturbations in DNA stacking.[15,17]

C. Photooxidants. We have focused on three different photooxidants to generate DNA damage via long-range charge transport, the two octahedral metallointercalators, $Rh(phi)_2bpy^{3+}$ and $Ru(phen)(bpy)(dppz)^{2+}$ (dppz, dipyridophenazine), and the organic intercalator, ethidium (Fig. 1.). Despite substantial differences in binding characteristics and photochemistry, all three intercalators have been shown to oxidatively damage 5′-GG-3′ doublets at long range with photoactivation.[11,18,20] Hence, long-range charge transport through the DNA to generate permanent base lesions appears to be a general chemical feature of the DNA base pair stack.

With each of these tethered intercalators, we have exploited photochemical reactions not only to oxidize the bases but also to mark the site of intercalation within the helix. These reactions can involve photocrosslinking, singlet oxygen sensitization, or a direct strand cleavage reaction. We have found that flexible tethers with linker lengths of at least six methylenes

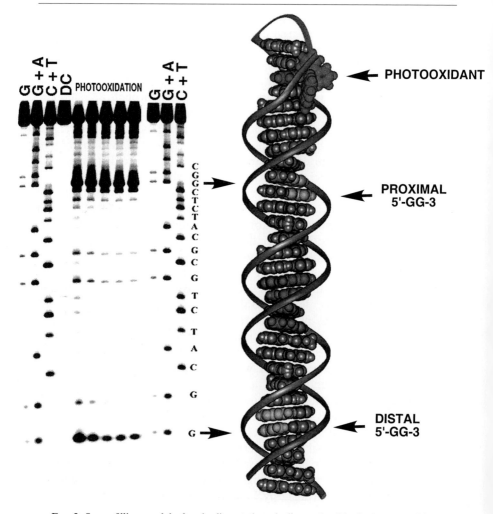

FIG. 2. Space-filling model of a rhodium-tethered oligonucleotide duplex assembly containing two guanine doublets and the corresponding 20% DPAGE autoradiogram showing damage at both proximal and distal 5'-GG-3' sites.

are needed to maintain deep intercalation without substantial constraints of the tether.[24] Thus experimental methods to establish the site of intercalation directly have been developed. Moreover, to demonstrate that oxidation does indeed occur over long range mediated by the DNA, it is

[24] S. O. Kelley and J. K. Barton, *Chem. Biol.* **5,** 413 (1998).

important to identify the binding site of a given photooxidant directly within the covalently tethered duplex assembly. Figure 3 illustrates the general strategy we employ.

Phi complexes of rhodium are powerful excited-state oxidants with an oxidation potential ~2.0 V versus NHE,[13] sufficiently strong to oxidize each of the bases. This rhodium photochemistry is initiated by irradiation at 365 nm. At higher energies of irradiation (313 nm), phi complexes of rhodium promote direct strand cleavage of DNA, to give products that are consistent with abstraction of the C-3' hydrogen atom from the DNA sugar at the site of intercalation.[12] This direct strand scission thus directly marks the binding site of the metal complex. Using this direct photocleavage reaction, we have found that tethered rhodium complexes intercalate 2 or 3 bp from the end of an oligonucleotide duplex.

Ground-state Ru(III) oxidants have an oxidation potential of ~1.6 V versus NHE; these potentials are sufficient to oxidize the purines within DNA.[18] Ru(III) oxidants, however, must be generated *in situ.* In the flash quench methodology that we employ, irradiation of Ru(II) with visible light (440 nm) in the presence of a nonintercalating, diffusible quencher yields Ru(III) by electron transfer.[18,25] $Co(NH_3)_5Cl^{2+}$, $Ru(NH_3)_6^{3+}$, and methyl viologen (MV^{2+}) have been used as the groove-bound quencher, with particular advantages for each. The yield of Ru(III) and thus oxidized guanine is greater with $Co(NH_3)_5Cl^{2+}$ and methyl viologen because back electron transfer from quencher to Ru(III) is disfavored; $Co(NH_3)_5Cl^+$ is unstable and falls apart to the aquo species, whereas MV^+ reacts with oxygen to form superoxide. Conversely, although the irradiation times are much longer using $Ru(NH_3)_6^{3+}$ because of the facility of the back reaction, the control samples and background are frequently cleaner. For these assemblies, the site of intercalation is determined by photoexciting the tethered ruthenium (II) intercalator in the *absence* of quencher. Without quencher the excited Ru(II) can sensitize the formation of small quantities of singlet oxygen that diffuse away from the metal center, oxidizing guanine bases in the *immediate* vicinity of the intercalation site. This ruthenium damage is the only visible damage when other quenchers are absent, making it relatively straightforward to distinguish the intercalation site from the 5'-GG-3' sites of oxidative base chemistry.

Tethered ethidium intercalators, when irradiated at 313 nm, are estimated to have an oxidation potential of ~2.0 V versus NHE[26] and can oxidatively damage 5'-GG-3' doublets at long range.[20] The short-range reaction that we have exploited to establish the ethidium intercalation site

[25] E. D. A. Stemp, M. R. Arkin, and J. K. Barton, *J. Am. Chem. Soc.* **119,** 2921 (1997).
[26] A. Waleh, B. Hudson, and G. Loew, *Biopolymers* **15,** 1637 (1976).

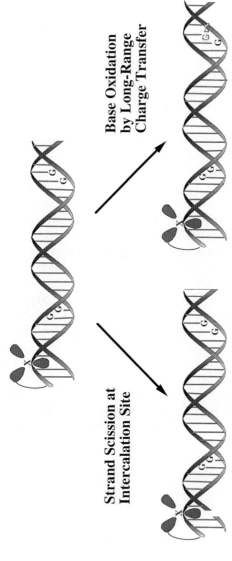

FIG. 3. To demonstrate that base oxidation occurs over a long distance, it is important to establish the binding site of a given photooxidant within the duplex assembly to which it is tethered. Strand scission at the intercalation site and base oxidation by long-range charge transfer are therefore employed as complementary reactions to mark and distinguish the site of binding from that of oxidation. In the case of phi complexes of rhodium, irradiation at 313 nm generates direct strand breaks as a result of reaction with the sugar moiety. When the ruthenium complex is irradiated at 440 nm in the absence of a quencher, singlet oxygen is formed at the intercalation site. This singlet oxygen reaction leads to piperidine-sensitive base damage in the immediate vicinity of the metal complex, allowing the intercalation site to be pinpointed. Ethidium' directly crosslinks the DNA at the intercalation site when irradiated at 313 nm, generating strand breaks even in the absence of piperidine. In contrast, to generate oxidized guanine (right), rhodium is irradiated at 365 nm, ethidium' is irradiated at 313 nm, and ruthenium is irradiated at 440 nm in the presence of a quencher; these photoexcitations lead to oxidative damage to 5'-GG-3' sites from a distance, which is revealed by treatment with hot aqueous piperidine.

has been a photocrosslinking reaction. Thus, in addition to oxidizing DNA bases by electron transfer, modified ethidium intercalators can also cross-link DNA directly when irradiated at 313 nm, generating direct strand breaks. Because these ethidium irradiations require high-energy light, however, it is critical when using this organic intercalator to include control reactions conducted in the absence of intercalator but with irradiation.

The choice of tethered oxidant therefore depends on the constraints of the particular experiment. Metallointercalators, for example, bind poorly to A-form nucleic acids, whereas ethidium, which binds from the minor groove, binds both A and B-form duplexes.[27] Ruthenium photochemistry is certainly the most efficient reaction and involves only visible light irradiation, a clear advantage. In generating ruthenium(III), however, an added reagent, the groove-bound quencher, is required, and the presence of this diffusible species may be detrimental to other components of the system being tested. Nonetheless, the tethered assemblies already prepared seem to offer sufficient flexibility in designing assays of long-range oxidative DNA damage under a variety of conditions.

II. Synthesis and Characterization of Assemblies.

A. Synthetic Methods for the Strand with Appended Photooxidant. Figure 4 shows our approach for functionalizing a DNA oligonucleotide with a tethered photooxidant. We have found the greatest reliability in coupling the photooxidant to the oligonucleotide on a solid support.[28] To tether covalently to the ends of the DNA strand, the photooxidants, properly functionalized to contain carboxylic acid groups, are first activated as N-hydroxysuccinic esters and then are condensed with aminoalkylated oligo-deoxyribonucleotides. The photooxidant is conjugated either to the 3' or to the 5' end of an oligonucleotide using solid-phase synthesis followed by deprotection of the DNA bases and cleavage of the DNA from the control pore-glass (CPG) resin. The virtue of the solid-phase coupling is that each step in the reaction is driven to completion by employing a large excess of reagent, which can be removed easily by thoroughly rinsing the resin with solvent. These solid-phase coupling reactions are carried out in peptide reaction vessels fitted with glass frits and stopcocks, which allow for simple bench-top manipulation and isolation of the CPG-tethered intermediates and products. A frequently used tethered photooxidant within our group is a derivative of $[Rh(phi)_2bpy]^{3+}$, which contains a carboxylic acid handle for site-specific DNA modification, $[Rh(phi)_2bpy']^{3+}$ (bpy', 4'-methylbipyridine-4-butyric acid). These covalent tethering reactions are also adapted

[27] T. W. Johann and J. K. Barton, *Phil. Trans. Royal Soc.* (*London*) **354,** 299 (1996).
[28] R. E. Holmlin, P. J. Dandliker, and J. K. Barton, *Bioconj. Chem.* **10,** 1122 (1999).

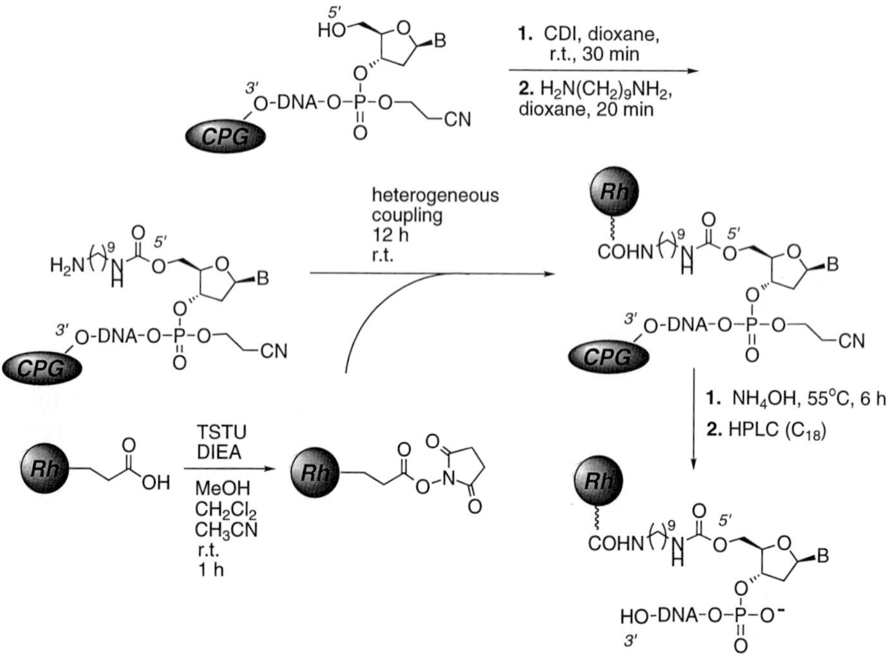

FIG. 4. Scheme for the covalent attachment of a carboxylic acid functionalized photooxidant [in this case, $Rh(phi)_2bpy'^{3+}$] to the 5' end of an oligonucleotide using solid-phase synthesis.

easily to complexes of ruthenium, ethidium, and similar molecules with appended carboxylate functionalities.[28] Syntheses of the functionalized intercalators are described elsewhere.[18,29,30]

In the procedure we have adopted, DNA is first prepared using standard methodology[31] on the resin and the aminoalkylated linker is added to the free 5'-hydroxyl on the DNA strand. We have synthesized DNA by phosphoramidite chemistry on 1-μmol scale synthesis columns (Applied Biosystems automated DNA synthesizer, 2000 Å CPG). After completion of the synthesis, the DNA remains attached to the resin and contains a free 5'-OH terminus. The CPG-linked oligonucleotide (2 μmol) is transferred to a peptide reaction vessel with a coarse glass frit in the bottom and a stopcock to control suction or positive pressure flow (Ar or N_2) through the reaction

[29] S. O. Kelley, R. E. Holmlin, E. D. A. Stemp, and J. K. Barton, *J. Am. Chem. Soc.* **119,** 9861 (1997).

[30] A. Sitlani and J. K. Barton, *Biochemistry* **33,** 12100 (1994).

[31] M. H. Caruthers, A. D. Barone, S. L. Beaucage, D. R. Dodds, E. H. Fisher, L. J. McBride, M. Matteucci, Z. Stabinsky, and J. Y. Tang, *Methods Enzymol.* **154,** 287 (1987).

vessel. Aminoacylation of the free 5′-hydroxyl with carbonyldiimidazole follows the method of Wachter and co-workers.[32] To the activated CPG-bound DNA is then added a solution of diaminononane (32 mg in 1 ml of a 9 : 1 dioxane : H_2O mixture) to form the 5′-aminoalkylated product. This reaction is allowed to proceed for 20–25 min under nitrogen or argon, followed by prompt removal of the reagent and rigorous washing of the aminoalkylated DNA with dioxane followed by methanol. Significantly, 1,9-diaminononane is sufficiently nucleophilic so as to affect release of the oligonucleotide from the CPG if allowed to react for longer than 25 min.[28] The DNA, now bearing the primary amine-functionalized C9 linkage, is ready for conjugation to the photooxidant.

Coupling of the photooxidant to the 5′-amino terminus of the DNA closely follows the methods described by Bannwarth and Knorr.[33] To activate the carboxylic acid terminus of the rhodium complex, 10 μmol Rh(phi)$_2$bpy′$^{3+}$ is added to 20 μmol of $N,N,N′,N′$-tetramethyl-O-(N-succinimidyl) uronium tetrafluoroborate (TSTU, 8 mg) and 20 μmol $N,N′$-diisopropylethylamine (DIEA, 6 μl) in 0.5 ml of 1 : 1 : 1 $CH_3CN/CH_2Cl_2/MeOH$. The solution is stirred at 25° for 1 hr, after which time the activation of the rhodium complex can be checked by alumina TLC. Following activation, 2 μmol of aminoalkylated DNA on the CPG resin is added directly to the Rh complex solution with an additional 20 μl (100 μmol) of DIEA, and the rhodium-CPG slurry is stirred at room temperature for 10–12 hr. The functionalized CPG-bound metal–DNA conjugates are isolated by filtration on a medium grade glass-fritted filter and washed extensively with distilled H_2O, CH_3OH, ethyl acetate, and CH_2Cl_2. The resulting conjugates are then cleaved from the CPG, and the DNA is deprotected by heating the functionalized resin at 55–60° in 2 ml concentrated NH_4OH. After a 6-hr incubation, the DNA–NH_4OH solution is cooled, decanted, and dried *in vacuo*. The crude reaction mixture is then purified by reversed-phase HPLC (see Section II,B).

We have also prepared DNA containing oxidants tethered to the 3′ end. In this case, DNA oligonucleotides are synthesized by standard phosphoramidite chemistry using 3′-amino-modifier DNA synthesis columns bearing an Fmoc-protected amine as the 3′ base (C7 CPG 500 Å, Glen Research). Following DNA synthesis (with the terminal DMT group on), the Fmoc group is removed from the 3′ base by incubation of the resin-bound DNA in 20% piperidine in dimethylformamide (DMF) for 30 min at ambient temperature. The piperidine solution is removed and the CPG is then washed repeatedly with DMF (5 × 1 ml), CH_3CN (5 × 1 ml), and

---

[32] L. Wachter, J.-A. Jablonski, and K. L. Ramachandran, *Nucleic Acids Res.* **14**, 7985 (1986).
[33] W. Bannwarth and R. Knorr, *Tetrahedron Lett.* **32**, 1157 (1991).

CH_2Cl_2 (5 × 1 ml) and dried *in vacuo*. Methodology for the subsequent photooxidant coupling to the 3′ free amine is identical to the 5′-coupling methodology. Following photooxidant coupling, dimethoxytrityl cleavage from the 5′ hydroxyl is best achieved on an automated DNA synthesizer. The synthesizer can be programmed to cleave the trityl group using an 18-sec wash of trichloroacetic acid and several washes with acetonitrile. After the DMT group is removed, the 3′-modified DNA is ready for deprotection and cleavage from the CPG resin in NH_4OH as described for the 5′-modified material.

   B. HPLC Purification. Reversed phase HPLC offers a very powerful tool by which to purify DNA conjugates in that it provides sufficient resolution to separate modified oligonucleotides both from unmodified DNA and from excess untethered photooxidant. Monitoring the HPLC separation using a photodiode array detector makes it possible to determine which components of the resulting product mixtures contain both DNA and the photooxidant.

   Product mixtures are separated using gradient chromatography over C-18 300-Å columns with an aqueous phase 50 mM in NH_4OAc, pH 7, and with CH_3CN as the organic solvent. Gradients usually start out at 5% CH_3CN and go to 13% over 30 min, during which time the unmodified DNA and then the metallointercalator–DNA conjugates elute. The gradient continues to 50% CH_3CN over a 45-min time interval to elute the untethered metal complex. These times and percentages vary somewhat based on DNA length, sequence, and identity of the tethered oxidant. In general, we prefer to use a flow rate that is as high as possible for the particular column and HPLC specifications to enhance the separation.

   Each of the intercalators contains an intense visible transition that is useful for separation and characterization (Fig. 5). In the case of $[Rh(phi)_2bpy]^{3+}$-tethered DNAs, the intense metal complex absorption at 390 nm, coupled with a DNA absorbance at 260 nm allows for the straightforward identification of those fractions containing functionalized DNA. Ethidium-tethered oligonucleotides are differentiated from unmodified DNA by their intense absorption at 480 nm, whereas untethered ethidium lacks the nucleic acid absorbance peak around 260 nm. Similarly, the ruthenium conjugates can be identified by characteristic absorbances at 260 nm (DNA) and at 440 nm (metal). In fact, we have found that purification by HPLC not only allows routine separation of metal-modified DNA from untethered DNA and free metal complex, but also permits separation of the different diastereomers resulting from coupling of each metal complex isomer to the single-stranded DNA. For DNA coupled to the rhodium complex, two diastereomers arise from coupling of the metal enantiomers (Δ and Λ), whereas for DNA tethered to tris-heteroleptic metal complexes

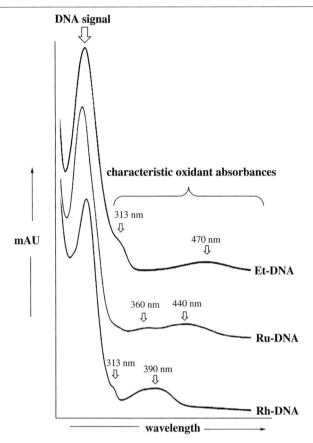

Fɪɢ. 5. UV-visible absorbance spectra of isolated oligonucleotides containing tethered Rh(phi)$_2$bpy$'^{3+}$, Ru(phen)(bpy$'$)(dppz)$^{2+}$, or ethidium. Wavelengths for visible and near-UV absorbance maxima of the photooxidants are indicated above the peaks.

such as Ru(phen)(bpy$'$)(dppz)$^{2+}$, four isomers are obtained. Unlike Rh and Ru complexes, the ethidium moiety is achiral and thus chromatographic analysis of Et-DNA conjugates yields only a single peak.

   C. *Characterization of DNA–Photooxidant Conjugate.* The sheer size of most DNA–oxidant conjugates precludes some characterization techniques traditionally associated with the synthesis of small molecules. We have found, however, that mass spectral data on conjugates up to about 10 kDa can be obtained by MALDI-TOF mass spectral analysis and that slightly smaller conjugates (5–7 kDa) can be characterized by electrospray ionization (ESI) mass spectral analysis. Both methods are amenable to use with conjugates directly after HPLC purification using NH$_4$OAc buffers. A complementary method of characterization is UV-visible absorbance spectroscopy, as each of the photooxidant–oligonucleotide conjugates displays a characteristic spectrum (as described earlier.) Along these lines, circular dichroism spectroscopy (CD) allows facile assignment of metal complex stereochemistry, which is useful, as preferential binding of the Δ-isomer of octahedral metallointercalators over the corresponding Λ-isomer to right-handed DNA is a critical consideration.

   D. *Preparation of Complementary Oligonucleotides.* Complementary oligonucleotides are synthesized on 1 μM scale (1000 Å CPG) using standard phosphoramidite chemistry.[31] The deprotected single-standed oligonucleotides are twice purified by reversed phase HPLC, once before and once after the removal of the DMT group. Alternatively, the detritylated oligonucleotide can be gel purified in a 18% denaturing polyacrylamide gel. The bands are identified by UV shadowing and are excised, and the DNA is eluted into buffered solution using the "crush and soak" method.[34] Urea and other contaminants can then be separated from the DNA by ethanol precipitation or size exclusion chromatography over Sephadex G-25 or G-50 resin. After purification, all DNA samples are resuspended in a dilute buffer solution with a pH ideally between 7.5 and 8.5. Approximately 100 pmol of purified complementary oligonucleotide is labeled enzymatically at the 5' or 3' end with ^{32}P according to standard techniques.[34] The radiolabeled oligonucleotides are then incubated in 100 μl of 10% piperidine at 90° for 30 min to cleave any damaged or depurinated oligonucleotides. Inclusion of this purification step cleans up the dark control lanes and improves the signal-to-noise ratio dramatically in later assays for oxidative damage. The samples are then dried *in vacuo* with centrifugation, and complete removal of residual salts, DTT, free ATP, and enzyme is achieved by DPAGE. The desired DNA band is revealed by autoradiography and excised, and the DNA is extracted into buffer by the "crush and soak" method.[34] Urea and other contaminants are removed using a Nensorb C-18 cartridge, and the amount of radioactivity is quantitated by liquid scintillation counting.

   E. *Preparation of Long Metal-Containing Oligonucleotides by Enzymatic Ligation (≥30 Nucleotides in Length).* Photooxidant-tethered oligonucleotides longer than approximately 35 bp cannot be synthesized directly in good yield and in pure form, but it is frequently interesting to examine oxidative damage on much longer DNA duplexes. Fortunately, long photooxidant-tethered duplexes can be synthesized in pieces and then ligated

---

[34] J. Sambrook, E. F. Fritsch, and T. Maniatis, "Molecular Cloning: A Laboratory Manual," 2nd Ed. Cold Spring Harbor Laboratory, Cold Spring Harbor, NY, 1989.

together (Fig. 6). For example, we have synthesized a 63-bp duplex assembly with an appended metallointercalator by ligating together a metallointercalator-tethered 17-bp oligonucleotide to a 46-bp oligonucleotide using T4 DNA ligase.[14] First the 17-mer and a 5′-phosphorylated 46-mer were each annealed to complementary oligonucleotides to create two duplexes with matching 6-bp sticky ends; then the two duplexes were mixed together with 10,000 units of high-concentration T4 DNA ligase (New England Biolabs) and the solution was incubated at 14° overnight. Note that ligase works quite well in T4 DNA ligase buffer prepared without DTT; this may be an issue with respect to thiol reactions with the photooxidant. The final solution contained 5 nmol of the metallated duplex and 6 nmol of the phosphorylated duplex in a 350-μl volume. The metal-conjugated 63-mer can be separated from the smaller pieces (including the complementary pieces, which cannot be ligated together without a 5′ phosphate terminus) by electrophoresis under denaturing conditions in a 14% polyacrylamide gel. The band containing the longest oligonucleotide was identified by UV shadowing, ex-

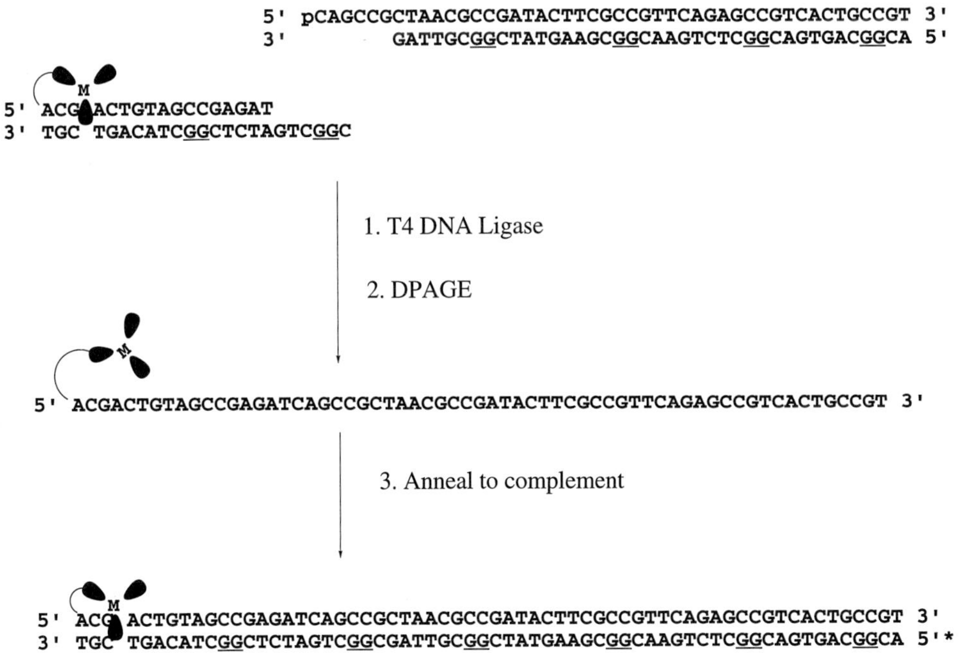

FIG. 6. Scheme for the synthesis of long photooxidant-tethered oligonucleotides from smaller pieces using T4 DNA ligase. [Modified from M. E. Núñez, D. B. Hall, and J. K. Barton, *Chem. Biol.* **6**, 85 (1999).]

tracted from the gel, and ethanol precipitated. Other photooxidant-tethered long duplexes can be prepared in the same manner.

III. Establishing Long-Range Oxidative DNA Damage

A. Solution Conditions. Although the exact solution conditions can vary somewhat from experiment to experiment, samples should contain between 1–10 μM of each DNA strand, 20–50 mM buffer, and 10 mM NaCl. The choice of buffers is fairly flexible, as Tris–HCl, Tris–acetate, phosphate, cacodylate, MES, citrate–acetate, and ammonium acetate buffers have all been shown to be generally suitable for carrying out experiments on long-range oxidative DNA damage. However, volatile buffers such as ammonium acetate are not recommended because the pH may change over the course of irradiation if the samples are not fully sealed. The ideal pH clearly depends on the nature of the photooxidant and the system being tested, but a pH above 7 is preferred to prevent spontaneous depurination.

B. Photolysis. Samples of photooxidant-tethered oligonucleotide duplex assemblies are irradiated by a 1000-W Hg-Xe arc lamp equipped with a monochromator and infrared UV filters. The 20- to 30-μl samples are irradiated in open 1.7-ml siliconized centrifuge tubes with the open end of the tube facing the light source and the solution placed exactly in the focal point. The arc lamp/monochromator system has the advantage that the excitation wavelength can be changed easily during the course of or between experiments. Alternatively, irradiations of Ru(phen)(bpy′)(dppz)$^{2+}$-tethered oligonucleotide duplex assemblies can be performed using a He/Cd laser (Liconix, Sunnyvale, CA, Model 4200 NB, 442 nm, 22 mW) with a 450-nm bandpass filter, with the open reaction tube placed directly in the line of the laser light. Other light sources can also be used, including transilluminators and sunlight, assuming that they initiate the desired photochemistry *and* do not induce direct damage on the DNA. Because of the possibility of direct DNA damage, the use of UV filters is generally recommended. Moreover, it is critical that "light-control" samples, containing the reaction components irradiated in the absence of the photooxidant, be examined for each set of experiments to demonstrate that the products are generated by the photooxidant and not by direct interaction of high-energy light with DNA.

Sample irradiation times are dictated by the strength of the light source, the geometry of the irradiation setup, and, if applicable, the quencher concentration. Preliminary experiments should be performed to determine the irradiation time needed to promote cleavage of between 5 and 10% of the duplexes. This level is usually appropriate to maintain single-hit conditions.

C. Analysis of Damage by Gel Electrophoresis. Photooxidative damage within the DNA duplex assemblies is revealed, either enzymatically or chemically, by strand cleavage at the site of base oxidation. Damage, for example, is detected readily by digestion with the DNA repair enzyme formamidopyrimidine-glycosylase (FPG). This enzyme recognizes a wide assortment of base damage lesions within duplex B-form DNA.[35] Importantly, the repair process involves sequential removal of the oxidized base followed by scission of the phosphodiester backbone, resulting from base-catalyzed elimination of both 3′ and 5′ phosphoryl moieties on the transient abasic sugar. As a result, the 5′-labeled DNA fragment that results bears a 3′ phosphate, which comigrates with the DNA markers obtained from Maxam–Gilbert chemical strand degradation.

FPG digestion of photooxidized DNA is optimal with a DNA duplex concentration of 50 nM. The buffer conditions used for photoinduced long-range charge migration experiments are flexible, but the most successful FPG analyses have been conducted on samples whose irradiation buffer is 20 mM Tris (pH 7.4), 5% glycerol, 8 mM spermidine, 75 mM glutamic acid, 0.1 mM EDTA, 4 mM MgCl$_2$, 13.5 μM (in nucleotides) poly (dG-dC), 0.01% Nonidet P-40, 5 μg/ml bovine serum albumin (BSA), and 20 mM KCl. To 2 μl of irradiated DNA is added 7 μl double-distilled H$_2$O and 1 μl FPG (ranging in concentration from 80 to 4 ng/μl stock FPG). Incubation of the 10-μl samples at 37° for 1 hr is followed immediately by the addition of 10 μl DPAGE loading dye and heating to 90° for 1–3 min prior to electrophoresis.

Chemical cleavage of DNA at sites of oxidative damage occurs on incubation of the damaged DNA in 10% piperidine. A major oxidative product of the reaction characterized by HPLC is 8-oxo-G, an important oxidative lesion within the cell.[11] Whether this lesion is piperidine sensitive is still under debate,[36] but a family of guanosine-derived oxidative lesions in addition to 8-oxo-G has been characterized after riboflavin-sensitized DNA oxidation.[37] These include deoxyribo-2,2-diaminooxazolone and deoxyribo-2-aminoimidazolone; both lesions were found to be alkaline labile.

Experimentally, piperidine digestion is accomplished by incubating the damaged DNA in 10% piperidine in a final volume of 100 μl at 90° for 30 min. The samples are dried thoroughly under vacuum with several washes of distilled water. Samples are then analyzed on an 18% denaturing polyacrylamide sequencing gel along with reference samples containing purine-specific and pyrimidine-specific Maxam–Gilbert sequencing reactions.[34]

[35] S. S. David and S. D. Williams, *Chem. Rev.* **98,** 1221 (1998).
[36] C. J. Burrows and J. G. Muller, *Chem. Rev.* **98,** 1109 (1998).
[37] K. Kino, I. Saito, and H. Sugiyama, *J. Am. Chem. Soc.* **120,** 7373 (1998).

Without drying, the gel is imaged and analyzed by phosphorimagery or by autoradiography and densitometry. The amount of oxidation at individual bases is determined by measuring the intensity of the band corresponding to that base as a fraction of the intensity of the whole lane. The fractional intensity of the corresponding band in the control lane is subtracted out to account for background damage. Alternatively, the ratio of oxidation at two sites in the same sample can be measured as the ratio of the intensities of the two bands.

D. Interduplex Control Reactions. In order to demonstrate that long-range charge transfer is responsible for guanine base damage, it is important to confirm that reactions occur intraduplex, e.g., that the photooxidant generates damage only on the duplex to which it is tethered.[11,14,18] Irradiation of samples containing a mixture of photooxidant-tethered but unlabeled duplexes and labeled duplexes lacking the tethered oxidant provides a useful control for interduplex reactions. In one 1.7-ml centrifuge tube, 2.5 μM duplex DNA bearing a radioactive label but no photooxidant is annealed in the appropriate buffers and salts; in a separate reaction tube, 2.5 μM duplex DNA bearing a photooxidant but no radioactive label is annealed with the appropriate buffers and salts. After the reactions have cooled thoroughly, the two reactions are mixed at a 1 : 1 ratio and irradiated. If the reaction occurs intraduplex, the electrophoretic analysis will reveal little detectable damage on the labeled strand. This will confirm not only that the photooxidant is not intercalating into the radiolabeled nontethered duplex, but also that long-range guanine damage is not generated by direct interaction of the radiolabeled DNA with light. Additionally, such an outcome refutes the argument that oxidative guanine damage arises by the interaction of guanine with singlet oxygen or other species that might diffuse to an alternate duplex.

Alternatively, the intraduplex nature of the reaction can be confirmed by showing that guanine oxidation occurs independently of the concentration of photooxidant. If the guanine oxidation reaction is a bimolecular process, the yield of oxidized guanine should increase substantially as the concentration of duplex with photooxidant is increased, but if the reaction occurs only in an intraduplex process, the yield of oxidized guanine should not change regardless of the photooxidant-tethered duplex concentration.

IV. Oxidative Repair of Thymine Dimer Lesions

The primary photochemical lesion in DNA is the thymine cyclobutane dimer lesion, generated by a [2 + 2] cycloaddition between adjacent thymines on the same strand of DNA.[38] In *Escherichia coli* this lesion is reductively repaired by the enzyme photolyase. The photolyase enzyme utilizes a flavin cofactor to add an electron to the dimer, causing it to revert to the original pair of thymine bases.[39,40] Interestingly, model studies showed that the thymine dimer lesion could be repaired both oxidatively and re-ductively,[41–44] hinting that this lesion could be an excellent target for DNA-mediated base oxidation chemistry. In studies of DNA assemblies con-taining a site-specifically incorporated thymine dimer, we then demon-strated that the rhodium intercalator excited with visible light (365 and 400 nm) could repair the dimer; moreover, it could do so at long range.[22]

In order to study the long-range oxidative repair of *cis-syn* thymine dimer lesions through the DNA base stack, we constructed duplex assem-blies containing a tethered rhodium intercalator at the end of one oligonu-cleotide strand and containing a thymine dimer lesion on the complemen-tary strand (Fig. 7). Single-stranded oligonucleotides containing a thymine dimer lesion were synthesized first by standard phosphoramidite chemistry as unmodified oligonucleotides with a 5'-TT-3' site and no other adjacent pyrimidine bases. Following deprotection and HPLC purification of the oligonucleotide, the thymine dimer lesions were formed photochemically in the presence of the triplet photosensitizer acetophenone.[45,46] Aqueous solutions (1 ml) of the oligonucleotide ($\sim$200 μM) and acetophenone (25 mM) were degassed repeatedly by the freeze/pump/thaw method and then were irradiated *in vacuo* at 330 nm for approximately 6 hr. The oligonucleo-tides were purified by high-temperature HPLC (65°) as described pre-viously. On a very slow gradient (a change of 7 to 11% acetonitrile over 30 min) the oligonucleotide containing the thymine dimer elutes $\sim$2 min before the unmodified starting material. Between one-third and one-half of 16-bp oligonucleotides can be converted to dimer-containing strands by this method. The presence of the dimer can be confirmed by treatment

[38] A. Sancar, *Annu. Rev. Biochem.* **65,** 43 (1996).

[39] T. P. Begley, *Acc. Chem. Res.* **27,** 394 (1994).

[40] P. F. Hellis, R. F. Hartman, and S. D. Rose, *Chem. Soc. Rev.* **24,** 289 (1995).

[41] C. Helene and M. Charlier, *Photochem. Photobiol.* **25,** 429 (1997).

[42] J. R. Jacobsen, A. G. Cochran, J. C. Stephans, D. S. King, and P. G. Schultz, *J. Am. Chem. Soc.* **117,** 5435 (1995).

[43] R. Epple, E.-U. Wallenborn, and T. Carrell, *J. Am. Chem. Soc.* **119,** 7440 (1997).

[44] S.-T. Kim and S. D. Rose, *Photochem. Photobiol.* **52,** 789 (1990).

[45] S. K. Banerjee, A. Borden, R. B. Christensen, J. E. LeClerc, and C. W. Lawrence, *J. Bacteriol.* **172,** 2105 (1990).

[46] S. K. Banerjee, R. B. Christensen, C. W. Lawrence, and J. E. LeClerc, *Proc. Natl. Acad. Sci. U.S.A.* **85,** 8141 (1988).

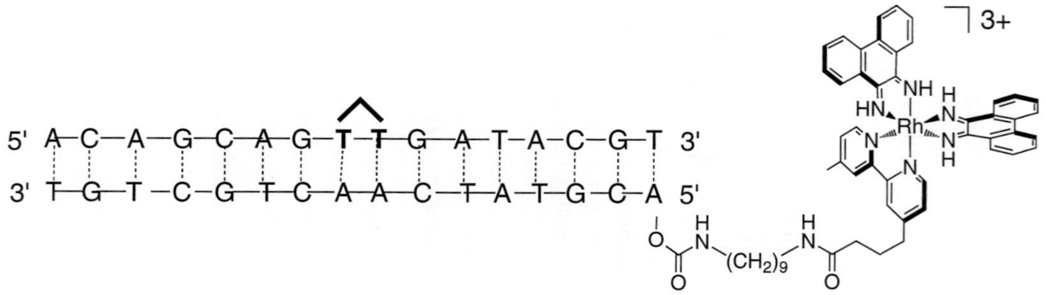

FIG. 7. Schematic of a rhodium-tethered duplex assembly containing a thymine dimer incorporated site-specifically. The oligonucleotide duplex assembly consists of two complementary strands, with a rhodium complex tethered covalently to the end of the sugar phosphate backbone of one stand and with a thymine dimer lesion present on its complement.

with 254-nm light, which causes spontaneous reversion to the unmodified strand, which can then be identified by HPLC and electrospray mass spectrometry. The presence of the dimer can also be confirmed by digestion with T4 DNA polymerase, which possesses a $3'-5'$ exonuclease activity. The T4 DNA polymerase digests the oligonucleotide from the $3'$ end until it reaches the dimer, generating short strands, which can be identified by gel electrophoresis if labeled radioactively. After purification, the dimer-containing strand is annealed to an equimolar amount of its rhodium-tethered complement by lineargradient cooling from $90°$.

Assays for the repair of the thymine dimer lesion are conducted similarly after irradiation using high-temperature HPLC. These duplexes, in the presence of rhodium (either tethered or untethered), are irradiated for 1–3 hr at 365 nm, and the products of the reaction are analyzed by HPLC under the same conditions as for purification. The oligonucleotide strands separate, and the dimer-containing strand, the repaired strand, and the rhodium strand all elute with distinct retention times. Importantly, we have found that the dimer strand is converted to the repaired strand without the formation of other products.

Using this methodology, we discovered that this dimer repair is relatively insensitive to distance, but it is sensitive to the integrity of the base stack, as the addition of unpaired base bulges to one strand decreases the efficiency of repair.[22] The dimer repair reaction is also relatively insensitive to the concentration of duplex, indicating that the reaction is not interduplex. Clearly, then, this oxidation occurs through the DNA π stack.

Furthermore, we observed that ground-state Ru(III) generated *in situ* cannot repair thymine dimer lesions.[47] This is understandable because the reduction potential of the ruthenium complex ($\approx +1.6$ V vs NHE) is significantly lower than that of rhodium ($\approx +2.0$ V) and therefore should be insufficient to oxidize the dimer ($+2.0$ V). Finally, thymine dimer repair does compete with 5'-GG-3' damage, which is surprising when we consider that the 5'-GG-3' site should be favored by at least 0.7 V. The competition between these two oxidizable sites indicates that *both* kinetic and thermodynamic effects must be involved in the trapping of the electronic hole in the helix.

Other photooxidants that damage DNA have not yet been shown to promote DNA repair oxidatively. Clearly, a very potent photooxidant is required. Not surprisingly, such reagents would also be expected to promote oxidative damage to the DNA.

[47] P. J. Dandliker, M. E. Nunez, and J. K. Barton, *Biochemistry* **37**, 6491 (1998).

Acknowledgments

We are grateful to the NIH (GM49216) for their financial support of this research. We also thank the American Cancer Society for a postdoctoral fellowship (to S.R.R.) and the Howard Hughes Medical Institute for a predoctoral fellowship (to M.E.N.). Most of all we thank the many talented scientists whose work contributed to that described here, especially Drs. D. B. Hall, M. R. Arkin, R. E. Holmlin, and P. J. Dandliker.

[17] Cholesterol Photodynamic Oxidation by Ultraviolet Irradiation and Cholesterol Ozonization by Ozone Exposure

By KYOICHI OSADA and ALEX SEVANIAN

Introduction

Cholesterol (3β-hydroxy-5-cholestene), which is found abundantly in cell membranes of various tissues or plasma lipoproteins, is as susceptible to peroxidative modification as other unsaturated lipids when exposed to exciting light of suitable wavelength and molecular oxygen. Figure 1 shows type I and type II pathways of photodynamic cholesterol peroxidation.[1] Cholesterol hydroperoxide products, including 3β-hydroxy-5α-cholest-6-ene-5-hydroperoxide (5α-OOH), 3β-hydroxycholest-4-ene-6α-hydroperoxide (6α-OOH), and 3β-hydroxycholest-4-ene-6β-hydroperoxide (6β-OOH), are formed in singlet excited oxygen ($^1O_2$)-mediated reactions (type II photooxidation). Among the hydroperoxides, 5α-OOH is generated at the highest rate in most reaction systems.[2] However, epimeric 3β-hydroxy-cholest-5-ene-7α-hydroperoxide (7α-OOH) and 3β-hydroxycholest-5-ene-7β-hydroperoxide (7β-OOH) are the most prominent hydroperoxide products in free radical-mediated reactions derived from ground state oxygen ($^3O_2$, type I photooxidation), although these can arise via allylic rearrangement of 5α-OOH in low polarity solvents such as chroloform.[3] These hydroperoxides produce deleterious effects on membranes, altering their structure and function.[4] Moreover, biological activities of cholesterol ozonization products are not completely elucidated, despite the fact that ozone

---

[1] A. W. Girotti, *J. Photochem. Photobiol.* **13,** 105 (1992).
[2] R. Langlois, H. Ali, N. Brasseur, J. R. Wagner, and J. E. van Lier, *Photochem. Photobiol.* **44,** 117 (1986).
[3] M. J. Kulig and L. L. Smith, *J. Org. Chem.* **38,** 3639 (1973).
[4] L. L. Smith, *J. Steroid Biochem.* **26,** 259 (1987).

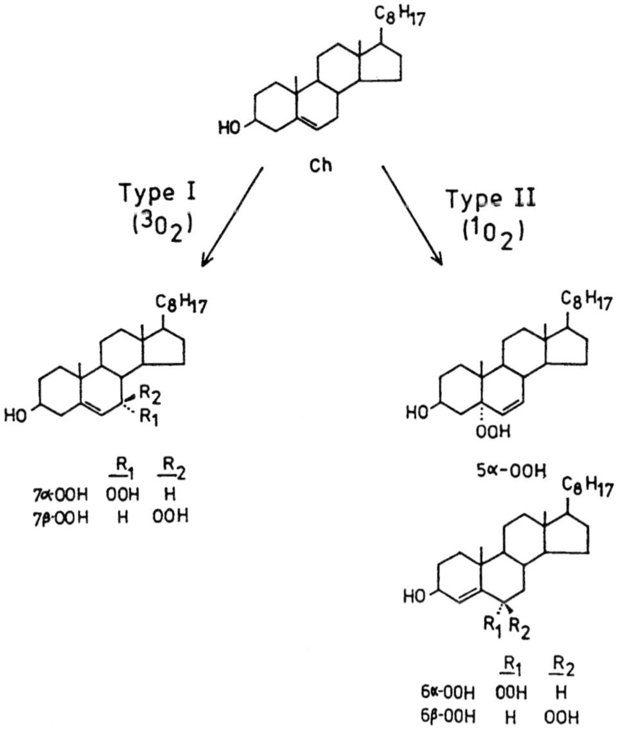

FIG. 1. Cholesterol hydroperoxides generated by type I and II photochemistry.[1]

toxicity is of great potential importance in our biosphere. Some cholesterol ozonization products, including 5,6α-epoxy-5α-cholestan-3β-ol (5α-epox), 5,6β-epoxy-5β-cholestan-3β-ol, 5ξ,6ξ-epidioxy-5,6-secocholestan-3β,5ξ,6ξ-triol, 3β-hydroxy-5-oxo-5,6-secocholestan-6-al, and 3β,10-dihydroxy-5,6:5,10-disecocholestan-5-oic acid lactone(5 → 10), were identified by GC/MS analysis (Fig. 2),[5] with the major ozonization product being 3β-hydroxy-5-oxo-5,6-secocholestan-6-al.[6] Both cholesterol hydroperoxide and ozonization products have been hard to identify directly; however, developments in ultrasensitive detectors or new methods for chromatographic analysis have overcome the problems of detection and quantification. The first half of this article focuses on the formation of cholesterol hydroperoxide and ozonization products by ultraviolet A (UVA, 315–400 nm) and ozone exposures.

[5] J. Gumulka and L. L. Smith, *J. Am. Chem. Soc.* **105**, 1972 (1983).
[6] K. Jaworski and L. L. Smith, *J. Org. Chem.* **53**, 545 (1988).

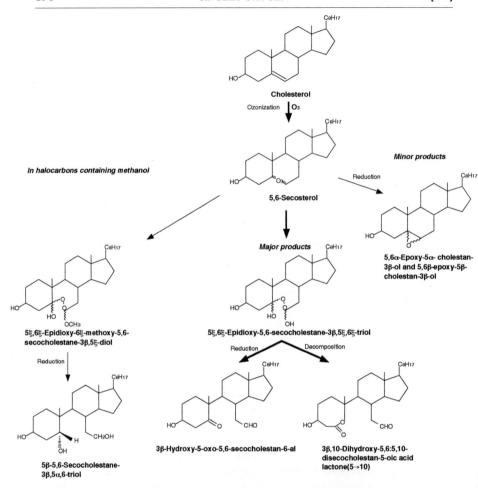

FIG. 2. Structure of cholesterol ozonization products.[6]

The formation of cholesterol hydroperoxides in human skin after sunlight exposure has been long identified. Ultraviolet light (UVL)-exposed human skin was shown to contain several cholesterol-derived photooxidation products,[7] one being 5α-epox, a rearrangement product of (5α-OOH).[8] Black and Chan[9] conducted a series of studies on hairless mice receiving chronic UVL radiation and found that 5α-epox levels gradually rose and

[7] H. S. Black and W. B. Lo, *Nature* **234,** 306 (1971).
[8] L. L. Smith, *in* "Cholesterol Autoxidation." Plenum Press, New York, 1981.
[9] H. S. Black and J. T. Chan, *Oncology* **33,** 119 (1976).

reached peak concentrations after 4 weeks. The pattern of 5α-epox levels in skin and tumor incidence and multiplicity were highly correlated. A similar relationship was reported for the levels of epimeric 7-OOH following UVB irradiation of rat skin.[10] Further studies revealed that the extent of lipid peroxidation and 5α-epox formation and tumor incidence were directly related to the extent of dietary corn oil intake and inversely related to the level of antioxidant supplementation (d).[11] A relationship between UVL-induced skin cancers and peroxidation of polyunsaturated fatty acids is now widely recognized.[12] Formation of 7-OOH as well as 5α-epox and other cholesterol oxidation products takes place during the peroxidation of tissue lipids, reflecting the progression of lipid-free radical propagation reactions.[13] However, the implication of cholesterol oxidation products as skin carcinogens, particularly 5α-epox, remains unsubstantiated, despite the evidence that this compound possesses mutagenic[14] and carcinogenic[15] properties. A reasonable assumption at this time is that 5α-epox levels represent the processes of UVL-induced oxidative processes, specifically involving lipid peroxidation, that at least in part contribute to skin cancer.

Production and Measurement of Cholesterol Oxidation Products by UVA Exposure

Photoperoxidation Reactions of Cholesterol in Cell-Free Model Systems

Large unilamellar liposomes (~100 nm) have been used to study cholesterol photooxidation reactions in model membranes. The vesicles are prepared by lipid extrusion according to the method of Korytowski et al.[16] using a lipid mixture consisting of dimyristoyl phosphatidylcholine (DMPC), cholesterol, and dicetylphosphate (DCP) in chloroform that is dried under nitrogen and then under vacuum. After hydration of the lipid film in phosphate-buffered saline (PBS), followed by five cycles of freezing and thawing, the multilamellar vesicles are passed through two 0.1-μm polycarbonate filters in an Extruder apparatus. The resulting vesicles con-

[10] N. Ozawa, S. Yamazaki, K. Chiba, and M. Tateishi, in "Oxygen Radicals" (K. Yagi, M. Kondo, E. Niki, and T. Yoshikawa, eds.), p. 323. Pergamon Press, New York, 1992.

[11] H. S. Black, W. A. Lenger, J. Gerguis, and J. I. Thornby, Cancer Res. **45,** 6254 (1985).

[12] V. E. Reeve, M. Matheson, G. E. Greenoak, P. J. Canfield, C. Boehm-Wilcox, and C. H. Gallagher, Photochem. Photobiol. **48,** 689 (1988).

[13] J. E. van Lier, in "Biological Activities of Oxygenated Sterols" (J. P. Beck and A. Crastes de Paulet, eds.), p. 15. INSERM, Paris, 1988.

[14] A. Sevanian and A. R. Peterson, Food Chem. Toxicol. **24,** 1103 (1986).

[15] F. Bischoff, Adv. Lipid Res. **7,** 165 (1969).

[16] W. Korytowski, G. J. Bachowski, and A. W. Girotti, Anal. Biochem. **197,** 149 (1991).

taining 2.5 mM DMPC, 2 mM cholesterol, and 0.25 mM DCP in the bulk phase are kept under argon at 4° until use. Isolated membranes from human erythrocytes (unsealed white ghosts) have also been studied that are prepared by conventional hypotonic lysis, stored refrigerated under argon, and used within 1 week.

Liposomal membranes (5 mM total lipid) or erythrocyte ghosts (1 mg protein/ml) in PBS are irradiated in the presence of a sensitizer such as aluminum phthalocyanine tetrasulfonate (10–40 μM) in a thermostated beaker using a quartzhalogen lamp (90-W rating) positioned above the beaker.[17] Incoming light is passed through a windowpane glass filter, which effectively removes all wavelengths below 305 nm (UVB). The fluence rate at the suspension surface is set at ~30 mW/cm^2. The reaction temperature is maintained at 37° for liposomes and 4° for ghosts. During irradiation, membrane suspensions are stirred continuously at a slow rate to ensure uniform exposure to light and oxygen.

In general, sensitizers of the flavin, ketone, or quinone family are believed to be intrinsically more disposed to free radical photochemistry (type I photooxidation) than porphyrin, thiazine, or xanthene sensitizers.[1] Therefore, protoporphyrin IX, uroporphyrin I, purified hematoporphyrin derivative, merocyanine 540, and aluminum phthalocyanine tetrasulfonate are used to generate 5α-OOH preferentially through the reactivity of $^1O_2$ with cholesterol (type II photooxidation).[1]

Analysis of Photooxidized Cholesterol: HPLC-EC Analysis

Lipid hydroperoxides are separated and quantitated by HPLC-EC(Hg) detection (Fig. 3).[17] HPLC analysis is performed using a mercury cathode electrochemical detector and a C-18 Ultrasphere column (4.6 × 70 mm; 3 μm particles) with a C-18 guard cartridge (4.6 × 5 mm; 5-μm particles) (Beckman Instruments). The deoxygenated mobile phase, consisting of methanol, acetonitrile, isopropanol, and aqueous solution containing 10 mM ammonium acetate and 1 mM sodium perchlorate (72:11:8:9, v/v), is delivered isocratically at a flow rate of 1.5 ml/min. The column effluent is passed first through a UV/visible detector at 205 nm and then through the EC detector. The typical setting for the mercury cathodes is −150 mV vs As/AgCl reference.

HPLC-Chemiluminescence Analysis

Extracted lipids are applied to Bond Elut NH$_2$ extraction columns after the addition of synthesized β-sitosterol hydroperoxide as the internal stan-

[17] P. G. Geiger, W. Korytowski, F. Lin, and A. W. Girotti, *Free Radic. Biol. Med.* **23,** 57 (1997).

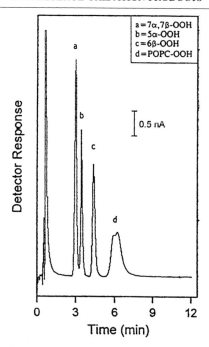

FIG. 3. HPLC-EC(Hg) profile of lipid hydroperoxide standards. Hydroperoxide assignments, retention times, and amounts injected are as follows: (a) unresolved $7\alpha,7\beta$-OOH (3.0 min, 135 pmol), (b) 5α-OOH (3.5 min, 90 pmol), (c) 6β-OOH (4.6 min, 110 pmol), and (d) POPC-OOH (6.2 min, 200 pmol). Full-scale detector sensitivity was 5.0 nA.[9]

dard. Following the procedure of Ozawa *et al.,*[18] the sterol fraction is eluted by ethyl acetate after washing with chloroform. The eluted fraction is injected onto an HPLC equipped with a normal phase silica column (4.6 × 150 mm), and the hexane/isopropanol (100:5, v/v) mobile phase is eluted at 2 ml/min. The cholesterol hydroperoxide-containing fraction is collected and applied to an HPLC equipped with a chemiluminescence detector (wavelength range 300–600 nm; chemiluminescence reagent: 10 mg/ml isoluminol, and 20 mg/ml microperoxidase, in 50 mM borate buffer, pH 9.5; flow rate: 1.2 ml/min) and a chiral phase column (Chiralcel OD, 4.6 × 250 mm, Daicel Chemical Industries). The mobile phase for separation is a mixture of methanol and water (85:15, v/v) at a flow rate of 1.0 ml/min.

[18] N. Ozawa, S. Yamazaki, K. Chiba, H. Aoyama, H. Tomizawa, M. Tateishi, and T. Watabe, *Biochem. Biophys. Res. Commun.* **178,** 242 (1991).

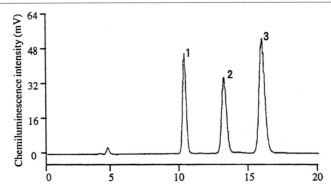

FIG. 4. Chromatograms of authentic standard 7α-OOH (peak 1, 50 pmol), 7β-OOH (peak 2, 250 pmol), and β-sitosterol hydroperoxide (peak 3, 500 pmol) by HPLC with a chemiluminescence detector.[10]

Figure 4 shows the chromatograph of 7α- and 7β-OOHs and β-sitosterol hydroperoxide. This approach was used to measure the formation of epimeric 7-OOH in the skin of aging rats and after UVB exposure.[18,19]

Production and Measurement of Cholesterol Ozonization Products

Ozonization of Cholesterol in Liposomes

A solution of 10 mg cholesterol, 20 mg dipalmitoyl phosphatidylcholine, and 20 mg egg PC containing approximately 50% unsaturated PC in chloroform is dried by evaporation at ambient temperature in a 25-ml round-bottom flask as described by Pryor *et al.*[20] A suspension of liposomes is cooled in an ice bath and sonicated under a blanket of argon to give unilamellar liposomes in 20 ml of 50 mM phosphate buffer (pH 7.4). The liposome solution (5 ml) is stirred and exposed for 2 hr to 5 ppm ozone in 0.6 ml/min air. After the addition of 10 ml of 0.118 M potassium chloride solution, the lipids are extracted and the reaction products are separated on a silica gel column packed with 0.4 g silica gel, with the cholesterol ozonization products being eluted by 10 ml ethyl acetate. The extract is reduced with 1.0 mg Zn powder in 1.0 ml acetic acid for 2 hr. The reduction mixture is diluted with 10 ml of water and extracted with dichloromethane. Then, 0.2 ml of 0.01 M 2,4-dinitrophenylhydrazine (DNPH) is added in ethanol. *Note:* a 0.01 M solution of DNPH is prepared by dissolving

[19] S. Yamazaki, N. Ozawa, A. Hiratsuka, and T. Watabe, *Free Radic. Biol. Med.,* in press.
[20] W. A. Pryor, K. Wang, and E. Bermudez, *Biochem. Biophys. Res. Commun.* **188,** 618 (1992).

1.42 g of 70% DNPH with 4 ml concentrated HCl in sufficient ethanol in a 500-ml volumetric flask. Thereafter, 1.0 ml methanol is added to the extract. The solution is bubbled with nitrogen for 5 min and stirred for 2 hr. The DNPH derivative of the cholesterol ozonization product is purified by thin-layer chromatography using ethyl acetate/hexane (2:3, v/v) and applied to an HPLC for further analysis as described later.

HPLC Analysis of Cholesterol Ozonization Products

The HPLC analysis of DNPH derivatives of cholesterol ozonization products is performed according to the method of Pryor *et al.*[20] The method utilizes a S5 ODS2 column (25 cm × 4.6 mm, Phase Sep Inc.) and methanol/water (7:3, v/v) as the mobile phase, with the eluent being monitored at 360 nm. Figure 5 shows a chromatograph of DNPH derivatization products of 3β-hydroxy-5-oxo-5,6-secocholestan-6-al. The DNPH derivatization of 3β-hydroxy-5-oxo-5,6-secocholestan-6-al results in the isolation of *trans* (peaks 1) and *cis* (peak 2) derivatives; however, the formation of 3,5-dihydroxy-β-norcholestane-6-carboxyaldehyde (peak 3), which is a

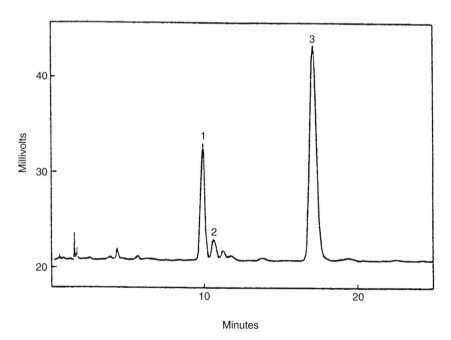

FIG. 5. HPLC separation of DNPH-3β-hydroxy-5-oxo-5,6-secocholestan-6-al. The elution times for peaks 1, 2, and 3 are 9.9, 10.6, and 17.2 min, respectively. Peaks before 3 min are due to solvent impurities. Small peaks between peaks 2 and 3 were not identified.[12]

rearrangement product of peaks 1 and 2, is also observed depending on the concentration of acid used and reaction time for derivatization.[21]

Application of HPLC/MS Techniques

Cholesterol oxidation products such as the isomeric 7-OOH and derived products have very commonly been analyzed by HPLC/MS techniques. Using a thermospray interface under isocratic conditions, an effective separation of some products was obtained.[22] This approach is hampered by limitations in the proportion of ammonium acetate required for the thermospray interface, severely affecting sample detection. Due to the ease with which these latter compounds are analyzed by GC/MS, measurement by this HPLC/MS approach has been limited. However, as is usually observed under thermospray conditions,[23] fragmentation processes are minimal and much of the total ion current is due to molecular species or protonated and NH_4^+ adducts,[24] which have been well described for cholesterol oxidation products.[25]

Under particle beam conditions, EI spectra are obtained easily and in principle a higher sensitivity can be obtained due to the higher production of ions.[22] HPLC analysis by means of a particle beam interface utilized an Altex-Ultrasphere-Si 250 column (250 × 4.6 mm) and a mobile phase composed of hexane/isopropanol (96:4, v/v) at a flow rate of 0.4 ml/min. The particle beam interface was set with 30 psi of helium nebulizing gas and with a desolvation chamber temperature of 45 EC. The EI source temperature was set at 200 EC, with a source housing pressure of 2.6 × 10^{-5} torr. Measurements were performed under normal EI conditions (70 eV and 300 FA) and the analyzer was set for positive ion detection. Under these conditions, isomeric 7-hydroxy cholesterols and 7-OOH are separated readily.

Acknowledgments

The authors acknowledge the support of Grant ES03466 (A.S.) and the Merck/Banyu Research Fellowship (K.O.) during the preparation of this manuscript.

[21] K. Wang, E. Bermudez, and W. A. Pryor, *Steroids* **58**, 225 (1993).

[22] A. Sevanian, R. Seraglia, P. Traldi, P. Rossato, F. Ursini, and H. N. Hodis, *Free Radic. Biol. Med.* **17**, 397 (1994).

[23] R. B. Voyksner and C. A. Haney, *Anal. Chem.* **57**, 991 (1985).

[24] F. W. McLafferty, Interpretation of Mass Spectra, 3rd Ed., p. 180. University Science Books, Mill Valley, CA, 1980.

[25] Y. Y. Lin and L. L. Smith, *Biomed. Mass Spectrom.* **5**, 604 (1978).

[18] Bactericidal and Virucidal Activities of Singlet Oxygen Generated by Thermolysis of Naphthalene Endoperoxides

By CORINNE PELLIEUX, ANNY DEWILDE, CHRISTEL PIERLOT, and JEAN-MARIE AUBRY

Introduction

Photodynamic antimicrobial chemotherapy (PACT), involving photosensitizers and visible or ultraviolet light, has been proposed in the treatment of locally occurring infection,[1] especially of caries and periodontal diseases,[2] oral candidiosis,[3] and infected wounds.[4] However, PACT also appears very promising in the disinfection of cellular blood products[5,6] because it causes much less damage to hematopoietic cells than physical or chemical processes of decontamination. Phenothiazines, merocyanine 540, hypericin, porphyrins, and the closely related chlorines, phthalocyanines, and psoralens have been shown to photoinactivate viruses and bacteria.[1,5,6] Except psoralen derivatives, which intercalate between nucleic acid bases, the other photosensitizers are believed to act mainly through singlet molecular oxygen ($^1O_2$, $^1\Delta_g$, abbreviated later as $^1O_2$) as microbial inactivation is inhibited by the addition of $^1O_2$ quenchers, such as β-carotene or sodium azide,[7–14] and is enhanced by D_2O in which the lifetime of $^1O_2$ is

[1] M. Wainwright, *J. Antimicrob. Chemother.* **42,** 13 (1998).

[2] M. Wilson, *Int. Dent. J.* **44,** 181 (1994).

[3] M. Wilson and N. Mia, *Lasers Med. Sci.* **9,** 105 (1994).

[4] M. Szpakowski, J. Reiss, A. Graczyk, S. Szmigielski, K. Lanockiand, and J. Grzybowski, *Int. J. Antimicrob. Agents* **8,** 23 (1997).

[5] E. Ben-Hur and B. Horowitz, *AIDS* **10,** 1183 (1996).

[6] L. Corash, *Vox Sang.* **70**(Suppl. 8), 9 (1996).

[7] Y. Wakayama, M. Takagi, and K. Yano, *Photochem. Photobiol.* **32,** 601 (1980).

[8] Z. Malik, J. Hanania, and Y. Nitza, *J. Photochem. Photobiol. B Biol.* **5,** 281 (1990).

[9] M. Bhatti, A. Mac Robert, S. Meghji, B. Henderson, and M. Wilson, *Photochem. Photobiol.* **68,** 370 (1998).

[10] W. Snipes, G. Keller, J. Woog, T. Vichroy, H. Deering, and A. Keith, *Photochem. Photobiol.* **29,** 785 (1979).

[11] S. Rywkin, L. Lenny, J. Goldstein, N. E. Geacintov, H. Margolis-Nunno, and B. Horowitz, *Photochem. Photobiol.* **56,** 463 (1992).

[12] J. Lenard, A. Rabson, and R. Vanderoef, *Proc. Natl. Acad. Sci. U.S.A.* **90,** 158 (1993).

[13] K. Müller-Breitkreutz and H. Mohr, *Infusionsther. Transfusionsmed.* **22**(Supp. 1), 8 (1995).

increased.[9,11-16] However, during photodynamic processes, various reactive oxygen derivatives, such as hydroxyl radical OH·, superoxide radical anion $O_2^{·-}$, peroxy radical ROO· (type I mechanism), and $^1O_2$ (type II mechanism), are produced simultaneously and it is therefore difficult to assess the role of each species unambiguously in the resulting biological effects. Thus, researchers have sought methods to generate pure $^1O_2$ in biological media.

In the "separated surface sensitizer" system proposed by Midden and Kaplan,[17] the photosensitizer is immobilized on a glass plate placed a short distance above the substrate solution and thus is physically separated from the substrate by a thin air layer. This method prevents type I photosensitized processes, which require a direct interaction between the substrate and the photosensitizer. Therefore, biological activity can be attributed unambiguously to $^1O_2$. By using this system, Dahl et al.[18,19] and Valduga et al.[20] have shown that gram-negative bacteria are much more resistant to the action of pure $^1O_2$ than gram-positive bacteria, with the outer wall of the gram-negative bacteria exerting a protective effect against $^1O_2$.

Water-soluble naphthalene endoperoxides have also been used as alternative sources of $^1O_2$ in biological systems.[21-23] These compounds are obtained from 1,4-disubstituted naphthalenes by reaction with $^1O_2$ at low temperature (0–5°). When incubated at 37°, they release molecular oxygen quantitatively, half of which is in a singlet state [Eq. (1)].[22,24] Thus, the cumulative amount of $^1O_2$ generated during the thermolysis is about half the starting concentration of the endoperoxide. The use of this nonphotochemical method appears both simpler and cleaner than photosensitization as it does not require a specifically designed system of irradiation and

[14] H. Abe, K. Ikebuchi, S. J. Wagner, M. Kuwabara, N. Kamo, and S. Sckiguchi, *Photochem. Photobiol.* **66,** 204 (1997).

[15] J. M. O'Brien, D. K. Gaffney, T. P. Wang, and F. Sieber, *Blood* **80,** 277 (1992).

[16] K. Müller-Breitkreutz, H. Mohr, K. Briviba, and H. Sies, *J. Photochem. Photobiol. B Biol.* **30,** 63 (1995).

[17] W. R. Midden and M. L. Kaplan, *J. Am. Chem. Soc.* **103,** 4129 (1983).

[18] T. A. Dahl, W. R. Midden, and P. E. Hartman, *Photochem. Photobiol.* **46,** 345 (1987).

[19] T. A. Dahl, W. R. Midden, and P. E. Hartman, *J. Bacteriol.* **171,** 2188 (1989).

[20] G. Valduga, G. Bertoloni, E. Reddi, and G. Jori, *J. Photochem. Photobiol. B Biol.* **21,** 81 (1993).

[21] A. W. M. Nieuwint, J. M. Aubry, F. Arwert, H. Kortbeek, S. Herzbzerg, and H. Joenje, *Free Radic. Res. Commun.* **1,** 1 (1985).

[22] P. Di Mascio and H. Sies, *J. Am. Chem. Soc.* **111,** 2909 (1989).

[23] J. M. Aubry, *in* "Membrane Lipid Oxidation" (C. Vigo-Pelfrey, ed.), p. 65. CRC Press, Boca Raton, FL, 1991.

[24] C. Pierlot, S. Hajjam, C. Barthelemy, and J. M. Aubry, *J. Photochem. Photobiol B Biol.* **36,** 31 (1996).

generates *known quantities* of $^1O_2$ by simple thermolysis of the naphthalene endoperoxides.

Up until now, only two anionic naphthalene derivatives, the endoperoxides of sodium 4-methyl-1-naphthalenepropanoate $MNPO_2$ and of disodium 1,4-naphthalenedipropanoate $NDPO_2$ **1a**, have been used to assess the antimicrobial activity of pure $^1O_2$. The first one kills the wild strain of the bacterium *Escherichia coli*,[25,26] but is less effective against lycopene-producing strains, suggesting that lycopene protects *E. coli* against $^1O_2$ toxicity by scavenging $^1O_2$.[26] The second one is able to inactivate efficiently extracellular enveloped viruses,[16,27] but has no effect on intracellular viruses,[27] probably because it is precluded from penetrating through cell membranes due to its negative charge. This article compares the *in vitro* bactericidal and virucidal activities of $NDPO_2$ to those of a new carrier $DHPNO_2$, the endoperoxide of *N,N'*-di(2,3-dihydroxypropyl)-1,4-naphthalenedipropanamide **1b**, which has been specifically designed to convey $^1O_2$ through lipid membranes thanks to its nonionic structure.[28]

Materials and Methods

Singlet Oxygen Carriers

Naphthalene derivatives and their endoperoxides are prepared and analyzed according to the procedures reported in article [1], this volume.

[25] T. Nagano, T. Tanaka, H. Mizuki, and M. Hirobe, *Chem. Pharm. Bull.* **42**, 884 (1994).
[26] M. Nakano, Y. Kambayashi, H. Tatsuzawa, T. Komiyama, and K. Fujimori, *FEBS Lett.* **432**, 9 (1998).
[27] A. Dewilde, C. Pellieux, S. Hajjam, P. Wattré, C. Pierlot, D. Hober, and J. M. Aubry, *J. Photochem. Photobiol. B Biol.* **36**, 23 (1996).
[28] A. Dewilde, C. Pellieux, P. Wattré, C. Pierlot, and J. M. Aubry, *Biol. Chem.* **379**, 1377 (1998).

TABLE I
BACTERIAL AND VIRUS STRAINS USED

	Species	Strain
Bacteria		
Gram negative	*Escherichia coli*	ATCC[a] 10536
	Pseudomonas aeruginosa	PIC[b] A 22
Gram positive	*Staphylococcus aureus*	ATCC 9144
	Enterococcus faecium	ATCC 10541
Viruses		
Enveloped double-stranded DNA	Human cytomegalovirus	AD 169 strain
Enveloped double-stranded RNA	Human immunodeficiency virus type 1	HIV-1 Lai strain
Nonenveloped double-stranded DNA	Adenovirus	Local laboratory strain

[a] American Type Culture Collection.
[b] Pasteur Institute Collection.

Bacterial and Virus Strains

Bacterial and virus strains used in this study are listed in Table I.

Growth of Bacteria

Bacteria are grown in nutrient broth 1.3% (Sanofi, Diagnostics Pasteur, France) to logarithmic phase at 37° for 4 hr. Then they are harvested by centrifuging at 800g for 10 min, washed twice with phosphate-buffered saline (PBS), pH 7.2, and, finally, resuspended in PBS to obtain a concentration of about 10^8 cells/ml.

Preparation of Virus Suspensions

Types of cells and media used to cultivate viruses are listed in Table II. Extracellular viruses are harvested from supernatants of infected cul-

TABLE II
CELLS AND MEDIA USED TO CULTIVATE VIRUSES

Virus	Cell	Medium
HCMV[a]	MRC5	MEM[b] supplemented with 10% FCS[c], 2 mM L-glutamine, and 1% nonessential amino acids
HIV-1	MT-4	RPMI 1640 supplemented with 2 mM L-glutamine and 10% FCS
ADV[d]	VERO	MEM supplemented with 5% FCS and 2 mM L-glutamine

[a] Human cytomegalovirus.
[b] Minimum essential medium.
[c] Fetal calf serum.
[d] Adenovirus.

tures. Contaminating cells are removed by low-speed centrifugation (800g, 10 min) followed by filtration through a 45-μm polysulfone filter.

Intracellular HCMV is obtained by incubating MRC5 cells in a 25-cm^2 tissue culture flask for 4 hr at 37° with 3 ml of a suspension of extracellular HCMV of high titer (>10^6/ml), treating them with 3 ml of trypsin-versene (Eurobio Laboratory, France) for 5 min at 37°, and washing them carefully with culture medium supplemented with 10% fetal calf serum (FCS) to remove free viruses. Intracellular HIV-1 is prepared by using the same method, except that nonattached infected MT-4 cells are simply collected by centrifugation and washed three times with culture medium.

Singlet Oxygen Carrier Exposures

For an unambiguous interpretation of results, it is essential to test separately (i) the reduced forms of the carriers NDP and DHPN, (ii) the oxidized forms, i.e., the endoperoxides NDPO$_2$ and DHPNO$_2$, and (iii) the oxidized forms deoxygenated previously by thermolysis for 150 min at 37°. The first experiment is performed to determine the intrinsic antimicrobial activity of the carrier, whereas the last test is carried out to check that the endoperoxides are free of impurities that might be toxic for cells and microorganisms.

All our experiments show a similar and low toxicity of both reduced and preheated oxidized forms of the carriers against microorganisms. Therefore, data corresponding to the reduced forms are omitted in the figures. The antimicrobial activity of pure $^1O_2$ is deduced from the comparison of the results obtained with the endoperoxides and those given by the preheated oxidized forms of the carriers.

So, bacterial or viral suspensions are treated with increasing concentrations (0 to 40 mM) of each form of $^1O_2$ carriers for 150 min at 37°. Then, surviving microorganisms are titrated by suitable methods.

Bacterial Survival Assays

After incubation with $^1O_2$ carriers, bacteria are serially 10-fold diluted with PBS. Each dilution is plated in duplicate on plate count agar (PCA) (Sanofi, Diagnostics Pasteur, France). After 48 hr incubation of the plates at 37°, the colony-forming units per milliliter (cfu/ml) are counted. Each experiment must be performed at least twice.

Virus Survival Assays

After incubation with $^1O_2$ carriers, surviving viruses are titrated in permissive cells. All titer estimates are based on mean counts of three replicative assays. Each experiment must be performed at least twice.

ADV and HCMV infectivity is determined by counting infectious centers stained with mouse monoclonal antibodies: briefly, 200 μl of a serial 10-fold dilution of viral suspensions is inoculated on VERO or MRC5 cells into 96-well plates. After incubation for 3 days at 37°, cells are fixed with methanol and stained for ADV with an antinuclear antigen present in all subtypes of ADV (H-60, Argene Biosoft, France) and for HCMV with an anti-immediate early antigen (E 13, Argene Biosoft, France) (50 μl of a 1/200 dilution of antibody in PBS, pH 7.2, per well). After incubation for 30 min at 37° and washings, peroxidase-labeled antimouse IgG (Sanofi, Diagnostics Pasteur, France) (1/200 in PBS, 50 μl per well) and then the enzyme substrate solution (3,3'-diaminobenzidine tetrahydrochloride 0.7 mg/ml in PBS with 0.5% H_2O_2 110 volume, 50 μl per well) are added successively. After 15 min, the enzymatic reaction is stopped by washings, and stained foci are enumerated using an epimicroscope. Results are expressed in infectious centers (IC) per milliliter.

In the case of HIV-1, 10-fold dilutions of residual viruses are inoculated on MT-4 cells into 48-well plates. The culture is inspected on day 10 under the microscope for a cytopathic effect (syncytium formation), and the median tissue culture infectious dose ($TCID_{50}$) is calculated by the Reed–Muench method.[29]

Cytotoxicity Studies

For a correct interpretation of the results of the virus titrations, it is necessary to check that neither the reduced forms nor the preheated forms of the naphthalene carriers are toxic for the cells used in titration experiments. Performed monolayers of VERO cells, MRC5 cells, MT-4 cells are exposed to various concentrations (0–10 mM) of the reduced and the preheated oxidized forms of the carriers. After washings with fresh medium, the cell viability is determined by the colorimetric MTT (tetrazolium) assay as described by Mosmann.[30]

Bactericidal Activity of Pure $^1O_2$ Generated by Naphthalene Endoperoxides

As shown in Table III and in Fig. 1, which represents the survival curves of the gram-positive bacterium *Staphylococcus aureus* (Fig. 1A) and the gram-negative bacterium *E. coli* (Fig. 1B) incubated for 150 min at 37° in the presence of oxidized and preheated oxidized forms of naphthalene $^1O_2$

[29] L. J. Reed and H. Muench, *Am. J. Hyg.* **27**, 493 (1938).
[30] T. Mosmann, *J. Immunol. Methods* **65**, 55 (1983).

TABLE III
INACTIVATION RATES[a] (Log$_{10}$) OF GRAM-NEGATIVE AND GRAM-POSITIVE BACTERIA
TREATED FOR 150 MIN WITH 40 mM OF $^1O_2$ CARRIERS AT 37°

	NDP or H-NDPO$_2$	NDPO$_2$	DHPN or H-DHPNO$_2$	DHPNO$_2$
Gram-negative bacteria				
Escherichia coli	0.3	1.2	0.3	2.3
Pseudomonas aeruginosa	0.4	1.2	0.3	2.0
Gram-positive bacteria				
Enterococcus faecium	2.7	3.3	0.9	4.3
Staphylococcus aureus	0.7	2.2	0.9	3.4

[a] The initial titer of bacteria was 10^8 cfu/ml, i.e., 8 log$_{10}$.

carriers, gram-positive bacteria were more susceptible to the action of naphthalene endoperoxides than gram-negative bacteria, as the decrease of gram-positive bacteria titer after exposure to 40 mM of the endoperoxides NDPO$_2$ and DHPNO$_2$ is about 10 times higher than the decrease of gram-negative bacteria titer. However, DHPNO$_2$ was more efficient than NDPO$_2$, and the inactivation rate measured after treatment by DHPNO$_2$ is 10 times higher than the one induced by NDPO$_2$.

The bactericidal activity of both endoperoxides is actually due to the action of $^1O_2$, as preheated endoperoxides and reduced carriers were almost inefficient except in the case of Enterococcus, which suffered a 2.7 log$_{10}$ reduction of infectivity when exposed to NDP and preheated NDPO$_2$ (Table III). This example shows that determination of the antimicrobial activity of reduced forms and preheated oxidized forms of $^1O_2$ carriers is essential in making an accurate interpretation of data.

Virucidal Activity of Pure $^1O_2$ Generated by Naphthalene Endoperoxides

Cytotoxicity of Carriers

Under our experimental conditions, reduced forms and preheated oxidized forms of the carriers are not cytotoxic for VERO, MRC5, and MT-4 cells and thus do not perturb virus infectivity assays.

Inactivation of Enveloped Viruses

Figure 2 shows the survival curves of extracellular HCMV incubated with increasing concentrations of the oxidized forms and the preheated

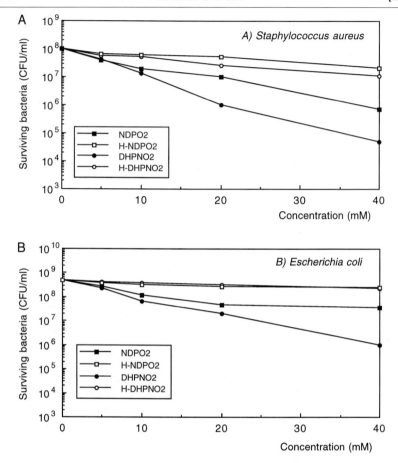

FIG. 1. Survival curves of a gram-positive bacterium (*Staphylococcus aureus*) (A) and a gram-negative bacterium (*Escherichia coli*) (B) treated for 150 min at 37° with $NDPO_2$ or preheated $NDPO_2$ (H-$NDPO_2$) and with $DHPNO_2$ and preheated $DHPNO_2$ (H-$DHPNO_2$). Each graph point is a mean of duplicate assays. For each compound concentration, at least two independent experiments were performed.

oxidized forms of both naphthalene endoperoxides. Similar curves were obtained with extracellular HIV-1. $DHPNO_2$ is slightly more effective than $NDPO_2$, as the complete inactivation of the extracellular viruses (i.e., a titer reduction greater than 5.3 $\log_{10}$) required 3.7 and 4.5 mM of $DHPNO_2$ and $NDPO_2$, respectively. The virucidal activity of both endoperoxides is actually due to $^1O_2$, as preheated oxidized forms H-$NDPO_2$ and H-$DHPNO_2$ were almost inefficient.

Figure 3 shows the survival curves of intracellular HCMV treated by

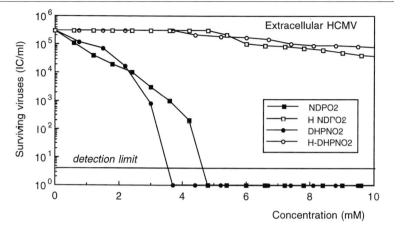

FIG. 2. Survival curves of extracellular HCMV treated for 150 min at 37° with $NDPO_2$ or preheated $NDPO_2$ (H-$NDPO_2$) and with $DHPNO_2$ and preheated $DHPNO_2$ (H-$DHPNO_2$). Each graph point is a mean of triplicate assays. For each compound concentration, at least two independent experiments were performed.

increasing concentrations of $NDPO_2$, $DHPNO_2$, and preheated $DHPNO_2$ (H-$DHPNO_2$). Similar curves were obtained with HIV-1. Whereas $NDPO_2$ and preheated $DHPNO_2$ were ineffective against intracellular viruses, $DHPNO_2$ was able to inactivate them efficiently, and 3.7 mM of this endo-

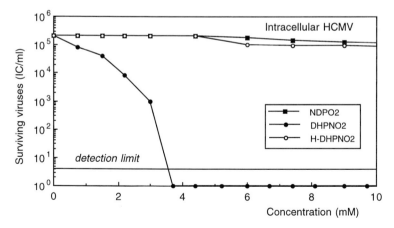

FIG. 3. Survival curves of intracellular HCMV treated for 150 min at 37° with $NDPO_2$, $DHPNO_2$, and preheated $DHPNO_2$ (H-$DHPNO_2$). Each graph point is a mean of triplicate assays. For each compound concentration, at least two independent experiments were performed.

peroxide reduced the infectivity of the intracellular viruses by more than five orders of magnitude under our experimental conditions. These results are particularly relevant, as several viruses (including HIV and HCMV) are carried by cells in blood.

Inactivation of Nonenveloped Viruses

As shown in Fig. 4, $NDPO_2$ was totally ineffective against the nonenveloped adenovirus, whereas $DHPNO_2$ was able to inactivate it efficiently. This activity is actually due to the action of $^1O_2$, as preheated $DHPNO_2$ had no effect on ADV infectivity. However, effective concentrations were 7.5 times higher than those required to kill enveloped viruses, and complete inactivation of the ADV (i.e., a titer reduction greater than 5.7 $\log_{10}$ under our experimental conditions) was obtained with 27.5 mM of $DHPNO_2$. The unexpected effectiveness of $DHPNO_2$ against nonenveloped viruses could be explained not only by its lipophily but also by the chemical similarity between its hydrophilic groups and amino acids. It is likely that this peculiar structure favors its interaction with the proteins of the viral capsid.

Conclusion

Water-soluble naphthalene endoperoxides are simple and powerful tools to assess the activity of a known quantity of pure $^1O_2$ against biological

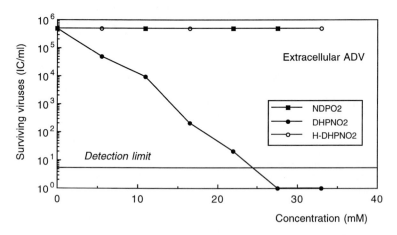

FIG. 4. Survival curves of extracellular adenovirus treated for 150 min at 37° with $NDPO_2$, $DHPNO_2$, and preheated $DHPNO_2$ (H-$DHPNO_2$). Each graph point is a mean of triplicate assays. For each compound concentration, at least two independent experiments were performed.

targets such as bacteria or viruses. Contrary to the well-known $NDPO_2$, the new nonionic carrier $DHPNO_2$ appears to be very effective in inactivating both nonenveloped viruses and intracellular enveloped viruses. It is also much more efficient than $NDPO_2$ in killing bacteria. These differences can be related to the structure of the endoperoxides. The anionic carrier $NDPO_2$ has no affinity for lipophilic membranes or negatively charged sites. Therefore, it releases $^1O_2$ randomly in the aqueous medium. Thanks to its nonionic structure, $DHPNO_2$ is able to penetrate into cell membranes where it generates $^1O_2$ close to the intracellular targets. Hence, it might be a useful tool in investigating the *in vitro* activity of pure $^1O_2$ on any intracellular target, such as DNA or mitochondria.

[19] Inactivation of Viruses in Human Plasma

By H. MOHR

Introduction

Human fresh frozen plasma (FFP) is widely used for therapeutic purposes and has a high safety margin based on two complementary approaches:

1. It is a single donor preparation isolated from donor blood or obtained by plasmapheresis. Each donation or plasma unit is tested for the absence of markers of human immunodeficiency virus (HIV) and the hepatitis viruses B and C (HBV, HCV), respectively.[1,2]
2. It is quarantined for up to 6 months in the frozen state. The plasma is not used before the respective donor or his new donation after this time period is retested and found to be negative for the viral markers mentioned.

Nevertheless, there is a slight, but significant, residual risk of viral contamination as a result of donations in the preseroconversion window period of infectivity.[3]

The donor may also be infected with a virus that is not tested for, e.g.,

---

[1] P. V. Holland, *N. Engl. J. Med.* **334,** 1734 (1996).
[2] W. K. Roth, M. Weber, and E. Seifried, *Lancet* **353,** 359 (1999).
[3] G. N. Vyas, *Transfusion* **35,** 367 (1995).

METHODS IN ENZYMOLOGY, VOL. 319   0076-6879/00 $30.00

parvovirus B19.[4] This justifies incorporating a step to ensure the viral safety of FFP.

For plasma pool products, heat treatment in the liquid or freeze-dried state or treatment with detergents in combination with organic solvents is routinely used.[5-7] These procedures are, however, too labor and time consuming to process numerous single donor units of FFP. In this case, a realistic approach is photodynamic treatment.

Principle of Photodynamic Virus Inactivation

The principle of photodynamic virus inactivation of FFP is as follows: The plasma is mixed with a photoactive compound that has a preference for viral structures, e.g., the nucleic acid or membrane components. The mixture is then illuminated with light of an appropriate wavelength, i.e., in the adsorption maximum of the photosensitizer. This activates the photosensitizer, and in the presence of oxygen, activated species of oxygen are generated, mainly the singlet form. Virus inactivation is accomplished by oxidative damage of the viral structures mentioned. In general, lipid-enveloped viruses are more susceptible to photodynamic treatment than nonenveloped viruses[8-10] (Table I).

Methylene Blue for Photodynamic Treatment of Plasma

Human plasma has a high light absorption of between 300 and approximately 450 nm, but above 550–600 nm it is almost negligible (Fig. 1). Accordingly, the photosensitizer used for viral decontamination needs to have its own absorption above 550–600 nm. This property is fulfilled by methylene blue (MB, Fig. 2), whose adsorption maximum is about 660 nm (Fig. 1). An additional advantage is that MB is relatively nontoxic and has been used therapeutically for many years, e.g., in the treatment of methemoglobinemia and depression.[11-13] Tolerated doses are in the range

[4] C. Wakamatsu, F. Takahura, E. Kojima, Y. Kiriyama, N. Goto, K. Matsumoto, M. Oyama, H. Sato, K. Okochi, and Y. Maeda, Vox Sang. 76, 14 (1999).

[5] P. Murphy, T. Nowak, S. M. Lemon, and J. Hilfenhaus, J. Med. Virol. 41, 61 (1993).

[6] P. Roberts and P. Feldman, Vox Sang. 73, 189 (1997).

[7] B. Horowitz, M. E. Wiebe, A. Lippin, and M. H. Stryker, Transfusion 25, 516 (1988).

[8] C. W. Hiatt, in "Concepts in Radiation Cell Biology," p. 57. Academic Press, New York, 1972.

[9] E. Ben-Hur and B. Horowitz, Photochem. Photobiol. 62, 383 (1995).

[10] R. Santus, P. Grellier, J. Schrével, J.-C. Mazière, and J.-F. Stoltz, Clin. Haemorheol. 18, 299 (1998).

[11] R. M. Devine, J. A. van Heerden, C. S. Grant, and J. J. Muir, Surgery 94, 916 (1983).

TABLE I
Lipid-Enveloped and Noneveloped Viruses Tested in Plasma for Their Sensitivity
to MB/Light Treatment

Name	Family	Charac-teristics	Inactivation Rate (log$_{10}$-steps)	Illumination time (min)
Enveloped Viruses				
HIV-1			≥ 6.32 **	10
HIV-2	Retro	ssRNA	≥ 3.81 *	15
SIV			≥ 6.26 *	15
Herpes Simplex			≥ 5.50	60
Bovine Herpes	Herpes	dsDNA	≥ 8.11 *	30
Suid Herpes Type 1			4.43 *	60
Sindbis	Toga	ssRNA	≥ 9.73	5
Semliki Forrest			≥ 8.77 *	15
West Nile	Flavi	ssRNA	≥ 4.39 *	2
Classical Swine Fever		ssRNA	≥ 3.20 *	<1
Bovine Viral Diarrhea	Pesti		≥ 5.63 *	2
			≥ 6.41 *	5
Vesicular	Rhabdo	ssRNA	≥ 4.89 *	60
Influenza	Orthomyxo	ssRNA	5.10	60
Non enveloped viruses				
EMC			0	60
Polio	Picorna	ssRNA	0	120
Hepatitis A			0	60
Porcine Parvo	Parvo	ssDNA	0	60
Adeno	Adeno	dsDNA	4.00	120
Calici	Calici	ssRNA	> 3.90 *	5
SV 40	Papova	dsDNA	> 4.00	30
Reo-3	Reo	dsRNA	3.80	30

* tested at production conditions
** 76 ml in cell culture flask

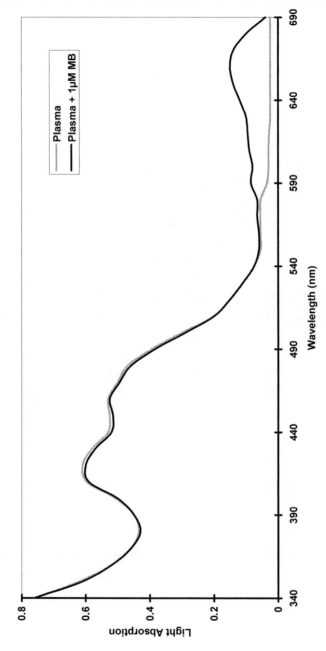

Fig. 1. Light absorption of human plasma in the absence and the presence of MB.

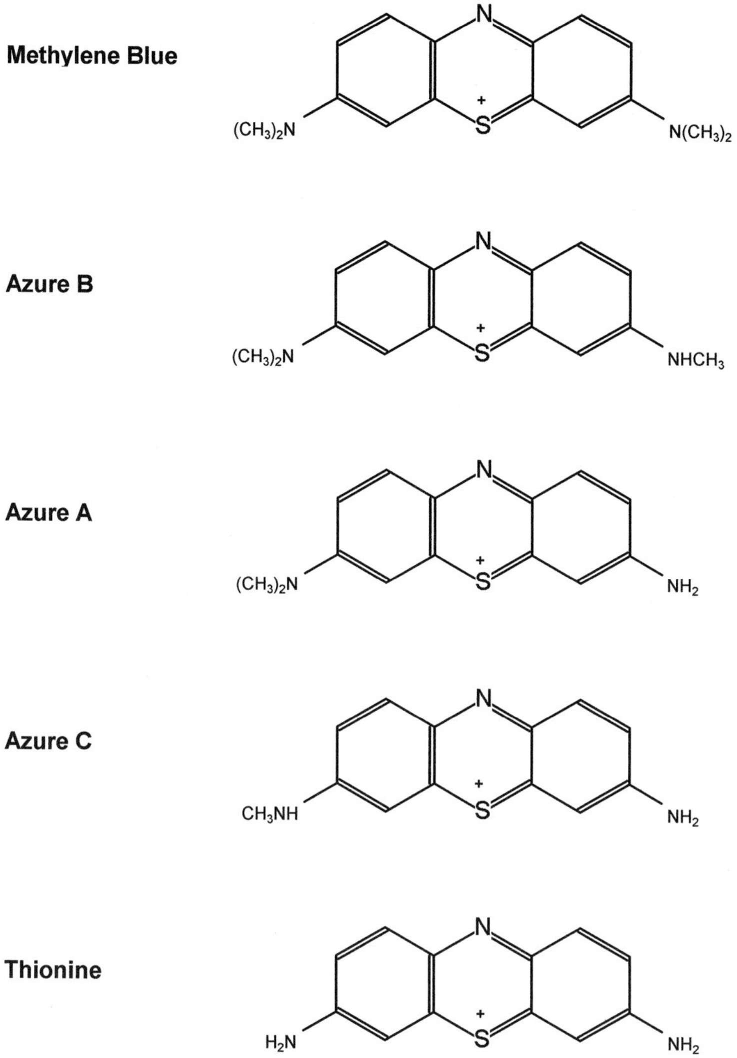

FIG. 2. Structures of MB and related phenothiazine dyes.

of 1–2 mg/kg body weight per day. For plasma treatment, a concentration of 1 μM is sufficient, i.e., approximately 300 μg/liter.[14,15]

Another important aspect is that during illumination photoproducts are formed from the photosensitizer itself. In the case of MB they are well defined. The photoproducts are the demethylated analogs of MB, i.e., azure A, B, C, and thionine, respectively (Fig. 2), whose photodynamic and other properties are similar to those of MB itself.

The virus inactivation properties of MB and other phenothiazine dyes in combination with visible light have been known for many years. Their use for the photodynamic treatment of sewage, tap water, vaccines, and blood plasma was proposed.[8,16–20] To date, however, only the treatment of single units of human plasma has been fully developed.

Production of MB Plasma

The procedure for manufacturing MB plasma is shown in Fig. 3. It is manufactured from fresh frozen plasma isolated from donor blood within 6–8 hr after donation. Alternatively, the plasma is obtained by plasmapheresis. The plasma units (approximately 300 ml) are shock-frozen and stored at $\geq -30°$ until further processing. They are thawed under gentle shaking in a water bath at 27–30° within 15–20 min. This is followed by visual inspection. Units that are hemalytic of lipemic are discarded for cosmetic reasons, as the plasma appears "dirty."

The freezing/thawing cycle is important because this destroys contaminating leukocytes that might harbor cell-associated viruses such as HIV-1 or provirus integrated into the cellular genome. Freezing/thawing makes these structures accessible to MB/light. As an alternative, leukocytes can be depleted from the plasma by filtration through membrane filters.[21–24] It

[12] J. W. Harvey and A. S. Keitt, *Br. J. Haematol.* **54,** 29 (1983).

[13] G. J. Naylor, B. Martin, S. E. Hopwood, and Y. Watson, *Biol. Psychiat.* **21,** 915 (1986).

[14] B. Lambrecht, H. Mohr, J. Knüver-Hopf, and H. Schmitt, *Vox Sang.* **60,** 207 (1991).

[15] H. Mohr, B. Bachmann, A. Klein-Struckmeier, and B. Lambrecht, *Photochem. Photobiol.* **65,** 441 (1997).

[16] J. R. Perdrau and F. R. S. Todd, *Proc. Roy. Soc. London* **112,** 288 (1933).

[17] C. W. Hiatt, E. Kaufman, J. J. Helprin, and S. Baron, *J. Immunol.* **84,** 480 (1960).

[18] C. Wallis and C. Melnick, *Viology* **23,** 520 (1964).

[19] G. S. L. Yen and E. H. Simon, *J. Gen. Virol.* **41,** 273 (1978).

[20] J. A. Badylak, G. Scherba, and D. P. Gustafson, *J. Clin. Microbiol.* **17,** 374 (1983).

[21] B. Rawal, B. T. S. Yen, G. N. Vyas, and M. Busch, *Vox Sang.* **60,** 214 (1991).

[22] D. Zucker-Franklin and B. A. Pancake, *Transfusion* **38,** 317 (1998).

[23] J. I. Willis, J. A. G. Lown, M. C. Simpson, and W. N. Erber, *Transfusion* **38,** 645 (1998).

[24] J. R. Rider, M. A. Winter, J.-M. Payrat, J.-M. Mathias, and D. H. Pamphilon, *Vox Sang.* **74,** 209 (1998).

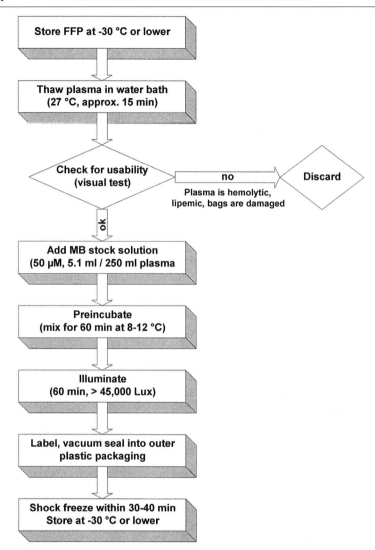

FIG. 3. Production procedure for MB/light-treated fresh plasma.

is the advantage of filtration that the freezing/thawing step can be omitted. After visual inspection the appropriate amount of a 50 μM stock solution of MB is added to the plasma to achieve a final dye concentration of 1 μM. When the plasma volume is about 250 ml, approximately 5.5 ml has to be added. The stock solution is in another plastic bag, which is positioned above the plasma-containing bag. Using a sterile docking system, the tubings of both bags are connected (Fig. 4). Flow of the dye solution into plasma is by gravity.

The plasma unit is placed on a computer-assisted balance, and the addition of dye solution is stopped when the required amount has been added to the plasma.

The manual addition of MB is cumbersome. In two alternative procedures, a fixed amount of dye is in the plasma bag system before isolation of the plasma, either in the form of a solution or as a pill, which is placed in the tubing between the blood bag and the plasma bag. When the plasma is transferred into its container, the dye-containing pill is dissolved simultaneously and mixed with the plasma. Because the plasma volume is variable in these two procedures, variations of the MB concentrations between 0.8 and 1.2 μM must be tolerated.

In the next production step, the MB-containing plasma units are rotated overhead in the dark in a thrombocyte rotator at $10–12°$ for $60–65$ min. This ensures complete mixing of dye and plasma, while at same time lowering the unit temperature to approximately $18°$. This is an advantage because the plasma temperature increases by $5–7°$ during illumination. Thus, precooling prevents the plasma from overheating.

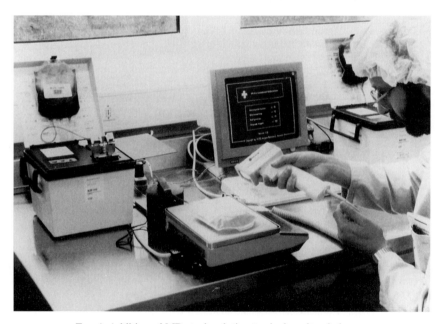

Fig. 4. Addition of MB stock solution to single units of plasma.

The following production step comprises the actual photodynamic treatment. The illumination device is a rectangular box with a glass plate on top. It is equipped with eight fluorescent tubes, e.g., Philips TL-M 115 W (33 RS) emitting white light and arranged horizontally below the glass plate. The device is air-cooled to remove excess heat. The plasma units to be treated are on the glass plate, with the labeled side up (Fig. 5). Up to 21 units can be illuminated at the same time. Minimal light intensity is 45,000 Lux. This is checked at different positions of the illumination area before and during each treatment cycle. Illumination time is 1 hr. After illumination the plasma units are labeled, sealed into outer plastic bags, and deep-frozen again at $\leq -30°$.

Alternative light sources may also be used that are equipped with LEDs emitting red light with a wavelength in the range of the absorption maximum of MB (i.e., approximately 660 nm) or low-pressure sodium lamps emitting high-intensity yellow light at 595 nm. It has been demonstrated that illumination time in both cases may be shortened considerably to achieve complete virus inactivation while at the same time better preserving the activities of plasma proteins.[15]

The efficacy of the procedure was proved using a variety of different lipid-enveloped and nonenveloped model viruses (Table I). These included those virus types that are recommended by CPMP guidelines 268/95 ("Note

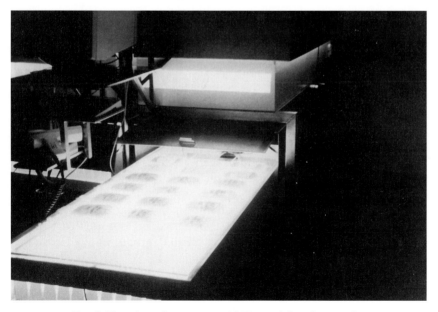

FIG. 5. Photodynamic treatment of MB containing plasma units.

for guidance on virus validation studies: the design, contribution, and interpretation of studies validating the inactivation and removal of viruses") and 269/95 ("Note for guidance on plasma-derived medicinal products"). These guidelines note which viruses should be used in validation studies on the inactivation and removal of viruses and how this should be done. It is necessary to test for HIV, a model for hepatitis C virus (HCV), for nonenveloped viruses and enveloped DNA viruses such as herpes viruses.

Because there is no cell culture system to assay HCV, flavi, toga, and pestiviruses may be used as models. It is evident from Table I that these viruses especially are very sensitive to MB/light treatment and are inactivated within less than 1–5 min. Polymerase chain reaction studies using HIV-infected plasma also indicate that HCV is probably effectively inactivated.[25]

[25] K. Müller-Breitkreutz and H. Mohr, *J. Med. Virol.* **56,** 239 (1998).

[20] 3-(4'-Methyl-1'-Naphthyl)propionic Acid, 1',4'-Endoperoxide for Dioxygenation of Squalene and for Bacterial Killing

By Minoru Nakano, Yasuhiro Kambayashi, and Hidetaka Tatsuzawa

Introduction

In recent years, considerable interest has been focused on the reaction of singlet oxygen ($^1O_2$; $^1\Delta_g$) with organic compounds and its toxicity toward living cells. Known $^1O_2$ generating systems, such as $NaOCl$-H_2O_2,[1] myeloperoxidase–hydrogen peroxide (H_2O_2)–halide ions,[2] and light-photosensitizer-O_2[3] systems contain reactive oxygen species other than $^1O_2$ and often generate free radicals, and therefore such systems cannot serve as pure $^1O_2$ generating reactions.

In the light of the aforementioned information, Saito *et al.*[4] synthesized a novel water-soluble naphthalene endoperoxide [3-(4'-methyl-1'-naphthyl)propionic acid, 1',4'-endoperoxide; NEPO], producing $^1O_2$ thermolyt-

[1] T. Kajiwara and D. K. Kearms, *J. Am. Chem. Soc.* **15,** 5886 (1976).
[2] J. R. Kanofsky, *Chem.-Biol. Interact.* **70,** 1 (1989).
[3] R. C. Straight and J. D. Spikes, *in* "Singlet O_2," p. 91. CRC Press, Boca Raton, FL, 1985.
[4] I. Saito, T. Matsuura, and K. Inoue, *J. Am. Chem. Soc.* **103,** 188 (1981).

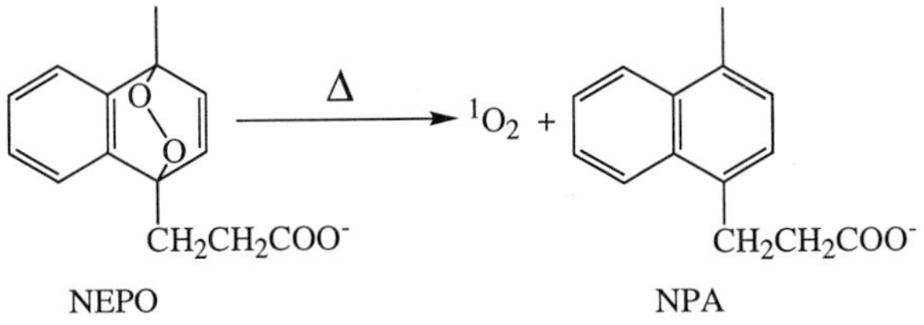

FIG. 1. Decomposition of NEPO to yield $^1O_2$ and NPA.

ically, not photochemically. However, little is known about the kinetic aspects of NEPO decomposition to produce $^1O_2$ in organic solvents or in aqueous media. The aim of this article is to clarify the stoichiometry of NEPO thermolysis with quantification of $^1O_2$ during incubation of NEPO in an organic or aqueous solution. On the basis of stoichiometric data obtained, reactivity of $^1O_2$ toward lipids in organic solvents, and its toxicity on two *Escherichia coli* strains and on human endothelial cells in aqueous media were examined with NEPO as a pure $^1O_2$ yielding compound.

Stoichiometry and Kinetics of NEPO Thermolysis

NEPO, 50–100 μM in a 3-ml cell, which should be shielded to avoid evaporation of solvent, is incubated at 37° and the absorbance of 3-(4′-methyl-1′-naphthyl)propionic acid (NPA) at 288 nm is recorded continuously during the time quoted. NEPO provides $^1O_2$ ($^1\Delta_g$) quantitatively at 37° in 0.1 M sodium phosphate buffer (pH 7.2), 0.1 M acetate buffer (pH 4.5), chloroform, or methanol, through the retro-Diels-Alder reaction shown in Fig. 1. Because NEPO has no absorption band at 288 nm, the non-zero-intercept (A_0) is attributable to NPA formation during the preparation of the solution and the thermoequilibrium in the cell. The experimental parameters fit well with a first-order kinetic equation, consistent with those reported by Saito *et al.*[4] and Fujimori *et al.*[5] The $^1O_2$ flux from NEPO and the total amount of $^1O_2$ are derived from Eqs. (1) and (2), respectively.

$$v = k[\text{NEPO}]_0, \text{M/s} \tag{1}$$
$$[^1O_2] = [\text{NEPO}]_0\{1 - \exp^{-kt}\}, \text{M} \tag{2}$$

The initial concentration of NEPO, $[\text{NEPO}]_0$, can be calculated from A_{max} and A_0 using Eq. (3):

$$[\text{NEPO}]_0 = (A_{max} - A_0)/\text{molar extinction coefficient } (\varepsilon) \text{ of NPA, M} \tag{3}$$

[5] K. Fujimori, T. Komiyama, H. Tabata, T. Nojima, K. Ishiguro, Y. Sawaki, H. Tatsuzawa, and M. Nakano, *Photochem. Photobiol.* **68**, 143 (1998).

TABLE I
DATA FOR THERMOLYSIS OF NEPO

Solvent	$k \times 10^4$ (sec^{-1})	ε at 288 nm (M^{-1} cm^{-1})	Half-life of NEPO (min)	Time for A_{max} measured (hr)
Sodium phosphate (0.1 M, pH 7.2)	4.16	7000	27.8	4
Acetate buffer (0.1 M, pH 4.5)	4.65	6890	24.8	4
Methanol	1.64	7100	70.4	8
Chloroform	1.55	7240	74.5	8

where A_{max} and A_0 are absorbance ($A_{288 \, nm}$) at 4 or 8 hr and 0 time, respectively. The parameters obtained in this study are shown in Table I and can be used in Eqs. (1)–(3). The yield of $^1O_2$ in the thermal decomposition of NEPO at 37° in water at pH 7.2 was found to be 93 ± 3%.[5] Other investigators have reported a maximum yield of 80% $^1O_2$ after thermal decomposition of NEPO.[6]

Dioxygenation of Lipids in Organic Solvents

1-Palmitoyl-2-oleoylphosphatidylcholine (POPC), 1-palmitoyl-2-linoleoylphosphatidylcholine (PLPC), 1-palmitoyl-2-arachidonoylphosphatidylcholine (PAPC), microperoxidase-11, and isoluminol are from Sigma (St. Louis, MO). Squalene (SQ) is from Tokyo Kasei (Tokyo). All solvents are of highest grade commercially available. POPC, PLPC, PAPC, and SQ are used after purification using an HPLC equipped with a UV detector monitored at 215 nm using a CAPCELLPAK C$_{18}$ column (20 × 250 mm, 5 μm; Shiseido, Tokyo) and methanol containing 0.02% triethylamine (methanol for purification of SQ) as an eluent (flow rate; 10.0 ml/min). The concentrations of POPC, PLPC, and PAPC are determined using a Phospholipid B test (Wako, Osaka, Japan), which is a choline-containing phospholipid-specific analytical kit. The concentration of SQ is determined by weighing the dried sample and from its molecular weight. POPC, PLPC, and PAPC are dissolved in methanol. SQ is dissolved in chloroform. They are stored at −80° prior to use.

Lipid samples are dissolved in chloroform or methanol and then the solvent is removed under reduced pressure. The residue obtained is redissolved in the reaction solvent to obtain an adequate concentration of lipid. One micromolar lipid is incubated with or without 1, 2, or 3 mM NEPO at 37° under aerobic conditions. The concentration of lipid hydroperoxide,

[6] J. R. Wagner, P. A. Mochinik, R. Stocker, H. Sies, and B. N. Ames, *J. Biol. Chem.* **268,** 18502 (1993).

the primary oxidation product of lipid, is measured by using a hydroperoxide-specific, isoluminol chemiluminescence assay.[7] The formation of phosphatidylcholine hydroperoxide (PC-OOH) is monitored by HPLC equipped with a silica gel column (4.6 × 250 mm, 5 μm; Supelco) using methanol/40 mM monobasic sodium phosphate (9:1, v/v) as the mobile phase (flow rate; 1.0 ml/min) and 1 mM isoluminol, 5 mg/liter microperoxidase-11 in methanol/0.1 M borate buffer, pH 10 (1:1, v/v), as the chemiluminescence reagent (flow rate; 1.5 ml/min). The formation of SQ hydroperoxide (SQ-OOH) is monitored by HPLC equipped with a CAPCELLPAK C_{18} column (4.6 × 250 mm, 5 μm; Shiseido) using methanol as the mobile phase (2.0 ml/min) and 1 mM isoluminol, 5 mg/liter microperoxidase-11 in methanol/0.1 M borate buffer, pH 10 (1:1, v/v), as the chemiluminescence reagent (1.5 ml/min). Fifty microliters of reaction mixture is injected onto the HPLC. Results are shown as mean ± standard deviation of three independent experiments and are analyzed statistically by Bonferroni comparisons.

When 1 mM SQ was incubated with 3 mM NEPO in chloroform at 37° for 90 min under aerobic conditions, chromatograms showed three SQ hydroperoxides as emission peaks. However, immediately after NEPO addition, no hydroperoxide formation was observed. These peak intensities correspond to monohydroperoxide, dihydroperoxide, and trihydroperoxide.[8] Because there is no conjugated diene in SQ, these hydroperoxides should have been found via $^1O_2$-mediated oxidation (ene reaction) of SQ. In this study, only mono-SQ-OOH was taken as the major dioxygenated product of SQ. When lipids were oxidized in chloroform instead of methanol, comparable results were obtained, i.e., the number of enes in lipids influences their oxidizability of $^1O_2$ (POPC < PLPC < PAPC < SQ). However, the amount of lipid hydroperoxide formed is much higher in chloroform than in methanol. For example, the 256 μM $^1O_2$ exposure for 30 min in methanol gave 0.018 ± 0.003 μM POPC-OOH, 0.031 ± 0.004 μM PLPC-OOH, 0.121 ± 0.047 μM PAPC-OOH, and 1.727 ± 0.237 μM SQ-OOH, whereas the 243 μM $^1O_2$ exposure for 30 min in chloroform gave 0.561 ± 0.174 μM POPC-OOH, 1.780 ± 0.876 μM PLPC-OOH, 5.083 ± 2.348 μM PAPC-OOH, and 20.200 ± 6.8151 μM SQ-OOH. This may be due to fact that the lifetime of $^1O_2$ is longer in chloroform (60 μsec) than in methanol (5–11.4 μsec).[9]

Of all lipids used in this study, SQ was the most susceptible to oxidation

[7] Y. Yamamoto, M. H. Brodsky, J. C. Baker, and B. N. Ames, *Anal. Biochem.* **160,** 7 (1987).
[8] Y. Kohno, O. Sakamoto, T. Nakamura, and T. Miyazawa, *J. Jpn. Oil. Chem. Soc.* (*Yukagaku*) **42,** 204 (1993).
[9] D. Bellus, *Adv. Photochem.* **11,** 105 (1979).

by $^1O_2$. However, the oxidizability of SQ cannot be explained only by the number of double bonds in the molecule. The oxidizability of SQ by $^1O_2$ may be attributable to the small ionization potential of SQ,[10] which is a measure of electron-donating capacity.

Cytotoxicity Studies on Human Umbilical Vein Endothelial Cells and Viability of Escherichia coli Strains in Aqueous Medium

Deep-frozen normal human umbilical vein endothelial cells (HUVEC) are from Kurabo (Osaka, Japan) and subcultured according to the protocol of the Kurabo cell culture kit when subconfluent monolayers are obtained. The medium, reagents, and buffer used are of Kurabo's cell culture kit. The frozen sample of HUVEC in a test tube is thawed in 50% ethanol at 37° and is immediately suspended in Humedia-EG2 culture medium. The inoculated viable cell number is determined using the trypan blue exclusion test.[11] The cells are incubated at 37° in 25-cm^2 tissue culture flasks (Corning Glass Works, Corning, NY) in a humidified atmosphere containing 5% CO_2 in air. At subconfluent monolayer stage, the cells are harvested according to Kurabo's cell culture kit protocol. The harvested cells are resuspended in HuMedia-EG2 medium at approximately 1×10^5 cells/ml. Then 250-μl aliquots of cell suspensions are transferred into the wells of a flat-bottom 24-well microtiter tissue culture plate (Corning) and incubated for 24 hr at 37° under 5% CO_2 in air. These cell plates are used for the cytotoxicity assay of $^1O_2$. Cytotoxicity of $^1O_2$ on HUVEC is measured by the lactate dehydrogenase (LDH) release assay by determining the decrease in the number of viable cells after exposure to $^1O_2$. After a 24-hr incubation, the culture medium is removed by pipette, attached cells are washed twice with phosphate-buffered saline without Ca^{2+} and Mg^{2+} (PBS) at pH 7.3, and 180 μl of PBS is added to each well. Twenty microliters of 1 mM NEPO buffer solution is added to each well and incubated at 37°. After the desired time period, the cell-free upper layer (buffer) is carefully removed from each well. LDH released in the buffer is measured using an LDH cytotoxic test kit (Wako). The viability of cells in each well is also determined by the trypan blue exclusion method.[11] To determine maximum and negative control cell membrane damage, cells are incubated with 0.2% Tween 20 (Sigma) in PBS or in PBS alone.

Escherichia coli (E. Coli) mutants carrying lycopene-producing genes (pACCCRT-EIB, Misawa et al.[12]) are used to determine the bactericidal

[10] H. Koizumi, Chem. Phys. Lett. **219**, 137 (1994).

[11] H. M. Evans and W. Schulmann, Science **39**, 443 (1914).

[12] N. Misawa, M. Nakagawa, K. Kobayashi, S. Yamano, Y. Izawa, K. Nakamura, and K. Harashima, J. Bacteriol. **172**, 6704 (1990).

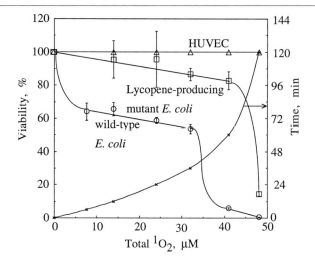

FIG. 2. Viability of wild-type *E. coli* ($n = 12$), a lycopene-producing mutant *E. coli* strain ($n = 12$), or human umbilical vein endothelial cell (HUVEC, $n = 4$) in exposure to $^1O_2$ produced from the thermolysis of NEPO at 37°. Two strains of *E. coli* were exposed to $^1O_2$ at pH 4.5 and HUVEC was exposed to $^1O_2$ at pH 7.3.

effect of NEPO. The amount of lycopene in lycopene-producing *E. coli* is estimated to be about 0.02 fmol/cell. Wild-type *E. coli* and a lycopene-producing one are cultured in LB medium[13] containing ampicillin (150 μg/ml) and/or chloramphenicol (30 μg/ml) at 27° with shaking (80 shakes/min) in test tubes. Growth is monitored by measuring the optical density of the test at 600 nm. Cells at the early stationary phase are harvested, washed with, and suspended in the minimal medium,[14] without vitamin B_{12}. The cell suspension are dispensed in polypropylene vials (17×51 mm diameter) and mixed with the $^1O_2$-generating medium. The reaction mixture contains 10^8 cells/ml, 0.1 m*M* NEPO, and 0.1 *M* acetate buffer, pH 4.5, or phosphate buffer, pH 7.4, in a total volume of 2 ml. The reaction is initiated by the addition of 200 μl of NEPO in water at 25° and is maintained at 37°. After the specified time period, samples are taken and washed, and viable cells are counted by spreading in triplicate on an LB agar plate after appropriate dilution. Colonies are counted after a 24-hr aerobic incubation at 37°.

When wild-type *E. coli* were exposed to 7 μ*M* $^1O_2$ derived from NEPO for 5 min at 37°, 34% of bacteria were killed (Fig. 2). Under identical

[13] J. Sambrook, E. F. Fritsch, and T. Maniatis, "Molecular Cloning," 2nd Ed. Cold Spring Harbor Laboratory Press, Cold Spring Harbor, NY, 1989.
[14] H. M. Hassan and I. Fridovich, *J. Biol. Chem.* **253**, 8143 (1978).

conditions, only 3% of lycopene-producing *E. coli* were killed, suggesting that lycopene protected *E. coli* against $^1O_2$, toxicity by scavenging $^1O_2$. The viability of *E. coli* at pH 4.5 was essentially the same as at pH 7.3. However, under our experimental conditions at pH 7.3, no toxic effect by $^1O_2$ on HUVEC was observed (Fig. 2). When wild-type *E. coli* were exposed to 1 m*M* NEPO for 3 min at pH 4.5 and at a given temperature, *E. coli* viabilities at 25 and 37° were 97 ± 5.4% (mean ± SEM, *n* = 6) and 53 ± 2.6%, respectively. The total amount of $^1O_2$ generated by thermal decomposition of NEPO for 3 min at 25° was negligible, but was 10 μM at 37°. Further, exposure of wild-type *E. coli* to 0.1 m*M* (or 1 m*M*) NPA for 12 or 60 min, instead of NEPO, had no effect on viability. These observations suggest that bacterial killing is via $^1O_2$ and not by a direct action of NEPO. We have found that electron transport systems located on the *E. coli* surface membrane are very vulnerable to short-lived $^1O_2$.[15] Based on these data, we believe that the resistance of HUVEC to $^1O_2$ toxicity under our experimental conditions may reflect the absence of oxidatively vulnerable electron transport systems in the HUVEC plasma membrane.

---

[15] H. Tatsuzawa, T. Maruyama, N. Misawa, K. Fujimori, K. Hori, Y. Sano, Y. Kambayashi, and M. Nakano, *FEBS Lett.* **439**, 329 (1998).

[21] Biological Singlet Oxygen Quenchers Assessed by Monomol Light Emission

By Karlis Briviba and Helmut Sies

Introduction

Physical techniques for the detection of singlet oxygen are based on monomol or dimol emission. This article reports on applications using an approach to determine the singlet oxygen ($^1O_2$) quenching constant by measuring steady-state monomol photoemission of singlet oxygen at 1270 nm [reaction (1)] detected by a germanium diode photodetector.

$$^1O_2 \rightarrow O_2 + h\nu \,(1270 \,\text{nm}) \tag{1}$$

$^1O_2$ was generated by thermodecomposition of the endoperoxide of disodium 3,3′-(1,4-naphthylidene)dipropionate (NDPO$_2$).[1]

---

[1] P. Di Mascio and H. Sies, *J. Am. Chem. Soc.* **111**, 2909 (1989).

Overall singlet oxygen quenching constants—the sum of the rate constants for chemical reaction and physical quenching—can be calculated from the decrease in singlet oxygen photoemission.[2] Time-resolved luminescence spectroscopy allows determination of the effect of tested compounds on the lifetime of singlet oxygen generated photochemically by pulse irradiation of a photosensitizer.[3,4]

Dimol emission [reaction (2)] of singlet oxygen at 634 and 703 nm can be detected by a red-sensitive photomultiplier.[5,6]

$$^1O_2 + {}^1O_2 \rightarrow 2\,O_2 + h\nu\,(634, 703\text{ nm}) \tag{2}$$

However, using monomol emission for the determination of quenching constants of singlet oxygen can be advantageous when tested quenchers absorb dimol emission, and some other reactive species, such as $NO_2{}^{\cdot}$, can generate light in the red spectral region.

A number of chemical competition reactions, such as oxidation of rubrene, 2-methyl-2-pentene, crocin, or some anthracene derivatives, have also been used for the determination of rate constants for singlet oxygen quenching.[7–10]

Assay

The infrared emission of $^1O_2$ is measured with a liquid nitrogen-cooled germanium photodiode detector (Model EO-81L7, North Coast Scientific Co., Santa Rosa, CA), sensitive in the spectral region from 800 to 1800 nm with a detector area of 0.25 cm^2 and a sapphire window as described in Sundquist et al.[2] A band-pass filter at 1270 ± 10 nm half-bandwidth (Spectrogon UK Ltd., Glenrothes, UK) is used. The diode signal is amplified (Model 5205, EG&G, Brookdeal Electronics Princeton Applied Research, Bracknell, Berkshire, UK) and documented with a recorder. The assay is carried out in a cuvette placed in a thermostated holder. After recording the baseline with the assay solvent, endoperoxide is added to the cuvette and photoemission is followed until maximum is reached; immediately

[2] A. R. Sundquist, K. Briviba, and H. Sies, Methods Enzymol. 234, 384 (1994).
[3] I. M. Byteva and K. I. Salokhiddinov, Opt. Spektrosk. 49, 707 (1979).
[4] T. Gensch and S. E. Braslavsky, J. Phys. Chem. B 101, 101 (1997).
[5] A. U. Khan and M. Kasha, J. Chem. Phys. 39, 2105 (1963).
[6] E. Cadenas and H. Sies, Methods Enzymol. 105, 221 (1984).
[7] C. S. Foote and R. W. Denny, J. Am. Chem. Soc. 90, 6233 (1968).
[8] W. Bors, C. Michel, and M. Saran, Biochim. Biophys. Acta 796, 312 (1984).
[9] M. Botsivali and D. F. Evans, J. Chem. Soc. Chem. Commun. 1114 (1979).
[10] B. M. Monroe, in "Singlet O₂" (A. A. Frimer, ed.), p. 177. CRC Press, Boca Raton, FL, 1985.

thereafter the tested compound is added and the resulting level of photo-emission is recorded. $NDPO_2$ as a water-soluble singlet oxygen source is suitable for determining the quenching constant for hydrophilic compounds; $DMNO_2$ should be used for hydrophobic compounds (see Ref. 11).

Singlet oxygen quenching constants ($k_r + k_q$), the sum of the rate constants for chemical reaction and physical quenching, are calculated according to Stern–Volmer plots, from $S_0/S = 1 + (k_r + k_q) * [Q] * \tau$, where S_0 and S are photoemission (1270 nm) intensities in the absence and presence of quenchers, respectively; [Q] is the quencher concentration; and τ is the lifetime of singlet oxygen.

Biological Singlet Oxygen Quenchers

Singlet oxygen formed by photoexcitation in skin on ultraviolet A exposure or in dark reactions by stimulated phagocytes has damaging effects on biomolecules and exerts genotoxic and cytotoxic effects (for reviews, see Refs 12 and 13). The reactivity of singlet oxygen can also be considered beneficial for the organism because of its toxicity to viruses[14] and bacteria.

Biological protection against singlet oxygen is afforded by quenchers such as carotenoids and tocopherols due to efficient quenching of singlet oxygen; singlet oxygen rate constants for carotenoids are in the range of 10^9–10^{10} M^{-1} sec^{-1}, and for α-tocopherol 3×10^8 M^{-1} sec^{-1}.[1,9,15] Multiplication of the concentration of a given compound with the corresponding rate constant for the reaction of singlet oxygen with that compound yields the rate of disappearance of singlet oxygen. This approach can be used to roughly estimate the biological relevance of the reaction of reactive species with their biological quenchers. In human plasma proteins react with singlet oxygen (Table I). However, in LDL lipophilic singlet oxygen scavengers such as carotenoids and α-tocopherol quench about 60% of singlet oxygen.[16] So far it is not clear which biological antioxidants are responsible for the defense against singlet oxygen in the aqueous phase because most plasma proteins can be damaged by singlet oxygen and are targets rather than antioxidants.

[11] C. Pierlot, J. M. Aubry, K. Briviba, H. Sies, and P. Di Mascio, Methods Enzymol. **319**, [3], 2000 (this volume).

[12] H. Sies and C. F. M. Menck, Mutat. Res. **275**, 367 (1992).

[13] K. Briviba, L. O. Klotz, and H. Sies, Biol. Chem. **378**, 1259 (1997).

[14] K. Müller-Breitkreutz, H. Mohr, K. Briviba, and H. Sies, J. Photochem. Photobiol. B Biol. **30**, 63 (1995).

[15] S. Kaiser, P. DiMascio, M. E. Murphy, and H. Sies, Arch. Biochem. Biophys. **277**, 101 (1990).

[16] J. R. Wagner, P. A. Motchnik, R. Stocker, H. Sies, and B. N. Ames, J. Biol. Chem. **268**, 18502 (1993).

TABLE I

RATE CONSTANTS FOR THE REACTION OF SINGLET OXYGEN WITH SOME BIOLOGICAL MOLECULES AND
RATES OF DISAPPEARANCE OF SINGLET OXYGEN IN THE PRESENCE OF THESE COMPOUNDS AT
IN VIVO CONCENTRATIONS

	Second-order rate constant $(M^{-1} \text{ sec}^{-1})$	Ref.	*In vivo* concentration (M)	Ref.	Rate of disappearance of $^1O_2$ $[k \times 10^3 \text{ (sec}^{-1})]$
Biological quenchers					
Human plasma		14			830[a]
Protein	1.1×10^4 (g/liter)$^{-1}$ sec^{-1}	21	60 g/liter		660
Plasma thiols	$8.3 \times 10^{6\,b}$	22	0.5×10^{-3}	23	4.2
α-Tocopherol	2.3×10^8	15	27×10^{-6}		6.2
Plasma carotenoids[c]	1×10^{10}	2	1.6×10^{-6}	24	16
Glutathione	2.4×10^6	22	$10 \times 10^{-3\,d}$	25	24
			$2 \times 10^{-6\,e}$	26	4.8×10^{-3}
Ascorbate	8.3×10^6	27	1.4×10^{-2}	28	116
			7×10^{-5}	23	0.6
			Concentrations for cell culture experiments (M)		
Exogenous $^1O_2$ quenchers					
Imidazole	4×10^7	17	40×10^{-3}		1,600
Azide	5×10^8	17	40×10^{-3}		20,000

[a] Calculated from data in Ref. 14.
[b] Rate constant for cysteine.
[c] The following carotenoids were taken into account for the calculation of the average singlet oxygen rate constant and *in vivo* concentration of plasma carotenoids: α-carotene, β-carotene, lycopene, lutein, and zeaxanthin.
[d] Intracellular.
[e] Plasma.

For cell culture experiments, imidazole and sodium azide are employed as quenchers of $^1O_2$. Reaction rate constants of imidazole and azide with $^1O_2$ are 4×10^7 and 5×10^8 M^{-1} sec^{-1}, respectively,[17] and rather high concentrations of these quenchers (20–40 mM) have to be used in biological systems (Table I). This apparently high concentration of azide was not toxic for cultured human skin fibroblasts.[18] Fibroblasts can tolerate even 100 mM azide for 2 hr.[19] Azide alone must also be tested for its effects on the assay system used because a number of enzymes can be inhibited. Because imidazole and azide do not react selectively with $^1O_2$, but also with hydroxyl radicals, an involvement of the latter in the investigated

[17] F. Wilkinson, W. P. Helman, and A. B. Ross, *J. Phys. Chem. Ref. Data* **24**, 663 (1995).
[18] K. Scharffetter-Kochanek, M. Wlaschek, K. Briviba, and H. Sies, *FEBS Lett.* **331**, 304 (1993).
[19] R. M. Tyrrell and M. Pidoux, *Photochem. Photobiol.* **49**, 407 (1989).

effect has to be tested. Hydroxyl radical scavengers such as mannitol can be used in cell culture experiments.[20]

Acknowledgment

Supported by the Deutsche Forschungsgemeinschaft, SFB 503, Project B1.

[20] L. O. Klotz, C. Pellieux, K. Briviba, C. Pierlot, J. M. Aubry, and H. Sies, *Eur. J. Biochem.* **260,** 917 (1999).
[21] J. R. Kanofsky, *Photochem. Photobiol.* **51,** 299 (1990).
[22] T. P. Devasagayam, A. R. Sundquist, P. Di Mascio, S. Kaiser, and H. Sies, *J. Photochem. Photobiol B* **9,** 105 (1991).
[23] R. Stocker and B. Frei, *in* "Oxidative Stress: Oxidants and Antioxidants" (H. Sies, ed.), p. 213. Academic Press, London, 1991.
[24] P. Di Mascio, M. E. Murphy, and H. Sies, *Am. J. Clin. Nutr.* **53,** 194S (1991).
[25] A. Wahlländer, S. Soboll, and H. Sies, *FEBS Lett.* **97,** 138 (1979).
[26] A. Wendel and P. Cikryt, *FEBS Lett.* **120,** 209 (1980).
[27] P. T. Chou and A. U. Khan, *Biochem. Biophys. Res. Commun.* **115,** 932 (1983).
[28] P. W. Washko, Y. Wang, and M. Levine, *J. Biol. Chem.* **268,** 15531 (1993).

[22] Synthetic Singlet Oxygen Quenchers

By Stefan Beutner, Britta Bloedorn, Thomas Hoffmann, and Hans-Dieter Martin

The quenching of singlet oxygen $^1O_2(^1\Delta_g)$ by synthetic compounds is described. The following classes of π- or n-electron containing substrates are considered: olefins, aromatic and heteroaromatic compounds, amines, sulfur compounds, phenols, metal chelates, nonaromatic heterocycles, and indigoids.

Introduction

Singlet oxygen is quenched by or reacts with many organic and bioorganic molecules, which possess, in most cases, reactive π electrons or n lone pairs of sufficiently low ionization energy. The quenching process can be physical (the quencher enters a vibrational or an electronic excited state) or chemical (the quencher combines with oxygen or is oxidized by oxygen) in nature. Physical quenching may occur via triplet energy transfer or by simple catalysis of the singlet oxygen $(^1O_2) \rightarrow$ ground state oxygen $(^3O_2)$ transition via spin-orbit coupling. Charge transfer interactions are important during the quenching process. In many reactions, intermediates are in-

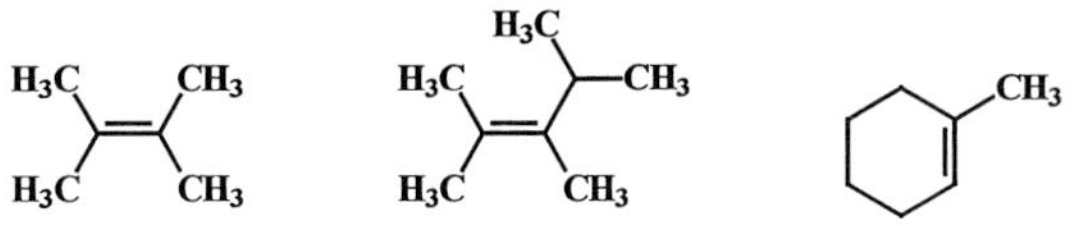

FIG. 1. Structures and $^1O_2$ quenching rates k_q (10^6 M^{-1} sec^{-1}) of selected olefins.

volved, e.g., exciplexes, diradicals, or zwitterions. Rate constants for quenching can be as high as 10^{10} to 10^{11} M^{-1} sec^{-1} (diffusion-controlled reactions) but are generally significantly less, 10^4 to 10^9 M^{-1} sec^{-1}.[1–3]

Olefins

Olefinic compounds, both conjugated and nonconjugated, show a wide variety of interactions with $^1O_2$. The most important among them are addition to C=C double bonds in many different mechanistic and structural ways (e.g., 2 + 1, 2 + 2, and 2 + 4 cycloadditions, ene-type reactions), charge transfer interactions, and triplet energy transfer.

Nonconjugated Olefins

Singlet oxygen attacks as an electrophilic reagent. The electron demand is given by the Hammett reaction constant ρ, which, for α,α,β-trimethylstyrene, has a negative value (-0.92).[4] A lower ionization energy increases the reactivity to $^1O_2$, which is often brought about by an increasing number of alkyl groups, whereas the opposite effect is shown by sterical hindering. With increasing sterical hindering the quenching efficiency decreases. This influence can be seen comparing the quenching rates k_q of the olefins 1–3 (Fig. 1).[2]

It has been shown that in the reaction of 1 with $^1O_2$ there is a rapid, reversible exciplex formation in the preequilibrium situation.[5] Enamines,

[1] A. A. Gorman and M. A. J. Rogers, in "Handbook of Organic Photochemistry" (J. C. Scaiano, ed.), p. 229. CRC Press, Boca Raton, FL, 1989.

[2] B. M. Monroe, in "Singlet O2" (A. A. Frimer, ed.), Vol. I, p. 177. CRC Press, Boca Raton, FL, 1985.

[3] R. V. Bensasson, E. J. Land, and T. G. Truscott, "Excited States and Free Radicals in Biology and Medicine." Oxford Univ. Press, Oxford, 1993.

[4] C. S. Foote and R. W. Denny, J. Am. Chem. Soc. 93, 5162 (1971).

[5] A. A. Gorman, I. Hamblett, C. Lambert, B. Spencer, and M. C. Standen, J. Am. Chem. Soc. 110, 8053 (1988).

as electron-rich olefins, can produce unstable dioxetanes that decompose to give carbonyl triplets.[6]

Conjugated Olefins and Aromatic or Heteroaromatic Compounds

Conjugated, cyclic dienes form 1,4-endoperoxides. However, allylic hydroperoxides, 1,2-dioxetane-derived products, endoperoxides, or mixtures arise from acyclic dienes. A recent example is the new colorless (transparency in the near UV) trap, 1,3-cyclohexadiene-1,4-diethanoate **4**, for the measurement of photogenerated $^1O_2$ in aqueous solution (Fig. 2).[7] There are some advantages compared with similar traps as **5**, **6**, **7a**, **8a**, **9**, and other "dienes."[7] The cyclohexadiene **4** interacts with $^1O_2$ via pure chemical quenching with a rate constant $k_q = 2.6 \times 10^7 \, M^{-1} \, \text{sec}^{-1}$.

In polymeric material, $^1O_2$ may play a role in the photodegradation of the macromolecule or of additives dissolved in the matrix. Aromatic compounds are also used as additives to remove $^1O_2$. Their quenching abilities in amorphous polymer glasses were determined.[8] For very efficient quenchers, k_q approaches k_{diff}. At this limit, a change from a liquid solvent to a polymer glass, with the concomitant reduction in diffusion coefficients, should result in a decrease in k_q, e.g., **7c** with $k_q = 1.3 \times 10^8 \, M^{-1} \, \text{sec}^{-1}$ in liquid toluene and $k_q = 2.4 \times 10^7 \, M^{-1} \, \text{sec}^{-1}$ in polystyrene glass.[8] For poor quenchers, the same change results in an increase in k_q because "cage escape" channels are more pronounced in a liquid solvent.

A new class of $^1O_2$ quenchers are amino acid-pyrrole N-conjugates synthesized from tyrosine, histidine, and glutathione, e.g., **11** and **12** (Fig. 3). They very effectively quench $^1O_2$ at rates ranging from 10^8 to $10^9 \, M^{-1} \, \text{sec}^{-1}$ in methanol. Their quenching ability compares favorably with that of natural antioxidants such as vitamins E and C.[9] Another useful factor consists in their rather good solubility in a variety of solvents such as water and ethanol. These antioxidants could provide protection against tissue damage caused by oxidative stress.

The rate constants k_q ($10^8 \, M^{-1} \, \text{sec}^{-1}$) for other aromatic heterocycles are similar or lower: furan, 0.11 (toluene)[10]; indole, < 0.01 (toluene)[10]; and imidazole, 0.40 (water).[11] The reaction mechanism shown in Fig. 4 accounts

[6] T. Schulte-Herbrüggen and E. Cadenas, *Photobiochem, Photobiophys.* **10**, 35 (1985).

[7] V. Nardello, N. Azaroual, I. Cervoise, G. Vermeersch, and J. M. Aubry, *Tetrahedron* **52**, 2031 (1996).

[8] P. R. Ogilby, M. Kristiansen, D. O. Martire, R. D. Scurlock, V. L. Taylor, and R. L. Clough, *Adv. Chem. Ser.* **249**, 113 (1996).

[9] J. M. Gaullier, M. Bazin, A. Valla, M. Giraud, and R. Santus, *J. Photochem. Photobiol. B Biol.* **30**, 195 (1995).

[10] A. A. Gorman, G. Lovering, and M. A. J. Rogers, *J. Am. Chem. Soc.* **101**, 3050 (1979).

[11] I. Kraljic and V. A. Sharpatyi, *Photochem. Photobiol.* **28**, 583 (1978).

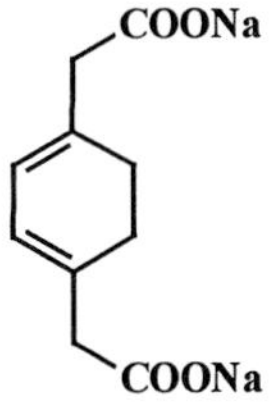

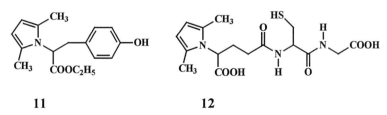

4

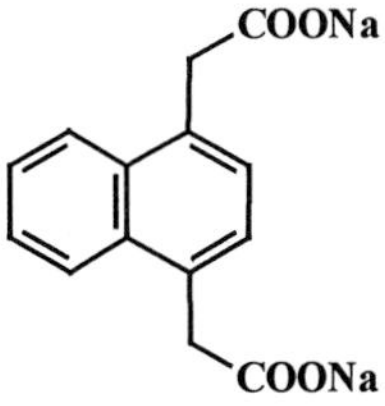

5

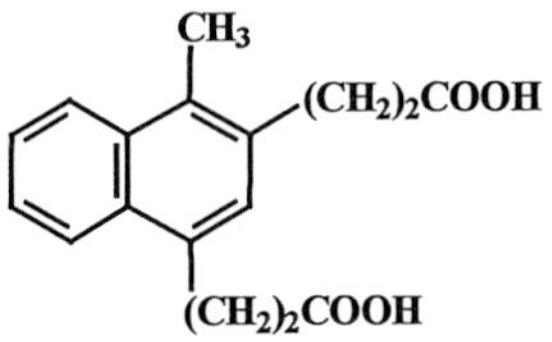

6

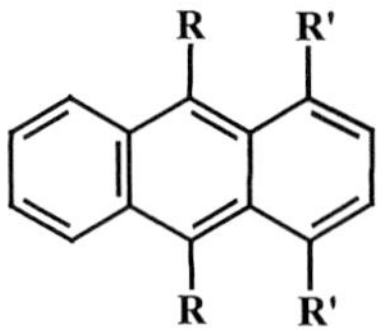

7

a: R = (CH$_2$)COONa, R' = H
b: R = Ph, R' = H
c: R = Ph, R' = OCH$_3$

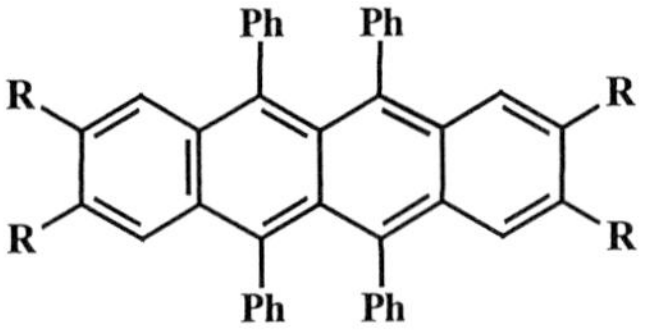

8

a: R = COOK
b: R = H

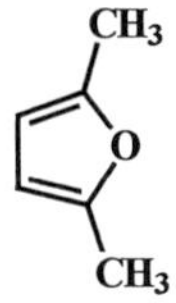

9 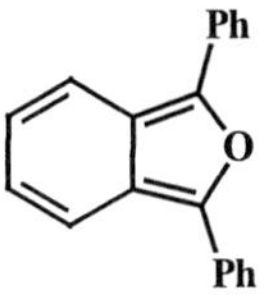   **10**

Fig. 2. Structures of endoperoxide-forming compounds.

11   **12**

Fig. 3. Structures of amino acid-pyrrole *N*-conjugates used as $^1O_2$ quenchers.

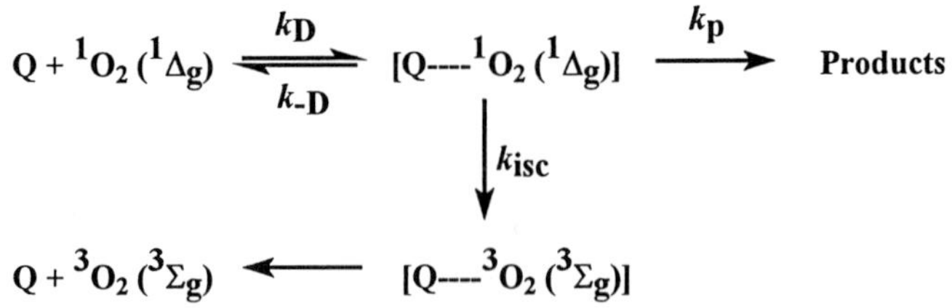

$$Q + {}^1O_2\,({}^1\Delta_g) \;\underset{k_{-D}}{\overset{k_D}{\rightleftharpoons}}\; [Q\text{----}{}^1O_2\,({}^1\Delta_g)] \;\xrightarrow{k_p}\; \text{Products}$$

$$\downarrow k_{isc}$$

$$Q + {}^3O_2\,({}^3\Sigma_g) \;\longleftarrow\; [Q\text{----}{}^3O_2\,({}^3\Sigma_g)]$$

$$k_q = \frac{k_D\,(k_{isc} + k_p)}{k_{-D} + (k_{isc} + k_p)}$$

Fast quenchers:
$$k_{-D} \ll (k_{isc} + k_p) \Rightarrow k_q \approx k_D \qquad\qquad \textbf{(diffusion limit)}$$

Slow quenchers:
$$k_{-D} \gg (k_{isc} + k_p) \Rightarrow k_q \approx \frac{k_D\,(k_{isc} + k_p)}{k_{-D}} \quad \textbf{(preequilibrium limit)}$$

FIG. 4. Mechanism and kinetics for the reactions of ${}^1O_2$ with a quencher Q.

for many experimental facts, especially for the temperature dependence of k_q.[5] For fast quenchers, a diffusion-limited reaction is obtained and for slow quenchers a preequilibrium limit holds.

Comparison of the temperature dependence of the deactivation of ${}^1O_2$ by the quenchers 1,3-diphenylisobenzofuran **10** (Fig. 2) and β-carotene **13b** (Fig. 5) is instructive.[5] An Arrhenius plot for the quenching by bicycle **10** shows two linear regions. At -30 to $-40°$ the plot undergoes inversion from the diffusion limit to the preequilibrium limit. The slope of the linear region at low temperature is identical to that of **13b**, indicating the diffusion control.

Polyenes and carotenoids are the best known among the compounds that quench ${}^1O_2$ by efficient energy transfer[2,3]: ${}^1O_2 + {}^1Q \rightarrow {}^3O_2 + {}^3Q$. A large number of modified, synthetic analogs and derivatives have been synthesized to prepare even better quenchers than the natural carotenoids β-carotene **13b**, canthaxanthin **14b**, or astaxanthin **15b** (Fig. 5).[12] Although in these cases the triplet energy of the polyene is the determinant, color can give an indication of the quenching efficiency: deeply colored, magenta,

[12] D. Baltschun, S. Beutner, K. Briviba, H. D. Martin, J. Paust, M. Peters, S. Röver, H. Sies, W. Stahl, A. Steigel, and F. Stenhorst, *Liebigs Ann./Recueil* 1887 (1997).

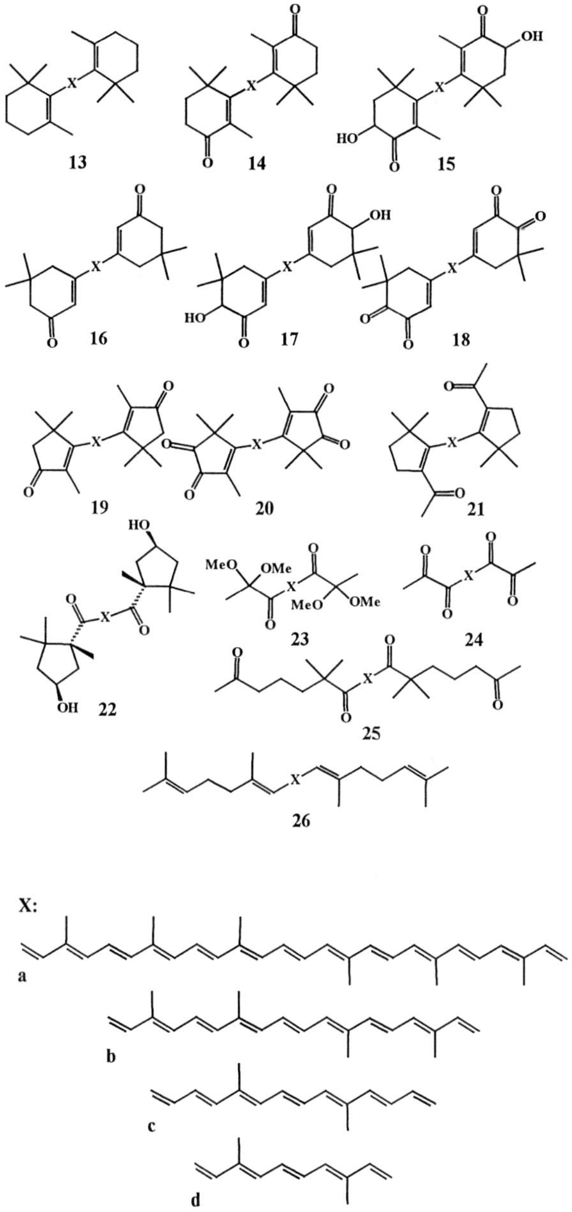

FIG. 5. Structures of polyenes and carotenoids used as $^1O_2$ quenchers.

TABLE I

SECOND-ORDER RATE CONSTANTS k_q FOR QUENCHING OF $^1O_2$ IN EtOH/CHCl$_3$/D$_2$O (50:50:1) AND $\pi\pi^*$ ABSORPTION DATA IN CHCl$_3$ FOR POLYENES AND CAROTENOIDS AT 37°

Compound	k_q (10^9 M^{-1} sec^{-1})	log (k_q)	λ_{max} (nm)	E(S) (10^3 cm^{-1})
16a	13.8	10.1	550	18.18
13a	13.3	10.1	516	19.38
15a	13.0	10.1	528	18.94
19b	12.7	10.1	505	19.80
18b	12.7	10.1	552	18.12
17b	12.6	10.1	519	19.27
17a	12.3	10.1	554	18.05
20b	12.1	10.1	584	17.12
16b	12.0	10.1	512	19.53
24b	11.7	10.1	532	18.80
23b	11.1	10.0	499	20.04
21b	11.1	10.0	505	19.80
14b	10.2	10.0	486	20.57
24c	10.2	10.0	488	20.49
15b	9.0	10.0	488	20.49
22b	9.0	10.0	491	20.37
26b	8.8	9.9	484	20.66
13b	8.4	9.9	460	21.74
25b	8.4	9.9	487	20.53
23c	3.0	9.5	447	22.37
22c	1.6	9.2	449	22.27
24d	0.1	8.0	442	22.62

purple, and blue polyenes (Table I) turn out to be excellent quenchers at the diffusion limit, e.g., **20b**, **16a**, **18b**, and **17a**.

Amines, Sulfur Compounds, and Phenols

The reduction potential for the couple $[^1O_2(^1\Delta_g)/O_2^-]$ is E $=$ +0.647 V (NHE). Singlet oxygen is therefore a better electron acceptor than ground state oxygen.[3] A charge transfer mechanism was suggested for the quenching of $^1O_2$ by amines and sulfides.[13] A logarithmic dependence of $^1O_2$ quenching rate constants on the ionization energy of aliphatic amines is known. Quenching in acetonitrile and benzene occurs via reversible charge transfer with the formation of an exciplex with a partial electron transfer (0.2

---

[13] R. H. Young and R. L. Martin, *J. Am. Chem. Soc.* **94,** 5183 (1972).

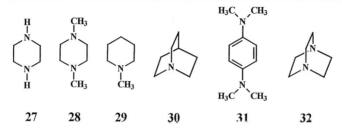

27   28   29   30   31   32

FIG. 6. Structures of amines used as $^1O_2$ quenchers.

esu).[14] The amines **27–32** have been determined in solvents with different polarity in dependence on pressure up to 1200 or 1600 bar (Fig. 6).[15]

The results are consistent with the assumption that quenching by amines occurs via polar exciplexes with singlet and triplet multiplicity. Only partial charge transfer takes place in the quenching process. TMPD **31**, $k_q = 1.1 \times 10^9 \ M^{-1} \ sec^{-1}$ in acetonitrile, is an exceptional case, in which full electron transfer (i.e., formation of superoxide and TMPD$^+$) was observed in D$_2$O.[14]

DABCO **32** quenches in polystyrene glasses with $k_q = 9 \times 10^7 \ M^{-1} \ sec^{-1}$.[8] The temperature dependence of k_q for quenching of $^1O_2$ in toluene by DABCO shows that at temperatures above $-20°$ inversion to a preequilibrium limited reaction takes place.[5]

Thioethers and thioketones react with $^1O_2$ to produce, depending on the structure and reaction conditions, sulfoxides, sulfones, and ketones, respectively, as well as other products (Fig. 7).[2,16,17,18]

Disulfides yield thiol sulfinates and thiol sulfonates. In addition, disulfides have the unique capability of deactivating $^1O_2$ by two different physical quenching mechanisms (Fig. 7).[19] The total quenching may be described by the reactions shown in Fig. 8. An estimate suggests the following proportions: chemical quenching (B) 4.0%, physical quenching (A) 0.2%, and physical quenching (via persulfoxide, B) 95.8%. The results demonstrate that disulfides in extended conformations (dihedral angle $>$RSSR $\approx 90°$) are ineffective as biological $^1O_2$ quenchers. Exposed disulfides ($>$RSSR $\leq 30°$), however, are physical quenchers with considerable rate constants.[19]

Time-resolved studies of $^1O_2$ quenching by phenols in aqueous solution

[14] A. P. Darmanyan, W. S. Jenks, and P. Jardon, *J. Phys. Chem.* **102**, 7420 (1998).
[15] M. Hild and H. D. Bauer, *Ber. Bunsenges. Phys. Chem.* **100**, 1210 (1996).
[16] B. M. Monroe, *Photochem. Photobiol.* **29**, 761 (1979).
[17] C. S. Foote and M. L. Kacher, *Photochem. Photobiol.* **29**, 765 (1979).
[18] T. Y. Ching, C. S. Foote, and G. G. Geller, *Photochem. Photobiol.* **20**, 511 (1974).
[19] E. L. Clennan, D. Wang, C. Clifton, and M. F. Chen, *J. Am. Chem. Soc.* **119**, 9081 (1997).

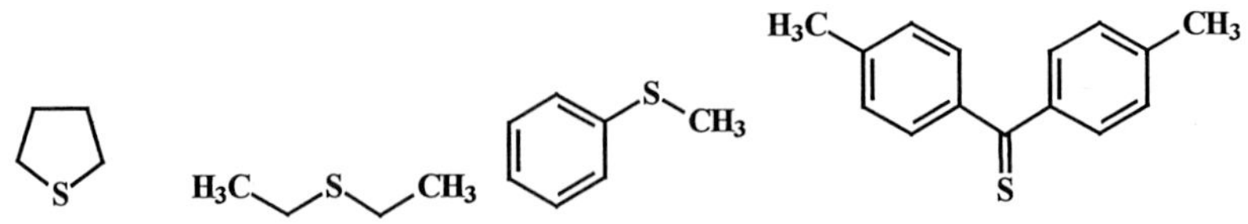

33 (4.2; $CHCl_3$) **34** (1.7; CH_3OH) **35** (0.2; CH_3OH) **36** (0.4; $CHCl_3$)

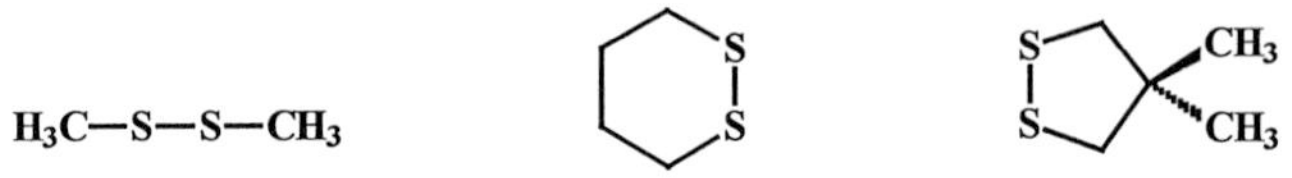

37 (0.042; acetone) **38** (0.014; acetone) **39** (14.400; acetone)

FIG. 7. Structures and $^1O_2$ quenching rates k_q (10^7 M^{-1} sec^{-1}; solvent) of sulfides and disulfides.

at different pH values show that the reaction occurs via a charge transfer from the phenolate PhO^-.[3,20] Phenols are also excellent chain-breaking antioxidants (e.g., α-tocopherol **40**, trolox c **41**, dihydrobenzofuran **42**)[21,22] and good $^1O_2$ quenchers (Fig. 9).[1,23] The protecting effect of phenols on the photofading of cyanine dyes has been described previously.[24]

Metal Chelates

Square-planar, diamagnetic nickel(II) complexes are very efficient quenchers of $^1O_2$, whereas paramagnetic complexes are somewhat less efficient quenchers.[2] They are used commercially as photostabilizers in polymers and against the fading of dyes by oxidation processes (Fig. 10, **46–49**).

In contrast to other $^1O_2$ quenchers, such as carotenoids or aromatic amines, many nickel chelates possess an extraordinarily high UV stability and a remarkably good thermal permanence. Also, they exhibit significant $^1O_2$ quenching ability even in the solid state.[25]

[20] R. Scurlock, M. Rougee, and R. V. Bensasson, *Free Radic. Res. Commun.* **8,** 251 (1990).
[21] G. W. Burton, L. Hughes, and K. U. Ingold, *J. Am. Chem. Soc.* **105,** 5950 (1983).
[22] M. J. Thomas and B. H. J. Bielski, *J. Am. Chem. Soc.* **111,** 3315 (1989).
[23] C. S. Foote and M. J. Thomas, *Photochem. Photobiol.* **27,** 683 (1978).
[24] Z. Wanxue, C. Ping, Z. Deshui, T. Okazaki, and M. Hayami, *Sci. China B* **38,** 1298 (1995).
[25] D. J. Carlsson, T. Suprunchuk, and D. M. Wiles, *Can. J. Chem.* **52,** 3728 (1974).

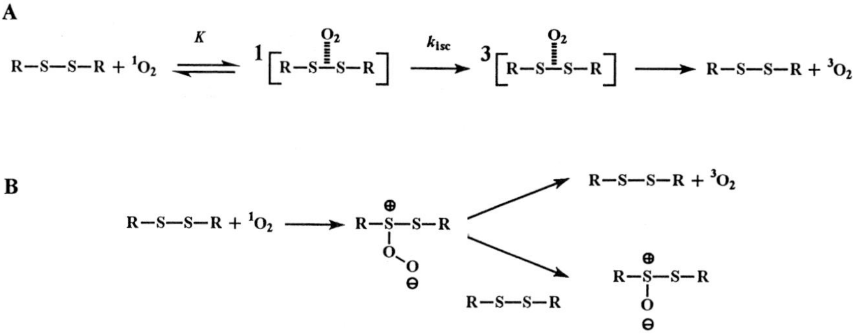

A

$$R-S-S-R + {}^1O_2 \underset{K}{\rightleftharpoons} {}^1\left[R-S\overset{\overset{O_2}{\|}}{-}S-R\right] \xrightarrow{k_{isc}} {}^3\left[R-S\overset{\overset{O_2}{\|}}{-}S-R\right] \longrightarrow R-S-S-R + {}^3O_2$$

B

$$R-S-S-R + {}^1O_2 \longrightarrow \underset{\underset{\underset{\ominus}{O}}{O}}{\overset{\oplus}{R-S-S-R}}$$

with products:

R—S—S—R + $^3O_2$

$\underset{\underset{\ominus}{O}}{\overset{\oplus}{R-S-S-R}}$

R—S—S—R

FIG. 8. Reactions for the quenching of $^1O_2$ by dialkyl disulfides.

Singlet oxygen quenchers have also become important in the optical data storage field. The dithiolene nickel complexes **49**, for example, have been applied to optical DRAW (Direct Reading After Writing) disks as inhibitors of both laser- and sunlight-induced fading of cyanine dyes contained in the recording layer. Generally, dithiolene nickel complexes are

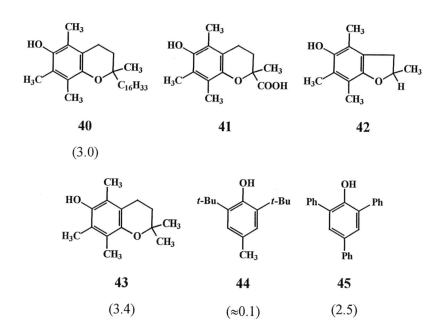

40

(3.0)

41

42

43

(3.4)

44

($\approx$0.1)

45

(2.5)

FIG. 9. Structures and $^1O_2$ quenching rates ($10^8\ M^{-1}\ sec^{-1}$) in methanol of selected phenolic compounds.

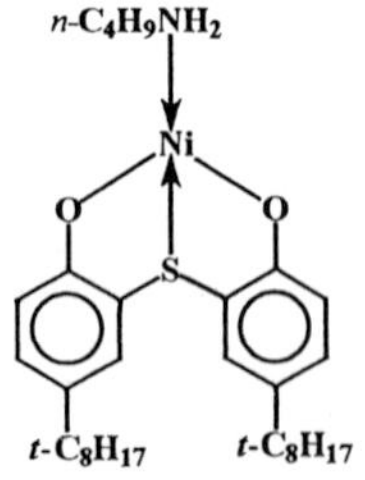

46

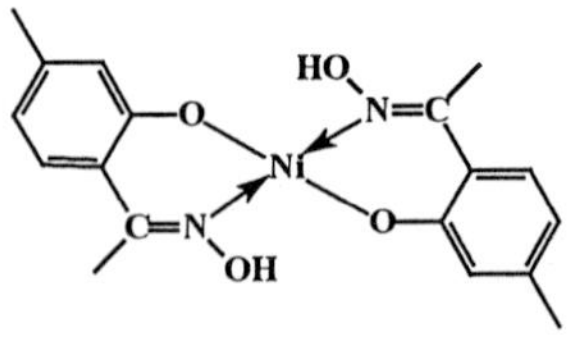

47

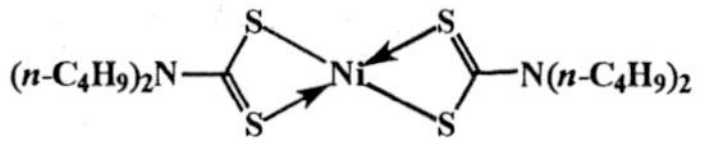

48

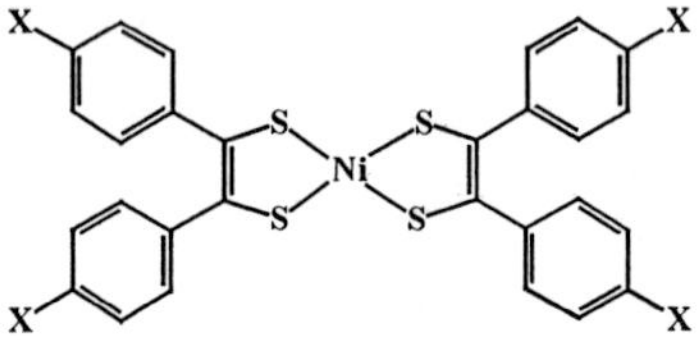

49

a: X = H
b: X = OCH_3
c: X = CF_3
d: X = Cl

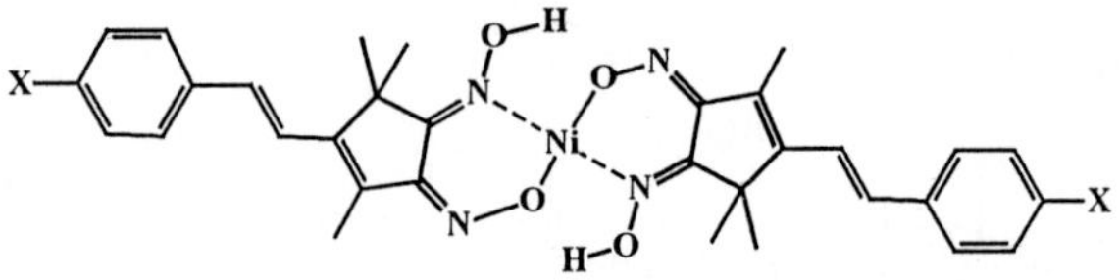

50

a: X = OCH_3
b: X = NO_2
c: X = $N(CH_3)_2$

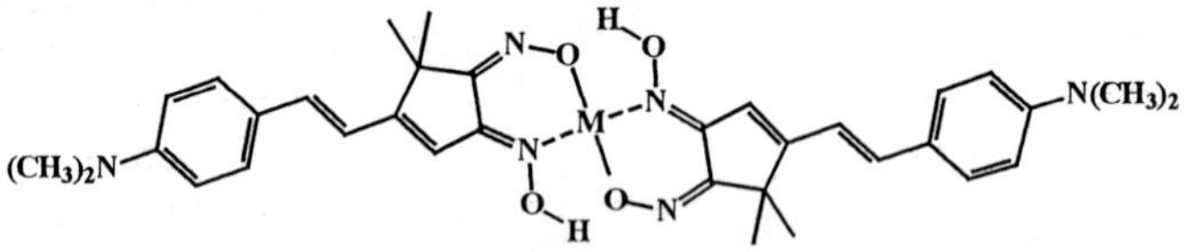

51

a: M = Ni
b: M = Pd
c: free ligand

FIG. 10. Structures of metal chelate $^1O_2$ quenchers.

TABLE II
QUENCHING CONSTANTS FOR TYPICAL METAL CHELATE $^1O_2$ QUENCHERS

Compound	k_q (10^9 M^{-1} sec^{-1})	Solvent	Ref.
46	0.17	Chloroform	28
47	3.00	Benzene	29
48	8.10	Chloroform	28
49a	12.00	Acetonitrile/benzene (4:1)	30
49b	10.10	Acetonitrile/benzene (4:1)	30
49c	13.50	Acetonitrile/benzene (4:1)	30
49d	11.70	Acetonitrile/benzene (4:1)	30
50a	1.00	Chloroform	27
50b	1.08	Chloroform	27
50c	0.84	Chloroform	27
51a	1.05	Chloroform	27
51b	0.02	Chloroform	27
51c	0.01	Chloroform	27

added as antioxidants (10–15, wt %) or their reduced form is used as a counteranion of cyanine dyes in DRAW disks.[26]

The quenching efficiency of several nickel-dioximato chelates **50** and **51a**, an analogous Pd-complex **51b**, and a free ligand **51c** have been investigated, all of them deriving from the end group of the blue carotenoid violerythrin **20b**[27] (Table II).[28–30]

Although a very large quantity of nickel complexes has been studied, their mechanism of $^1O_2$ quenching is yet unknown. However, quenching seems to be a physical process as no indication of a chemical reaction between nickel chelates and $^1O_2$ has been found. In fact, exposure of chelate solutions to $^1O_2$ merely leads to an insignificant change in their UV spectra.[27,31,32]

The high quenching rates (cf. Table II), which for some sulfur-coordinated nickel chelates reach the limit of diffusional control with a k_q of 1.1×10^{10} M^{-1} sec^{-1} in chloroform, indicate an electronic energy transfer

[26] J. Fabian, M. Matsuoka, and H. Nakazumi, *Chem. Rev.* **92,** 1197 (1992).

[27] B. Bloedorn, Diplomarbeit Universitaet Duesseldorf, 1999.

[28] B. M. Monroe and J. J. Mrowea, *J. Phys. Chem.* **83,** 591 (1979).

[29] A. Farmilo and F. Wilkinson, *Photochem. Photobiol.* **18,** 447 (1973).

[30] T. Kitao, H. Nakazumi, H. Shiozaki, and Y. Takamura, *Bull. Chem. Soc. Jpn.* **63,** 2653 (1990).

[31] J. Flood, K. E. Russell, and J. K. S. Wan, *Macromolecules* **6,** 669 (1973).

[32] D. J. Carlsson, G. D. Mendenhall, T. Suprunchuk, and D. M. Wiles, *J. Am. Chem. Soc.* **94,** 8960 (1972).

mechanism, as this type of reaction seems to be somewhat more efficient than the charge-transfer mechanism with a k_q often lower than $10^9 M^{-1}$ sec^{-1}.[30]

Studies on substituted nickel chelates gave very small positive Hammett ρ values: 0.14 for k_q of **49**[30] and 0.06 for k_q of **51**.[27] Thus the formation of a charge-transfer complex with $^1O_2$ being the acceptor is not likely.

Concerning the relationship between the ligand structure as well as the nature of metal and the $^1O_2$ quenching ability, several conclusions can be drawn. Because free ligands most often fail to quench $^1O_2$, the quenching efficiency seems to depend on the complex as a whole, cf. **51a** and **51c**.[32] However, k_q is influenced extensively by the nature of the central atom. So while keeping the ligand constant and changing the central metal atom, the quenching efficiency decreases in the order of Ni > Co ≫ Pd, cf. **51a** and **51b**.[27,29,33]

The magnetism and geometry of the chelate appear to be very important factors for $^1O_2$ quenching efficiency. Diamagnetic nickel chelates are excellent $^1O_2$ quenchers with quenching rates varying from diffusional-controlled ($1.1 \times 10^{10} M^{-1}$ sec^{-1}) to a tenth of this value. Paramagnetic nickel complexes, however, give $^1O_2$ quenching rates of 100 to 1000 times smaller, although the unpaired electrons should support intersystem crossing.[28]

The correlation between k_q and the magnetic properties does not necessarily mean that magnetism is the k_q-influencing measure, rather the geometry of the complex seems to be the rate-constant controlling factor. In diamagnetic nickel complexes the coordinating atoms are arranged in a square planar geometry so that $^1O_2$ has a practically unhindered approach from above or below this plane to the nickel atom and can interact easily with an orbital perpendicular to this plane. In the case of paramagnetic nickel complexes, this approach of the nickel atom is blocked by the octahedral or tetrahedral arrangement of the ligands.[28]

Remember that in other classes of $^1O_2$ quenchers, such as olefins,[34] amines,[35] and sulfides,[16] the quenching rate constant is also very sensitive to steric effects.

The relative position of a diamagnetic nickel chelate within this group depends on the nature of the ligands' coordinating atoms. The quenching rate constant k_q decreases in the order $S_4 > N_4 > N_2O_2$, which would indicate that the rate constant increases as the polarizability of the ligands

[33] W. A. Henderson and A. Zweig, *J. Polym. Sci. Polym. Chem. Ed.* **13**, 717 (1975).
[34] S. H. Liu, B. M. Monroe, and F. G. Moses, *Mol. Photochem.* **1**, 245 (1969).
[35] B. M. Monroe, *J. Phys. Chem.* **82**, 15 (1978).

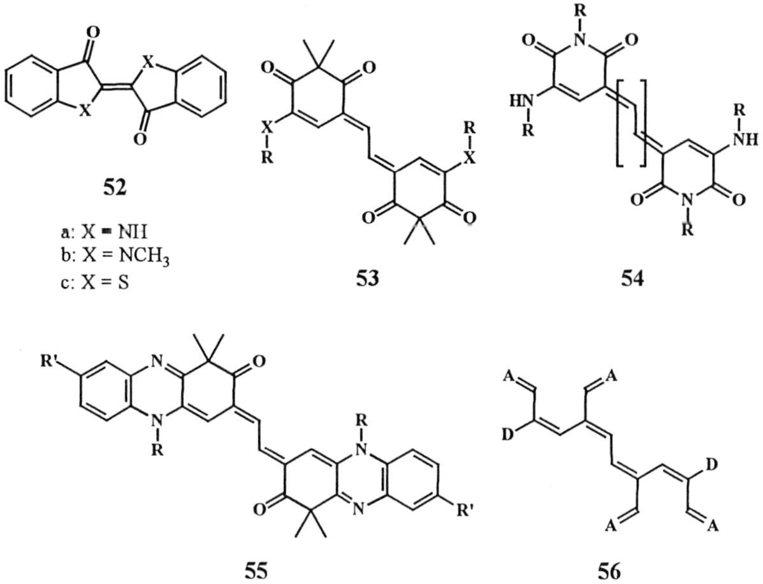

FIG. 11. Structures of indigoid $^1O_2$ quenchers.

increases.[28] The triplet energies decrease in the same order. Diamagnetic complexes with sulfur donor ligands give the highest k_q and show very low-lying triplet energies. This observation corresponds to an energy transfer mechanism.

Nonaromatic Heterocycles and Indigoids

The nature of the heteroatom and the chemical structure influence strongly the quenching ability. Many important molecules have been in vestigated: chlorophyll, amino acids, bilirubins, and pyrimidine fungicides.[2]

Many indigoid dyes were studied with the remarkable result that most but not all members of this class of chromophores quench $^1O_2$ at the diffusion limit. Indigoids are special chromophores, called H-chromophores, which contain an efficient combination of two merocyanine moieties. The parent compound, although not the smallest member of this family, is Indigo 52a. More elaborate examples are 53, 54, and 55 (Fig. 11).[36,37]

[36] I. A. Hernandez Blanco, Diplomarbeit Universitaet Duesseldorf, 1998.
[37] Th. Hoffmann, Diplomarbeit Universitaet Duesseldorf, 1998.

The indigoidine or "bacterial indigo" chromophore **54**, n = O, R = H, is found in *Pseudomonas indigofera*. The chromophores **53–55** may be reduced to the structure **56**, which indicates the special nature of these indigoid π systems.

Because Indigo **52a** is a sparingly soluble compound with a high tendency to form aggregates, the determination of quenching constants is subject to considerable experimental error; k_q is likely to be higher than 1.6×10^8 M^{-1} sec^{-1} in CHCl$_3$.[37] These problems can be avoided by measuring the readily soluble derivative **52b**: $k_q = 7.9 \times 10^9 \, M^{-1}$ sec^{-1} in CHCl$_3$.[37] Thioindigo **52c**, however, is a very inefficient quencher with $k_q < 10^6 \, M^{-1}$ sec^{-1}, which is similar to the low efficiencies of other sulfur compounds (vide supra).

The indigoids **53–55** seem to follow regularities similar to carotenoids: the deeper the color and the more bathochromic the absorption wavelength, the more efficient is the quenching ability, up to the diffusion limit (Table III). This might indicate an energy transfer mechanism.

TABLE III

SECOND-ORDER RATE CONSTANTS k_q FOR QUENCHING OF $^1O_2$ IN CHCl$_3$ AT 37° AND $\pi\pi^*$ ABSORPTION DATA IN CHCl$_3$ FOR INDIGOID COMPOUNDS

Compound	k_q ($10^9 \, M^{-1}$ sec^{-1})	log (k_q)	λ_{max} (nm)	E(S) (10^3 cm^{-1})
54e, n = 0, R = *n*-Bu	10.77	10.0	633	15.80
55b, R' = H, R = Ph	10.77	10.0	648	15.43
53e, X = NH, R = Ph	9.74	10.0	562	17.79
54a, n = 1, R = *n*-Bu	9.74	10.0	596	16.78
54c, n = 0, R = *n*-Pr	9.74	10.0	631	15.85
54d, n = 0, R = *i*-Bu	9.74	10.0	633	15.80
55a, R' = H, R = Me	9.74	10.0	642	15.58
55c, R' = H, R = *p*-MeO-C$_6$H$_4$	8.82	9.9	650	15.39
55e, R' = OMe, R = Et	8.82	9.9	682	14.66
53f, X = NH, R = *n*-Bu	7.98	9.9	538	18.59
55d, R' = *n*-Bu, R = Et	7.98	9.9	664	15.06
53g, X = NH, R = *m*-CF$_3$-C$_6$H$_4$	7.22	9.9	546	18.32
54b, n = 1, R = *i*-Bu	7.22	9.9	599	16.69
53a, X = O, R = H	0.29	8.5	439	22.78
53d, X = O, R = Me	0.01	7.0	431	23.20
53b, X = O, R = Ac	<0.01	<6.5	387	25.84
53c, X = O, R = Bz	<0.01	<6.5	390	25.64

Acknowledgments

The pleasant and fruitful collaboration with Professor Sies, University of Duesseldorf, is very much appreciated. Support of our own work by the Fonds der Chemischen Industrie, the Deutsche Forschungsgemeinschaft, Bayer AG, BASF AG, and Henkel KGaA is gratefully acknowledged.

Section II

Ultraviolet A

A. Dosimetry
Article 23

B. Biological Responses: Signaling
Articles 24 through 32

C. Biological Responses: Protocarcinogenesis, Photoaging, and Photoallergic Reactions
Articles 33 through 36

D. Oxidative Damage Markers
Articles 37 through 40

E. Protection
Articles 41 through 44

[23] Dosimetry of Ultraviolet A Radiation

By Brian L. Diffey

Dosimetry is the science of radiation measurement. There are two principal reasons why ultraviolet A (UV-A) radiation should be measured: to allow consistent exposure of cells and tissues over many months and years within a local laboratory, and to allow the results of irradiations made in different laboratories to be published and compared.

It is important to distinguish between these two objectives. The first requires *precision,* or reproducibility. The dosimeter is used as a monitor to give a reference measurement and so it needs to be stable. *Accuracy,* i.e., absolute calibration against some accepted standard, is not essential. The second objective requires both precision and accuracy. Here the dosimeter must not only be stable from one day to the next, but also the display (in, say, milliwatts per square centimetre) must be traceable to absolute standards. Although electrooptical technology has improved over the years, resulting in the availability of versatile and precise ultraviolet measurement equipment, these improvements have not always been accompanied by improved accuracy due to misunderstandings about calibration.

Radiometric Terms and Units

In photobiological dosimetry it is customary to use the terminology of radiometry rather than that of photometry, as photometry is based on visible light measurements that simulate the human eye's photopic response curve and, strictly speaking, a source that emits only UV-A has zero intensity in photometric terms.[1]

Common radiometric terminology is given in Table I. Terms relating to a beam of radiation passing through space are the "radiant energy" and "radiant flux." Terms relating to a source of radiation are the "radiation intensity" and "radiance." The term "irradiance," which is the most commonly used term in photobiology, relates to the object (e.g., target molecules) struck by the radiation. The time integral of the irradiance is strictly termed the "radiant exposure," but is sometimes expressed as "exposure dose" or even more loosely as "dose."

---

[1] A. W. S. Tarrant, *in* "Radiation Measurement in Photobiology" (B. L. Diffey, ed.), p. 1. Academic Press, London, 1989.

TABLE I
RADIOMETRIC TERMS AND UNITS

Term	Unit	Symbol
Wavelength	nm	λ
Radiant energy	J	Q
Radiant flux	W	Φ
Radiant intensity	Wsr^{-1}	I
Radiance	$Wm^{-2}\ sr^{-1}$	L
Irradiance	Wm^{-2}	E
Radiant exposure	Jm^{-2}	H

Radiometric Calculations

Time, Dose, and Irradiance. The most frequent radiometric calculation is to determine the time to irradiate a sample in order to achieve a certain dose (in J/m^2) when the dosimeter indicates an irradiance in W/m^2. The relationship among these three quantities (time, dose, and irradiance) is simply:

$$\text{exposure time (seconds)} = \frac{\text{dose (J/m}^2\text{)}}{\text{measured irradiance (W/m}^2\text{)}} \tag{1}$$

Inverse Square Law and Cosine Laws of Irradiation. The irradiance E on a surface of area dA subtending a solid angle $d\Omega$ at a point source of radiant intensity I is

$$E = I\frac{d\Omega}{dA} \tag{2}$$

From geometrical considerations $d\Omega = dA'/d^2$, where d is the distance from the source to the surface and dA' is the projection of the area dA on the plane perpendicular to the direction of the source, given by

$$dA' = dA \cos\theta \tag{3}$$

where θ is the angle between the direction of the source and the perpendicular to the surface. Eliminating $d\Omega$ and dA from Eq. (2) leads to

$$E = I \cos\theta/d^2 \tag{4}$$

Equation (4) is the symbolic expression of the two fundamental laws of radiometry and may be formally stated as follows. The irradiance of an elementary surface due to a point source of radiation is proportional to the radiant intensity of the source in the direction of that surface and to the cosine of the angle between this direction and the normal to the surface, and it is inversely proportional to the square of the distance between the surface and the source.

Surface Irradiation. When the surface cannot be regarded as small in comparison with its distance from the source, the irradiance varies over the surface. For the simple case of a circular disk of radius a irradiated by a uniform point source of radiant intensity I placed at a distance d from the disk along the axis of the latter, the irradiance at any point P on the disk at a distance r from the center is[2]

$$\frac{I \cos \theta}{(r^2 + d^2)} \tag{5}$$

where θ is the angle of incidence at P. Now because

$$\cos \theta = \frac{d}{(r^2 + d^2)^{1/2}} \tag{6}$$

expression (5) reduces to

$$I \cos^3 \theta / d^2 \tag{7}$$

This is sometimes referred to as the *cosine-cubed* law. It follows that the average irradiance is

$$(1/\pi a^2) \int_0^a (I \cos^3 \theta / d^2)\, 2\pi r\, dr \tag{8}$$

which equals

$$(2I/a^2) \left[1 - \frac{d}{(a^2 + d^2)^{1/2}} \right] \tag{9}$$

Because fluorescent lamps are used commonly in photobiology as sources of UV-A, the irradiance at a point distance d from the center of a lamp of breadth $2a$, length $2l$, and radiance L, is given by[2]

$$(2aL/d) \left[\tan^{-1}(l/d) + \frac{ld}{(l^2 + d^2)} \right] \tag{10}$$

Volume Irradiation. In many instances in cellular photobiology, the samples to be irradiated are in suspension in a liquid. For example, a liquid suspension of single cells may absorb some of the radiation so that cells near the back of the cuvette receive less radiation than those near the

[2] J. W. T. Walsh, "Photometry." Academic Press, New York, 1978.

front.[3,4] Even if the sample is stirred vigorously, the average irradiance per cell will be lower than the incident irradiance. For a parallel beam of monochromatic radiation larger than the cross-sectional area of the suspension perpendicular to the direction of the propagation of the beam, Morowitz[5] has shown that the average irradiance per cell, $\bar{E}$ is related to the incident irradiance E by

$$\bar{E} = E[1 - \exp(-2.3A)]/2.3A \tag{11}$$

where A is the absorbance of the sample at the irradiating wavelength. The factor 2.3 approximates $\log_e 10$ as the absorbance A is related to the fraction of incident radiation transmitted T by

$$A = -\log_{10} T \tag{12}$$

Equation (11) has been modified by Johns,[6] who considered the case of a converging beam of radiation on the cuvette. Jagger *et al.*[7] discuss the errors that can arise in the estimation of $\bar{E}$ by Eq. (11) if radiation scattering by suspensions of small cells is not taken into account.

Detection of UV-A Radiation

Techniques for the measurement of UV-A fall into three classes: physical, chemical, and biological. In general, physical devices measure power, whereas chemical and biological systems measure energy. In this article, UV-A measurement will be described by reference only to physical detectors.

Spectroradiometry

It is common practice to talk loosely of *UV-A lamps.* However, such a label does not characterize adequately ultraviolet lamps, as nearly all UV-A lamps will emit not only UV-A but often UV-B and visible light, and sometimes UV-C and infrared radiation. The only correct way to specify the nature of the emitted radiation is by reference to the spectral power distribution. This is a graph (or table) that indicates the radiated power as a function of wavelength. Data are obtained by a technique known

[3] J. Jagger, *in* "The Science of Photobiology" (K. C. Smith, ed.), p. 1. Plenum, New York, 1977.
[4] W. Harm, "Biological Effects of Ultraviolet Radiation." Cambridge Univ. Press, Cambridge, 1980.
[5] H. J. Morowitz, *Science* **111**, 229 (1950).
[6] H. E. Johns, *Photochem. Photobiol.* **8**, 547 (1968).
[7] J. Jagger, T. Fossum, and S. McCaul, *Photochem. Photobiol.* **21**, 379 (1975).

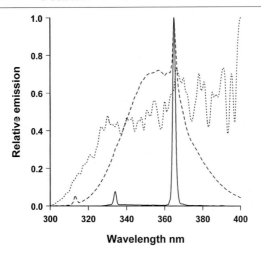

Fig. 1. Spectral power distribution of ultraviolet radiation from an optically filtered medium-pressure mercury arc lamp (solid line), a UV-A fluorescent lamp (dashed line), and sunlight (dotted line).

as spectroradiometry.[8] Figure 1 shows the spectral power distribution of three different sources commonly used by researchers requiring UV-A irradiation.

Components of a Spectroradiometer

The three basic requirements of a spectrometer system are the *input optics,* designed to conduct the radiation from the source into the *monochromator,* which disperses the radiation onto a *detector.*

Input Optics. The spectral transmission characteristics of monochromators depend on the angular distribution and polarization of the incident radiation as well as the position of the beam on the entrance slit. For measurement of spectral irradiance, particularly from extended sources such as linear arrays of fluorescent lamps or daylight, direct irradiation of the entrance slit should be avoided. Two types of input optics are available to ensure that the radiation from different source configurations is depolarized and follows the same optical path through the system: the integrating sphere or the diffuser. Both of these types of input optics produce a cosine-weighted response (see later), as the radiance of the source as measured through the entrance aperture varies as the cosine of the angle of incidence.

[8] P. Gibson and B. L. Diffey, *in* "Radiation Measurement in Photobiology" (B. L. Diffey, ed.), p. 71. Academic Press, London, 1989.

Monochromator. A blazed ruled diffraction grating is normally preferred to a prism as the dispersion element in the monochromator used in a spectroradiometer, mainly because of better stray radiation characteristics. High-performance spectroradiometers, used for determining low UV spectral irradiances in the presence of high irradiances at longer wavelengths, demand extremely low stray radiation levels. Such systems may incorporate a double monochromator, i.e., two single ruled grating monochromators in tandem.

Detector. Photomultiplier tubes, incorporating a photocathode with an appropriate spectral response, are normally the detectors of choice in spectroradiometers. However, if radiation intensity is not a problem, solid-state photodiodes may be used, as they require simpler and less expensive electronic circuitry.

Calibration of Spectroradiometers

It is important that spectroradiometers are calibrated over the wavelength range of interest using standard lamps.[9] A tungsten filament lamp operating at a color temperature of about 3000 K can be used as a standard lamp for the spectral interval 250–2500 nm, although workers concerned solely with the ultraviolet region (200–400 nm) may prefer to use a deuterium lamp.

Sources of Error in Spectroradiometry

Accurate spectroradiometry requires careful attention to detail. Factors that can affect accuracy include wavelength calibration, bandwidth, stray radiation, polarization, angular dependence, linearity, and calibration sources.[10]

Commercial Spectroradiometers

Modern spectroradiometers (e.g., Model OL754; Optonic Laboratories, Inc., Orlando, FL) incorporate a number of features, such as automated computer control of data collection and display; wavelength accuracies of typically ±0.2 nm over the spectral range 200 to 1600 nm; low stray light rejection level of 1×10^{-8} at 285 nm using a double holographic grating monochromator in combination with an order blocking filter wheel; high sensitivty and wide dynamic range; and user selectable bandwidths.

The just-described type of spectroradiometer operates by stepwise scan-

[9] T. M. Goodman, *in* "Radiation Measurement in Photobiology" (B. L. Diffey, ed.), p. 47. Academic Press, London, 1989.
[10] J. R. Moore, *Light. Res. Tech.* **12,** 213, (1980).

ning through the required wavelength range at scan speeds of 0.1 to 2 nm per second. By using a diode array as the detector in conjunction with a single grating spectrograph, instantaneous spectral power distributions can be obtained in much more compact and portable systems (e.g., Solatell; 4D Controls Ltd., Redruth, Cornwall, UK). What such a device gains in speed, cost, and portability, it loses in performance in terms of stray light rejection, which is typically at a level of no better than 1×10^{-4} at 285 nm. This is particularly important in the spectroradiometry of solar UV-B (wavelengths less than 315 nm), but may not be problem in the spectral characterization of UV-A lamps in studies where the investigator believes the small UV-B (and possibly UV-C) component is of no biological significance.

UV-A Radiometry

Although spectroradiometry is the fundamental way to characterize the radiant emission from a UV-A source, routine measurement of radiation output is normally achieved using UV-A radiometers. UV-A radiometers generally combine a detector (such as a vacuum phototube or a solid-state photodiode[11]) with a wavelength-selective device (such as a color glass filter or interference filter) and suitable input optics [such as a quartz hemispherical diffuser or polytetrafluoroethylene (PTFE) window]. A typical commercial UV-A radiometer is shown in Fig. 2.

Spectral Sensitivity of UV-A Radiometers

In order to meet the criterion for an ideal UV-A radiometer the sensor should have a uniform spectral response from 315 to 400 nm (the UV-A wave band) with zero response outside this interval. In other words, the electrical output from the sensor should depend only on the total power within the UV-A wave band received by the sensor and not on how the power is distributed with respect to wavelength. In practice, no such sensor exists with this ideal spectral response. All UV-A radiometers that combine a photodetector with an optical filter have a nonlinear spectral sensitivity of the form shown in Fig. 3.

Angular Response of UV-A Radiometers

UV-A radiometers are often used to measure the irradiance from extended sources of radiation such as linear fluorescent lamps or the sky. In

[11] A. D. Wilson, *in* "Radiation Measurement in Photobiology" (B. L. Diffey, ed.), p. 23. Academic Press, London, 1989.

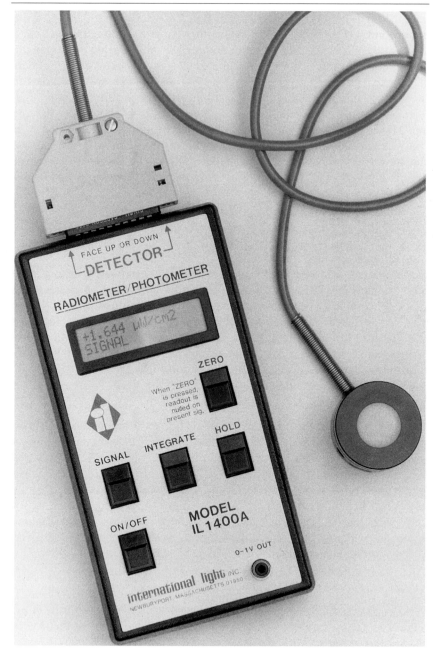

FIG. 2. A commercially available UV-A radiometer (IL1400A radiometer with SEL033/ UVA/TD detector; International Light Inc., Newburyport).

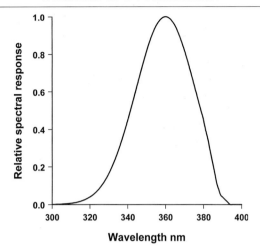

FIG. 3. Spectral sensitivity of a typical UV-A radiometer.

these instances it is important that the sensor "sees" radiation coming from all parts of the source and does not have a limited field of view, i.e., the sensor is not collimated. Furthermore, the sensor should have a cosine-weighted response,[12] which can be best explained by reference to Fig. 4.

Consider a parallel flux (Φ) of UV-A normally incident on a surface of area A. The irradiance on the surface is Φ/A. If, however, the radiation reaches the surface inclined at an angle θ to the normal, the same flux now covers a larger area equal to $A/\cos \theta$. Thus, the irradiance is $\Phi \cos \theta/A$. In other words, the irradiance has decreased by a factor of $\cos \theta$ relative to the same beam of UV-A normally incident on the surface. This means that a sensor that is measuring irradiance on a plane must weight the incident flux by the cosine of the angle between the incoming radiation and the normal to the surface; this property is called a *cosine-weighted* response.

In practice, it is very difficult to achieve a cosine-weighted response, but sensors incorporating a PTFE or quartz input optic, as illustrated in Fig. 2, can get very close and only diverge significantly from a cosine-weighted response at angles θ exceeding 70° from the normal.

Calibration of UV-A Radiometers

Because of the nonlinear spectral response of UV-A radiometers, accurate estimates of UV-A irradiance can only be made if the radiometer is

[12] M. G. Holmes, *in* "Radiation Measurement in Photobiology" (B. L. Diffey, ed.), p. 85. Academic Press, London, 1989.

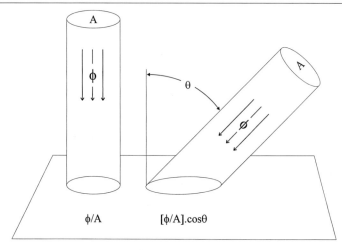

FIG. 4. The cosine effect. A parallel flux (Φ) of cross-sectional area A normally incident on a surface results in an irradiance of Φ/A. When the beam reaches the surface inclined at an angle θ to the normal, the irradiance is $[\Phi/A] \cdot \cos \theta$.

calibrated using a UV-A source with the same spectral emission as the source that is to be routinely monitored. If this is not the case, inaccuracies in UV-A irradiance will result. For example, suppose a UV-A radiometer is calibrated using the optically filtered mercury arc lamp with the spectrum shown by the solid line in Fig. 1. The spectral power distribution of this source is confined largely to a narrow band of radiation centered at 365 nm (one of the characteristic lines present in the mercury spectrum) and is very different from that of the fluorescent UV-A lamp shown by the broken line in Fig. 1. The consequence is that a radiometer calibrated in this way and then used to determine the irradiance from UV-A fluorescent lamps will underestimate the true UV-A irradiance by about 22%.

The technique recommended by the author[13] for calibrating UV-A radiometers intended to monitor the output from UV-A fluorescent lamps is as follows. Using a spectroradiometer with a cosine-weighted input optic, measure the spectral irradiance from 315 to 400 nm in 1-nm steps at 20 cm from a bank of at least four UV-A fluorescent lamps (60 cm or greater in length) having the same spectrum as the lamps to be monitored and after switching on the lamps for a sufficient period of time to allow the output to stabilize. Sum the spectral irradiance across the wave band 315–400 nm to determine the UV-A irradiance. Remove the spectroradiometer, place

[13] B. L. Diffey, *in* "Handbook of Noninvasive Methods and the Skin" (J. Serup and G. B. E. Jemec, eds.), p. 619. CRC Press, Boca Raton, FL, 1995.

the entrance aperture of the sensor at the same point as the input optic of the spectroradiometer, and adjust the meter display so that it reads the UV-A irradiance.

It is important that users of spectroradiometers have their own standard lamps (either deuterium or tungsten) calibrated regularly by standards laboratories so that these can be used to provide an absolute spectral sensitivity calibration of the spectroradiometer.

Radiometer Stability

It should be remembered that the sensitivity of all radiometers will change with time; frequent exposure to high-intensity sources of optical radiation will accelerate this change. For this reason it is always a sound policy to acquire two radiometers, preferably of the same type, one of which has a calibration traceable to a national standards laboratory.[14] This radiometer should be reserved solely for intercomparisons with the other radiometer(s) used for routine purposes. A measurement of the same source is made with each radiometer and a ratio calculated. It is the stability of this ratio over a period of months and years that indicates long-term stability and good precision.

[14] L. A. Mackenzie, *Photoderm.* **2,** 86 (1985).

[24] Radiation-Induced Signal Transduction

By Axel Knebel, Frank D. Böhmer, and Peter Herrlich

Cells depend for almost all their life reactions, from survival to apoptosis, from circadian rhythm to changes in behavior and differentiation, on extracellular cues. Much information has accumulated on the mechanisms of cell–cell communication and on the network of signaling inside the cell. We now know how hormones act and how cytokines address the cellular program of gene expression, and we know some of the components of the signal transduction network from the cell membrane to the cytoskeleton and to the nucleus. There are, nevertheless, still many gaps in our understanding. One research field that may contribute to a better understanding of such pathways concerns cellular responses to light, to other types of radiation, and to nonnatural chemicals. This article addresses some of the methodology that has been employed to study radiation-induced signaling and which led to the recognition of radiation-induced pathways and of some of the primary target molecules.

Background

To elicit a biological response, radiation needs to be absorbed by cellular molecules. The nature of the absorbing molecule depends on the wavelength of radiation and almost any constituent, from lipids, proteins, nucleic acids to water, can be a radiation target. The nature of the target molecule will in turn define the structural consequence on absorption of radiation energy.

Studies in the 1930s found radiation a powerful agent for the generation of mutations, which are the result of a primary damaging event in DNA. The DNA complement is present only once or twice per cell in haploid and diploid organisms, respectively, and is, because of its singularity, particularly vulnerable. Nonrepaired damage to DNA can be converted into a mutation at a given rate, depending on a number of parameters. If these are constant, the risk of mutagenesis will increase with dose of radiation, giving rise to a linear dose–response curve within a certain range of doses. The central role of genes magnifies the initial radiation-induced structural change in that mutations can create enormous phenotypic consequences.

For most cellular molecules, destruction of single copies will go by unnoticed. The discovery that most qualities of radiation as well as various toxic chemical agents can induce impressive, if not massive, activations of signal transduction and subsequent changes in macromolecule synthesis and structure in all biological cells examined has therefore created a puzzle. Intracellular communication as well as the regulation of macromolecular structure and gene expression are highly ordered processes. They depend on reversible functional changes of regulatory proteins whose actions are interdigitated and ordered into cascades of dependencies. For each situation a new balance of positive and negative regulatory activities needs to be established. To exemplify, receptor tyrosine kinases represent sensors built into the plasma membrane of eukaryotic cells and are primed to respond by autophosphorylation to their specific ligand, which may be presented by a neighboring cell or as a soluble factor. Receptor tyrosine kinases act as homo- or heterodimers (or oligomers) such that the subunits can cross-phosphorylate each other. At phosphorylated receptor domains, complexes of several proteins assemble. Either their interactions with regulatory switches activate a cascade of protein kinases or they acquire themselves catalytic activities, which then trigger the release of intracellular message molecule, e.g., the low molecular weight inositol phosphates or the STAT proteins that are released from the receptor-associated complexes and become nuclear transcription factors. At each level of these signaling pathways, negatively regulating enzymes control the intensity and duration of the signal. Autophosphorylated receptor tyrosine kinases, for instance, are dephosphorylated rapidly by receptor-directed tyrosine phosphatase(s). A second mechanism of inactivation of the receptor tyrosine kinases is their clustering and internalization.

The study of stress-induced signal transduction needs to establish links between primary target molecules, which are most likely damaged by the stress agent such as radiation, and signaling cascades. One idea has been that the cells possess specialized sensors for toxic agents, e.g., a sensor that reacts to ultraviolet A (UVA) or a sensor that is primed to react to an alkylating event. Such sensors would be connected to the physiologic communication network and, just like physiologic sensors, receptor tyrosine kinases, send signals to various parts of the cell. More recently, several laboratories discovered another possibility: the inactivation of negatively acting regulatory components, which shifts the balance toward increased signaling. As it turns out, the two hypotheses—a sensor or the inactivation of a protein—are not mutually exclusive. Once signal transduction pathways are turned on, the cellular fate depends on the signaling state of the complete network within the cell. Choices between survival pathways and apoptosis or cellular differentiation are not yet well understood, irrespective of the source of stimulation.

We will subsequently attempt at summarizing what is now known about radiation-induced signal transduction.

Transcriptional Arrest at DNA Damage Sites as a Signal Generator

The originally envisaged radiation target, DNA, is not only the substrate of mutagenesis, but also appears to generate signals that act on regulatory proteins, particularly transcription factors. It is not yet resolved how DNA damage is converted into a signal, but several suggestions have been published: protein kinases with the specific property to recognize DNA lesions, e.g., DNA strand breaks, and to be activated by this interaction (e.g., reviewed by Morgan and Kastan[1]); and transcriptional arrest at DNA lesions, which may lead to the recruitment of signaling components.[2a,2b] This latter possibility has been suggested by experiments with DNA repair-deficient mutant cells. Transcription-coupled repair removes signal-generating lesions much more readily than repair in nontranscribed regions of the genome. Arrest of transcription with subsequent signaling can also be accomplished in the absence of DNA lesions, e.g., by elongation inhibitors. Transcriptional arrest-dependent signaling addresses the level of the tran-

[1] S. E. Morgan and M. B. Kastan, *Adv. Cancer Res.* **71,** 1 (1997).

[2a] C. Blattner, E. Tobiasch, M. Litfin, H. J. Rahmsdorf, and P. Herrlich, *Oncogene* **18,** 1723 (1999).

[2b] C. Blattner, K. Bender, P. Herrlich, and H. J. Rahmsdorf, *Oncogene* **16,** 2827 (1998).

scription factor p53 (by preventing its degradation) and the release of cytokines.[3]

Signal Generation by Block of Translational Elongation

Ultraviolet-induced lesions in the S/R loop and of domains V and VI of ribosomal 28S RNA, as well as the binding of inhibitors to this RNA sequence, appear to strongly activate Jun N-terminal kinase.[4] Although the intermediate steps are not known, a process coupled to translational elongation can obviously feed into the protein kinase cascade, which links to processes in the nucleus.

Protein Tyrosine Phosphatases as Primary Targets of Radiation-Induced Signaling

Stimulated by the findings that nucleus-free cytoblasts can be induced by radiation to activate the transcription factor NF-κB and Jun-N-terminal kinase[5] and that induction by UVC of c-fos transcription requires Ras-Raf[6] (why should a DNA lesion-dependent process involve components at the inner surface of the plasma membrane?), radiation targets in the plasma membrane have been considered. Radiation of different wavelengths activates effectively several membrane receptors in a ligand-independent fashion. For the tumor necrosis factor α (TNFα) receptor, clustering has been the readout; autophosphorylation of epidermal growth factor (EGF), platelet derived growth factor (PDGF), and insulin receptors and the association of the receptors with the cytoplasmic signaling complexes, as another end point, have been observed in response to several types of radiation.[7] Chemical cross-linkers, which efficiently fix PDGF receptor dimers on treatment of cells or cell membrane preparations with PDGF, cannot do so on UV-activated receptors, despite the fact that UV also causes mutual phosphorylation of receptor subunits, which must be in close proximity to each other.[8] This led to the suggestion that receptor subunits preexist, at least transiently,

[3] K. Bender, M. Göttlicher, S. Whiteside, H. J. Rahmsdorf, and P. Herrlich, *EMBO J.* **17,** 5170 (1998).

[4] M. S. Iordanov, D. Pribnow, J. L. Magun, T.-H. Dinh, J. A. Pearson, and B. E. Magun, *J. Biol. Chem.* **273,** 15794 (1998).

[5] Y. Devary, C. Rosette, J. A. DiDonato, and M. Karin, *Science* **261,** 1442 (1993).

[6] A. Radler-Pohl, C. Sachsenmaier, S. Gebel, H.-P. Auer, J. T. Bruder, U. Rapp, P. Angel, H. J. Rahmsdorf, and P. Herrlich, *EMBO J.* **12,** 1005 (1993).

[7] P. Herrlich, C. Blattner, A. Knebel, K. Bender, and H. J. Rahmsdorf, *Biol. Chem.* **378,** 1217 (1997).

[8] S. Gross, A. Knebel, T. Tenev, A. Neininger, M. Gaestel, P. Herrlich, and F.-D. Böhmer, *J. Biol. Chem.* **274,** 26378 (1999).

as dimers, even in the absence of ligand, and that their activation requires a mechanism in addition to dimerization.

Because many agents of totally different mechanisms of action, from ionizing radiation to metal toxins, can activate receptor tyrosine kinase activities, the idea that these agents indeed caused gain of functions, such as an elevation of catalytic function, became unlikely. Studies of the lifetime of the receptor phosphorylation state of receptors then identified the (major) mechanism of radiation action.[9] Protein tyrosine phosphatases associated with receptor tyrosine kinases (as well as cytoplasmic dual-specificity phosphatases, e.g., MKP-1; Cavigelli *et al.*[10]) are inactivated by radiation as well as by chemical agents. The result is an increase of the phosphorylation level of their substrates and elevated signal transduction. As a common mechanism the oxidation of a catalytic cysteine in the active center of tyrosine phosphatases is proposed.[8] Thus, the primary radiation event must be the generation of reactive oxidative intermediates. These could be oxidative derivatives of chromophores absorbing the radiation, or could be oxygen intermediates produced by energy absorption of a chromophore. This is where the two hypotheses—sensor or inactivation of a negative regulator—may fuse. Tyrosine phosphatases may well be subjected to redox regulation under physiologic conditions, and radiation influences this redox system through a "sensor" generating oxidative potential.

Threshold Hypothesis

We hypothesize that there is a decisive difference between radiation-induced mutagenesis and radiation-induced signal transduction and gene expression. While a single DNA lesion suffices to produce a risk for a mutation, a single activated signaling component is not likely to cause a biological response. Despite the fact that magnification mechanisms are built into the signaling cascades, a "minimal" dose of activation will be required. Thus dose–response curves should indicate thresholds below which no response can be seen.

Subsequent sections will describe the methodology that has led to the recognition of radiation-induced signal transduction.

Inducers of Oxidative Stress

Ultraviolet Radiation

Ultraviolet radiation is formally subdivided into UVC (100–280 nm), UVB (280–320 nm), and UVA (320–400 nm). A very narrow 256-nm UVC

[9] A. Knebel, H. J. Rahmsdorf, A. Ullrich, and P. Herrlich, *EMBO J.* **15,** 5314 (1996).
[10] M. Cavigelli, W. W. Li, A. Lin, B. Su, K. Yoshioka, and M. Karin, *EMBO J.* **15,** 6269 (1996).

peak is emitted by a G15T8-10R tube. An UG-5 filter glass (Philips) should be used to eliminate visible light. UVB is emitted by a TL20/12 tube (Philips). Visible bands can be filtered out by the UG-5 filter. However, 30% of the emitted energy under these conditions still is UVA. UVA irradiation is emitted by F15T8-BL tubes (NIS, Japan) in combination with an UVC filter (UG-1, Philips). This combination nevertheless produces 1–2% contaminating UVB. UVC (256 nm) is absorbed by a large variety of molecules. The most efficient absorber is probably DNA, but proteins and some amino acids (Phe, Cys, Trp) also absorb UVC. It is a common practice to irradiate monolayers of cells that had been rinsed with phosphate-buffered saline (PBS), either without a fluid layer or with PBS only.

Oxidizing Agents

Hydrogen peroxide (H_2O_2) is used in concentrations of 100 μM to 10 mM. For "trapping" experiments, 5 mM H_2O_2 were used. Potassium permanganate ($KMnO_4$) is effective in concentrations between 0.5 and 5 mM. Shift into an oxidative condition can also be achieved by treating cells with ionizing radiation (γ rays: 1 to 50 γ ray) or by depletion of cellular glutathione. The latter is achieved by the inhibition of synthesis of reduced glutathione (GSH), an important cellular defense molecule against oxidation. Treatment of cells with low millimolar concentrations of buthionine sulfoximine (BSO), a reversible inhibitor of γ-glutamylcysteine synthetase, reduces the amount of cellular GSH by at least 50% within 2 to 6 hr.[11]

Ionizing Radiation: Cobalt Source, 1–50 Gy

Organometal compounds, e.g., tributyl-tin, interact with cellular thiol groups, mimicking oxidative stress.[12] Tributyl-tin at 1 to 5 μM causes apoptosis in cultured cells.

Inhibitors of Oxidative Stress (Antioxidants)

Several thiol-containing substances can be used to protect proteins or cells from oxidative damage. Because the level of cellular GSH is already in the range of 5 mM, GSH or synthetic thiol reagents, e.g., N-acetylcysteine (NAC), are usually only effective in concentrations above 5 mM. However, GSH does not penetrate the cellular membrane easily and some of the effects of GSH are explained by its capacity to directly inactivate oxidants in the medium. NAC, however, is in part lipophilic and penetrates readily

---

[11] A. Meister, *Pharmacol. Ther.* **51,** 155 (1991).

[12] E. Corsini, B. Viviani, M. Marinovich, and C. L. Galli, *Toxicol. Appl. Pharmacol.* **145,** 74 (1997).

into cells. If antioxidants are employed to investigate the effect of ultraviolet radiation, it should be noted that they may absorb UV. For instance, UVC is absorbed by GSH or, less efficiently, by NAC. Therefore, the UV spectrum of the medium with or without antioxidant should be compared. Strongly reducing agents such as 2-mercaptoethanol or dithiol agents such as dithiothreitol (DTT) or dimercaptosuccinic acid are toxic in millimolar concentrations.

Assays for Signaling Components

Phosphorylation of EGF Receptor and Assembly of Signaling Complexes[8,9,13,14]

Buffers

Ca^{2+}/Mg^{2+}-free PBS: 137 mM NaCl, 2.7 mM KCl, 6.5 mM $Na_2HPO_4 \times 2 H_2O$, 1.5 mM KH_2PO_4, pH 7.5

Coimmunoprecipitation (Co-IP) buffer: 30 mM Tris, pH 7.4, 150 mM NaCl, 1 mM EDTA, 1% Triton X-100, 10 mM NaF, 5 mM Na_3VO_4, 10 μg/ml aprotinin, 10 μg/ml leupeptin, 1 mM phenylmethylsulfonyl fluoride (PMSF)

IP wash buffer III: 25 mM Tris, pH 7.5, 1 mM EDTA

1$\times$ sodium dodecyl sulfate (SDS) sample buffer: 10% (v/v) glycerol, 2% SDS, 0.08 M Tris, pH 6.8, 2% 2-mercaptoethanol, 0.02% bromphenol blue

Cells (1.5 $\times$ 10^6) expressing the EGF receptor (A431, HeLa, Rat-1-HER) are incubated for 36 hr on 10-cm cell culture dishes in Dulbecco's modified Eagle's medium (DMEM), supplemented with 10% fetal calf serum (FCS) and antibiotics. They are incubated further in 5 ml serum-free DMEM for 6 hr. The cells may be treated with 100 μM to 1 mM Na_3VO_4 for 1 hr to boost the effects of ultraviolet radiation on receptor phosphorylation. The cells are then treated with 10 ng/ml EGF or with other stimuli of choice. Usually, treatment for 2 to 5 min is sufficient to achieve maximal receptor activation with a given dose and stimulus. Sensitive assays detect receptor phosphorylation as early as 10 sec after treatment. The cell culture medium is aspirated and the cells are washed with 5 ml ice cold Ca^{2+}/Mg^{2+}-free PBS. The cells are lysed in 800 μl Co-IP buffer and incubated for 10

[13] P. J. Coffer, B. M. Burgering, M. P. Peppelenbosch, J. L. Bos, and W. Kruijer, *Oncogene* **11,** 561 (1995).
[14] C. Sachsenmaier, A. Radler-Pohl, R. Zinck, A. Nordheim, P. Herrlich, and H. J. Rahmsdorf, *Cell* **78,** 963 (1994).

min. Nuclei are sedimented at 13,000 rpm and 4° for 15 min and the supernatant is transferred into a new tube. A 50-μl aliquot is taken to measure total protein. The main fraction is supplemented with 5–10 μg of an EGF receptor antibody that recognizes the extracellular domain (e.g., clone 528, Calbiochem) and 30 μl protein G$^+$ agarose (Calbiochem). The EGF receptor is immunoprecipitated at 4° for 2 hr. The agarose is washed four times with lysis buffer and once with IP wash buffer III. Fifty microliters of 1× SDS sample buffer is added, and agarose beads containing immune precipitates are incubated at 96° for 5 min. The agarose is spun down, vortexed, centrifuged again, and incubated at 96° for another 3 min. This is to assure maximal protein recovery from the agarose. After another spin the solubilized protein is loaded without transferring the agarose onto the gel and is resolved by 7.5% SDS–polyacrylamide electrophoresis (PAGE). The gel is analyzed by Western blotting, using antibodies against phosphotyrosine, EGF receptor, Shc, Grb-2, or Sos. If coprecipitated Grb-2 is to be determined, a 7.5 to 15% SDS–polyacrylamide (PAA) gradient gel should be used because of the low molecular weight of Grb-2 (23,000).

Activation of Mitogen-Activated Protein Kinases and Stress-Activated Protein Kinases

The determination of Raf activation, MEK activation, and Erk activation was described thoroughly before in this series.[15] Several methods on how to measure activation of Jun-N-terminal kinase have been described in Westwick and Brenner.[16] More recently, phosphospecific antibodies against a growing number of protein kinases such as Erk, JNK, and p38 have become available. With their use the phosphorylation state of several protein kinases can be investigated with remarkable ease. Other activity determinations may also be employed.

Assays to measure the activation of p21 Ras have been described previously.[17–21]

[15] D. R. Alessi, P. Cohen, A. Ashworth, S. Cowley, S. J. Leevers, and C. J. Marshall, *Methods Enzymol.* **255,** 279 (1995).

[16] J. K. Westwick and D. A. Brenner, *Methods Enzymol.* **255,** 342 (1995).

[17] G. Bollag and F. McCormick, *Methods Enzymol.* **255,** 161 (1995).

[18] A. M. de Vries-Smits, L. van der Voorn, J. Downward, and J. L. Bos, *Methods Enzymol.* **255,** 156 (1995).

[19] J. B. Gibbs, *Methods Enzymol.* **255,** 118 (1995).

[20] T. Satoh and Y. Kaziro, *Methods Enzymol.* **255,** 149 (1995).

[21] A. J. Self and A. Hall, *Methods Enzymol.* **256,** 67 (1995).

Membrane Preparation and Dephosphorylation Reaction

This assay measures EGF receptor dephosphorylation in a cell-free system.[8,9]

Partial Purification of A431 Cell Membranes

Buffers

Hypotonic buffer: 20 mM Tris, pH 7.5, 10 mM NaCl, 10 μg/ml aprotinin (Sigma), 10 μg/ml leupeptin (Sigma), 1 mM PMSF (Sigma)

Dephosphorylation buffer: 20 mM Tris, pH 7.5, 10 mM NaCl, 0.5% Triton X-100, 10 μg/ml aprotinin, 10 μg/ml leupeptin, 1 mM PMSF

Ten confluent 15-cm dishes of A431 (about 10^7 cells per plate) are treated for 1 hr with freshly prepared 1 mM Na$_3$VO$_4$ (Sigma). Na$_3$VO$_4$ acts as a reversible inhibitor of tyrosine phosphatases and serves here to prevent EGF receptor dephosphorylation during the membrane preparation. EGF receptor tyrosine phosphorylation is induced by treatment with 50 ng/ml human recombinant EGF (Sigma) for 5 min. Cells are subsequently washed twice with ice cold Ca^{2+}/Mg^{2+}-free PBS, collected 5 ml per plate using a rubber policeman scraper, and sedimented by centrifugation at 330g and 4° for 5 min. The cells are resuspended in 8 ml hypotonic buffer and incubated on ice for 30 min. The suspension is then transferred into a cold 10-ml tightly fitting dounce homogenizer and the cells are disrupted with 50 strokes. Intact cells, nuclei, and large cellular organelles are sedimented by centrifugation at 2000g and 4° for 5 min. The turbid supernatant is aliquoted in six 1.5-ml polyallomer tubes and centrifuged at 100,000g and 4° for 25 min. The membrane sediment is resuspended and lysed in 250 μl dephosphorylation buffer per tube and pooled. It is crucial to keep the preparation cold to prevent uncontrolled dephosphorylation. These membrane vesicle preparations contain both EGFR and associated protein tyrosine phosphatase.

Membrane Dephosphorylation Assay

Buffer

1.5 M NAC stock solution (titrated with NaOH to pH 7.5)

Tyrosine phosphatases are very sensitive to oxidation. Therefore, the A431 cell membrane preparations are supplemented with 1–20 mM NAC. In a pilot experiment the concentration of thiol reagent to obtain maximal phosphatase activity should be tested. It is expected to be around 5 mM NAC. The dephosphorylation is started by transfer of the reaction mix to 37°. Aliquots of 100 μl are withdrawn, e.g., at 0, 0.5, 1, and 3 min, and boiled after the addition of 25 μl 5× sample buffer. The samples are resolved by 7.5% SDS–PAGE, and the presence of tyrosine phosphate is determined by Western blotting using the PY20 antibody (Transduction Laboratories, Exeter, UK).

To determine the effect of radiation or of inhibitors on the dephosphorylation reaction, 450-μl aliquots of the membrane preparation are treated with 200 μM H_2O_2, 5 mM iodoacetamide, or 100–5000 J/m^2 ultraviolet light and incubated on ice for 5 min. Exposure to ultraviolet light can be performed by pipetting the solubilized membranes onto a small piece of Parafilm and recollecting it after irradiation. A part of the solution may be not recoverable from the Parafilm. Therefore, a greater volume should be used and the appropriate control without irradiation is particularly important. Then the dephosphorylation reaction is started and measured as described previously.

Membrane Mix Experiment

Membrane preparations can serve as sources of substrate or phosphatase separately. To this end, a membrane preparation obtained from EGF-treated cells is deprived of phosphatase activity by covalent interaction with an inhibitor. A membrane preparation from nontreated cells (and thus devoid of phosphorylated substrate) is used as phosphatase preparation. Upon mixing the two types of the Triton X-100 solubilized membranes, phosphatases get access to the EGF receptor.

The procedure is performed in detail as follows: "Substrate membranes" are prepared from cells that had been treated with 1 mM Na_3VO_4 for 1 hr, 10 mM iodoacetamide for 30 min, and 50 ng/ml EGF for 10 min. Substrate membranes will not contain any tyrosine phosphatase activity because of covalent alkylation by iodoacetamide and therefore will not catalyze EGF receptor dephosphorylation. In parallel, "phosphatase membranes" are prepared from nontreated cells. This preparation contains full protein tyrosine phosphatase activity and no detectable amount of tyrosine phosphorylated EGF receptor. Phosphatase membranes can be treated, e.g., by radiation *in vitro*, to determine an effect of radiation on phosphatase, and subsequently be mixed with substrate membranes to examine the kinetics of dephosphorylation of its EGF receptors. The dephosphorylation is measured as described earlier.

Dephosphorylation *in Vivo* (*Cell Culture*)

Determination of EGF Receptor Dephosphorylation Kinetics

EGF receptor tyrosine phosphorylation is controlled stringently by protein tyrosine phosphatases. Dephosphorylation of EGF receptor *in vivo* as described later is a matter of fractions of a minute. Therefore, accurate timing is essential.

Cells (10^5) overexpressing EGF receptor (A431, Rat-1-HER, or transfected NIH3T3) are cultivated in full medium on 3-cm cell culture dishes for 48 hr and further incubated in serum-free medium for 6 hr. One plate is used for each time point. Freshly prepared Na_3VO_4 (100 μM) may be added for 60 min in order to slow down receptor dephosphorylation, and EGF receptor tyrosine phosphorylation is induced by the addition of 5 ng/ml EGF. Further autophosphorylation is stopped by the addition of 100 nM of the specific EGF receptor kinase inhibitor AG1478 (see also section on the use of Specific Inhibitors of Signaling Proteins). If an inhibitor of tyrosine phosphatase is to be tested, cells are treated either at this same time or prior to AG1478 (e.g., 2 mM iodoacetamide for 20 min or 1 mM H_2O_2 for 5 min). The timing after the addition of AG1478 is critical! Fifteen- and 30-sec time points are advisable. The medium is removed thoroughly and cells are lysed in 250 μl 2$\times$ SDS sample buffer. The cell extracts are transferred to Eppendorf tubes and sonified to disrupt the DNA. Aliquots of fifty microliters are boiled and loaded on SDS–PAGE and analyzed by phosphotyrosine Western blot.

Determination of PDGF Receptor Dephosphorylation Kinetics

The assay is essentially the same as for the EGF receptor. PDGF receptor dephosphorylation is nevertheless much slower and sensitive to changes in the pH of the culture medium.[22] If the medium changes to a more basic pH due to the loss of CO_2 during incubation outside an incubator, dephosphorylation of the PDGF receptor is already inhibited markedly. Therefore, cells should be kept under 5% CO_2 or be cultured with HEPES buffer. NIH3T3 cells overexpressing the PDGF-β receptor are pretreated with 10 ng/ml PDGF-BB for 10 min. PDGF receptor kinase activity is blocked by the addition of 50 μM AG1296, and dephosphorylation is determined as described earlier for the EGF receptor. Time points of 2, 5, 15, and 30 min may suffice.

---

[22] F.-D. Böhmer, S.-A. Böhmer, and C. H. Heldin, *FEBS Lett.* **331**, 276 (1993).

PTP Assays *in Vitro*

Inhibition of individual cellular PTPs may be visualized *in vitro* by assaying immunoprecipitated PTPs with artificial substrates. A very sensitive assay is the dephosphorylation of ^{32}P-phosphorylated Raytide (trade name of Oncogene Science). This tyrosine-containing, gastrin-related peptide (the sequence has not been disclosed by the manufacturer) can be phosphorylated very well by a variety of tyrosine kinases, including Src kinase, Abl kinase, or EGF receptor kinase. We use the assay method described by Pot *et al.*[23] A technical problem associated with the detection of stress-induced PTP inactivation is created by the fact that the inactivation by treatment with reducing agents seems to be, at least partially, reversible. However, PTP assays have to be performed under reducing conditions to enable reproducible measurements. Thus, part of the stress-imposed PTP inactivation may escape detection with such *in vitro* assays.

Labeling of Raytide

> Buffer A: 50 mM HEPES, pH 7.5; 0.1 mM EDTA; 0.015% Brij 35
> Buffer B: Buffer A prepared with 0.1 mg/ml BSA and 0.2% 2-mercaptoethanol
> 10 mM MgCl$_2$
> 30% (v/v) acetic acid
> ATP, 10 mM, stored in aliquots at $-20°$
> Assay buffer: 40 mM MES, pH 6.0, 20 mM DTT
> Stop solution: 0.9 M HCl, 90 mM sodium pyrophosphate, 2 mM NaH$_2$PO$_4$, 4%
> Norit A charcoal (v/v)
> Raytide EL (Oncogene Science, Uniondale) dissolved at 1 mg/ml in buffer A and stored in 10-μl aliquots at $-80°$
> Src (p60c-src) kinase (UBI; Lake Placid): many other tyrosine kinases may work similarly well. Kinase is stored in aliquots at $-80°$. Even under these conditions, the enzyme will gradually lose some activity on prolonged storage.

Labeling Reaction and Purification

To prepare the ATP solution, combine 4.5 μl 1 mM ATP (it is very important to include unlabeled ATP to obtain a significant stoichiometry of Raytide labeling and, subsequently, have a sufficient substrate concentra-

---

[23] D. A. Pot, T. A. Woodford, E. Remboutsik, R. S. Haun, and J. E. Dixon, *J. Biol. Chem.* **266,** 19688 (1991).

tion for the PTP assay), 7.0 μl buffer B, 0.5 μl MgCl$_2$, and 3 μl [γ^{32}P]ATP (ca. 30 μCi). Mix 2.5–10 units Src kinase, 10 μl buffer B, 10 μl Raytide solution, and 10 μl ATP solution. Incubate at 30° for 2–3 hr.

Run the mixture over a 0.5-ml column of Dowex AG 1 × 8, equilibrated previously in 30% acetic acid. Collect 0.1- to 0.2-ml fractions and monitor individual fractions with a Geiger counter to detect the peak of ^{32}P-Raytide eluting in the void volume. Pool the two to three peak fractions (usually 0.5 ml in total), measure the aliquot, and store the pool at −80°. This solution, typically with 5–25 × 10^4 cpm/μl, can be used for about 4 weeks. On the day of the experiment, a suitable aliquot is dried in a Speed-Vac and restored in assay buffer. It may require 30 min with occasional vigorous vortexing to dissolve most of the radioactive peptide. Centrifuge and transfer supernatant to a new tube.

PTP Assay

The PTP of interest is immunoprecipitated from lysates of appropriately treated cells using protein A-Sepharose to collect the immunocomplexes. It is important that no PTP inhibitors (e.g., vanadate, zinc acetate, molybdate) be present in the lysis buffer or immunoprecipation reagents! The washed beads are suspended in a small volume of assay buffer, and in a first round of assays, different amounts of a control sample are subjected to the PTP assay to determine the range of linearity. This is a critical preliminary experiment, as frequently immunoprecipitated PTP activities are very high and the ^{32}P-Raytide in the assay is easily completely dephosphorylated.

Assay Procedure. The release of ^{32}P-phosphate is determined. The remaining ^{32}P-Raytide is removed by trapping with charcoal. Five microliters of suspended immunoprecipitate (appropriately diluted as determined in a preliminary experiment), 40 μl assay buffer, and 5 μl ^{32}P-Raytide (3–5 × 10^4 cpm) are incubated at 30° for 30 min. To stop the reaction, 0.75 ml of ice-cold stop solution is added (suspend charcoal before pipetting!). The samples are vortexed and centrifuged at 17,000 rpm at room temperature for 5 min, and 0.4 ml of the supernatant is subjected to liquid scintillation counting.

Use of Specific Inhibitors of Signaling Proteins

Because dysregulation of signal transduction accompanies many diseases, attempts at isolating specific inhibitors are abundant. Specific inhibitors have become an important research tool. In contrast to a dominant-negative mutant, which is introduced by transfection, acute blocks of signal transduction are achieved with inhibitors. The specificity of inhibitors, however, remains a permanent problem. It is very difficult to prove whether an inhibitor is really specific. Taking into consideration that not all kinases have yet been cloned and can therefore not be tested, specificity is always uncertain. Inhibitor studies should therefore be complemented by data with dominant-negative mutants, refractority studies,[14] and analysis of the complete pathway. With the current state of the art, involvement of a particular enzyme in a particular signal transduction pathway cannot be proved beyond doubt. Of course, inhibitors still have their place in guiding the further planning of experiments, and they have their place in the treatment of disease.

The utility of protein kinase inhibitors can be illustrated with the commonly used compounds described in this section. All kinase inhibitors are soluble in DMSO. 1000× stock solutions are prepared. In comparing inhibitor studies using purified enzyme with *in vivo* applications, we note that hydrophobic inhibitors accumulate in cellular membranes. Membranous inhibitor concentrations may therefore be considerably higher than those in the cytosol or medium.

AG1478, a quinazoline derivative,[24] was developed as a very potent and selective inhibitor of EGF receptor tyrosine kinase activity. (Similar compounds have been developed by other laboratories: PD153035 = AG1517.[25]) Despite the fact that AG1478 competes with ATP, 50% inhibition of EGF receptor kinase activity is reached at only 3 nM. AG1478 is amazingly selective in that even other members of the EGF receptor family are not inhibited significantly at doses below 50 μM.[24] AG1478 can penetrate the cellular membrane extraordinarily fast and blocks EGF receptor kinase activity in less than 30 sec. Standard concentrations used are 40–100 nM. It is therefore a useful tool, not only to investigate the role of the EGF receptor as a mediator of signaling,[26] but also to study its dephosphorylation.[9,27]

AG1295 and AG1296 are quinoxaline derivatives and are structurally related to AG1478.[24,28] They potently inhibit the PDGF receptor at an IC$_{50}$ of about 1 μM. Nevertheless, they do not inhibit the EGF receptor or other tyrosine kinases significantly at doses below 100 μM. AG1296 has been used to investigate the role of the PDGF receptor and to analyze PDGF

[24] A. Levitzki and A. Gazit, *Science* **267,** 1782 (1995).
[25] D. W. Fry and A. J. Bridges, *Curr. Opin. Biotechnol.* **6,** 662 (1995).
[26] H. Daub, F. U. Weiss, C. Wallasch, and A. Ullrich, *Nature* **379,** 557 (1996).
[27] F.-D. Böhmer, A. Böhmer, A. Obermeier, and A. Ullrich, *Anal. Biochem.* **228,** 267 (1995).
[28] M. Kovalenko, A. Gazit, A. Böhmer, C. Rorsman, L. Rönnstrand, C. H. Heldin, J. Waltenberger, F. D. Böhmer, and A. Levitzki, *Cancer Res.* **54,** 6106 (1994).

receptor dephosphorylation.[9,27,29,30] It is sufficient to treat cells with 20 μM AG1296 for 5 min to completely inhibit subsequent PDGF receptor activation.

GF 109203x (Bisindolylmaleimide I)[31] is a frequently used relatively selective inhibitor of PKC isoforms α, β, γ, δ, and ε; 500 nM GF 109203x inhibits Erk1,2 activation completely on treatment of cells with 10 ng/ml TPA (phorbol-12-myristate-13-acetate). However, Erk1,2 activation induced by 50 ng/ml TPA is not inhibited completely. GF 109203x and another frequently used PKC inhibitor, RO318220, are both equipotent inhibitors of p90^{RSK} = MAPKAP kinase 1β and of p70^{S6kinase},[32] which implies that the ATP-binding pockets of these kinases are structured similarly.

PD98059 inhibits MEK1 (= MAP kinase kinase 1) activation induced by many agents such as growth factors, mitogens, stress, and cytokines. The IC$_{50}$ is between 1 and 50 μM in cell culture experiments, depending on the stimulus and its dose. PD98059 is poorly soluble in aqueous solutions and, if applied at concentrations above 50 μM, it should be given in two aliquots with a 10-min interval to allow resorption by the cells. PD98059 was assumed to be particularly MEK1 specific, as it binds to the inactive form of MEK1 and prevents its activation by Raf1 in a noncompetitive manner.[33] Reports have cast doubts on the specificity. PD98059 appears to inhibit the binding of TCDD to the arylhydrocarbon receptor[34] and to inhibit cyclooxygenase activity,[35] a finding with some relevance, as MEK1 controls the activation of PLA$_2$ and the release of the cyclooxygenase substrate arachidonic acid.

U0126, a novel putatively selective MEK1 inhibitor,[36] possesses a 100-fold higher affinity to inactive MEK1 than PD98059. It is a noncompetitive inhibitor as well and inhibits basal MEK1 activity 100-fold better than activated MEK. U0126 is structurally unrelated to PD98059. It is too early

[29] F.-D. Böhmer and S.-A. Böhmer, FEBS Lett. 391, 219 (1996).

[30] A. Herrlich, H. Daub, A. Knebel, P. Herrlich, A. Ullrich, G. Schultz, and T. Gudermann, Proc. Natl. Acad. Sci. U.S.A. 95, 8985 (1998).

[31] D. Toullec, P. Pianetti, H. Coste, P. Bellevergue, F. Loriolle, L. Duhamel, D. Charon, and J. Kirilovsky, J. Biol. Chem. 266, 15771 (1991).

[32] D. R. Alessi, FEBS Lett. 402, 121 (1997).

[33] D. R. Alessi, A. Cuenda, P. Cohen, D. T. Dudley, and A. R. Saltiel, J. Biol. Chem. 270, 27489 (1995).

[34] J. Reiners, Jr., J. Y. Lee, R. E. Clift, D. T. Dudley, and S. P. Myrand, Mol. Pharmacol. 53, 438 (1998).

[35] A. G. Börsch-Haubold, S. Pasquet, and S. P. Watson, J. Biol. Chem. 273, 28766 (1998).

[36] M. F. Favata, K. Y. Horiuchi, E. J. Manos, A. J. Daulerio, D. A. Stradley, W. S. Feeser, D. E. van Dyk, W. J. Pitts, R. A. Earl, F. Hobbs, R. A. Copeland, R. L. Magolda, P. A. Scherle, and J. M. Trzaskos, J. Biol. Chem. 273, 18623 (1998).

to comment on its specificity for MEK1. *In vivo*, 20 μM is used to inhibit completely TPA-induced transcription of a 2× TRE-luciferase construct.

SB203580 inactivates the p38 isoforms p38 and p38β2 (SAPK2a, SAPK2b) at 50 and 600 nM *in vitro*.[37,38] SB203580 also addresses JNK2β with half-maximal inhibition at 5 μM and JNK2α at 20 μM (tested with purified kinases).[39] Furthermore, cyclooxygenase and thromboxane synthase are inhibited directly at low micromolar concentrations.[35] *In vivo* concentrations used to block p38 are 10–25 μM.

Use of Dominant-Negative Mutants

To investigate whether a particular enzymatic reaction is a necessary step in a signaling cascade in living cells, a mutant form of the enzyme can be introduced that still serves as a substrate for the upstream signaling molecules, but is incapable of transmitting the signal further (e.g., catalytically inactive). The dominant-negative enzyme must be expressed in excess over the endogenous enzyme. The use of dominant-negative mutants suffers from disadvantages different than those discussed for chemical inhibitors. No acute inhibition can be obtained. At least the time required for synthesis needs to pass (exception fusion constructs with the ligand-binding domain of the estrogen receptor). Thus new equilibrium states are established. Also, the mechanism of dominant-negative action cannot be predicted easily. The mutant protein may compete either with downstream or with upstream interaction partners. Interaction partners may yet be unknown or vary with growth conditions. Because dominant-negative mutants are introduced by the expression of transiently transfected plasmids, it is often necessary to cotransfect a reporter construct. This construct may encode a tagged version of a downstream component of the signaling cascade or a responsive promoter that controls the expression of a reporter enzyme that does not exist naturally in the investigated cell line, such as luciferase or chloramphenicol transferase. The experimental setup depends obviously very much on the mutant and the reporter used. The following protocol describes the inhibition of activation of a hemagglutinin-tagged version of Erk-2 by a dominant-negative EGF receptor.

[37] A. Cuenda, J. Rouse, Y. N. Doza, R. Meier, T. F. Gallagher, P. R. Young, and J. C. Lee, *FEBS Lett.* **364,** 229 (1995).

[38] Eyers, M. Craxton, N. Morrice, P. Cohen, and M. Goedert, *Chem. Biol.* **6,** 321 (1998).

[39] A. J. Whitmarsh, S. H. Yang, M. S. Su, A. D. Sharrocks, and R. J. Davis, *Mol. Cell. Biol.* **17,** 2360 (1997).

Dominant-Negative EGF Receptor

The most efficient dominant-negative mutant of EGF receptor is a deletion mutant lacking both the tyrosine kinase domain and the phosphorylation sites at the C terminus. Because EGF receptor acts as a homo- or heterodimer, it is not sufficient to delete only the kinase domain or the phosphorylation sites because such mutants can be phosphorylated by the wild-type partner or phosphorylate their partner, respectively, thereby creating still functional signaling complexes. They would only inhibit EGF receptor signaling effectively when their copy number is well in excess of the endogenous receptor. The plasmid pRK5-NA8 encodes a C-terminal deletion mutant of EGF receptor,[40] which still dimerizes and can therefore inactivate wild-type EGF receptor under stoichiometric conditions. Furthermore, it can heterodimerize with and thereby inhibit ErbB2 and ErbB3, EGF receptor family members that are naturally involved in EGF receptor signaling.

pRK5-NA8 is cotransfected with pcDNA3-HA-Erk (M. J. Weber, University of Virginia). The measurement of HA-Erk-2 activity gives a direct readout of EGF receptor inhibition. Alternatively, a 5× TRE promoter[41] controlling the synthesis of chloramphinecol transferase (CAT) or luciferase can be used as reporter.

Experimental Details: Cotransfection of Constructs Expressing Dominant-Negative EGF Receptor and Hemagglutinin-Tagged ERK2

Buffer

TBS: Prepare 500 ml 10× TBS (250 mM Tris, pH 7.4, 1.37 M NaCl, 50 mM KCl, 7 mM CaCl$_2$, 5 mM MgCl$_2$) and autoclave. Sterile filtrate 10 ml of 0.5 M Na$_2$HPO$_2$. Add 50 ml of 10× TBS and 600 μl of Na$_2$HPO$_4$ to 450 ml of deionized, autoclaved water to get TBS.

Lysis buffer: 30 mM Tris, pH 7.4, 150 mM NaCl, 1 mM EDTA, 1% Triton X-100, 10 mM NaF, 3 mM Na$_3$VO$_4$, 5 mM sodium pyrophosphate, 5 mM β-glycerophosphate, 10 μg/ml aprotinin, 10 μg/μl leupeptin, 1 mM PMSF

IP wash buffer I: Lysis buffer supplemented with 0.5 M NaCl

IP wash buffer II: 30 mM Tris, pH 7.4, 150 mM NaCl, 1 mM EDTA, 1% Triton X-100

IP wash buffer III: 25 mM Tris, pH 7.5, 1 mM EDTA

[40] N. Redemann, B. Holzmann, T. von Röden, E. F. Wagner, J. Schlessinger, and A. Ullrich, *Mol. Cell. Biol.* **12**, 491 (1992).

[41] C. Jonat, H. J. Rahmsdorf, K.-K. Park, A. C. B. Cato, S. Gebel, H. Ponta, and P. Herrlich, *Cell* **62**, 1189 (1987).

DEAE-Dextran (500,000 Da) (Sigma), 10 mg/ml in water, autoclave

Chloroquine, 100 mM in water, sterilize

1× MBP phosphorylation buffer: 25 mM Tris, pH 7.5, 5 mM 2-mercaptoethanol, 10 mM Mg^{2+}-acetate, 0.1 mM [γ-^{32}P]ATP (5 μCi per reaction), 20 μg/reaction myelin basic protein (MBP)

HeLa tk$^-$ cells (5 × 10^5) are plated evenly onto 6-cm cell culture dishes in DMEM containing 10% fetal calf serum (antibiotics in the medium do not interfere with the transfection) and allowed to settle and grow for 16 hr. To mix the transfection solution, 2 μg pRK5-NA8 or the empty control vector, respectively, and 1 μg pcDNA3-HA-Erk2 are diluted in 950 μl TBS. Fifty microliters of DEAE-Dextran solution is added and gently mixed (do not vortex). The cells are washed twice with TBS and incubated with 1 ml transfection solution for 30 min. Gentle movement assures even distribution. The cells are incubated further with an additional 3 ml of serum-free DMEM. After 2 hr the medium is supplemented with 100 μM chloroquine. Sixty minutes later the transfection medium is removed, and the cells are washed once with TBS, incubated in 3 ml 10% DMSO for 1 min, washed twice with TBS, and left in DMEM containing 10% FCS for 24 hr. The cells can now be treated with 20 ng/ml EGF or other stimuli for 20 min. The medium is removed thoroughly, and the cells are lysed on the plate using 600 μl lysis buffer (keep on ice for 10 min to lyse efficiently). Nuclei are sedimented in a microcentrifuge at 13,000 rpm and 4° for 15 min. The supernatant is transferred into a fresh tube. A 50-μl aliquot is taken for the analysis of protein concentration by conveniently probing for the HA-tag in a Western blot analysis. The main fraction is incubated with 5 μg of an antibody against the hemagglutinin tag (clone 12CA5) and 40 μl of protein A–Sepharose for 2 hr. The Sepharose is sedimented and washed with 600 μl IP wash buffer I, three times with IP wash buffer II, and two times with IP wash buffer III. After removing all wash buffer from the Sepharose with a 0.25-mm needle, 50 μl MPB-phosphorylation buffer is added and an immune complex kinase reaction is allowed to procede under agitation at 30° for 15 min (for a detailed description, see Alessi et al.[15]). The reaction can be stopped either by the addition of 15 μl 5× SDS sample buffer and loaded on a 15% SDS–PAA gel for evaluation or by pipetting 40 μl onto a 2 × 2-cm piece of Whatman p81 paper, washing five times for 2 min in 0.5% orthophosphoric acid and once in acetone, and counting the radioactivity of this paper.

[25] Signaling Pathways Leading to Nuclear Factor-κB Activation

By NANXIN LI and MICHAEL KARIN

Introduction

Exposure of cells to long wavelength ultraviolet radiation A (320–400 nm) stimulates signaling pathways that activate transcription factors and lead to induction of specific gene expression. These transcription factors include activator protein (AP)-1, AP-2, and nuclear factor (NF)-κB.[1,2] It has been shown that singlet oxygen mediates effects of UVA on gene expression and NF-κB activity can be induced on treatment with singlet oxygen generated photochemically by rose bengal or methylene blue plus light.[3]

The Rel/NF-κB family of transcriptional factors regulates the expression of numerous cellular and viral genes that play important roles in immune and stress responses, in inflammation, and in apoptosis.[4–7] The NF-κB proteins are activated by a variety of stimuli, ranging from cytokines, bacterial and viral infections to various forms of radiation. They are kept in the cytoplasm of nonstimulated cells through interaction with inhibitory proteins, IκBs. In response to pro-inflammatory cytokines, such as tumor necrosis factor (TNF) and interleukin 1 (IL-1), the IκBs are phosphorylated rapidly at two specific serine residues located at their N-terminal region (e.g., Ser-32 and Ser-36 for IκBα) and then undergo ubiquitination and proteolysis by the 26S proteasome, resulting in release and translocation of NF-κBs to the nucleus, where they activate the transcription of specific target genes. Several components of a protein kinase complex that mediates IκB phosphorylation and whose activity is stimulated by TNF or IL-1 were purified and cloned molecularly. Two of them, IKKα and IKKβ (IκB kinase α and β), contain an N-terminal protein kinase domain and a C-terminal regulatory domain with several

[1] K. Briviba, L. O. Klotz, and H. Sies, *Biol. Chem.* **378,** 1259 (1997).
[2] S. Grether-Beck, R. Buettner, and J. Krutmann, *Biol. Chem.* **378,** 1231 (1997).
[3] B. Piret, S. Legrand-Poels, C. Sappey, and J. Piette, *Eur. J. Biochem.* **228,** 447 (1995).
[4] P. A. Baeuerle and T. Henkel, *Annu. Rev. Immunol.* **12,** 141 (1994).
[5] P. A. Baeuerle and D. Baltimore, *Cell* **87,** 13 (1996).
[6] P. J. Barnes and M. Karin, *N. Engl. J. Med.* **336,** 1066 (1997).
[7] V. R. Baichwal and P. A. Baeuerle, *Curr. Biol.* **7,** R94 (1997).

protein interaction motifs.[8,9] IKKγ/NEMO is a regulatory subunit of the complex that is necessary for the stimulation of IKK activity and NF-κB activation.[10,11] In certain cases, however, NF-κB activation does not seem to involve IκB phosphorylation by IKK or even IκB degradation. Short-wavelength UV (UV-C) light activates NF-κB in certain cell types concomitantly with IκBα degradation. Pretreatment of cells with proteasome inhibitors blocked IκBα degradation and NF-κB activation induced by UV radiation, indicating that IκBα degradation is required.[12] However, neither IKK activation nor phosphorylation of IκBα on Ser-32 and Ser-36 was observed to occur after UVC irradiation.[12,13] Similar to UV-C radiation, treatment of cells with amino acid analogs also activates NF-κB through proteasome-mediated IκBα degradation without apparent phosphorylation at Ser-32 and Ser-36.[14] A different pathway leads to NF-κB activation when cells are treated with tyrosine phosphatase inhibitors (e.g., pervanadate) or in response to reoxygenation of hypoxic cells.[15,16] On these treatments, NF-κB is activated through tyrosine phosphorylation of IκBα without its degradation. The phosphorylation site was identified as Tyr-42, and this site is present only in IκBα. It was shown that the tyrosine phosphorylation of IκBα led to its dissociation from NF-κB.[15]

This article focuses on studying signaling pathways triggered by a stimulus (e.g., UVA) that ultimately lead to NF-κB activation.

Preparation of Cell Extracts

Nuclear Extracts

At various intervals after treatment, cells (4×10^6) are washed once with ice-cold phosphate-buffered saline (PBS) and scraped into 1 ml of hypotonic buffer containing 10 mM HEPES, pH 7.6, 10 mM KCl, 1.5 mM MgCl$_2$, 0.5 mM dithiothreitol (DTT), and 0.5 mM phenylmethylsulfonyl

[8] J. A. DiDonato, M. Hayakawa, D. M. Rothwarf, E. Zandi, and M. Karin, *Nature* **388,** 548 (1997).

[9] E. Zandi, D. M. Rothwarf, M. Delhase, M. Hayakawa, and M. Karin, *Cell* **91,** 243 (1997).

[10] S. Yamaoka, G. Courtois, C. Bessia, S. T. Whiteside, R. Weil, F. Agou, H. E. Kirk, R. J. Kay, and A. Israel, *Cell* **93,** 1231 (1998).

[11] D. M. Rothwarf, E. Zandi, G. Natoli, and M. Karin, *Nature* **395,** 297 (1998).

[12] N. Li and M. Karin, *Proc. Natl. Acad. Sci. U.S.A.* **95,** 13012 (1998).

[13] K. Bender, M. Gottlicher, S. Whiteside, H. J. Rahmsdorf, and P. Herrlich, *EMBO J.* **17,** 5170 (1998).

[14] C. Kretz-Remy, E. E. Bates, and A. P. Arrigo, *J. Biol. Chem.* **273,** 3180 (1998).

[15] V. Imbert, R. A. Rupec, A. Livolsi, H. L. Pahl, E. B. Traenckner, C. Mueller-Dieckmann, D. Farahifar, B. Rossi, P. Auberger, P. A. Baeuerle, and J. F. Peyron, *Cell* **86,** 787 (1996).

[16] S. Singh, B. G. Darnay, and B. B. Aggarwal, *J. Biol. Chem.* **271,** 31049 (1996).

fluoride (PMSF). After centrifugation at 10,000 cpm, the supernatants are discarded and cell pellets are resuspended in hypotonic buffer containing 0.1% NP-40 and incubated on ice for 15 min. After centrifugation, the supernatants are removed carefully and the pellets are resuspended in 30 μl of nuclear extract buffer (20 mM HEPES, pH 7.6, 420 mM NaCl, 1.5 mM MgCl$_2$, 0.2 mM EDTA, 1 mM DTT, 0.5 mM PMSF, 20% glycerol) and incubated on ice for 15 min. The volume should be kept as small as possible because the protein concentration in the nuclear extracts should be kept relatively high (higher than 1 μg/μl). After centrifugation at 13,000 rpm for 10 min, the supernatants are transferred carefully to clean tubes and used as nuclear extracts.

Whole Cell Extracts

After treatment, cells (3 × 10^6) are washed twice with ice-cold PBS and scraped into 300 μl of cell lysis buffer containing 20 mM HEPES, pH 7.6, 150 mM NaCl, 1.5 mM MgCl$_2$, 0.2 mM EDTA, 1% Triton X-100, 10% glycerol, 1 mM DTT, 1 mM PMSF, 10 μg/ml aprotinin, 5 μg/ml leupeptin, 10 mM p-nitrophenyl phosphate (pNPP), 100 mM β-glycerophosphate, 1 mM sodium fluoride, and 0.1 μM sodium orthovanadate. After incubation on ice for 10 min, cells are centrifuged at 13,000 rpm for 10 min and the resulting supernatants are transferred to fresh tubes and used as whole cell extracts. DTT, proteinase inhibitors, and phosphatase inhibitors should be added fresh.

Lysates for Luciferase Assay

After treatment, cells (0.5–1.0 × 10^6) are washed twice with PBS and scraped into 100–200 μl of luciferase lysis buffer (100 mM Tris, pH 7.8, 100 mM magnesium acetate, 1 mM EDTA, 1 mM DTT, 1% Triton X-100). After centrifugation at 13,000 rpm for 10 min, the supernatants are transferred into fresh tubes and used to measure luciferase activity.

Measurement of NF-κB Activation

NF-κB activation may be assayed by measuring its DNA-binding activity, its nuclear translocation, and its transcriptional activation activity.

Electrophoretic Mobility Shift Assay (EMSA)

NF-κB DNA-binding activity is measured by EMSA using a ^{32}P-labeled oligonucleotide probe containing a NF-κB consensus sequence. Cell extracts are prepared at various intervals after treatment. Five micrograms of nuclear extracts of 10 μg of whole cell extracts is incubated at room temperature with the oligonucleotide (5000–10,000 cpm) in the presence of 3 μg of poly(dI-dC) (Pharmacia) and 5 μl of 3× EMSA reaction buffer [30 mM Tris, pH 7.6, 300 mM NaCl, 3 mM EDTA, 15 mM DTT, 30% glycerol, 0.3 mg/ml bovine serum albumin (BSA)]. Water is added to make the total incubation volume 15 μl. After 30 min, the reactions are loaded onto 5% nondenaturing polyacrylamide gel and separated by electrophoresis. Addition of poly(dI-dC) helps to reduce nonspecific binding, especially if using nuclear extracts. We have found that in measuring induction of NF-κB DNA-binding activity by TNF, UVC, and ionizing irradiation in HeLa cells there is no difference between using nuclear extracts and whole cell extracts.[12]

Detection of NF-κB Nuclear Translocation by Indirect Immunofluorescence

Cells are grown on glass coverslips. After treatment, cells are washed once with PBS and fixed in PBS containing 3% formaldehyde and 2% sucrose for 10 min. After washing with PBS, cells are permeablized with PBS containing 0.2% Triton X-100 for 10 min and blocked with PBS containing 2% BSA, 2% goat serum, and 0.1% Triton X-100 for 1 hr at room temperature. Samples are then washed with PBS containing 0.2% Triton X-100 and incubated for 1 hr with polyclonal anti-p65 (RelA) (sc-372, Santa Cruz) (1:60 dilution) in PBS containing 0.1% Triton X-100 and 2% BSA. After washing five times with PBS plus 0.2% Triton X-100, samples are incubated with secondary antibody rhodamine-conjugated goat antirabbit IgG (1:200 dilution) (DAKO) for 1 hr. After washing five times with PBS plus 0.2% Triton X-100, samples are mounted with Vectashield mounting solution (Vector) and viewed by an epifluorescence-equipped microscope. If a treatment induces nuclear translocation of NF-κB (p65 RelA), enhanced nuclear staining with anti-p65 should be detected.

Luciferase Reporter Gene Assay

To determine transcriptional activity of NF-κB, cells are transfected with the plasmid carrying the luciferase reporter gene whose expression is driven by a minimal promoter containing κB sites together with a control plasmid carrying β-galatosidase whose expression is not regulated by NF-κB. This control plasmid allows normalizing variability in transfection efficiencies. Because this assay measures reporter gene expression at the protein level, a certain interval should be given to allow accumulation of the protein in cells. We typically wait for 6 hr after the maximal induction of

NF-κB DNA-binding activity to harvest cells. After treatment, cells are harvested in the lysis buffer for luciferase assay. The amount of extracts required for the assay may be optimized. For this, 100 μl of luciferase reaction mixture (100 mM Tris, pH 7.8, 10 mM magnesium acetate, 1 mM EDTA) containing 66 μM luciferin (D-luciferin-potassium, Analytical Luminescence Laboratory) and 2 mM ATP are incubated with extracts and luciferase activity is measured by the luminometer. β-Galactosidase activity is measured by incubating extracts with 300 μl of the β-Gal reaction mixture [0.061 M Na$_2$HPO$_4$, 0.039 M NaH$_2$PO$_4$, pH 7.5, 1 mM MgCl$_2$, 50 mM 2-mercaptoethanol, 0.15% o-nitrophenyl β,D-galactopyranoside (ONPG)] at 37°. Once the color of the mixture turns yellow, the reaction is stopped by adding 150 μl of 1 M Na$_2$CO$_3$ and then light absorption is measured at 420 nm.

IκBα Degradation and Phosphorylation

Degradation and phosphorylation of IκBα in response to various stimuli can be examined by Western blotting. Whole cell extracts are prepared at various time points after treatment of cells with an NF-κB-activating agent. Equal amounts of extracts (usually 20 μg/lane) are separated by 12% SDS–PAGE and then transferred to a nitrocellulose membrane. The membrane is then blocked with 2% BSA plus 2% milk in PBS containing 0.1% Triton X-100 for 1 hr at room temperature or overnight at 4°. After brief washing with PBS plus 0.2% Triton X-100, the membrane is incubated with anti-IκBα (sc-371, Santa Cruz) (1 : 1000 dilution) in PBS containing 2% BSA and 0.1% Triton X-100 for 1 hr at room temperature. After washing four times (10 min each time) with PBS plus 0.2% Triton X-100 the membrane is incubated with horseradish peroxidase (HRP)-conjugated secondary antibody and then visualized with enhanced chemiluminescence (ECL) detection. Degradation of IκBα protein can be determined by disappearance of the IκBα band and its phosphorylation by retardation of the electrophoretic mobility of the IκBα band. The mobility retardation of IαBα is caused by its phosphorylation at Ser-32 and Ser-36. To detect mobility retardation, easily, it is sometime necessary to run the SDS gel for a long time and to pretreat cells with proteasome inhibitors to allow the accumulation of phosphorylated IκBα, which otherwise would be degraded quickly.

To determine whether IκBα degradation is a prerequisite for NF-κB activation, cells are treated with proteasome inhibitors prior to exposure to an NF-κB-activating agent. Three types of proteasome inhibitors may be used: peptide aldehydes such as N-Ac-Leu-Leu-norLeucinal (AcLLnL, Calpain inhibitor I) or MG132, peptide vinyl sulfones such as Cbz-Leu-

Leu-Leu-vinyl sulfone (Z-L3vs), and lactacystin (Calbiochem).[17–19] The effective concentrations of these inhibitors that are needed may vary according to the cell type being examined. For HeLa cells, typically 50–100 μM of AcLLnL or MG132, 20–50 μM of ZL3vs, or 10 μM of lactacystin is added to the cultures 45–60 min prior to other treatments. NF-κB activation and IκBα degradation/phosphorylation are then determined by EMSA and Western blotting, respectively. If induction of NF-κB DNA-binding activity and IκBα degradation is blocked by the pretreatment of proteasome inhibitors, then the primary mechanism through which the inducer under examination activates NF-κB activation is based on proteasome-mediated IκBα degradation. If mobility retardation of IκBα is detected by Western blotting, it can be considered that the inducer is able to induce phosphorylation of IκBα.

Measurement of IKK Activity

If an NF-κB activating agent induces IκBα phosphorylation and degradation, its ability to activate IKK should be examined in order to determine whether it operates with this almost universal NF-κB activation pathway.

IKK Immune-Complex Kinase Assay

To determine whether IKK activity is induced by a given stimulus, whole cell extracts (see earlier discussion) are prepared at various intervals after stimulation of cells. Two hundred micrograms of extract protein is incubated with 40 μl of 50% slurry of protein A–Sepharose beads (in PBS or in whole cell lysis buffer) together with 1 μl of anti-IKKα antibody (PharMingen) for 2 hr at 4°. The mixture should be rotated, and additional whole cell lysis buffer may be added to allow good rotation of the mixture. The anti-IKKα antibody precipitates the entire IKK complex, which includes IKKα and IKKβ subunits. The immunoprecipitates are collected by gentle and brief centrifugation and washed twice with cold whole cell lysis buffer and washed twice with cold IKK kinase buffer (20 mM HEPES, pH 7.6, 10 mM MgCl$_2$, 1 mM DTT, 10 mM PNPP, 100 mM β-glycerol-3-phosphate, 25 μM sodium orthovanadate, and 10 μM ATP). The supernatants are removed carefully and the immunoprecipitates are resuspended

[17] K. L. Rock, C. Gramm, L. Rothstein, K. Clark, R. Stein, L. Dick, D. Hwang, and A. L. Goldberg, *Cell* **78,** 761 (1994).
[18] M. Bogyo, J. S. McMaster, M. Gaczynska, D. Tortorella, A. L. Goldberg, and H. Ploegh, *Proc. Natl. Acad. Sci. U.S.A.* **94,** 6629 (1997).
[19] G. Fenteany, R. F. Standaert, W. S. Lane, S. Choi, E. J. Corey, and S. L. Schreiber, *Science* **268,** 726 (1995).

and incubated with 20 μl of kinase buffer containing 5 μCi of ^{32}P-γ-ATP and 2 μg of GST-IκBα(1–54) used as a substrate at 30° for 30 min. The reactions are stopped by the addition of SDS sample buffer and analyzed by SDS–PAGE (12%). The IKK kinase activity is determined by the amount of radioactive ^{32}P incorporated into the GST-IκBα(1–54) fusion protein.

Preparation of GST-IκBα(1–54) Fusion Protein in Escherichia coli

The cDNA fragment containing the first 54 amino acids of IκBα is ligated in frame to the 3' of the GST sequence in pGEX expression vector (Pharmacia). Bacteria are transformed by the resulting plasmid, and GST-IκBα(1–54) protein expression is induced by isopropyl-β-thiogalactopyranoside (IPTG) according to standard procedure.[20] Bacteria are then collected by centrifugation and resuspended in lysis buffer [PBS containing 0.3 M $(NH_4)_2SO_4$, 1 mM EDTA, 0.5% Triton X-100, 5 mM DTT, and proteinase inhibitors]. Lysozyme is added to a final concentration of 1 mg/ml. Tubes are immersed in ice and bacteria are lysed by sonication. The lysates are centrifuged at 12,000 rpm for 20 min and the supernatants are removed and incubated with 50% slurry of glutathione–agarose beads (Pharmacia) for 1 hr at 4°. The beads are washed twice with the lysis buffer containing 1 mM PMSF, twice with the lysis buffer without 0.3 M $(NH_4)_2SO_4$, and once with PBS. After removal of PBS, the GST-IκBα(1–54) protein is eluted from the beads by incubating with two bed volumes of elution buffer (75 mM Tris, pH 7.8, 20 mM glutathione) for 1 hr at 4°. After centrifugation, the supernatants are collected and the beads are subjected to one more elution. The supernatants are combined and dialyzed against 2 liters of 20 mM Tris, pH 8.0, 100 mM NaCl, 0.2 mM EDTA, 10% glycerol, and 10 mM β-glycerophosphate. The quality and quantity of GST-IκBα(1–54) protein should be checked by SDS–PAGE and the protein stored in aliquots at −80°.

[20] F. Ausubel, R. Brent, and R. E. Kingston, et al. (eds.), "Current Protocols in Molecular Biology." Wiley, New York, 1991.

[26] Gene Regulation by Ultraviolet A Radiation and Singlet Oxygen

By SUSANNE GRETHER-BECK and JEAN KRUTMANN

Introduction

Human skin is increasigly exposed to long-wave ultraviolet (UV) radiation that is UVA radiation in the range of 340–400 nm. Sources for UVA radiation exposure are solar radiation and artificial sources used for cosmetic or phototherapeutic purposes. In order to understand the effects of UVA radiation on human skin, it is critical to study the mechanisms by which UVA radiation affects gene expression in human skin cells, particularly epidermal keratinocytes, which constitute the primary cellular target for UVA radiation. Until very recently, studies addressing UV radiation-induced gene regulation in human cells have employed nonskin cells and have almost exclusively focused on short (UVC) and middle (UVB) wave UV radiation. Studies on UVA radiation-induced gene expression in human keratinocytes, however, are important because there is increasing evidence that the photobiological and molecular mechanisms that mediate UVA radiation-induced gene regulation differ from those identified previously for UVC and UVB radiation-induced gene expression.[1] The specific signal transduction pathway being induced by UVA radiation may additionally depend on the cell type studied and may, e.g., vary for epidermal keratinocytes as compared with dermal fibroblasts.

By focusing on the regulation of the human intercellular adhesion molecule-1 (ICAM-1) gene in primary human epidermal keratinocytes, it has been demonstrated that UVA radiation-induced gene expression depends critically on the activation of transcription factor AP-2.[2] Interestingly, UVA radiation-induced AP-2 activation and subsequent induction of ICAM-1 transcription were found to be mediated through the generation of singlet oxygen, thus implying that UVA radiation and singlet oxygen-induced gene expression in human epithelial cells may share a common signal transduction pathway. These studies have been made possible by the availability of a variety of techniques that allow the study of gene expression in UVA-irradiated cultured human keratinocytes at different levels of expression, including transcription factor activation, mRNA, and protein expression

[1] S. Grether-Beck, R. Buettner, and J. Krutmann, *Biol. Chem.* **378,** 1231 (1997).
[2] S. Grether-Beck, S. Olaizola-Horn, H. Schmitt, M. Grewe, A. Jahnke, J. P. Johnson, K. Briviba, H. Sies, and J. Krutmann, *Proc. Natl. Acad. Sci. U.S.A.* **93,** 14586 (1996).

under various conditions that allow the monitoring of the role of singlet oxygen in this system. These techniques and assay system will be reviewed and their relevance for studying UVA radiation-induced gene expression in human keratinocytes will be discussed.

Ultraviolet A Irradiation of Human Epithelial Cells

For UV radiation, cells have been grown on polystyrene cell culture dishes (10 or 15 cm diameter) or multiple well plates (6 or 12 well) of different size according to the experiment to be done. To avoid phototoxic reactions that might result from absorption of UV radiation by components such as amino acids, vitamins, phenol red pH indicator of the cell culture medium the latter was replaced by phosphate-buffered saline (PBS) during irradiation. The PBS chosen should not contain calcium or magnesium in order to prevent the differentiation of normal human keratinocytes. Depending on the irradiation device, it is important to keep in mind that the volume of PBS chosen to be on the opened cell culture dish or multiple well plate has to be large enough to avoid desiccation of the monolayer cell culture on the surface of the cell culture dish/multiple well plate. Moreover, it is critical to control the increase of temperature due to the irradiation device using a thermometer. If a temperature rise is a problem, the cell culture dishes/multiple well plates can be put on refrigerant packs. Keratinocytes, e.g., long-term cultured normal human keratinocytes from human foreskin or cell lines such as KB cells or HaCaT cells, are irradiated routinely with 30 J/cm^2 UVA1. The unirradiated control cells are kept under identical conditions as irradiated samples. The so-called "sham irradiation" of cells is done by incubation in the presence of PBS for the time the irradiated cells are kept under PBS. Moreover, unirradiated cells also get fresh medium identical to irradiated ones to exclude effects from the change of media.

Irradiation Device, Dosimetry, and Irradiation Technique

Cells are exposed to UVA1 irradiation using a Sellamed irradiation device (Sellamed 12000, Sellas GmbH, Dr. Sellmaier, Wuppertal, Germany), which emits wavelenghts between 340 and 450 nm with a broad peak between 360 and 400 nm (Fig. 1). The UVA emission can be determined using a UVAMETER (Mutzhas, Munich, Germany) and has been found to be 150 mW/cm^2 at a tube to target distance of 30 cm. The irradiation device used does not emit significant amounts of UV radiation at wavelenghts shorter than 340 nm and thus irradiation-induced effects cannot be attributed to contaminating UVC or UVB radiation.

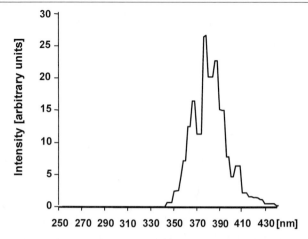

FIG. 1. Spectral analysis of a UVA1 source (SellasUVA1).

Keratinocytes are irradiated routinely with a dose of 30 J/cm^2 UVA1. This dose has to be adapted for other cell types such as fibroblasts or dendritic cells, which are more sensitive to UVA radiation. To test viability of the cells, trypan blue exclusion or propidium iodine staining can be used.[3]

Singlet Oxygen Quenchers, Deuterium Oxide, and Singlet Oxygen
    Generating System

Although there is increasing acceptance of reactive oxygen intermediates as universal and important biochemical intermediates, conclusive evidence of their involvement is lacking in many instances. The latter is a consequence of the difficulties in detecting and measuring reactive oxygen intermediates that generally have lifetimes measured in microseconds and which are extremely difficult to measure per se. These include radicals such as superoxide anion free radical (O_2^-), hydroxyl radical (HO·), lipid (L), and other peroxy radicals (LOO· and XOO·), nitrous oxide (NO·), peroxynitrate ($^-OONO_2$), peroxynitrite (^-OONO), hydrogen peroxide (H_2O_2), lipid peroxide (LOOH), hypochlorous acid (HOCl), other N-chloramine compounds, and singlet oxygen ($^1O_2$).[4] Therefore, the indirect approach of using quenchers or enhancers of lifetime of the species to be investigated is widely used and accepted. Unfortunately, the specificity of these quenchers

[3] R. Baserga, "Cell Growth and Division: A Practical Approach." IRL Press, Oxford, 1992.
[4] N. A. Punchard and F. J. Kelly, "Free Radicals: A Practical Approach." IRL Press, Oxford, 1996.

might pose a problem, and usually testing of a set of quenchers with specificity for different types of reactive oxygen intermediates is recommended.

In our hands, UVA effects are mainly found to involve the generation of singlet oxygen, and therefore mainly singlet oxygen quenchers will be discussed. These quenchers effectively inhibited UVA radiation-induced effects. In addition, the involvement of a certain species can be further proved by the use of a system to generate the reactive oxygen intermediate of interest independent from UVA irradiation.

Vitamin E is used as singlet oxygen quencher.[5] The cells have to be preincubated with vitamin E in order to achieve integration of this quencher into membranes. As α-tocopherol is quite insoluble in cell culture media, we used the esterified α-tocopheryl succinate, which can be dissolved easily in ethanol. Alternatively, one can use α-tocopheryl acetate or phosphate. α-Tocopheryl succinate was added to cell cultures 24 hr before irradiation at a concentration of 25 μM. As keratinocytes contain the corresponding esterases, they are able to hydrolyze the α-tocopheryl succinate to have the α-tocopherol integrated into the membranes.[6] As a general rule, some of the solvents used to dissolve the stimulus or quencher/inhibitor may affect the readout system by themselves. An excellent example is dimethyl sulfoxide (DMSO), which is not only a solvent but also an effective free radical scavenger.[7]

Sodium azide (NaN$_3$) is a well-known specific quencher for singlet oxygen.[8] NaN$_3$ is dissolved in phosphate-buffered saline (PBS; 50 mM) and applied to cell culture during irradiation only. Sodium azide is highly water soluble and will quench singlet oxygen in aqueous solutions such as cell culture medium. In addition, it can function as an inhibitor for complex IV in the mitochondrial respiratory chain.[9] The concentration used in these experiments during the short period of irradiation does not influence cell viability or induce apoptosis.

To make sure that no other reactive oxygen intermediates are involved, a panel of other quenchers known to be specific for other reactive oxygen intermediates, such as mannitol (hydroxyl quencher) or catalase (which breaks down H$_2$O$_2$ to oxygen and water) or superoxide dismutase (which converts O$_2^-$ into H$_2$O$_2$), has been used. None of these reagents has inhibited UVA radiation effects.

[5] P. DiMascio, S. Kaiser, and H. Sies, *Arch. Biochem. Biophys.* **274,** 532 (1989).

[6] Z. Nabi, A. Tavakkol, N. Soliman, and T. G. Polefka, *J. Invest. Dermatol.* **110,** A1239 (1998).

[7] S. P. Tam, X. Zhang, C. Cuthbert, Z. Wang, and T. Ellis, *J. Lipid. Res.* **38,** 2090 (1997).

[8] G. Deby-Dupont, C. Deby, A. Mouithys-Mickalad, M. Hoebeke, M. Mathy-Hartert, L. Jadoul, A. Vandenberghe, and M. Lamy, *Biochim. Biophys. Acta* **1379,** 61 (1998).

[9] B. van de Water, J. P. Zoeteweij, H. J. de-Bont, and J. F. Nagelkerke, *Mol. Pharmacol.* **48,** 928 (1995).

As an enhancer for singlet oxygen, half-lifetime deuterium oxide[10] (e.g., Sigma D-4501) is used instead of water. For our purpose, we have irradiated cells in PBS dissolved in D_2O (99.9 atom% ^2H) rather than H_2O. PBS is prepared as a 20× stock solution: 4.00 g/liter KCl, 4.00 g/liter KH_2PO_4, 160.00 g/liter NaCl, and 43.2 g/liter $Na_2HPO_4 \cdot 7H_2O$ in H_2O. Therefore, the final concentration of D_2O is 95% (v/v) in PBS/D_2O. Under these conditions, UVA radiation effects have been enhanced.

Singlet oxygen has been generated by the thermal decomposition of the endoperoxide of the disodium salt of 3,3'-(1,4-naphthylidene)dipropionate ($NDPO_2$),[11] 1 mM in PBS, for 1 hr in the dark at 37° (e.g., CO_2 incubator for stimulation of cells) and yielded excited singlet molecular oxygen and 3,3'-(1,4-naphthylidene)dipropionate (NDP). This singlet oxygen-generating system[11] has been shown to be well suited to cell culture experiments because it is water soluble and nontoxic for up to 40 mM for 1 hr of incubation. Infrared emission of singlet oxygen has been measured with a liquid nitrogen-cooled germanium photodiode detector (Model EO-817L, North Coast Scientific, Santa Rosa, CA). The rate of singlet oxygen generation has been monitored by the formation of NDP. Fifteen minutes after the addition of 1 mM $NDPO_2$, the rate of singlet oxygen generation has been 3 μM per minute. In order to exclude effects of the heterocyclic molecule on gene expression, control cells have been stimulated with NDP, which can be generated by thermal decomposition from the same batch of $NDPO_2$. NDP has been generated by leaving an aliquot of $NDPO_2$ at room temperature for a week. By using this singlet oxygen-generating system it has been possible to mimic UVA radiation-induced effects in human keratinocytes.

Materials

Long-term cultured human keratinocytes have been prepared from human neonatal foreskin or mamma reduction plastic surgery and cultured under serum-free conditions in media such as Keratinocyte-SFM (Life Technologies, GIBCO/BRL, Karlsruhe, Germany), Keratinocyte-Medium Kit (Sigma, Deisenhofen, Germany), or MCDB 153 supplemented with trace elements, ethanolamine, phosphoethanolamine, hydrocortisone, and epidermal growth factor and bovine pituitary extract (Biochrom KG seromed, Berlin, Germany). As an alternative to primary human keratinocytes, for selected experiments, keratinocyte-like cell lines have been employed. The epidermoid carcinoma cell line KB KB is provided by ATCC (Rock-

[10] L. O. Klotz, K. Briviba, and H. Sies, *FEBS Lett.* **408,** 289 (1997).
[11] P. DiMascio and H. Sies, *J. Am. Chem. Soc.* **111,** 2909 (1989).

ville, MD) as CCL-17. The cells are cultured in Eagle's MEM with nonessential amino acids and Earle's BSS (buffered salt solution) in 10% fetal bovine serum (FBS). The keratinocyte cell line HaCaT[12] is cultured in Dulbecco's MEM supplemented with sodium pyruvate (0.11 g/liter) and 10% FBS.

Assessment of Gene Expression

Ultraviolet A radiation-induced gene expression can be studied at the mRNA and protein level. To assess modulation of protein expression, Western blotting or FACS analysis is appropriate. The latter can be used if the cells of interest are not available in large amounts, such as primary cells.

Protein Expression: FACS Analysis

Collected and washed cells (1×10^6) have been incubated with a $1:40$ dilution of MoAb 84H10 (anti-ICAM-1) (Dianova, Hamburg, Germany) or the equivalent IgG1 isotype control antibody (Sigma, Deisenhofen, Germany) for 30 min at $4°$, washed three times in PBS containing 0.05% sodium azide, resuspended, and incubated with a $1:20$ dilution of goat antimouse FITC-$(Fab')_2$ for 30 min at $4°$. Subsequently, cells have been washed three times and either fixed in paraformaldehyde or analyzed within 4 days. FACS analysis is performed using a FACS SCAN (Becton-Dickinson, Mountain View, CA). Prior to analysis, cells have been routinely stained with propidium iodine to exclude dead cells, and usually 5000 cells per sample are assesed. Data can be given either as histograms of fluorescence intensity versus cell number or as the percentage of ICAM-1 reactive cells in comparison to unstimulated control cells.

mRNA Expression: Differential RT-PCR

In general, gene expression can be analyzed at the mRNA level Northern blotting combined with hybridization techniques or by reverse transcriptase polymerase chain reaction (RT-PCR) if the amounts of RNA are very small.

Keratinocytes have been grown on 10-cm cell culture dishes. For harvest, cells are washed with PBS, snap frozen with liquid nitrogen, and stored at $-80°$. Total RNA has been isolated using RNeasy total RNA kits (Qiagen, Hilden, Germany). Induction of gene expression has been measured by differential RT-PCR using the Geneamp RNA PCR core kit (Perkin-Elmer

[12] P. Boukamp, R. T. Petrussevska, D. Breitkreutz, J. Hornung, A. Markham, and N. E. Fusenig, *J. Cell Biol.* **106,** 761 (1988).

Applied Biosystems, Weiterstadt, Germany). The primers for quantifying gene expression are deduced from the corresponding cDNA sequence of interest (e.g., ICAM-1) obtained from databases such as EMBL or Gen-Bank. Computer programs such as PCRPLAN included in PC/GENE are useful for primer design because they reject primers complementary to another region of the template. RT-PCR products can be analyzed on a 1% agarose gel containing 0.5 (μg/liter ethidium bromide. Specificity of the PCR products can be determined by restriction analysis. Selection of an appropriate restriction enzyme for digestion of the amplificates is possible using a restriction site analysis program, e.g., RESTRI as included in PC/GENE. It may be useful to purify the PCR products prior to restriction analysis using the Qiaquick PCR purification kit (Qiagen, Hilden, Germany), which allows for appropriate restriction buffer conditions. Semi-quantification of gene induction has been determined by differential RT-PCR using housekeeping genes, such as β-actin or glyceraldehyde-3-phosphate dehydrogenase (GAPDH), which are known to remain non modulated under the conditions assessed.[13] Subsequently, amplification products have been separated by HPLC and quantified using an online UV photometer as described.

Reporter Gene Assay

For identification of transcriptional regulatory proteins, homology searches of corresponding consensus sequences in the genome are informative for a very preliminary analysis. To determine the functional relevance of such a site for UVA radiation-induced gene expression, it is necessary to study the binding of trancription factors to this site (gel electrophoresis mobility shift assay) as well as its role in promotor activation (transient transfection with reporter gene).

Transfection assays have been performed on multiple well plates (12 well). Plasmids for transfection are prepared by anion-exchange chromatography, e.g., Qiagen plasmid kit (Qiagen, Hilden, Germany), by silica gel, e.g., Wizard Maxipreps DNA or $CsCl_2$. Transfection of primary keratinocytes is still a problem, although there are a variety of protocols and reagents for transfection available commercially. In our hands, the transfection of primary human keratinocytes or KB cells has been mediated best by Polybrene (Aldrich, Deisenhofen, Germany) as described[14] or by Superfect (Qiagen, Hilden, Germany), respectively. The principle of both types of

[13] M. Berneburg, N. Gattermann, H. Stege, M. Grewe, K. Vogelsang, T. Ruzicka, and J. Krutmann, *Photochem. Photobiol.* **66,** 271 (1997).
[14] Ch-K. Jiang, D. Connolly, and M. Blumenberg, *J. Invest. Dermatol.* **94,** 969 (1991).

transfection reagents is their polycationic nature, which assembles the negatively charged plasmid DNA into compact structures that bind to receptors
on the surface of eukaryotic cells. These complexes are internalized via
lysosomal pathways. A commercially available luciferase reporter gene
system (Promega, Madison, WI) has been used. The firefly luciferase reporter gene has been widely applied in both transient and stable transfection
protocols in eukaryotic cells to measure the transcriptional regulation of
DNA elements. The firefly luciferase is a 62,000-Da protein, which is active
as a monomer and does not require subsequent processing for its activity.
This enzyme produces light by the ATP-dependent oxidation of luciferin.
In the presence of excess substrate, light output from samples containing
firefly luciferase becomes proportional to enzyme concentration in the
assay, thus allowing for its quantification.[15] Conventional luciferase assays
generate a flash of light that decays rapidly after the enzyme and substrate
are mixed, therefore requiring automated injection luminometers for the
measurement of photon production. Stabilized forms of luciferase are also
available that can be detected in scintillation counters. This may be interesting if the amount of gene expression is expected to be high because scintillation counters have a less sensitive detection limit in comparison to luminometers, which can measure as little as 10^{-20} mol of luciferase. Firefly
luciferase has the highest efficiency of any known bioluminescence reaction
with a quantum yield of 0.88. Luciferase assays are linear over at least eight
orders of magnitude. Sensitivity of the reporter gene to be cloned behind
the regulatory sequence of interest is an important issue. In general, luciferase assays are thought to be 1000-fold more sensitive than [^{14}C]chloram-
phenicol acyltransferase assay.

For transfection of cells (triplicates of 2×10^5 cells), 20 μg of DNA is
necessary to transfect primary human keratinocytes (1.5 μg for KB cells).
Transfection efficiency has been monitored by cotransfecting cells with the
simian virus 40 promoter and enhancing region β-galactosidase control
plasmid (Promega) using the β-galactosidase assay system Galacto-Light
(Tropix, Bedford, MD). This detection kit excludes the detection of endogenous β-galactosidase activity according to a shift in pH by adding a reaction
buffer. Routinely, 10 μg of DNA of this control plasmid is cotransfected.
Use of a plasmid containing green fluorescent protein [plasmid pGreen
Lantern (Life Technologies, Karlsruhe, Germany)] as a reporter gene could
avoid the problem of endogenous enzyme activity. This reporter gene can
be detected by fluorescent microscopy, FACS, or fluorescent plate readers.
A convenient way to combine the speed the sensitivity of two luciferase
reporter enzymes into an integrated single-tube, dual reporter assay for

---

[15] A. R. Brasier and D. Ron, *Methods Enzymol.* **216,** 384 (1992).

internal normalization of gene expression gene is Dual-Luciferase (Promega, Mannheim, Germany). The firefly luciferase reporter activity is detected first; addition of a quencher for firefly luciferase and concomitant activation of *Renilla* luciferase as transfection control allow the completion of both assays in about 30 sec using a dual auto injector system.

Promoter activation has been expressed as mean ±SD of relative specific luciferase activity, which was based on protein content. The latter has been determined according to Bradford.[16] There are several protein assays based on methods available, such as Bio-Rad protein assay (Bio-Rad Laboratories GmbH, Munich, Germany) or Micro BCA protein assay reagent (Pierce Europe b.v., Ba oud-Beijerland, Holland). Note that the commercially available lysis buffers contain detergents to lyse the cells. These detergents may falsify the protein microassay. It may therefore be necessary to add an aliquot of the lysis buffer to the different dilutions of the protein standard to control for a potential effect of the detergent on the quality of the assay. By employing these techniques it has been possible to demonstrate that the AP-2 site serves as the UVA radiation-responsive as well as the singlet oxygen-responsive element of the human ICAM-1 promotor.

Gel Electrophoresis Mobility Shift Assays

The gel electrophoresis mobility shift assay is simple and straightforward. An end-labeled, double-stranded DNA oligonucleotide in the range of 20 to 30 bp containing the promoter site of interest is mixed with a nuclear extract and then subjected to a native gel electrophoresis on a 5% polyacrylamide gel with 1× TBE buffer.[17] Nuclear proteins that bind to the promoter site cause a retardation in migration of the radiolabeled oligonucleotide. The specificity of the binding is proved by competition experiments with access of specific, i.e., identical, but unlabeled oligonucleotide. Unspecific protein binding to any DNA fragment can be competed with addition of a synthetic poly(dI · dC) × (dI · dC) (Amersham Pharmacia Biotech, Freiburg, Germany). To identify the protein binding to the fragment, antibodies raised to specific DNA-binding proteins can be added to the binding reactions. The addition of antibodies leads either to a supershift of the fragment, which now binds the transcription factor and the antibody directed against it, or to the disappearence of the shifted band due to the antibody binding to a domain of the transcription factor that is required for DNA binding. In this context it is important to note that the antibody of choice should bind to the protein in its native configuration. For supershift

[16] M. M. Bradford, *Anal. Biochem.* **72,** 248 (1976).
[17] M. M. Garner and A. Revzin, "Gel Electrophoresis: A Practical Approach," p. 202. IRL Press, Oxford, 1992.

assay experiments, samples are preincubated for a period of time with an amount of antiserum against the transcription factor of interest before the addition of the radiolabeled fragment. Time and temperature of preincubation have to be determined for each extract/antibody system and range from 4° up to 37° and between 15 min and 1 hr.

Cells for the preparation of nuclear extracts have been cultured on 15-cm cell culture dishes. For harvest, cells have been washed with PBS at room temperature. Ten milliliters of ice-cold PBS is added and cells are scraped off the plate using a rubber policeman. In order to prepare extracts, cells are washed with cold PBS. All subsequent steps are performed at 4°. Cells are suspended in five packed cell pellet volumes of buffer A [10 mM HEPES, pH 7.9; 1.5 mM MgCl$_2$, 10 mM KCl, and 0.5 mM dithiothreitol (DTT)] and allowed to stand for 10 min. Cells have been collected by centrifugation [10 min, 460g ($=$ 1500 rpm) Heraeus Megafuge 3.0R, Kendro Laboratories] and suspended in two packed cell pellet volumes of buffer A and lysed with 10 strokes of a Dounce tissue homogenizer (Teflon B type pestle, Wheaton). The homogenate is centrifuged for 10 min at 818g (2000 rpm Megafuge 3.0R) to pellet nuclei. The nuclei pellet is transferred in Corex glass tubes and is centrifuged for 20 min at 25,000g (Sorvall SS34 rotor) to remove residual cytoplasmic material. The crude nuclei pellet is resuspended in 0.5 ml of buffer C [20 mM HEPES, pH 7.9, 25% (v/v) glycerol, 0.42 M NaCl, 1.5 mM MgCl$_2$, 0.2 mM EDTA, 0.5 mM phenylmethylsulfonyl fluoride (PMSF) and 0.5 mM DTT] and homogenized with a Dounce tissue homogenizer with a Teflon pestle (Wheaton) using 10 strokes. The resulting suspension is kept on ice for 30 min and centrifuged for 30 min at 25,000g (Sorvall SS34 rotor). The clear supernatant is dialyzed overnight against 50 volumes of buffer D [20 mM HEPES, pH 7.9; 20% (v/v) glycerol, 0.1 M KCl, 0.2 mM EDTA, 0.5 mM PMSF, and 0.5 mM DTT]. To avoid protein degradation it might be necessary to add proteinase inhibitors. In the original paper, PMSF is recommended. Meanwhile, there are many more proteinase inhibitors available, which may be added as a cocktail, e.g., Complete (Roche Molecular Biochemicals, Mannheim, Germany).

After dialysis, the protein content of the nuclear extract is determined according to Bradford and aliquots are frozen in liquid nitrogen. If the nature of the transcription factor to be analyzed is unknown, freezing/thawing cycles that might influence DNA binding of the protein should be avoided. Electromobility shift assays have been performed according to Goodwin.[18] The consensus sequences for transcription factor binding sites

---

[18] G. H. Goodwin, "Gel Electrophoresis: A Practical Approach," p. 225. IRL Press, Oxford, 1992.

have been taken from published sequences or deduced from the corresponding promoter using analysis programs such as PC/Gene or Omiga.

Using these methods, it has been demonstrated that transcription factor AP-2 is activated in human keratinocytes by UVA radiation as well as singlet oxygen. The activation pattern is biphasic with an early maximum at 2 hr after stimulation and a second later peak at 24 to 48 hr. This biphasic activation pattern strongly resembles the biphasic activation of UVA-inducible genes and, in general, might be highly characteristic for UVA radiation-induced gene transcription, as it has been observed for every UVA-inducible gene that has been studied thus far.

Concluding Remarks

The methods described in this article have made it possible to partially determine the signal transduction pathway responsible for increased gene expression in UVA-irradiated human keratinocytes. A central observation has been that the generation of singlet oxygen is critically involved in UVA radiation-induced gene expression. Ongoing studies are currently directed at answering the question of which mechanism singlet oxygen is capable of activating transcription factor AP-2. These studies will not only help to better understand the biological consequences that UVA radiation exposure might exert on human skin, but they will also improve our knowledge about the regulation and control of singlet oxygen-mediated effects in human skin cells. Finally, it can be expected that the previously unrecognized role of AP-2 in mediating oxidative stress responses will lead to a better understanding of the mechanisms by which the activity of this well-known and very important transcription factor is being regulated.

[27] Role for Singlet Oxygen in Biological Effects of Ultraviolet A Radiation

By REX M. TYRRELL

A series of observations in the early 1970s established that the cytotoxic action of ultraviolet A (UVA) radiation on both bacteria[1] and mammalian cells[2] is dependent on oxygen, so active oxygen species are certainly involved in cell killing by long wavelength radiation. Furthermore, most

---

[1] R. B. Webb, *Photochem. Photobiol. Rev.* **2,** 169 (1977).
[2] H. J. Danpure and R. M. Tyrrell, *Photochem. Photobiol.* **23,** 171 (1976).

UVA-mediated biological events ranging from DNA damage to skin erythemas[3] are dependent on oxygen. There is substantial *in vitro* evidence predicting that ultraviolet A (UVA 320–400 nm) radiation will generate several types of active oxygen intermediates in cells (see Ref. 4 for review). Active oxygen intermediates generated by UVA radiation therefore clearly have critical biological consequences.

Various studies have attempted to define the nature of the active oxygen intermediates involved in UVA effects. From a series of experiments with different modifying agents, it was concluded that singlet oxygen was involved in the inactivation of cultured human skin fibroblasts that had been irradiated with near-monochromatic UV or near-visible radiation.[5] Some years later it was shown that UVA activation of the heme oxygenase 1 gene, which is the strongest UVA-activable response observed in eukaryotic cells,[6] is also apparently mediated by singlet oxygen generated intracellularly by UVA radiation.[7] Shortly afterward it was shown that the collagenase gene (known to be UVA inducible) could be activated by singlet oxygen liberated by thermal decomposition of a thermolabile endoperoxide.[8] Since then it has been shown that singlet oxygen is probably involved in other examples of UVA-inducible gene expression.[9,10] Not unexpectedly, singlet oxygen has now been implicated in UVA-mediated signal transduction pathways, including p38 and c-jun-N-terminal kinases but not extracellular regulated kinases[11]; UVA also activates certain transcription factors.[10,12] Finally, singlet oxygen has been implicated in UVA induction of oxidative DNA damage.[13]

The approaches that implicate singlet oxygen in UVA-mediated events are necessarily indirect but generally include several lines of supporting evidence showing that singlet oxygen can be a mediator of the event concerned.

[3] M. Auletta, R. W. Gange, O. Tan, and E. Mazinger, *J. Invest. Dermatol.* **86,** 649 (1986).

[4] R. M. Tyrrell, in "Oxidative Stress: Oxidants and Antioxidants" (H. Sies, ed.), p. 57. Academic Press, London, 1991.

[5] R. M. Tyrrell and M. Pidoux, *Photochem. Photobiol.* **49,** 407 (1989).

[6] S. M. Keyse and R. M. Tyrrell, *Proc. Natl. Acad. Sci. U.S.A.* **86,** 99 (1989).

[7] S. Basu-Modak and R. M. Tyrrell, *Cancer Res.* **53,** 4505 (1993).

[8] K. Scharffetter-Kochanek, M. Wlaschek, K. Briviba, and H. Sies, *FEBS Lett.* **331,** 304 (1993).

[9] M. Wlaschek, J. Wenk, P. Brenneisen, K. Briviba, A. Schwarz, H. Sies, and K. Scharffetter-Kochanek, *FEBS Lett.* **413,** 239 (1997).

[10] S. Grether-Beck, R. Buettner, and J. Krutmann, *Proc. Natl. Acad. Sci. U.S.A.* **93,** 14586 (1996).

[11] L.-O. Klotz, C. Perlieux, K. Briviba, C. Pierlot, J.-M. Aubry, and H. Sies, *Eur. J. Biochem.* **260,** 917 (1999).

[12] O. Reelfs, Ph.D. thesis, University of Lausanne, Switzerland, 1999.

[13] E. Kvam and R. M. Tyrrell, *Carcinogenesis* **18,** 2379 (1997).

Evidence for the Role of Singlet Oxygen in a Biological Event:
  Methods

Deuterum Oxide

The lifetime of singlet oxygen is enhanced 10-fold when the species is generated in deuterum oxide rather than water. The simplest test of singlet oxygen involvement is therefore to make up the buffer solutions so that the experiments will be performed in this medium rather than water. In practice, deuterium oxide buffer solution is prepared by dissolving 1 phosphate buffer saline (PBS) tablet (e.g., Oxoid, Basingstoke, UK) in the appropriate volume of 100% D_2O. For monolayer cultures of human fibroblasts, cells are seeded (5×10^5/10-cm dish) in 10 ml of medium 3 days prior to the experiment. Prior to irradiation, the culture medium is removed and kept aside. The cell monolayer is then rinsed with 5 ml PBS (in D_2O supplemented with Ca^{2+}/Mg^{2+}, 0.01% each) and then added to each dish, which are then incubated at 37° for 15 min. Control dishes are incubated for the same time in 5 ml of PBS (made up with H_2O containing 0.01% Ca^{2+}/Mg^{2+}). After pretreatment, cells are irradiated with different doses (0–1 MJ/m^2) of UVA using a broad-spectrum lamp with maximum energy between 360 and 420 nm (UVASUN 3000, Mutzhas, Munich, Germany or a Sellas 4000, Germany) at a distance that does not cause heating during irradiation. After irradiation, the PBS is aspirated, cells are washed with fresh PBS, and the original medium is added back. The cells are then assayed for colony-forming ability or specific mRNA accumulation. Using this approach, it was shown that both the cytotoxic effect of UVA on cultured human fibroblasts[5] and UVA activation of the HO-1 gene[7] are enhanced more than 2-fold by deuterium oxide. The biological enhancement factors (around two) are much lower than would be predicted from the *in vitro* lifetime experiments because components of the cellular milieu will rapidly quench singlet oxygen and counter the deuterum oxide effect. In fact, this quenching would be expected to eliminate all traces of singlet oxygen so that the net increase in biological action is evidence that singlet oxygen is generated by an endogenous chromophore in close proximity to, or even an integral part of, the critical target. The procedure does assume that isotope effects are not significant.

Singlet Oxygen Quenchers

Sodium azide and histidine are both commonly used to quench singlet oxygen effects. Because azide also has many biological effects, including inhibition of RNA polymerase, it is not an ideal modifier to use. However, histidine is not only a good singlet oxygen quencher but has a negligible interaction with superoxide anion so that if a positive quenching effect is shown with this compound then a role for superoxide anion in the effect is unlikely. The chemicals (100 mM sodium azide, 10 mM L-histidine) are dissolved in PBS as described earlier and incubated with cells for 15 min prior to irradiation.

Carotenoids, such as β-carotene and lycopene, both present at significant levels in the diet, are extremely effective physical scavengers of singlet oxygen. However, they have not been widely used as modifying agents. This is partly because they are extremely lipophilic and require appropriate solubilization to deliver them effectively to cells. In addition, they are oxidized readily, which further complicates experimental procedures. Some of them (e.g., β-carotene) form retinoid-like metabolites, which are active in signal transduction, a property that can add further complexity to the biological results observed.

Modifiers of Other Active Oxygen Intermediates

Because effects of deuterum oxide and the modulation of biological events by singlet oxygen scavengers are not in themselves absolutely diagnostic for singlet oxygen, it is useful to use an array of agents believed to interact with other intermediates (e.g., mannitol in the molar range or 4% dimethyl sulfoxide, both of which will quench hydroxyl radicals). Other approaches that have been used include the use of cell lines overexpressing antioxidant enzymes (e.g., catalase, superoxide dismutase, gluthathione peroxidase) and test if the biological events are modified. Inhibitors of these enzymes (e.g., aminotriazole for catalase) may also be used. A lack of effect of these interventions will strengthen the case for the role of singlet oxygen. If modifications are observed by pathways that should not influence singlet oxygen-mediated events, then the role for singlet oxygen becomes less clear.

Generators of Singlet Oxygen

If a biological effect resulting from a given treatment (e.g., UVA irradiation) is believed to act via singlet oxygen, then it is pertinent to examine whether other procedures known to generate singlet oxygen can mimic the effects. Several compounds (including dyes such as methylene blue, rose bengal, and phenol red) interact with light to generate singlet oxygen in a type II photodynamic reaction. Endogenous chromophores also undergo such reactions, an important example being the immediate heme precursor, protoporphyrin IX. When irradiated with light, exogenous protoporphyrin IX or rose bengal can both generate singlet oxygen and strongly activate the HO-1 gene.[14] However, because these reactions are generally carried

---

[14] S. Ryter and R. M. Tyrrell, *Free Radic. Bio. Med.* **24,** 1 (1998).

out in solution, other intermediates, such as hydroxyl radical, can also be generated in type I reactions and complicate interpretation. A simpler approach to generate pure singlet oxygen is to irradiate a sensitizer (such as rose bengal) attached to beads or to a membrane in close vicinity to the target cells but with an air interface.[15]

Thermal degradation of unstable peroxides such as the 1,4-endoperoxide of disodium 3,3'-(1,4-naphthylidene) has also proved to be a suitable way to generate pure singlet oxygen for biological experiments.[16] In this way, it was shown that collagenase (a UVA-activable gene) is induced by singlet oxygen[16] and supports the idea that the UVA activation of collagenase is via singlet oxygen.[17] However, it should be noted that, for certain effects, singlet oxygen must be generated intracellularly so that interpretation of results showing a lack of effect of an exogenous generator must take this into account.

A Role for Iron

When evidence for a role of iron in oxidative reactions has been obtained, it has often been concluded that Fenton chemistry is involved and that hydroxyl radicals are implicated in the response. However, iron also acts as a catalyst for the lipid chain peroxidation reaction. Because singlet oxygen may initiate lipid peroxidation events and because oxidative membrane damage has been linked to both UVA-induced cell death and early steps in signaling pathways, an influence of iron on UVA effects may well be related to events quite independent of Fenton chemistry. Certainly iron can influence UVA activation of the HO-1 gene,[18] but detailed investigations have not been undertaken.

There are several approaches that can be undertaken to understand the role of iron. Cellular iron may be depleted with iron chelators with a variety of selected properties or, alternatively, cells can be iron loaded. Two convenient iron-loading treatments involve either overnight incubation of the cells with ferric citrate (10–100 μM) or hemin treatment (5–100 μM) for 2–4 hr. Because iron that becomes available within the cells will be stored in the iron storage protein, ferritin, within a few hours the effects should be analyzed fairly soon after the addition of iron to avoid negative effects due to natural sequestration.

[15] M. R. Midden and S. Y. Wang, *J. Am. Chem. Soc.* **105**, 4129 (1983).

[16] K. Briviba, L. O. Klotz, and H. Sies, *Biol. Chem.* **378**, 1259 (1997).

[17] K. Scharffetter-Kochanek, M. Wlaschek, A. Hogg, K. Bolsen, A. Schothorst, G. Goerz, T. Krieg, and G. Flewig, *Arch. Dermatol. Res.* **283**, 506 (1991).

[18] S. M. Keyse and R. M. Tyrrell, *Carcinogenesis* **11**, 787 (1990).

Two very sensitive methods have become available for estimating the presence of free (labile) intracellular iron. The first relies on the measurement of the level of activation of iron regulatory protein (IRP) in the cytoplasmic extract of treated cells.[19,20] This protein binds to specific elements within mRNA molecules (IREs) in response to low iron levels and controls levels of both ferritin and transferrin receptors. Extremely sensitive estimates of free iron can be made by monitoring changes in IRP/RNA binding by a bandshift assay that uses a radio labeled probe containing the specific IRE from H-chain ferritin. Cytoplasmic extracts of cells to be examined are mixed with the RNA probe and are run on acrylamide gels to separate the bound complex. Levels of the complex estimated by autoradiography will provide information about the availability of intracellular free iron. Alternatively, the level of aconitase activity in the cytosolic fractions of cells, devoid of mitochondria, can provide an estimate of intracellular free iron, as in response to high iron, IRP-1 protein becomes inactive as an IRE/binding protein and instead functions as a cytosolic aconitase.

Because IRP binding can be oxidatively modified independently of iron levels, it is useful to use an independent method to indicate free intracellular iron. A popular new approach is the method developed by Breuer and co-workers,[21] which is based on total dequenching of the sector of calcein (a divalent fluorescent chelator) fluorescence that is quenched by bound iron. In this method, cells loaded with calcein are transferred to a spectrofluorimeter and the level of intracellular calcein-bound iron is determined by the increase in fluorescence produced by the addition of the highly permeable iron chelator salicaldehyde isonicotinoyl hydrazone. To establish the relationship between fluorescence change and intracellular chelatable iron concentrations, calibrations are undertaken using various concentrations of ferrous ammonium sulfate ($0.1-1~\mu M$) in the presence of ionophore A23187 ($10~\mu M$) and the corresponding change in fluorescence is monitored. We have used the latter technique to follow the increased levels of free, chelatable iron following UVA irradiation of primary human skin fibroblasts, FEK4.[22]

In summary, we have described a series of experimental approaches that together have provided strong evidence that singlet oxygen is involved in several UVA-mediated events. A likely target for singlet oxygen is the membrane, and the peroxidation of the lipid that ensues may be the trigger

[19] R. D. Klausner, T. A. Rouault, and J. B. Harford, *Cell* **72**, 19 (1993).
[20] C. Pourzand, O. Reelfs, E. Kvam, and R. M. Tyrrell, *J. Invest. Dermatol.* **112**, 419 (1999).
[21] W. Breuer, S. Epsztejn, and Z. I. Cabantchik, *FEBS Lett.* **382**, 304 (1996).
[22] C. Pourzand, R. D. Watkin, J. E. Brown, and R. M. Tyrrell, *Proc. Natl. Acad. Sci. U.S.A.* **96**, 6751 (1999).

for both cell death and events that modulate gene expression. We therefore also include new approaches to measure labile iron pools, as this element may act in concert with singlet oxygen to catalyze lipid peroxidation chain reactions. These methods should be generally applicable to a study of the role of single oxygen and iron in the biological events that result from treatments that induce cellular stress.

Acknowledgments

The research of the author has been supported by the Association for International Cancer Research (UK) and the U.K. Department of Health (Contract No. 121/6378).

[28] Ultraviolet A-1 Irradiation as a Tool to Study the Pathogenesis of Atopic Dermatitis

By JEAN KRUTMANN

Introduction

Ultraviolet A (UVA) radiation is widely used for the treatment of skin diseases, including atopic dermatitis.[1] Although the beneficial effects of phototherapy for patients with atopic dermatitis have been appreciated for decades, it has been only recently that the mechanisms by which UVA irradiation improves skin symptoms in these patients have been analyzed. These studies have been made possible through the availability of *in situ* techniques that allow detection and monitoring of immunomodulatory consequences in lesional skin of patients and their subsequent correlation with the course of atopic dermatitis. From these studies a novel concept concerning the pathogenesis of eczematous skin lesions in atopic dermatitis patients has emerged.[2] It is now generally accepted that cytokines derived from skin-infiltrating T-helper cells are responsible for and contribute to the generation and maintenance of eczematous skin lesions in atopic dermatitis patients. In addition, different T-helper cell cytokines are relevant for different stages of atopic dermatitis.

---

[1] J. Krutmann, *in* "Fitzpatrick's Dermatology in General Medicine" (I. M. Freedberg, A. Z. Eisen, K. Wolff, K. F. Austen, L. A. Goldsmith, S. I. Katz, and T. B. Fitzpatrick, eds.), 5th Ed., p. 2870. McGraw-Hill, New York, 1999.
[2] M. Grewe, C. A. F. M. Bruijnzeel-Koomen, E. Schöpf, T. Thepen, A. G. Langeveld-Wildschut, T. Ruzicka, and J. Krutmann, *Imunol. Today* **19,** 359 (1998).

This article describes highly sensitive techniques to semiquantitatively assess the expression of T-helper cell-derived, proinflammatory cytokines in lesional skin of patients with atopic dermatitis at the mRNA and protein level. Results obtained by employing these methods have suggested that skin-infiltrating T-helper cells are peferential targets for UVA radiation phototherapy.[3] We have therefore subsequently developed a double-labeling immuofluorescence technique to examine the effects of UVA phototherapy on the number and morphology of T cells present in eczematous lesions of atopic dermatitis patients.

Ultraviolet A Irradiation of Human Skin

Phototherapeutic modalities that may be used to treat patients with atopic dermatitis include broadband UVB, 311-nm- or narrowband-UVB, UVA/UVB, and UVA-1 (340–400 nm) phototherapy.[4] Within recent years it has been realized that the efficacy of phototherapy for atopic dermatitis is greatest in UVA range. As a result, UVA-1 phototherapy is currently being regarded as the photherapeutic modality of choice for this indication.[5–7] In UVA-1 phototherapy, the almost complete lack of irradiation shorter than 340 nm, i.e., in the UVA-2 (320–340 nm) as well as UVB (280–320 nm) range, allows exposure of human skin to single doses of up to 130 J/cm^2 UVA-1 (= high-dose UVA-1 phototherapy).

For UVA-1 irradiation, a metal halogenid lamp with an appropriate filtering and cooling system is employed. In our studies, a UVA-1 Sellamed System Dr. Sellmaier (Sellas GmbH, Gevelsberg, Germany) irradiation device has been used. The emission is filtered with a UVA-1 filter (Sellas GmbH, Gevelsberg, Germany) and an infrared absorbing KG1 filter (Schott, Mainz, Germany). As a consequence, only UV wavelengths greater than 340 nm and smaller than 450 nm are emitted. The UVA-1 irradiance needs to be measured at regular intervals (1× daily) with a UVAMETER (Mutzhas, Munich, Germany) and a IL 1700 photometer (International Light, Newburyport, MA). In order to achieve optimal therapeutic effects, irradiance should be in the range of 70–90 mW/cm^2 at body distance. Similar irradiation devices with comparable emission spectra and irradiation intensities can be obtained from Schulze & Böhm, Cologne, Germany.

[3] J. Krutmann and A. Morita, *J. Invest. Dermatol. Symp. Proceed.* **4,** 70 (1999).

[4] J. Krutmann, *Dermatol. Ther.* **1,** 24 (1996).

[5] J. Krutmann, W. Czech, T. Diepgen, R. Niedner, A. Kapp, and E. Schöpf, *J. Am. Acad. Dermatol.* **26,** 225 (1992).

[6] J. Krutmann and E. Schöpf, *Acta Derm. Venereol.* (*Stockh.*) **176,** 120 (1992).

[7] J. Krutmann, T. Diepgen, T. A. Luger, S. Grabbe, H. Meffert, N. Soennichsen, W. Czech, A. Kapp, H. Stege, M. Grewe, and E. Schöpf, *J. Am. Acad. Dermatol.* **38,** 589 (1998).

To rule out sensitivity to UVA-1 irradiation, all patients are phototested before whole body exposure with increasing doses of UVA-1 radiation (0 to 130 J/cm^2) with a UVASUN 5000 (Mutzhas, Munich, Germany) or a Sellamed System Dr. Sellmaier (Sellas GmbH, Gevelsberg, Germany) partial body irradiation device. For each treatment, the patient's whole body is exposed to 130 J/cm^2 (= high-dose UVA-1 phototherapy) per body half from a Sellamed irradiation device (Sellas GmbH, Gevelsberg, Germany). Because this irradiation device allows exposure only from the top, patients have to turn from back to front every 10 min during the irradiation. For the treatment of atopic dermatitis, the total number of exposures is limited to 15. Patients are irradiated on a daily basis five times per week.

RT-PCR Based Semiquantitative Assessment of T-Cell Cytokine Expression in Human Skin

This technique allows the monitoring of cytokine mRNA expression in 4-mm punch biopsies that are taken from eczematous skin lesions in patients with atopic dermatitis before, during, and after phototherapy.[8,9] The specimens are washed twice in 0.9% NaCl for 20 sec to remove blood and then snap-frozen in liquid nitrogen. Frozen biopsy specimens are grounded to powder and total RNA is extracted by a modified chloroform/phenol method.[10] Total RNA is washed and pelleted three times in 70% ethanol and then routinely taken up in 200 μl aqua bidest.

In order to semiquantitatively assess cytokine mRNA expression in these samples, a differential RT-PCR method can be employed.[8–13] For this purpose, total RNA is reverse transcribed using mouse maloney leukemia virus reverse transcriptase and an oligo-dT$_{18}$ primer. Identical amounts of cDNA are subjected to increasing cycle numbers of PCR to obtain the linear amplification range, and then increasing amounts of cDNA (for up to 64-fold of the starting amount) are subjected to PCR of a given cycle number within the linear range to exclude the possibility that increased amounts of specific DNA lead to disturbance of the linearity in PCR ampli-

[8] M. Grewe, K. Gyufko, E. Schöpf, and J. Krutmann, *Lancet* **343**, 25 (1994).

[9] M. Grewe, S. Walter, K. Gyufko, W. Czech, E. Schöpf, and J. Krutmann, *J. Invest. Dermatol.* **105**, 407 (1997).

[10] P. Chomzynski and N. Sacchi, *Anal. Biochem.* **162**, 156 (1987).

[11] H. P. Henninger, R. Hoffmann, M. Grewe, A. Schulze-Specking, and K. Decker, *Biochem. Biophys. Hoppe-Seyler* **374**, 625 (1993).

[12] S. Grether-Beck, S. Olaizola-Horn, H. Schmitt, M. Grewe, A. Jahnke, J. P. Johnson, K. Briviba, H. Sies, and J. Krutmann, *Proc. Natl. Acad. Sci. U.S.A.* **93**, 14586 (1996).

[13] M. Grewe, K. Gyufko, A. Budnik, M. Berneburg, T. Ruzicka, and J. Krutmann, *J. Invest. Dermatol.* **107**, 865 (1996).

fication. For the cytokine primers used in our studies (Table I), amplification was found to be linear for up to 31 cycles and for up to 26 cycles for the β-actin (= housekeeping gene) primers.[9] These "housekeeping" gene primers are being used for estimation of similar amounts of cDNA used for PCR. Accordingly, relative amounts of cDNA of each sample to be inserted into PCR can be calculated to yield similar amounts of β-actin PCR products, and this is confirmed by repeating RT-PCR for β-actin with the calculated amount of cDNA. A sample (1 of 200) of total cDNA is routinely subjected to 24 PCR cycles (which is in the linear amplification range for cycle numbers) using a primer pair for β-actin and is then subjected directly to ion-exchange chromatography. The negatively charged DNA fragments are bound to an ion-exchange column and subsequently exposed to an increasing NaCl concentration.[11] This leads to the elution according to the amount of negative charges (length) of the DNA fragments. For quantification, these fractions are then evaluated by an on-line UV spectro-photometer (Gynkotek, Germering, Germany) at a wavelength of 260 nm. Data can be obtained as histograms of absorption (DNA concentration) on the y axis and elution time (DNA amount) on the x axis for each PCR product. Values for areas under the curve of each PCR product can be used for further calculations.[11]

To investigate cytokine mRNA expression, five times more cDNA is inserted than for β-actin PCR. PCR is carried out with 28 cycles, which is in the linear amplication range of all cytokine cDNAs. Paralell to each cytokine PCR, a PCR for β-actin is also performed. Furthermore, each PCR of each sample for each cytokine should be carried out at least two times. Products are quantified by ion-exchange chromatography as de-

TABLE I
PRIMER PAIRS SPECIFIC FOR CYTOKINES AND β-ACTIN

Cytokine	Sequence 5'–3'
Interferon-γ	AGTTATATCTTGGCTTTTCA
	ACCGAATAATTAGTCAGCTT
Interleukin-2	ACTCACCAGGATGCTCACAT
	AGGTAATCCATCTGTTCAGA
Interleukin-4	CTTCCCCCTCTGTTCTTCCT
	TTCCTGCCGAGCCGTTTCAG
Interleukin-12 p35	ACCCAGGAATGTTTCCCATGC
	TCTGTCAATAGTCACTGCCCG
Interleukin-12 p40	AAAGGAGGCGAGGTTCTAAGCC
	TTTGCGGCAGATGACCGTGG
β-Actin	GTGGGGCGCCCCAGGCACCA
	CTCCTTAATGTCACGCACGATTTC

scribed earlier. To ensure identity of products it is recommended that their chromatogram peaks be collected and digested with an appropriate restriction endonuclease. Fragments are then subsequently visualized on agarose gel by ethidium bromide staining. Lengths of the restriction fragments are compared to those deduced from published mRNA sequences of the respective cytokines, e.g., by using PC/GENE software (Intelligenetics Inc., Mountain View, CA).

By employing this technique, we have been able to demonstrate that in lesional atopic skin, the T-helper-1-cell-derived cytokine interferon (IFN)-γ is overexpressed.[8] Successful UVA-1 phototherapy significantly and specifically reduced *in situ* expression of IFN-γ to background levels. This indicated that skin-infiltrating, IFN-γ-producing T-helper cells are targets for UVA-1 phototherapy. The predominance of IFN-γ production at this later stage of the disease is in contrast to the initiation phase of atopic dermatitis, which includes the time point at which an inhalant allergen, a given patient is sensitized to, penetrates into the skin until the development of clinically apparent eczema.[9] During this early period, IFN-γ mRNA expression is down-regulated below constitutive levels, whereas, at the same time, the T-helper-2-cell-derived cytokine interleukin (IL)-4 is overexpressed. At later time points, however (which is when the patient is undergoing phototherapy because of preexisting eczema), this early Th-2 response is switched into a Th-1 response that is dominated by IFN-γ. By employing the techniques described earlier we have been able to provide evidence that this switch is being mediated by the cytokine IL-12.[2]

This two-phase model of the pathogenesis of eczematous lesions in patients with atopic dermatitis has been corroborated and extended at the protein level in studies employing a double-labeling technique that allows the simultaneous immunohistochemical detection of T-cell-derived cytokines and T-cell surface markers.[14]

Assessment of UVA-1 Radiation Effects on Morphology and Number of T Cells

In order to assess the effects of UVA-1 phototherapy on the morphology and number of skin-infiltrating T cells, sequential biopsies can be taken in patients with atopic dermatitis from chronic, lichenified eczematous skin lesions present in the flexural ceases of their elbows before and after the 1st, 2nd, 3rd, 4th, and 10th UVA1 radiation exposure. Cryostat sections are prepared and fixed in chilled acetone for 10 min. These specimens can

---

[14] T. Thepen, E. G. Langeveld-Wildschut, I. C. Bihari, and C. A. F. M. Bruynzeel-Koomen, *J. Allergy Clin. Immunol.* **97**, 828 (1996).

then be processed further in order to simultaneously detect T-helper cells and cells undergoing apoptosis (programmed cell death).[15] Identification of T-helper cells is achieved by immunofluorescence microscopy employing a monoclonal antibody directed against the CD4 antigen being expressed by human T-helper cells. For the *in situ* detection of apoptosis in CD4$^+$ T cells, the TUNEL method can be employed. This method is based on the labeling of free DNA 3'-OH termini, which are labeled with fluorescein-labeled nucleotides in the presence of a terminal deoxynucleotidyltransferase for 1 hr at 37°. Prior to labeling it is necessary to permeablize the cells, e.g., by treatment of cryostat sections with 0.1% sodium citrate and 0.1% Triton X-100 (Boehringer Mannheim, Mannheim, Germany). Apoptotic cells will give a green fluorescence, whereas CD4$^+$ T cells will show up in red, if the anti-CD4 monoclonal antibody is rhodamin conjugated (e.g., monoclonal antibody MT310; mIgG1; Dako Diagnostika, Hamburg, Germany). As a consequence, double-positive, i.e., apoptotic, CD4$^+$ T cells can be identified as orange cells. For semiquantitative assessment of apoptotic T-helper cells it is recommended that three serial sections per specimen be analyzed by counting the number of double-positive cells in three high-power ($\times$200) view fields using a fluorescence microscope.[15]

By employing this technique we have been able to demonstrate that UVA-1 phototherapy is highly effective in inducing apoptosis in skin-infiltrating T-helper cells in patients with atopic eczema. As a result of apoptosis, T cells are being depleted from eczematous skin. This is followed by significant clinical improvement of the patients' skin.[15]

Conclusion

The availability of the highly sensitive *in situ* techniques described previously has provided the methodological basis for monitoring immunological changes being induced by UVA-1 phototherapy in the skin of patients with atopic dermatitis. As a consequence, it is now generally appreciated that the induction of apoptosis in skin-infiltrating T cells is a central mechanism operative in high-dose UVA-1 phototherapy.[15–17] A clinical result of this observation is the successful use of UVA-1 phototherapy for the treatment of other T-cell-mediated skin diseases, such as cutaneous T-cell lymphoma (CTCL) or mycosis fungoides.[18] It has also been learned that

[15] A. Morita, T. Werfel, H. Stege, C. Ahrens, K. Karmann, M. Grewe, S. Grether-Beck, T. Ruzicka, A. Kapp, L.-O. Klotz, and J. Krutmann, *J. Exp. Med.* **186,** 1763 (1997).
[16] J. Krutmann, *J. Photochem. Photobiol. B Biol.* **44,** 159 (1998).
[17] D. E. Godar, *J. Invest. Dermatol.* **112,** 3 (1999).
[18] H. Plettenberg, H. Stege, M. Megahed, T. Ruzicka, Y. Hosokawa, T. Tsuji, A. Morita, and J. Krutmann, *J. Am. Acad. Dermatol.* **41,** 47 (1999).

induction of apoptosis represents a general mechanism by which UV photo-therapy of T-cell-mediated skin diseases works.[16] Successful UVB therapy of patients with psoriasis was also found to result from the induction of apoptosis in skin-infiltrating T cells.[19,20] Similarly, the fact that 311-nm UVB phototherapy, which uses UVB radiation in the less erythemogenic (>300 nm) range, is superior to broadband UVB phototherapy, which contains relatively large amounts of erythemogenic UVB radiation below 300 nm, has been attributed to the observation that 311-nm UVB is more efficient in inducing T-cell apoptosis in the dermal compartment of psoriatic skin as compared with broadband UVB irradiation.[18]

In conclusion, it can be expected that the continuing use of the methods described in this article will not only help improve existing and develop new phototherapeutic modalities, but will also provide new insight into the pathogenesis of skin diseases that respond to phototherapy.

Acknowledgment

Our studies were supported by the Deutsche Forschungsgemeinschaft, SFB 503, Project B2 and by the Alexander-von-Humboldt-Foundation.

[19] J. G. Krueger, J. T. Wolfe, R. T. Nabeja, V. P. Vallat, P. Gilleaudeau, N. S. Heftler, L. M. Austin, and A. R. Gottlieb, *J. Exp. Med.* **182,** 2057 (1995).
[20] M. Ozawa, K. Ferenci, T. Kikuchi, I. Cardinale, L. M. Austin, T. R. Coven, L. H. Burack, and J. G. Krueger, *J. Exp. Med.* **189,** 711 (1999).

[29] Ultraviolet A Radiation-Induced Apoptosis

By AKIMICHI MORITA and JEAN KRUTMANN

Introduction

Ultraviolet (UV) radiation has been used for decades in the treatment of patients with a variety of T-cell-mediated skin diseases, including atopic dermatitis and psoriasis. Among the different phototherapeutic modalities currently available, recent interest has focused on the use of irradiation devices that allow exposure of human skin to wavelengths in the longer UVA range, i.e., UVA-1 (340–400 nm).[1] The almost complete absence of erythemogenic wavelengths shorter than 340 nm allows exposure of human

[1] J. Krutmann, *in* "Fitzpatrick's Dermatology in General Medicine" (I. W. Freedberg, A. Z. Eisen, K. Wolff, K. F. Austen, L. A. Goldsmith, S. I. Katz, and T. B. Fitzpatrick eds.), 5th Ed., p. 2870. McGraw-Hill, New York, 1999.

skin to single doses of up to 130 J/cm^2.[2] These clinical developments have stimulated studies on the mechanisms by which UVA phototherapy works.[3] From these studies it has been learned that UVA phototherapy improves symptoms in patients with T-cell-mediated skin diseases by killing skin-infiltrating T cells through the initiation of apoptosis.[4] These studies have been made possible through a unique combination of *in vivo* methods that allow the monitoring of apoptosis in skin-infiltrating T cells *in situ* in the skin of patients undergoing UVA phototherapy with *in vitro* techniques that mimic the therapeutic situation as closely as possible and have helped greatly in unraveling the photobiological and molecular mechanisms by which UVA radiation induces T-cell apoptosis. This article describes these methods and discusses their relevance to UVA radiation-induced T-cell apoptosis.

Ultraviolet A Irradiation of Human T Cells

Ultraviolet A radiation applied during phototherapy can reach the lower levels of the human dermis. In order to test whether phototherapy-induced apoptosis in skin-infiltrating T cells results from direct effects, we have performed *in vitro* studies under conditions closely resembling the therapeutic situation. The UVA source employed exhibited an emission spectrum identical to that used for phototherapy, and the *in vitro* UVA radiation doses were equivalent to those achieved during phototherapy in the dermal compartment of human skin.

In Vitro Ultraviolet A Irradiation of Human T Cells

Human cultured or freshly isolated T cells are harvested and resuspended in phosphate-buffered saline (PBS) or nonsupplemented RPMI 1640 medium without phenol red (Biochrom, Berlin, Germany) in appropriately sized dishes for UVA irradiation. The suspended cells must not overlay each other, as otherwise a homogenous irradiation will not be achieved. For this purpose, the cell suspension in the dish should be observed under a microscope to make sure that cells are single suspended and do not overlay. For most applications, e.g., analysis of apoptosis, surface protein expression, or mRNA expression, one million cells/sample is enough. Six- or 12-well flat-bottom tissue culture plates are recommended. The concentration of cell suspensions should be $3-5 \times 10^5$ cells/ml. For UVA irradia-

---

[2] J. Krutmann, *Dermatol. Ther.* **1**, 24 (1996).

[3] J. Krutmann and A. Morita, *J. Invest. Dermatol. Symp. Proceed.* **4**, 70 (1999).

[4] A. Morita, T. Werfel, H. Stege, C. Ahrens, K. Karmann, M. Grewe, S. Grether-Beck, T. Ruzicka, A. Kapp, L. O. Klotz, H. Sies, and J. Krutmann, *J. Exp. Med.* **186**, 1763 (1997).

tion, plastic lids are removed and cells are exposed. Cells are harvested and subjected to apoptosis detection systems as described in the following sections.

Irradiation Device and Dosimetry. UVASUN 5000 BIOMED irradiation devices (Mutzhas, Munich, Germany) are used for UVA radiation. The emission is filtered with UVACRYL (Mutzhas, Munich, Germany) and UG1 (Schott Glaswerke, Munich, Germany) and consists exclusively of wavelengths greater than 340 nm. The UVA output is determined with a UVAMETER (Mutzhas, Munich, Germany) or an IL 1700 photometer (International Light, Newburyport, MA) equipped with an IL SED 038 photodetector. It is approximately 100 mW/cm^2 at target distance of 30 cm from a tube.

Singlet Oxygen Quencher and Deuterium Oxide Treatment, and Singlet Oxygen Generation. Reagents capable of quenching (sodium azide) or enhancing (deuterium oxide) singlet oxygen effects have have been found to modulate UVA radiation-induced human T-cell apoptosis. The singlet oxygen-mediated T-cell apoptosis can be inhibited by coincubation with a substance that quenches the effect of singlet oxygen such as sodium azide (NaN$_3$). NaN$_3$ is resolved in PBS and is present at a concentration of 50 mM only during irradiation of cells. The presence of NaN$_3$ leads to a dose-dependent reduction of UVA radiation-induced apoptosis. However, concentrations >100 mM NaN$_3$ have cytotoxic effects on T cells. However, deuterium oxide (D$_2$O) enhances the half-life of singlet oxygen. Coincubation with D$_2$O during irradiation leads to an increase of UVA radiation-induced apoptosis as compared to UVA irradiation alone. Deuterium oxide (99.9 atom % D) is used at a final concentration of 90% in PBS only during the irradiation.

Singlet oxygen is generated by thermal decomposition of the endoperoxide of the disodium salt of 3,3'-(1,4-naphthylidene)dipropionate (NDPO$_2$), 1 mM in PBS, for 1 hr in the dark at 37°, yielding excited singlet molecular oxygen and 3,3'-(1,4-naphthylidene)dipropionate (NDP). This singlet oxygen-generating system has been shown previously to be well suited for use in cell cultures, as it is water soluble and nontoxic up to 40 mM for 1-hr incubation.[5] Infrared emission of singlet oxygen is measured with a liquid nitrogen-cooled germanium photodiode detector (Model EO-817L, North Coast Scientific Co., Santa Rosa, CA). The rate of singlet oxygen generation is monitored by the formation of NDP. Fifteen minutes after the addition of 1 mM NDPO$_2$, the rate of singlet oxygen generation is 3 μM/min. As controls, cells should be stimulated with NDP, which is generated by thermal decomposition from the same batch of NDPO$_2$ used. Incubation of T cells

[5] P. DiMascio and H. Sies, *J. Am. Chem. Soc.* **111**, 2909 (1989).

with 30 mM NDPO$_2$ in PBS for 1 hr in the dark at 37° instead of UV irradiation leads to the induction of apoptosis at the same time kinetics.

Material

Atopen-Specific Human T-Helper Cells from Atopic Dermatitis

Human atopen-specific T-helper cell lines have been used for *in vitro* studies. These T-cell lines are specific for *Dermatophagoides pteronyssinus* (Dp) antigen and have been generated from lesional atopic skin by limiting dilution cloning in the presence of interleukin-2 and an extract of Dp allergen as antigen at a concentration of 3 mg/ml.[6] The cells are expanded in 24-well flat-bottom tissue culture plates with allogeneic, irradiated (50 Gy) peripheral blood mononuclear cells (PBMC) as feeder cells (1 × 10^6 cells/well) in the presence of PHA (10 μg/ml; Gibco Life Technologies, Rockville, MD) and rh IL-2 (20 U/ml; Boehringer Mannheim, Mannheim, Germany). Culture medium (Iscove's medium, Biochrom, Berlin, Germany) contains 4% heat-inactivated human AB serum (PAA Laboratories, Cölbe, Germany), 2 mM glutamine, 50 μg/ml of gentamycin, 100 μg/ml of penicillin/streptomycin, and nonessential amino acids (Biochrom, Berlin, Germany).

Freshly Isolated T Cells

Human PBMC are isolated using Ficoll–Paque density gradient (Amersham Pharmacia Biotech, Uppsala, Sweden). CD4$^+$ T cells are selected positively by a flow cytometry sorting (FACSCalibur System, Becton-Dickinson, San Jose, CA) or a magnetic cell-sorting system (Miltenyi Biotec, Bergisch, Germany). The isolated CD4$^+$ T cells are washed with PBS and subjected to irradiation.

UVA Phototherapy for Atopic Dermatitis

For UVA phototherapy, the patient's whole body is exposed to 130 J/cm^2 UVA-1 radiation from a UVASUN 30,000 BIOMED (Mutzhas, Munich, Germany). The emission is filtered with UVACRYL (Mutzhas, Munich, Germany) and UG1 (Schott, Munich, Germany) filters and consists exclusively of wavelengths greater than 340 nm. The irradiance at the body site is approximately 75 mW/cm^2 for UVA, as determined with a UVAMETER (Mutzhas, Munich, Germany) and an IL 1700 photometer (International Light, Newburyport, MA) equipped with an IL SED 038

[6] T. Werfel, A. Morita, M. Grewe, H. Renz, U. Wahn, and J. Krutmann, *J. Invest. Dermatol.* **107**, 871 (1996).

photodetector. UVA phototherapy is conducted as a monotherapy; exposures are given daily for 10 consecutive days. UVA phototherapy leads to a significant improvement of skin symptoms, as has been assessed by a clinical scoring system (total score before UVA phototherapy: 65.4 ± 6.2; total score after UVA phototherapy: 18.2 ± 3.4; $p < 0.001$).[4,7]

Assessment of Apoptosis

In Vitro Detection of Apoptosis

Two distinct modes of cell death, apoptosis and necrosis, can be distinguished based on differences in morphological and molecular changes. Cells undergoing apoptosis show a characteristic pattern of structural changes in nucleus and cytoplasm, including blebbing of the membrane and nuclear disintegration. The nuclear disintegration is associated with DNA cleavage to oligonucleosomal length DNA fragments by endogenous endonuclease. This process can be detected by agarose gels as "DNA laddering." This method is commonly used for the detection of apoptosis. The other methods as described in the following section are also used for the detection of apoptosis at the single cell level.

DNA Laddering. In human T cells, apoptosis is detected by DNA laddering. To examine DNA fragmentation, cells are collected and resuspended in 10 mM Tris–HCl (pH 7.5), 1 mM EDTA, 0.15 M NaCl, 1% SDS, 0.2 mg/ml proteinase K, and 0.5 mg/ml of RNase (Sigma Chemicals, St. Louis, MO). After a 1-hr incubation at 65°, DNA is extracted with phenol and chloroform, precipitated with ethanol, and dissolved in 10 mM Tris–HCl (pH 7.5) containing 1 mM EDTA. Extracted DNA is loaded on a 2% agarose gel and visualized by staining with ethidium bromide.

TUNEL Assay. In human T cells, apoptosis is detected by employing the TUNEL (TdT-mediated dUTP nick end labeling) technique. The TUNEL assay kit is available from several companies (e.g., Boehringer Mannheim, Mannheim, Germany, and MBL, Nagoya, Japan). The irradiated cells are harvested and fixed in a freshly prepared paraformaldehyde solution (4% in PBS, pH 7.4) for 10 min. Following permeabilization with 0.1% sodium citrate and 0.1% Triton X-100 (Boehringer Mannheim, Mannheim, Germany), free DNA 3'-OH termini are labeled with fluorescein-labeled nucleotides in the presence of a terminal deoxynucleotidyltransferase (TdT) for 1 hr at 37°. Cells are washed and analyzed by flow cytometry using a FACScan (Becton-Dickinson, San Jose, CA) counting 1×10^4 cells per sample. Incubation of dUTP without TdT serves as the negative control

[7] J. Krutmann, W. Czech, T. Diepgen, R. Niedner, A. Kapp, and E. Schopf, *J. Am. Acad. Dermatol.* **26**, 225 (1992).

population and data are given as percentage positive (= apoptotic) cells. In atopen-specific human T-helper cells, *in vitro* UVA irradiation induced apoptosis in a dose- and time-dependent manner. Significant apoptosis is already detectable 4 hr after exposure, reaching a maximum 24 hr after irradiation with 30 J/cm^2 UVA radiation.

Effects on Mitochondria. UVA radiation induces mitochondrial transmembrane potential changes ($\Delta\Psi m$), which is followed by the induction of apoptosis.[8] $\Delta\Psi m$ is measured using a specific fluorescent probe, 5,5',6,6'-tetrachloro-1,1',3,3'-tetraethylbenzimidazolylcarbocyanine iodide, JC-1 (Molecular Probes, Eugene, OR). The probe is incorporated and forms aggregates in mitochondriae (red). After UVA irradiation, cells are harvested and incubated with 10 μM of JC-1 for 15 min at 37°. The cells are washed and suspended with an appropriate volume of PBS to be subjected to FACS analysis. After UVA irradiation, JC-1 does not aggregate and exists as a monomer (green) because of a low membrane potential. Another fluorescent probe, rhodamine 123 (Molecular Probes, Eugene, OR), can also be used for a similar assay.

Other Methods for Detection of Apoptosis. During the early process of apoptosis, cells expose phosphatidylserine (PS) on the surface of the membrane. Annexin V binds to PS residues on the membrane. Annexin V is available with several fluorescence makers, e.g., fluorescein isothiocyanate (FITC), green fluorescence protein (GFP), Cy3, which are readily detectable by FACS analysis. After UVA irradiation, cells are harvested and incubated with the labeled annexin V for 15 min at room temperature. The cells are washed and suspended in an appropriate volume of PBS to be subjected to FACS analysis. Optionally, propidium iodide (PI) can be coincubated with annexin V to detect the membrane permeability. During the late stage of apoptosis or in necrotic cells, the cell membrane loses its integrity and becomes leaky. Early apoptotic cells showing annexin V$^+$ PI$^-$ can be distinguished from late apoptotic and necrotic cells. This method can identify earlier stages of apoptosis than can assays based on DNA fragmentation.

In Situ Detection of Apoptosis in CD4$^+$ T Cells

Patients should not have been treated with any systemic or topical agent 4 weeks prior to the start of UVA phototherapy to avoid other therapeutic effects on the skin. In order to assess whether UVA phototherapy of atopic dermatitis patients is accompanied by the induction of apoptosis in skin-infiltrating T cells, sequential biopsies are taken from eczematous skin. Skin specimens are analyzed for apoptotic cells using the TUNEL assay,

---

[8] D. E. Godar, *J. Invest. Dermatol.* **112,** 3 (1999).

followed by an anti-CD4 staining to detect T-helper cells. Before therapy, numerous CD4+ cells are present intradermally in eczematous skin, as reported previously. After the first UVA radiation exposure, CD4+ apoptotic cells are detected. During subsequent UVA treatments, the number of double-positive cells is increased further, whereas the total number of CD4+ cells decreases. After 10 exposures, the total number of intradermally located, CD4+ T cells is diminished significantly, and remaining cells almost exclusively show signs of apoptosis.

Double-Staining Technique. Sequential biopsies are taken in each patient from eczematous skin lesions present in the flexural creases of the elbows before the 1st and after the 1st, 2nd, 3rd, 4th, and 10th UVA radiation exposure. Each skin specimen is embedded in OCT compound (Miles, Elkard, IN), frozen in liquid nitrogen, and stored at $-80°$. For analysis of apoptotic, CD4+ T cells, a double-staining method employing the TUNEL technique, followed by staining with a fluorescence-labeled anti-CD4 monoclonal antibody, is used. Six-micrometer cryostat sections are prepared and fixed in chilled acetone for 10 min. Following permeabilization with 0.1% sodium citrate and 0.1% Triton X-100 (Boehringer Mannheim, Mannheim, Germany), free DNA 3'-OH termini are labeled with fluorescein-labeled nucleotides in the presence of a terminal deoxynucleotidyltransferase (TdT) for 1 hr at $37°$. Sections are subsequently stained with RPE-conjugated mouse antihuman CD4 monoclonal antibody MT310 (mIgG1, Dako Diagnostika, Hamburg, Germany) for 10 min, washed three times with PBS, and then immediately examined by fluorescence microscopy using an Axioplan microscope (Zeiss, Jena, Germany).

Conclusion

The methods described in this article have made it possible to confirm that UVA radiation-induced apoptosis in skin-infiltrating T cells is the major mechanism of action underlying UVA phototherapy of patients with atopic dermatitis. These findings indicate that UVA phototherapy might be used effectively to treat other T-cell-mediated skin diseases as well. Very recent studies have indeed indicated that UVA phototherapy may be used effectively to treat patients with cutaneous T-cell lymphoma.[9] In addition, the identification of singlet oxygen as a major mediator of UVA radiation-induced T-cell apoptosis suggests that singlet oxygen generation in human skin might constitute a previously unrecognized phototherapeutic principle and that it might be of interest to test other wavelengths known

[9] H. Plettenberg, H. Stege, M. Megahed, T. Ruzicka, Y. Hosokawa, T. Tsuji, A. Morita, and J. Krutmann, *J. Am. Acad. Dermatol.* **41,** 47 (1999).

to be capable of singlet oxygen generation within the UV, but possibly also the visible range for their phototherapeutic potential.

Acknowledgments

Our studies were supported by Grant-In-Aid for Scientific Research from the Ministry of Education, Science and Culture of Japan (10670801, AM), a grant from the Deutsche Forschungsgemeinschaft, SFB 503, Teilproject B2, and a grant from the Alexander-von-Humboldt-Foundation.

[30] Singlet Oxygen-Triggered Immediate Preprogrammed Apoptosis

By DIANNE E. GODAR

Introduction

The two distinct morphological forms of cell death are apoptosis and necrosis. These morphological distinctions can be seen using light, fluorescent, or electron microscopy. The salient morphological features of apoptosis include cell shrinkage, plasma membrane "blebbing," vacuolization, and chromatin digestion and condensation along the nuclear membrane. In the final stages, the cell fragments into membrane-bound vesicles called apoptotic bodies. *In vivo,* these apoptotic cells and bodies are removed by phagocytosis; however, *in vitro* they are not removed and consequently undergo "secondary necrosis." During secondary necrosis, lysosomes rupture and release hydrolytic enzymes into the cytoplasm, causing further destruction of internal components, including the plasma membrane, which results in cell swelling and lysis. When the term apoptosis was originally coined in 1972,[1] it only referred to the morphological changes characteristic of this form of cell death, as the underlying biochemical mechanisms were not yet known. Since then the terms apoptosis and programmed cell death (PCD)[2] have been used synonymously, even though some mechanisms of apoptosis are found to proceed independent of transcription and translation,[3,4] a biochemical mechanism referred to as preprogrammed cell death (prePCD).[4] Thus, the term apoptosis only describes the morphological changes that occur during this mode of cell death,

[1] J. F. R. Kerr, A. H. Wyllie, and A. R. Currie, *Br. J. Cancer* **26,** 239 (1972).
[2] R. A. Lockshin and C. M. Williams, *J. Insect. Physiol.* **10,** 643 (1964).
[3] S. Martin, S. Lennon, A. Bonham, and T. Cotter, *J. Immunol,* **145,** 1859 (1990).
[4] D. E. Godar, *Photochem. Photobiol.* **63,** 825 (1996).

whereas the terms prePCD and PCD segregate the general underlying biochemical mechanisms that drive these changes.

Reactive oxygen species (ROS) cause damage to a variety of cellular components that can initiate different apoptotic mechanisms simultaneously, complicating the interpretation of results. ROS cause damage to mitochondrial membranes, triggering immediate prePCD apoptosis (T $\leq$ 30 min), damage to receptors initiating intermediate prePCD apoptosis (T $>$ 0.5 hr $\leq$ 4 hr), and damage to DNA, inducing delayed PCD apoptosis (T $\gg$ 4 hr).[5] Both UVA1 (340–400 nm) radiation and photodynamic therapy (PDT; photosensitizer and visible light $\geq$400 nm) mediate primarily singlet oxygen damage to membranes. Singlet oxygen damage to mitochondrial membranes causes immediate depolarization of the inner transmembrane potential and the immediate appearance of apoptotic cells. Unlike most initiators of apoptosis, singlet oxygen triggers the immediate onset of apoptotic morphology because mitochondria are downstream of all initiating signaling events, such as gene activation and the "initiator" caspase cascade (e.g., caspase 8). When apoptosis is triggered at the mitochondria, either apoptosis-initiating factor (AIF) or cytochrome c is released, which activates the "executioner" caspase cascade (e.g., caspase 3) that causes morphological changes. Thus, morphological characteristics associated with apoptosis occur very rapidly ($\leq$1 hr) when triggered by singlet oxygen damage and is consequently referred to as "immediate" apoptosis.[6]

For several reasons, it is particularly challenging to detect and accurately quantify apoptotic cells and changes that occur during apoptosis following singlet oxygen damage. In addition to initiating different apoptotic mechanisms simultaneously, singlet oxygen damage causes cells to undergo apoptosis so rapidly that significant percentages of cells can be in the final stages within minutes after exposure (some exposure times are 60 min). In the final stages of apoptosis, the plasma membranes become permeable to dyes such as propidium iodide so that these cells appear necrotic. In addition, secondary necrosis quickly follows singlet oxygen-triggered apoptosis, which further complicates the interpretation of results. Moreover, singlet oxygen damage to plasma membranes renders these cells particularly sensitive to common technical manipulations, such as vortexing, centrifuging, washing, pipetting, fixing, and even vital dye staining methods. Some routine methods cannot be used at all or must be modified to optimize the yield by protecting the cells from further damage and loss during processing. A couple of examples of routine techniques that cannot be used at all are cytospins and vital dye staining. Cytospins cannot be used even if the vessels

[5] D. E. Godar, *J. Invest. Dermatol.* **112,** 3 (1999).
[6] D. E. Godar, S. A. Miller, and D. P. Thomas, *Cell Death Different.* **1,** 59 (1994).

and slides are precoated with serum/protein and the centrifugal force and time are decreased as much as possible because singlet oxygen damage to plasma membranes causes them to rupture during this procedure and the remaining cells appear necrotic after fixing and staining (unpublished observations). Simple vital dye stains, such as trypan blue, cannot be relied on to quantify apoptotic cells because they swell and give misleading data.[7] In general, the most reliable techniques include those that have the fewest processing steps or those that can be modified to afford adequate protection of these cells during the entire technical procedure to increase the yield and maintain the true identity of all cell populations.

The methods described in this article allow good recovery and separate analysis of both apoptotic and morphologically normal cell populations following singlet oxygen damage by UVA1 radiation or PDT. These methods are divided into three sections. The first section describes general modifications to various common processing techniques that maximize the yield of all cell populations to assure that accurate and reliable results are obtained. It also describes irradiation procedures and postexposure treatments. The second section describes some specific ways to determine if a new system creates singlet oxygen and if this damage causes cells to die by immediate ($T < 30$ min) prePCD apoptosis or by some other mechanism of apoptosis. The third section describes a flow cytometry procedure that allows separate analysis and quantification of apoptotic and morphologically normal cells and changes that occur to the mitochondria in either population.

General Modifications of Methods

The delicate nature of singlet oxygen-damaged cells requires some general modifications to all procedures in order to obtain maximum recovery of all cell populations. These modifications will minimize erroneous results created by technical procedures that cause membrane damage before and/ or after exposure.

Centrifuging

The time and force of centrifugation must both be decreased, e.g., the time should be reduced to 5–7 min and the rpm should not exceed 1200–1400 for most centrifuges. For microcentrifuges with a set speed (10,000 rpm), use for only 30–45 sec, unless it is a variable speed model, and then set it no higher than eight (8000 rpm) for no longer than 1 min.

[7] J. Z. Beer, K. M. Olvey, S. A. Miller, D. P. Thomas, and D. E. Godar, *Photochem. Photobiol.* **58,** 676 (1993).

Washing

Procedures that require washing the cells, other than annexin V binding to phosphatidylserine residues (because proteins interfere with binding), should contain at least 1% serum or protein (protease-free bovine serum albumin) in all wash buffers. The number of washes should be reduced as much as possible whenever possible (e.g., three washes can usually be reduced to two) and can sometimes be completely eliminated (see Section VI,B).

Suspending Cells

To suspend cells after centrifuging and aspirating, either "finger flick" the cell pellet on the bottom of the centrifuge tube or tap the bottom of the tube a few quick times on the countertop. Alternatively, a quick (2–3 sec) vortex at a moderate setting (half maximum speed) may be used, but only on the pelleted cells. Never add buffer or media first and then vortex. For homogeneous suspension of cell pellets, it is important to centrifuge them briefly and lightly as just described. Check for "clumps" of cells and, if present, mix by inversion and "finger flick" the bottom of the tube a few times while upside down. Pipetting the solution gently once or twice (without bubble formation) is better for larger volumes. If clumps of cells are found, the procedure that was used should be modified by decreasing the centrifugal force and/or time and possibly altering the suspending technique.

Vortexing

Never vortex singlet oxygen-damaged cells for more than 3 sec, at a high setting, or in solution. Use quick vortex times (≤3 sec) at half the highest setting unless the cells are being fixed (one-third highest setting), and always vortex the cells in pellet form rather than in solution.

Pipetting

It is important to pipette without too much force and without forming many bubbles during any processing step, especially when suspending cells after irradiation. Bubbles cause oxidative damage to plasma membranes, increasing the percentage of apoptotic cells, causes loss of cells through lysis,[7] and also increases the percentages of necrotic cells.

Irradiation Procedures

Irradiation of cells in phosphate-buffered saline (PBS) is preferred to media for three reasons: (1) it does not absorb wavelengths above 200 nm;

(2) it has no components that can be oxidized like the essential nutrients present in media (nutrient deprivation can cause apoptosis as well); and (3) it cannot produce toxic products. If media must be present during exposure, then a control for the exposed media must be included and added to cells that were not exposed. In addition, the time of irradiation must be increased somewhat to account for the radiation absorbed by the media, if irradiated from above. Note that the time the cells are in PBS should not exceed 2 hr or the cells will start to undergo apoptosis due to nutrient deprivation.

Cells in PBS can be irradiated in uncovered cell culture dishes, e.g., 5 ml of $6–10 \times 10^5$ suspended cells/ml in 60-mm dishes or 2 ml in 35-mm dishes. Alternatively, flasks (T-25 with 5 ml) or covered dishes can be used to prevent evaporation during irradiation. However, this system should only be used for UVA/UVB and visible light exposures because UVC is absorbed completely by polystyrene and other plastics. If a covered system is chosen, then transmission of the light through the plastic must be assessed and the time of irradiation increased to obtain the correct dose. In addition, dishes no smaller than 35 mm should be used, e.g., 24- or 96-well plates, because edging effects of the light occur and alter the effective dose received by the cells. For ionizing radiation procedures, dishes, flasks, or centrifuge tubes may be used for irradiation of cells in PBS.

The dose rate of some UVA-emitting devices, e.g., black light bulbs, is too low to produce a reasonable dose in 2 hr. Although biological affects can be observed using these lower dose-rate bulbs, they are mainly due to the significant amounts of UVA2 (320–340 nm) and UVB (290–320 nm) emitted by these sources. To observe UVA1-triggered immediate apoptosis, the UVA-emitting source should have a dose rate ≥ 25 W/m^2 and emit very little UVA2 and virtually no UVB.

The temperature must be controlled and monitored carefully during UVA/visible exposures[8] because almost all high-dose rate UVA/visible light sources (except lasers) emit significant amounts of infrared radiation. The best control of temperature can be achieved with a circulating water bath connected to a black cooling plate (or block). Alternatively, one can use a long shallow dish (to reduce reflected light) with a shallow pool of water (2–3 mm depth), some ice, and possibly salt. Whatever system is chosen, the temperature must be monitored during the entire course of exposure to assure that the highest dose does not create temperatures above 30° for more than a few minutes. For when temperatures exceed 30°, the enzymes involved in apoptosis are more active and higher percentages of

[8] D. E. Godar and J. Z. Beer, in "Biological Responses to Ultraviolet A Radiation" (F. Urbach, ed.), p. 65. Valdenmar, Overland, KS, 1992.

apoptotic cells are obtained immediately after exposure. In addition, cellular damage from IR can also occur and/or heat shock proteins can be induced at higher temperatures, either of which may interfere with the interpretation of results.

Radiation Sources

1. UVA1 Source. UVASUN 2000, 3000, or 5000 sunlamps (Mutzhas, Munich, Germany, and similar high-output UVA sources are available through other companies) emit wavelengths $\geq$340 nm and have an emission peak around 365 nm. The UVA1 irradiance at the sample level (43–71 cm) should be about 200 W/m^2. The emission spectrum of the UVASUN 3000 sunlamp,[6] the instrument used for routine dosimetric measurements, and the calibration of the detector have been described previously.[9]

2. Visible Light Source. UVASUN sunlamps can be equipped with a UF3 cutoff filter (Read Plastics, Rockville, MD) to yield wavelengths $\geq$400 nm.

3. UVAR Source. The UVAR photoactivation chamber (Therakos, West Chester, PA) has an irradiance of about 50 W/m^2 through the plastic top of a T-25 flask at the sample level (16 cm). This source primarily emits UVA, but also emits biologically significant amounts of UVA2 and UVB wavelengths.

4. UVB Source. The FS20 sunlamp (Westinghouse) has an emission peak near 313 nm. An U340 filter (Hoya Optics, Fremont, CA) may be used to eliminate all the UVC and to reduce the visible and UVA wavelengths so that the UVB wavelengths are predominate. Using this filter, UVB irradiance at the sample level (29 cm) is about 0.48 W/m^2 (UVA irradiance is about 0.32 W/m^2).

5. X-Ray Source. A 120-kV, constant potential, 3-mm aluminum half-value thickness X-ray source can be used at a target-to-source distance of 70 cm. If the dose rate is 0.81 Gy/min, then 12.35 min of exposure will generate 10 Gy of ionizing radiation. A 20-cm-thick lucite sheet should be used to backscatter the beam in air for even dosimetry.

Irradiating Suspension Cells

Cells, which may be obtained from the American Type Tissue Collection, are grown to 3–5×10^5 cells/ml in complete CO_2-independent medium (GIBCO) containing 10% serum, 4 mM glutamine, and sometimes antibiotics. This medium is recommended for all procedures because it does not cause a significant pH change when processing the cells. The cells are

[9] D. E. Godar, D. P. Thomas, S. A. Miller, and W. Lee, *Photochem. Photobiol.* **57,** 1018 (1993).

harvested and concentrated by centrifugation (see Section II), aspirated, suspended by vortexing the pellet, and then diluted in PBS (without serum or protein) using about half the original volume. The final concentration of cells should be between 6 and 10×10^5 cells/ml, about twice the original, but should never exceed 1×10^6 cells/ml to ensure even dosimetry (by reducing shielding effects from "piled-up" cells). Count these cells to be sure they are not too dense and to later determine the recovery after removal from the dishes.

Irradiating Adherent Cells

Cells should be grown to 70–80% confluency and no higher to assure even dosimetry. These cells can be irradiated in either the attached or the detached (suspended) state in PBS. If the later is chosen, then the cell concentration should not exceed 10^6 cells/ml.

Postexposure Treatment

Suspension Cells

Immediately following exposure, an equal volume of complete medium is added to either the suspension cells or the adherent cells that were irradiated in the unattached state. Cells are then removed by gently pipetting the media over the dish or back of the flask once or twice with minimum bubble production and mild force using a spiraling or back and forth motion, respectively.[7] Tip the dish at an angle (30° increasing to 45°) while drawing up the solution to minimize bubble formation. Check the dishes visually for clarity by holding them up to the light under the biological hood and verify that most of the cells have been removed using an inverted microscope. Put these cells in either T-25 flasks or centrifuge tubes for observation at different times postexposure or processing for different end points, respectively. In addition, count the cells to determine the recovery, which should be ≥90%. After monitoring the cells over a short time course (<24 hr), more medium may be added for overnight growth so that the concentration does not exceed $2–3 \times 10^5$ cells/ml.

Adherent Cells

For postexposure times other than immediately after exposure, add an equal volume of medium and incubate at 37°. For the time point immediately after exposure, do not aspirate the PBS off the adherent cells because they detach when undergoing apoptosis and are lost during this process. Instead, remove the PBS containing cells from the dishes or flasks by pipetting and place this solution in a centrifuge tube (15 ml). Remove the remaining attached cells using trypsin-EDTA (1 ml for 35-mm dish and 3 ml for 60-mm) at room temperature for 5–10 min with an occasional gentle swirling motion. When detached, add an equal volume or more of complete media and gently collect the cells by pipetting and then pool these cells with the PBS cells already in the centrifuge tube. Harvest the cells by centrifugation and suspend cells in fresh 37° complete media as described in Section II. For dose–response curves and time courses of adherent cell types, a different dish or flask of cells must be used for each dose and postexposure observation time.

Determining which ROS Initiates which General Mechanism of Apoptosis

An extensive list of biologically suitable reagents used for ROS studies is given elsewhere in this volume. The few, but most reliable, reagents used in ROS studies in cell culture are given here as a first approach for determining if a given system may involve the production of singlet oxygen.[5] In addition, some alternate approaches are described for obtaining evidence that suggests singlet oxygen is involved and for determining which general apoptotic mechanism a given system initiates. It is important to note that externally added enzymes, such as catalase and superoxide dismutase, can only decrease plasma membrane damage caused by hydroxyl and/or superoxide anion radicals, respectively, because enzymes cannot enter the cell in active form. Furthermore, externally added oxidants, such as H_2O_2, affect the plasma membrane lipids and proteins to a greater extent than internal components.

Immediate apoptosis can be triggered by singlet oxygen- or superoxide anion-mediated damage.[5] In either case, the appearance of a significant number of apoptotic cells can be observed immediately after exposure (real time ≤ 15 min). Dose–response curves (0–1000 kJ/m^2) with clonogenic survival data (platting for reproductive capacity) should be obtained first, and the cells should be monitored over a time course beginning immediately after exposure and extending to at least 48 hr.[4] Cyclosporin A (CsA) is used to distinguish between singlet oxygen- and superoxide anion-triggered immediate apoptotic mechanisms because it only inhibits singlet oxygen-triggered immediate apoptosis.[5] In addition, immediate and intermediate apoptotic mechanisms cannot be inhibited with transcription or translation inhibitors because both immediate apoptosis and intermediate apoptosis are prePCD mechanisms. Only delayed apoptosis is inhibited significantly by transcription or translation inhibitors because DNA damage-induced apoptosis is a PCD mechanism that partially relies on the synthesis of protein.[4]

In Vitro Evidence for Singlet Oxygen

For singlet oxygen studies, one sample of cells is suspended in PBS and maintained in the dark (sham exposed), whereas another sample is exposed and compared to other exposed samples that have added reagents. One sample should be suspended in PBS containing an agent that extends the lifetime of singlet oxygen, such as deuterium oxide ($10\times$ PBS may be diluted into D_2O, 90%, or powdered PBS may be dissolved into D_2O, 100%, either solution must be sterile filtered before use). Another sample should contain an agent that quenches singlet oxygen, such as sodium azide (50–100 mM final concentration in PBS).[5] The control for mitochondrial depolarization and uncoupling of oxidation from phosphorylation is sodium cyanide (1 mM), which does not quench or enhance the lifetime of singlet oxygen. However, these reagents should be removed immediately after exposure for three reasons: (1) they can induce, promote, or inhibit an apoptotic mechanism; (2) they can alter mitochondrial function (e.g., azide, deuterium oxide, and cyanide); and (3) they can interfere with certain cellular processes under study and give misleading results. For example, azide also uncouples oxidation from phosphorylation, which decreases intracellular ATP levels and consequently decreases the activity of energy-dependent enzymes. To remove reagents after irradiation, cells are harvested by centrifugation as described in Section II. Other chemicals, such as water-soluble antioxidant glutathione (10 mM),[5] can also be added to the PBS during irradiation to obtain information as to whether ROS are involved at all. The hydrophobic ROS quencher, vitamin E, can determine if oxidative damage to membranes is important.[10] However, because vitamin E (type VI, Sigma) is an oil, it must be emulsified first and then preincubated for 2–24 hr with the cells (with slow rocking) to allow incorporation into the membranes. Furthermore, a variety of agents may be used to test for hydroxyl radicals, such as ethanol, dimethyl sulfoxide, histidine, and mannitol, but care should be exercised because most of these reagents also quench singlet oxygen. A good handbook to consult for designing reactive oxygen studies is "Free Radicals in Biology and Medicine."[11]

Figure 1 shows that UVA1 radiation results in the immediate appearance of apoptotic cells (0 in the figure). The presence of a lifetime extender of singlet oxygen (100% D_2O) increased the percentage of apoptotic cells

---

[10] D. E. Godar and A. D. Lucas, *Photochem Photobiol.* **62**, 108 (1995).
[11] B. Halliwell and J. M. C. Gutteridge, "Free Radicals in Biology and Medicine." Clarendon Press, Oxford, 1991.

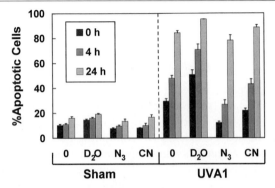

FIG. 1. UVA1 mediates singlet oxygen damage that results in the immediate appearance of apoptotic cells. The percentages of apoptotic cells in sham (left side) and UVA1 (800 kJ/ m^2; right side) samples at 0, 4, and 24 hr after exposure are shown. O, PBS; D$_2$O, deuterium oxide (100%); N$_3$, sodium azide (50 mM); CN, sodium cyanide (1 mM). Note that D$_2$O increased the percentages of apoptotic cells, whereas N$_3$ decreased the percentages of apoptotic cells and CN had little affect. Results represent means from three separate experiments and error bars are standard deviations (SD).

significantly, whereas a quencher of singlet oxygen (50 mM N$_3$) decreased the percentage of apoptotic cells significantly. In addition, the control for uncoupling oxidation from phosphorylation, sodium cyanide (1 mM), shows that it has little effect on the percentages of apoptotic cells. These results suggest that UVA1 mediates singlet oxygen damage, which triggers immediate apoptosis. The sham controls are shown on the left-hand side of Fig. 1 and the UVA1-exposed cells are shown on the right-hand side of Fig. 1. In both cases the reagents were removed immediately after exposure. The percentages of apoptotic cells were obtained using a flow cytometry procedure described in Section VI,A, but may also be obtained using other methods of detection (see Section VII).

Systems Known to Generate Singlet Oxygen, Superoxide Anions, and/or Hydroxyl Radicals, and Systems That Do Not Involve ROS in the Initiation of Apoptosis

Other systems that are known to generate singlet oxygen, superoxide anions, and/or hydroxyl radicals or do not create any reactive oxygen species may be used to reveal which ROS, if any, and which general mechanism of apoptosis a given system produces. For example, singlet oxygen may be generated by photosensitizers exposed to visible light, i.e., PDT. This can be done using natural precursors, such as δ-aminolevulinic acid (ALA), to increase intracellular levels of a photosensitizer, such as protoporphyrin

IX,[12] or by adding a photosensitizing dye, such as rose bengal.[13] Superoxide anions may be generated using pyruvate (10 mM) with UVA1 radiation,[5] vitamin K_3, or paraquat in the dark.[11] Hydroxyl radicals (and superoxide anions) may be generated using high doses of UVB.[14] High doses of UVB[15] and UVC[16] radiation are also reported to cause cross-linking of the Fas receptor on the plasma membrane. Thus, systems that initiate apoptosis through receptor cross-linking should also be used, such as anti-Fas IgM antibodies, to determine if intermediate apoptosis is initiated. In addition, some systems also cause some DNA damage, such as UVA1 radiation, and others almost exclusively cause DNA damage, such as X-rays, and consequently the delayed apoptotic mechanism. Furthermore, some systems exclusively cause DNA damage without any ROS, such as etoposide. Thus, additional systems should be included that are known to cause DNA damage without producing appreciable ROS, e.g., moderate doses of UVB/ UVC,[10] PUVA,[17] X-rays,[5,18] or etoposide.[19]

One approach for generating singlet oxygen inside cells is to increase the intracellular concentration of the photosensitizer protoporphyrin IX using ALA, which is a natural biochemical precusor of protoporphyrin IX, but is not a photosensitizer. Cells are preincubated at 37°, in complete medium with 0.25–1 mM ALA for 2–4 hr, and no longer,[5] to prevent overproduction and delocalization of protophorpyrin IX to the plasma membrane, which causes excessive damage, resulting in necrosis. The increase in protoporphyrin IX may be monitored by fluorescence emission >650 nm (FL3 channel in flow cytometry).[12] The stock solution of ALA is 1 M in DMSO. To generate singlet oxygen, the cells are processed as described and irradiated with at least 100 kJ/m^2 visible light. A higher dose may be used, but it alone should not induce appreciable apoptosis ($\leq$20%). Two sham-exposed controls must be included: one with only ALA and no light and the other with only visible light and no ALA.

Another approach for generating singlet oxygen inside cells uses photosensitizing dyes, such as rose bengal.[5,13] Cells are preincubated with 10–100

[12] E. A. Hryhorenko, K. Rittenhouse-Diakun, and N. S. Harvey et al., Photochem. Photobiol. 67, 565 (1998).

[13] C. R. Lambert and I. E. Kochevar, J. Am. Chem. Soc. 118, 3297 (1996).

[14] A. Gorman, A. McGowan, and T. G. Cotter, FEBS Lett. 404, 27 (1997).

[15] Y. Aragane, D. Kulms, D. Metze, G. Wilkes, B. Poppelmann, T. A. Luger, and T. Schwarz, J. Cell Biol. 140, 171 (1998).

[16] A. Rehemtulla, C. A. Hamilton, A. M. Chinnaiyan, and V. M. Dixit, J. Biol. Chem. 272, 25783 (1997).

[17] B. R. Vowels, E. K. Yoo, and F. P. Gasparro, Photochem. Photobiol. 63, 572 (1996).

[18] J. G. Peak and M. J. Peak, Mutat. Res. 246, 187 (1991).

[19] S. Kasibhatla, T. Brunner, L. Genestier, F. Echeveri, A. Mahboubi, and D. R. Green, Mol. Cell. 1, 543 (1998).

μM rose bengal (Sigma) for 1–2 hr, or overnight, in complete medium, processed as described, and irradiated with about 100 kJ/m^2 or more of visible light. The stock solution of rose bengal can be 1–10 mM in PBS. Two sham-exposed controls must be included: one with only rose bengal and no light and the other with only the light and no dye.

To generate superoxide anions, vitamin K$_3$ (Sigma) is incubated at different concentrations (10 μM–1 mM) with the cells in complete medium in the dark.[5,11] The stock solution is 100 mM in DMSO.

To generate hydroxyl radicals, use high doses of UVB (>99.99% clonogenic killing; see Section II).

To initiate apoptosis via a mechanism that does not involve ROS, but does involve receptor cross-linking, anti-Fas monoclonal IgM antibody, CH11 (Medical & Biological Laboratories, Nagoya, Japan), can be incubated with the cells at the recommended concentration (100 ng/ml) or higher in fresh complete medium.

To generate primarily DNA damage with minimal ROS production, one may use low doses of UVB or UVC radiation (<99% clonogenic killing), X-rays (10 Gy; see Section II,B), chemicals, such as etoposide (5 μM), or PUVA (see later).

To generate DNA damage-induced delayed apoptosis by PUVA,[17] cells are preincubated in complete medium with 100–300 ng/ml (200 ng/ml is shown in Fig. 2) 8-methoxypsoralen (8-MOP) for 15–20 min at 37° before processing for irradiation with 10–30 kJ/m^2 (30 kJ/m^2 is shown in Fig. 2) of UVAR from the photoactivation chamber (see Section II). The stock solution of 8-MOP is 200 μg/ml (or 1000 times higher than that used) in ethanol or DMSO. Two sham-exposed samples must be run: one for 8-MOP alone and the other for UVAR alone. The latter will have some immediate apoptosis due to the UVA present in this source, which may be subtracted from the PUVA values to obtain only the effect of PUVA. Alternatively, the UVAR dose may be reduced (e.g., 10 kJ/m^2) and the photosensitizer increased (e.g., 300–400 ng/ml).

Figure 2 shows a time course (0–24 hr) after exposure of cells to a variety of different initiators of apoptosis. Note that only singlet oxygen-generating systems, such as protoporphyrin IX (ALA in Fig. 3) and rose bengal, trigger immediate apoptosis, as does UVA1 radiation. Vitamin K and high doses of UVB initiated an intermediate mechanism, whereas moderate dose UVB, PUVA, and X-rays all initiated delayed apoptosis. Collectively, the results in Fig. 1 and 2 show that only single oxygen damage leads to the immediate appearance of apoptotic cells unlike damage produced by other ROS (intermediate and delayed) or damage to DNA that does not involve ROS (delayed).

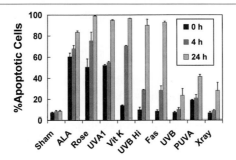

FIG. 2. Various initiators of apoptosis. Cells were treated as described in Section V,B. All sham-exposed controls were within 5% of the PBS sham, except the UVAR control for PUVA, which had the same value as PUVA immediately and at 4 hr after exposure. Sham, PBS; ALA, δ-aminolevulenic acid (0.5 mM, 2-hr preincubation) and 100 kJ/m^2 visible light, Rose, rose bengal (50 μM, 1-hr preincubation) and 100 kJ/m^2 of visible light; UVA1, UVA1 radiation (800 kJ/m^2; 95% clonogenic killing); Vit K, vitamin K$_3$ (100 μM) in the dark; UVB Hi, high-dose UVB (500 J/m^2; >99.99% clonogenic killing); Fas, anti-Fas IgM antibody (CH11, 100 ng/ml); UVB, moderate-dose UVB (100 J/m^2; 95% clonogenic killing); PUVA, 200 ng/ml 8-MOP and 30 kJ/m^2 of UVAR; and Xray, X-rays (10 Gy; >99.99% clonogenic killing). Note that only the singlet oxygen-generating systems, ALA and rose bengal, caused immediate apoptosis like UVA1 radiation. Results represent means of three separate experiments and error bars are SD.

Specifically Inhibiting Singlet Oxygen-Triggered Immediate Apoptosis

CsA only inhibits singlet oxygen-triggered immediate apoptosis. It blocks the release of apoptosis initiating factor (AIF), a 50-kDa protease, from the mitochondrial megachannel.[20] A final concentration of 10–100 μg/ml (8–80 μM) CsA is preincubated with the cells at 37° in complete medium for 30–45 min, and no longer, because it is only effective for short periods of time (<2 hr). In addition, 10–100 μg/ml of CsA is present during and after exposure. The stock solution is 10 mg/ml in ethanol or DMSO. The sham-exposed cell sample and other samples that did not receive CsA had the same final percentage of solvent (1% here) before, during, and after irradiation because ethanol and DMSO can quench hydroxyl radicals. Furthermore, ethanol leads to increases in NADH levels through the action of alcohol dehydrogenase, which inhibits superoxide anion-triggered immediate apoptosis by UVA1 radiation (400 kJ/m^2) combined with pyruvate (10 mM).[5]

[20] G. Kroemer, N. Zamzami, and S. A. Susin, *Immunol. Today.* **18,** 44 (1997).

Figure 3 shows that 100 μg/ml (about 80 μM) CsA only inhibits singlet oxygen (ALA PDT and UVA1 radiation)-triggered immediate apoptosis, whereas it has little effect on superoxide anion-triggered (vitamin K_3) or receptor-initiated (anti-Fas antibody) intermediate apoptosis. Although not shown in Fig. 3, CsA also significantly inhibits singlet oxygen-triggered immediate apoptosis generated by rose bengal and visible light. In contrast, it has no inhibitory effect on superoxide anion-triggered immediate apoptosis generated by either UVA1 radiation combined with pyruvate or high concentrations of vitamin K_3 (1 mM) in the dark.[5] Furthermore, CsA does not inhibit any intermediate or delayed apoptotic mechanism and can increase the percentages of these cells. Note that concentrations of CsA as low as 8 μM may be used to inhibit singlet oxygen-triggered immediate apoptosis.

Assessing Preprogrammed and Programmed Cell Death Mechanisms of Apoptosis

Preprogrammed cell death is an apoptotic mechanism that relies solely on proteins that are synthesized constitutively,[4] whereas programmed cell death is an apoptotic mechanism that also relies on the synthesis of new

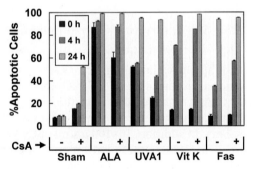

FIG. 3. CsA inhibits only singlet oxygen-triggered immediate apoptosis. All samples had the same final concentration of ethanol present (1%). The ALA and visible light sham-exposed controls were within 5% of that shown as the Sham($-$). Sham, sham-exposed PBS cells without ($-$) CsA and with ($+$) CsA (100 μg/ml); ALA, δ-aminolevulenic acid (1 mM, 2-hr preincubation) and 100 kJ/m^2 of visible light without ($-$) CsA and with ($+$) CsA (100 μg/ml); UVA1, UVA1 radiation (800 kJ/m^2) without ($-$) CsA and with ($+$) CsA; Vit K, vitamin K_3 (100 μM) in the dark without ($-$) CsA and with ($+$) CsA (100 μg/ml); Fas, anti-Fas IgM antibody (CH11, 100 ng/ml) without ($-$) CsA and with ($+$) CsA (100 μg/ml). Note that only singlet oxygen-triggered immediate apoptosis (ALA PDT and UVA1 radiation) is inhibited by CsA, whereas superoxide anion-/hydroxyl radical or receptor-initiated intermediate apoptosis (vitamin K_3 or anti-Fas antibody, respectively) are not affected. Results represent means of three separate experiments and error bars are SD.

protein.[2] The best way to determine if an apoptotic mechanism is prePCD or PCD is to inhibit translation with cycloheximide (CHX) because it does not inhibit aminopeptidases like puromycin.[21] However, great care must be taken not to inhibit translation much more than 50% or else the cells start to undergo apoptosis (after 4 hr) from the presence of the inhibitor.[4,5] This general concept applies to all other inhibitors as well.

Some preliminary testing should be done to find a suitable concentration of CHX that inhibits translation by about 50%, while it does not cause appreciable apoptosis ($\leq$15%) in the controls when present for 48 hr.[4] Low concentrations around 0.1 μg/ml are effective on leukemia and lymphoma cells and are a good starting point for finding the correct concentration for other cell types. The correct concentration should be verified by quantifying the amount of [^{35}S]methionine incorporated into newly synthesized proteins.[4] When the concentration is approximately correct, CHX-treated cells will only incorporate about half the amount of [^{35}S]methionine into trichloroacetic acid-precipitable protein as the controls. In addition, the effect of CHX on translation should be monitored before every experiment by [^{35}S] methionine labeling. Be sure to add some fresh media, with or without CHX, for overnight growth. Note that some adherent cells detach during this procedure, which may be irradiated in the detached state. Alternatively, shorter preincubation (4 hr) with CHX can be used for adherent cell types.

Once the correct concentration of CHX has been determined, it is preincubated with the cells for at least 4 hr, or overnight (12–24 hr), for irradiation by UVA1, PDT, or X-rays. However, if PUVA, UVB, or UVC exposures will be used, the CHX should be added immediately after exposure because it absorbs short wavelength UV. Use a UVA1 or PDT dose that gives 30–50% apoptosis at almost zero time postexposure so that significant differences may be observed. Other apoptotic mechanisms will require longer postexposure observation times (e.g., 4–48 hr). In addition, include at least one positive control for intermediate apoptosis, such as high doses of UVB (>99.99% clonogenic death; 500 kJ/m^2), anti-Fas IgM antibody (CH 11 at 100 ng/ml), and/or staurosporine (1 μM), and at least one positive control for delayed apoptosis, such as moderate dose UVB (90–99% clonogenic death; 100 kJ/m^2), etoposide (5 μM), PUVA (see Section IV,B), or X-rays (10 Gy; Section II,B).

Figure 4 shows that neither the UVA1-triggered immediate nor the high-dose UVB-initiated intermediate apoptotic mechanisms are inhibited significantly by CHX (0.08–0.1 μg/ml). However, after 24 hr, a difference is observed, which suggests that both UVA1 radiation and high doses of

[21] D. B. Constam, A. R. Tobler, A. Rensing-Ehl, I. Kemler, L. B. Hersh, and A. Fontana, *J. Biol. Chem.* **270,** 26931 (1995).

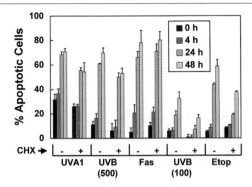

Fig. 4. Preprogrammed and programmed cell death mechanisms of apoptosis determined by inhibiting translation with cycloheximide (CHX). Cells were pretreated without (−) or with (+) CHX (0.1 μg/ml) overnight before exposure to UVA1 (800 kJ/m^2), Fas (anti-Fas IgM antibodies; CH11, 100 ng/ml), or Etop (etoposide, 5 μM). CHX was added immediately after UVB exposure (100 and 500 J/m^2) because it absorbs UVB (and UVC) radiation. Similar results for anti-Fas antibodies and etoposide were obtained if the CHX was added immediately after the addition of either reagent. Results are also shown for 48 hr because the UVB-induced delayed apoptotic mechanism, like X-rays, requires more time to display a significant difference. Results represent means of three separate experiments and error bars are SD.

UVB radiation also initiate a delayed apoptotic mechanism by causing DNA damage. Intermediate apoptosis initiated by anti-Fas IgM antibodies is also a prePCD mechanism of apoptosis because it is not inhibited by CHX. Moderate doses of UVB (100 J/m^2) or etoposide (5 μM) cause DNA damage and are positive controls for PCD because they induce a delayed apoptotic mechanism (24–48 hr) that is inhibited by CHX to the same extent as protein synthesis (about 50%). Note that the UVB-induced delayed apoptotic mechanism requires at least 48 hr to display a significant difference.

Flow Cytometry Procedures

The flow cytometry method described here relies on the differences in size between apoptotic and morphologically normal cells. This procedure is recommended for regularly shaped cells (round), such as most lymphomas and leukemias, and some adherent cell types, such as L929. However, it is difficult to separate apoptotic and morphologically normal cell populations of irregularly shaped cells, such as the human T-helper cell lymphoma, H9, or some large adherent cell types (that must also be analyzed shortly after preparation to minimize "clumping"). In addition, some inducers of apoptosis, such as staurosporine, etoposide, and anti-Fas-initiated apoptosis, cause a diverse distribution of shapes and sizes of cells undergoing apoptosis that makes the analysis tricky and less reliable. For these other cell types and inducers of apoptosis, another method for detecting apoptosis, such as Annexin V binding of phosphatidylserine residues expressed on the outer leaflet of the plasma membrane and/or TUNEL labeling of digested DNA, must also be used to confirm these flow cytometry data. These other methods will not be discussed here because kits with detailed instructions are available from a variety of companies.

Analyzing Apoptotic and Morphologically Normal Cells Separately

Flow cytometry can be performed using a single beam 488-nm excitation source, e.g., FACscan (Becton-Dickinson, San Jose CA). Apoptotic and morphologically normal cell percentages are determined using a flow cytometric method described previously.[22] At least 10,000 cells are collected and analyzed using suitable software, such as LYSYS II or CellQuest, according to the side light-scattering (SSC) profile (proportional to cellular granularity) versus the forward light-scattering (FSC) profile (proportional to cellular cross-sectional area or size). Depending on the inducer, the apoptotic cells can be more granular, but are always smaller, than the normal cell population. The FSC and SSC settings must be adjusted for the different inducers of apoptosis and cell types to achieve a distinct separation between these two populations of cells. This can be accomplished by first centering (between 400 and 600 on a linear scale) the sham-exposed cells on the FSC (size) and then changing the SSC (granularity) and FSC settings by comparing both sham and exposed populations until adequate separation is achieved between them. The morphologically normal cell population may have to be adjusted up field in some cases (centered around 600), especially for large adherent cells and irregularly shaped cells, such as H9 lymphomas. Once the optimum settings are determined, data are collected in the linear mode for both SSC and FSC. For analysis, the region marker cutoffs between populations are best determined by comparing both sham and exposed (highest dose) populations in adjacent windows. This method can be verified either by sorting these two populations and analyzing the different end points associated with apoptosis or by analyzing different end points, such as Annexin V or TUNEL.

An example of a SSC (granularity) versus FSC (size) profile is shown in Fig. 5A as a dot plot with polygon region markers separating apoptotic and morphologically normal cell populations after UVA1 irradiation of Jurkat T-helper cells.

[22] C. Dive, C. D. Gregory, D. J. Phipps, D. L. Evans, A. E. Milner, and A. H. Wyllie, *Biochim. Biophys. Acta* **1133,** 275 (1992).

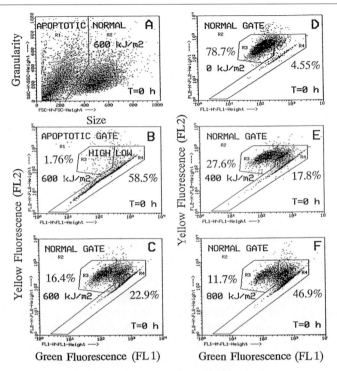

Fig. 5. UVA1 irradiation of transformed human T-helper cells (Jurkat) immediately ($T \le 20$ min) reduces the mitochondrial transmembrane potential ($\Delta\Psi_m$) in a dose-dependent manner. The $\Delta\Psi_m$ was monitored using a specific probe, JC-1, as described in Section VI,B. (A) The side (granularity) versus the forward (side) light-scatter profile of a 600-kJ/m^2 UVA1 exposed sample immediately after exposure: the apoptotic cells are smaller (downfield) than the normal population as displayed by the region markers in the dot plot. (B) The $\Delta\Psi_m$ of only the apoptotic cells in a yellow/orange (FL2) versus green (FL1) fluorescent profile after gating on this population in the side versus the forward light-scatter profile as shown in A. Apoptotic cells display a low $\Delta\Psi_m$ as indicated in the polygon region marker. (C) Morphologically normal cells, after gating from A, immediately following 600 kJ/m^2 UVA1 radiation: these cells show both low and high $\Delta\Psi_m$. (D) Almost all of the morphologically normal sham-exposed cells have a high $\Delta\Psi_m$. (E and F) The $\Delta\Psi_m$ of morphologically normal cells immediately after 400 or 800 kJ/m^2 of UVA1 radiation, respectively. Similar results were obtained in three separate experiments.

Analysis of Intracellular Organelles: Mitochondria

Various flow cytometry grade fluorescent dyes are now available to explore changes to different organelles and endpoints, such as pH, Ca^{2+}, intracellular GSH, and increases in ROS, to mention but a few (Molecular Probes, Eugene, OR). Fluorescent dyes that emit primarily in one fluorescent channel can be used so that double labeling allows another end point to be monitored simultaneously in another fluorescent channel. Alternatively, confocal microscopy may be used to monitor mitochondrial or other organelle changes. Fluorescent microscope pictures of these cells can be obtained to verify and document the organelle changes.

Dramatic changes in the mitochondrial transmembrane potential ($\Delta\Psi_m$) can be measured using a specific fluorescent probe, such as 5,5′,6,6′-tetrachloro-1,1′,3,3′-tetraethylbenzimidazolcarbocyanine iodide (JC-1; Molecular Probes, Eugene, OR).[23] The cells are incubated at 37° for about 10 min with JC-1 at a final concentration of 1 μg/ml.[5] The cells are not washed to remove the unbound dye after incubation because a threshold setting of 52 does not allow the flow cytometer to count the free fluorescent particles as data. The JC-1 stock solution is 1 mg/ml in DMSO, so that 1 μl is added to 1 ml of cells. Stock solutions (10×; 10 mg/ml) may be stored at −20° for extended periods of time (1 year or more), and the diluted solution (1 mg/ml), protected from light, is stable for up to a month or more at room temperature. The FACScan settings can vary dramatically between instruments, cell types, and procedures. The following initial settings are recommended for leukemia and lymphoma cells: FL1 (530 ± 15 nm) in log data collection mode with a PMT setting around 425 and FL2 (585 ± 20 nm) in log data collection mode with a PMT setting around 310. Although compensation is not shown in Fig. 5 because it is not necessary for accurate quantification, the different populations may be separated further for better qualitative presentation by adjusting the compensation: FL1-FL2% is recommended at 1.2–1.4 and FL2-FL1% is recommended at 22.4–23.2 for Jurkat T cells.

Once suitable settings are obtained, data are collected from 10,000 cells and a separate analysis of each population is achieved by gating on that cell population in the FSC versus SSC profile as shown in Fig. 5A. Each population is then analyzed for mitochondrial changes ($\Delta\Psi_m$) using two-dimensional dot plots of the yellow/orange (FL2) versus the green (FL1) fluorescent profiles as shown in Figs. 5B–F. Figure 5B shows a UVA1-exposed sample where all the apoptotic cells display a low mitochondrial transmembrane potential ($\Delta\Psi_m$), whereas Fig. 5C shows that morphologically normal cells have both low and high $\Delta\Psi_m$'s. Figure 5D shows a sham-exposed sample where almost all of the morphologically normal cells have a high $\Delta\Psi_m$, but as the dose of UVA1 radiation increases, the percentages of cells with a high $\Delta\Psi_m$ decreases (compare Figs. 5D, 5E, 5C, and 5F). Figure 6 shows that UVA1 radiation causes a dose- and time-dependent decrease of the $\Delta\Psi_m$ in the morphologically normal cell population. Unlike

---

[23] S. Salvioli, A. Ardizzoni, C. Franceschi, and A. Cossarizza, *FEBS Lett.* **411**, 77 (1997).

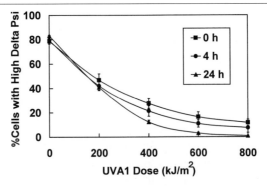

Fig. 6. UVA1-mediated singlet oxygen damage to mitochondrial membranes reduces the inner transmembrane potential ($\Delta\Psi_m$) in a dose- and time-dependent manner. UVA1 radiation causes a dose- and time-dependent drop in the $\Delta\Psi_m$ of morphologically normal cells after gating on this population as shown in Fig. 5A. Results represent means from three separate experiments and error bars are the SD.

other inducers of apoptosis, singlet oxygen causes a drop in the $\Delta\Psi_m$ before the cells display typical morphological characteristics of apoptosis. For most other apoptotic mechanisms, the $\Delta\Psi_m$ drops while the cells are decreasing in size (or after). Compare $\Delta\Psi_m$ changes over a time course using other inducers of apoptosis.

Discussion

Unlike other initiators of apoptosis, singlet oxygen causes damage to a variety of cellular components, particularly membranes. Further technical damage to the plasma membranes can give misleading results because more cells undergo apoptosis and some appear necrotic or lyse during processing. To assure that reliable results are obtained, many common methods must be modified so that the cells do not acquire more plasma membrane damage from these technical procedures.

Both UVA1 radiation[24] and PDT[25] mediate the production of singlet oxygen.[5] Although some reagents, such as deuterium oxide or azide (Fig. 1), and other approaches (Fig. 2) can suggest that singlet oxygen may be involved in a given system, only CsA has been found to specifically inhibit singlet oxygen-triggered immediate apoptosis (Fig. 3).[5] Immediate apoptosis occurs very rapidly (T ≤ 30 min) and is a prePCD mechanism of apoptosis because it does not rely on the synthesis of new proteins to

[24] A. Morita, T. Werfel, and H. Stege et al., J. Exp. Med. **186**, 1763 (1997).
[25] D. Kessel and Y. Luo, J. Photobiol. Photochem. **42**, 89 (1998).

proceed (Fig. 4). Singlet oxygen causes damage to a variety of cellular components, but immediate prePCD apoptosis is triggered primarily by mitochondrial membrane damage. This damage causes the mitochondrial transmembrane potential to immediately depolarize (Figs. 5 and 6). Once depolarized, the mitochondrial megachannel opens at the so-called "S", or sulfhydryl sensitive, site[26] and apoptosis-initiating factor (AIF) is released, unless CsA is present.[20] Unlike singlet oxygen-triggered immediate apoptosis, superoxide anion (1 mM vitamin K_3)-triggered immediate apoptosis and superoxide anion- or hydroxyl radical-initiated intermediate apoptosis cannot be inhibited with CsA. Moreover, other systems that initiate apoptosis cannot be inhibited by CsA, such as receptor (e.g., anti-Fas antibody, CH 11)-initiated intermediate apoptosis or DNA damage-induced delayed apoptosis (e.g., etoposide). CsA does not inhibit other mechanisms of apoptosis because they do not rely on the release of AIF, but rather rely on the release of cytochrome c[27] after Bax forms a pore[28] at the "P," or pyrimidine nucleotide-sensitive, site of the mitochondrial megachannel.[26] Antioxidants, such as glutathione,[5] or some antioxidant proteins associated with the outer mitochondrial membrane, such as Bcl-2, can inhibit UVA1-[29] and PDT-mediated[30] immediate apoptosis. Bcl-2, and family members, can also inhibit many other mechanisms of apoptosis. Details of the different apoptotic mechanisms, i.e., immediate prePCD, intermediate prePCD, and delayed PCD are discussed elsewhere.[31]

Flow cytometry can be used to determine the order of biochemical events that occur before, during, and/or after the apoptotic morphology occurs, i.e., cell shrinkage. Flow cytometry allows separate analysis of the apoptotic and morphologically normal cells while analyzing various end points. This method relies on the difference in size between apoptotic and morphologically normal cells that is determined by forward (size) versus side (granularity) light-scattering profiles. Unlike other methods that ana-lyze the entire population of cells and average end points associated with apoptosis, this method segregates events that occur before, during, and after cell shrinkage, the salient morphological feature of apoptotic cells. A

[26] P. Costantini, B. V. Chernyak, V. Petronilli, and P. Bernardi, *J. Biol. Chem.* **271,** 6746 (1996).

[27] E. Bossy-Wetzel, D. D. Newmeyer, and D. R. Green, *EMBO J.* **17,** 37 (1998).

[28] J. M. Jurgensmeier, Z. Xie, Q. Deveraux, L. Ellerby, D. Bredesen, and J. C. Reed, *Proc. Natl. Acad. Sci. U.S.A.* **95,** 4997 (1998).

[29] C. Pourzand, G. Rossier, O. Reelfs, C. Borner, and R. M. Tyrrell, *Cancer Res.* **57,** 14051 (1997).

[30] J. He, M. L. Agarwal, H. E. Larkin, L. R. Friedman, L. Y. Xue, and N. L. Oleinick, *Photochem. Photobiol.* **64,** 845 (1996).

[31] D. E. Godar, *J. Invest. Dermatol. Symp. Proc.* **4,** 17 (1999).

variety of flow cytometry methods[32] allow different cellular changes to be monitored, such as external and internal antigens, pH, Ca^{2+}, and organelles, like mitochondria, while they occur in normal and apoptotic populations. Because singlet oxygen causes damage to mitochondrial membranes resulting in immediate depolarization of the inner mitochondrial transmembrane potential, measuring mitochondrial transmembrane potential gives a powerful tool to help distinguish singlet oxygen-triggered immediate apoptosis from other apoptotic mechanisms. In addition, the order of signaling events that occur before and after mitochondrial transmembrane depolarization, such as gene induction or activation of the different caspase cascades, can be determined. Thus, biochemical events that initiate the apoptotic morphology can be segregated from those that execute morphological changes. In addition, other flow cytometry methods for detecting apoptosis can help confirm the order of events. For example, Annexin V can determine if an event occurs before or after phosphatidylserine residues are expressed on the outer leaflet of the plasma membrane or TUNEL can determine if an event occurs before or after DNA fragmentation.[33]

These methods will help investigators determine which ROS, if any, are responsible for initiating which apoptotic mechanism, i.e., immediate prePCD, intermediate prePCD, or delayed PCD. In addition, the different biochemical changes, such as gene activation, signal transduction, and caspase activation, may be analyzed separately in apoptotic and morphologically normal cells to help determine the order of events. This is particularly important for singlet oxygen-triggered immediate apoptosis because it occurs so rapidly, especially to cancer cells. Since the therapeutic relevance of UVA1-mediated singlet oxygen-triggered immediate apoptosis has recently been noted,[24] understanding how this mechanism works may help improve future clinical applications of UVA1 phototherapy.

[32] Z. Darzynkiewicz and H. A. Crissman (eds.), *in* "Methods in Cell Biology," Vol. 33. Academic Press, New York, 1990.
[33] T. G. Cotter and S. J. Martin (eds.), "Techniques in Apoptosis: A Users Guide." Portland Press, London, 1996.

[31] Determination of DNA Damage, Peroxide Generation, Mitochondrial Membrane Potential, and Caspase-3 Activity during Ultraviolet A-Induced Apoptosis

By SAEKO TADA-OIKAWA, SHINJI OIKAWA, and SHOSUKE KAWANISHI

Introduction

Solar ultraviolet (UV) light is a well-known carcinogen. UVB (280–320 nm) has been shown to cause direct photoactivation of the DNA molecule to generate, mainly, pyrimidine photoproducts, leading to mutation and carcinogenesis.[1] In addition to UVB, UVA (320–380 nm) has been implicated in multistage photocarcinogenesis.[2,3] In UVA wavelengths, it is assumed that cellular DNA damage is produced indirectly through photosensitized reactions mediated by photosensitizers, as UVA can hardly be absorbed by the DNA. An excited endogenous photosensitizer molecule could react with DNA through direct electron transfer (type I) or through generation of reactive oxygen species such as singlet oxygen ($^1O_2$) and superoxide anion radical (type II).[4–6] Thus, UVA, as well as UVB, may play an important role in the induction of carcinogenesis.

Apoptosis contributes to the pathogenesis of a number of human diseases, including cancer. DNA damage can induce death by apoptosis,[7] which is characterized by cell shrinkage, chromatin condensation, and formation of oligonucleosome-length fragments of DNA (DNA ladder).[7,8] UVA induces apoptosis in human and rodent cell lines.[9–11] Roles for mitochondrial alterations and caspase activation in the regulation of apoptotic process have

[1] IARC, *in* "IARC Monographs on the Evaluation of Carcinogenic Risks to Humans, Solar and UV Radiation," Vol. 55. IARC, Lyon, 1992.

[2] R. B. Setlow, E. Grist, K. Thompson, and A. D. Woodhead, *Proc. Natl. Acad. Sci. U.S.A.* **90,** 6666 (1993).

[3] A. de Laat, J. C. van der Leun, and F. R. de Gruijl, *Carcinogenesis* **18,** 1013 (1997).

[4] K. Ito, S. Inoue, K. Yamamoto, and S. Kawanishi, *J. Biol. Chem.* **268,** 13221 (1993).

[5] K. Ito and S. Kawanishi, *Biochemistry* **36,** 1774 (1997).

[6] K. Ito and S. Kawanishi, *Biol. Chem.* **378,** 1307 (1997).

[7] C. B. Thompson, *Science* **267,** 1456 (1995).

[8] A. H. Wyllie, *Nature* **284,** 555 (1980).

[9] S. Tada-Oikawa, S. Oikawa, and S. Kawanishi, *Biochem. Biophys. Res. Commun.* **247,** 693 (1998).

[10] E. G. Dianne and D. L. Anne, *Photochem. Photobiol.* **62,** 108 (1995).

[11] C. Pourzand, G. Rossier, O. Reelfs, C. Borner, and R. M. Tyrrell, *Cancer Res.* **57,** 1405 (1997).

drawn much interest.[12–14] A decrease in mitochondrial membrane potential ($\Delta\Psi$m), causing opening of the permeability transition pore (PTP), has been implicated as a critical effector of apoptosis in a variety of cells.[12,13] Subsequently, cytochrome c released from mitochondria to cytosol has been shown to initiate the activation of caspase-3.[15] The activated caspase-3 can cleave an inhibitor of caspase-activated deoxyribonuclease (ICAD)[16] and lead to DNA ladder formation.

This article focuses on a methodological approach to study roles of DNA damage, peroxide generation, change of $\Delta\Psi$m, and caspase-3 activity in a UVA-induced apoptotic pathway.

Materials and Methods

Materials

Nuclease P_1 obtained from Yamasa Shoyu Co. (Chiba, Japan) and bacterial alkaline phosphatase (Sigma, St. Louis, MO) are used for digesting DNA to deoxynucleosides. Ribonuclease A from Sigma and proteinase K from Merck (Darmstadt, Germany) are used for digesting cellular RNA and proteins, respectively. A caspase-3 substrate Ac-Asp-Glu-Val-Asp-7-amino-4-trifluoromethyl coumarin (Ac-DEVD-AFC) is from Enzyme System Products (Livermore, CA). 2′,7′-Dichlorofluorescin diacetate (DCFH-DA) for detection of cellular peroxide is from Molecular Probes, Inc. (Eugene, OR). $\Delta\Psi$m-sensitive probes, 3,3′-dihexyloxacarbocyanine iodide [$DiOC_6(3)$] and 5,5′,6,6′-tetrachloro-1,1′,3,3′-tetraethylbenzimidazolylcarbocyanine iodide (JC-1), are also obtained from Molecular Probes, Inc.

Cell Culture and UVA Irradiation

The human myelomonocytic leukemia cell line, HL-60, is grown in RPMI 1640 supplemented with 6% (v/v) fetal calf serum (FCS) at 37° under 5% (v/v) CO_2 in a humidified atmosphere. Exponentially growing cells are used throughout all experiments at a concentration of 1×10^6 cells/ml in freshly prepared phenol red-free RPMI 1640 (Sigma) containing 6% (v/v) FCS. The cells are irradiated with five 8-W UV lamps (365 nm, UVP, Inc., Model TDM-20, San Gabriel, CA) placed at a distance of 8–9 cm. The

[12] C. N. Kim, X. Wang, Y. Huang, A. M. Ibrado, L. Liu, G. Fang, and K. Bhalla, *Cancer Res.* **57,** 3115 (1997).
[13] G. Kroemer, N. Zamzami, and S. A. Susin, *Immunol. Today* **18,** 44 (1997).
[14] D. R. Green and J. C. Reed, *Science* **281,** 1309 (1998).
[15] X. Liu, C. N. Kim, J. Yang, R. Jemmerson, and X. Wang, *Cell* **86,** 147 (1996).
[16] M. Enari, H. Sakahira, H. Yokoyama, K. Okawa, A. Iwamatsu, and S. Nagata, *Nature* **391,** 43 (1998).

cells are protected from direct sunlight. After UVA irradiation, cells are harvested by centrifugation at 200g for 5 min after incubation at 37° for various time and washed with phosphate-buffered saline (PBS) three times.

Measurement of 8-oxodG in Cultured Cells Irradiated with UVA

The cells (3.0 × 10^6 cells) are irradiated with UVA and then incubated continuously at 37° for various times. DNA is extracted by using a DNA Extractor WB Kit (Wako Pure Chemical Industries, Ltd., Osaka, Japan) in ambient atmosphere, or DNA extraction and digestion are carried out inside an anaerobic incubator EAN-140 (TABAI Espec, Japan).[17] The content of 8-oxodG is determined by enzyme digestion of DNA using an HPLC-ECD.[4,18] The DNA is dissolved in 20 mM acetate buffer (pH 5.0) and digested to deoxynucleosides by incubation first with 8 units of nuclease P$_1$ at 37° for 30 min and then with 0.6 units of bacterial alkaline phosphatase at 37° for 1 hr in 0.1 M Tris–HCl (pH 7.5). The resulting deoxynucleoside mixture is injected into an HPLC apparatus (LC-10AD, Shimadzu, Kyoto, Japan) equipped with both a UV detector (SPD-6A, Shimadzu) and an electrochemical detector (Coulochem II Model 5200A, ESA, Bedford, MA): column, Tosoh TSK-GEL (ODS-80TS, 4.6 × 250, Tosoh, Tokyo, Japan); eluent, 8% aqueous methanol containing 10 mM sodium dihydrogen phosphate; flow rate, 1 ml/min. The molar ratio of 8-oxodG to deoxyguanosine (dG) in each DNA sample is measured based on the peak height of authentic 8-oxodG with the electrochemical detector and the UV absorbance at 254 nm of dG.

After irradiating with 10 J/cm^2 of 365-nm light, the formation of 8-oxodG in cells increased in a time-dependent manner (Fig. 1). The result suggests that UVA induces oxidative DNA damage via immediate and delayed mechanisms.

Detection of Cellular DNA Damage Induced by UVA Irradiation

After UVA irradiation, the cells (2.5 × 10^6 cells) are incubated continuously at 37° for various times. The cells are resuspended in 75 μl of PBS and the cell suspension is mixed with an equal volume of 1.3% (w/v) low melting point agarose (Bio-Rad Laboratories, Hercules, CA), poured into sample holders (Bio-Rad) and allowed to harden at 4° for 30 min. Upon solidification, the agarose plugs are transferred to 2 ml of Tris/EDTA/2-mercaptoethanol buffer [0.5 M EDTA (pH 8.0), 10 mM Tris–HCl (pH 8.0),

---

[17] M. Nakajima, T. Takeuchi, and K. Morimoto, *Carcinogenesis* **17,** 787 (1996).
[18] H. Kasai, S. Nishimura, Y. Kurokawa, and Y. Hayashi, *Carcinogenesis* **8,** 1959 (1987).

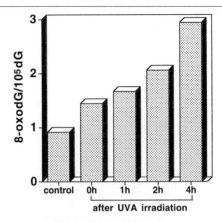

FIG. 1. Time course of 8-oxodG production in cultured cells irradiated with UVA. HL-60 cells (3×10^6 cells) were irradiated with 10 J/cm^2 of 365-nm light. After irradiation, cells were incubated at 37° for the indicated time, and extracted DNA was subjected to enzyme digestion and analyzed by an HPLC-ECD.

and 7.5% (v/v) 2-mercaptoethanol] and incubated at 37° overnight, followed by incubation in 1 ml of NDS buffer [0.5 M EDTA (pH 8.0), 10 mM Tris–HCl (pH 8.0), 1% (w/v) N-laurylsarcosine, and 1 mg/ml proteinase K] at 50° for 24 hr. Electrophoresis is performed in 0.5× TBE buffer [45 mM Tris, 45 mM boric acid, and 1.25 mM EDTA, (pH 8.3)] by a CHEF-DRII pulsed field electrophoresis system (Bio-Rad) at 200 V at 14°. Switch time is 60 sec for 15 hr followed by 90 sec for 9 hr. The DNA in the gels is visualized in ethidium bromide (0.5 mg/liter).

After irradiating with 10 J/cm^2 of 365-nm light, DNA fragments corresponding to 1–2 Mb were observed slightly at 0–2 hr after UVA irradiation and approximately 50-kb fragments were observed at 1–3 hr (Fig. 2A). Then, 1- to 2-Mb DNA fragments disappeared at 3 hr (Fig. 2A), suggesting that the 1- to 2-Mb fragments are cleaved into approximate 50-kb DNA fragments.

Detection of DNA Ladder Formation Induced by UVA Irradiation

For detection of DNA ladder formation, the cells (2.5×10^6 cells) are disrupted in 1 ml of cytoplasm extraction buffer [10 mM Tris (pH 7.6), 0.15 M NaCl, 5 mM MgCl$_2$, and 0.5% (v/v) Triton X-100]. After centrifugation at 1000g for 5 min at 4°, intact nuclei are pelleted. Nuclei are then treated with 200 μl of lysis buffer [10 mM Tris (pH 7.6), 0.4 M NaCl, 1 mM EDTA, and 1% (v/v) Triton X-100] for 10 min and centrifuged at 12,000g at 4° to separate the nucleoplasm from high molecular weight chromatin. The

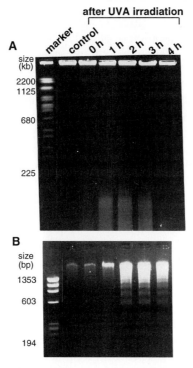

Fig. 2. DNA damage and DNA ladder formation in cultured cells irradiated with UVA. HL-60 cells (2.5×10^6 cells) were irradiated with 10 J/cm^2 of 365-nm light and incubated at 37° for the indicated time. (A) Cells were prepared as agarose plugs, lysed, and subjected to pulsed field gel electrophoresis through a 1% agarose gel. The gel was stained in ethidium bromide. Marker lane: size marker DNA (*Saccharomyces cerevisiae*). (B) Cells were lysed, and DNA was extracted and analyzed by conventional electrophoresis. Marker lane: size marker DNA (ΦX174/*Hae*III digest). Control lane: no UVA irradiation.

nucleoplasm is incubated with 0.2 mg/ml heat-treated ribonuclease A for 1 hr at room temperature, subsequently with 100 μg/ml proteinase K for 2 hr at 37°. The DNA is extracted with phenol/chloroform and subsequently with water-saturated ether, and precipitated with 2.5 volumes of ethanol at −90° for 30 min or at −20° overnight. The pellet is dissolved in 25 μl of TE buffer [10 mM Tris–HCl (pH 8.0) and 1 mM EDTA]. The DNA is electrophoresed on a 1.4% agarose gel containing 0.375 μg/ml ethidium bromide in 0.5× TBE buffer.

After irradiating with 10 J/cm^2 of 365-nm light, the DNA ladder formation, which is associated with apoptosis, was observed at 2–4 hr (Fig. 2B). These results suggest that UVA causes DNA strand breakage in cells, resulting in internucleosomal DNA fragmentation and apoptosis.

Flow Cytometric Detection of Peroxide

The intracellular generation of peroxide is examined using DCFH-DA, which is deacetylated in cells to a nonfluorescent compound, 2',7'-dichloro-fluorescin, which can be oxidized to a highly fluorescent compound, 2',7'-dichlorofluorescein (DCF).[19,20] After UVA irradiation, to evaluate cellular peroxide level, the cells (2×10^6 cells) are treated with 5 μM DCFH-DA in culture medium (from a solution of 1 mM DCFH-DA in ethanol) for 30 min at 37°.[21] After treatment with DCFH-DA, the cells are washed twice with PBS, suspended in 2 ml of PBS, and analyzed with a flow cytometer (FACScan, Becton-Dickinson, San Jose, CA) equipped with a single 488-nm argon laser. The filter in front of the fluorescence 1 (FL1) photomultiplier (PMT) transmits at 530 nm and has a bandwidth of 30 nm, and the filter used in the FL2 channel transmits at 585 nm and has a bandwidth of 42 nm. DCF data are recorded using FL1 PMT. The 10,000 cells per sample are acquired in histograms using a data analysis program CELL Quest (Becton-Dickinson). Dead cells and debris are excluded from the analysis by electronic gating of forward and side scatter measurements (Fig. 3A).

As shown in Fig. 3, peroxide generation was observed immediately after 5 J/cm^2 of UVA irradiation in HL-60 cells. This result suggests that hydrogen peroxide (H_2O_2) generated by the type II (reactive oxygen species) mechanism can participate significantly in DNA damage by UVA. This may be supported by reports showing that UVA irradiation mediates the formation of H_2O_2.[22] Alternatively, UVA may induce DNA damage by the type I (electron transfer) mechanism[4–6] and DNA damage induces H_2O_2 generation.[23]

Flow Cytometric Measurement of $\Delta\Psi m$

To assess $\Delta\Psi m$ in intact cells, lipophilic cations such as cyanine dyes and rhodamine 123 have been used to probe $\Delta\Psi m$. This section describes the measurements of $\Delta\Psi m$ using two cyanine dyes, DiOC$_6$(3) and JC-1.

Measurement of $\Delta\Psi m$ Using DiOC$_6$(3). The change of mitochondrial membrane potential ($\Delta\Psi m$) is studied using DiOC$_6$(3). DiOC$_6$(3) incorpo-

[19] D. A. Bass, J. W. Parce, L. R. Dechatelet, P. Szejda, M. C. Seeds, and M. Thomas, *J. Immunol.* **130,** 1910 (1983).

[20] P. Ubezio and F. Civoli, *Free Radic. Biol. Med.* **16,** 509 (1994).

[21] Y. Hiraku and S. Kawanishi, *Cancer Res.* **56,** 5172 (1996).

[22] X. Zhang, B. S. Rosenstein, Y. Wang, M. Lebwohl, and H. Wei, *Free Radic. Biol. Med.* **23,** 980 (1997).

[23] S. Tada-Oikawa, S. Oikawa, M. Kawanishi, M. Yamada, and S. Kawanishi, *FEBS Lett.* **442,** 65 (1999).

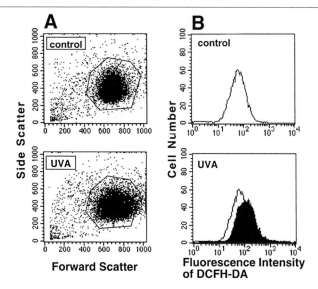

FIG. 3. Generation of peroxide in cultured cells irradiated with UVA. HL-60 cells (2.0 × 10^6 cells) were irradiated with 5 J/cm^2 of 365-nm light. After irradiation, cells were treated continuously with 5 μM DCFH-DA for 30 min at 37°. Cells were analyzed with a flow cytometer (FACScan). (A) Forward and side scatters were gated on the major population of normal-sized cells. (B) The horizontal axis shows the relative fluorescence intensity of DCFH-DA, and the vertical axis shows the cell number. Distribution of fluorescence intensity of control is also shown (open peaks). Control: no UVA irradiation.

rates into mitochondria in strict nonlinear dependence of $\Delta\Psi m$ and emits exclusively within the spectrum of green light.[24] After UVA irradiation and incubation, the cells (2 × 10^6 cells) are treated continuously with 40 nM DiOC$_6$(3) [from a solution of 40 μM DiOC$_6$(3) in dimethyl sulfoxide (DMSO)] in culture medium for 15 min at 37°.[25] After the treatment with DiOC$_6$(3), cells are washed twice with PBS and suspended in 2 ml of PBS and then are analyzed with a flow cytometer. DiOC$_6$(3) data are recorded using FL1 PMT, and 10,000 cells per sample are acquired in histograms using the data analysis program CELL Quest as described previously.

Figure 4 shows flow cytometric distribution of cells irradiated with 10 J/cm^2 365-nm light, subsequently treated with DiOC$_6$(3). UVA irradiation shows a decrease in fluorescence intensity of DiOC$_6$(3) in a time-dependent manner (Fig. 4), suggesting that $\Delta\Psi m$ is decreased by UVA exposure.

[24] P. X. Petit, J. E. O'Connor, D. Grunwald, and S. C. Brown, *Eur. J. Biochem.* **194,** 389 (1990).
[25] N. Zamzami, P. Marchetti, M. Castedo, C. Zanin, J. L. Vayssiere, P. X. Petit, and G. Kroemer, *J. Exp. Med.* **181,** 1661 (1995).

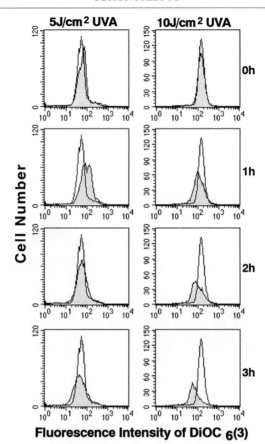

FIG. 4. The loss of $\Delta\Psi m$ detected by $DiOC_6(3)$ in cultured cells irradiated with UVA. HL-60 cells (2.0×10^6 cells) were irradiated with 5 and 10 J/cm^2 of 365-nm light and incubated for 0–4 hr at 37°. After incubation, cells were treated with 40 nM $DiOC_6(3)$ for 15 min at 37°. Cells were analyzed with a flow cytometer (FACScan). The horizontal axis shows the relative fluorescence intensity, and the vertical axis shows the cell number. Distribution of fluorescence intensity of control is also shown in 0–4 hr (open peaks). Control: no UVA irradiation.

Measurement of $\Delta\Psi m$ Using JC-1. $\Delta\Psi m$ variations are also analyzed with another cyanine, JC-1, whose monomer emits at 520 nm (green) after excitation at 490 nm, when the membrane potential in mitochondria is highly negative. Depending on the membrane potential, JC-1 is able to form "J aggregates" that are associated with a large shift in emission (585 nm).[26] After UVA irradiation and incubation, the cells (2×10^6 cells) are

[26] S. T. Smiley, M. Reers, C. Mottola-Hartshorn, M. Lin, A. Chen, T. W. Smith, G. D. Steele, Jr., and L. B. Chen, *Proc. Natl. Acad. Sci. U.S.A.* **88,** 3671 (1991).

resuspended in PBS containing 5 μg/ml JC-1 (from a solution of 2.5 mg/ml JC-1 in DMSO) and incubated for 20 min at 37°. After treatment with JC-1, the cells are washed twice with PBS, suspended in 2 ml of PBS, and analyzed with a flow cytometer. JC-1 fluorescence is analyzed on FL1 and FL2 channels for the detection of the fluorescence of the dye monomer and J aggregates, respectively. For the FACScan acquisition, PMT values are 480 and 310 V, respectively, for FL1 and FL2; the FL1-FL2 compensa-

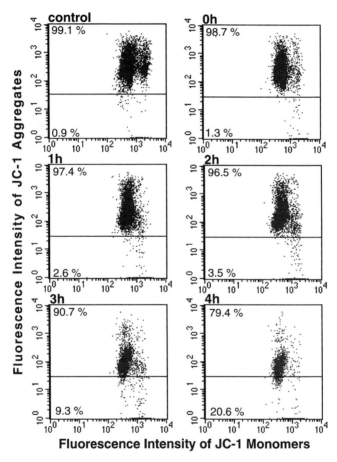

FIG. 5. The loss of $\Delta\Psi$m detected by JC-1 in cultured cells irradiated with UVA. HL-60 cells (2.0×10^6 cells) were irradiated with 10 J/cm^2 of 365-nm light and incubated for 0–4 hr at 37°. After incubation, cells were treated with 5 μg/ml JC-1 for 20 min at 37°. Cells were analyzed with a flow cytometer (FACScan). The horizontal axis shows fluorescence intensity of JC-1 monomers, and the vertical axis shows that of JC-1 aggregates. Percentages refer to the proportion of cells found in the respective window. Control: no UVA irradiation.

tion was 2.0–4%, and FL2-FL1 compensation is 8.0–9.5%. The 10,000 cells per sample are acquired in a dot plot using the data analysis program CELL Quest. Dead cells and debris are excluded from the analysis by electronic gating of forward and side scatter measurements.

Figure 5 shows flow cytometric distribution of cells irradiated with 10 J/cm² of 365-nm light, subsequently treated with JC-1. UVA irradiation has shown a decrease in fluorescence intensity of J aggregates in a time-dependent manner (Fig. 5), suggesting that ΔΨm is decreased by UVA exposure.

DiOC₆(3) emits one spectrum of green light and is convenient to measure ΔΨm.[24] JC-1 is a fluorescent probe characterized by two emission peaks (520 and 585 nm with excitation at 490 nm) corresponding to monomer and aggregate forms of the dye.[26] Although DiOC₆(3) is easy to assess as it has only one emission peak, JC-1 may be a more useful reagent for specifically identifying ΔΨm than DiOC₆(3).

Measurement of Activity of Caspase-3 in Cultured Cells Irradiated with UVA

After UVA irradiation, the cells are incubated continuously at 37° for 0–4 hr. Cells (5.0 × 10⁶ cells) are washed with PBS twice and suspended

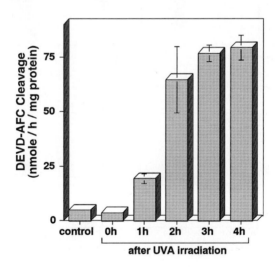

FIG. 6. Measurement of DEVD-AFC proteolysis catalyzed by a UVA-induced apoptotic extract. HL-60 cells were irradiated with 5 J/cm² of 365-nm light and incubated for indicated times. After incubation, samples were assessed for DEVD-specific protease activation. Reactions were initiated by the addition of 20 μM Ac-DEVD-AFC. Product formation (protease activity) was monitored by measuring the fluorescence (excitation at 400 nm, emission at 505 nm). Results are expressed as means ± SD of values obtained from four experiments.

in 500 μl of assay buffer [0.1 M HEPES (pH 7.4), 2 mM dithiothreitol, 0.1% (w/v) CHAPS, and 1% (w/v) sucrose]. Cell suspensions are frozen in liquid N_2 and thawed at 37° until the ice has melted in each tube. The freeze–thaw procedure is repeated three times and cell lysates are centrifuged at 18,500g for 20 min, and the supernatant is obtained. Fifty microliters of apoptotic extract is diluted with 450 μl of freshly prepared assay buffer. The reaction is initiated by the addition of 20 μM enzyme substrate Ac-DEVD-AFC at 37° for 10 min. The level of released 7-amino-4-trifluoromethyl coumarin (AFC) is measured using a Shimadzu RF-5300PC spectrofluorophotometer with excitation at 400 nm and emission at 505 nm with a standard curve of AFC. The protein content of cell extracts was determined using a dye (Bio-Rad Protein Assay, Bio-Rad).

At 2–4 hr after irradiation with 5 J/cm^2 of 365-nm light, the activity of caspase-3 was increased in a time-dependent manner (Fig. 6). This result suggests that caspase-3 is involved in UVA-induced apoptosis in HL-60 cells.

Conclusion

This article described techniques useful in the analysis of DNA damage, peroxide generation, change of $\Delta\Psi$m, and caspase-3 activity. Time course studies using these methods have revealed that UVA irradiation induces oxidative DNA damage and peroxide generation, followed by the loss of $\Delta\Psi$m and subsequent activation of caspase-3, resulting in apoptosis. In addition to the mitochondria-mediated pathway, lysosome might be involved in the apoptotic pathway.[27,28] Although cultured skin-derived cells should be used instead of the human leukemic cell line for studying the influence of UV light, epithelial cell lines show the resistance to UVA-mediated apoptosis. This study has suggested that UVA induces DNA damage by the type I mechanism[4–6] and the minor type II (superoxide anion radical) mechanism. In addition, several reports have shown that UVA irradiation mediates the formation singlet oxygen via the major type II mechanism[29] and leads to damage that kills cells by initiating apoptosis.[30]

[27] R. Ishisaka, T. Utsumi, M. Yabuki, T. Kanno, T. Furuno, M. Inoue, and K. Utsumi, *FEBS Lett.* **435**, 233 (1998).
[28] K. Roberg and K. Öllinger, *Am. J. Pathol.* **152**, 1151 (1998).
[29] S. Kawanishi, S. Inoue, S. Sano, and H. Aiba, *J. Biol. Chem.* **261**, 6090 (1986).
[30] A. Morita, T. Werfel, H. Stege, C. Ahrens, K. Karman, M. Grewe, S. Grether-Beck, T. Ruzicka, A. Kapp, L. O. Klotz, H. Sies, and J. Krutmann, *J. Exp. Med.* **186**, 1763 (1997).

Thus, several potential mechanisms of apoptosis may exist. Further analysis is necessary to elucidate various types of apoptotic mechanisms.

Acknowledgment

This work was supported by Grants-in-Aid for Scientific Research from the Ministry of Education, Science, and Culture of Japan.

[32] Mechanism of Photodynamic Therapy-Induced Cell Death

By NIHAL AHMAD and HASAN MUKHTAR

Photodynamic therapy (PDT) is a promising new therapeutic procedure for the management of a variety of solid tumors and also possesses a potential against many nonmalignant diseases. PDT depends on the uptake of a photosensitizing compound by the diseased tissue to be treated, followed by its irradiation with visible light of an appropriate wavelength that correlates with the absorption spectrum of the photosensitizing compound, in the presence of molecular oxygen. A proper understanding of the mechanism of PDT-mediated cancer cell kill that may result in improving the efficacy of this treatment modality is far from complete. It is believed that PDT, by generating reactive oxygen species (ROS), results in oxidative damage to the cellular organelles. This damage triggers a combination of biochemical, genetic, and molecular events, which have a common cell destruction goal. Necrosis and apoptosis have been shown to contribute for cell kill by PDT. In this article, an attempt has been made to provide general information on PDT and a brief account of the mechanisms responsible for PDT-mediated cell death.

Background Information

Photodynamic therapy of cancer is a modality based on photochemical sensitization, which results in tumor destruction. PDT requires three components for its action: a photosensitizing compound, light, and oxygen, which individually are nontoxic but tumoricidal in combination.[1] The concept of photochemical sensitization is not new as the healing potential of light was first reported by the Greeks. Herodotus, the great greek historian

[1] H. I. Pass, *J. Natl. Cancer Inst.* **85,** 443 (1993).

and geographer, in fact, is considered the father of "heliotherapy."[1] The first published information on photodynamic action dates back to the year 1888 when it was reported that the toxicity of quinine and cinchonamine for enzymes, plants, and frog eggs was greater in light than in darkness.[2] More than a decade later, in 1900, it was demonstrated that the toxic response of acridine towards *Paramecium* increased with increasing sunlight.[2] The first oncological use of PDT was reported in 1903 when eosin, in combination with light, was used to treat cancer.[1] In following years, many compounds were recognized to promote photochemical toxicity.[1] In the 1940s, the crude preparation of hematoporphyrin was demonstrated to be retained preferentially in malignant tissues as compared to normal ones.[1] This was followed by a number of studies aimed at improving the tumor-localizing potential by synthesizing a porphyrin mixture, viz. "hematoporphyrin" derivative (HPD), which resulted in a novel method for the detection of a tumor.[1] In 1975, Kelly *et al.*[3] were successful in treating a patient with recurrent bladder cancer with a combination of HPD (injected intravenously) and light (delivered through a fiber-optic device). This probably started the modern era of PDT, which received great impetus from the studies of Dougherty and colleagues.[2,4] The detailed history of PDT is beyond the scope of this article and has been reviewed.[1,2]

Clinical Use of Photodynamic Therepy

Many preclinical and clinical studies conducted around the world, in the last two decades, have established that PDT is a promising therapeutic approach for the management of a variety of solid tumors.[1,2,4] The treatment relies on the selective uptake of a porphyrin-based photosensitizing chemical by the tumor relative to the surrounding normal tissue, followed by light irradiation in visible or near infrared region that is typically derived through a laser.[1,2,4] The United States Food and Drug Administration (US-FDA) approved PDT (with Photofrin as a photosensitizer) as a therapeutic modality in 1995 for the treatment of selected patients with advanced esophageal cancer.[4] In January 1998, the application of PDT for the treatment of microinvasive lung cancer was also approved by the USFDA.[4] The most common site for which PDT has been widely employed is the skin.[1,2] PDT has been used for the treatment of basal cell and squamous cell carcinomas, breast cancer metastatic to the skin, Kaposi's sarcoma, and

[2] N. L. Oleinick and H. H. Evans, *Radiat. Res.* **150**(Suppl.), S146 (1998).

[3] J. F. Kelley, M. E. Snell, and M. C. Berenbaum, *Br. J. Cancer* **31**, 237 (1975).

[4] T. J. Dougherty, C. J. Gomer, B. W. Henderson, G. Jori, D. Kessel, M. Korbelik, J. Moan, and Q. Peng, *J. Natl. Cancer Inst.* **90**, 889 (1998).

cutaneous T-cell lymphoma. PDT has also been employed for the treatment of cancers of the head and neck, eye, brain, lung, and ovary.[1,2] Other preneoplastic lesions for which PDT has been investigated include actinic keratoses, benign prostatic hyperplasia, and Barrett's esophagus.[1,2,4] The efficacy of PDT is also being tested against certain nonmalignant conditions such as vascular restenosis, macular degeneration, atherosclerosis, and rheumatoid arthritis.[1,2,4] The efficacy of PDT against a variety of other cancers and noncancerous conditions is also being investigated.

Mechanism of PDT: The Photochemical Process

The response of PDT depends on a combination of photosensitizer, light, and oxygen. Most of the studies conducted on PDT have employed Photofrin as a photosensitizer. Despite its proven efficacy, Photofrin possesses certain limitations: (i) it is a complex mixture, (ii) it utilizes light in the 625- to 630-nm wavelength range of the absorption spectrum, which has limited penetration in the tissue, and (iii) it causes severe and persistent cutaneous photosensitivity. For this reason, many second-generation photosensitizers, such as benzoporphyrin derivatives, tin etiopurpurin, lutetium texaphyrin, and phthalocyanines, have been synthesized and studied, some of which have been found to be effective in cell culture systems and in animal models. In recent years, the concept of "endogenous compounds as a photosensitizer" is being widely appreciated.[2] In this regard, 5-aminolevulinic acid (ALA) has also been widely studied because ALA application results in the increased production of protoporphyrin IX (PPIX), which is the penultimate product of the heme biosynthetic pathway and is an effective photosensitizer.[2]

The mechanism of photosensitization includes the absorption of a photon by the photosensitizer, which results in an excited singlet state and then to an excited triplet state.[2,5] The triplet state may either (i) transfer an electron to a biomolecule such as a membrane lipid, which, in the presence of oxygen, generates free radicals and triggers a chain reaction, or (ii) transfer the electron to molecular oxygen in the ground state, resulting in the formation of a reactive oxygen species (ROS), including singlet oxygen ($^1O_2$). These ROS are believed to be responsible for the cytotoxic effects of PDT.[1,2,4,5]

Mechanism of PDT: Tumor Response

The response of PDT is generally rapid as it can cause the ablation of a tumor within a few days of treatment. Based on the results of many

[5] M. Ochsner, *J. Photochem. Photobiol.* **39**, 1 (1997).

animal studies, Oleinick and Evans[2] have suggested three types of mechanisms involved in PDT-mediated tumor ablation: (i) PDT may directly damage the malignant cells of the tumor; (ii) PDT may cause changes in the tumor vasculature affecting blood supply and may also result in vasculture collapse and/or vascular leakage; and (iii) PDT may cause a release of cytokines and other molecules from treated cells, which produce an inflammatory response and recruit additional phagocytic cells to eliminate tumor cells. The mechanism of the PDT response is probably dependent on the irradiation protocol, the photosensitizer being employed, and the type of tumor being treated.[1,2,5]

Mechanism of PDT: Apoptosis vs Necrosis

Studies have suggested the involvement of apoptosis as well as necrosis during the PDT response.[1,2,4,5] In recent years, apoptosis has become an important concept in biomedical research as the life span of cancer cells as well as normal cells within a living system in regarded to be affected by the rate of apoptosis. In addition, apoptosis, different from necrosis, is a discrete way of cell death and is regarded an ideal way of cell elimination. Many *in vitro* as well as *in vivo* studies have shown the involvement of apotosis during PDT.[2] The involvement of apotosis during PDT was first demonstrated by Agarwal *et al.*[6] This study investigated the mode of cell death of two strains of mouse lymphoma L5178Y cells (LY-R and LY-S strains). Both cell strains were demonstrated to undergo rapid apoptosis, as shown by DNA fragmentation, following chloroaluminum phthalocyanine PDT. The involvement of apoptosis in *in vivo* situations was first demonstrated by an earlier study[7] from this laboratory in which PDT was found to result in apoptosis in radiation-induced fibrosarcoma (RIF-1) tumors grown in C3H/HeN mice, as evident from DNA fragmentation, increased formation of apoptotic bodies, and condensation of chromatin material around the periphery of the nucleus as early as 1 hr post-PDT. The extent of these changes was found to increase during the later stages of tumor ablation. This study suggested that apoptosis is an early event during the *in vivo* tumor shrinkage following PDT. In a subsequent *in vivo* study[8] from this laboratory, PDT was found to result in early apoptosis, as evident by DNA fragmentation, appearance of apoptotic bodies, and direct immu-

[6] M. L. Agarwal, M. E. Clay, E. J. Harvey, H. H. Evans, A. R. Antunez, and N. L. Oleinick, *Cancer Res.* **51,** 5993 (1991).
[7] S. I. Zaidi, N. L. Oleinick, M. T. Zaim, and H. Mukhtar, *Photochem. Photobiol.* **58,** 771 (1993).
[8] R. Agarwal, N. J. Korman, R. R. Mohan, D. K. Feyes, S. Jawed, M. T. Zaim, and H. Mukhtar, *Photochem. Photobiol.* **63,** 547 (1996).

noperoxidase detection of digoxigenin-labeled genomic DNA during the ablation of chemically induced squamous papillomas in SENCAR mice.

Mechanism of PDT-Mediated Necrosis

As stated earlier, the mechanism of the PDT response depends on the irradiation protocol, the photosensitizer being employed, and the type of tumor being treated. With some exceptions, in general, hydrophobic photosensitizers attack the tumors by direct interaction, whereas hydrophilic photosensitizers damage blood vessels, which interrupts the supply of oxygen and other nutrients, thereby killing tumor cells.[5] It has also been shown that the mechanism of tumor destruction, at least in part, depends on the availability of oxygen to the tumor.[5] The drug delivery system can change the mechanism of action entirely.[5] The timing of irradiation has also been shown to affect the mechanism of PDT action.[5]

Therefore, it is not possible to suggest a universal mechanism for PDT-mediated necrosis. In an earlier review, Ochsner[5] has presented a reasonable picture of the processes that are involved during the necrosis by PDT. Figure 1 summarizes the series of events during necrosis by PDT. For a detailed account, Ochsner review and the references therein should be consulted.[5]

Mechanism of PDT-Mediated Apoptosis

The mechanism of PDT-mediated apoptosis has received increasing attention in the recent past. Many studies have attempted to determine the mechanism of PDT-mediated apoptosis at biochemical as well as genetic/molecular levels. Looking at all the available information, no universal mechanism of apoptosis by PDT exists. PDT-mediated apoptosis may follow different mechanism(s) depending on (i) the types of cells being treated, (ii) the type of photosensitizer being used, (iii) the light delivery protocols employed, and (iv) the time lag between photosensitizer and light treatment.[1,2,4,5] The existence of multiple pathways for one treatment protocol is also an intriguing possibility. However, the major efforts in this direction are summarized next.

Involvement of Phospholipase Activation and Intracellular Ca^{2+} in PDT-Mediated Apoptosis

Agarwal et al.[9] demonstrated that aluminum phthalocyanine PDT results in extensive DNA fragmentation within 1–2 hr following PDT in

[9] M. L. Agarwal, H. E. Larkin, S. I. Zaidi, H. Mukhtar, and N. L. Oleinick, *Cancer Res.* **53**, 5897 (1993).

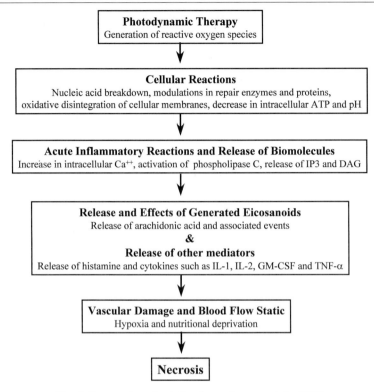

FIG. 1. Proposed mechanistic scheme for necrosis by PDT.

mouse lymphoma cell L5178Y. This study demonstrated that PDT resulted in (i) rapid (<1 min) activation of phospholipase C, (ii) breakdown of membrane phosphoinositides, and (iii) rapid release of Ca^{2+} from intracellular pools. This study further demonstrated that a phospholipase C inhibitor, U73122, blocked the rapid transient increases in both inositol-1,4,5-trisphosphate and intracellular Ca^{2+} levels as well as the subsequent fragmentation of nuclear DNA. In addition, p-bromphenacyl bromide, an inhibitor of phospholipase A_2, was also found to block DNA fragmentation. PDT also stimulated the release of arachidonic acid by the phospholipase A_2-dependent breakdown of membrane phospholipids. This study suggested that the PDT-mediated damage to cell membranes might mimic natural stimuli of phospholipases and initiate apoptosis in L5178Y cells.

In another study, Tajiri et al.[10] demonstrated that the changes in intra-

---

[10] H. Tajiri, A. Hayakawa, Y. Matsumoto, I. Yokoyama, and S. Yoshida, *Cancer Lett.* **128**, 205 (1998).

cellular Ca^{2+} concentrations are correlated to Photofrin PDT-induced apoptosis in human squamous cell carcinoma cells (HSC-2). This study showed extensive DNA fragmentation within 2 hr of PDT, as evident from flow cytometric analysis, and (ii) a slight change in the intracellular Ca^{2+} concentration occuring at 30 min after PDT and a subsequent marked increase at 1–2 hr after PDT as shown by confocal laser scanning microscopy. These findings further indicated that an increase in the intracellular Ca^{2+} concentration may play an important role in the induction of PDT-induced apoptosis.

Serine/Threonine Dephosphorylation during PDT-Mediated Apoptosis

Luo and Kessel[11] demonstrated the induction of PDT-mediated apotosis in murine leukemia cells as early as 30 min after irradiation. Apoptosis was found to be inhibited by concurrent exposure of the cells to a serine/threonine phosphatase inhibitor calyculin A and to be promoted by a serine/threonine kinase inhibitor. These results suggested that, probably at late stage, PDT-mediated apoptosis requires serine/threonine dephosphorylation.[11]

Involvement of bcl2 during PDT Response

In a study by He et al.[12] employing a pair of Chinese hamster ovary cell lines that differ from one another by a transfected human bcl2 gene, the ability of this antiapoptotic gene to modulate PDT-induced apoptosis and overall cell killing was investigated. It is important to emphasize that the study employed a silicon phthalocyanine compound [$HOSiPcOSi(CH_3)_2$ $(CH_2)_3 N(CH_3)_2$] termed "Pc4", a photosensitizer developed at Case Western Reserve University.[13] Based on encouraging experimental results and preclinical pharmacokinetics, efficacy, and toxicology studies completed by the "Drug Decision Network" of the National Cancer Institute, Pc4 has been recommended for clinical testing. Data from this study indicated that parental cells displayed a high incidence of apoptosis following PDT, whereas bcl2-expressing cells demonstrated a much lower incidence of apoptosis as assessed by (i) DNA fragmentation by gel electrophoresis, (ii) flow cytometry of cells labeled with fluorescently tagged dUTP by terminal deoxynucleotidyltransferase, and (iii) fluorescence microscopy of acridine

---

[11] Y. Luo and D. Kessel, *Biochem. Biophys. Res. Commun.* **221,** 72 (1996).

[12] J. He, M. L. Agarwal, H. E. Larkin, L. R. Friedman, L. Y. Xue, and N. L. Oleinick, *Photochem. Photobiol.* **64,** 845 (1996).

[13] N. L. Oleinick, A. R. Antunez, M. E. Clay, B. D. Rihter, and M. E. Kenney, *Photochem. Photobiol.* **57,** 242 (1993).

orange-stained cells. Further, clonogenic assays demonstrated that bc12 was able to inhibit the overall cell killing as well. Thus, this study suggested an involvement of bc12 in PDT-mediated apoptosis.[11]

Role of Wild-Type P53 in PDT

The tumor suppressor gene p53 is regarded as a key element in maintaining a balance between cell growth and cell death in the living system.[14] p53, in response to DNA damage, triggers a variety of regulatory events that limit the proliferation of damaged cells. Because of these facts and also because the loss of p53 function is shown to be correlated with decreased sensitivity to chemotherapy and radiation therapy in a variety of human tumors in a study by Fisher *et al.*,[15] the efficacy of tin ethyl etiopurpurin (SnET2)-PDT, or Photofrin PDT, was investigated in human promyelocytic leukemia HL60 cells exhibiting wild-type p53, mutated p53, or deleted p53 expression. This study demonstrated that HL60 cells expressing wild-type p53 were more sensitive to Photofrin and SnET2 PDT, as well as to UVC irradiation, when compared to HL60 cells exhibiting deleted or mutated p53 phenotypes. The authors of this study suggested that (i) photosensitivity is increased in HL60 cells expressing wild-type p53 and that (ii) PDT-mediated oxidative stress can induce apoptosis through a p53-independent mechanism in HL60 cells.

Caspase Activation in PDT-Mediated Apoptosis

Because caspases have been shown to play crucial role in apoptosis, Granville *et al.*[16] demonstrated that PDT (with benzoporphyrin derivative monoacid ring A, BPD MA, as a photosensitizer) of human promylocytic leukemia HL-60 cells results in (i) cleavage and subsequent activation of caspase-3, but not caspase-1, and (ii) cleavage of poly(ADP-ribose) polymerase (PARP) and catalytic subunit of DNA-dependent protein kinase (DNA PKc$_S$). Further, the general caspase inhibitor Z-Asp-2,6 dichlorobenzoyloxymethylketone (Z-Asp-DCB) was found to block PARP cleavage and DNA fragmentation, whereas the serine protease inhibitor 3,4-dichloroisocoumarin (DCI) and *N*-tosyllysylchloromethylketone (TLCK) were able to block caspase-3 cleavage and DNA fragmentation. This study

[14] M. L. Agarwal, W. R. Taylor, M. V. Chernov, O. B. Chernova, and G. R. Stark, *J. Biol. Chem.* **273**, 1 (1998).

[15] A. M. Fisher, K. Danenberg, D. Banerjee, J. R. Bertino, P. Danenberg, and C. J. Gomer, *Photochem. Photobiol.* **66**, 265 (1997).

[16] D. J. Granville, J. G. Levy, and D. W. C. Hunt, *Cell Death Differ.* **4**, 623 (1997).

demonstrated, for the first time, the involvement of caspases in PDT-mediated apoptosis.[16] Granville *et al.,*[17] in a followup study, demonstrated that the overexpression of the Bcl-X(L) gene in HL-60 cells prevented BPD MA PDT-induced caspase activation, DFF cleavage, and DNA fragmentation. These results also demonstrated an example of chemotherapeutic drug-induced activation of DFF and its regulation by Bcl-X(L).

In another study, He *et al.*[18] assessed the ability of Pc4 PDT to activate cellular caspases during rapid apoptosis in murine lymphoma L5178Y-R cells. This study demonstrated a rapid dose-dependent apoptosis by Pc4 PDT as evident from the DNA ladder assay. Western blot analysis revealed cleavage of the normally ~116-kDa PARP into ~90-kDa fragments at approximately the same time as the earliest DNA fragmentation. This analysis also revealed that levels of a poly(ADP-ribosylated) 100-kDa protein, identified tentatively as topoisomerase I, were maintained in cells after total cleavage of PARP. Further, caspase-3 (and/or caspase-7) activity, as measured in cell lysates with the fluorogenic substrate DEVD-AMC, was found to be elevated almost immediately after PDT. PDT-induced apoptosis and PARP cleavage were shown to be inhibited by the irreversible caspase inhibitor, benzyloxycarbonyl-Val-Ala-Asp(*O*-methyl)fluoromethylketone, whereas the inactive peptide analog, benzyloxycarbonyl-Phe-Ala-fluoromethylketone, was without effect. These results further strengthened the hypothesis that PDT-induced apoptosis is mediated by the activation of caspases.

Because it has been shown for other proapoptotic stimuli that the integral endoplasmic reticulum protein Bap31 is cleaved by caspases 1 and 8 to generate a 20-kDa Bap31 cleavage fragment that can induce apoptosis, Granville *et al.*[19] conducted a study to investigate whether Bap31 cleavage and generation of p20 are early events in PDT-induced apoptosis. They specifically evaluated the effect of BPD MA PDT on (i) the mitochondrial release of cytochrome c, (ii) the involvement of different caspases, and (iii) the status of several known caspase substrates, including Bap31, in HeLa cells. This study demonstrated that BPD MA PDT results in (i) an immediate release of cytochrome c into the cytosol, (ii) activation of caspases 3, 6, 7, and 8 within 1–2 hr post-PDT, and (iii) cleavage of Bap31 at 2–3 hr post-PDT. Further, it was found that the caspase-3 inhibitor DEVD-fmk blocked caspase-8 and Bap31 cleavage, suggesting that caspase-8 and Bap31

[17] D. J. Granville, H. Jiang, M. T. An, J. G. Levy, B. M. McManus, and D. W. Hunt, *FEBS Lett.* **422,** 151 (1998).

[18] J. He, C. M. Whitacre, L. Y. Xue, N. A. Berger, and N. L. Oleinick, *Cancer Res.* **58,** 940 (1998).

[19] D. J. Granville, C. M. Carthy, H. Jiang, G. C. Shore, B. M. McManus, and D. W. Hunt, *FEBS Lett.* **437,** 5 (1998).

processing occurs downstream of caspase-3 activation in PDT-induced apoptosis. These results demonstrated that mitochondrial cytochrome c release into the cytoplasm is a primary response of PDT preceded by caspase activation and cleavage of Bap31.

Ceramide Generation during PDT-Mediated Apoptosis

Because ceramide is a second messenger that has been associated with stress-induced apoptosis, Separovic et al.[20] investigated a possible link among PDT, ceramide, and apoptosis in L5178Y-R cells. This study demonstrated that the cells undergo rapid apoptosis, initiated within 30 min of PDT. With a dose of PDT that produces a 99.9% loss of clonogenicity, L5178Y-R cells responded with an increased production of ceramide that reached a maximum level in 60 min. For a constant light fluence and varying concentrations of the photosensitizer Pc4, the ED_{50} for ceramide generation (46 nM) was similar to the LD_{50} for clonogenic cell death (40 nM). Further, when cells were exposed to exogenous N-acetylsphingosine (10 μM), the apoptotic response was found to be delayed, possibly due to an induction of apoptosis via a different mechanism. This study suggested that the PDT-stimulated increase in synthesis of ceramide in LY-R cells may be related to PDT-induced apoptosis.

Extending the just-described study, Separovic et al.[21] demonstrated that Pc4 PDT results in a rapid induction of apoptosis (starting at 1 hr post-PDT) in mouse lymphoma L5178Y-R cells, human leukemia (U937) cells, and Chinese hamster ovary (CHO) cells. However, similar treatment did not cause apoptosis in mouse radiation-induced fibrosarcoma (RIF-1) cells even at 24 hr post-PDT. Further, it was shown that the same doses of PDT result in a time-dependent ceramide accumulation in L5178Y-R, U937, and CHO cell lines. Results demonstrated a significant increase in ceramide levels within 1 and 10 min in U937 and CHO cells, respectively. In RIF-1 cells, however, elevated ceramide production was measured in only 30-min post-PDT cells. In addition, exogenous N-acetylsphingosine was found to mimic PDT-induced apoptosis in U937 and CHO cells. Based on this study, it was suggested that ceramide accumulation is associated with PDT-induced apoptosis and photocytotoxicity.[21]

Cell Cycle-Mediated Apoptosis by PDT

A growing body of evidence indicates that cell cycle perturbations may be related to apoptosis because certain cell cycle regulatory molecules such

[20] D. Separovic, J. He, and N. L. Oleinick, Cancer Res. **57,** 1717 (1997).
[21] D. Separovic, K. J. Mann, and N. L. Oleinick, Photochem. Photobiol. **68,** 101 (1998).

as the tumor suppressor protein p53 and the cyclin kinase inhibitor (cki). WAF1/CIP1/p21 is shown to be associated with apoptosis and cell cycle arrest.[14] Studies have demonstrated the implication of WAF1/CIP/p21 induction through a p53-dependent as well as a p53-independent mechanism during apoptosis and cell cycle deregulation at the $G_1 \rightarrow S$ phase transition checkpoint.[14,22] WAF1/CIP1/p21 is known to control the cell cycle by regulating the cyclin-dependent kinase (cdk)–cyclin complexes in response to various physiological processes, such as the cellular response to DNA damage, senescence, growth inhibitory signals, and tumor suppression.[14,22]

Because of these facts, we investigated the involvement of cell cycle regulatory events during Pc4 PDT-mediated apoptosis in human epidermoid carcinoma cell A431. In a recently published study,[23] we demonstrated that PDT of A431 cells resulted in apoptosis, inhibition of cell growth, and G_0–G_1 phase arrest of the cell cycle in a time-dependent fashion. Further, Western blot analysis showed that PDT resulted in an induction of the cyclin kinase inhibitor WAF1/CIP1/p21 and a downregulation of cyclins (cyclin D1 and E) and cyclin-dependent kinases (cdk2 and cdk6). PDT was also found to result in a decrease in kinase activities associated with all the cdks and cyclins examined. We further demonstrated that PDT also resulted in increased binding of cyclin D1 and cdk6 with WAF1/CIP1/p21 and decreased binding of cyclin D1 with cdk2 and cdk6. Based on this study, we suggested that the PDT-mediated induction of WAF1/CIP1/p21 results in an imposition of the artificial checkpoint at the $G_1 \rightarrow S$ transition, which, through inhibition in the cdk2, cdk6, cyclin D1, and cyclin E, causes a G_0–G_1 phase cell cycle arrest. This cell cycle arrest is an irreversible process and the cells, unable to repair these damages, ultimately suffer an apoptotic cell death. However, the involvement of two independent mechanisms for apoptosis and cell cycle arrest cannot be ruled out. This study, for the first time, established the involvement of cell cycle deregulation by a cki–cyclin–cdk network during PDT-mediated apoptosis.

Extending our study, showing that PDT results in an induction of WAF1/CIP1/p21, which through downregulating cyclin (E and D1)/cyclin-dependent kinases (cdk2 and cdk6) results in a G_0/G_1 phase arrest of the cell cycle and apoptosis in A431 cells, we investigated the involvement of pRb-E2F/DP machinery during Pc4 PDT-mediated cell cycle deregulation and apoptosis.[24] This study was based on the rationale that retinoblastoma (pRb) and the E2F family transcription factors are important proteins that

[22] C. J. Sherr, *Cell* **73,** 1059 (1993).

[23] N. Ahmad, D. K. Feyes, R. Agarwal, and H. Mukhtar, *Proc. Natl. Acad. Sci. U.S.A.* **95,** 6977 (1998).

[24] N. Ahmad, S. Gupta, and H. Mukhtar, *Oncogene* **18,** 1891 (1999).

regulate the $G_1 \rightarrow S$ transition in the cell cycle. This study[24] demonstrated that Pc4 PDT resulted in a significant decrease in the hyperphosphorylated form of pRb at 3, 6, and 12 hr post-PDT, with a relative increase in hypophosphorylated pRb. The study further revealed that a PDT-caused decrease in phosphorylation of pRb occurs at serine-780. ELISA data demonstrated that PDT results in a time-dependent accumulation of hypophosphorylated pRb. Further, PDT was found to result in downregulation of the protein expression E2F family transcription factors (E2F1 through 5) and their heterodimeric partners DP1 and DP2. This study provided evidence for the involvement of pRb-E2F/DP machinery in PDT-mediated cell cycle arrest leading to apoptosis.[24]

In another study from our laboratory,[25] we investigated the involvement of nitric oxide (NO), which is considered to be involved in a variety of physiological and pathological processes, including cell cycle regulation and apoptosis. In this study we employed two different cell lines—PDT apoptosis-resistant radiation-induced fibrosarcoma (RIF-1) cells and PDT apoptosis-sensitive human epidermoid carcinoma (A431) cells. This study demonstrated that Pc4 PDT resulted in a rapid increase in nitrite production in A431 cells starting as early as 15 sec following PDT and showed a progressive increase up to 15 min post-PDT. RIF-1 cells, however, did not show an increase in nitrite production. PDT was also found to result in a rapid (as early as 15 sec post-PDT) increase in the constitutive form of NO synthase as shown by Western blot analysis and the [^{3}H]L-citrulline assay. RIF-1 cells did not show any change in protein expression or enzyme activity following the same treatment. This study suggested the involvement of NO during PDT-mediated apoptosis.[25]

All three studies just discussed are related to each other, and in Fig. 2 we have tried to piece together the information from these studies and from our existing source of knowledge to provide a composite scheme for the mechanism of PDT-mediated cell cycle arrest and apoptosis. It is important to emphasize here that this composite scheme may still not be complete because of the possibility of additional or complimentary mechanisms that may be operative during the PDT response.

Mechanism Of PDT-Mediated Apoptosis Under *In Vivo* Situations

Although the occurrence of apoptosis as a mechanism of cell death has been shown in both *in vitro*[1,2,4,9,23] and in vivo[1,2,4,7,8] situations, the molecular mechanisms responsible for PDT-mediated apoptosis in *in vivo* situations have not been investigated in detail. To our knowledge, the only such

[25] S. Gupta, N. Ahmad, and H. Mukhtar, *Cancer Res.* **58,** 1785 (1998).

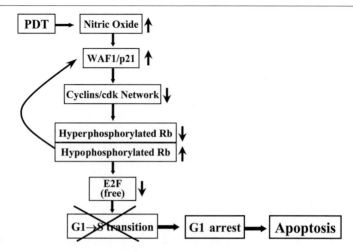

FIG. 2. Proposed mechanistic scheme for cell cycle-mediated apoptosis by PDT. Up arrows (↑) show an increase whereas the down arrows (↓) show a decrease.

investigations stem from our laboratory. We have shown that Pc4 PDT of ovarian epithelial carcinoma (OVCAR-3) tumors implanted in athymic nude mice results in PARP cleavage and induction of p21/WAF1/CIP1.[26] Because the validation of *in vitro* observations to *in vivo* situations is regarded as a key factor for the application of any therapeutic approach to humans, this study may be important. We are extending our efforts in this direction.

Other Mechanisms of Growth Inhibitory/Antiproliferative Effects of PDT

Some studies have attempted to deliniate the signaling events, which may not be linked directly to apoptosis, but contribute to the growth inhibitory/antiproliferative potential of PDT. A summary of these studies is as follows.

Involvement of Glucose-Regulated Protein in PDT-Response

Gomer *et al.*[27] examined porphyrin-photosensitizing parameters associated with the induction of the glucose-regulated family of stress proteins

[26] V. C. Colussi, D. K. Feyes, J. W. Mulvihill, Y.-S. Li, M. E. Kenney, C. A. Elmets, N. L. Oleinick, and H. Mukhtar, *Photochem. Photobiol.* **69,** 236 (1999).
[27] C. J. Gomer, A. Ferrario, N. Rucker, S. Wong, and A. S. Lee, *Cancer Res.* **51,** 6574 (1991).

(GRPs). Increased levels of mRNA encoding GRPs as well as increases in GRP protein synthesis were found to occur in mouse radiation-induced fibrosarcoma cells exposed to a 16-hr porphyrin incubation prior to light exposure. A short (1 hr) porphyrin incubation prior to light treatment, however, caused only minimal increases in GRP mRNA levels or GRP protein synthesis. In this study the relationship between GRP levels and PDT sensitivity was also sought in cells pretreated with the calcium iono- phore A-23187 in order to overexpress GRPs prior to photosensitization. Resistance to PDT was observed in cells overexpressing GRPs only with an extended porphyrin incubation protocol, and this response was found to be independent of the cellular porphyrin uptake. In additional experi- ments, a transient elevation of GRP mRNA levels was observed in trans- planted mouse mammary carcinomas following *in vivo* PDT treatments. These results suggested that (i) specific targets of oxidative damage were correlated with PDT-mediated GRP induction, which may be a useful *in vivo* biochemical marker of PDT-mediated injury, and (ii) GRPs may play a role in modulating sensitivity to cellular stresses, including certain types of oxidative injury.

Involvement of Nuclear Transcription Factor Nuclear Factor Kappa B in PDT Response

Ryter and Gomer[28] demonstrated that photofrin II PDT enhances the DNA-binding activity of nuclear factor (NF) κB, which is a nuclear tran- scription factor regarded as playing a key role in the transcription of a variety of a genes known to be involved in cellular proliferation and apoptosis. Photosensitization with photofrin II was found to result in a significant increase in the NF-κB-binding activity in mouse leukemia L1210 cells.

Involvement of Early Response Genes c-fos, c-jun, c-myc, and egr-1

Luna *et al.*[29] demonstrated that PDT-mediated oxidative stress induced the early response genes c-fos, c-jun, c-myc, and egr-1 in murine radiation- induced fibrosarcoma cells. This study showed that the incubation of cells with some porphyrin-based photosensitizers in the dark also increase the mRNA levels of these genes. Using nuclear runoff experiments it was shown that the induction of c-fos mRNA is controlled, in part, at the transcriptional level. Also, a chloramphenicol acetyltransferase reporter construct containing the major c-fos transcriptional response elements was found to be inducible by porphyrin and PDT. This study further examined

[28] S. W. Ryter and C. J. Gomer, *Photochem. Photobiol.* **58,** 753 (1993).
[29] M. C. Luna, S. Wong, and C. J. Gomer, *Cancer Res.* **54,** 1374 (1994).

the signal transduction pathways associated with PDT-mediated c-fos activation by treating the cells with protein kinase inhibitors. Staurosporine and 1-(5-isoquinolinesulfonyl)-2-methylpiperazine were found to inhibit PDT-mediated c-fos activation, whereas N-(2-guanidinoethyl)-5-isoquinoline sulfonamide was without any effect. Quinacrine, known to inhibit phospholipase activity, was found to block PDT-induced c-fos mRNA. This study suggested that the oxidative stress caused by PDT acts through the protein kinase-mediated signal transduction pathway(s) to activate early response genes, which may be involved in the biological response of PDT.

Involvement of α_2-Macroglobulin Receptor/Low-Density Lipoprotein Receptor-Related Protein in PDT Response

Employing mRNA differential display, Luna et al.[30] evaluated parental and photodynamic therapy (PDT)-resistant mouse, radiation-induced fibrosarcoma cell lines to identify unique transcripts. One transcript that was differentially expressed in parental cells, but not in PDT-resistant cells, was detected. This transcript was cloned and sequenced to be identified as an α_2-macroglobulin receptor/low-density lipoprotein receptor-related protein (α_2-MR/LRP). This was also confirmed by Northern and Western blot analyses. The functionality of this receptor was evaluated by exposing cells to Pseudomonas exotoxin A, and α_2-MR/LRP was found to be responsible for Pseudomonas exotoxin A internalization. The binding and endocytosis of activated α_2-macroglobulin and lipoproteins by α_2-MR/LRP were shown to be consistent with the uptake and localization of photosensitizers. This study suggested that α_2-MR/LRP plays a role in the PDT response by modulating photosensitizer uptake and/or subcellular localization.

Stimulation of Stress-Activated Protein Kinase and p38 HOG1 Kinase by PDT

Tao et al.[31] investigated activation of the members of the mitogen-activated protein kinase family following BPD MA PDT in a naturally transformed murine keratinocyte cell line, Pam 212. PDT was found to result in rapid activation of stress-activated protein kinase (SAPK) and p38 HOGI in a dose- and time-dependent fashion. However, PDT did not cause any significant activation of extracellularly regulated kinase (ERK) 1 and ERK2. The known scavengers of reactive oxygen species (ROS), viz. L-histidine and N-acetyl-L-cysteine, showed a significant inhibitory effect

[30] M. C. Luna, A. Ferrario, N. Rucker, and C. J. Gomer, Cancer Res. **55**, 1820 (1995).
[31] J. Tao, J. S. Sanghera, S. L. Pelech, G. Wong, and J. G. Levy, J. Biol. Chem. **271**, 27107 (1996).

on PDT-mediated SAPK and p38 HOGI activation, indicating that PDT-induced SAPK and p38 HOG1 activation may be partially mediated by ROS.

Induction of Heat Shock Proteins by PDT

Gomer *et al.*[32] demonstrated that PDT-mediated oxidative stress can activate the heat shock factor as well as increase HSP-70 mRNA and protein levels in mouse RIF-1 cells. This study further showed that this response may vary for the different photosensitizers as treatments using chlorin (mono-L-aspartylchlorin-e⁶)- or purpurin (tin etiopurpurin)-based photosensitizers induced HSP-70 expression, whereas the porphyrin (Photofrin)-based photosensitizer was without effect. This response was associated with the concomitant induction of thermotolerance in PDT-treated cells. The interesting observation of this study was that the *in vivo* PDT of RIF-1 tumors was found to result in HSP-70 induction with all photosensitizers, including Photofrin.

Tyrosine Phosphorylation by PDT

Xue *et al.*[33] investigated the role of tyrosine phosphorylation during Pc4 PDT of L5178Y-R cells. This study showed a dramatic increase in tyrosine phosphorylation of ~80 and ~55-kDa proteins due to Pc4 PDT. The increase was found to be dose dependent, occurring as early as 20 sec post-PDT. This response was amplified by the protein tyrosine phosphatase inhibitor sodium orthovanadate ($NaVO_4$). The vanadate treatment, however, was not found to have any influence on the overall photocytotoxicity or on the rate of apoptotic DNA fragmentation. The ~80-kDa phosphorylated protein was identified as HS1, a substrate of nonreceptor-type protein tyrosine kinases. The protein tyrosine kinase inhibitor genistein diminished tyrosine phosphorylation of this and other proteins and protentiated cell killing by PDT, but did not affect the rate of apoptosis. This study suggested that Pc4 PDT causes tyrosine phosphorylation of proteins, which is a stress response that may be protecting the cells from the lethal effects of PDT without altering the mechanism.

In another study, Granville *et al.*[34] demonstrated that BPD-MA PDT, which induces apoptosis rapidly in murine P815 mastocytoma cells, results in tyrosine phosphorylation of the ~200-kDa protein during or immediately

[32] C. J. Gomer, S. W. Ryter, A. Ferrario, N. Rucker, S. Wong, and A. M. Fisher, *Cancer Res.* **56,** 2355 (1996).

[33] L. Y. Xue, J. He, and N. L. Oleinick, *Photochem. Photobiol.* **66,** 105 (1997).

[34] D. J. Granville, J. G. Levy, and D. W. Hunt, *Photochem. Photobiol.* **67,** 358 (1998).

following PDT. Increased tyrosine phosphorylation of a ~15-kDa protein was evident 15 min postirradiation. Also, treatment with the protein kinase inhibitor staurosporine resulted in the prevention of tyrosine phosphorylation of the ~200-kDa species, but did not affect tyrosine phosphorylation of the ~15-kDa protein or the level of DNA fragmentation due to PDT. The authors suggested that the protein tyrosine phosphorylation observed in this study might not be related directly to the induction of the apoptotic cell death pathway.

Involvement of c-Jun N-Terminal Kinase (JNK) and p38 Mitogen-Activated Protein Kinase (MAPK) in PDT Response

Klotz et al.[35] demonstrated that 5-aminolevulinate (ALA) PDT of Ha-CaT cells resulted in a significant elevation of cellular c-Jun N-terminal kinase activity and phosphorylation of p38 mitogen-activated protein kinase. The phosphorylation of p38 was also observed in the human melanoma cell lines Bro and SkMel-23. This study suggested that the activation of JNK and p38 may be involved in the biological response of ALA PDT of human skin cells.

Conclusion

Understanding the mechanism(s) involved in the biological response of PDT is important because the efforts in this direction may result in designing better approaches for an improved PDT response. These efforts may also be useful in broadening the application of PDT. Although significant advances have been made, knowledge of the mechanism(s) of PDT-mediated cell death is far from complete. Concerted investigations are warranted. Also, there is a need for extensive effort directed to the validation of *in vitro* observations to *in vivo* situations. This will not only help in improving PDT response, but in developing novel approaches to improve its efficacy.

[35] L. O. Klotz, C. Fritsch, K. Briviba, N. Tsacmacidis, F. Schliess, and H. Sies, *Cancer Res.* **58,** 4297 (1998).

[33] Photocarcinogenesis: UVA vs UVB

By Frank R. de Gruijl

Introduction

The sun emits ultraviolet (UV) radiation extending from the UVA band (315–400 nm) through the UVB (280–315 nm) down to the high-energy UVC band (190–280) and vacuum UV (below 190 nm). The UV radiation is very (photo-) chemically active. Conjugated bonds in organic molecules absorb radiation of wavelengths around and below 200 nm, and in linear repeats or in ring structures the absorption bands of these bonds shift to higher wavelengths (250–300 nm and higher, e.g., the broad absorption up to 550 nm by β-carotene with its extended linear repeats and terminal rings). After absorbing UV radiation, a molecule may become altered (damaged) and/or affect (damage) other molecules, e.g., by producing reactive oxygen species (ROS). Thus, UV radiation, especially in the short wavelength range, forms a direct threat to the stability of unprotected organic molecules, which are essential to life on earth: foremost, DNA with aromatic rings in all of its bases (absorption maximum around 260 nm).

Besides enabling us to breath, the production of oxygen is a very fortunate evolutionary event: by absorbing the most energetic UV radiation (wavelengths below 240 nm) in the outer reaches of our atmosphere and by forming ozone in the process, oxygen protects the earth's surface from the most damaging solar UV radiation. The rare stratospheric ozone absorbs most of the radiation with wavelengths below 310 nm. UV radiation below 290 nm is virtually undetectable at ground level. However, the fraction of UVB radiation that reaches us can still be absorbed by DNA and proteins and is enough to damage and kill unprotected cells (mainly through DNA damage). Hence, organisms on the earth's surface have to be well adapted, e.g., by forming UV-absorbing surface layers (skin and fur), by antioxidant defences to quench UV-generated ROS, by efficiently repairing damaged DNA, and by repairing or replacing damaged cells. Human skin—UV radiation does not penetrate any deeper—shows all of these adaptive features. Nevertheless, the skin's defenses are not perfect and the continuous UV challenge leads to deterioration ("photoaging") in the long run and to an increasing likelihood of developing skin cancer.

Carcinogenesis

According to current contention, cancer arises as a consequence of a combination of disturbances in signal pathways that control cell cycling and differentiation. A straightforward way to introduce such disturbances persistently is to mutate genes whose products are crucial in these signal pathways. Thus, the central idea in cancer research is that a cancer develops as a clonal expansion of a cell with mutations in a combination of regulatory genes (actually, this concept of "disturbed signaling" was largely inferred from the finding that certain genes were consistently mutated or deleted in cancers, and from subsequently tracing the functions of these genes). These genes are subdivided into two categories: (proto-) oncogenes, which by mutation or amplification contribute actively to carcinogenesis, and tumor suppressor genes, which by mutation become dysfunctional or which are completely lost and thus fail to repress the carcinogenic progression.

There are, of course, other ways to interfere with these signaling pathways. Viruses may introduce oncogenes or gene products that inactivate tumor suppressor proteins. Exogenous chemicals or physical agents may directly affect the cascades of molecules that transfer signals. The latter epigenetical challenges are transient and need to be applied repetitively to contribute to a persistent development of a tumor.

DNA Damage, Mutations, and Action Spectra

From the general outline in the previous section it is immediately clear that DNA damage, causing mutations in genes, can play a pivotal role in carcinogenesis. UVB and UVC radiations are very genotoxic through direct absorption by DNA. The damage is mainly targeted at dipyrimidine sites where the pyrimidines become dimerized either by a cyclobutane ring at the 5 and 6 positions of the neighboring bases (forming a cyclobutane pyrimidine dimer, CPD) or by binding the 6th position in one base to the 4th in the next base (forming a 6-4 pyrimidine-pyrimidone photoproduct, 6-4PP). The wavelength dependencies (action spectra) of inducing CPD and 6-4PP run parallel[1] and follow the DNA absorption spectrum[2] (measured up to 313 nm). Action spectra of mutation induction (acquired thioguanine or ouabain resistance) and cell killing run parallel to the induction of CPD in human fibroblasts up to 313 nm, but at higher wavelengths in the UVA band mutation and cell killing become more frequent per induced CPD

[1] T. Matsunaga, K. Hieda, and O. Nikaido, *Photochem. Photobiol.* **54,** 403 (1991).
[2] B. S. Rosenstein and D. L. Mitchell, *Photochem. Photobiol.* **45,** 775 (1987).

lesion than in the UVB band[3] (well over 10 times more frequent at 365 nm; at 385 nm and higher wavelengths CPD are virtually undetectable[4]). The latter result indicates that DNA damage other than CPD and 6-4PP becomes more important in the UVA band. In accord with this inference, Applegate et al.[5] found that glutathione depletion caused a threefold increase in mutations at 365 nm, but had no effect at 313 nm where mutations arise far more efficiently, mainly through the direct induction of CPD and 6-4 PP. This effect of glutathione indicates that ROS such as hydrogen peroxides and hydroxyl radicals are important in the induction of mutations by UVA radiation. These ROS are generated by an endogenous photosensitizer (not DNA but, for example, porphirines or NADH) that becomes excited by absorbing UVA radiation, and then reacts with oxygen (a "type II reaction"; in a "type I reaction" the sensitizer reacts directly with the target molecule, e.g., DNA). Singlet oxygen can also be generated by UVA radiation through this sensitizing mechanism (see [27], this volume). The DNA damage caused by various ROS types appear to converge on commonly inducing 8-oxo-7,8-dihydroxyguanine (8-oxoG). The measured induction of this DNA lesion by UVA radiation is therefore not informative on which type of ROS was responsible; other types of damage also need to be considered.[6] UVA radiation also induces DNA strand breaks and DNA–protein cross-links, but action spectra for these lesions differ from that of 8-oxoG, which indicates that relative contributions from various sensitizers and types of ROS shift with wavelength. These dissimilarities in action spectra were clearly demonstrated in a study by Kielbassa et al.,[7] who used a selection of endonucleases to detect various types of photolesions in DNA from UV-exposed Chinese hamster ovary (CHO) cells (see Table I for an excerpt; also see [39], this volume). They found that the wavelength dependencies of 8-oxoG (i.e., Fpg-sensitive sites) and single-strand breaks (SSB) run fairly parallel to that of CPD (i.e., T4endoV-sensitive sites) from 290 to 310 nm. Over this wavelength interval the lesion frequency (presented per 20 kJ/m^2 of UV exposure) drops by a factor of 10^2, but CPD is about 10^3 times more frequent than the other two types of lesions. For higher wavelengths the frequency of CPD continues to drop, albeit more slowly from 330 nm (to 380 nm by about a factor of 6), whereas

[3] I. C. Enninga, R. T. L. Groenendijk, A. R. Filon, A. A. van Zeeland, and J. W. I. M. Simons, Carcinogenesis 7, 1829 (1986).
[4] H. Hacham, S. E. Freeman, R. W. Gange, D. J. Maytum, J. C. Sutherland, and B. M. Sutherland, Photochem. Photobiol. 52, 893 (1990).
[5] L. A. Applegate, D. Lautier, E. Frenk, and R. M. Tyrrell, Carcinogenesis 13, 1557 (1992).
[6] J. Cadet, M. Berger, T. Douki, B. Morin, S. Raoul, J.-L. Ravanat, and S. Spinelli, Biol. Chem. 378, 1275 (1997).
[7] C. Kielbassa, L. Roza, and Bernd Epe, Carcinogenesis 18, 811 (1997).

TABLE I

WAVELENGTH DEPENDENCIES (ACTION SPECTRA) OF A NUMBER OF
CPD, SSB, AND 8-OXOG PER 10^6 bp PER 20 kJ/m^2 [a]

	Wavelength			
	290 nm	310 nm	345 nm	425 nm
CPD (T4endoV)[b]	6×10^3	7×10^1	2×10^{-1}	1×10^{-2}
SSB	2×10^1	9×10^{-2}	1×10^{-2}	1×10^{-2}
8-oxoG (Fpg)[b]	1×10^1	2×10^{-1}	2×10^{-2}	1×10^{-1}

[a] From C. Kielbassa, L. Roza, and B. Epe, *Carcinogenesis* **18,** 811 (1997).

[b] The actual endonuclease to detect lesions is given between parentheses.

SSB level off to a broad plateau from 330 up to 420 nm, and 8-oxoG show a minimum around 350 nm and a secondary broad maximum in the violet around 420 nm.

8-OxoG is known to form a miscoding lesion, which can cause a G to T transversion after replication of the damaged DNA.[8] Using this mutation as a specific signature of UVA has thus far not been particularly successful.[9] However, without the Fpg gene product (formamidopyrimidine glycosylase that removes 8-oxoG), *LacZ* in plasmid retrieved from *Escherichia coli* shows a 16-fold increase in G to T transversions after UVA/B irradiation (>290 nm).[10] As discussed by Ito and Kawanishi,[11] the high number of T to G transversions in the *aprt* locus of CHO cells after UVA irradiation[12] may result from the combined effect of Cu(II) and hydrogen peroxide on genomic DNA, but it may also be attributable to mispairing of A to 8-oxoG[13] that is formed in the nucleoside pool (e.g., by singlet oxygen). The same types of reactions could also explain the frequent mutations in A:T base pairs in a human cell line.[14]

[8] S. Shibutani, M. Takeshita, and A. P. Grollman, *Nature* **349,** 431 (1991).
[9] J. M. T. de Laat and F. R. de Gruijl, in "Cancer Surveys" (M. L. Kripke, ed.), Vol. 26, p. 173. Imperial Cancer Research Fund, London, 1996.
[10] C. M. Palmer, D. K. Serafini, and H. E. Schellhorn, *Photochem. Photobiol.* **65,** 543 (1997).
[11] K. Ito and S. Kawanishi, *Biol. Chem.* **378,** 1307 (1997).
[12] E. A. Drobestky, J. Turcotte, and A. Chateauneuf, *Proc. Natl. Acad Sci. U.S.A.* **92,** 2350 (1995).
[13] V. I. Polter, N. V. Shulyupina, and V. I. Bruskov, *J. Mol. Recogn.* **3,** 45 (1990).
[14] C. Robert, B. Muel, A. Benoit, L. Dubertret, A. Sarasin, and A. Stary, *J. Invest. Dermatol.* **106,** 721 (1996).

Cellular Responses and Epigenetical Effects

Cell killing is directly proportional to the mutation rate over the UVB range, but the number of mutations per lethal event drops toward longer wavelengths in the UVA range.[3] This could indicate that damage other than genotoxic damage, e.g., lipid peroxidation at the cell membrane, contributes more to the cell death under long wave UVA irradiation. The time kinetics for the induction of apoptosis ("programmed cell death") differs dramatically between UVA and UVB radiations: it evolves in a matter of hours after long wave UVA exposure, but is much delayed after UVB exposure, up to and over 24 hr (see [30], this volume).[15] Similarly, the course of the cell cycle arrest, specifically in the S phase, differs: the DNA synthesis is lowered immediately with UVA exposure and returns to normal after about 8 hr, whereas it appears to reach a minimum around 6 hr after UVB exposure and normalizes after about 24 hr.[16] The induced cell cycle arrest is believed to buy the cell time to repair the DNA damage adequately before proceeding with DNA synthesis. The repair of 6-4PP takes several hours, and of CPD over 24 hr, but the repair of UVA-induced SSB is virtually completed within half an hour.[17] The $p53$ tumor suppressor gene appears to play an important role in the response to genotoxic damage by regulating DNA repair, by blocking damaged cells in the late G_1-phase (no progression to DNA synthesis) and G_2/M phase (no cell division), and by forcing overly damaged cells into apoptosis. P53 protein levels are elevated after long wave UVA irradiation as well as after UVB irradiation.[18]

Cellular responses to UV radiation do not originate exclusively from DNA damage: a clear example is the activation of the transcription factor NF-κB in relation to peroxidation of membrane lipids by UVA radiation, which can also be induced in enucleated cells.[19] Thus, like ROS, UVA radiation may affect signal pathways, which may stimulate tumor development independently of DNA damage (i.e., not causing irreversible gene mutations); a process traditionally referred to as "tumor promotion." Many chemical "tumor promotors," such as phorbol esters, are almost exclusively membrane active agents. Like the classical tumor promotor 12-O-tetradecanoylphorbol-13-acetate (TPA), UVA radiation causes arachidonic acid re-

[15] D. E. Godar and A. D. Lucas, *Photochem. Photobiol.* **57,** 62, 108 (1995).
[16] A. de Laat, M. van Tilurg, J. C. van der Leun, W. A. van Vloten, and F. R. de Gruijl, *Photochem. Photobiol.* **63,** 492 (1996).
[17] L. Roza, G. P. van der Schans, and P. H. M. Lohman, *Mutat. Res.* **146,** 89 (1985).
[18] J. M. T. de Laat, E. D. Kroon, and F. R. de Gruijl, *Photochem. Photobiol.* **65,** 730 (1997).
[19] G. T. Vile, A. Tanew-Illitschew, and R. M. Tyrrell, *Photochem. Photobiol.* **62,** 436 (1995).

lease from membrane lipids,[20] inflammation and hyperplasia through eicosanoids, inhibition of ligation of the epidermal growth factor to its receptor, and an increase in cellular PKC.[21,22]

Skin Cancer

Like UVB radiation, UVA radiation proved to be a complete carcinogen in experiments in which skin carcinomas developed in (hairless) mice under chronic exposure.[23] Unlike UVB radiation, UVA radiation initially induces mainly benign papillomas, which increase at a slow rate and are at a later stage outnumbered rapidly by carcinomas that increase at a much higher rate.[24] Chronic UVB exposure mainly induces a rapidly increasing number of squamous cell carcinomas (SCC) and actinic keratoses as precursors, with intermixed rare occurrences of benign papillomas.[25] As pointed out by Marks and Fuerstenberger,[26] the chronic application of a (chemical) "initiator" of tumor development (generally a mutagenic agent) will lead to carcinomas, whereas the repeated application of a "tumor promotor" (after a single application of an "initiator") will generate benign papillomas; the promoting action appears to be related to the release of ROS. In this perspective, UVB radiation appears to exert predominantly a "tumor-initiating" effect, whereas "tumor promotion" is relatively more pronounced with UVA radiation.

The wavelength dependence (the "SCUP-m action spectrum") of murine SCC has been determined.[27] It shows a maximum at 293 nm (at shorter wavelengths the penetration of radiation into skin diminishes dramatically), and from 300 to 320 nm it drops by a factor of 10^3, following the induction of CPD. Above 340 nm it appears to level off, with a minimum occurring around 350 nm, which is very much reminiscent of the action spectrum of 8-oxoG (see previous section). Replacing a portion of UVB irradiation by equally carcinogenic UVA irradiation lowers the level of cyclobutane

[20] D. L. Hanson and V. A. DeLeo, *Photochem. Photobiol.* **49,** 423 (1989).

[21] M. S. Matsui and V. A. DeLeo, *Cancer Cells* **3,** 8 (1991).

[22] M. S. Matsui, N. Wang, D. MacFarlane, V. A. DeLeo, *Photochem. Photobiol.* **59,** 53 (1994).

[23] H. van Weelden, F. R. de Gruijl, and J. C. van der Leun, *in* "The Biological Effects of UVA Radiation" (F. Urbach and R. W. Gange, eds.), p. 137. Praeger, New York, 1986.

[24] G. Kelfkens, F. R. de Gruijl, and J. C. van der Leun, *Carcinogenesis* **12,** 1377 (1991).

[25] F. R. de Gruijl, J. B. van der Meer, and J. C. van der Leun, *Photochem. Photobiol.* **37,** 53 (1983).

[26] F. Marks and G. Fuerstenberger, *in* "Oxidative Stress" (H. Sies, ed.), p. 437. Academic Press, London, 1985.

[27] F. R. de Gruijl, H. J. C. M. Sterenborg, P. D. Forbes, R. E. Davies, G. Kelfkens, H. van Weelden, H. Slaper, and J. C. van der Leun, *Cancer Res.* **53,** 53 (1993).

thymine dimers,[28] which indicates that DNA lesions other than CPD and 6-4PP contribute to UVA carcinogenesis. Curiously enough, the action spectrum of the induction of melanoma (malignant tumors originating from pigment cells) in fish shows a much smaller difference between UVB and UVA.[29] The authors speculate that this melanoma action spectrum indicates that UV absorption by the melanin pigment may actually contribute to the development of melonomas, probably by generating ROS.

As found in UV-related human SCC,[30] a high percentage (50–70%) of experimentally UVB-induced murine SCC and precursors carried point mutations in the *p53* tumor suppressor gene at dipyrimidinic sites, typical UVB targets.[31,32] To investigate whether there was a shift in the types of mutations going from UVB to long wave UVA radiation, the *p53* genes from UVA-induced carcinomas were sequenced. Surprisingly, less than 15% of these UVA-SCC appeared to bear *p53* point mutations and almost all of them occurred at a UVB hot spot (codon 267).[33] Apparently, UVA carcinogenesis does not markedly involve the induction of mutations in the *p53* gene, and other oncogenic routes play a more important role. The frequent occurrence of papillomas appears to underline this inference. Papillomas also occur more frequently in UVB-irradiated transgenic mice with a defect in the XP-A DNA repair gene. These papillomas carry mutations in codon 12 of the *Ha-ras* oncogene,[34] a gene that is not found mutated in UVB-SCC in wild-type mice.[31,34] Although this still has to be verified for UVA-induced papillomas, codon 12 of *Ha-ras* (with a CCT UVB target in the transcribed strand that is repaired efficiently in wild types in contrast to the nontranscribed strand)[34] may become a mutational target going from UVB to UVA radiation because of a shift in photoproducts from the transcribed to the nontranscribed DNA strand. Instead of, or next to, a "tumor-promoting" effect of UVA radiation, such a shift in mutational targets may explain the high frequency of papillomas under UVA irradiation.

[28] R. J. W. Berg, F. R. de Gruijl, and J. C. van der Leun, *Cancer Res.* **53,** 4212 (1993).

[29] R. B. Setlow, E. Grist, K. Thompson, and A. P. Woodhead, *Proc. Natl. Acad. Sci. U.S.A.* **90,** 6666 (1993).

[30] D. E. Brash, J. A. Rudolph, J. A. Simon, A. Lin, G. J. McKenna, H. P. Baden, A. J. Halperin, and J. A. Ponten, *Proc. Natl. Acad. Sci. U.S.A.* **88,** 10124 (1991).

[31] H. J. van Kranen, F. R. de Gruijl, A. de Vries, Y. Sontag, P. W. Wester, H. C. M. Senden, E. Rozenmuller, and C. F. van Kreijl, *Carcinogenesis* **16,** 1141 (1995).

[32] N. Dumaz, H. J. van Kranen, A. de Vries, R. J. W. Berg, P. W. Wester, C. F. van Kreijl, A. Sarasin, L. Daya-Grosjean, and F. R. de Gruijl, *Carcinogenesis* **18,** 897 (1997).

[33] H. J. van Kranen, A. de Laat, J. van de Ven, P. W. Wester, A. de Vries, R. J. W. Berg, C. F. van Kreijl, and F. R. de Gruijl, *Cancer Res.* **57,** 1238 (1997).

[34] A. de Vries, R. J. W. Berg, S. Wijnhoven, A. Westerman, P. W. Wester, C. F. van Kreijl, P. J. A. Capel, F. R. de Gruijl, H. J. van Kranen, and H. van Steeg, *Oncogene* **16,** 2205 (1998).

Conclusion

Like UVB radiation, albeit differently and less efficiently, UVA radiation is clearly mutagenic and carcinogenic. In contrast to UVB radiation, the fundamental photochemistry and premutagenic DNA lesions are still not well determined for UVA radiation. Ten to 20% of sunburn and carcinogenic UV doses from sunlight stems from the UVA band.[35] UVA radiation is therefore an important toxic factor in our natural environment.

Summary

Ultraviolet B and A radiations (respective wavelength ranges 280–315 and 315–400 nm) are present in sunlight at ground level. The ultraviolet radiation does not penetrate any deeper than the skin and has been associated with various types of human skin cancers. The carcinogenicity of UVB radiation is well established experimentally and, to a large extent, understood as a process of direct photochemical damage to DNA from which gene mutations arise. Although UVA is generally far less carcinogenic than UVB radiation, it is present more abundantly in sunlight than UVB radiation (>20 times radiant energy) and can, therefore, contribute appreciably to the carcinogenicity of sunlight. In contrast to UVB, UVA radiation is hardly absorbed by DNA. Hence, the absorption by other molecules (endogenous photosensitizers) becomes more important, thus radicals and, more specifically, reactive oxygen species can be generated that can damage DNA, membranes, and other cellular constituents. These photochemical differences between UVA and UVB radiations are reflected in differences in cellular responses and carcinogenesis.

[35] G. Kelfkens, F. R. de Gruijl, and J. C. van der Leun, *Photochem. Photobiol.* **52,** 819 (1990).

[34] Photoaging-Associated Large-Scale Deletions of Mitochondrial DNA

By Mark Berneburg and Jean Krutmann

Introduction

One of the main functions of mitochondria is to supply the cell with energy via oxidative phosphorylation, which is carried out by five protein complexes located in the inner mitochondrial membrane. A proton gradient is generated in a multistep process, involving several redox reactions, resulting in the generation of ATP from ADP and organophosphate. Erroneous oxidative phosphorylation can lead to the generation of reactive oxygen species (ROS), which is why mitochondria are the main site of ROS turnover in the cell. Oxidative phosphorylation is encoded by both nuclear and mitochondrial (mt) DNA.

The human mitochondrial DNA is a 16569-bp, circular double-stranded molecule of known sequence.[1] Mutations in mitochondrial DNA have been found in degenerative diseases and are thought to play a key role in the normal aging process[2] as well as photoaging of the skin.[3–5]

The mutation frequency of mitochondrial DNA is much higher than that seen in the nuclear genome, partly due to the lack of protective histones and the absence of nucleotide excision repair,[6] but also due to the deleterious effects of ROS.[7] Reactive oxygen species can generate 8-hydroxy(deoxy)guanosine (8-OH-Guo) in mtDNA directly,[8] which can lead to strand breaks.[9] However, they can also lead to large-scale deletions of the mtDNA via an indirect mechanism involving strand breaks during replication of the mtDNA.[10,11] The analysis of large-scale deletions via singlet oxygen and their detection *in vivo* and *in vitro* will be discussed in this article.

Detection of Large-Scale Mitochondrial Deletions by Polymerase Chain Reaction (PCR)

Polymerase chain reaction is a quick and highly sensitive method for selective amplification of DNA. Because mtDNA is of low abundance in the cell, the application of PCR facilitated the investigation of mtDNA mutations greatly.

[1] S. Anderson, *Nature* **290,** 457 (1981).
[2] D. C. Wallace, *Science* **256,** 628 (1992).
[3] J.-H. Yang, H.-C. Lee, and Y.-H. Wei, *Arch. Dermatol. Res.* **287,** 641 (1995).
[4] M. Berneburg *et al., Photochem. Photobiol.* **66,** 271 (1997).
[5] M. A. Birch-Machin, M. Tindall, R. Turner, F. Haldane, and J. L. Rees, *J. Invest. Dermatol.* **110,** 149 (1998).
[6] D. A. Clayton, J. N. Doda, and E. C. Friedberg, *Proc. Natl. Acad. Sci. U.S.A.* **71,** 2777 (1974).
[7] F. M. Yakes and B. Van Houten, *Proc. Natl. Acad. Sci. U.S.A.* **94,** 514 (1997).
[8] T. P. Devasagayam, S. Steenken, M. S. Obendorf, W. A. Schulz, and H. Sies, *Biochemistry* **30,** 6283 (1991).
[9] H. Sies, W. A. Schulz, and S. Steenken, *J. Photochem. Photobiol. B* **32,** 97 (1996).
[10] J. M. Shoffner *et al., Proc. Natl. Acad. Sci. U.S.A.* **86,** 7952 (1986)
[11] M. Berneburg, S. Grether-Beck, V. Kürten, T. Ruzicka, K. Briviba, H. Sies, and J. Krutmann, *J. Biol. Chem.* **274,** 15345 (1999).

Extraction of Total Cellular DNA

Conventional DNA extraction methods using phenol/chloroform, ethanol, and cesium chloride gradients may exclude small cytoplasmic circular DNA, resulting in a low yield of mtDNA. Extraction kits designed to yield DNA of around 5–50 kbp, which avoid phenol/chloroform and cesium chloride, give better results and have the added advantage of removing inhibitory substances such as hemoglobin. The QIA Amp Tissue kit (Qiagen, Hilden Germany) gives good yields for cell cultures (fibroblasts and keratinocytes) and tissues (skin, blood). Extractions of total cellular DNA should yield above 0.1 μg, as lower concentrations can lead to false-negative results.

Strategy to Detect Large mtDNA Deletions

For detection of large mtDNA deletions, oligonucleotide primers are designed to anneal outside the deletion (Fig. 1). Thus, amplification of a large PCR fragment can be expected for undeleted mtDNA, whereas the existence of a deleted mtDNA molecule reduces the distance between primers, leading to the amplification of a correspondingly shorter PCR product. Because mitochondrial DNA molecules shortened by deletions are amplified more efficiently than undeleted mtDNA, this strategy results in selective amplification of deleted mtDNA.

Short-Cycle PCR

To generate large PCR products, the amplification time during PCR cycles needs to be sufficiently long. By keeping the extension time of the PCR short it is possible to manipulate the selectivity of the reaction to

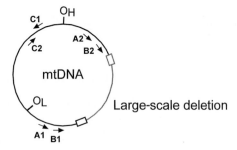

FIG. 1. Primer positioning for detection of large-scale mtDNA deletions. Primers A and B anneal outside the deletion and primers C in a nonmutated area. Gray line denotes the deletion and boxes denote direct repeats surrounding it. O_H, origin of heavy-strand replication; O_L, origin of light-strand replication.

1st PCR   2nd PCR

                           (primer shift)

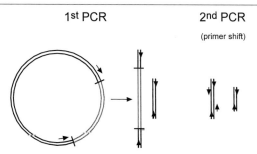

FIG. 2. Experimental design for nested PCR.

preferentially reveal mtDNA deletions. Most mtDNA deletions are between 3 and 10 kbp long. If the amplification time is designed too short for the amplification of the wild-type mtDNA, the deleted molecule is amplified selectively.[12]

Nested PCR

As discussed later in this section, different tissues require different detection methods. For a specimen with high abundance of mutated mtDNA, primary PCR is often sufficient. However, tissues with a low content of the deletion of interest may require a more sensitive PCR method, which uses nested primers.[13] In this method, a second PCR reaction is carried out, using the initial amplification product as a template. A second set of primers is designed to anneal inside the initial PCR fragment (Fig. 2). Nested PCR can be carried out in several ways.

1. A small volume of the primary PCR can be used directly in the secondary PCR (i.e., 2 μl from a 100-μl PCR assay). This is the simplest way of carrying out nested PCR. Because nucleotides in PCR are not limiting the reaction and because overabundance of nucleotides may inhibit PCR efficiency, for nested PCR, it is recommended that the nucleotide concentration be one-tenth of the first PCR (i.e., 40 μM instead of 400 μM in primary PCR).
2. Before transferring a primary PCR product, it can be purified by column purification (Geneclean or Qiagen columns). This ensures removal of inhibitory substances.

[12] G. A. Cortopassi, D. Shibata, N. W. Soong, and N. Arnheim, *Proc. Natl. Acad. Sci. U.S.A.* **89,** 7370 (1992).
[13] S. Ikebe *et al., Biochem. Biophys. Res. Commun.* **170,** 1044 (1990).

TABLE I
OLIGONUCLEOTIDE PRIMERS TO DETECT THE MOST PREVALENT LARGE-SCALE mtDNA DELETIONS[a]

Name	Sequence 5'–3'	Annealing site	Annealing temperature (°C)	Site of Δ	Size of Δ
L116	AACTCAAAGGACCTGGCGGT	1161–1180	50	1836–5447	3610 bp
L173	TTCATGATTTGAGAAGCCTT	1731–1750	55	1836–5447	3610 bp
H594	AACCCCCTGAAGCTTCACCG	5941–5960	55	1836–5447	3610 bp
H617	CGTGAAATCAATATCCCGCA	6171–6190	50	1836–5447	3610 bp
L729	GCAGTAATATTAATAATTTTCATG	7293–7316	58	8469–13447	4977 bp
L790	TGAACCTACGAGTACACCGA	7901–7920	68	8469–13447	4977 bp
H1390	CTAGGGTAGAATCCGAGTATGTTG	13905–13928	58	8469–13447	4977 bp
H1363	GGGGAAGCGAGGTTGACCTG	13631–13650	68	8469–13447	4977 gp
L825	GCCCGTATTTACCCTATAGC	8251–8270	55	8648–16085	7436 bp
L834	ACCAACACCTCTTTACAGTG	8344–8363	56	8648–16085	7436 bp
H1620	GTTGAGGGTTGATTGCTGTAC	16208–16228	56	8648–16085	7436 bp
H1641	TGCGGGATATTGATTTCACG	16411–16430	55	8648–16085	7436 bp
L219	ATGCTTGTAGGACATAATAA	219–238	55	N/A	N/A
H447	AGTGGGAGGGGAAAATAATA	447–466	55	N/A	N/A

[a]Annealing temperatures are values, which gave the best results in our PCR assays. Amplifiable deletions are as cited in Kogelnik et al.[14] Abbreviations: Δ, deletion; L, light strand; H, heavy strand.

3. Before transfer, the primary PCR is separated electrophretically, and the area of the gel corresponding to the PCR product is excised and processed by a purification technique such as Geneclean before the subsequent nested PCR. This procedure is preferable because it both eliminates any inhibitory substances and results in the transfer of less nonspecific products to the secondary PCR reaction.

Oligonucleotide Primers

The human mitochondrial genome is sequenced completely[1] and oligonucleotide primers can be chosen to fit experimental designs. Table I shows primers that amplify the most frequent large-scale deletions, as well as a control region of the mtDNA in which no large-scale deletions have thus far been reported.[14] The primer sequence is given in a 5' to 3' direction together with site, length, and annealing temperature of primer. Primer concentrations around 1 μM produce good results.

[14] A. M. Kogelnik, M. T. Lott, M. D. Brown, S. B. Navathe, and D. C. Wallace, *Nucleic Acids Res.* **26,** 112 (1998).

Quantification of mtDNA Deletions

A PCR fragment from a nonmutated region of the mtDNA can be used as an internal standard for quantification. This fragment should be small so that it will amplify from both normal and mutated mtDNA molecules in order to represent the entire population of mtDNA molecules. In this way, PCR products representing deleted mtDNA molecules can be normalized to the total amount of mitochondria simply by comparing the quantity of deleted mtDNA with that from the internal standard.

The number of DNA copies per PCR cycle reaches a plateau toward the end of the reaction. Thus, for quantification it is important that products are produced only during the linear phase of the PCR. This needs to be determined empirically for each primer pair. For this, different concentrations of the template DNA need to be amplified with increasing cycle numbers.

The obtained PCR products can be quantitated semiquantitatively by optically comparing dilutions of the PCR with known standards.[3] Furthermore, electrophoretically separated products can be visualized and quantified with a phosphorimager.[5] The most sensitive method is the quantification of PCR products via chromatographic separation and subsequent detection with an on-line ultraviolet spectrophotometer (Gynkotek, Germering, Germany). This is especially useful for results where signal intensity is low.

Quantification of PCR Products by Ion-Exchange Chromatography[15]

PCR products for normal and mutated mtDNA are taken up by a sampling robot (20 μl per sample). The negatively charged DNA fragments are then bound to an ion-exchange column and subsequently exposed to an increasing NaCl concentration. This leads to elution according to the total amount of negative charge associated with the DNA fragments (a measure of length). These fractions are then evaluated by an on-line ultraviolet spectrophotometer at a wavelength of 260 nm. The results can be expressed graphically by plotting ultraviolet (UV) absorption (DNA concentration) on the y axis and elution time (DNA amount) on the x axis for each PCR product. Values for the area under the curve of each PCR product can then be used for quantitative comparison.

Material

Each eukaryotic cell contains several hundred mtDNA molecules with normal and mutated molecules coexisting in varying ratios.[16] This so-called

[15] M. Grewe, K. Gyufko, E. Schöpf, and J. Krutmann, *Lancet* **343**, 25 (1994).
[16] C. Richter, *Int. J. Biochem. Cell Biol.* **27**, 647 (1995).

TABLE II
LEVEL OF LARGE-SCALE DELETIONS IN DIFFERENT HUMAN TISSUES

Tissue	Tissue type	Level of deletion
Muscle Brain	Nonreplicating	High
Liver Kidney Spleen Prostate Fatty (adipose) tissue	Mixed	Intermediate
Skin Bone marrow blood	Replicating	Low

heteroplasmy influences the phenotype of the cell and must be taken into consideration during experimental design. Furthermore, the majority of DNA in a cell is nuclear and mtDNA is only a small and varying fraction of this. Additionally, selection mechanisms have been described that influence the ratio of mutated and wild-type mtDNA molecules on the molecular[2] and cellular level.[17]

In Vivo

Large-scale deletions of mtDNA have been detected in many human tissues. Postreplicative tissues, such as muscle and nerve, contain the highest levels of large-scale deletions of the mtDNA (Table II). A second group of tissues, such as liver, kidney, and spleen, exhibit an intermediate level of deletions. Replicative tissues, such as blood and skin, contain the lowest levels of deletions. Highly sensitive methods need to be employed to detect deletions of mtDNA in such tissues.

Normal human skin is among tissues with a low content of large-scale mtDNA deletions. It has been shown, however, that sun-exposed skin has a higher content of deletions than sun-protected skin,[3–5] indicating that mutations of mtDNA may play a role in the process of photoaging. Furthermore, it has been shown that singlet oxygen can induce the most prevalent deletion of the mtDNA.[11] To investigate mtDNA mutations in human skin, punch biopsies are taken from the gluteal region (sun protected) as well

[17] T. Bourgeron, D. Chretien, A. Rotig, A. Munnich, and P. Rustin, *J. Biol. Chem.* **268,** 19369 (1993).

as from the neck and forearm (sun exposed). Specimens (4–6 mm diameter) are subjected directly to extraction of total cellular DNA or can be first snap frozen in liqid nitrogen and stored at −80° until use. For extraction of DNA from skin, incubation times with proteinase K may have to be increased from the normal 1–2 hr up to 6 hr to break down architectural structures of the skin.

Blood samples can also be employed for the detection of large mitochondrial deletions. However, anticoagulants should not contain heparin as this may inhibit subsequent PCR reactions.

In Vitro

Deletions of the mtDNA can be detected in cell culture. The conditions of culture, treatment, and assay of choice for the detection of mutations vary from cell to cell.

Normal Human Fibroblasts

Fibroblasts are involved in photoaging—changes of the skin after chronic exposure to ultraviolet light—in which reactive oxygen species play a critical role.[18] It has been shown that the most prevalent large-scale mtDNA deletion, also called common deletion, can be detected in normal human fibroblasts and that the abundance of this deletion increases after repetitive irradiation with UVA light. Evidence suggests that this increase is due to the effects of singlet oxygen.[11] There is no nucleotide excision repair in mitochondria.[6] However, mtDNA is repaired readily after ROS-induced damage after 4 hr.[19] In order to shift the steady state in cells between ROS-induced damage and ongoing repair toward the damage side, they either have to be exposed to ROS-generating systems continuously or at least in shorter intervals than the time necessary to remove generated damage. Furthermore, when irradiating cells repetitively, the lethal effects of UV light have to be taken into account.

Figure 3 shows the effect on cell viability of different UVA doses. In this experiment, cells were exposed to doses of 0, 4, 8, and 16 J/cm^2 of UVA three times daily. Irradiation with 16 J/cm^2 three times daily induced significant cell death, whereas 8 and 4 J/cm^2 three times daily had little effect. To induce the common deletion, normal human fibroblasts can be irradiated with 8 J/cm^2 at 4-hr intervals. Irradiation with 4 J/cm^2 also induced the common deletion, but the total time needed for induction is

---

[18] K. Scharffetter-Kochanek, *Adv. Pharmacol.* **38,** 639 (1997).
[19] W. J. Driggers, S. P. LeDoux, and G. L. Wilson, *J. Biol. Chem.* **268,** 22042 (1993).

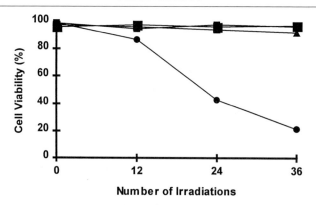

Fig. 3. Viability of normal human fibroblasts after repetitive UVA irradiation. Normal human fibroblasts were irradiated for 12, 24, and 36 times at 0 (◆), 4 (■), 8 (▲), and 16 (●) J/cm^2. After each interval, cell viability was assessed by trypan blue exclusion before and after irradiation and compared to sham-irradiated cells.

longer. These doses are of physiological relevance as they are similar to those administered to human skin during the course of a 15- to 30-min exposure on a summer day at noon with a northern latitude of 30–35°.[20] Furthermore, repetitive irradiation of cells is comparable to the type of sun exposure received by tourists on summer holidays.

Culture and Medium

Normal human fibroblasts are grown in 10-cm petri dishes in Eagle's minimum essential medium (Life Technologies GmbH, Eggenstein, Germany) containing 15% fetal calf serum (Greiner, Frickenhausen, Germany), 0.1% L-glutamine, 2.5% $NaHCO_3$, and 1% streptomycin/amphotericin B in a humidified atmosphere containing 5% CO_2. For irradiation, the medium is removed and replaced with phosphate-buffered saline (PBS).

UVA Irradiation

Lids of petri dishes are removed and cells are irradiated with a UVA-SUN 5000 Biomed irradiation device (Mutzhas, Munich, Germany) whose emission is filtered with UVACRYL (Mutzhas) and UG1 (Schott Glaswerke, Munich) so that it consists of wavelengths greater than 340 nm. Irradiation is carried out three times daily for four consecutive days and cells are checked for viability by trypan blue exclusion. To investigate the

[20] J. E. Frederick and A. D. Alberts, "Biological Responses to Ultraviolet A Radiation" (F. Urbach, ed.). Valdenmar, Overland Park, KS, 1992.

effects of repetitive irradiation longer than 4 days, cells are subsequently aliquoted 1 : 1 with one aliquot stored at $-80°$ until extraction of DNA and the other aliquot is plated to a new 10-cm culture dish for ongoing culture and irradiation.

In vitro irradiation of normal human fibroblasts with this regimen induces the common deletion to detectable levels after 24 irradiations with a maximal induction after 36 exposures.

Coincubation with Deuterium Oxide (D_2O)

Deuterium oxide enhances the half-life of singlet oxygen. Coincubation of fibroblasts with D_2O during irradiation leads to a small but consistent increase of the common deletion when compared to UV irradiation alone. The high toxicity of D_2O, which is observed when cells are exposed repeatedly, requires that cells are only exposed to D_2O during irradiation with D_2O (99.9 atom % ^2H) at a final concentration of 95%.

Incubation with NDPO$_2$

Incubation of cells with a singlet oxygen-generating system can also induce large-scale deletions of the mtDNA. The endoperoxide of the disodium salt of 3,3'-(1,4-naphthylidene)dipropionate (NDPO$_2$) generates singlet oxygen through thermal decomposition. Its generation, biochemical characteristics, and applications are discussed in more detail in a different section of this volume. This singlet oxygen-generating system is well suited for cell culture applications due to its good water solubility and absence of toxic effects at concentrations up to 40 mM for a 1-hr incubation period. Incubation of fibroblasts with 0.3 mM NDPO$_2$ in PBS for 1 hr in the dark at $37°$, instead of UV irradiation, leads to the induction of the common deletion with kinetics comparable to UV irradiation.

Coincubation with Quenchers of Singlet Oxygen

The singlet oxygen-mediated induction of the common deletion can be inhibited by coincubation of fibroblasts with substances that quench the effect of singlet oxygen such as sodium azide (NaN$_3$) and vitamin E (α-tocopherol succinate). Sodium azide should be applied in concentrations of 20 to 50 μM. The presence of NaN$_3$ leads to a dose-dependent reduction of the irradiation-induced generation of the common deletion. To ensure uptake of vitamin E into cells it has to be in culture 24 hr prior to the first irradiation. Concentrations of 25 μM vitamin E show good results in our system.

Keratinocytes

It has been shown *in vivo* that the epidermis has a lower content of the common deletion than dermis.[5] This may be due to the fact that keratinocytes, the predominant cell type in the epidermis, have a higher efficiency in quenching ROS when compared with fibroblasts, which prevail in the dermis. Nevertheless, the repetitive irradiation of keratinocytes does induce the common deletion, but it does require higher doses or longer irradiation periods.

Summary

Heteroplasmy, replicative segregation, low copy numbers of mtDNA, and selection mechanisms at the molecular and cellular level are all factors that determine requirements toward the experimental design for the detection and the quantification of mtDNA mutations. The short half-life and low stability of ROS further increase the technical demands. However, the continuous improvement of techniques has given us more insight into the interactions between ROS and mtDNA, both at the level of endogenous ROS produced by the normal mitochondrial metabolism and exogenous sources of ROS, such as singlet oxygen, which can result from treatments such as UVA exposure.

[35] Role of Activated Oxygen Species in Photodynamic Therapy

By Wesley M. Sharman, Cynthia M. Allen, and Johan E. van Lier

Introduction

Photodynamic therapy (PDT) finds its roots at the turn of the century when a young medical student found that acridine orange killed paramecia on exposure to sunlight.[1] Subsequently, a plethora of information concerning the lethal effects of the combination of photosensitizers and light, both *in vitro* and *in vivo,* has been documented. Several reviews have been published outlining the clinical aspects of PDT and the preclinical and clinical studies predominantly accomplished within the past 25-year pe-

---

[1] O. Raab, *Z. Biol.* **34,** 524 (1900).

riod.[2–7] This chapter outlines the basic principles of PDT using first- and second-generation photosensitizers. Attention will be given to their mode of action, either via singlet oxygen or other reactive oxygen species.

Conventional cancer therapies include radiation and chemotherapies, surgery, and a combination of any or all of the just-described therapies. The treatments themselves have important side effects, even life-threatening. Consequently, the development of an effective treatment that is more selective for diseased tissue is of utmost importance.

Photodynamic therapy offers an alternative, less invasive treatment for such illnesses as psoriasis and several types of cancers. It involves the use of three basic components. First, a photosensitizer, a light-absorbing molecule that is activated by the second element, light of a corresponding wavelength. Third, by definition, molecular oxygen is consumed during the photochemical reaction to produce cytotoxic agents, thus destroying neoplastic tissue.

The traditional treatment protocol used in photodynamic therapy involves the intravenous injection of a photosensitizer. The photosensitizer is distributed rapidly throughout the body and experiences a differential uptake and/or retention time in tumor tissue such that, typically 48–72 hr postinjection, there is a marked increase in the photosensitizer concentration in the tumor as compared to surrounding normal tissue.[2] Several explanations have been proposed as to why there is more or less selective uptake by the tumor of the photosensitizer. They include lower intratumoral pH, increased phagocytosis, increased permeability of the tumor vasculature, as well as reduced lymphatic drainage, and an increased number of receptors on the cell membrane for cellular proteins (i.e., lipoproteins), which are able to target the photosensitizer.[3,8,9]

Unlike laser surgery or psoralen ultraviolet (UV-A) treatment, PDT employs light with wavelengths typically 600–800 nm and therefore not toxic as such. The photosensitizer alone is not able to generate cytotoxic agents, hence following administration of the photosensitizer and localiza-

[2] D. Phillips, *Progr. React. Kinet.* **22,** 175 (1997).

[3] G. I. Stables and D. V. Ash, *Cancer Treat. Rev.* **21,** 311 (1995).

[4] T. J. Dougherty, C. J. Gomer, B. W. Henderson, G. Jori, D. Kessel, M. Korbelik, J. Moan, and Q. Peng, *J. Natl. Cancer Inst.* **90,** 889 (1998).

[5] W. M. Sharman, C. M. Allen, and J. E. van Lier, *Drug Discovery Today* **4,** 507 (1999).

[6] D. Wöhrle, A. Hirth, T. Bogdahn-Rai, G. Schnurpfeil, and M. Shapova, *Russian Chem. Bull.* **47,** 807 (1998).

[7] C. Ell, *Endoscopy* **30,** 408 (1998).

[8] G. Jori, in "CRC Handbook of Organic Photochemistry and Photobiology" (W. H. Horspool, ed.), p. 1379. CRC Press, Boca Raton, FL, 1995.

[9] J. Moan, Q. Peng, R. Sorensen, V. Iani, and J. M. Nesland, *Endoscopy* **30,** 387 (1998).

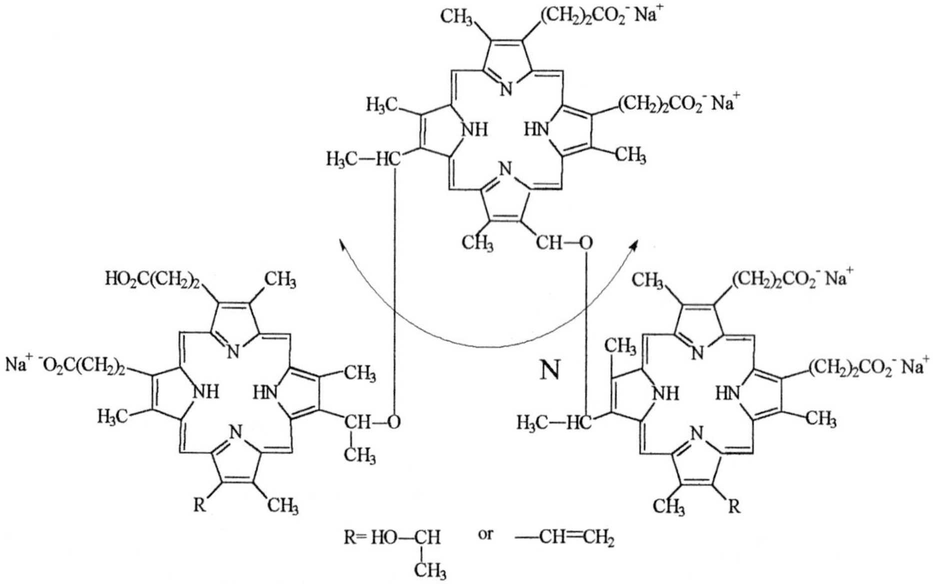

FIG. 1. Photofrin II structure where N is from 0 to 7 repeating units.

tion in the malignant tumor, the beam of light can be directed to the tumor, increasing the selectivity of the treatment. This is advantageous as it spares normal tissue, which is not the case with conventional therapies. PDT offers several advantages. It can be used in conjunction with other therapies, there is regeneration of healthy tissue following treatment, and, due to photosensitizer fluorescence, *in vivo* detection is possible, providing a means of monitoring photosensitizer concentration and location.

Photosensitizers

There are hundreds of naturally occurring and synthetic dyes that can function as photosensitizers. The most commonly used are first-generation hematoporphyrin derivatives, most notably Photofrin II (PII) (Fig. 1). Photofrin II is obtained by treating hematoporphyrin (Hp) with 5% sulfuric acid in acetic acid at room temperature.[10] Subsequently, the mixture is treated with an aqueous base and then neutralized, yielding a complex mixture of dimers and oligomers. The active compound is either an ether or an ester derivative. Purification of the most active compounds via HPLC leads to Photofrin II. PII has been accepted in several countries, including

[10] R. Bonnett, *Chem. Soc. Rev.* 19 (1995).

Canada, France, Germany, Japan, the Netherlands, and the United States.[4] It is used for the treatment of advanced stage eosophageal cancer, both early and advanced stage lung cancer, and gastric cancers, as well as cervical cancer and dysplasia. In addition, it is used for the prophylactic treatment of bladder cancer (see Dougherty et al.[4] for a complete review of clinical trials). Treatments of various other cancers are being thoroughly investigated using PII in the hopes of licensing. Among them are early stage eosophageal cancer, head and neck cancers, and superficial bladder cancer.[4]

Regardless of the success of PII, it has many important drawbacks.[2,11] The first being it is not a pure substance but a poorly reproducible mixture that varies with different preparations and storage times. It has poor absorption at 630 nm ($\varepsilon \approx 10^3 \, M^{-1} \, cm^{-1}$). Tissue transmittance of light at this wavelength is minimal, thus limiting the treatment to tumors at a depth of 5 mm or less. There is a nonspecific accumulation of PII in various organs with only 0.1–3% of the injected dose retained by the malignant tissue. Finally, PII is retained by cutaneous tissue for up to 2 months post-PDT, causing skin photosensitivity; the patient must therefore avoid bright sunlight.[6,10] With that said, the search for the ideal photosensitizer has led to second-generation photosensitizers in clinical trials.

Most importantly, the photosensitizer should be nontoxic with minimal systemic photosensitivity, especially cutaneous. Ideally, the photosensitizer should absorb in the red or near infrared (600–1000 nm) spectrum, thus having enough energy to produce long-lived triplet states so as to generate cytotoxic species usually believed to be singlet oxygen (see later). Light at a longer wavelength also allows for almost double the tissue penetration depth as compared to PII. There should be selective retention in tumor tissue as compared to adjacent healthy tissue. For convenience, the photosensitizer should fluoresce to allow for visualization. Using fluorescent microscopy, the dye can be followed to determine subcellular targets. Mitochondria, lysosomes, plasma membrane, tumor cell nuclei, and tumor vasculature have all been identified as important PDT targets, depending on which photosensitizer is employed. Thus fluorescence is a useful tool in drug development. Finally, a defined chemical composition is beneficial, preferably water soluble, to avoid the need for emulsifiers and organic solvents for formulation and delivery.[10,12]

Various second-generation photosensitizers (Fig. 2) have entered into phase I/II and II/III clinical trials. One such sensitizer is the very active

[11] J. V. Moore, C. M. L. West, and C. Whitehurst, *Phys. Med. Biol.* **42,** 913 (1997).
[12] A. J. MacRoberts, S. G. Bown, and D. Phillips, *in* "Photosensitising Compounds: Their Chemistry, Biology and Clinical Use" (G. Bock and S. Harnett, eds.), p. 4. Wiley, Chichester, 1989.

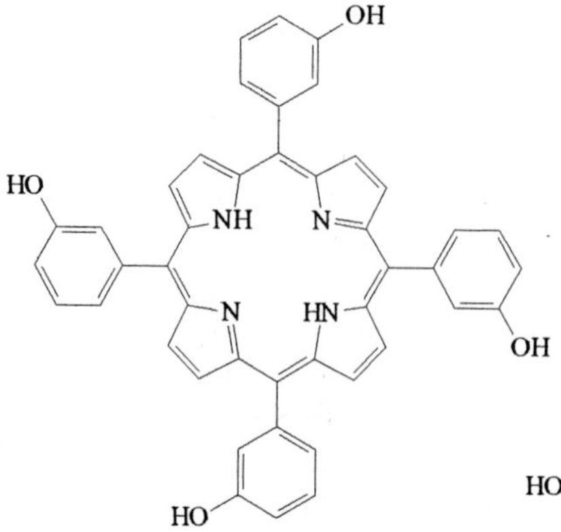

m-THPC

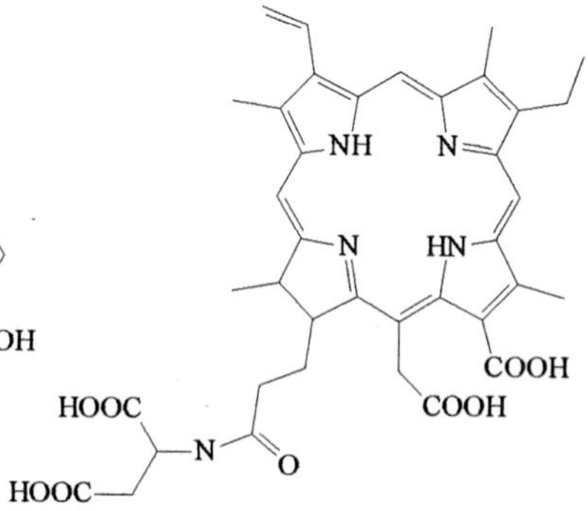

mono-L-aspartylchlorin e6

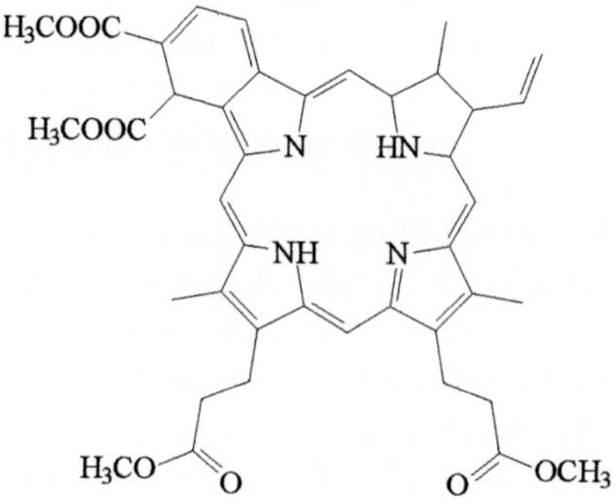

BPD-MA

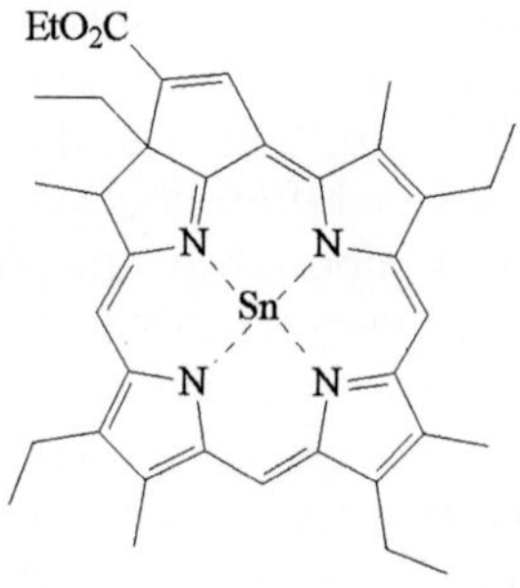

SnET2

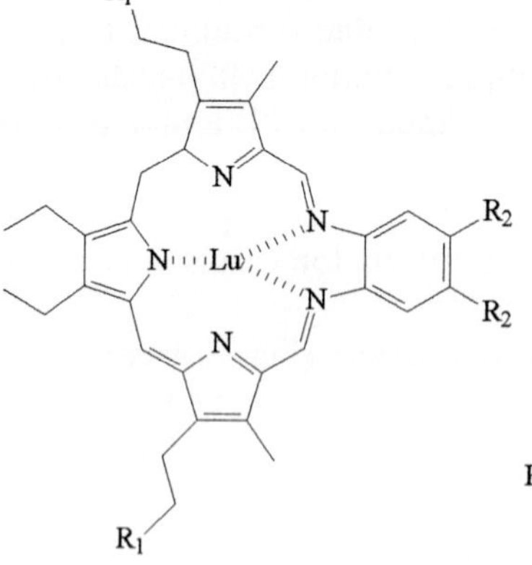

Lu-tex

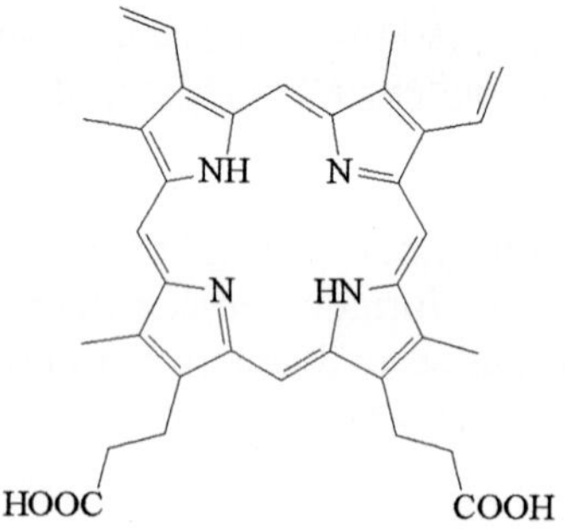

Protoporphyrin IX

FIG. 2. Second-generation photosensitizers.

chlorin, tetra(m-hydroxyphenyl) chlorin(mTHPC), which has the greatest potential. It is now in Phase III clinical trials for head and neck cancers under the commercial name Temoporfin. It is activated at a number of wavelengths, 514 and 652 nm. It is very potent at a dose of 0.1 mg/kg of body weight and light doses as low as 10 J cm^{-2} for superficial esophagus cancers and Barrett's esophagus. As it is so potent at 652 nm, there is the option of illuminating at 514 nm where the product has a smaller extinction coefficient yet efficacy is not sacrificed. A potential drawback is that mTHPC has prolonged skin photosensitivity of about 6 weeks.[10,13]

Subsequently, mono-L-aspartyl chlorin e6 (Npe6) is undergoing phase I/II trials for endobronchial lung cancer in Japan. Npe6 is very hydrophilic and is therefore cleared rapidly from the blood. It is only effective as a photosensitizer if it is irradiated within 2 hr of injection when serum levels are at their peak. Limited skin photosensitivity is experienced.[2,4]

Benzoporphyrin derivative monoacid ring A (BPD-MA), under the trade name Verteporfin, has been in clinical trials for cutaneous tumors such as metastatic breast cancer and basal cell carcinoma. It is a lipophilic photosensitizer with a strong absorption maximum at 690 nm. Because BPD-MA localizes rapidly in neoplastic tissue, light is administered 3 hr postinjection on the tumor as opposed to 48–72 hr postinjection as is the case for PII. With its rapid body clearance, BPD-MA causes mild cutaneous photosensitivity for only 3–5 days.[4,14,15] The same drug, under the trade name Visudyne, has been accepted in Switzerland for the treatment of wet age related macular degeneration (AMD).[5]

Tin etiopurpurin (SnET2) is undergoing phase II trials in the United States for cutaneous breast cancer and Kaposi's sarcoma for HIV patients,[16] whereas lutetium texaphrin (Lu-tex) is in phase II/III trials for cancer. Lu-tex is one of the sensitizers with higher tumor-to-tissue ratios.[4]

A novel PDT method is the employment of endogenous photosensitization. 5-Aminolevulinic acid (ALA) is a precursor in the heme pathway.[17,18] ALA formation is the rate-limiting step in the formation of protoporphyrin IX (PpIX) and is formed from glycine and succinyl-CoA. Excess exogenous ALA can cause an accumulation of PpIX in the tissue where ALA was applied. It has been suggested that cells with higher turnover rates produce

[13] M. C. Berenbaum, R. Bonnett, E. B. Chevretton, S. L. Akande-Adebakin, and M. Ruston, *Lasers. Med. Sci.* **8**, 235 (1993).

[14] A. M. Richter, B. Kelly, J. Chow, D. J. Liu, G. M. N. Towers, D. Dolphin, and J. G. Levy, *J. Natl. Cancer. Inst.* **79**, 1327 (1987).

[15] E. Sternberg and D. Dolphin, *Photodyn. Newslett.* **7**, 4 (1993).

[16] N. Razum, A. Synder, and D. Dorion, *SPIE Conf. Proc.* **2675**, 43 (1996).

[17] J. C. Kennedy, R. H. Pottier, and D. C. Pross, *J. Photochem. Photobiol. B Biol.* **6**, 143 (1990).

[18] J. C. Kennedy, S. L. Marcus, and R. H. Pottier, *J. Clin. Laser Med. Surg.* **14**, 289 (1996).

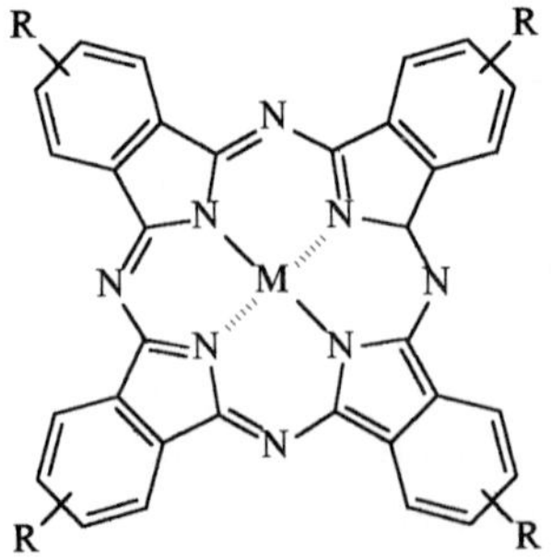

FIG. 3. Phthalocyanine structure.

more PpIX possibly due to decreased ferrochelatase activity. In addition, it has been postulated that tumor cells require more iron, thus limiting the amount available to proceed with the pathway to heme. The topical application of ALA to actinic keratoses, squamous cell carcinoma, and superficial basal cell carcinoma induces a favorable biological response postillumination. ALA or its methyl ester gives excellent results when applied topically, yet shows systemic toxicity when administered orally or intravenously. It is a noninvasive and convenient treatment and merits further investigation for any diseases dealing with epithelial surfaces, oral, vaginal, rectal, gastric, respiratory mucosa, and so on.

Phthalocyanines (Pc) (Fig. 3) are another second-generation class of photosensitizers.[19–21] They are azoporphyrin derivatives with four pyrrole subunits fused together with nitrogen atoms. Four benzo rings on the pyrrole units extend the macrocycle. These modifications lead to enhanced absorption in the far-red region of the spectrum as compared to PII. In addition, metallo-Pc can be prepared by chelating one of several possible metal cations with the four central benzisoindole nitrogens to form stable complexes that are purified easily. Pc can be modified either by substitution on the benzene rings, e.g., the addition of sulfonato and phosphorous groups, or by substituting the axial ligand. Pc have attractive photophysical properties. They absorb strongly ($\varepsilon = 10^5 \, M^{-1} \, cm^{-1}$) in the far red (680 nm) where tissue penetration is optimal. These photochemically stable

[19] J. E. van Lier and J. D. Spikes, in "Photosensitising Compounds: Their Chemistry, Biology and Clinical Use" (G. Bock and S. Harnett, eds.), p. 17. Wiley, Chichester, 1989.
[20] B. Paquette and J. E. van Lier, in "Photodynamic Therapy: Basic Principles and Clinical Applications" (B. W. Henderson and T. J. Dougherty, eds.), p. 145. Dekker, New York, 1991.
[21] I. Rosenthal, in "Phthalocyanines, Properties and Applications" (C. C. Leznoff and A. B. P. Lever, eds.), p. 481. VCH, Weinheim New York, 1996.

compounds have been used for a number of *in vitro* and *in vivo* studies to evaluate their potential usefulness as anticancer agents. Both sulfonated aluminum and zinc phthalocyanine are in phase I/II clinical trials for early tracheobronchial, esophagus, and digestive tract cancers.[22]

Mode of Action

Typically the photosensitizer is administered systemically and is allowed to localize in the tumor. At this time, the photosensitizer is illuminated with light of the appropriate wavelength, exciting the photosensitizer to a higher energy state, ultimately leading to the production of cytotoxic species, resulting in cell death and tumor necrosis. The underlying mechanisms behind the cytotoxic effects displayed during photodynamic therapy on the cellular level are described schematically as

Sen. $\rightarrow$ Sen.* $\rightarrow$ Cytotoxic Agents $\rightarrow$ Biological Damage $\rightarrow$ Cell Death

While the first and last steps are indeed well known, as it is clear that excitation of the photosensitizer will ultimately lead to cell death, the intervening steps are not so clearly understood and are most often assumed relying on conjecture and indirect evidence. However, the overall mechanism involved in photodynamic therapy can be divided easily into two distinct and well-defined steps. The photophysical and photochemical properties of the photosensitizer and its ability to generate cytotoxic agents govern the first. The second results from the biological response of the cell toward the cytotoxic agents produced.

The initial photophysical processes experienced by the photosensitizer on illumination have been examined extensively for a wide range of potential compounds. Upon illumination with light of the appropriate energy, the ground state sensitizer (S_0) is promoted to its short-lived excited singlet state (S_1) (Fig. 4). The lifetime of this excited singlet state (τ_s) is generally in the nanosecond range,[2] which is far too short to allow for significant interaction with the surrounding molecules. As such, it is generally accepted that photodynamic damage induced by the excited singlet state of the sensitizer is negligible. The excited singlet state can dissipate its energy via radiative emission of its excitation energy (fluorescence) or nonradiative decay (internal conversion). Internal conversion entails the loss of energy via collisions with solvent molecules, resulting in the generation of heat. It has been suggested that photothermal effects caused by internal conversion are one of the most important mechanisms for photosensitized cell killing.[23]

[22] U. Isele, P. van Hoogevest, H. Leuenberger, H.-G. Capraro, and K. Schieweck, *SPIE Conf. Proc.* **2078,** 397 (1994).

[23] G. Jori and J. D. Spikes, *J. Photochem. Photobiol. B Biol.* **6,** 93 (1990).

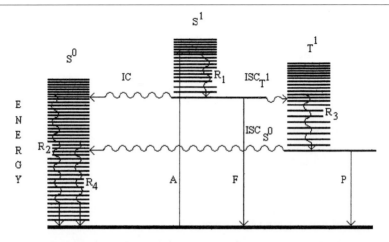

FIG. 4. Jablonski diagram illustrating some of the physical processes that can occur after a molecule absorbs a photon, excited state levels, and transitions. S^0 is the ground electronic state of the molecule. S^1 and T^1 are the lowest excited singlet and triplet states, respectively. Straight arrows represent processes involving photons, and wavy arrows represent radiationless transitions. A, absorption; F, fluorescence; P, phosphorescence; IC, internal conversion; ISC, intersystem crossing; R, vibrational and rotational relaxation. From D. Phillips, *Progr. React. Kinet.* **22,** 175 (1997).

For instance, Davila and Harriman[24] estimated that illumination of a cell stained with Merocyanine 540, a polymethine dye, could increase the internal temperature of the cell by about 12°/min provided the cell membrane functioned as an adiabatic sink. However, photosensitizer fluorescence can be used to monitor compound distribution in tissues both *in vitro* and *in vivo*.

Despite these important factors, the most important method of dissipating excited photosensitizer singlet state energy in terms of photodynamic therapy is the nonradiative intersystem crossing to populate the much longer-lived triplet state (T_1). Lifetimes for longer-lived triplet species are typically in the micro- to millisecond range, as the $T_1 \rightarrow S_0$ transition is spin-forbidden.[25] This allows for sufficient time for interaction between the excited photosensitizer and surrounding molecules. Accordingly, it is believed that excited triplet states of the photosensitizer are responsible for the generation of the cytotoxic species produced during PDT. In fact, Takemura *et al.*[26] have shown that the phototoxic effects of a given porphyrin are enhanced significantly as its triplet state quantum yield and lifetime

[24] J. Davila and A. Harriman, *Photochem. Photobiol.* **54,** 1 (1991).

[25] M. Ochsner, *J. Photochem. Photobiol. B Biol.* **39,** 1 (1997).

[26] T. Takemura, N. Ohta, S. Nakajima, and I. Sakata, *Photochem. Photobiol.* **50,** 339 (1989).

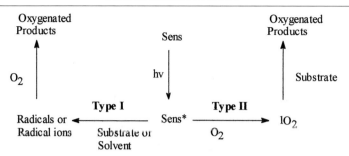

FIG. 5. Diagrammatic presentation of type I and type II photosensitized oxidation reactions. From C. S. Foote, *Photochem. Photobiol.* **54,** 659 (1991).

are increased. In addition, phthalocyanines containing paramagnetic central metal ions (Cu^{2+}, Fe^{2+}, Ni^{2+}), which shorten the lifetime of the triplet state greatly, are far less effective photosensitizers as compared to phthalocyanines chelating diamagnetic metal ions (Al^{3+}, Ga^{3+}, Zn^{2+}), whose triplet lifetimes are much longer.[19] Finally, it has been well established that the addition of heavy atoms such as bromine or chlorine to a photosensitizer improves it photosensitizing activity by improving triplet state quantum yields.[27] For instance, the triplet yield of rhodamine dyes has been enhanced greatly by the addition of bromine to the chromophore.[28] This is due to the internal heavy atom effect, which increases spin-orbital coupling and facilitates intersystem crossing, thus allowing otherwise forbidden changes in the spin state ($S_1 \rightarrow T_1$). Typical photosensitizers being examined for use in photodynamic therapy have triplet state quantum yields (ϕ_T) of 0.2 to 0.7,[9] whereas triplet state lifetimes (τ_T) greater than 500 nsec[25] are generally considered a prerequisite for efficient photosensitization.

Once produced, the excited triplet state can lose its excitation energy via radiative triplet-singlet emission known as phosphorescence. Of more importance for PDT, however, is the quenching of the excited triplet state, which turns out to be the process that generates the majority of the cytotoxic agents needed to induce a biological effect. The quenching mechanism of the T_1 state of the photosensitizer can be distinguished as occurring via a type I or a type II mechanism (Fig. 5).[29] A type I mechanism involves hydrogen atom extraction or electron transfer reactions between the excited state of the sensitizer and some substrate, biological, solvent, or another

[27] M. Wainwright, *Chem. Soc. Rev.* 351 (1996).
[28] P. Pal, H. Zeng, G. Durocher, D. Girard, T. Li, A. K. Gupta, R. Giasson, L. Blanchard, L. Gaboury, A. Balassy, C. Turmel, A. Laperrière, and L. Villeneuve, *Photochem. Photobiol.* **63,** 161 (1996).
[29] C. S. Foote, *Photochem. Photobiol.* **54,** 659 (1991).

photosensitizer, to yield radicals and radical ions. These radical species are highly reactive and can interact readily with molecular oxygen to either generate reactive oxygen species such as superoxide anion or fix the damage so it is irrepairable. These reactions cause the formation of oxidative damage and eventually lead to the cytotoxic effects seen during photodynamic therapy. The type II mechanism, however, results from an energy transfer from the triplet state of the photosensitizer to ground state molecular oxygen, leading to the generation of an excited state of oxygen known as singlet oxygen. Due to its high reactivity, singlet oxygen can react with a large number of biological substrates, causing oxidative damage and cell death.

Role of Oxygen

Ever present in either type I or type II photodestruction is the role of molecular oxygen. It has been well established that the presence of oxygen is an absolute requirement for the photoinactivation of cells via photodynamic therapy as anoxic conditions totally abolish PDT-mediated cellular inactivation. For PhotofrinII photosensitization of cells *in vitro,* full effects were observed at around 40 torr oxygen tension with half-values at about 8 torr.[30,31] In addition, PDT-dependent oxygen consumption rates for PhotofrinII have been estimated to be as high as 6–9 μM/sec using an incident light intensity of 50 mW/cm^2.[32] As such, it is clear that oxygen plays a vital and absolute role in photodynamic therapy and is in fact one of its key limiting factors. For instance, tumor cells are poorly supplied with blood, leading to local areas of hypoxia. In addition, during irradiation, oxygen levels within a tumor will be affected by PDT-induced vascular damage and even by the production of reactive oxygen species by PDT. These effects can greatly limit the potential of PDT against solid tumors where regions of hypoxia become important.[4,33]

Type II: Singlet Oxygen

Until recently, it has been universally accepted that the type II mechanism predominates in photodynamic therapy. As such, the single most important cytotoxic agent generated during PDT is viewed as singlet oxy-

[30] B. W. Henderson and V. H. Fingar, *Cancer Res.* **47,** 3110 (1987).

[31] J. D. Chapman, C. C. Stobbe, and M. R. Arnfield, *Radiat. Res.* **126,** 73 (1991).

[32] T. H. Foster, R. S. Murant, R. G. Bryant, R. S. Knox, S. L. Gibson, and R. Hilf, *Radiat. Res.* **126,** 296 (1991).

[33] J. Fuchs and J. Thiele, *Free Radic. Biol. Med.* **24,** 835 (1998).

gen. Singlet oxygen is produced during PDT via a triplet–triplet annihilation reaction between ground state molecular oxygen (which is in a triplet state) and the excited triplet state of the photosensitizer. Such an energy transfer reaction requires a collision between the two species involved, thus the need for a long triplet state lifetime for the photosensitizer and a minimum oxygen concentration in the tissue. It should be noted that, with certain sensitizers, singlet oxygen can be produced via an energy transfer reaction between the S_1 state of the sensitizer and molecular oxygen.[25] This is possible when the S_1–T_1 energy gap is large enough and the sensitizer has a singlet state lifetime that is sufficiently long to allow ample bimolecular collisions with molecular oxygen.

Two excited singlet S_1 states of molecular oxygen exist.[34] The $^1\Delta_g$ state has an energy level of 94 kJ/mol above the ground state, whereas the $^1\Sigma_g^+$ state is excited by 156.7 kJ/mol. The higher energy $^1\Sigma_g^+$ state of singlet oxygen has a very short lifetime (20 psec in methanol[35]) and is quenched rapidly to yield the $^1\Delta_g$ singlet state in a spin-allowed process. Considering this, the $^1\Delta_g$ singlet state of molecular oxygen is assumed to be the only singlet oxygen species involved in the photodynamic effect and any further reference to singlet oxygen refers to the $^1\Delta_g$ state.

The two highest energy electrons of the $^1\Delta_g$ state are paired and in the same orbital, leading to zwitterionic reactivity.[34] Such an electronic configuration lends a relatively high reactivity to singlet oxygen toward a number of biological substrates. The major chemical entities that constitute biological matter are water, amino acids, pyrimidine and purine bases, and phospholipids. Except for water, all these constituents are sensitive to being damaged oxidatively by singlet oxygen. Presumably such chemical modifications can lead to biological lesions and possibly cell death. Singlet oxygen adds readily to unsaturated carbon–carbon bonds of biomolecules via a 1:4 addition to eventually yield hydroperoxides among primary oxidation products.[36] The reaction of singlet oxygen with membrane lipids, proteins, or nucleic acids can lead to disruption of the cell membrane, the lose of functionality of vital proteins, and irreparable DNA damage. Any or all of these biological lesions can lead to cell death.

Singlet oxygen was first proposed as the cytotoxic agent responsible for photoinactivation of tumor cells in 1976 by Weishaupt et al.[37] Despite this, direct evidence of the involvement of singlet oxygen in PDT has been

[34] J. E. van Lier, in "Photobiological Techniques" (D. P. Valenzeno et al., eds.), p. 85. Plenum Press, New York, 1991.

[35] R. Schmidt and M. Bodescheim, J. Phys. Chem. 98, 2874 (1994).

[36] A. Singh, Can. J. Physiol. Pharmacol. 60, 1330 (1982).

[37] K. R. Weishaupt, C. J. Gomer, and T. J. Dougherty, Cancer. Res. 36, 2326 (1976).

illusive. However, extensive indirect evidence is available that backs up the hypothesis of the involvement of singlet oxygen in PDT-induced biological damage.

One of the main problems associated with the detection of singlet oxygen in biological systems is its high reactivity, which results in extremely short singlet oxygen lifetimes. Singlet oxygen is known to have a lifetime of about 3 μsec in water.[38,39] In the meantime, due to its high reactivity with biological substrates, the lifetime of singlet oxygen in cells has been estimated to be in the region of 200 nsec[40] In cellular systems, saturated with oxygen, the rate of singlet oxygen generation in the cell interior has been assessed to be in the range of 2×10^5 to 4×10^5 sec^{-1}.[41] As such, in the best possible conditions (oxygen saturation), the decay of singlet oxygen will occur at a rate that is 5–10 times more rapid than the rate of formation. The situation is even more precarious in biological systems equilibrated with air only. The resulting fivefold decrease in oxygen concentration reduces the rate of generation of singlet oxygen to about 1×10^4 to 4×10^4 sec^{-1}. Clearly, this results in an infinitesimally small amount of singlet oxygen being available for detection.

The other consequence of the extremely short lifetime of singlet oxygen in cells is its short diffusion range, which has been predicted to be limited to approximately 45 nm.[42] This is in excellent agreement with experimental results where singlet oxygen generated by hematoporphyrin outside the cell wall of *Escherichia coli* could not induce DNA strand breaks within the bacteria.[43] Thus, because the diameter of human cells ranges from 10 to 100 μm, the primary site of singlet oxygen generation will determine what subcellular target will be attacked. Hence, the importance of the subcellular distribution of the photosensitizer is obvious in determining the efficiency of the compound as a therapeutic agent.

The only true method of detecting singlet oxygen directly is by means of its monomol luminescence at approximately 1270 nm.[44,45] Singlet oxygen also has a dimol luminescence emission at 610 nm,[46] and while this emission has proven an effective tool in the identification of the $^1\Delta_g$ state, its produc-

[38] S. Y. Egorov, V. F. Kamalov, N. I. Korroteev, A. A. Krasnovsky, B. N. Toleutaev, and S. V. Zinukov, *Chem. Phys. Lett.* **163,** 421 (1989).

[39] A. A. Krasnovsky, Jr., *Chem. Phys. Lett.* **81,** 443 (1981).

[40] A. Baker and J. R. Kanofsky, *Photochem. Photobiol.* **55,** 523 (1992).

[41] A. A. Gorman and M. A. J. Rodgers, *J. Photochem. Photobiol. B Biol.* **14,** 159 (1992).

[42] J. Moan and E. Boye, *Photobiochem. Photobiophys.* **2,** 301 (1981).

[43] J. Moan, *J. Photochem. Photobiol. B Biol.* **6,** 343 (1990).

[44] J. G. Parker, *IEEE Circuits Devices January,* 10 (1987).

[45] E. Cadenas and H. Seis, *Methods Enzymol.* **105,** 221 (1984).

[46] A. A. Krasnovsky, Jr., and K. V. Neverov, *Chem. Phys. Lett.* **167,** 591 (1990).

tion requires a bimolecular reaction between a pair of singlet oxygen species. Obviously, this is extremely disadvantageous at the micromolar levels seen in PDT. Furthermore, the emission at 610 nm is clearly far from ideal as most photosensitizers absorb in that range of the electromagnetic spectrum.

While the demonstration of the intermediacy of singlet oxygen in reactions carried out in homogeneous environments using the monomol luminescence is a relatively easy procedure, the molecular complexity and heterogeneity of biological systems have made such demonstrations both difficult and ambiguous. The ability of a given photosensitizer to generate singlet oxygen can be determined easily and is in fact often considered to be one of the first parameters to be determined in order to evaluate the potential of the sensitizer in PDT. Photofrin has a singlet oxygen quantum yield of 0.87,[47] whereas second-generation photosensitizers have similar or better quantum yields.

While the triplet excited state of the photosensitizer can be detected easily by laser flash photolysis techniques for a sensitizer localized intracellularly, there is no such detection of singlet oxygen luminescence in intact cells.[48] However, an improvement in techniques has allowed for the detection of singlet oxygen luminescence in a variety of cell suspensions. For instance, Girotti and associates have detected singlet oxygen luminescence from porphyrins bound to erythrocyte ghosts and other membranes.[49–51] Singlet oxygen luminescence in the case of porphyrin sensitization was also reported by Bohm et al.[52] using Fourier transform techniques for porphyrins bound to membrane surfaces. Yeast cells sensitized with tetra-(p-phenosulfide) porphyrin also displayed singlet oxygen luminescence using flash photolysis techniques.[38,53] Kanofsky[54] and Baker and Kanofsky[55,56] have detected singlet oxygen in sensitized suspensions of red blood cell ghosts and L1210 leukemia cell using 5-(N-hexadecanoyl)aminoeosin via its emission at 1270 nm. Lifetimes determined in all cases suggested that the vast major-

[47] A. Blum and L. I. Grossweiner, *Photochem. Photobiol.* **41,** 27 (1985).

[48] P. A. Firey, T. W. Jones, G. Jori, and M. A. J. Rodgers, *Photochem. Photobiol.* **48,** 357 (1988).

[49] R. D. Hall and A. W. Girotti, *Photochem. Photobiol.* **45,** 83S (1987).

[50] J. P. Thomas, R. D. Hall, and A. W. Girotti, *Cancer Lett.* **35,** 295 (1987).

[51] J. P. Thomas and A. W. Girotti, *Photochem. Photobiol.* **47,** 79S (1988).

[52] F. Bohm, G. Marston, T. G. Truscott, and R. P. Wayne, *J. Chem. Soc. Farad. Trans.* **90,** 2453 (1994).

[53] S. Y. Egorov, S. V. Zinukov, V. F. Kamalov, N. I. Koroteev, A. A. Kransnovskii, and B. N. Toleutaev, *Opt. Spectros. (USSR)* **65,** 530 (1988).

[54] J. R. Kanofsky, *Photochem. Photobiol.* **53,** 93 (1991).

[55] A. Baker and J. R. Kanofsky, *Photochem. Photobiol.* **57,** 720 (1993).

[56] A. Baker and J. R. Kanofsky, *Arch. Biochem. Biophys.* **286,** 70 (1991).

ity of the detected luminescence was due to singlet oxygen that had diffused into the buffer.[54–57]

Considerations such as these, along with a number of other theoretical considerations of the *in vivo* singlet oxygen detection problem, i.e., short lifetime of singlet oxygen in cells, rapid rate of singlet oxygen consumption, and low luminescent quantum yields of singlet oxygen (only 200 singlet oxygen molecules per billion undergo radiative decay[41]), suggest that it is unlikely that luminescence emission will be detected emitting from within cells. However, extensive work using indirect means of evaluating the presence of singlet oxygen in biological systems has been used to verify the hypothesis of the involvement of singlet oxygen in PDT. For instance, it has been shown that photooxidation of cholesterol by hematoporphyrin or tetrasulfonated aluminum phthalocyanine yields characteristic reaction products due to singlet oxygen oxidation, 3β-hydroxy-5α-hydroperoxy-cholest-6-ene, 3β-hydroxy-6α-hydroperoxycholest-4-ene, and 3β-hydroxy-6α-hydroperoxycholest-4-ene.[58–61] In addition, typical singlet oxygen products are acquired on sensitization of guanosine (4-hydroxy-8-oxo derivatives).[62] Cholesterol and guanosine both give complex mixtures of species when oxidized by radicals (type I mechanism), mainly leading to epimeric 7-hydroperoxyl derivatives from cholesterol and imidazole ring-opening products from guanosine. In the case of cholesterol, reduction of the products using sodium borohydride yields the corresponding stable hydroxyl analogs, which can be identified readily using chromatographic procedures. This leads to a simple diagnostic test to distinguish between the two reaction pathways in homogeneous systems.[34] However, in complex biological milieu, the ensuing lipid peroxidation of cholesterol results in the formation of complex mixtures of oxysterols, which obscure the initial products.[63] In addition, the 5-α-hydroperoxyl derivative is slowly converted via an intramolecular rearrangement to the 7-α-hydroperoxide, which subsequently epimerizes to the 7-β-hydroperoxyl derivative via a dissociative radical mechanism.[64] Accordingly, although the presence of the 5-α-hydro-

[57] Y. Fu and J. R. Kanofsky, *Photochem. Photobiol.* **62,** 692 (1995).

[58] M. J. Kulig and L. L. Smith, *J. Org. Chem.* **38,** 3639 (1973).

[59] W. Korytowski, G. J. Bachowski, and A. W. Girotti, *Photochem. Photobiol.* **56,** 1 (1992).

[60] P. G. Geiger, W. Korytowski, and A. W. Girotti, *Photochem. Photbiol.* **62,** 580 (1995).

[61] R. Langlois, H. Ali, N. Brasseur, J. R. Wagner, and J. E. van Lier, *Photochem. Photobiol.* **44,** 117 (1986).

[62] J. L. Ravanat, M. Berger, F. Benard, R. Langlois, R. Ouellet, J. E. van Lier, and J. Cadet, *Photochem. Photobiol.* **55,** 809 (1992).

[63] L. L. Smith, "Cholesterol Autooxidation." Plenum Press, New York, 1981.

[64] A. L. J. Beckwith, A. G. Davies, I. G. E. Davison, A. Maccoll, and M. H. Mruzek, *J. Chem. Soc. Perkin Trans.* **II,** 815 (1989).

peroxyl derivative of cholesterol is unambiguous evidence for a type I mechanism, its absence or the presence of 7-hydroperoxycholesterols alone does not rule out the involvement of singlet oxygen in the photochemical process. Despite these problems involving biological systems, studies using [^{14}C]cholesterol clearly showed the intermediacy of singlet oxygen in unilamellar phospholipid vesicles, in ghost erythrocytes, and in L1210 leukemia cells.[59]

Routinely, the responsibility of singlet oxygen in the photodamage caused during PDT has been indicated by the inhibition of the biological effect using competitive quenchers of singlet oxygen. Numerous compounds exist that can react competitively with singlet oxygen, thus providing protection against its cytotoxic effects. Among these are sodium azide, histidine, 2,5-dimethylfuran, β-carotene, and 1,4-diazabicyclo[2,2,2]octane (DABCO).[36,41,65–67] In addition, compounds such as 1,3-diphenylisobenzofuran and 9,10-diphenylanthracene have been used to determine singlet oxygen quantum yields in heterogeneous environments by following their photooxidation via fluorescence.[68,69] These quenchers work via different mechanisms. β-Carotene involves an electronic energy transfer[70] whereas amines such as DABCO and sodium azide react by charge transfer quenching.[71] Others, such as 1,3-diphenylisobenzofuran and 9,10-diphenylanthracene, undergo chemical reactions with singlet oxygen to form peroxy derivatives.[41]

Quenchers have routinely been used to show not only the intermediacy of singlet oxygen in PDT, but to also distinguish between the reaction mechanism involved in a given system. For instance, PDT-induced damage to cytochrome P450 and associated monooxygenases (aryl hydrocarbon hydroxylase, 7-ethoxycoumarin-O-deethylase, and 7-ethoxyresorufin-O-deethylase), along with lipid peroxidation by chloroaluminium phthalocyanine tetrasulfonate, was studied by Agarwal et al.[66,67] in hepatic microsomes. It was determined that among quenchers of reactive oxygen species, only those of singlet oxygen, such as sodium azide, histidine, and 2,5-dimethylfuran, afforded substantial protection toward the photodestruction of cytochrome P450 and associated monooxygenase activities along with photooxi-

[65] B. W. Henderson and A. C. Miller, *Radiat. Res.* **108**, 196 (1986).

[66] R. Agarwal, S. I. A. Zaidi, M. Athar, D. R. Bickers, and H. Mukhtar, *Arch. Biochem. Biophys.* **294**, 30 (1992).

[67] R. Agarwal, M. Athar, D. R. Bickers, and H. Mukhtar, *Biochem. Biophys. Res. Commun.* **173**, 34 (1990).

[68] E. Gross, B. Ehrenberg, and F. M. Johnson, *Photochem. Photobiol.* **57**, 808 (1993).

[69] C. Hadjur, A. Jeunet, and P. Jordon, *J. Photochem. Photobiol. B Biol.* **26**, 67 (1994).

[70] A. Farmillo and F. Wilkenson, *Photochem. Photobiol.* **18**, 809 (1973).

[71] C. S. Foote, T. T. Fujimoto, and Y. C. Chang, *Tetrahedron Lett.* **13**, 45 (1972).

dation of lipids. As such, it was suggested that PDT-induced damage was due to the production of singlet oxygen. Similar results were obtained for mammalian cells using phthalocyanines as the photosensitizer.[72]

Unfortunately, the use of quenchers is not entirely specific for singlet oxygen. The quenchers used are typically systems of low oxidation potential and are almost certainly capable of interacting with other reactive oxygen species produced during irradiation of the photosensitizer. For instance, sodium azide, an effective quencher of singlet oxygen, is also known to react with hydroxyl radicals, but at a much slower rate, and inhibition of lipid peroxidation by sodium azide has been found to occur at a rate constant that is 50 times lower than expected.[59] This could be explained by the limited access of the quencher to the singlet oxygen generated in the membrane. However, it could also indicate the intermediacy of the hydroxyl radical. Tryptophan is also a popular quencher of singlet oxygen, but has been shown to react under certain circumstances via a mixed type I/type II mechanism.[73] Only careful identification of the reaction products can clearly identify the reaction mechanism involved in a given case.[61]

A second diagnostic test for the involvement of singlet oxygen depends on the truly amazing and totally characteristic lifetime variation in different solvents exhibited by singlet oxygen. As was stated previously, singlet oxygen has a lifetime in water of about 3 μsec. This increases to 65 μsec in heavy water (D_2O).[41,74] The result of changing from water to heavy water leads to an increased lifetime for singlet oxygen and presumably an increase in the biological effect, which has been used to verify the presence of singlet oxygen during PDT. However, while large increases in the rate of photooxidation have been seen, the results are not unequivocal. For instance, the quantum yield for the triplet state formation of the photosensitizer is greater in D_2O, which will favor both type I and type II reaction pathways.[2,75] In addition, the structural conformation of proteins may not be comparable in H_2O and D_2O environments, thus making the protein more susceptible to oxidative damage, be it via a type I or type II mechanism.[2] The replacement of water with heavy water may also hamper the biological processes involved in recovery from sublethal damage and would therefore potentiate cell killing.[76] Finally, D_2O may be expected to have little or no effect on the lifetime of singlet oxygen that is generated with

[72] J. Y. Chen, X. Rong, S. M. Chen, F. D. Lu, K. T. Chen, and H. X. Cai, *Cancer Biochem. Biophys.* **12,** 103 (1991).

[73] M. Shopova and T. Gantchev, *J. Photochem. Photobiol. B Biol.* **6,** 49 (1990).

[74] A. A. Gorman and M. A. J. Rodgers, *in* "Handbook of Organic Photochemistry" (J. C. Scaiano, ed.), p. 229. CRC Press, Boca Raton, FL, 1989.

[75] I. Rosenthal and E. Ben-Hur, *Int. J. Radiat. Biol.* **67,** 85 (1995).

[76] E. Ben-Hur and I. Rosenthal, *Radiat. Res.* **103,** 403 (1985).

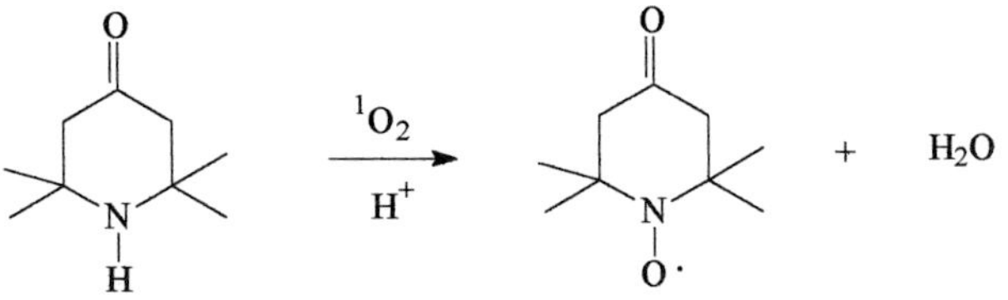

FIG. 6. The reaction of TEMP with singlet oxygen. From A. Viola, A. Jeunet, R. Decreau, M. Chanon, and M. Julliard, *Free Radic. Res.* **28,** 517 (1998).

a lipid membrane or another hydrophobic environment where water does not have access. Thus, while the use of D_2O can be used to show the involvement of singlet oxygen, it is not unequivocal proof of which mechanism is involved.

One final indirect method for the detection of singlet oxygen in a biological system is the use of electron paramagnetic resonance (EPR).[69,77–83] Although singlet oxygen is not radical in nature and will not give a signal in the EPR spectrum, it has been shown to react readily with the spin trap 2,2,6,6-tetramethyl-4-piperidone (TEMP).[77] Such a reaction leads to the formation of 2,2,6,6-tetramethyl-4-piperidone-*N*-oxyl radical (TEMPO) (Fig. 6), which gives a clear and easily identifiable signal in the EPR spectrum. To ensure that the production of TEMPO was due to the reaction of TEMP with singlet oxygen, various quenchers and the effects of D_2O were used. EPR has been used to identify singlet oxygen production in liposomes,[79] in human erythrocytes,[80] and in bronchial epithelial cells[83] and has proved to be a useful technique. Of particular interest is the reported feasibility of *in vivo* skin EPR spectroscopy and imaging, which might provide for the detection of singlet oxygen production *in vivo*.[84,85]

[77] A. Viola, A. Jeunet, R. Decreau, M. Chanon, and M. Julliard, *Free Radic. Res.* **28,** 517 (1998).

[78] C. Hadjur, G. Wagnières, P. Monnier, and H. van den Bergh, *Photochem. Photobiol.* **65,** 818 (1997).

[79] C. Hadjur, G. Wagnières, F. Ihringer, P. Monnier, and H. van den Bergh, *J. Photochem. Photobiol. B Biol.* **38,** 196 (1997).

[80] M. Hoebeke, H. J. Schuitmaker, L. E. Jannink, T. M. A. R. Dubbelman, A. Jakobs, and A. Van de Vorst, *Photochem. Photobiol.* **66,** 502 (1997).

[81] Y. Y. He, J. Y. An, and L. J. Jiang, *Int. J. Radiat. Res.* **74,** 647 (1998).

[82] Y.-Z. Hu and L.-J. Jiang, *J. Photochem. Photobiol. B Biol.* **33,** 51 (1996).

[83] A. C. Nye, G. M. Rosen, E. W. Gabrielson, J. F. W. Keana, and V. S. Prabhu, *Biochim. Biophys. Acta* **928,** 1 (1987).

[84] J. Fuchs, N. Groth, T. Herrling, and L. Packer, *Methods Enzymol.* **203,** 140 (1994).

[85] T. E. Herrling, N. Groth, and J. Fuchs, *Appl. Magn. Reson.* **11,** 471 (1996).

Type I: Radical and Other Reactive Oxygen Species

While the techniques described earlier have surely identified singlet oxygen as an important reactive oxygen species implicated in photodynamic therapy, it is not the only reactive oxygen species formed during PDT. Reactive oxygen species such as superoxide anion, hydrogen peroxide, and hydroxyl radicals can be formed easily via type I processes. In general, excitation of the photosensitizer to its excited triplet state and the resulting promotion of an electron from an occupied to an unoccupied orbital lead to the formation of a reducing electron and an oxidizing hole. As such, the triplet state of the photosensitizer is oxidized and reduced more easily as compared to the ground state molecule. The excited triplet state can react with another photosensitizer in the ground state to form a radical anion and radical cation pair. Furthermore, the excited triplet state can react easily with electron-donating molecules to form the sensitizer anion radical. In fact, numerous molecules capable of donating electrons to the triplet state of the sensitizer exist in biological matter. They include NADH, vitamin C, cysteine, methionine, tyrosine, uracil, and guanine, among many others.[86] The sensitizer anion radical can react with molecular oxygen in an electron-exchange reaction, leading to the formation of superoxide anion ($\cdot O_2^-$). It should be noted that superoxide anion can also be formed theoretically via an electron transfer reaction between molecular oxygen and the triplet state of the photosensitizer. However, this process has been shown to be thermodynamically unfavorable as compared to the energy transfer reaction that forms singlet oxygen.[87]

Superoxide anion can react with biological substrates either by electron transfer or oxidation reactions and can interact directly with a number of cellular structures such as polyunsaturated fatty acids, alcohols, amino acids, and proteins. It has been shown that the superoxide anion can inactivate several enzymes as well. In addition, superoxide anion reacts rapidly with ascorbic acid and α-tocopherol, while one of the most important biologically relevant reactions is that of superoxide anion with sulfhydryl compounds, leading to the formation of RS $\cdot$ radicals and hydrogen peroxide.[36] Overall, however, the reactivity of the superoxide anion is rather limited. Its most important role in the induction of biological damage induced by PDT is in the generation of the highly reactive hydroxyl radical via the Fenton reaction (Fig. 7). The hydroxyl radical can readily add to double bonds or abstract hydrogen atoms, resulting in the formation of secondary radicals,

[86] J. Davila and A. Harriman, *Photochem. Photobiol.* **50,** 29 (1989).
[87] P. C. C. Lee and M. A. J. Rodgers, *Photochem. Photobiol.* **45,** 79 (1987).

$$2 O_2^- + 2 H^+ \rightarrow O_2 + H_2O_2$$

$$Fe^{3+} + O_2^- \rightarrow Fe^{2+} + O_2$$

$$Fe^{2+} + H_2O_2 \rightarrow Fe^{3+} + OH^- + OH^-$$

FIG. 7. Fenton reaction.

ultimately leading to chain reactions such as those implicated in lipid peroxidation.[88]

Reactive oxygen species such as superoxide anion, hydrogen peroxide, and hydroxyl radical have been identified in photodynamic therapy using a number of indirect methods. Superoxide anion, for instance, has been implicated in photodynamic-induced damage using the quenching effects of superoxide dismutase (SOD). However, superoxide dismutase is too big to enter intact cells and, therefore, SOD is less useful in cellular systems.[89] In addition, the dismutation of superoxide anion by SOD leads to the formation of hydrogen peroxide, which can also induce a biological effect.[36,41,65-67] Another quencher of superoxide anion, p-benzoquinone, has also been used, primarily to distinguish between effects due to superoxide anion and the other reactive oxygen species.[90] The reduction of ferricytochrome c to ferrocytochrome c has been used to quantify the amount of superoxide anion formed during PDT.[77] Quenching has also been used to examine the presence of hydrogen peroxide (using catalase) and hydroxyl radical (using sodium benzoate, mannitol and ethanol) with varying results.[30,36,66,67] As was mentioned previously, quenching experiences seemed to identify singlet oxygen as the only reactive species involved in the photodestruction of cytochrome P450 activity. However, the use of quenchers in studying photodynamic cell killing of EMT6 and CHO cells by Photofrin reaffirmed the predominant role of singlet oxygen in PDT, but also indicated some involvement of free radical species such as hydroxyl radical.[65] In addition, using quenching experiments and studying the effects of D_2O, it was determined that the inactivation of catalase in erythrocytes and K562 leukemia cells using tetrasulfonated metallophthalocyanines involved a mixed type I/type II mechanism.[91] Similar studies indicated a mixed type I/type II mechanism in the photoinactivation of Chinese hamster lung

[88] A. W. Girotti, *Photochem. Photobiol.* **51,** 497 (1991).
[89] I. A. Menon, S. D. Persad, and H. F. Haberman, *Clin. Biochem.* **22,** 197 (1989).
[90] W. Bors, M. Saran, E. Lengfelder, R. Spöttl, and C. Michel, *Curr. Top. Radiat. Res.* **9,** 247 (1974).
[91] T. G. Gantchev and J. E. van Lier, *Photochem. Photobiol.* **62,** 123 (1995).

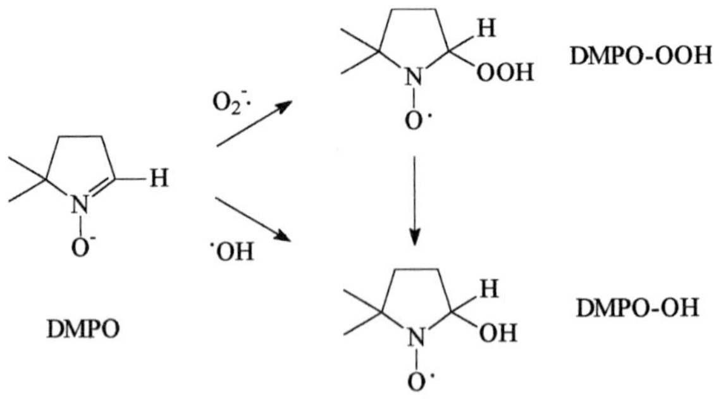

Fig. 8. Formation of DMPO spin adducts of superoxide anion and hydroxyl radical. From A. Viola, A. Jeunet, R. Decreau, M. Chanon, and M. Julliard, *Free Radic. Res.* **28,** 517 (1998).

fibroblasts.[92] Clearly, quenching experiments such as these seem to show that the mechanism behind photodynamic therapy is not quite so straightforward and most likely involves a combination of type I and type II mechanisms.

The presence of reactive oxygen species such as superoxide anion and hydrogen peroxide has also been examined using flow cytometry.[93] Using fluorescence probes such as hydroethidine (which reacts with superoxide anion to give ethidium bromide, which emits a red fluorescence) and dihydrorhodamine 123 (which reacts with hydrogen peroxide to give rhodamine 123, which emits a green fluorescence), it was determined that the ALA-induced photodestruction of primary human skin fibroblast correlates with intracellular superoxide anion production but does not correlate with intracellular hydrogen peroxide production. This apparent discrepancy was explained by the fact that the flow cytometric assay used focused on early stages of cell death and that hydrogen peroxide can diffuse readily out of the cell and might not be seen by the intracellular method used.

One of the most important methods for detecting radicals in biological systems is electron paramagnetic resonance and it has been used extensively in the case of photodynamic therapy. Superoxide anion, for instance, can react with the spin trap 5,5-dimethyl-1-pyrolidine-1-oxide (DMPO) (Fig. 8) to give the superoxide spin trap adduct DMPO-OOH.[77] However, DMPO-OOH is relatively unstable, especially in the presence of transition metals, and decomposes rapidly into various species, including DMPO-OH, the spin trap adduct of the hydroxyl radical. Despite this, the EPR signal due

---

[92] M. O. K. Obochi, R. W. Boyle, S. D. Watts, and J. E. van Lier, *Photochem. Photobiol.* **57S,** 12S (1993).

[93] Y. Gilaberte, D. Pereboom, F. J. Carapeto, and J. O. Alda, *Photodermatol. Photoimmunol. Photomed.* **13,** 43 (1997).

to DMPO-OOH can be seen when desferrioxamine is used to chelate the iron ions present in solution and this technique has been used to show the presence of superoxide anion during the photosensitization of numerous photosensitizers.[69] The same goes for the hydroxyl radical, which has been shown to be present in a number of photodynamic systems. In one very interesting EPR study, Viola *et al.*[77] demonstrated that a series of phthalocyanines generated both type I and type II reactive oxygen species readily in a homogeneous solution of DMF with type II products predominating.[77] In a membrane model, those phthalocyanines tested having fixed axial ligands were unchanged, with type II products predominating. However, in the same membrane model, type I products became important for those phthalocyanines without axial ligands, whereas the type II pathway was shown to be negligible. Results such as these clearly show the importance of environment on the mechanism involved in PDT.

In addition to the spin trapping of superoxide anion and hydroxyl radical, spin adducts of the sensitizer anion radical have been detected in certain systems,[69,81,82] clearly identifying this species as the precursor to the formation of the other reactive oxygen species. Furthermore, carbon-centered radicals have been seen, most likely being due to the formation of radicals on biological targets.[94,95] Finally, studies have shown that the radical cation of the photosensitizer may be involved in the photosensitized decomposition of biological peroxides, resulting in the generation of peroxyl radicals.[96]

Type I versus Type II

Although the factors that govern the competition between type I and type II processes are reasonably well understood, the complexity of the biological environment, as well as uncertainty concerning the localization and binding of the sensitizer to tissue and cell constituents, combined with fluctuations in oxygen concentrations in tissues and even in cellular compartments make it impossible to predict which type of reaction mechanism will prevail during PDT. It has been well established, for instance, that type II processes predominate in oxygenated systems, whereas type I reactions prevail under hypoxic conditions.[33] At low oxygen concentrations, type I processes have been demonstrated to contribute significantly to the photooxidation of membrane components and amino acids using flash

[94] T. G. Gantchev, *Cancer Biochem. Biophys.* **13,** 103 (1992).

[95] T. G. Gantchev, I. J. Urumov, D. J. Hunting, and J. E. van Lier, *Int. J. Radiat. Biol.* **65,** 289 (1994).

[96] T. G. Gantchev, J. Lusztyk, and J. E. van Lier, unpublished results.

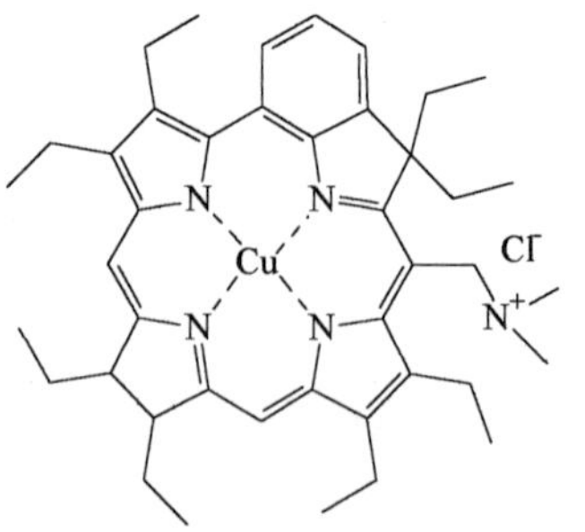

FIG. 9. Copper(II)-α-meso-N,N-dimethyloctaethylbenzochlorin imminium chloride. From D. Skalkos, J. A. Hampton, R. W. Keck, M. Wagoner, and S. H. Selman, *Photochem. Photobiol.* **59,** 175 (1994).

photolysis studies with tetrasulfonated chlorogallium(III) phthalocyanine.[97] In fact, the consumption of oxygen during PDT has been shown to change the mechanism of action from type II to type I as the oxygen concentration decreases below a critical level and tumor vascularitity is damaged.[33,81,98] For instance, electrochemical measurements have demonstrated that the number of cells with a regular intracellular oxygen concentration is obviously reduced following the irradiation of photosensitizer-loaded cells.[99] Moreover, it has been demonstrated clearly that type I processes are favored in more polar environments where the high dielectric constant of the medium should stabilize the radical pair that is generated. In the meantime, type II reactions should predominate in more lipophilic environments where the lifetime of singlet oxygen is much longer.[25] However, despite this, it has been shown that even in relatively polar media, singlet oxygen mechanisms are often more efficient. Results by Rossi *et al.*[100] indicated that indole derivatives appear to be photooxidized largely by a type I mechanism involving electron transfer from the triple state of hematoporphyrin to the indole moiety in water. Type II processes only become important as the environment becomes more lipophilic. Despite this, the situation is not quite so clear, as at low trytophan concentrations, singlet oxygen has been shown to be able to compete with radical-type reactions even in highly polar media.[73] Finally, binding of the sensitizer to cellular components should favor type I hydrogen abstraction or electron transfer reactions,

[97] G. Ferraudi, G. A. Argüello, H. Ali, and J. E. van Lier, *Photochem. Photobiol.* **47,** 657 (1988).
[98] B. W. Henderson and V. H. Fingar, *Photochem. Photobiol.* **49,** 299 (1989).
[99] F. W. Hetzel and M. Chopp, *SPIE Conf. Proc.* **1065,** 41 (1989).
[100] E. Rossi, A van de Vorst, and G. Jori, *Photochem. Photobiol.* **34,** 447 (1981).

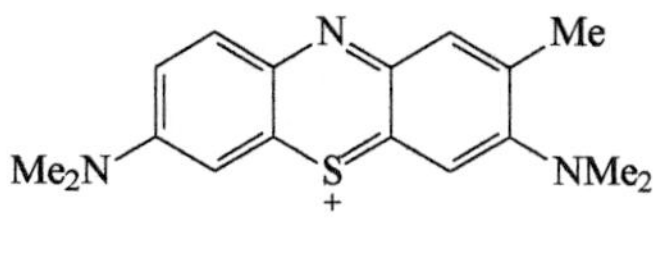

FIG. 10. Toluidine blue.

which will not be evident under *in vitro* conditions. An interesting example of the importance of type I mechanisms involves the recently studied photosensitizer copper(II)-α-meso-N,N-dimethyloctaethylbenzochlorin iminium chloride (Fig. 9).[101] This sensitizer has a triplet lifetime of less than 20 nsec, which is far too short to allow efficient energy transfer to oxygen in order to form singlet oxygen. Despite this, this compound has been demonstrated to be an effective photosensitizer against urothelial tumors in rats. Using superoxide dismutase and catalase, it was shown that this sensitizer induces its damage via reactive oxygen species other than singlet oxygen. It was determined that the close proximity and/or binding of this compound to important biological molecules and the rapid time scale of electron transfer reactions (less than 10 psec[102]) most likely promoted a type I mechanism and helped explain the usefulness of this compound as a PDT agent.

Further complicating the identification of the processes responsible in a given system is the finding that the mechanism involved may depend on the cell type.[2] It has been demonstrate that killing of bacteria depended on whether the bacteria were gram positive or gram negative. By adding toluidine blue (Fig. 10) to Sepharose beads, the effects of type I cytotoxic agents could be ignored as cells could not penetrate the bead. Type I cytotoxic agents could not escape it. Using this, it was determined that singlet oxygen was responsible for the cell killing of *Streptococcus mutans,* a gram-positive bacteria. However, gram-negative bacteria such as *Porphyromonas gingivalis* and *Escherichia coli* along with a yeast, *Candida albicans,* were not killed using the beads, while toluidine blue is capable of inactivating these bacteria by itself. From this, it was concluded that the action of type I free radicals was vital for the cell-killing ability of toluidine blue toward these cells. In addition, it was determined that the kinetics of cell killing of these cells by toluidine blue was binomial, suggesting a two-step mechanism whereby free radicals caused disruption of the cell wall, which allowed access to the cell interior to toludine blue and the cytotoxic

[101] D. Skalkos, J. A. Hampton, R. W. Keck, M. Wagoner, and S. H. Selman, *Photochem. Photobiol.* **59,** 175 (1994).
[102] F. R. Hopf and D. G. Whitten, *in* "The Porphyrins" (D. Dolphin, ed.), Vol. II, p. 161. Academic Press, New York, 1978.

effects of singlet oxygen. Such cell specificity is backed up by Rywkin *et al.*,[103] who, using various quenchers, established that virus kill in red blood cell concentrates operated through a type II mechanism, whereas both type I and type II mechanisms contributed to red blood cell damage.[103]

Conclusion

Our knowledge of photodynamic therapy has increased substantially. So much so that it is now an approved treatment for various cancers and subsequent uses for PDT are being determined regularly. Although it is well known that PDT is very toxic under specific circumstances and conditions, there is much to learn about its exact mode of action. For overall photodamage, the initial reaction is of less importance as both type I and type II reactions lead to similar oxidative damage and can lead to comparable radical chain reactions in the presence of oxygen. Overall, the effect of either type I or type II reactions is the production of oxidative damage within the cell, which will ultimately lead to cell death. Where distinguishing between these two reaction pathways becomes important is in order to thoroughly understand how PDT actually works so that the modulation of its effect can be achieved to maximize the biological effect. It is highly unlikely that a given photosensitizer or a given PDT mechanism is ideal for every application. It is only in completely understanding the processes involved that ideal compounds and conditions can be determined for each use.

Acknowledgments

Financial support by the Medical Research Council of Canada is gratefully acknowledged. W.M.S. and C.M.A. are recipients of a doctoral stipend from the Minister of Education of Québec. J.E.v.L. is the holder of the Jeanne and J.-Louis Lévesque Chair in Radiobiology.

[103] S. Rywkin, L. Lenny, J. Goldstein, N. E. Geacintov, H. Margolis-Nunno, and B. Horowitz, *Photochem. Photobiol.* **56,** 463 (1992).

[36] Gas Chromatography–Mass Spectrometry Analysis of DNA: Optimization of Protocols for Isolation and Analysis of DNA from Human Blood

By ALMAS REHMAN, ANDREW JENNER, and BARRY HALLIWELL

Introduction

Biomolecules are susceptible to damage by a variety of reactive oxygen (ROS), chlorine (RCS), and nitrogen species (RNS) that are constantly produced *in vivo* by various mechanisms, including aerobic metabolism, inflammatory responses, and exposure to toxins or ionizing radiation.[1] Many of these species are capable of causing damage to DNA, culminating in chemical changes to both pyrimidine and purine bases,[2-4] as well as a variety of other lesions, such as single- and double-strand breaks, abasic sites, DNA–protein cross-links, and modified sugars.[5,6] The frequency of these lesions and the effectiveness of repair systems appear to be important factors in the development of mutations and carcinogenic states, as well as other degenerative diseases.[3,7-9] The accurate assessment of oxidative DNA damage in human samples may therefore be a valuable marker of the risk of occurrence of carcinogenic events as well as the overall level of oxidative stress within the body[3,10] and how it may be influenced by diet.

Gas chromatography–mass spectrometry (GC/MS) is becoming widely used to quantify DNA damage because of its ability to identify a wide range of DNA base products in a single analysis.[2,10-12] The identification of the pattern of DNA base damage and the alternative products that arise

[1] B. Halliwell and J. M. C. Gutteridge, "Free Radicals in Biology and Medicine," 3rd Ed. Oxford Univ. Press, Oxford, 1999.
[2] M. Dizdaroglu, *Methods Enzymol.* **193**, 842 (1990).
[3] R. A. Floyd, *Carcinogenesis* **11**, 1447 (1990).
[4] H. Wiseman and B. Halliwell, *Biochem. J.* **313**, 17 (1996).
[5] C. von Sonntag, *in* "The Chemical Basis of Radiation Biology." Taylor and Francis, London, 1987.
[6] Z. Nackerdien, G. Rao, M. A. Cacciuttolo, E. Gajewski, and M. Dizdaroglu, *Biochemistry* **30**, 4872 (1991).
[7] B. N. Ames, M. K. Shigenaga, and T. M. Hagen, *Proc. Natl. Acad. Sci. U.S.A.* **90**, 7915 (1993).
[8] L. A. Loeb, *Cancer Res.* **49**, 5489 (1989).
[9] B. Singer and B. Hang, *Chem. Res. Toxicol.* **10**, 713 (1997).
[10] B. Halliwell, *Free Radic. Res.* **29**, 469 (1998).
[11] B. Halliwell and M. Dizdaroglu, *Free Radic. Res. Commun.* **16**, 75 (1992).
[12] M. Dizdaroglu, *FEBS Lett.* **315**, 1 (1993).

from the same DNA base radical intermediate under different chemical environments (e.g., from the 8-hydroxyguanine radical) can help identify the damaging species and its mechanism of action, e.g., the patterns of damage to DNA by $^1O_2$,[13] OH·,[1,2,14] ONOO$^-$,[15] and HOCl[16] are markedly different. Indeed, measurement of only a single oxidation product can give misleading results regarding the extent of DNA damage, as illustrated by studies in Parkinson's disease[17] and in healthy human volunteer trials investigating dietary ascorbate supplementation.[18]

Baseline levels of DNA oxidation in healthy individuals have yet to be established unequivocally. Most study has been devoted to the quantitation of 8-hydroxyguanine, either as the nucleoside (8-hydroxy-2'-deoxyguanosine) after enzymatic hydrolysis of DNA[3,19] or as the base after acid hydrolysis.[2,12,20] Levels of 8-hydroxyguanine measured in cellular DNA are often, but not always, greater when measured by GC/MS compared to other methods, such as high-performance liquid chromatography with electrochemical detection (HPLC-ECD) (reviewed in Refs. 11 and 21).

The accurate measurement of DNA base damage in human tissue is vital as the values obtained may be related to cancer risk and may also be used in human dietary studies to test agents with the potential to decrease oxidative DNA damage. However, at present, there is a large variation in the literature for values reported as estimates of the steady-state level of 8-hydroxy-2'-deoxyguanosine and 8-hydroxyguanine in human material.[11,21–24] These differences can arise from the methods employed to measure DNA damage. In addition sample preparation can lead to further oxidative damage. Due to the large excess of undamaged compared to damaged bases in cellular DNA, as little as 0.0001% artifactual base oxida-

[13] T. P. A. Devasagayam, S. Steenken, M. S. W. Obendorf, W. A. Schulz, and H. Sies, *Biochemistry* **30**, 6283 (1991).

[14] O. I. Aruoma, B. Halliwell, and M. Dizdaroglu, *J. Biol. Chem.* **264**, 13024 (1989).

[15] J. P. E. Spencer, J. Wong, A. Jenner, O. I. Aruoma, C. E. Cross, and B. Halliwell, *Chem. Res. Toxicol.* **9**, 1152 (1995).

[16] M. Whiteman, A. Jenner, and B. Halliwell, *Chem. Res. Toxicol.* **10**, 1240 (1997).

[17] Z. I. Alam, A. Jenner, A. J. Lees, N. Cairns, C. D. Marsden, P. Jenner, and B. Halliwell, *J. Neurochem.* **69**, 1196 (1997).

[18] I. D. Podmore, H. R. Griffiths, K. E. Herbert, N. Mistry, P. Mistry, and J. Lunec, *Nature* **392**, 559 (1998).

[19] M. K. Shigenaga and B. N. Ames, *Free Radic. Biol. Med.* **10**, 211 (1991).

[20] H. Kaur and B. Halliwell, *Biochem. J.* **318**, 21 (1996).

[21] J. Lunec, *Free Radic. Res.* **29**, 601 (1998).

[22] H. J. Helbock, K. B. Beckman, M. K. Shigenaga, P. B. Walter, A. A. Woodall, H. C. Yeo, and B. N. Ames, *Proc. Natl. Acad. Sci. U.S.A.* **95**, 288 (1998).

[23] M. Dizdaroglu, *Free Radic. Res.* **29**, 551 (1998).

[24] C. M. Gedik, S. G. Wood, and A. R. Collins, *Free Radic. Res.* **29**, 609 (1998).

tion will confound the detection of small changes in levels of oxidized DNA bases.

The most practical tissue to examine in human supplementation studies or in ongoing disease processes is blood, which can be obtained easily at several time points in a study. We have used GC/MS to determine oxidative DNA damage in human blood from healthy individuals before and after dietary supplementation.[25,26] Results from our study and those of others illustrate another value of GC/MS in that changes in base damage products other than 8-hydroxyguanine occur frequently.[18,25–27] This article describes (i) optimized procedures for the isolation and handling of DNA from human blood so as to maximize yield and purity and minimize further oxidative damage prior to GC/MS analysis and (ii) a simple and reliable derivatization procedure for GC/MS for analysis of multiple DNA base damage products, which appears to avoid artifactual oxidation problems.

Materials

All chemicals are of the highest quality available from Sigma Chemical Co. (Poole, Dorset, UK), BDH Chemical Co. (Gillingham, Dorset, UK), and/or Aldrich Chemical Co. (Milwaukee, WI).

Ribonucleases A (RNase A from bovine pancreas, molecular biology grade) and T_1 (RNase T_1 from *Aspergillus oryzae*), RNA (diethylaminoethanol salt, type IX from torula yeast), 6-azathymine, 2,6-diaminopurine, 8-bromoadenine, 5-hydroxyuracil (isobarbituric acid), 4,6-diamino-5-formamidopyrimidine (FAPy-adenine), 2,5,6-triamino-4-hydroxypyrimidine, 5-(hydroxymethyl)uracil, 5-chlorouracil, and guanase are from Sigma Chemical Co. Nonidet P-40 is from BDH Chemical Co. 8-Hydroxyguanine and ethanethiol are from Aldrich Chemical Co. Regenerated cellulose tubular membrane (Cellusep T_1) with a nominal relative molecular mass cutoff (RMMCO) of 3500, silylation grade acetonitrile, and bis(trimethylsilyl)trifluoroacetamide (BSTFA) (containing 1% trimethylchlorosilane, TMCS) are from Pierce Chemical Co. (Rockford, IL).

8-Hydroxyadenine and 2,6-diamino-4-hydroxy-5-formamidopyrimidine (FAPy-guanine) are synthesized by, respectively, treatment of 8-bromoadenine with concentrated formic acid (95%) at 150° for 45 min with purification by crystallization from water[28] and treatment of 2,5,6-triamino-4-hydroxy-

[25] A. Rehman, C. S. Collis, M. Yang, M. Kelly, A. T. Diplock, B. Halliwell, and C. Rice-Evans, *Biochem. Biophys. Res. Commun.* **246,** 293 (1998).
[26] B. Halliwell, *Mut. Res.* **443,** 37 (1999).
[27] H. Favre, M. Mousseau, J. Cadet *et al., Free Radic. Res.* **28,** 377 (1998).
[28] M. Dizdaroglu and D. S. Bertgold, *Anal. Biochem.* **156,** 182 (1986).

pyrimidine with concentrated formic acid with purification by crystallization from water.[29]

Thymine glycol is synthesized by reaction of 5-methyluracil with OsO_4 for 1 hr at 60°, and excess OsO_4 is removed by freeze drying.[30] The purity of standards (>99%) is assessed by mass spectrometry.

2-Hydroxyadenine, 5-hydroxycytosine, and 5-hydroxy-5-methyl-hydantoin are kind gifts from Dr. M. Dizdaroglu (National Institute of Standards and Technology, Gaithersburg, MD).

Distilled water passed through a purification system (Elga, High Wycombe, Bucks, UK) is used for all purposes.

Extraction of DNA: Optimization of the Protocol for DNA Isolation from Human Blood for Analysis of Base Damage by GC/MS

Blood Preparation

Venous blood is collected into lithium heparin- or EDTA-coated tubes and is mixed well before aliquoting into 5 ml volumes. Five milliliters of blood provides 50–100 μg of DNA, which is convenient for GC/MS analysis. Often plasma is required for other analytical studies, and this is prepared by centrifugation at 2000 rpm (823g) for 20 min at 0–4°. Following careful removal of the plasma supernatant, the red cell pellet and the undisturbed buffy coat interface containing the white blood cell layer can be stored for DNA base damage analysis.

DNA Extraction Protocols

We prefer to utilize a method that does not involve separation of cell types in order to maximize DNA yield and minimize phagocyte activation, which may lead to further damage to their own DNA and that of other blood cells such as lymphocytes.

We compared several reported methods for extracting DNA from whole blood and packed cells (fresh and frozen), measuring DNA quantity (A_{260} of 1.0 = 50 μg DNA/ml), quality [assessed by the 260-nm:280-nm absorbance ratio (A_{260}/A_{280} of 1.8 is good-quality DNA, A_{260}/A_{280} of >2 is poor-quality DNA:RNA contamination, A_{260}/A_{280} of 1.2 is too much protein)], and level of DNA base modification measured by GC/MS.

Methods tested included a variety of alternative procedures,[31] such

[29] L. F. Cavelieri and A. Bendich, *J. Am. Chem. Soc.* **72,** 2587 (1950).

[30] M. Dizdaroglu, E. Holwitt, and M. P. Hagen, *Biochem. J.* **235,** 531 (1986).

[31] Special issue on DNA damage: Measurement and Mechanism. *Free Radic. Res.* **29,** 461 (1998).

as inclusion[32,33] or noninclusion[34–36] of enzymatic protein digestion, and organic[32,33,36] or nonorganic[34,35] extraction procedures. The method selected[36] can be applied to whole blood and packed cells (fresh and frozen), and it avoids lengthy enzymatic protein digestion for 1.5–2 hr at 55°, possibly leading to temperature- and time-dependent DNA damage. It provides the best yield and quality of DNA from both fresh and frozen samples with no difference in base damage compared to other methods (Fig. 1). Subsequently, this method was modified further to improve the speed of isolation.

We find no differences in DNA base damage with or without the use of phenol. The use of phenol together with chloroform : isoamyl alcohol did not produce increased levels of DNA damage when compared to a nonorganic protocol (Fig. 1).[37,38] Some reports have claimed that phenol sensitizes DNA to oxidative damage,[39,40] but the extraction protocols in those studies used lengthy proteinase digestions at 55° and/or cell separation techniques. In all our experiments, phenol is kept at 0–4° in the dark and is used only while colorless. Variability between laboratories may relate to the quality and age of phenol used and its contamination with transition metal ions.

One striking finding is that chloroform : isoamyl alcohol alone does increase the DNA base damage[31] (Fig. 2), although some groups prefer this method, believing that phenol is the damaging agent.[41,42] It is possible that phenol may act as an antioxidant when used in combination with chloroform : isoamyl alcohol as chloroform : isoamyl alcohol did not cause extra damage in the presence of phenol.

RNA Contamination

It is likely that a small amount of RNA is coextracted with DNA. It could contribute to measured "DNA base damage" as 8-hydroxyguanosine and 8-hydroxydeoxyguanosine will both be converted to 8-OH guanine

[32] R. C. Gupta, *Proc. Natl. Acad. Sci. U.S.A.* **81**, 6943 (1984).
[33] R. A. McIndoe, M. S. Linhardt, and L. Hood, *BioTechniques* **19**, 30 (1995).
[34] D. K. Lahiri, S. Bye, J. I. Nurnberger, Jr., M. E. Hodes, and M. Crisp, *J. Biochem. Biophys. Method* **25**, 193 (1992).
[35] D. K. Lahiri and J. I. Nurnberger, Jr., *Nucleic Acids Res.* **19**, 5444 (1991).
[36] S. W. M. John, G. Weitzner, R. Rozen, and C. R. Scriver, *Nucleic Acids Res.* **19**, 408 (1991).
[37] G. Harris, S. Bashir and P. G. Winyard, *Carcinogenesis* **15**, 411 (1994).
[38] T. Hofer and L. Möller, *Chem. Res. Toxicol.* **11**, 882 (1998).
[39] H. G. Claycamp, *Carcinogenesis* **13**, 1289 (1992).
[40] M. T. V. Finnegan, K. E. Herbert, M. D. Evans, H. R. Griffiths, and J. Lunec, *Free Radic. Biol. Med.* **20**, 93 (1996).
[41] M. B. Johns, Jr., and J. E. Paulus-Thomas, *Anal. Biochem.* **180**, 276 (1989).
[42] K. Frenkel and C. B. Klein, *J. Chromatogr.* **618**, 289 (1993).

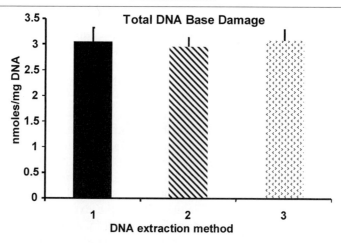

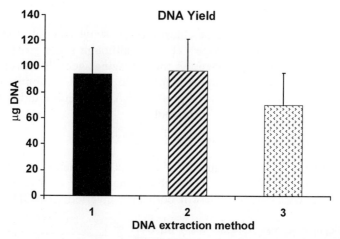

FIG. 1. Comparison of different DNA extraction methods using fresh whole blood by measuring total oxidative DNA base damage (sum of 5-Cl uracil, 5-OH Me hydantoin, 5-OH hydantoin, 5-OH uracil, 5-OH Me uracil, 5-OH cytosine, thymine glycol (*cis* and *trans*), FAPy adenine, 8-OH adenine, 2-OH adenine, FAPy guanine, and 8-OH guanine) and DNA yield. Method 1, nonenzymatic, phenol extraction protocol with chloroform:isoamyl alcohol; method 2, nonenzymatic, phenol extraction protocol with chloroform:isoamyl alcohol, but without additional chloroform:isoamyl alcohol and with only one phenol extraction; and method 3, nonorganic, nonenzymatic protocol. Results are mean ± SD, $n = 4$. There was no statistically significant difference using the unpaired t test with Bonferroni correction at the 95% confidence level.

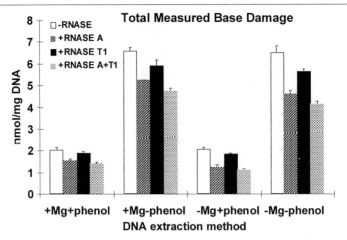

FIG. 2. Effects of omitting Mg^{2+} and phenol from the DNA extraction procedure on total measured base damage. Results are mean ± SD, $n = 3$. Statistical analysis was carried out using the paired t test with Bonferroni correction at the 95% confidence level. Results for total DNA base damage showed statistically significant differences for the following samples when compared with the sample with the same number: +Mg+phenol; −RNase = 1, 2, 3, 12, 13, +RNase A = 1, 16, 17, 18, +RNase T_1 = 2, 22, 23, +RNase A+T_1 = 3, 26, 27, 28; +Mg − phenol; −RNase = 4, 5, 6, 12, 14, +RNase A = 4, 16, 19, 20, +RNase T_1 = 5, 22, 24, +RNase A+T_1 = 6, 26, 29, 30; −Mg+phenol; −RNase = 7, 8, 9, 14, 15, +RNase A = 7, 17, 19, 21, +RNase T_1 = 8, 24, 25, +RNase A+T_1 = 9, 27, 29, 31; −Mg − phenol; −RNase = 10, 11, 13, 15, +RNase A = 10, 18, 20, 21, +RNase T_1 = 23, 25, +RNase A+T_1 = 11, 28, 30, 31.

during the formic acid hydrolysis step. This possible artifact is avoided by incorporating an RNase digestion step into the protocol (Fig. 2). However, Mg^{2+} must be removed from the buffer used during DNA extraction to suspend the nuclear pellet and into which RNase is to be added, as Mg^{2+} is known to inhibit RNase activity (Fig. 2). It is not necessary to remove Mg^{2+} from the initial digestion buffer as this is discarded, and the absence of Mg^{2+} here leads to poor yield and quality of DNA, although oxidative base damage is unaffected.

Two types of RNase are required to ensure complete digestion of all RNA (Fig. 2). RNase A specifically catalyzes the cleavage between the 3′ and 5′ positions of the ribose moieties in RNA with the formation of oligonucleotides terminating in 2′,3′-cyclic phosphate derivatives. RNase T_1 attacks guanine sites specifically to yield 3′-GMP and oligonucleotides with a 3′-GMP terminal group.[43] RNase A is added to a final concentration

[43] L. Z. Rubsam and D. S. Shewach, *J. Chromatogr. B* **702,** 61 (1997).

of 300 μg/ml and RNase T_1 to 150 units/ml, the samples are incubated at 37° for 30 min, and the protocol followed as before.

RNA is found to be digested both in the normal and the oxidized form (10 mg/ml RNA incubated with 50 mM phosphate buffer, 14 mM H_2O_2, 500 μM $CuCl_2$, and 5 mM ascorbate at 37° for 1 hr followed by dialysis) by both RNase A and T_1. The amount of RNA that could be digested (assessed by measuring the concentration of RNA, using the absorbance at 260 nm, after the cleaved bases are dialyzed out) by the selected concentrations of both types of enzyme is 300 μg, which is well in excess of the maximum amount that could be present in cells from 5 ml of blood (100–150 μg).

Studies with O_2 Exclusion and Addition of Antioxidants during DNA Isolation

The effects of excluding atmospheric oxygen during DNA extraction were examined: all solutions and samples were degassed by bubbling with helium and, subsequently, any air entering during their use was displaced with nitrogen by allowing a stream of the gas to enter the partially covered container for 30–120 sec (depending on the size of the container and the volume of liquid contained). We also examined the effect of extracting DNA at low temperature, samples kept on ice throughout the experiment, dialyzing at 0–4°, and incorporating antioxidants into every step of the extraction procedure. Experiments were performed with and without RNases, and DNA was analyzed with and without dialysis.

The exclusion of oxygen,[44] maintaining samples on ice, omitting dialysis, and the incorporation of several different antioxidants individually and in combination had no effect on levels of DNA base damage. Several compounds that have been reported previously to prevent DNA oxidation during extraction[38,45] were tested, including desferrioxamine (DFO) replacing EDTA (to examine the effects of producing a redox inactive iron complex as opposed to a redox active one[46]), DFO in combination with EDTA, butylated hydroxyanisole (BHA),[47,48] histidine,[49] reduced glutathi-

[44] M. Nakajima, T. Takeuchi, and K. Morimoto, *Carcinogenesis* **17,** 787 (1996).
[45] E. Kvam and R. M. Tyrrell, *Carcinogenesis* **18,** 2281 (1997).
[46] J. M. C. Gutteridge, R. Richmond, and B. Halliwell, *Biochem. J.* **184,** 469 (1979).
[47] S. J. Stohs, T. A. Lawson, L. Anderson, and E. Bueding, *Mech. Ageing Dev.* **37,** 137 (1986).
[48] P. A. E. L. Schilderman, E. Rhijnsburger, I. Zwingmann, and J. C. S. Kleinjans, *Carcinogenesis* **16,** 507 (1995).
[49] O. Cantoni and P. Giacomoni, *Genet. Pharmacol.* **29,** 513 (1997).

one,[50,51] 2-mercaptoethanol (ME), and 2,2,6,6-tetramethyl-1-piperidinyl-oxy, free radical (TEMPO).[38] These compounds were used at final concentrations ranging from 0.1 to 10 mM and did not affect DNA base damage levels measured either individually or in combination, although at high concentrations they decreased the quantity and quality of DNA recovered. In addition, none of these agents interferes with RNase activity. When samples are treated on ice, less DNA is extracted, indicating that the reagents may require higher temperatures for efficient extraction.

These data suggest that artifactual oxidation of DNA is not a problem with our extraction protocol. Even when a prooxidant system (Fe^{3+}-EDTA, H$_2$O$_2$, and ascorbate incubated at 37° for 1 hr to generate OH·) is added to whole blood or packed cells, no rise in oxidative DNA damage levels is detected. The high antioxidant activity of human blood[52] may serve to protect DNA during the early stages of the isolation process.

Blood Sample Storage

No differences in DNA quantity, quality, or levels of oxidative base damage are observed when whole blood or packed cells are stored at −20° for up to 1 year (Fig. 3) compared to either fresh samples or samples stored at −80° for the same length of time. Fresh or frozen blood (and whole blood or packed cells stored at 0 to 4° for 1 week) can be used without compromising DNA yield, quality, or levels of base damage, a finding supported in the literature by some workers[53] but not all.[54,55] Yields are greater when frozen samples are thawed gently overnight at 0–4°, compared to thawing quickly at room temperature. This may be due to reduced nuclear lysis during gentle thawing, which means that during the first step of extraction, less DNA is discarded with other cellular contents, whereas intact nuclei are spun down. Previous work in our laboratory has shown that DNA in solution can be stored at 0–4° for up to 1 week and at −20° either in solution or freeze dried indefinitely.

[50] N. Spear and S. D. Aust, *Arch. Biochem. Biophys.* **324,** 111 (1995).
[51] M. V. M. Lafleur, J. J. Hoorweg, H. Joenje, E. J. Westmijze, and J. Retèl, *Free Radic. Res.* **21,** 9 (1994).
[52] B. Halliwell and J. M. C. Gutteridge, *Arch. Biochem. Biophys.* **280,** 1 (1990).
[53] B. Towne and E. J. Devor, *Hum. Biol.* **62,** 301 (1990).
[54] K. S. Ross, N. E. Haites, and K. F. Kelly, *J. Med. Genet.* **27,** 569 (1990).
[55] L. Madisen, D. I. Hoar, C. D. Holroyd, M. Crisp, and M. E. Hodes, *Am. J. Med. Genet.* **27,** 379 (1987).

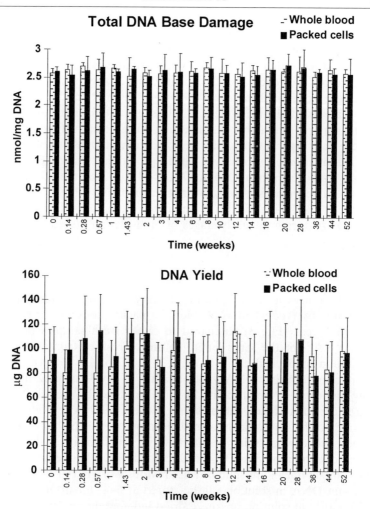

FIG. 3. Effects of storing blood at $-20°$ on DNA base damage and yield over a period of 1 year. Results are mean $\pm$ SD, $n = 3$. There was no statistically significant difference using the unpaired t test with Bonferroni correction at the 95% confidence level.

Optimized DNA Extraction Method

To 5 ml of anticoagulated blood or packed cells from this volume, solution 1 (10 mM Tris-base, 10 mM KCl, 10 mM MgCl$_2$, pH 7.6) is added to make the volume up to 10 ml in a 15-ml centrifuge tube, followed by 120 μl Nonidet P-40. The tubes are mixed well by inversion on a rotator for 15 min. The nuclear pellet is isolated by centrifugation at 2000 rpm ($823g$) for 10 min at room temperature and the supernatant is discarded.

The pellet is gently, but thoroughly, resuspended in 800 μl solution 2 (10 mM Tris-base, 10 mM KCl, 10 mM MgCl$_2$, 0.5 M NaCl, 0.5% SDS, 2 mM EDTA, pH 7.6). RNase A is added to a final concentration of 300 μg/ml and RNase T$_1$ to 150 units/ml, and the samples are incubated at 37° for 30 min. The suspension is vortexed with 400 μl phenol (saturated with 10 mM Tris–HCl and 1 mM EDTA, pH 8.0) and is then centrifuged at 4000 rpm ($3291g$) for 10 min at room temperature.

The upper (aqueous) phase is transferred to a clean tube, vortexed with 200 μl phenol and 200 μl chloroform : isoamyl alcohol (24 : 1), and centrifuged as before. The bottom (organic) layer is discarded and the remaining aqueous phase is vortexed with 700 μl chloroform : isoamyl alcohol (24 : 1) and centrifuged as before.

The upper (aqueous) phase is transferred to a clean tube with 0.1 volume (of the aqueous phase collected) of 3 M sodium acetate buffer, pH 6.0. Then 2.5 volumes (of the aqueous phase collected) of ice-cold ethanol is added and the DNA is precipitated by inverting the tube several times.

Samples are placed at -20° overnight to maximize DNA precipitation and are then centrifuged at 3500 rpm ($2520g$) for 10 min at room temperature. The supernatant is discarded and the pellet is washed in 1 ml 70% (v/v) ethanol and centrifuged as before. This process is repeated for the second wash. The pellet is then air dried for 15 min before dissolving in 2 ml distilled water by gently agitating the tube at room temperature.

DNA samples are dialyzed against distilled water for 18–20 hr using regenerated cellulose tubular membrane (Cellusep T$_1$) with a nominal RMMCO of 3500. The amount of DNA recovered for each sample is determined spectrophotometrically at 260 nm (A_{260} of 1.0 = 50 μg DNA/ ml), and the DNA quality is assessed by the ratio of the absorbance measured at 260 nm to that at 280 nm. The samples are then aliquoted into volumes containing 100 μg DNA, and 25 μl of 20 μM internal standard (a mixture of 6-azathymine and 2,6-diaminopurine) is added before lyophilization and acid hydrolysis.

Analysis of DNA Base Modification by GC/MS

Acid Hydrolysis

Release of normal and modified bases from DNA is achieved by acidic hydrolysis, which cleaves both phosphodiester bonds and the glycosidic bonds linking bases to deoxyribose. DNA samples (100 μg) containing the internal standards 6-azathymine and 2,6-diaminopurine (0.5 nmol) are hydrolyzed by the addition of 0.5 ml 60% (v/v) cold formic acid and heating at 150° for 45 min in an evacuated, sealed hydrolysis tube. Samples are cooled and lyophilized together with calibration standards that are prepared as mixtures of all the DNA base damage products to be measured (content range of each product is 0.02–1.0 nmol).

Derivatization

After acid hydrolysis of DNA, GC/MS requires a derivatization procedure in order to convert the polar nucleosides/bases and internal standards to volatile, thermally stable derivatives that possess characteristic mass spectra. Trimethylsilylation is the most common derivatization reaction used[2,6,12,14] and is often carried out at a high temperature (90–140°) to ensure optimal derivative formation. However, if oxygen is not removed completely during derivatization, artifactual oxidation of some undamaged bases can occur, leading to an overestimation of oxidative DNA damage.[56–59] We have determined that only 8-OH guanine, 8-OH adenine, 5-OH cytosine, and 5-(hydroxymethyl)uracil are subject to this artifact during derivatization of hydrolyzed DNA at high temperatures[60] (Fig. 4) and that the artifact is worsened by not purging with nitrogen prior to derivatization.[56–59] Indeed, studies failing to exclude air at all obtain the greatest artifacts.[56–58]

Artifacts due to oxidation of DNA bases can apparently be prevented by the removal of undamaged bases from hydrolyzed DNA prior to derivatization, using such techniques as HPLC-ECD[56–58] or immunoaffinity column prepurification.[56] However, these extra precautionary steps require more time, equipment, and expense, especially when all four bases need to be removed.

Our laboratory has adopted an alternative means of eliminating artifacts based on the derivatization of nitrogen-purged samples at room temperature, which minimizes artifactual oxidation, but necessitates longer time periods. Further investigation indicated that purging with nitrogen does not remove oxygen completely, as the complete prevention of artifactual guanine oxidation is only achieved by the addition of the reducing agent ethanethiol (Fig. 5). When this further antioxidant protection is provided

[56] J.-L. Ravanat, R. J. Turesky, E. Gremaud, L. J. Trudle, and R. H. Stadler, *Chem. Res. Toxicol.* **8,** 1039 (1995).
[57] T. Douki, T. Delatour, F. Bianchini, and J. Cadet, *Carcinogenesis* **17,** 347 (1996).
[58] J.-L. Ravanat, T. Douki, R. Turesky, and J. Cadet, *J. Chim. Phys. Physico-Chim. Biol.* **94,** 306 (1997).
[59] M. Hamberg and L. Y. Zhang, *Anal. Biochem.* **229,** 336 (1995).
[60] T. G. England, A. Jenner, O. I. Aruoma, and B. Halliwell, *Free. Radic. Res.* **29,** 321 (1998).

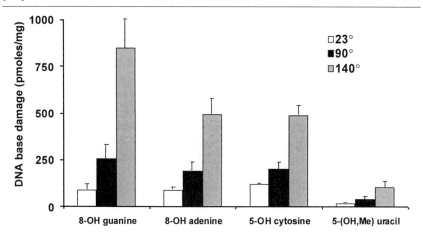

FIG. 4. Influence of derivatization reaction temperature on GC/MS analysis of DNA base damage in acid hydrolyzed calf thymus DNA. Samples were purged with nitrogen before derivatization at each temperature for 1 hr and then kept at 23° for 1 hr prior to GC/MS analysis. Results are mean ± SD, $n \geq 4$. In the absence of artifactual oxidation, levels of oxidized bases should be the same whatever the derivatization conditions, provided that appropriate standards are carried through the same protocols. For these four base oxidation products, levels increased with temperature, suggesting artifactual oxidation of the parent bases at high temperatures. Data published previously.[60]

by ethanethiol, levels of 8-OH guanine are identical in hydrolyzed DNA samples whether or not guanine is first removed by treatment with guanase (which should prevent possible artifactual 8-OH guanine formation) (Fig. 5). Such levels in calf thymus DNA are comparable to those measured by HPLC-ECD and GC/MS after HPLC prepurification.[56,58,61]

Because guanine is the most readily oxidizable DNA base, these derivatization conditions are also likely to prevent artifactual oxidation of adenine, thymine, and cytosine (Table I). Trifluoroacetic acid (TFA) has been reported to increase the solubility of guanine and 8-OH guanine in derivatization mixtures and promote the efficiency of room temperature derivatization,[59] but we have found that TFA interferes with the accurate gas chromatographic separation of pyrimidines after derivatization. However, ethanethiol has no detectable detrimental effect on the derivatization of other bases.

Using the following protocol, artifactual oxidation during derivatization can apparently be prevented. Levels of oxidized bases measured by GC/MS after this protocol are comparable with those measured by HPLC-ECD.[61]

[61] A. Jenner, T. G. England, O. I. Aruoma, and B. Halliwell, *Biochem. J.* **331,** 365 (1998).

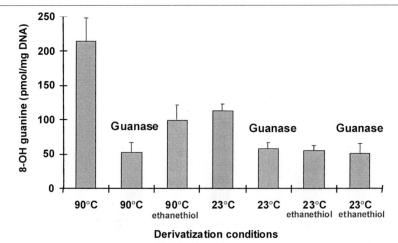

FIG. 5. Effect of temperature, ethanethiol, and guanine removal on GC/MS derivatization of hydrolyzed calf thymus DNA. Samples were purged with nitrogen before derivatization at each temperature for 1 hr and then kept at 23° for 1 hr prior to GC/MS analysis. Guanine was removed prior to derivatization by its enzymatic conversion to xanthine using guanase. Results are mean ± SD, $n \geq 4$. Data published previously.[61]

TABLE I

GC/MS MEASUREMENT OF OXIDIZED DNA BASES IN COMMERCIAL CALF THYMUS DNA[a,b]

Damaged DNA base (pmol mg^{-1} DNA)	90°	90° Ethanethiol	23°	23° Ethanethiol
8-OH guanine	212 ± 66	100 ± 22	114 ± 9	56 ± 8
5-OH Me hydantoin	39 ± 5	47 ± 2	41 ± 9	40 ± 8
5-OH hydantoin	27 ± 6	11 ± 1	26 ± 4	16 ± 3
5-OH uracil	39 ± 4	14 ± 3	34 ± 5	5 ± 1
5-(OH Me) uracil	37 ± 9	26 ± 3	19 ± 1	23 ± 4
5-OH cytosine	226 ± 9	117 ± 3	82 ± 18	57 ± 2
Thymine glycol	139 ± 12	105 ± 12	93 ± 25	92 ± 16
FAPy adenine	97 ± 4	93 ± 15	73 ± 6	71 ± 9
8-OH adenine	172 ± 20	120 ± 26	97 ± 7	88 ± 11
2-OH adenine	43 ± 12	42 ± 4	52 ± 4	43 ± 3
FAPy guanine	214 ± 13	123 ± 29	111 ± 6	122 ± 4

[a] Results are mean ± SD; $n > 4$.
[b] Data adapted from Jenner et al.[61]

Caution: Ethanethiol has a very unpleasant odor. A mixture of acetonitrile/ethanethiol (3:1) is freshly prepared. Samples and calibration standards are derivatized in poly(tetrafluoroethylene)-capped glass vials after purging with nitrogen. Seventy-five microliters of a BSTFA (+1% TMCS)/acetonitrile/ethanethiol (16:3:1, v/v) mixture is added and mixed well. Samples are derivatized at 23° for 2 hr.[16]

GC/MS Analysis

GC/MS analysis is essentially as described by Spencer *et al.*[62] and Jenner *et al.*[61] Derivatized samples are analyzed by a Hewlett-Packard 5971A mass selective detector interfaced with a Hewlett-Packard 5890II gas chromatograph and equipped with an automatic sampler and a computer workstation. The injection port and the GC/MS interface are kept at 250 and 290°, respectively. Separations are carried out on a fused silica capillary column (12 m × 0.2 mm i.d.) coated with cross-linked 5% phenylmethylsiloxane (film thickness 0.33 μm) (Hewlett-Packard Ltd., Stockport, Cheshire, UK).

Helium is the carrier gas with a flow rate of 0.93 ml/min. Derivatized samples (2 μl) are injected into the GC injection port using a split ratio of 8:1. The column temperature is increased from 125° to 175° at 8°/min after 2 min at 125°, then from 175 to 220° at 30°/min and held at 220° for 1 min, and finally from 220 to 290° at 40°/min and held at 290° for 2 min. Selected ion monitoring is performed using the electron ionization mode at 70 eV with the ion source maintained at 190°.

Quantitation of modified bases is achieved by relating the peak area of the compound with the internal standard peak area and applying the following formula:

$$\text{nmol base/mg DNA} = (A/A_{IS} \times [IS] \times (1/k) \times (1000/\mu\text{g DNA used})$$

where A is the peak area of product, A_{IS} is the peak area of the internal standard, [IS] is the concentration of the internal standard (0.5 nmol), and k is the relative molar response factor for each product calculated from the gradient of the calibration curve for each product.

It is crucial to obtain calibration values from authentic base damage products and internal standards that are hydrolyzed exactly as the experimental samples. MS detection allows stable isotopically enriched authentic compounds (at least 3 mass units different) to be added to the sample as an internal standard, which has essentially the same physical and chemical properties. This technique is often referred to as isotope-dilution mass

---

[62] J. P. E. Spencer, A. Jenner, K. Chimel, O. I. Aruoma, C. E. Cross, R. Wu, and B. Halliwell, *FEBS Lett.* **375,** 179 (1995).

spectrometry and ideally compensates for any loss of analyte during hydrolysis and GC/MS analysis.[12] It is the most accurate method for internal standardization and provides extra assurance of analyte identification. Stable isotope-labeled standards must be synthesized and purified from simpler starting materials that contain one or more heavier isotopes of H, C, or N.[59,63,64] Therefore isotopically labeled standards are more expensive to obtain and if unavailable, structurally similar, but different compounds must be used as internal standards.

We use 6-azathymine to quantitate the pyrimidine-derived base products and 2,6-diaminopurine for purines, which are both stable during hydrolysis and derivatization.[60,61] Levels of oxidized bases are similar to those reported by other laboratories using isotope dilution when similar GC/MS protocols are employed. In our laboratory, isotope dilution for GC/MS analysis of 8-OH guanine in hydrolyzed calf thymus DNA has a lower coefficient of variation, but does not result in significantly different levels compared to 2,6-diaminopurine internal standardization. Nevertheless, isotope dilution should preferably be incorporated into GC/MS analysis.

Conclusion

We have shown that 5 ml of whole blood, or packed cells from this volume, is required to extract 100 μg DNA, which allows easy analysis of base damage by GC/MS, and storing blood at $-20°$ for up to 1 year does not affect DNA damage. However, it is important to note that although the DNA extraction protocol followed in this study requires removal of RNA to prevent artifacts in DNA base damage determination, showing a lack of effect with all the other conditions tested, it is not necessary that every DNA extraction procedure follows this pattern. The optimal protocol may depend on sources of reagents, and the general level of contamination with prooxidants (e.g., transition metal ions in the water and reagents) in the laboratory concerned.

We have subsequently applied this protocol to the investigation of factors affecting oxidative DNA damage in the human body.[25,26,65] We found that steady-state levels of oxidative DNA damage varied widely among healthy individuals.[66] This may be related to a number of endogenous and exogenous factors, including genetic predisposition, exposure to environmental toxins, and antioxidant intake. Levels of DNA damage and

[63] C. J. LaFrancois, K. Yu, and L. C. Sowers, *Chem. Res. Toxicol.* **11,** 75 (1998).

[64] C. J. LaFrancois, K. Yu, and L. C. Sowers, *Chem. Res. Toxicol.* **11,** 786 (1998).

[65] A. Rehman, J. Nourooz-Zadeh, W. Möller, H. Tritschler, P. Pereira, and B. Halliwell, *FEBS Lett.* **448,** 120 (1999).

[66] B. Halliwell, *Free Radic. Res.* **31,** 261 (1999).

repair will also be influenced by endogenous rates of free radical production and level of antioxidant defenses, as well as the activity of repair enzymes and the ability to respond positively to increased levels of oxidative stress, including the absorption, processing, and transport of dietary antioxidants to the site of oxidative stress. Collectively, such factors will contribute to the overall levels of oxidative stress in the human body and thus DNA damage and, in the long term, perhaps susceptibility to cancer.

Acknowledgment

The authors are grateful to the Ministry of Agriculture, Fisheries, and Food for research support.

[37] Sequence Specificity of Ultraviolet A-Induced DNA Damage in the Presence of Photosensitizer

By Kimiko Ito and Shosuke Kawanishi

Introduction

Solar ultraviolet radiation in the ultraviolet A (UVA; 320–400 nm) region, as well as in the UVB (280–320 nm) region, is considered probably carcinogenic to humans.[1] Although the precise mechanism of UV carcinogenesis is not fully understood, it is generally accepted that UV-induced DNA damage plays an important role in tumor induction. In wavelengths shorter than about 320 nm, pyrimidine photoproducts generated by the direct absorption of UV light by the DNA molecule have been shown to be relevant to mutation and carcinogenesis.[2] Much less is known, however, of the UVA light-induced DNA lesions, which may be involved in the development of human skin cancer. It is assumed that UVA light could induce cellular DNA damage indirectly through photosensitized oxidation reaction mediated by endogenous or exogenous molecules (photosensitizers) by which UVA light is primarily absorbed.[3-5] The photosensitized

---

[1] IARC, in "IARC Monographs on the Evaluation of Carcinogenic Risks to Humans, Solar and UV Radiation," Vol. 55. IARC, Lyon, 1992.
[2] H. N. Ananthaswamy and W. E. Pierceall, *Photochem. Photobiol.* **52**, 1119 (1990).
[3] T. P. Coohill, M. J. Peak, and J. G. Peak, *Photochem. Photobiol.* **46**, 1043 (1987).
[4] I. E. Kochevar and D. A. Dunn, in "Bioorganic Photochemistry" (H. Morrison, ed.), p. 273. Wiley, New York, 1990.
[5] K. Ito and S. Kawanishi, *Biol. Chem.* **378**, 1307 (1997).

reaction may produce radical intermediates by interaction of an excited photosensitizer with a substrate by direct electron transfer (type I) or may give rise to reactive oxygen species such as singlet oxygen ($^1O_2$) and superoxide anion radical (type II).[6]

Mutations in *ras* oncogenes and in the *p53* tumor suppressor gene have been analyzed in human skin cancers.[2,7,8] It is generally believed that dipyrimidine sequences where cyclobutane pyrimidine dimers and pyrimidine(6-4)pyrimidone photoadducts can be formed are sunlight-induced mutation hot spots. In agreement with this belief, mutations detected in skin tumors were predominantly C → T and CC → TT transitions at dipyrimidine sites. Nonetheless, it seems possible that DNA lesions produced via photosensitized oxidation reactions may lead to mutations at another DNA sequence, which could be involved in UV carcinogenesis. It is of great interest, therefore, to investigate the distribution of the photosensitized DNA lesions within defined DNA sequences and the mechanism involved in their formation. This article describes a DNA-sequencing method that has been used in our works to analyze the UVA-induced DNA damage in human c-Ha-*ras*-1 protooncogene in the presence of a photosensitizer. This method is an application of the Maxam–Gilbert sequencing procedure.[9] The reaction mechanism of the photosensitizer with DNA has been examined by the electron spin resonance (ESR) spin destruction method.[10,11]

Methods

Subcloning of AvaI Fragments of c-Ha-ras-1 Protooncogene

Plasmid pbcNI, which carries a 6.6-kb *Bam*HI chromosomal DNA segment containing human c-Ha-*ras*-1 protooncogene, can be obtained from American Type Culture Collection. The plasmid is digested with *Bst*EII and *Ava*I, and the resulting DNA fragments are fractionated by electophoresis on a 2% agarose gel to produce two *Ava*I fragments; a 602-bp *Ava*I fragment (*Ava*I 1645–*Ava*I 2246) and a 435-bp *Ava*I fragment (*Ava*I 2247–*Ava*I 2681). The two DNA fragments are ligated separately into pUC18 plasmids, which are then transferred to *Escherichia coli* MC1061.

[6] C. S. Foote, *Photochem. Photobiol.* **54**, 659 (1991).
[7] W. E. Pierceal, T. Mukhopadhyary, L. H., Goldberg, and H. N. Ananthaswamy, *Mol. Carcinogen.* **4**, 445 (1991).
[8] A. Ziegler, D. J. Leffell, S. Kunala, H. W. Sharma, M. Gailani, J. A. Simon, A. J. Halperin, H. P. Baden, P. E. Shapiro, A. E. Bale, and D. E. Brash, *Proc. Natl. Acad. Sci. U.S.A.* **90**, 4216 (1993).
[9] A. M. Maxam and W. Gilbert, *Methods Enzymol.* **65**, 499 (1980).
[10] J. Moan, *Acta Chem. Scand. B* **34**, 519 (1980).
[11] K. Reszka and R. C. Sealy, *Photochem. Photobiol.* **39**, 293 (1984).

Preparation of ^{32}P 5' End-Labeled DNA Fragments

pUC18 plasmid (ca. 800 μg) carrying the cloned 602-bp fragment (or the 435-bp fragment) is digested with 150 units of *Ava*I in 300 μl buffer at 37° for approximately 20 hr. Subsequently, 6 μl of calf intestine phosphatase (1 unit/μl) is added and incubated at 37° for 15 min to remove the 5'-terminal phosphate. After dephosphorylation, the reaction solution is mixed with 60 μl of glycerol dye (30% glycerol, 0.25% bromophenol blue, and 0.25% xylene cyanol) and fractionated on a 1% agarose gel (Seakem ME agarose 50012) containing 0.5 mg/ml ethidium bromide. After electrophoresis at 75 mA for approximately 3 hr, the 602-bp (or 435-bp) *Ava*I fragment is recovered by extraction from the gel with Gilbert's buffer [0.5 M ammonium acetate, 10 mM magnesium acetate, 1 mM EDTA, and 0.1% sodium dodecyl sulfate (SDS)].

The resulting two *Ava*I fragments are end labeled at the 5' termini. Twenty-five picomoles of each fragment is incubated at 37° for 20 min with [γ-^{32}P]ATP (3.7 MBq) and 20 units of polynucleotide kinase in 50 μl buffer [50 mM Tris–HCl (pH 7.6), 10 mM MgCl$_2$, 5 mM dithiothreitol, 0.1 mM spermidine, and 0.1 mM EDTA]. The ^{32}P 5' end-labeled 602-bp *Ava*I fragment (*Ava*I* 1645–*Ava*I* 2246) is then digested with 70 units of *Xba*I at 37° for 60 min to produce two singly 5' end-labeled fragments: a 261-bp fragment (*Ava*I* 1645–*Xba*I 1905) and a 341-bp fragment (*Xba*I 1906–*Ava*I* 2246), which, after the addition of glycerol dye, are fractionated by electrophoresis on a 6% polyacrylamide gel (5.8% acrylamide and 0.2% bisacrylamide in TBE buffer) at 50 mA for approximately 2 hr. The two singly end-labeled fragments are recovered from the gel by extraction with ca. 1600 μl Gilbert's buffer and are precipitated by ethanol in 15–20 Eppendorf tubes before storing at −80°. Similarly, the ^{32}P 5' end-labeled 435-bp *Ava*I fragment (*Ava*I* 2247–*Ava*I* 2681) is digested with *Pst*I and is purified by electrophoresis on a 6% polyacrylamide gel to produce another two singly 5' end-labeled fragments: a 98-bp fragment (*Ava*I* 2247–*Pst*I 2344) and a 337-bp fragment (*Pst*I 2345–*Ava*I* 2681). The asterisk indicates ^{32}P labeling and the nucleotide numbering starts with the *Bam*HI site.[12] The 12th and 13th codons of human c-Ha-*ras*-1 protooncogene are located on the 261-bp fragment.

Detection of DNA Damage

Singly 5' end-labeled DNA (ca. 0.1 μM/base) is mixed with various concentrations of photosensitizer in 100 μl of 10 mM sodium phosphate

[12] D. J. Capon, E. Y. Chen, A. D. Levinson, P. H. Seeburg, and D. V. Goeddel, *Nature* **302,** 33 (1983).

buffer (pH 7.9) containing 2 μM/base of sonicated calf thymus DNA. For the experiment with single-stranded DNA, the labeled DNA is denatured at 90° for 5 min in the buffer containing calf thymus DNA before the addition of sensitizer. The mixture is irradiated in a 1.5-ml Eppendorf tube at ca. 2.7 mJ/cm^2/sec for various durations with UV lamps (UVP, Inc., CA), which have an emission peak at approximately 365 nm; a long-wave filter is used to reduce the light intensity at wavelengths shorter than 300 nm negligible small as compared with that of the 365-nm line. The fluence rate at 365 nm is measured using a UVX radiometer (UVP, Inc.). Irradiated DNA is recovered by ethanol precipitation, rinsed with 70% ethanol to remove the sensitizer, and dried under vacuum. To induce strand breaks at base alteration and liberation sites, irradiated DNA is treated further with freshly diluted 1 M piperidine at 90° for 20 min, and the DNA is again ethanol precipitated, washed, and dried. After piperidine treatment, or without piperidine treatment (for detection of directly produced strand brakes), samples are dissolved in 10 μl of loading buffer (80% formamide, 1 mM EDTA, 0.1% bromphenol blue, and 0.1% xylene cyanol), denatured at 90° for 2 min, chilled rapidly, and loaded immediately onto 12 × 16-cm 8% polyacrylamide gels (7.76% acrylamide, 0.24% bisacrylamide, and 8 M urea in TBE buffer). Electrophoresis is performed at 50 mA for approximately 40 min. Gels are dried on filter papers at 80° for 60–80 min, and autoradiograms are obtained by exposing X-ray films to the gels at −85° overnight. Samples are protected from direct sunlight throughout the experiment.

DNA Sequence Analysis

Singly 5′ end-labeled DNA (ca. 1 μM/base) is irradiated in the presence of a photosensitizer as described earlier. The concentration of the sensitizer and the irradiation dose should be chosen to introduce an average of one base modification per labeled fragment. The irradiated DNA is ethanol precipitated, washed twice with 70% ethanol, dried, and treated with piperidine as described earlier. In parallel, the same labeled fragment is also treated with the chemical reactions of the Maxam–Gilbert procedure to induce cleavage at adenine and guanine residues, and thymine and cytosine residues in the unirradiated DNA fragment.[9] The radioactivity of each sample is then measured and samples are dissolved in 1–20 μl of loading buffer to give approximately same counts per volume in each sample. One microliter of the sample is loaded onto 18 × 50-cm 8% polyacrylamide gels (7.76% acrylamide, 0.24% bisacrylamide, and 8 M urea in TBE buffer) immediately after denaturation at 90° for 2 min. Gels are run at 40 mA for approximately 2 hr at 60° using a DNA-sequencing system (LKB2010

Macrophor). After electrophoresis, the gels are treated as described earlier. The preferred cleavage sites are determined by comparison of the electrophoretic mobilities of the oligonucleotides produced by the photosensitized reaction with those produced by the Maxam–Gilbert procedure.[9] A laser densitometer (LKB 2222 UltroScan XL) is used for the measurement of the relative amounts of oligonucleotides from treated DNA fragments.

Results and Discussion

DNA Damage Induced by UVA in the Presence of Various Sensitizers

Figure 1A shows the autoradiogram of the double-stranded ^{32}P 5′ end-labeled 337-bp fragment exposed to UVA light with 0.05 mM riboflavin. The upper and lower bands in the control show single- and double-stranded

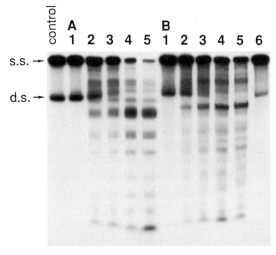

FIG. 1. Autoradiogram of ^{32}P-labeled DNA fragments exposed to UVA light in the presence of riboflavin or hematoporphyrin. The ^{32}P 5′ end-labeled 337-bp fragment was exposed to UVA light in the presence of 0.05 mM riboflavin (A) or 0.1 mM hematoporphyrin (B) in 100 μl of 10 mM sodium phosphate buffer (pH 7.9) containing 2 μM/base of sonicated calf thymus DNA. For the experiment with hematoporphyrin, the labeled DNA fragment was denatured by treatment at 90° for 5 min followed by quick chilling. After irradiation, the DNA fragments were treated with 1 M piperidine and subjected to electrophoresis on an 8% polyacrylamide–8 M urea gel (12 × 16 cm). The autoradiogram was obtained by exposing X-ray film to the gel. Single- and double-stranded intact DNA fragments are indicated as s.s. and d.s., respectively. Control, no photosensitizer, no irradiation; lane 1, 0 J/cm^2; lane 2, 1.65 J/cm^2; lane 3, 3.3 J/cm^2; lane 4, 6.6 J/cm^2; lane 5, 10 J/cm^2; and lane 6, 10 J/cm^2 without sensitizer. From K. Ito *et al., J. Biol. Chem.* **268,** 13221 (1993).

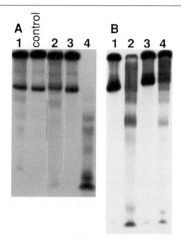

Fig. 2. Autoradiogram of ^{32}P-labeled DNA fragments exposed to UVA light in the presence of various photosensitizers. (A) The ^{32}P 5' end-labeled 337-bp fragment was exposed to 6.6 J/cm^2 of UVA light as described in the legend to Fig. 1 in the presence of indicated photosensitizers. Lane 1, 1 mM menadione; lane 2, 0.5 mM folic acid; lane 3, 1 mM NADH; and lane 4, 1 mM NADH + 20 μM Cu(II). (B) The 337-bp fragment was exposed to the indicated doses of UVA light as described in the legend to Fig. 1 in the presence of 0.5 mM NADH + 20 μM Cu(II). Lane 1, 0 J/cm^2; lane 2, 5 J/cm^2; lane 3, 5 J/cm^2 + 15 units of catalase; and lane 4, 5 J/cm^2 + 50 μM bathocuproine. After irradiation, the DNA fragments were treated as described in the legend to Fig. 1.

forms of intact DNA fragments, respectively. DNA photodegradation in the presence of riboflavin increased with radiation dose. In contrast, UVA radiation in the presence of 0.1 mM hematoporphyrin induced damage in single-stranded DNA (Fig. 1B), but not in double-stranded DNA.[13,14] DNA damage was not observed with UVA radiation alone (Fig. 1, lane 6).

Figure 2A shows that damage of double-stranded DNA is also induced by UVA irradiation in the presence of menadione (lane 1) and folic acid (lane 2). Although UVA light with NADH did not induce DNA damage in the absence of Cu(II) (lane 3), the addition of Cu(II) produced significant DNA damage (lane 4). Similarly, we have demonstrated that pterin derivatives[15] and nalidixic acid[16] induce DNA damage with exposure to UVA light.

Under our experimental conditions, the DNA damage induced by UVA light and riboflavin was not affected in D$_2$O,[14] which increases the lifetime of ^{1}O$_2$. In addition, ESR studies indicated that photoexcited riboflavin

[13] S. Kawanishi, S. Inoue, S. Sano, and H. Aiba, *J. Biol. Chem.* **261,** 6090 (1986).

[14] K. Ito, S. Inoue, K. Yamamoto, and S. Kawanishi, *J. Biol. Chem.* **268,** 13221 (1993).

[15] K. Ito and S. Kawanishi, *Biochemistry* **36,** 1774 (1997).

[16] Y. Hiraku, H. Ito, and S. Kawanishi, *Biochem. Biophys Res. Commun.* **251,** 466 (1997).

reacted specifically with dGMP to produce a riboflavin radical anion and a guanine radical cation.[14] From these results, it is considered that riboflavin-sensitized damage in duplex DNA is induced predominantly through the type I (electron transfer) mechanism. Similar results were obtained with pterins,[15] folic acid,[15] and nalidixic acid.[16] However, an enhancing effect of D_2O was significant in hematoporphyrin-mediated DNA damage. By ESR studies, no reactivity of hematoporphyrin radical with dGMP was observed.[14] These observations suggest that the damage of single-stranded DNA by UVA light with hematoporphyrin is due to photochemically generated $^1O_2$ (major type II). As for photoinduced DNA damage in the presence of NADH and Cu(II), the addition of both catalase and bathocuproine, a chelating agent for Cu(I), inhibited damage substantially (Fig. 2B), indicating that H_2O_2 and Cu(I) were involved. It is speculated therefore that photoexcited NADH reacts with oxygen to form superoxide anion radical (minor type II), which subsequently autodismutates to H_2O_2, and reactive species produced by the reaction of H_2O_2 with Cu(I) may cause DNA photodamage.[17]

Sequence Specificity of DNA Damage induced by Type I and Type II Photosensitization

Figure 3 shows the autoradiogram of a sequencing gel obtained by exposure of the ^{32}P 5′ end-labeled 341-bp fragment to UVA light in the presence of riboflavin or hematoporphyrin, followed by treatment with piperidine. For the measurement of relative intensity of DNA damage, autoradiograms were scanned with a laser densitometer (Figs. 4 and 5). As seen in Figs. 3–5, riboflavin induced cleavage in double-stranded DNA predominantly at the 5′ G of 5′-GG-3′ sequences. The preferred cleavage sites of 5′-GGG-3′ were the 5′ and central guanine. Although very weak cleavage at the 5′ G of 5′-GA-3′ sites occurred, no cleavage was found at other sequences, including a sequence in which guanine was located 5′ to a pyrimidine nucleoside. In contrast, hematoporphyrin induced cleavage at most guanine residues without sequence selectivity in denatured single-stranded DNA (Fig. 5A). In addition, it was found that when denatured DNA was irradiated by UVA light in the presence of riboflavin, cleavage occurred with less preference for consecutive guanines (Fig. 5C). The sequence-specific damage at 5′ G of 5′-GG-3′ sites was also observed, in our laboratory, with an exposure of double-stranded DNA to UVA light in the presence of pterin derivatives,[15] nalidixic acid,[16] folic acid,[5] and menadione,[18] all of which have been shown to induce DNA damage mainly

[17] K. Yamamoto and S. Kawanishi, *J. Biol. Chem.* **264,** 15435 (1989).
[18] K. Ito and S. Kawanishi, unpublished result (1996).

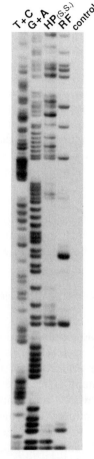

Fig. 3. Autoradiogram of a sequencing gel obtained by exposure of ^{32}P-labeled DNA fragments to UVA light in the presence of riboflavin or hematoporphyrin. The [32]P 5' end-labeled 341-bp fragment was exposed to 6.6 J/cm^2 of UVA light with 0.05 mM riboflavin (RF) and 0.1 mM hematoporphyrin (HP) as described in the legend to Fig. 1. For the experiment with denatured single-stranded DNA (S.S.), the 5' end-labeled DNA fragment was treated at 90° for 5 min and chilled quickly before the addition of sensitizer. After piperidine treatment, samples were electrophoresed on an 8% polyacrylamide–8 M urea gel (18 × 50 cm) using a DNA-sequencing system (LKB2010 Macrophor), and the autoradiogram was obtained by exposing X-ray film to the gel. The G + A and T + C lanes represent the patterns obtained for the same fragment after cleavage by the chemical methods of Maxam and Gilbert.[9]

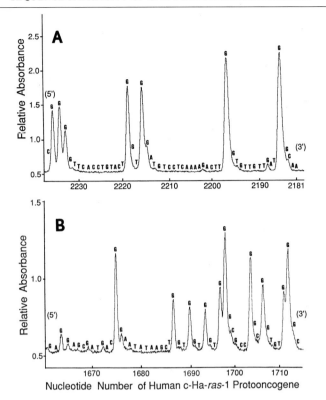

Fig. 4. Site specificity of DNA cleavage induced by UVA radiation in the presence of riboflavin. The ^{32}P 5′ end-labeled 341-bp fragment (A) and 261-bp fragment (B) were exposed to 6.6 J/cm^2 of UVA light with 0.05 mM riboflavin as described in the legend to Fig. 1. The autoradiogram was obtained as described in the legend to Fig. 3 and scanned with a laser densitometer. Horizontal axis: the nucleotide number of human c-Ha-*ras*-1 protooncogene starting with the *Bam*HI site. From K. Ito *et al.*, *J. Biol. Chem.* **268,** 13221 (1993).

through type I mechanism. These findings suggest that the type I-photosensitized reaction induces sequence-specific oxidation at consecutive guanines in duplex DNA, whereas the type II reaction ($^1O_2$) can damage most guanines without sequence selectivity. Interestingly, similar sequence-specific guanine damage has been also reported from other laboratories using various systems.[19–22] Of these, it is noteworthy that the oxidative damage at the 5′ G of 5′-GG-3′ sites is induced by a long-range electron transfer

[19] I. Saito, M. Takayama, and S. Kawanishi, *J. Am. Chem. Soc.* **117,** 5590 (1995).

[20] D. T. Breslin and G. B. Schuster, *J. Am. Chem. Soc.* **118,** 2311 (1996).

[21] D. B. Hall, R. E. Holmlin, and J. K. Barton, *Nature* **382,** 731 (1996).

[22] H. Sies, W. A. Schulz, and S. Steenken, *J. Photochem. Photobiol. B Biol.* **32,** 97 (1996).

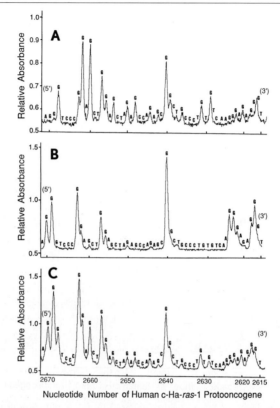

FIG. 5. Comparison of photoreactivity of DNA with riboflavin and hematoporphyrin. The ^{32}P 5' end-labeled 337-bp fragment was exposed to 6.6 J/cm^2 of UVA light with 0.1 mM hematoporphyrin (A) or 0.05 mM riboflavin (B and C) as described in the legend to Fig. 1. For the experiment with denatured DNA (A and C), the 5' end-labeled DNA fragment was treated at 90° for 5 min and chilled quickly before the addition of sensitizer. After the piperidine treatment, DNA fragments were analyzed by the method described in the legend to Fig. 4. From K. Ito *et al., J. Biol. Chem.* **268,** 13221 (1993).

over a distance of ca. 37 Å on irradiation of a metallointercalater attached covalently to DNA.[21] A theoretical calculation has been made to explain the sequence-specific oxidation at consecutive guanines in duplex DNA.[23]

In addition, we found that the content of 7,8-dihydro-8-oxo-2'-deoxy-guanosine (8-oxodG) in DNA, as determined by HPLC with electrochemical detection,[24] was increased significantly by photosensitized reactions with type I sensitizers.[14–16] The guanine radical cation generated in DNA via

[23] H. Sugiyama and I. Saito, *J. Am. Chem. Soc.* **118,** 7063 (1996).
[24] H. Kasai, S. Nishimura, Y. Kurokawa, and Y. Hayashi, *Carcinogenesis* **8,** 1959 (1987).

the type I reaction could therefore react with water to give the C-8 OH adduct radical of guanine, which leads to the formation of 8-oxodG through one-electron oxidation.[25,26] The increase of this premutagenic DNA base lesion has been reported in human skin cells following UVA exposure.[27] However, studies have shown that 8-oxodG lesions in DNA are stable on treatment with piperidine.[28,29] Therefore, it is suggested that, in addition to 8-oxodG, other guanine lesions that are piperidine labile should occur at the 5' G of 5'-GG-3' sequences. At present, it would appear that the likely candidates are 2,2-diamino-4-[(2-deoxy-*b*-D-*erythro*-pentofurano-syl)amino]-5(2*H*)-oxazolone (oxazolone) and its imidazole precursor.[30] It should be added that further oxidation of 8-oxodG may produce piperidine-labile lesions.[31]

Concluding Remarks

We have investigated the distribution of UVA-induced oxidative damage within isolated DNA of human c-Ha-*ras*-1 protooncogene using a DNA-sequencing method, an application of the Maxam–Gilbert procedure. The photosensitization mechanism involved in DNA damage induction, as well as the photooxidation products, was also studied. Results have demonstrated that different photosensitization mechanisms generate different patterns of the DNA damage at the DNA sequence level; the type I (electron transfer) reaction induces sequence-specific damage at guanine located 5' to guanine in double-stranded DNA, whereas the major type II (singlet oxygen) reaction damages most guanines without sequence selectivity. The method herein, with a minor modification, is applicable to the analysis of the distribution of DNA damage in the *p53* tumor suppressor gene.[32]

[25] S. Steenken, *Chem. Rev.* **89**, 503 (1989).

[26] H. Kasai, Z. Yamaizumi, M. Berger, and J. Cadet, *J. Am. Chem. Soc.* **114**, 9692 (1992).

[27] E. Kvam and R. M. Tyrrell, *Carcinogenesis* **18**, 2379 (1997).

[28] P. M. Cullis, M. E. Malone, and L. A. Merson-Davies, *J. Am. Chem. Soc.* **118**, 2775 (1996).

[29] D. Gasparutto, J.-L. Ravanat, O. Gerot, and J. Cadet, *J. Am. Chem. Soc.* **120**, 10286 (1998).

[30] J. Cadet, M. Berger, G. W. Buchko, P. C. Joshi, S. Raoul, and J.-L. Ravanat, *J. Am. Chem. Soc.* **116**, 30 (1994).

[31] W. Adam, C. R. Saha-Moller, and A. Schonberger, *J. Am. Chem. Soc.* **118**, 9233 (1996).

[32] N. Yamashita, M. Murata, S. Inoue, M. J. Burkitt, L. Milne, and S. Kawanishi, *Chem. Res. Toxicol.* **11**, 855 (1998).

[38] Protein Oxidative Damage

By EMILY SHACTER

Introduction

Proteins can undergo numerous covalent changes on exposure to oxidants.[1,2] Some of these changes result from direct attack of a free radical on the protein molecule whereas others are incurred indirectly such as through covalent attachment of oxidation by-products. One mode of direct oxidative attack on a protein derives from site-specific metal-catalyzed oxidation (MCO) in which the reduced form of a protein-bound transition metal (e.g., Fe^{2+}, Cu^+) reduces H_2O_2 to form a reactive intermediate (e.g., $OH \cdot$, ferryl radical) in the immediate proximity of amino acid side chains.[3,4] Radical-mediated oxidation induces the formation of amino acyl carbonyl groups (aldehydes and ketones) in amino acids, especially in Lys, Arg, Pro, Thr.[4,5] A metal-catalyzed attack on His and Tyr residues leads to the formation of oxo-His[6] and dityrosine,[7] respectively. The sulfhydryl moiety of Cys is highly prone to oxidative attack by several mechanisms, leading to the formation of disulfide bonds and thiyl radicals.[8,9] Similarly, numerous oxidative pathways lead to the oxidation of Met residues, leading to the formation of methionine sulfoxide.[10] Val, Leu, and Tyr residues are oxidized to hydroxy and hydroperoxy derivatives on exposure to γ-irradiation in O_2.[11] Reaction of myeloperoxidase-derived HOCl with Tyr, Trp, Lys, and Met residues leads to the formation of chlorotyrosine, chloramines, aldehydes, and methionine sulfoxide.[12–15] Peroxynitrite, generated from the

[1] R. T. Dean, S. Fu, R. Stocker, and M. J. Davies, *Biochem. J.* **324,** 1 (1997).
[2] B. S. Berlett and E. R. Stadtman, *J. Biol. Chem.* **272,** 20313 (1997).
[3] A. Samuni, J. Aronovitch, D. Godinger, M. Chevion, and G. Czapski, *Eur. J. Biochem.* **137,** 119 (1983).
[4] E. R. Stadtman, *Free Radic. Biol. Med.* **9,** 315 (1990).
[5] A. Amici, R. L. Levine, L. Tsai, and E. R. Stadtman, *J. Biol. Chem.* **264,** 3341 (1989).
[6] S. A. Lewisch and R. L. Levine, *Anal. Biochem.* **231,** 440 (1995).
[7] C. Giulivi and K. J. A. Davies, *Methods Enzymol.* **233,** 363 (1994).
[8] M-L. Hu, *Methods Enzymol.* **233,** 380 (1994).
[9] B. Kalyanaraman, *Biochem. Soc. Symp.* **61,** 55 (1995).
[10] W. Vogt, *Free Radic. Biol. Med.* **18,** 93 (1995).
[11] S. L. Fu and R. T. Dean, *Biochem. J.* **324,** 41 (1997).
[12] L. J. Hazell, J. J. M. van den Berg, and R. Stocker, *Biochem. J.* **302,** 297 (1994).
[13] C.-Y. Yang, Z.-W. Gu, H.-X. Yang, M. Yang, A. M. Gotto, and C. V. Smith, *Free Radic. Biol. Med.* **23,** 82 (1997).

interaction of nitric oxide and superoxide, causes nitrosylation of tyrosine residues and oxidative modification of other amino acid residues, including Cys, Trp, Met, and Phe,[16] but does not induce protein carbonyls (R. L. Levine, personal communication). Indirect modification of protein amino acyl side chains occurs through adduct formation. For example, lipid peroxidation breakdown products such as hydroxynonenal (HNE) and malondialdehyde (MDA) bind covalently to Lys, His, and Cys residues, leading to the addition of aldehyde moieties to the protein.[17,18] Glutathiolation of Cys residues[19] and addition of p-hydroxyphenylacetaldehyde to Lys residues[20] also occur under conditions of oxidative stress. Carboxymethyllysine represents a form of protein modification generated by oxidation products of sugars and lipids.[2,21] A summary of well-characterized protein amino acyl oxidative modifications is provided in Table I, and a list of assays commonly employed to measure these modifications is given in Table II. Note that ozone can induce a similar spectrum of protein oxidative modifications as MCO systems and has a particularly strong ability to induce protein carbonyl groups.[22,23] Singlet oxygen generated by exposure to UV light or by photochemical methods has a greater tendency to attack Trp, His, Tyr, Met, and Cys residues and a lesser capacity to induce carbonyls.[24-26]

This article focuses on the detection of carbonyl groups as markers of oxidative protein modification. Protein carbonyls become elevated under various conditions of oxidative stress and are generated readily under experimental conditions *in vitro*, such as through exposure to iron/ascorbate,[5]

[14] S. L. Hazen and J. W. Heinecke, *J. Clin. Invest.* **99**, 2075 (1997).

[15] A. J. Kettle, *FEBS Lett.* **379**, 103 (1996).

[16] H. Ischiropoulos and A. B. Al-Mehdi, *FEBS Lett.* **364**, 279 (1995).

[17] K. Uchida and E. R. Stadtman, *Methods Enzymol.* **233**, 371 (1994).

[18] J. R. Requena, M. X. Fu, M. U. Ahmed, A. J. Jenkins, T. J. Lyons, J. W. Baynes, and S. R. Thorpe, *Biochem. J.* **322**, 317 (1997).

[19] C.-K. Lii, Y.-C. Chai, W. Zhao, J. A. Thomas, and S. Hendrich, *Arch. Biochem. Biophys.* **308**, 231 (1994).

[20] S. L. Hazen, J. P. Gaut, F. F. Hsu, J. R. Crowley, A. d'Avignon, and J. W. Heinecke, *J. Biol. Chem.* **27**, 16990 (1997).

[21] T. P. Degenhardt, S. R. Thorpe, and J. W. Baynes, *Cell. Mol. Biol.* **44**, 1139 (1998).

[22] C. E. Cross, A. Z. Reznik, L. Packer, P. A. Davis, Y. J. Suzuki, and B. Halliwell, *Free Radic. Res. Commun.* **15**, 347 (1992).

[23] B. S. Berlett, R. L. Levine, and E. R. Stadtman, *J. Biol. Chem.* **271**, 4177 (1996).

[24] D. Balasubramanian, X. Du, and J. S. Zigler, *Photochem. Photobiol.* **52**, 761 (1990).

[25] H. R. Shen, J. D. Spikes, P. Kopecekova, and J. Kopecek, *J. Photochem. Photobiol. B* **34**, 203 (1996).

[26] M. L. Hu and A. L. Tappel, *Photochem. Photobiol.* **56**, 357 (1992).

TABLE I
EXAMPLE OF OXIDATIVE PROTEIN MODIFICATIONS

Modification	Amino acids modified	Oxidizing sources[a]
Disulfides, mixed disulfides (e.g., glutathiolation)	Cys	All, ONOO⁻ (with GSSG)
Methionine sulfoxide	Met	All, ONOO⁻
Carbonyls	All (especially Lys, Arg, Pro, Thr)	All
oxo-Histidine	His	γ-ray, MCO, $^1O_2$
Dityrosine	Tyr	γ-ray, MCO, $^1O_2$
Chlorotyrosine	Tyr	HOCl
Nitrotyrosine	Tyr	ONOO⁻
Tryptophanyl modifications [e.g., (N-formyl)kynurenine]	Trp	γ-ray, $^1O_2$, MCO, ozone
Hydro(pero)xy derivatives	Val, Leu, Tyr, Trp	γ-ray
Chloramines, deamination	Lys	HOCl
Lipid peroxidation adducts (MDA, HNE, acrolein)	Lys, Cys, His	γ-ray, MCO (not HOCl)
Glycoxidation adducts	Lys	Glucose
Amino acid oxidation adducts	Lys, Cys, His	HOCl
Cross-links, aggregates, fragments	several	All

[a] MCO, metal-catalyzed oxidation; all = γ-ray, MCO, HOCl, $^1O_2$, ozone.

ionizing radiation,[27] singlet oxygen,[28] and ozone.[22] Their chemical stability makes them suitable targets for laboratory measurement. Early techniques for the detection of protein carbonyls were based either on incorporation of tritium through reduction with tritiated borohydride[29] or on incorporation of a stable dinitrophenyl (DNP) adduct through reaction of the carbonyls with 2,4-dinitrophenylhydrazine (DNPH).[30] Because DNP absorbs light at 370 nm, the carbonyl groups can be measured spectrophotometrically. While reliable and quantitative, these two techniques suffer from three significant technical drawbacks: (a) They require the use of a substantial amount of purified protein (~1–2 mg); (b) the derivitizing reagents have to be removed prior to analysis in order to reduce background levels sufficiently to allow detection of protein-bound label (either H^3 or DNP); and (c) they are unable to differentiate oxidized from nonoxidized proteins in cell or tissue extracts.

[27] E. Shacter, J. A. Williams, and R. L. Levine, *Free Radic. Biol. Med.* **18,** 815 (1995).
[28] J. A. Silvester, G. S. Timmins, and M. J. Davies, *Arch. Biochem. Biophys.* **350,** 249 (1998).
[29] A. Lenz, U. Costabel, S. Shaltiel, and R. L. Levine, *Anal. Biochem.* **177,** 419 (1989).
[30] R. L. Levine, D. Garland, C. N. Oliver, A. Amici, I. Climent, A. Lenz, B. Ahn, S. Shaltiel, and E. R. Stadtman, *Methods Enzymol.* **186,** 464 (1990).

TABLE II
METHODS FOR DETECTION OF OXIDATIVE PROTEIN MODIFICATIONS[a]

Modification	Methods of detection	References[b]
Disulfides	SDS–gel electrophoresis ($\pm$2-ME); DTNB	41
Thiyl radicals	Electron spin resonance spectroscopy	9
Glutathiolation	RP-HPLC/mass spectroscopy; IEF; S^{35}-Cys/Chx/SDS–PAGE	19,42,43
Methionine sulfoxide	CNBr cleavage/amino acid analysis	44,45
Carbonyls	DNPH/Western blot/ELISA/ immunocytochemistry/HPLC/ A_{370}; reduction with Na-borotritide	This article and 30,35,36,46 29,47
2-oxo-His	Amino acid analysis	6
Dityrosine	Fluorescence; proteolysis or hydrolysis/HPLC	7,48
Chlorotyrosine	Hydrolysis/nitroso-napthol/HPLC; HBr hydrolysis-GC/MS	14,15
Nitrotyrosine	Immunoassay; hydrolysis/HLPC; HPLC/electrochemical detection	49–53
Tryptophanyl	Fluoroscence spectroscopy; amino acid analysis (alk. hydrolysis)	
Hydroperoxides	KI/I_3^-/NaBH$_4$/hydrolysis/OPA-HPLC	11
Lipid peroxidation adducts	NaBH$_4$/hydrolysis/OPA-HPLC; hydrolysis-GC/MS; DNPH; immunoassays	17,18,54,55
Amino acid oxidation adducts	Reduction/hydrolysis/NMR	20
Cross-links, aggregates, fragments	SDS–gel electrophoresis; HPLC	12,56,57

[a] DTNB, dithiobisnitrobenzoate; IEF, isoelectric focusing; DNPHI, dinitrophenylhydrazine; CNBr, cyanogen bromide; Chx, cyclohexamide; OPA, o-pthaldialdehyde.
[b] References for some methods currently in use. Other methods are also available.

Western Blot Immunoassay for Protein Carbonyls

Several years ago, we described a new method for detection of protein carbonyls.[31] The method takes advantage of the fact that DNP groups are immunogenic and that anti-DNP antibodies are available commercially. By derivatizing proteins with DNPH in an sodium dodecyl sulfate (SDS) buffer, oxidized protein can be detected by performing SDS–polyacrylamide gel electrophoresis (SDS–PAGE), electroblotting to nitrocellulose (Western blotting), and assaying with anti-DNP antibodies (immunoassay).

[31] E. Shacter, J. A. Williams, M. Lim, and R. L. Levine, Free Radic. Biol. Med. 17, 429 (1994).

This method has significant advantages in that it requires very little protein (as little as 50 ng of a 50-kDa protein oxidized to the extent of 0.3–0.5 mol carbonyl/mol protein) and it identifies which proteins in a complex mixture are oxidized and which are not.[31] The sensitivity of the method for detecting minor oxidized proteins in a cell can be enhanced greatly by preparing subcellular fractions prior to derivatization and performance of the Western blot immunoassay.[32,33] The carbohydrate groups in glycoproteins have no apparent effect in the assay and are not themselves readily oxidizable by metal-catalyzed systems.[34] Aldehydes introduced through the adduction of lipid peroxidation breakdown products (e.g., 4-hydroxynonenal, MDA, acrolein) can react with DNPH and thus can be detected by this method. These adducts are somewhat unstable and the extent to which they actually contribute to protein-associated carbonyl levels in tissue extracts remains to be established.

Reagents required for running the Western blot immunoassay are all available from commercial sources and the only special equipment employed is an SDS–PAGE/electroblotting apparatus, available in most biochemistry laboratories. Due to the nature of Western blotting, the assay is only semiquantitative. This drawback can be overcome by running size-exclusion HPLC instead of SDS–PAGE if there is enough sample available.[35] Alternatively, the quantitation of total DNPH-derivatized protein carbonyls can be achieved by use of an ELISA, as described by Buss et al.[36]

The actual procedure for running the Western blot immunoassay for protein carbonyls is very straightforward and there are few "tricks" required to ensure its success. Detailed procedures have been described in previous publications.[31,34,35,37] Directions and some useful pointers follow.

Derivitization Reagents

DNPH/TFA stock solution: 20 mM DNPH in 20% (v/v) trifluoroacetic acid (TFA). Dissolve the DNPH in 100% TFA and then dilute with

[32] L. J. Yan, R. L. Levine, and R. S. Sohal, *Proc. Natl. Acad. Sci. U.S.A.* **94**, 11168 (1997).

[33] L. J. Yan and R. S. Sohal, *Proc. Natl. Acad. Sci. U.S.A.* **95**, 12896 (1998).

[34] Y-J. Lee and E. Shacter, *Arch. Biochem. Biophys.* **321**, 175 (1995).

[35] R. L. Levine, J. Williams, E. R. Stadtman, and E. Shacter, *Methods Enzymol.* **233**, 346 (1994).

[36] H. Buss, T. P. Chan, K. B. Sluis, N. M. Domigan, and C. C. Winterbourn, *Free Radic. Biol. Med.* **23**, 361 (1997).

[37] E. Shacter, J. A. Williams, E. R. Stadtman, and R. L. Levine, *in* "Free Radicals: A Practical Approach" (N. Punchard and F. Kelly, eds.). Oxford Univ. Press, Oxford, 1994.

water. Note that most commercial sources of DNPH contain roughly 30% H_2O. For a 10-ml stock DNPH/TFA solution, dissolve 40 mg of dry DNPH (52 mg of the labeled "wet" powder) in 1 ml of 100% TFA and then dilute with 4 ml water.

20% TFA (v/v) (reagent control)
24% SDS in H_2O (24 g/100 ml)
2 M Tris base containing 30% glycerol
2-mercaptoethanol (2-ME), undiluted stock (14.4 M)

Samples

Any protein solution may be used, including purified proteins and cell and tissue extracts. Extracts containing SDS are most suitable as the derivitization reaction and gels contain SDS buffer.

Reaction with Dinitrophenylhydrazine

Derivitization is carried out in a solution containing 6% SDS/5 mM DNPH/5% TFA. The concentration of protein may be as high as 10 mg/ml; use a concentration that will be convenient for gel loading (e.g., to load 2–20 μl of sample per lane). Start the derivitization reaction by mixing 2 volumes of protein solution (e.g., 10 μl) with 1 volume of 24% SDS (e.g., 5 μl). Then add 1 volume of DNPH/TFA stock solution (e.g., 5 μl). Incubate at room temperature for 5–30 min. Longer derivitization times lead to a nonspecific incorporation of DNP, which will interfere with correct interpretation of the results. Immediately following derivitization, neutralize the sample and prepare for final gel loading by adding 1/3 volume (e.g., 6.7 μl) of Tris/glycerol ± 2-ME. The sample should turn from yellow to orange on neutralization. To run reducing gels, add 1 volume of 2-ME to 4 volumes of the Tris/glycerol solution. In most cases, heating is not performed. Load and run the SDS–PAGE gels promptly using the conditions described by Laemmli.[38] The total time elapsed between initiation of the derivitization reaction and running the gels should not take more than about 45–60 min for 20 samples.

---

[38] U. K. Laemmli, *Nature London* **227,** 680 (1970).

Electroblotting

Western transfer is carried out according to standard procedures.[39,40] In our hands, the lowest nonspecific immunoassay backgrounds are achieved with PVDF paper (e.g., Immobilon-P from Millipore). Free DNPH present in the samples will not interfere with immunodetection and does not contribute a background.

When nonreducing gels are run (i.e., no 2-ME is added to the samples), treat the gels with 2-ME prior to transfer. This reduction of the proteins facilitates immunodetection of DNP groups in proteins containing large numbers of disulfide bonds (e.g., albumin).[31] After electrophoresis, soak the gel in 1% 2-ME in Tris/glycine transfer buffer for 5 min at room temperature. Then, rinse the gel in transfer buffer for 5 min to remove excess 2-ME. Set up the transfer.

[39] H. Towbin, T. Staehelin, and J. Gordon, *Proc. Natl. Acad. Sci. U.S.A.* **76,** 4350 (1979).

[40] J. M. Gershoni and G. E. Palade, *Anal. Biochem.* **131,** 1 (1983).

[41] R. Radi, K. M. Bush, T. P. Cosgrove, and B. A. Freeman, *Arch. Biochem. Biophys.* **286,** 117 (1991).

[42] D. A. Davis, K. Dorsey, P. T. Wingfield, S. J. Stahl, J. Kaufman, H. M. Fales, and R. L. Levine, *Biochemistry* **35,** 2482 (1996).

[43] J. A. Thomas, Y.-C. Chai, and C.-H. Jung, *Methods Enzymol.* **233,** 385 (1994).

[44] K. L. Maier, A. G. Lenz, I. Beck-speier, and U. Costabel, *Methods Enzymol.* **251,** 455 (1995).

[45] R. L. Levine, L. Mosoni, B. S. Berlett, and E. R. Stadtman, *Proc. Natl. Acad. Sci. U.S.A.* **93,** 15036 (1996).

[46] M. A. Smith, L. M. Sayre, V. E. Anderson, P. L. R. Harris, M. F. Beal, N. Kowall, and G. Perry, *J. Histochem. Cytochem.* **46,** 731 (1998).

[47] L. J. Yan and R. S. Sohal, *Anal. Biochem.* **265,** 176 (1998).

[48] C. Leewenburgh, J. E. Rasmussen, F. F. Hsu, D. M. Mueller, S. Pennathur, and J. W. Heinecke, *J. Biol. Chem.* **272,** 3520 (1998).

[49] M. A. Smith, P. L. R. Harris, L. M. Sayre, J. S. Beckman, and G. Perry, *J. Neurosci.* **17,** 2653 (1997).

[50] Y. Z. Ye, M. Strong, Z. Q. Huang, and J. S. Beckman, *Methods Enzymol.* **269,** 201 (1996).

[51] H. Ischiropoulos, L. Zhu, J. Chen, M. Tsai, J. C. Martin, C. D. Smith, and J. S. Beckman, *Arch. Biochem. Biophys.* **298,** 431 (1992).

[52] J. P. Eiserich, C. E. Cross, A. D. Jones, B. Halliwell, and A. Van der Vliet, *J. Biol. Chem.* **271,** 19199 (1996).

[53] M. K. Shigenaga, H. H. Lee, B. C. Blount, E. T. Shigeno, H. Yip, and B. N. Ames, *Proc. Natl. Acad. Sci. U.S.A.* **94,** 3211 (1997).

[54] M. E. Rosenfeld, J. C. Khoo, E. Miller, S. Parthasarathy, W. Palinski, and J. L. Witztum, *J. Clin. Invest.* **87,** 90 (1991).

[55] J. G. Kim, F. Sabbagh, N. Santanam, J. N. Wilcox, R. M. Medford, and S. Parthasarathy, *Free Radic. Biol. Med.* **23,** 251 (1997).

[56] K. J. A. Davies and M. E. Delsignore, *J. Biol. Chem.* **262,** 9908 (1987).

[57] R. E. Pacifici and K. J. A. Davies, *Methods Enzymol.* **186,** 485 (1990).

Immunoassay Reagents

TBS: 20 mM Tris/500 mM NaCl

Tween TBS: TBS containing 0.05% (v/v) Tween 20

Tween TBSA: Tween TBS containing 1% bovine serum albumin (BSA)

Primary antibody solution: We employ a monoclonal anti-DNP antibody (a mouse IgE) marketed by Sigma (St. Louis, MO). Use at a 1:1000 dilution in Tween/TBSA. Other anti-DNP antibodies are also available commercially. These need to be tested to find the best dilution for obtaining high specificy with low backgrounds.

Secondary antibody solution: biotinylated rat antimouse IgE from Southern Biotechnology Assoc. (Birmingham, AL). Use at at a 1:4000 dilution in Tween TBSA.

Immunoassay Procedure

Block the blot for 30 min using either 3% BSA in TBS or 5% milk. Incubation with the primary antibody is generally overnight at 4° or for at least 3 hr at room temperature. After treating with the anti-DNP antibody, wash the blot well with Tween/TBS. Incubate with the secondary antibody for 1 hr at room temperature. Final detection is carried out using a biotin–avidin–peroxidase-conjugated system from Vector Laboratories and a chemiluminescence reagent (e.g., from NEN/Dupont or Amersham). Note that cell and tissue extracts often contain biotin-binding proteins that will react with biotin–avidin reagents. These proteins are detected by running the immunoassay in the absence of the primary (anti-DNP) antibody and can serve as convenient internal markers. If their presence interferes with the carbonyl assay, use a secondary antibody that is conjugated directly to peroxidase.

Staining of Total Protein Bands

After completion of the immunoassay, protein bands can be visualized by washing the blots extensively (>1 day with several buffer changes) in TBS and then staining with Amido black (0.1% in 40% methanol/10% acetic acid). Destain with 40% methanol/10% acetic acid. By staining the same blot as used for the immunoassay, bands detected by staining and by immunoassay can be lined up accurately. There is no need to run replicate gels in order to visualize protein bands.

Controls

Two negative controls should be run when new samples are tested.

Control for the Anti-DNP Reaction. This is accomplished by "derivatizing" with TFA without DNPH and performing the remainder of the protocol as usual. Samples treated in this manner should show no bands in the immunoassay. If any protein bands are seen, determine by process of elimination whether they derive from reactivity with the primary antibody, secondary antibody, or the biotin–avidin detection system.

Control for Specificity of the Secondary Antibody. In our experience, most problems with "nonspecific" background derive from problems with the secondary antibody. It is a good idea to test this periodically by running the immunoassay without the primary antibody step.

In addition, positive and negative control samples should be run in all gels. Any oxidatively modified protein and its native (or sodium borohydride-reduced) counterpart can be employed. We routinely use fibrinogen that has been oxidized by exposure either to iron/ascorbate or to ionizing radiation.[27] Protocols for preparing oxidatively modified proteins can be found in the literature.[5,27,31] Use a positive control of known carbonyl content (i.e., nmol/mg protein) to obtain an estimate of the extent of oxidation of positive protein bands. Use the negative control to gauge how long to do the chemiluminescence exposure (i.e., this band should be negative at all exposure times).

[39] DNA Damage Induced by Ultraviolet and Visible Light and Its Wavelength Dependence

By Christopher Kielbassa and Bernd Epe

Introduction

DNA damage induced by solar radiation in mammalian cells consists largely of two types of modification: pyrimidine dimers and oxidative modifications.[1] Pyrimidine dimers, which can be subdivided into cyclobutane pyrimidine dimers (CPDs) and (6-4) photoproducts, are the characteristic and most abundant modifications after direct excitation of DNA, although they can also be formed indirectly by energy transfer from other excited

---

[1] J. Cadet, M. Berger, T. Douki, B. Morin, S. Raoul, J.-L. Ravanat, and S. Spinelli, *Biol. Chem.* **378**, 1275 (1997).

molecules such as carbonyl compounds.[2] Oxidative DNA damage, which includes various pyrimidine and purine modifications, sites of base loss (AP sites), and strand breaks, is generated in only low yield after direct excitation of DNA (except at very short wavelengths, e.g., 193 nm[3]). It is, however, the most frequent type of DNA damage generated by so-called indirect mechanisms, which emanate from the excitation of cellular chromophores other than DNA. These chromophores (endogenous photosensitizers) can react directly with DNA (type I reaction) or give rise to the formation of reactive oxygen species (ROS) such as singlet oxygen ($^1O_2$) or superoxide (O_2^-).[4]

For the repair of both pyrimidine dimers and oxidative DNA modifications, many bacteria and eukaryotic cells contain special enzymes that specifically or selectively recognize these lesions.[5,6] Most of these enzymes are endonucleases, which convert the substrate DNA modifications into single-strand breaks (SSB). This can be exploited for a very sensitive quantification of substrate modifications in cellular DNA using assays originally developed to detect SSB. For this purpose, cellular DNA is incubated with a purified repair endonuclease prior to the analysis, which then yields the sum of "direct" SSB and endonuclease-sensitive modifications. The methodology has been worked out using the alkaline elution technique,[7,8] the DNA unwinding technique,[9] and the single cell gel electrophoresis (comet assay)[10] for the detection of endonuclease incisions and SSB. A great advantage of the approach is that several types of DNA modification can be determined in parallel when a set of repair endonucleases with different substrate specificities is used.

This article describes the application of this approach to determine the wavelength dependencies (action spectra) of the generation of both cyclobutane pyrimidine dimers and oxidative DNA modifications in the range between 290 and 500 nm in cultured mammalian cells by means of the alkaline elution technique. In addition, DNA damage profiles (ratios of various types of DNA modification) induced by various wavelength

[2] B. Epe, H. Henzl, W. Adam, and C. R. Saha-Möller, *Nucleic Acids Res.* **21,** 863 (1993).
[3] T. Melvin, S. M. T. Cunniffe, P. O'Neill, A. W. Parker, and T. Roldan-Arjona, *Nucleic Acids Res.* **26,** 4935 (1998).
[4] B. Epe, *in* "DNA and Free Radicals" (B. Halliwell and O. Aruoma, eds.), p. 41. Ellis Horwood, Chichester, 1993.
[5] E. Seeberg, L. Eide, and B. Bjørås, *Trends Biochem. Sci.* **20,** 391 (1995).
[6] R. P. Cunningham, *Mutat. Res.* **383,** 189 (1997).
[7] A. J. Fornace, Jr., *Mutat. Res.* **94,** 263 (1982).
[8] B. Epe and J. Hegler, *Methods Enzymol.* **234,** 122 (1994).
[9] A. Hartwig, H. Dally, and R. Schlepegrell, *Toxicology* **110,** 1 (1996).
[10] A. R. Collins, S. J. Duthie, and V. L. Dobson, *Carcinogenesis* **14,** 1733 (1993).

ranges and by natural sunlight are described. Scope and limitations of the technique are discussed.

Endonuclease Preparations

Substrate specificities of some endonucleases, according to presently available data, are summarized in Table I. An increasing number of these enzymes are available commercially. Enzyme preparations suitable for damage analysis can best be obtained from *Escherichia coli* strains carrying the cloned genes under the control of an inducible promoter. In these cases, a purification of the protein to homogeneity is not always necessary. However, it is helpful when the host strains are deficient in (some of) their own repair endonucleases to avoid contaminations with other endonuclease activities. Activities of the enzyme preparations should be tested against

TABLE I

REPAIR ENDONUCLEASES SUITABLE FOR ANALYSIS OF CYCLOBUTANE PYRIMIDINE DIMERS AND OXIDATIVE DNA MODIFICATIONS INDUCED BY UV AND VISIBLE LIGHT

Repair endonuclease	Recognition spectrum[a]	
	Site of base loss	Base modifications
Fpg protein	+	8-oxoG[b]; Fapy[c]
Endonuclease III	+	5,6-dihydropyrimidines; hyd[d]
T4 endonuclease V	+	CPD[e,f]
Exonuclease III	+	—
Endonuclease IV	+	—

[a] See R. P. Cunningham, *Mutat. Res,* **383,** 189 (1997); B. Demple and L. Harrison, *Annu. Rev. Biochem.* **63,** 915 (1994); A. Karakaja, P. Jaruga, V. A. Bohr, A. P. Grollman, and M. Dizdaroglu, *Nucleic Acids Res.* **25,** 474 (1997); and M. Häring, H. Rüdiger, B. Demple, S. Boiteux, and B. Epe, *Nucleic Acids Res.* **22,** 2010 (1994).

[b] 7,8-Dihydro-8-oxoguanine (8-hydroxyguanine).

[c] Formamidopyrimidines (imidazole ring-opened purines), e.g., 4,6-diamino-5-formanidopyrimidine (Fapy-A) and 2,6-diamino-4-hydroxy-5-formamidopyrimidine (Fapy-G). Additional substrates have been described [J. J. Jurado, M. Saparbaev, T. J. Matray, M. M. Greenberg, and J. Laval, *Biochemistry* **37,** 7757 (1998)].

[d] 5-Hydroxy-5-methylhydantoin and other ring-contracted and fragmented pyrimidines.

[e] Cyclobutane pyrimidine photodimers.

[f] At 100-fold higher concentration, Fapy-A was also found to be a substrate of T4 endonuclease V [M. Dizdaroglu, T. H. Zastawny, J. R. Carmical, and R. S. Lloyds, *Mutat. Res.* **362,** 1 (1996)].

various substrate and nonsubstrate modifications. This can be done by measuring saturation curves (enzyme concentration dependencies) for the incision reaction in cell-free supercoiled DNA containing suitable reference modifications in a relaxation assay as described previously.[8] According to our experience, enzyme concentrations that fully recognize the substrate modifications in the relaxation assay (and do not lead to an incision in DNA containing no or other subtrate modifications) are also suitable for the alkaline elution assay described.

Concentrated endonuclease preparations can be kept in a sterile buffer. Repeated freezing and thawing should be avoided. Frozen aliquots in BE_1 buffer (20mM Tris–HCl, pH 7.5, 100 mM NaCl, 1 mM EDTA) are stable for months at $-70°$.

Cells, Reagents, and Light Sources

AS52 Chinese hamster ovary cells[11] are cultured in Ham's F12 medium with 5% fetal calf serum,(FCS), penicillin (100 units/ml), and streptomycin (100 μg/ml). Spontaneously immortalized human keratinocytes (HaCaT cells)[12] were kindly provided by N. E. Fusenig, Heidelberg, Germany, and cultured in Dulbecco's modified Eagle medium with 10% FCS and antibiotics.

A 1000-W xenon arc lamp (976C-0010, Hanovia) equipped with a monochromator and/or cutoff filters (3mm; Schott, Mainz, Germany), which has a rather constant output between 250 and 800 nm, is used as the light source in the determination of action spectra. For irradiations with broad-spectrum UVB, a Philips TL20/12RS lamp (maximum emission at 306 nm) is used.

The dosimetry is performed with a calibrated dosimeter (Krochmann GmbH, Berlin, Germany) equipped with a GaP photoelement.

Irradiation of AS52 Cells at Various Wavelengths

AS52 cells are suspended at 3×10^6 cells in 3 ml phosphate-buffered saline (PBS) buffer (137 mM NaCl, 2.7 mM KCl, 8.3 mM NaH$_2$PO$_4$, 1.5 mM KH$_2$PO$_4$, 1 mM CaCl$_2$, 0.5 mM MgCl$_2$, pH 7.4). Irradiation of cells is carried out in a quartz tube (size 1.2×3.5 cm) on ice as shown in Fig. 1. The irradiated area is approximately 1.4×4 cm, which allows a dose rate of approximately 4 W/m^2/nm. The transmission of the cell suspension should be >50% at all wavelengths between 290 and 600 nm. A grid mono-

[11] K. R. Tindall and L. F. Stankowski, Jr., *Mutat. Res.* **220**, 241 (1989).
[12] P. Boukamp, R. T. Petrussevska, D. Breitkreutz, J. Hornung, A. Markham, and N. E. Fusenig, *J. Cell Biol.* **106**, 761 (1988).

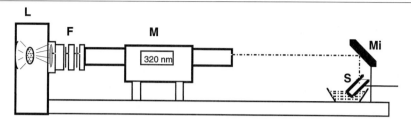

Fig. 1. Irradiation of cells with monochromatic light. L, xenon arc lamp; F, cutoff filters; M, grid monochromator (band width 8 nm); Mi, aluminum-coated mirror; S, sample holder with the suspended cells in a quartz tube in a water bath (0°).

chromator (bandwidth 8 nm) is used for irradiation at wavelengths between 290 and 330 nm. At longer wavelengths, cutoff filters (two or three in line) with half-maximal absorptions at 335, 360, 400, 420, 455, 475, and 495 nm are used to define wavelength intervals. In these cases, differences observed between experiments with cutoff filters at different wavelengths are determined and assigned to the mean wavelength of the interval. To ensure that wavelengths below the nominal setting of the monochromator (or below the wavelength intervals defined by the filters) do not contribute significantly to DNA damage, tests with additional cutoff filters, the specified half-maximal absorbance of which is ~10 nm below the monochromator setting (or at the lower end of the wavelength interval), are carried out: the damage is assigned to the monochromator setting (or to the mean wavelength of the interval defined by the cutoff filters) only if the additional filters has no effect. Otherwise, another cutoff filter of the same type is added and the experiment is repeated.

Irradiation times are chosen that generate an extent of damage that is in the detection range of the alkaline elution assay (0.05–1 modification/10^6 bp) and in the linear part of a dose–response curve. For the induction of oxidative base modifications sensitive to Fpg protein in AS52 cells in the visible range of the spectrum, the latter condition is only fulfilled for doses that generate less than 0.2 modifications per 10^6 bp. To avoid sedimentation during long irradiation times, cells in the quartz tube are resuspended every 5 min.

Irradiation of HaCaT Cells with Natural Sunlight

Exposures of HaCaT cells to natural sunlight are carried out in PBS buffer in open culture dishes at 0°. One minute illumination at 88,000 lux from clear sky during midday in Würzburg (50' latitude) in spring corresponds to approximately 1352 J/m^2 between 400 and 420 nm, equivalent to 8800 J/m^2 between 400 and 500 nm.

Quantification of DNA Modifications by Alkaline Elution

Principle

The alkaline elution assay follows essentially the protocol of Kohn *et al.*[13] Cells to be analyzed are lyzed on a membrane filter and all cell constituents other than DNA are eluted from the filter with the lysis solution. DNA remaining on the filter is then eluted under alkaline conditions (pH 12.2). When DNA contains SSB or when it is incised by a repair endonuclease in a preceding incubation, the elution is more rapid. Within a certain range, the elution rate is proportional to the average number of SSB (plus endonuclease-sensitive sites) per base pair. The assay is calibrated by means of ionizing radiation (6 Gy induce 1 SSB per 10^6 bp).

Procedure

The protocol has been described in detail previously[8,14] and is outlined here only briefly. Cells (1×10^6) are applied to a polycarbonate membrane filter (25 mm diameter, 2-μm pore size). Cells are washed twice with 3 ml ice-cold PBSG without Ca^{2+} and Mg^{2+}. Subsequently, a lyzing solution (0.1 M glycine, 0.02 M Na_2EDTA, 2% sodium dodecyl sulfate, pH 10.0) is pumped through the filter for 1 hr at 25°. The lysis buffer is washed out completely with $5\times$ 5 ml BE_1 buffer, pumped through the filter at maximum speed. Subsequently, 2×1 ml repair endonuclease solution is pumped through the filter, with the second milliliter for 30 min at 37°. To quantify direct SSB, this step is carried out without endonucleases. After washing with BE_1 buffer, 5 ml lyzing solution containing 500 μg/ml proteinase is passed through the filter at 25° for 30 min. After another washing with BE_1 DNA is eluted at 2.3 ml/hr with a solution of 20 mM EDTA (acid form) adjusted to pH 12.2 with tetraethylammonium hydroxide and collected for 10 hr in 4.6-ml fractions. DNA eluted and retained on the filter is quantified after neutralization by Hoechst 33258 fluorescence measurement (excitation at 360 nm; emission at 450 nm; final dye concentration 0.5 μM). The number of modifications per 10^6 bp is calculated from the slopes of the elution curves after subtraction of the slopes obtained with unirradiated control cells. Experiments without repair endonuclease (to obtain the number of SSB) and experiments with endonuclease (to obtain the sum endonuclease–sensitive modifications and SSB) are carried out in parallel.

[13] K. W. Kohn, L. C. Erickson, R. A. G. Ewing, and C. A. Friedman, *Biochemistry* **15**, 4629 (1976).

[14] M. Pflaum and B. Epe, *in* "Measuring *In vivo* Oxidative Damage—A Practical Approach" (J. Lunec and H. R. Griffiths, eds.), pp. 95–104. Wiley, Chichester, UK, 2000.

Wavelength dependencies (action spectra) determined with the alkaline elution technique for the induction of SSB and modifications sensitive to Fpg protein and T4 endonuclease V in AS52 cells are shown in Fig. 2. DNA damage profiles induced by various ranges of UVA plus visible light in AS52 cells and by natural sunlight and UVB in HaCaT cells are depicted in Fig. 3.

Discussion

In combination with purified repair endonucleases, the alkaline elution technique allows a quantification of both cyclobutane pyrimidine dimers and oxidative DNA modifications in cells exposed to relatively low doses of UVA, visible light, and natural sunlight. Action spectra (Fig. 2) and damage profiles (Fig. 3A) obtained indicate that pyrimidine dimers are detectable even after irradiation at wavelengths >360 nm. It is not clear whether the generation in this range of the spectrum is completely due to the very weak absorption of DNA at these wavelengths or whether energy transfer mechanisms from other cellular chromophores contribute as well.

Action spectra (Fig. 2) further demonstrate that oxidative (Fpg-sensitive) DNA base modifications in the cells are generated by two different mechanisms: at wavelengths up to ~340 nm they are generated in parallel with pyrimidine dimers as (relatively minor) by-products of the direct excitation of DNA and at longer wavelengths—with a maximum

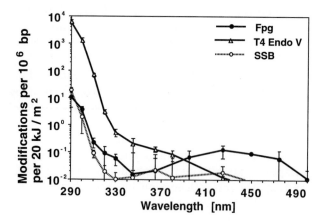

Fig. 2. Wavelength dependence of the induction of Fpg-sensitive modifications (●), T4 endonuclease V-sensitive modifications (△), and SSB (○) in AS52 cells calculated for a dose of 20 kJ/m². Data represent the means of three to seven independent experiments (±SD). Adapted from C. Kielbassa, L. Roza, and B. Epe, *Carcinogenesis* **18,** 811 (1997), with permission.

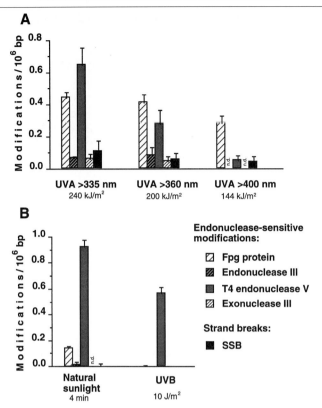

FIG. 3. (A) DNA damage profiles induced in AS52 (Chinese hamster ovary) cells by radiation from a xenon arc lamp at wavelengths above an indicated cutoff level. Doses refer to the wavelengths <500 nm. (B) Comparison of DNA damage profiles induced in HaCaT cells (immortalized human keratinocytes) by natural sunlight and by a UVB source (maximum at 312 nm). Columns indicate the number of various endonuclease-sensitive modifications (see Table I) and of (directly generated) SSB. For details, see C. Kielbassa, L. Roza, and B. Epe, *Carcinogenesis* **18,** 811 (1997) and M. Pflaum, C. Kielbassa, M. Garmyn, and B. Epe, *Mutat. Res.* **408,** 137 (1998).

between 400 and 450 nm—they are generated indirectly via excitation of other cellular chromophores, possibly porphyrins. Damage profiles induced at wavelengths >400 nm in AS52 cells (Fig. 3A) and in various other cell types[15] are dominated by base modifications sensitive to Fpg protein and closely resemble those induced by singlet oxygen and many exogenous photosensitizers under cell-free conditions and in cells.[16] For cell-free condi-

[15] M. Pflaum, C. Kielbassa, M. Garmyn, and B. Epe, *Mutat. Res.* **408,** 137 (1998).
[16] B. Epe, M. Pfaum, and S. Boiteux, *Mutat. Res.* **299,** 135 (1993).

tions, Fpg-sensitive modifications induced by singlet oxygen and several photosensitizers were shown to be mostly 8-hydroxyguanine (8-oxoG) by means of HPLC with electrochemical detection.[17]

For natural sunlight, the low quantum yield of indirect oxidative DNA damage is partly compensated by the much higher energy output in the visible range of the solar spectrum. Therefore, the exposure of human keratinocytes to natural sunlight generates endonuclease-sensitive oxidative base modifications in a yield of ~10% of that of pyrimidine dimers, i.e., in much higher yields than the exposure to UVB, as shown in Fig. 3B.

The high sensitivity of the alkaline elution technique is of particular importance for the detection of indirectly generated oxidative base modifications, as the generation was shown to be saturated in human keratinocytes and AS52 cells at levels around 0.2 modifications/10^6 bp, most probably due to photobleaching of the cellular chromophore responsible.[15,18] DNA modifications resulting from the excitation of this kind of chromophore may escape the detection by other techniques such as the quantification of 8-oxoG by HPLC with an electrochemical detector.[19]

Because of early saturation, the contribution of the oxidative base damage to the mutagenicity of natural sunlight, UVA, and visible light may be underestimated from experiments with high doses and dose rates, at which the induction of these modifications is saturated and is therefore much less relevant for the mutagenicity than at low doses and dose rates (which might be more relevant in human exposures).

A limitation of the endonuclease approach described here is that most of the repair endonucleases recognize more than one substrate (Table I). Because the enzymes frequently become more promiscuous at higher concentrations, determination of the appropriate enzyme concentrations in experiments with DNA containing various substrate modifications as described earlier are of high importance. Furthermore, experiments with additional endonucleases frequently allow to distinguish between different substrate lesions. Thus, the failure of endonuclease III and exonuclease III to recognize Fpg-sensitive modifications induced by UVA and visible light (Fig. 3) excludes that these are oxidized pyrimidine derivatives or sites of base loss. Similarly, a low recognition by Fpg protein excludes that a major part of the T4 endonuclease V-sensitive modifications are formamidopyrimidines, which have been described as additional substrates of T4 endonuclease V at very high concentrations (see footnote in Table I).

[17] M. Pflaum, O. Will, H.-C. Mahler, and B. Epe, *Free Radic. Res.* **29,** 585 (1998).
[18] C. Kielbassa, L. Roza, and B. Epe, *Carcinogenesis* **18,** 811 (1997).
[19] E. Kvam and R. Tyrrell, *Carcinogenesis* **18,** 2379 (1997).

In conclusion, the high sensitivity of the approach described allows one to obtain action spectra for the induction of several types of DNA modification by UVA and visible light in mammalian cells with a resonable resolution (~20 nm) at low doses. The damage profile observed after the exposure of mammalian cells to natural sunlight is consistent with action spectra obtained and the solar emission spectrum.

[40] Photoprotection of Skin against Ultraviolet A Damage

By HANS SCHAEFER, ALAIN CHARDON, and DOMINIQUE MOYAL

Introduction

Changes in lifestyle and the development of leisure activities and holiday habits have led to an increase in daily exposure of the skin to ultraviolet (UV) light, affecting an ever larger section of the population. At the same time, sun protective products have undergone improvements in their ability to protect skin against sunburn as reflected by continuous increases in sun protection factors (SPF).

Until not long ago UVA was perceived to be beneficial to the skin. Sun protection products were optimized according to their capacity to prevent acute sunburn, i.e., against the UVB-induced erythema. Only when the reasons for UV-mediated skin aging were investigated in greater depth was it realized that repetitive UVA exposure is damaging.

Thus, though there is a plethora of end points related to UVB damage, most of them being directly or indirectly linked to DNA damage, there are only few *specific* UVA-related phenomena in the skin that can be quantified.

In the field of photoprotection, considerable progress was made in the investigation of UVA damage when potent filter substances with maximum absorption in the range of longer than 320 nm became available. It became clear that some subchronic to chronic skin disorders are mainly due to UVA exposure (polymorphous light eruption being a typical and frequent example).

In relation to UV-mediated skin aging, attention should be paid to the difference in the depth of penetration into the skin of UVB versus UVA light: UVB is absorbed rather superficially, i.e., mostly in the horny layer and the epidermis, whereas UVA penetrates deep into the dermis, where it provokes the typical signs of skin aging, such as wrinkling, chronic thickening of the tissue, and dilation of small blood vessels (teleangiectasia).

METHODS IN ENZYMOLOGY, VOL. 319

In this article, emphasis is placed on those methods of quantification of the impact of UV light on the skin, which are sufficiently discriminative to distinguish between mere UVB and complete UVB + UVA photoprotection and which are of use in the assessment of the efficacy of photoprotection.

Solar Ultraviolet A

The maximal solar UVA irradiation at sea level, including the direct flux from the sun and air-scattered radiation, is 6.4 mW/cm², and maximal solar UVB irradiation is about 0.36 mW/cm² when the sun is close to its zenith. Under these conditions, UVA and UVB together represent about 5% of the total radiation energy received.[1] UVA radiation is less affected than UVB by such variables as sun elevation, altitude, or cloud cover (i.e., atmospheric filtration); its total flux is less attenuated in the morning and in the evening, when the sun is low.[2] Furthermore, the main portion of UVA radiation is not trapped by standard glass: automobile windows, verandas, conservatories, and windows in general fail to protect against UVA in contrast to UVB because the short wavelength cutoff of glass is about 320 nm.

Transmission[3]

The most obvious approach to quantify the capacity of any UV-shielding measure (shade, cloths, sunscreens) consists of measuring the transmission of UV light through the respective material by a photometer. Diffey and Robson[3] have described a standardized technique to quantify the spectral UV transmission through sunscreen products applied to a carrier matrix.

The validity of this and other methods depends critically on the specifications of the light source. This seemingly obvious caveat cannot be overemphasized: even today light sources are advertized to emit exclusively UVA, although a careful evaluation may reveal variable amounts of UVB. Furthermore, the yield and spectral distribution of light sources may not be stable over time. Thus every experiment in this field must be accompanied by measurement of the output and spectrum of the light source by a calibrated photometer.

The carrier matrix proposed by O'Neill[4] is Transpore TM tape (3M

[1] N. Robinson, "Solar Radiation." Elsevier Publishing, New York, 1966.
[2] B. L. Diffey and O. Larkö, *Photodermatology* **1**, 30 (1984).
[3] B. L. Diffey and J. Robson, *J. Soc. Cosmet. Chem.* **40**, 127 (1989).
[4] J. J. O'Neil, *J. Pharm. Sci.* **73**, 888 (1984).

Company, St. Paul, MN), which fulfills two requirements: It is translucent to UV light down to 290 nm and it has an irregular ("knobbled") surface and thereby mimics the surface of human skin after the application of sunscreen products.

The principle of the method is to measure the spectral transmission of ultraviolet radiation through the tape with and without sunscreen products applied. UV light sources, which provide a continuous spectral distribution between 290 and 400 nm, are used (e.g., unfiltered xenon arc lamps). The UV radiation is conducted by a liquid light guide to the input of a scanning spectroradiometer (e.g., Optronic Model 742, Optronic Labs. Inc., Orlando, FL). The bandwidth of the monochromator is set at 1.5 nm. The wavelength calibration can be achieved using a low-pressure mercury discharge lamp (lines at 253.7 and 435.8 nm). For input optics, a double ground quartz diffuser of 10 mm in diameter is used.

A piece of Transpore tape of 4×4 cm^2 is placed over the quartz input optics of the spectroradiometer. The intensity of the UV light transmitted through the tape is determined by recording the photocurrent in 5-nm steps from 290 to 400 nm. The tape sample is then placed onto a piece of stiff photographic film over a circular open aperture of 2 cm^2. An appropriate volume of sunscreen product is applied to achieve a final surface density of 2 μl/cm^2 by "spotting" the product at several locations over the application area of 16 cm^2. The sunscreen is spread to a uniform thickness using a gloved finger and rubbing for 10 sec with a light circular motion. The application volume is higher (ca. 3 μl/cm^2) than the final surface density, as water evaporates from the formulation during rubbing. The UV transmission is measured immediately after finishing the application.

The monochromatic protection factor at wavelength λ (nm) is calculated as the ratio of the photocurrent recorded at this wavelength in the absence of sunscreen to the corresponding photocurrent after the product has been applied. For each sunscreen product the transmittance is determined on three different pieces of tape, and mean monochromatic protection factors [PF(λ)] are calculated.

The sun protection factor (SPF) can be predicted from the transmission measurements according to the following formula:

$$\text{SPF} = \sum_{290}^{400} E(\lambda)\varepsilon(\lambda) \Big/ \sum_{290}^{400} E(\lambda)\varepsilon(\lambda)/\text{PF}(\lambda)$$

where $E(\lambda)$ is the spectral irradiance of terrestrial sunlight under defined conditions and $\varepsilon(\lambda)$ is the relative effectiveness of UV light at wavelength λ (nm) in producing delayed erythema in human skin (erythema action spectrum). Values of $E(\lambda)$ used in the following are chosen to represent

TABLE I

SOLAR SPECTRAL IRRADIANCE AND ACTION SPECTRUM FOR ERYTHEMA
IN HUMAN SKIN USED TO CALCULATE SUN PROTECTION FACTORS

Wavelength (nm)	Midday midsummer global irradiance at 40°N $(Wm^{-2}\ nm^{-1})$	Erythemal effectiveness (CIE 1987)
290	3.68×10^{-6}	1.0
300	1.28×10^{-2}	0.65
310	0.171	7.4×10^{-2}
320	0.398	8.6×10^{-3}
330	0.630	1.4×10^{-3}
340	0.68	9.7×10^{-4}
350	0.70	6.8×10^{-4}
360	0.73	4.8×10^{-4}
370	0.78	3.4×10^{-4}
380	0.83	2.4×10^{-4}
390	0.90	1.7×10^{-4}
400	0.97	1.2×10^{-4}

midday midsummer sunlight for southern Europe (latitude 40°N; solar zenith angle 20°; ozone layer thickness 0.305 cm). The erythema action spectrum has been adopted by the International Commission on Illumination (CIE) as a reference action spectrum.[5] Numerical values of $\varepsilon(\lambda)$ and $E(\lambda)$ are given in Table I.

Diffey and Larkö[2] have determined the monochromatic protection factors at 5-nm intervals for 14 different sunscreen products, including the standard reference formulations published by the American Food and Drug Administration (FDA) and the German standards organization (Deutsches Institut für Normung, DIN). (For more detail, see the original communications[2,3]).

Close agreement was observed between the *in vitro* protection factor and the published *in vivo* sun protection factor SPF (inhibition of erythema). The same approach can be followed for the investigation of the stability of sunscreens to UV light.[6]

The advantage of this method is that it allows the determination of transmission over the full UV spectrum and its relative simplicity. Furthermore, the method is particularly suited to assess sun protection factors of sunscreen combinations and finished cosmetic products. However, the method does not completely mimic the interaction of sunscreen products

[5] A. F. McKinlay and B. L. Diffey, *CIE J.* **6**, 17 (1987).
[6] B. L. Diffey, R. P. Stokes, S. Forestier, C. Mazilier, and A. Rougier, *Eur. J. Dermatol.* **7**, 226 (1997).

with the complex structure of the uppermost layer of the skin, the horny layer. For example, emulsions, which are a common basis for sunscreens, may break on the skin surface when rubbed on. These products may attain a specific distribution pattern on and within the upper layers of the horny layer that may affect their UV absorption capacity. These effects may only be measured *in vivo* (see later).

Persistent Pigment Darkening[7]

The UVA part of sun radiation has been shown to be biologically more active than previously thought and has become a genuine concern to the scientific community and health authorities, particularly due to its potential adverse effect on fair-skinned populations. As a result, protection from UVA has to be improved. The sunscreen industry has been challenged to meet the demand by developing and marketing new ingredients and appropriate formulations.

UVA radiation has weak erythemogenic activity. The UVA minimal erythema dose (UVA-MED) varies with skin type from 20 to 80 J/cm^2, whereas UVB-MED ranges from 20 to 70 mJ/cm^2.[8] However, taking into account the relative high level in the sun UVA flux at the zenith angle (about 16 times that of UVB), the effective erythemal potential of UVA amounts to about one-fifth of that of UVB (erythemal effectiveness: UVB 84%, UVA 16%, UVB:UVA ratio 5.2:1). This is not negligible: 2 hr of exposure of fair skin in the sun without protection may result in excess of one UVA-MED in fair skin. The SPF index remains the criterion of choice to substantiate sunscreen protection because it already takes into account the whole UV erythemal effect and more than 80% of the potential harmful effects. However, the contribution of UVA radiation to the erythemal index amounts to only about 15% of the total UV, whereas it represents more than 90% of the sun UV radiant energy. Thus, the SPF is not an accurate reflection of the protective capacity of a product, particularly for those filters designed to protect against UVA damage.

Persistent pigment darkening (PPD), or the stable portion of immediate pigment darkening (IPD), was proposed as an end point to measure UVA damage as it occurs quickly, is readily inducible, and is visible. This pigmentation occurs in the uppermost nonviable layer of the skin, the horny layer, and is due to a Maillard reaction, i.e., is unrelated to melanin production

---

[7] A. Chardon, D. Moyal, and C. Hourseau, *in* "Sunscreens: Development, Evaluation and Regulatory Aspects." (N. Lowe, N. Shath, and M. Pathak, eds.), 2nd Ed. Dekker, New York, 1997.

[8] B. Diffey, P. Farr, and A. Oakle, *Br. J. Dermatol.* **117**, 57 (1987).

by melanocytes in the living layers. To define accurate and reliable conditions of measurements, the characteristics of UVA-induced pigmentation were studied thoroughly in various skin types, using colorimetry as an objective tool to quantify and analyze the response.

For enhanced efficacy, absorption spectra of UVA filters used in sunscreen products should include the entire action spectrum of potential skin damage. However, the action spectrum of deep skin damage has not been fully established. To overcome this uncertainty, the widest UVA spectrum possible is taken into account.

Ultraviolet Immediate Pigmentation of Skin

Ultraviolet A doses required to induce the same minimal pigmenting effects on human skin with and without protection by the test sample of a sunscreen product are determined.

Time Course

When Caucasian volunteers with very light, light, and intermediate color categories of skin (skin types I to III) are exposed to progressive doses (1–28 J/cm^2) of UVA (340–400 nm) supplied by a filtered metal halide source, pigmentation of an essentially bluish-gray hue is formed after 2 hr. This pigmentation gradually loses its blue component and a gray residue persists. The pigmentation induced can be monitored visually and colorimetrically using the L*a*b* color space of a chromameter (Minolta chromameter, Minolta, Japan). This approach of quantitative assessment has been selected because it corresponds to human color perception.[9–11]

The bluish-gray skin coloration develops during exposure, reaching its maximum at the end of exposure. A partial fading occurs rapidly within 1 hr after the end of exposure, but for UVA doses higher than about 6–10 J/cm^2, a stable residual pigmentation remains, persisting for up to several days. The dose–response relationship is presented in Fig. 1. (For further details see Ref. 7.)

The color of this residual (blue-gray) pigmentation is different from that of a delayed pigmentation [tanning = neomelanization with typical brown (dark yellow) hue] because the latter occurs only with a delay of about 2–3 days.

[9] A. Chardon, D. Moyal, M. F. Bories, and C. Hourseau, *Cosmet. Toilet.* **108**, 79 (1993).
[10] W. Westerhos, B. A. van Hasselt, and A. Kammerijer, *Photodermatology* **3**, 310 (1986).
[11] J. Weatherall and B. Coombs, *J. Invest. Dermatol.* **99**, 468 (1992).

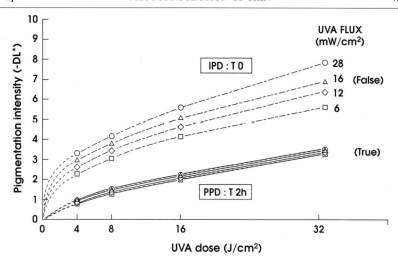

FIG. 1. Reciprocity of immediate pigment darkening (IPD) and persistent pigment darkening (PPD) versus UVA doses.

Ultraviolet A Source Definition and Efficacy Spectrum

Immediate pigment darkening and persistent pigment darkening responses are not specific to the UVA range of wavelengths; consequently, the emission spectrum of the UVA sources should be restricted to the UVA range with suitable filters. A xenon arc lamp, already used for SPF determination, filtered with a specific Schott WG335 3-mm short cutoff filter and a Schott UG11 1-mm filter to discard visible and infrared rays, is considered to be a valid source.[12] When the action spectrum of the PPD end point is multiplied by the emission spectrum of such a UVA source, the resulting efficacy spectrum is suitable for the UVA protection assessment of sunscreens, as it covers the whole UVA range.

Definition of a UVA Protection Factor Protocol

On the basis of an established data set for UVA-induced immediate skin pigmentation, the essential parameters of a suitable protocol for the UVA protection assessment of sunscreen products can be defined as follows.

Conditions concerning the test area (back), size of test sites, product application (2 mg/cm²; surface at least 35 cm²), selection of volunteers, inclusion and exclusion criteria, follow the same current standards as those for the SPF determination, unless otherwise specified. The PPD, which is

---

[12] C. Cole, *J. Am. Acad. Dermatol.* **30,** 729 (1994).

the lasting stable portion of the IPD skin response, is used as an end point. At least 10 (maximum 20) subjects with skin types II, III, and IV are used.

Six UVA doses ranging from 4 to about 30 J/cm^2 are applied to unprotected skin of the volunteer's back in a 25–50% geometric progression; a wide range of UVA doses allows the same dose range to be suitable for any skin type. On protected skin the UVA doses are multiplied by the expected UVA protection factor (UVA-PF) of the test product. This is to ascertain that a measurable pigmentation is produced despite the protective capacity of the filter substance. After exposure, the products, particularly those including pigments and dyes, are gently removed with a mild cleansing lotion; the volunteers are requested to rest for the time lag period before observation and to avoid rubbing the test area. Observation of the skin response is delayed for 3 ± 1 hr after the end of the last exposure.

Visual observation of UVA-induced progressive pigmentation is performed under standard room and illumination conditions. Minimal pigmenting doses (MPD) are determined in a simultaneous observation of protected (MDPp) and unprotected areas (MPDu) by paired matching of pigmentation thresholds, with interpolation between successive doses. Colorimetry, using a colorimeter (Minolta Chromameter, see earlier discussion), also serves to quantify pigmentation.

Ultraviolet A Protection Factor

The individual UVA protection factor (UVA-PFi) for a product is calculated as the ratio of MPDp to MPDu. The UVA-PF of the product is the arithmetic mean of the individual UVA-PFi of n volunteers; n may vary between 10 and 20, provided the standard error of the mean (SEM) is kept below 10% of the mean. The method is calibrated using neutral density physical filters as standards with respective absorbances giving nominal UVA-PF values of about 3 and 7.

The importance of the time lag between the end of the irradiation and the reading of the pigmentation is underlined by the fact that results at T_0 may have considerable interindividual variability, but tend to converge at 2 hr and beyond, which results in reduced variability. The pigmentation decreases slightly at 24 hr, whereas the variability tends to increase: the mean variability among volunteers was about 18% at 2 hr and 28% at 24 hr.

Ultraviolet A Filters

The dose effects of various UVA sunscreens in similar basic formulations were evaluated separately, according to the protocol defined earlier, and are presented in Fig. 2.

The vehicle as well as a pure UVB filter provided a very small UVA-

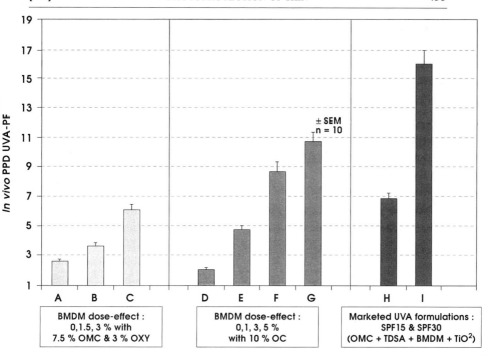

Fig. 2. *In vivo* UVA-PF of UVA-protective sunscreens: Increasing concentrations of various single UVA sunscreen formulations, tested with standard protocol based on PPD response, yield increasing UVA-PF values. The protective efficacy of the vehicle is not significant. Mexoryl SX and Parsol 1789 are significantly more efficient than oxybenzone. OC, octocrylene; OXY, oxybenzone; OMC, octyl methoxycinnamate (Parsol MCX); BMDM, butyl methoxydibenzoylmethane (parsol 1789, Avobenzone) DSA, terephthalylidene dicamphor sulfonic Acid (Mexoryl SX).

PF (about 1.3). Benzophenone-3 and Parsol 1789 provided good UVA protection, whereas the protection by Mexoryl SX was more pronounced.

With the same method, a strong increase in efficacy was found with various combinations of Mexoryl SX and Parsol 1789.

Chlorpromazine Induced Phototoxicity in Skin Equivalents[13]

Phototoxicity is a term used to describe all nonimmunologic photoinduced skin reactions. In most cases of phototoxicity, an endogenous or exogenous chemical (chromophore) absorbs light and transfers the energy to, or reacts in the excited state with, cellular components.

[13] C. C. Cohen, K. G. Dossou, A. Rougier and R. Roguet, *Toxicol. in Vitro* **8,** 669 (1994).

Chlorpromazine, a phenothiazine tranquilizer, is known to be phototoxic *in vivo*. This adverse reaction occurs through the absorption of UVA radiation by chlorpromazine metabolites.

Reconstructed Human Epidermis (Episkin)

Episkin is provided by Imedex (Chaponost, France) as a kit containing 12 units of epidermis. One unit consists of collagen matrix (types I and III) attached to the bottom of a plastic chamber that is coated with a thin layer of collagen IV. Human adult keratinocytes seeded on this dermal substitute are grown submerged for 3 days in the medium [DMEM/HAM F12 (3 : 1) + 10% fetal calf serum] and exposed at the air–liquid interface for 10 days to develop a fully differentiated epidermis.

UVA Radiation Source

The UVA source in this case was a 1000-W solar simulator Oriel (Model 68820) presenting a 102 × 102-mm size beam. In combination with additional filters—Oriel n° 81050 (= Schott UG11-1mm) and Oriel n° 81019 (= Schott WG 335-3 mm)—it emits in the 320- to 400-nm range with a maximum at 360 nm. Irradiation (13 mW/cm^2 sec) is monitored with a Osram Centra UV meter.

Tested Formulas

Two sunscreen formulas (Anthelios T, formula A; Anthelios L, formula B, both La Roche-Posay Pharm. Laboratories) were tested for their capacity to protect the epidermis from the phototoxic reaction. Formula A contains Mexoryl SX and titanium dioxide, and formula B contains Mexoryl SX, Parsol 1789, Eusolex 6300, and micronised titanium dioxide. Clinical testing had been performed previously to determine their respective UVB and UVA protective indexes, which are reported in Table II.

TABLE II
CLINICAL PROTECTIVE INDEXES

Anthelios[a]	SPF[b]	IPD[c]	PPD[d]
T	25	20	5
L	60	55	12

[a] La Roche-Posay/Pharmaceutical Laboratories.
[b] Sun protection factor.
[c] Immediate pigment darkening.
[d] Persistent pigment darkening.

Incubation

One hundred milliliters of an aqueous chlorpromazine solution are applied to the Episkin surface (1 cm^2), which is subsequently incubated at 37° under 5% CO_2 for 24 hr. Drying of the surface is achieved with a hair dryer, and 2 mg/cm^2 of each sunscreen formula is applied and incubated for 1 hr before UVA irradiation. Experiments are run in duplicate. After irradiation, the samples are rinsed with phosphate-buffered saline (PBS) and incubated overnight at 37° under 5% CO_2. Cell viability is assessed using the MTT conversion test described by Mosmann,[14] and the release of interleukin (IL)-1α into the culture medium is measured.

For the determination of cytotoxicity, the Episkin chambers are placed in 12-well culture dishes containing 2 ml MTT solution/well (0.3 mg MTT/ml test medium) and incubated at 37° for 3 hr. A punch biopsy (diameter 8 mm) is performed, and formazan precipitates are extracted with acidified isopropanol (0.04 N HCl); optical density is measured at 570 nm. Cytotoxicity is given as IC_{50} (concentration inhibiting 50% of MTT conversion by mitochondria).

IL-1α release into the culture medium is evaluated by an EIA kit (Enzyme Immuno-Assay) provided by Amersham International (Bucks, UK).

Results

Chlorpromazine Dose–Response and UVA Irradiation Dose

Figure 3 gives the phototoxic reaction relative to the chlorpromazine concentration. Chlorpromazine induced a cytotoxic reaction in Episkin at a IC_{50} (concentration killing 50% of the cells) of 97 nmol/epidermal equivalent. When Episkin treated with chlorpromazine was exposed to a UVA irradiation dose of 50 J/cm^2, IC_{50} dropped to 45 nmol/epidermal equivalent.

Thus a concentration of 50 nmol per epidermal equivalent was selected for use in the subsequent experiment. This concentration is not cytotoxic in the dark. However, in combination with 50 J/cm^2 UVA irradiation, it induced 50% cytotoxicity, revealing a phototoxic effect.

Efficacy of Sunscreen Formulas against Phototoxicity

The two sunscreen formulas A and B mentioned earlier were evaluated for their capacity to protect from the onset of the phototoxic reaction in

[14] T. Mosmann, *J. Immunol. Methods* **65,** 55 (1983).

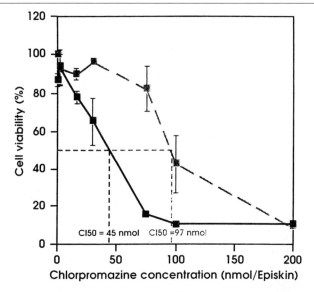

FIG. 3. Cytotoxicity and photocytotoxicity of UVA in reconstituted epidermis "Episkin." Solid line, UVA; dashed line, UVA + chlorpromazine (50 nmol); CI, cytotoxicity index.

comparison with their respective placebos. The UVA dose of 50 J/cm^2 induced a phototoxic reaction, resulting in a decrease of 50% in cell viability when chlorpromazine-incubated Episkin wells were exposed to the solar simulator either without any application or treated with 2 mg/cm^2 of placebos of A or B. The placebos allow phototoxic cellular damage to occur with no difference in cytotoxicity to nonirradiated and chlorpromazine-treated (50 nmol) specimens.

The two sunscreen formulas were evaluated against a 70-J/cm^2 UVA dose in an attempt to differentiate their protective potential against a UVA-induced chemical phototoxic reaction. As shown in Fig. 4, the phototoxicity was increased further in the irradiated Episkin control as well as for the placebos of A and B. Formula B proved to be photoprotective. Formula A was less effective (cell viability 65%).

Thus reconstructed skin or epidermis *in vitro* constitutes a useful tool to study the phototoxic potential of topically applied chemicals and to assess the efficacy of sunscreens, especially of formulations providing very high UVA protection. Though unproved, this approach can be reasonably proposed for the quantification of other photoprotective measures, such as protection by clothes, and window glass.

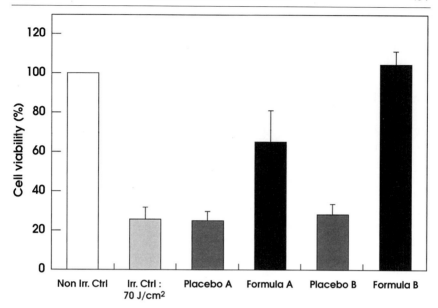

FIG. 4. Protection against photocytotoxicity in Episkin of UVA by sunscreen products. A, Anthelios T; B, Anthelios L, La Roche-Posay Laboratories.

Lipid Peroxidation

Lipid peroxidation[15] subsequent to the formation of free radicals on and in the skin by UVA is considered to be a major mechanism of cutaneous photodamage. Thus it is of interest to determine the photoprotection offered by sunscreens against lipid peroxidation. Skin, particularly epidermal equivalents grown in culture, lend themselves to such investigations,[16] as such specimens allow the application of sunscreen products under conditions approximating normal use, and lipid peroxides can be extracted from these equivalents after irradiation with and without sunscreen application.

Episkin kits (Imedex, Chaponost, France) are generally used. One kit consists of 12 wells containing 12 specimens of cultured epidermis, grown on a collagen I and III matrix coated with collagen IV. (For culture conditions see earlier discussion.)

A solar simulator can be used as the UVA source. In the following

[15] C. Cohen, R. Roguet, M. Cottin, M. H. Grandidier, E. Popovic, J. Leclaire, and A. Rougier, in "Protection of the Skin against Ultraviolet Radiations" (A. Rougier and H. Schaefer, eds.), p. 189. John Libbey Eurotext, Paris, 1998.
[16] L. Marrot, J. P. Belaidi, C. Chaubo, J. R. Meunier, P. Perez, and C. Agapakis-Caussé, *Eur. J. Dermatol.* **8,** 403 (1998).

example, a 1000-W Oriel solar simulator (Model 68820), providing a 102 × 102-mm beam, was used. In combination with the filters Oriel n° 81050 or equivalent Schott filter UG11-1 mm, and Oriel n° 81019 or equivalent Schott filter WG 335-3mm, a UVA range of 320–400 nm with a maximum at 360 nm is provided. It is recommended that the irradiation (13 mW/cm^2/sec) be monitered with an Osram Centra UV meter.

Typically, 2 mg/cm^2 of a given sunscreen product is applied in duplicate to the surface of the skin equivalents. The wells are then incubated for 1 hr at 37° under 5% CO_2 before irradiation. The samples are tested for lipid peroxidation immediately after UVA exposure without prior rinsing as the complete removal of the products from the surface is not possible. The specimens are crushed and extracted with isopropanol. Lipid peroxides are determined using the very sensitive K assay (LPO-CC, Kamiya Biomedical Company). Cumene hydroperoxide is used to establish a calibration curve. When the equivalents are depleted of glutathione by the glutathione synthetase inhibitor buthionine sulfoxamide (BSO), increased lipid peroxide formation is observed (Fig. 5), confirming the role of glutathione in the detoxification of lipid peroxides.

Protection against UVA-Induced Lipid Peroxidation

A typical study was performed using a noncytotoxic UVA exposure dose (50 J/cm^2) without BSO pretreatment. The protective effect against

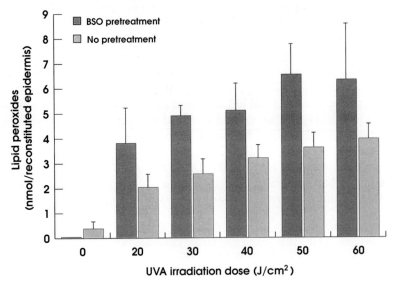

FIG. 5. Dose-related lipid peroxidation by UVA in reconstituted epidermis "Episkin."

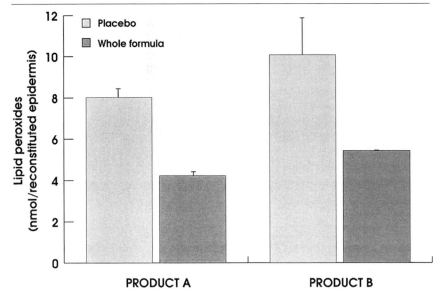

FIG. 6. Prevention of lipid peroxidation by UVA in reconstituted epidermis "Episkin" by sunscreen products versus placebo. A, Anthelios T, containing Mexoryl SX and titanium dioxide. B, Anthelios L, containing Eusolex 6300, Parsol 1789, Mexoryl SX, and titanium dioxide. (A and B, La Roche-Posay Laboratories).

lipid peroxidation of different sunscreen formulas was investigated. As shown in Fig. 6, treating the reconstructed epidermis with formula A or B resulted in a significant reduction (50%) in lipid peroxide production compared to treatment with their respective placebos. Because the placebos of formulas C and D were unavailable, such an evaluation on formulas C and D was not possible. Figure 7 clearly shows that the four formulas investigated were different with regard to their ability to protect the epidermis from the *in situ* production of lipid peroxides. As compared to the amount of lipid peroxides produced by nonprotected irradiated epidermis, significant lipid peroxide formation was induced by UVA when using formula C. No significant rise in lipid peroxides was found when using formulas B and D. Only limited protection was provided by formula A.

Immnosuppression Measured by the Delayed-Type Hypersensitivity (DTH) Reaction *in Vivo*[17]

Numerous studies have addressed the effects of UVB light on immune responses. For a report on different mechanisms and end points, see Ref.

[17] D. Moyal, *Eur. J. Dermatol.* **8,** 209 (1998).

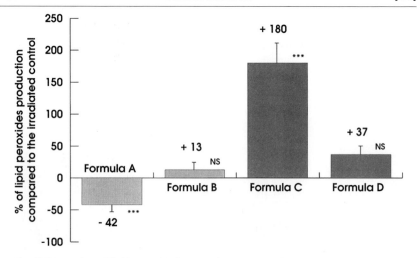

F<small>IG</small>. 7. Prevention of lipid peroxidation by UVA in reconstituted epidermis "Episkin" by four different sunscreen products. A and B as in Fig. 6, C containing Eusolex 232, Uvinyl T150, Parsol MCX, Parsol 1789, and titanium dioxide. D containing titanium dioxide and zinc oxide. ***Highly significant, $p > 0.0001$ (Student Newmann–Keuls test). NS, nonsignificant.

18. The following approach allows the assessment of photoprotection against immunosupression by total UV radiation as well as by UVA.

The DTH skin reaction is assessed *in vivo* by exposure to a cocktail of recall antigens. For this purpose, a multitest kit like the one provided by Institut Pasteur/Merieux, France, can be used. The seven test antigens are tetanus toxoid, diphtheria toxoid, *Streptococcus* antigen, tuberculin, *Candida albicans,* Trichophyton, and *Proteus* antigens. A 70% glycerol solution serves as the inactive control. Volunteers are recruited according to a study protocol approved by an ethics committee and upon signed informed consent. Typically their skin type is II to III and their age is between 18 and 40 years. Their initial DTH response (see later) should be greater than 5. Exclusion criteria include conditions or medications causing immunomodulation or risk of photosensitization. When a comparison of photoprotection to UVA versus UVA + UVB is envisaged, three groups can be constituted. Group 1 is not exposed to UV; this control group allows the variability of reactivity to the multitest without UV exposure to be assessed. Group 2 is exposed to complete UVB + UVA. Group 3 is exposed to UVA alone.

In a second step, the efficacy of two sunscreens in preventing the immu-

[18] T. H. Kim, S. E. Ullrich, H. N. Ananthaswamy, S. Zimmerman, and M. L. Kripke, *Photochem. Photobiol.* **68,** 738 (1998).

nosuppression induced by UVB + UVA versus UVA exposure in additional groups (4 and 5) can be compared. A control group 6 is exposed without sunscreen.

Areas of approximately 600 cm^2 on the middle of the back and on the abdomen are irradiated using a metal-halide Mutzhas lamp with a wide beam. Groups 2, 4, 5, and 6 are exposed to an emission spectrum ranging from 290 to 390 nm. Group 3 is exposed to a UVA emission spectrum ranging from 320 to 400 nm. Groups 2 and 6 are exposed on 10 separate occasions. The UV dose is increased progressively from 0.8 individual minimal erythema dose (MED) to 2 MED. The total dose is then 14.5 MED per body side, and the UVA dose delivered within this total UV dose is 75 J/cm^2. For groups 4 and 5, these UV doses are multiplied by 4.5, which corresponds to half the SPF of the sunscreen products. It is recommended to choose UV doses that do not exceed the protective efficacy of the sunscreens or other protective measures against UV erythema. At the last exposure, the UV dose can be 9 MED, which corresponds to the level of erythemal protection provided by the applied sunscreen with a SFP of 9. The total UV dose delivered in the given example was 58 individual MED and its UVA part corresponded to an average of 315 J/cm.^{2} For group 3 exposed to UVA alone, 12 exposures are performed. The UVA dose is increased progessively from 20 to 50 J/cm.^{2} The total UVA dose received is then 350 J/cm.^{2}

Sunscreen Products

To compare photoprotection against UVB versus complete UVA + UVB protection, two sunscreen products having the same SPF 9 but with different absorption spectra were prepared. The SPF of each sunscreen was determined using the Mutzhas lamp (290–390 nm). Sunscreen A was composed of two UVB filters (9% Uvinyl N539, 0.3% Eusolex 232) and two UVA filters, one absorbing short UVA (0.7% Mexoryl SX) and the other one absorbing long UVA (3% Parsol 1789). With such a composition, the absorption spectrum of sunscreen A covers the entire UV range and has a flat absorption spectrum profile. Sunscreen B is composed of the same UVB filters, but without UVA filter substances and with adjusted concentrations (10% Uvinyl N539, 1% Eusolex 232) to provide the same SPF as sunscreen A.

The efficacy of both sunscreens in the UVA range was assessed using the persistent pigment darkening method[7] (see earlier discussion). The UVA protection factors were, respectively, 9 for sunscreen A and 2 for sunscreen B. Sunscreens A and B were applied 15 min before each exposure at the rate of approximately 2 mg/cm^2.

Delayed-Type Hypersensitivity Tests

The test antigens are applied to the upper part of the back to one skin site before exposure and to two other sites 24 hr after the last UV exposure: a nonexposed skin site for the evaluation of systemic immunosuppression and an exposed skin site for the assessment of both local and systemic immunosuppression. Measurements of DTH test responses are made 48 hr after application of the test sample. The diameter of each positive test, identified as erythema accompanied by local induration, is measured in two directions at a 90° angle with a caliper. These diameters are then averaged. To obtain the cumulative diameter (total score) the average diameters are added for each subject.

Statistical Analysis

Differences between DTH responses after UV exposure in each group and differences between DTH responses between each group are analyzed by one-way analysis of variance (SPSS computer program).

Results

The results of the study are summarized in Table III. No significant variation was observed among the three multitests sites in control group 1 (not exposed to UV). A significant and equivalent decrease in the response to the antigens was observed in group 2 after application to exposed and nonexposed skin sites. In group 3, a significant and equivalent decrease in DTH responses was also observed. The decrease of the DTH response observed in the two groups exposed either to UVB + UVA or to UVA alone was similar on both the exposed and the nonexposed site (60 and 70%). In control group 6 exposed to UVB + UVA without sunscreen, the decrease in the DTH response on exposed and nonexposed skin sites was confirmed. In group 5, exposed after application of sunscreen B, the decrease in the response was also observed on the two skin sites. There was no significant difference between groups 6 and 5 with regard to exposed and nonexposed skin sites. In group 4, protected by sunscreen A, the DTH response was decreased slightly (-27%) on the exposed site and was unchanged on the nonexposed site.

Thus, the first part of the study showed that the response to DTH tests is reduced significantly by UVB + UVA and UVA irradiation alone. In both cases, immunosuppression is induced both locally and systemically. In the second part, group 4, having received sunscreen A, showed no significant decrease at the nonexposed site, which means that systemic immunosuppression was prevented. The observed decrease (-27%) on the

TABLE III

Delayed Hypersensitivity Test Responses: Intergroup Comparison of Total Scores[a]

Test site	Group 1 (n = 12) Control, nonexposed	Group 2 (n = 10) UVB + UVA	Group 3 (n = 11) UVA	Group 4 (n = 12) Exposed, with sunscreen A	Group 5 (n = 11) Exposed, with suncreeen B	Group 6 (n = 5) Control group UVA Exposed, without sunscreen
Pre-UV skin site S1	10 ± 2.9	9.2 ± 3.6	8.1 ± 2.1	13.1 ± 2.9	13.3 ± 3.3	13.8 ± 4.5
Post-UV skin site S2 (nonexposed)	12 ± 3.4	4.3 ± 2.5[b]	3.1 ± 1.7[b]	11.5 ± 2.4[d]	6.3 ± 1.9[c]	4.5 ± 2.3[c]
Post-UV skin site S3 (exposed)	9.8 ± 3.3	2.9 ± 1.8[b]	2.7 ± 1.3[b]	9.2 ± 1.9[c,d]	4.6 ± 2.2[c]	3.5 ± 2[c]

[a] SE.
[b] Significantly different from pre-UV and control group ($p < 0.05$).
[c] Significantly different from pre-UV ($p < 0.05$).
[d] Significantly different from group 6 and from group 5 ($p < 0.05$).

exposed site was significantly lower than in the control group, which reflects reduced local immunosuppression. Under UVB + UVA exposure, a UVB sunscreen failed to provide local and systemic immunoprotection.

These results confirm that UVA contributes significantly to inducing photoimmunosuppression. A sunscreen product providing an adequate protection in the UVA range reduces UV-induced local immunosuppression significantly and prevents systemic immunosuppression. Further, these results demonstrate that SPF, which is based on acute erythema as an end point, is not an appropriate indicator for *in vivo* assessment of the protective efficacy of sunscreen products against photoimmunosuppression.

Further End Points

The increased gene expression in mammalian cells on UV exposure, particularly to UVA, has been linked to singlet oxygen as the primary effector. Singlet oxygen appears to induce a signal transduction cascade, which depends on activation of the transcription factor AP-2.[19] In this context it is of particular interest that this activation, which mediates the expression of the human intercellular adhesion molecule 1 (ICAM-1) gene, is specific for UVA radiation.[20] In the future, this may provide a UVA-specific assay of photodamage and photoprotection.

Further end points,[16] which could not be presented here in detail, are as follows: UVA-induced p53 expression *in vivo*[21]; supercoiled circular plasmid DNA for the detection of structural alterations; cytotoxicity and genotoxicity in *Saccharomyces cerevisiae;* comet assay for the determination of DNA damage and repair in human keratinocytes; and induction of pigmentation in human melanocytes.

Conclusion

Because UVA is recognized to be an important, if not predominant, factor in chronic light damage in our Western society, it was of particular interest to describe several methods based on physical as well as biological end points that are sensitive enough to quantify the efficacy of photoprotective measures against UVA. This article presented only those methods that have proved to be of practical use in the research and development of

[19] S. Grether-Beck, S. Olaizola-Horn, H. Schmitt, M. Grewe, A. Jahnke, J. P. Johnson, K. Briviba, H. Sies, and J. Krutmann, *Proc. Natl. Acad. Sci. U.S.A.* **93,** 14586 (1996).

[20] S. Grether-Beck, R. Buettner, and Jean Krutmann, *Biol. Chem.* **378,** 1231 (1997).

[21] S. Séité, D. Moyal, M. P. Verdier, C. Hourseau, and A. Fourtanier, "Photodermatology, Photoimmunology and Photomedicine," in press.

sunscreen substances and preparations. One may therefore assume that in the future these methods will be suited for broader investigations in UVA protection, including that provided by textiles, glass, and shade.

[41] Topically Applied Antioxidants in Skin Protection

By Franz Stäb, Rainer Wolber, Thomas Blatt, Reza Keyhani, and Gerhard Sauermann

Introduction

Reactive oxygen species (ROS) and free radicals can be generated in skin by ultraviolet (UV) irradiation, ultrasonication, toxic or allergic chemical noxes, or even during normal metabolic processes of cells.[1,2] It has been assumed that more than 10^{10} ROS are generated per day and per cell under normal physiological conditions,[3] which have to be dealt with by the endogenous antioxidant system. Oxidative stress arises when the balance between antioxidant and prooxidant processes drifts in favor of a prooxidant status.[4] Some of the general events in the early phase of oxidative stress response in skin are depletion of endogenous intra- and intercellular antioxidants,[5,6] the enhancement of the intracellular hydroperoxide level,[7] and the induction of specific signal transduction pathways.[1,8]

In the early phase of the diverse cascades of oxidative stress-induced signal transduction pathways, specific proteins become phosphorylated enzymatically by kinases and/or dephosphorylated by phosphatases, leading finally to the activation of specific signal transduction factors (e.g., NFkB, AP-1, AP-2, JNK), which are involved in the expression of specific genes.[1,9] The resulting gene products can modulate, e.g., inflammatory, immune

[1] J. Fuchs, *Free Radic. Biol. Med.* **25,** 848 (1998).
[2] A. K. Campbell (ed.), "Chemiluminescence, Principles and Applications in Biology and Medicine." Ellis Horwood, Chichester, 1988.
[3] B. Chance, H. Sies, and A. Boveris, *Physiol. Rev.* **59,** 527 (1979).
[4] H. Sies (ed.), "Oxidative Stress: Oxidants and Antioxidants." Academic Press, San Diego, 1991.
[5] J. Fuchs and L. Packer (eds.), "Oxidative Stress in Dermatology." Dekker, New York, 1993.
[6] Y. Sagara, R. Dargusch, D. Chambers, J. Davis, D. Schubert, and P. Maher, *Free Radic. Biol. Med.* **24,** 1375 (1998).
[7] A. R. Rosenkranz, S. Schmaldienst, K. M. Stuhlmeier, W. Chen, W. Knapp, and G. J. Zlabinger, *J. Immunol. Methods* **156,** 39 (1992).
[8] H. Sies (ed.), *Adv. Pharmacol.* **38** (1997).
[9] S. W. Ryter and R. M. Tyrrell, *Free Radic. Biol. Med.* **24,** 1520 (1998).

suppressive, or apoptotic processes in skin.[1,8] The intracellular and extracellular structures are normally protected against oxidative stress by endogenous enzymatic and nonenzymatic antioxidant and repair systems linked closely to intracellular energy metabolism.[10–12] Vitamins C and E, as well as α-lipoic acid, uric acid, or glutathione, are examples of well-known endogenous nonenzymatic antioxidants taken up orally with food or synthesized endogenously. Catalase and superoxide dismutases, as well as enzymes of the glutathione redox cycle (glutathione peroxidase and reductase), are representatives of the enzymatic antioxidant system in skin.[1,13] It is evident that oxidative stress plays a central role in skin aging and in the pathogenesis of many skin diseases.[5,8,14] Therefore, it appears obvious to apply antioxidants topically, which might counteract the early phase of an oxidative stress response in skin cells to maintain the homeostasis of the endogenous redox system. However, there is poor evidence in the literature, as yet, that topically applied antioxidants have a proven *in vivo* antioxidant efficacy in human skin.[5,15–17]

The aim of this article is to present an example of a screening strategy for the successful selection of *in vivo* functional topical antioxidants using special *in vitro, ex vivo,* and *in vivo* methods. The screening protocol used is subdivided into three steps. In the primary step, the efficacy of antioxidants is evaluated *in vitro* in cultures of primary skin cells, and the biocompatibility of the selected efficient antioxidants is proved *in vitro* and *in vivo*. In the second step, the antioxidant efficacy is tested *ex vivo* in a clinical pilot study using biopsies from *in vivo*-pretreated skin sites. In the third step, the *in vivo* efficacy of preselected antioxidants has been confirmed in clinical studies on panels of human volunteers.

[10] B. C. Ames, M. K. Shigenaga, and T. M. Hagen, *Proc. Natl. Acad. Sci. U.S.A.* **90,** 7915 (1993).
[11] C. Richter, in "Molecular Aspects of Aging" (K. Esser, G. Martin, and J. Wiley, eds.), p. 99, 1995.
[12] M. Podda, M. G. Traber, C. Weber, L.-J. Yan, and L. Packer, *Free Radic. Bio. Med.* **24,** 55 (1998).
[13] Y. Shindo, E. Witt, D. Han, W. Epstein, and L. Packer. *J. Invest. Dermatol.* **102,** 122 (1994).
[14] I. Hadshiew, F. Stäb, S. Untiedt, K. Bohnsack, F. Rippke, and E. Hölzle, *Dermatology* **195,** 362 (1997).
[15] F. Bonnia and L. Montenegro, *SÖFW J.* **122,** 684 (1998).
[16] J. Fuchs, N. Groth, and T. Herrling, *Free Radic. Biol. Med.* **24,** 643 (1998).
[17] L. Packer, *Methods Enzymol.* **300** (1999).

In Vitro Prescreening of Antioxidants

Based on the results of previous *in vitro* and *in vivo* efficacy studies on antioxidant preparations[14,18] and data from the literature,[1,19–21] three key parameters of the early afferent phase of oxidative stress response in primary human skin cells were selected for the first step of efficacy screening. The aim of the selection of these parameters was to focus very closely on the primary effects of oxidative stress in cells to identify antioxidants counteracting the oxidative stress-induced intracellular cascade as early as possible.

The selected *in vitro* parameters are (a) the intracellular level of thiols (THS), mainly represented by glutathione, used as an index of the intracellular antioxidant status; (b) the degree of intracellular phosphotyrosine (PT) to evaluate the overall activation status of the early afferent part of the oxidative stress-induced signal transduction pathways; and (c) the mitochondrial membrane potential (MMP) in cells as a sensitive marker for the integrity of mitochondria, important for the cellular energy supply.

In Vitro Methods

Primary human epidermal cells (HEC) and primary human dermal cells (HDC) are prepared enzymatically from skin biopsies using a standardized dispase (Boehringer Mannheim) and/or trypsin (Sigma) digestion technique.[22] Epidermal cells are cultured in K-SFM (Life Technologies) and dermal cells in DMEM (Life Technologies), respectively. After the incubation of primary cells in 384-well microtiter plates (5×10^3 cells/well) with putative antioxidants (i.e., 50–450 μM, for 24 hr), the supernatants are removed by washing in phosphate-buffered saline (PBS), and the oxidative stress experiments are started in PBS. The chemical induction of oxidative stress is conducted using H_2O_2 (0.25–2 mM, for 30 min). For the physical induction of oxidative stress, a dose of 20 J UVA/cm^2 (Dermalight 2020, Hönle) is applied.

Immediately after exposure to H_2O_2, the cells are washed in PBS and incubated with the specific fluorescent probes corresponding to the method

[18] U. Kiistala, *J. Invest. Dermatol.* **50,** 129 (1968).

[19] I. A. Cotgreave, M. Weis, M. Berggren, M. S. Sandy, and P. W. Moldeus, *J. Biochem. Methods* **16,** 247 (1988).

[20] G. Papadopoulus, G. V. Z. Dedoussis, G. Spanakos, A. D. Gritzapis, C. N. Baxevanis, and M. Papamichail, *J. Immunol. Methods* **177,** 101 (1994).

[21] Cossariza, M. Baccarani-Contri, G. Kalashnikova, and C. Franceschi, *Biochem. Biophys. Res. Commun.* **197,** 40 (1993).

[22] F. Stäb, G. Lanzendörfer, U. Schönrock, and H. Wenck, *SÖFW J.* **124,** 10, 604 (1998).

used for the detection of the three test parameters (THS, PT, and MMP). The spectrometric detection of intracellular fluorescence is carried out in black 384-well microtiter plates (Cliniplates 384, Labsystems) for 30 min after the induction of oxidative stress, using the fluorescence spectrometer system Fluoroscan Ascent (Labsystems).

For the detection of intracellular THS, the probe monobromobimane (mbrb; Molecular Probes, Inc.) is used, which penetrates readily into living cells and becomes fluorescent by a specific irreversible interaction with intracellular thiol groups.[19] After washing in PBS, stressed cells are incubated with 60 μM mbrb for 10 min at 37° in 384-well microtiter plates. mbrb fluorescence is measured spectrometrically (Ex 390 nm/Em 460 nm).

The intracellular level of PT is determined by applying a specific immunofluorescence technique. After washing in PBS, cells are pretreated with methanol/acetone (1 : 1) at −20° for 10 min, washed again, and preincubated with 1% bovine serum albumin in PBS for 20 min at 4° in 384-well microtiter plates. Finally, intracellular PT are detected fluorometrically (Ex 485 nm, Em 530 nm) using a FITC-labeled specific primary anti-PT antibody (clone PT-66, Sigma).

The mitochondrial membrane potential is measured in cultures of primary skin cells using the styryl dye JC-1 (Molecular Probes, Inc.). Fluorescent probe JC-1 is a dual-emission potential-sensitive probe exhibiting a broad excitation spectrum and emits a green fluorescence at low MMP (Em 530 nm) and a red fluorescence (Em 590 nm) at high MMP. JC-1 can be used specifically to quantify alterations in MMP in living cells.[21] After stress experiments are conducted, cells are washed and incubated with JC-1 (10 μg/ml) for 20 min, and the ratio between green and red fluorescence is determined spectrometrically in 384-well microtiter plates.

In addition to these *in vitro* assays, the probe ethidium homodimer-1 (Molecular Probes, Inc.) is used routinely in our *in vitro* screening system for evaluating membrane integrity[20] and the probe Calcein-AM (Molecular Probes, Inc.) for measuring the intracellular esterase activity and vitality of cells[20] as internal standards and controls to calculate, e.g., the cytotoxic or mitogenic potential of test substances applied in *in vitro* assay systems.

Statistics

In every test cycle, unstressed controls (without H_2O_2 or UVA) as well as untreated controls (without antioxidant) are used per microtiter plate. Data from *in vitro* studies are calculated by normalizing the mean values of five corresponding wells on the respective control values ($= 100\%$). The statistical calculations are made using the student's t test.

In Vitro Results

Depending on the quality and quantity of the oxidative stressor applied, as well as the cell type and the assay system used for *in vitro* studies on antioxidant efficacy, antioxidants are more or less effective *in vitro*. If very low stressor concentrations (i.e., <0.1 mM H_2O_2n) are applied in the *in vitro* assay systems used, we found that even antioxidants with very poor *in vivo* efficacy (e.g., vitamine E acetate, magnesium ascorbyl phosphate) can be effective *in vitro* (data not shown). However, the use of low stressor concentrations (<0.1 mM H_2O_2) has proved not to be suitable for a relevant prediction of the efficacy potential *in vivo* in human skin. Therefore, the *in vitro* prescreening assays are conducted at rather high doses of oxidative stress (i.e., $0.25–2$ mM H_2O_2 or 20 J UVA/cm^2) to separate highly efficient antioxidants from moderately efficient ones.

For evaluation of the H_2O_2-induced oxidative stress response, the levels of intracellular thiols and phosphotyrosines have been found to be the most sensitive and suitable parameters in the *in vitro* screening system used (Fig. 1), whereas for UVA-induced oxidative stress experiments, the parameters intracellular thiol level and mitochondrial membrane potential appear to be the most suitable ones for reproducibility and sensitivity (Fig. 2).

Furthermore, we found that the sensitivity of primary human dermal cells against H_2O_2- or UVA-induced oxidative stress has been significantly higher than that of primary human epidermal cells (data not shown). These findings correlate closely with a significantly higher antioxidant status *in vivo* described for human epidermis compared with the dermis.[12]

Using the *in vitro* screening protocol mentioned earlier, we have been able to identify flavonoid α-glucosyl rutin (AGR; Fig. 3) and ubiquinone 10 (Q10) as being significantly efficient in the protection against oxidative stress *in vitro* (Figs. 4 and 5). However, the efficacy pattern of AGR and Q10 appears to differ for the intracellular thiol level (THS), mitochondrial membrane potential, and phosphorylation of intracellular tyrosines (PT) parameters.

In comparison with equimolar ratios of vitamin E acetate or magnesium ascorbyl phosphate as examples of widely used antioxidants in topical preparations, AGR and Q10 protect the primary human epidermis as well as dermis cells against oxidative stress in our *in vitro* assay system. Interestingly, α-glucosyl rutin has proved to be significantly more effective in the inhibition of oxidative stress-induced depletion and in bolstering of intracel-

a

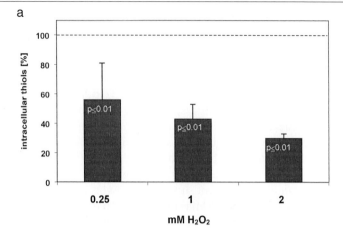

b

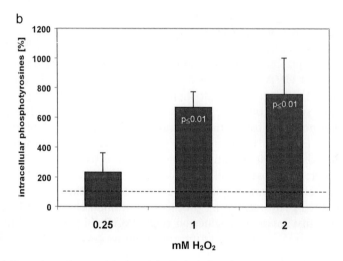

Fig. 1. Dose-dependent modulation of the H_2O_2-induced stress response in primary human epidermis cells. Intracellular parameters measured: (a) level of thiols and (b) level of phospho-tyrosines. Data are normalized on unstressed controls (---).

lular thiols (Fig. 4a) as well as in protection against stress-induced tyrosine phosphorylation (Fig. 4b), whereas Q10 seems to be more effective in the protection of MMP against UVA-induced damage (Fig. 5).

These *in vitro* findings indicate a different mode of action of AGR and Q10, depending on specificity in radical scavenging as well as solubility and localization within the intracellular compartments. In contrast to rutin or

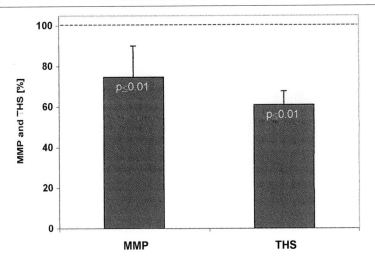

FIG. 2. Modulation of the UVA-induced stress response (20 J UVA/cm^2) in primary human epidermis cells. Intracellular parameters measured: level of thiols (THS) and mitochondrial membrane potential (MMP). Data are normalized on unstressed controls (---).

ubiquinone 10, α-glucosyl rutin is a highly water-soluble flavonoid with the basic structure of quercetin (Fig. 3). Additionally, applied ubiquinone 10 has to be reduced intracellularly to ubiquinol to become effective as an anti-oxidant.

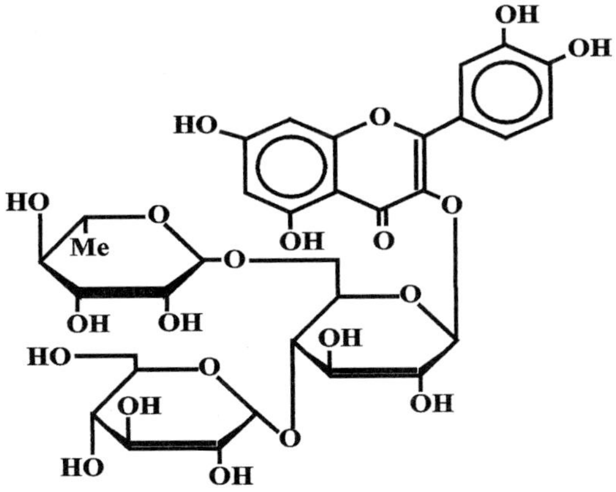

FIG. 3. Chemical structure of flavonoid α-glucosyl rutin.

a

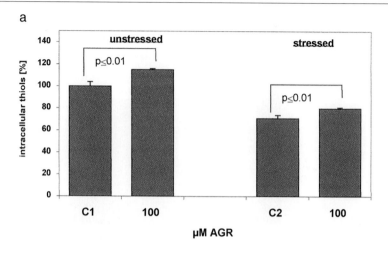

b

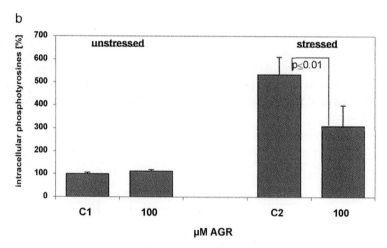

FIG. 4. Dose-dependent efficacy of α-glucosyl rutin (100 μM AGR) in the protection of primary human epidermis cells against H_2O_2-induced oxidative damage. Data of stressed (1 mM H_2O_2, 30 min) and unstressed HEC cultures. Intracellular parameters measured: (a) level of thiols and (b) level of phosphotyrosines. Data are normalized on the correspondent untreated controls (C1, unstressed; C2, stressed control without AGR).

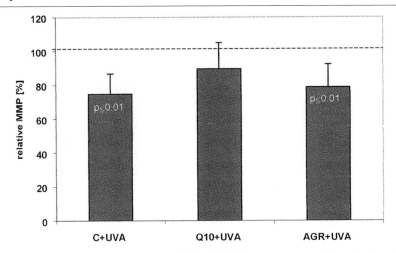

Fig. 5. Efficacy of α-glucosyl rutin (AGR) and Q10 in the protection of the mitochondrial membrane potential (MMP) in primary human epidermis cells (HEC) against UVA-induced stress (20 J UVA/cm^2). HEC were pretreated *in vitro* using 100 μM AGR (for 24 hr) or 50 μM Q10 (for 24 hr), respectively. Data are normalized on correspondent unstressed HEC cultures. C, control without AGR or Q10.

Ex Vivo/in Vitro Screening of Antioxidants

The second step in the screening protocol includes clinical pilot studies on rather small panels of human volunteers ($n = 6–10$) to preevaluate the efficacy potential *in vivo* of the antioxidants selected *in vitro* in the first screening step. For *ex vivo/in vitro* studies the MMP and intracellular thiol level parameters are selected as the most sensitive and suitable ones of the *in vitro* parameters used. Due to the limitation of freshly prepared primary skin cells available from skin biopsies, we have investigated only these two parameters routinely.

Ex Vivo/in Vitro Methods

In clinical pilot studies, the volunteers apply a verum preparation (antioxidant in a basic formulation) and the corresponding vehicle (placebo) contralaterally onto randomized skin sites (50 mg/25 cm^2) of the volar forearms twice a day for 7 days. Biopsies are taken from pretreated skin sites on day 8, 14 hr after the last application. Primary HEC are harvested from epidermal biopsies using the suction blister technique.[18] For studies on HDC, punch biopsies are taken. Suspensions of HDC and HEC are prepared enzymatically from the biopsies using the dispase and trypsin

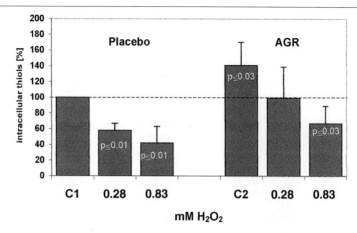

FIG. 6. *Ex vivo/in vitro* studies on efficacy of topically applied α-glucosyl rutin (AGR) in the protection of intracellular thiol levels in freshly isolated human epidermis cells (HEC) against depletion *in vitro* by H_2O_2-induced stress (0.28 and 0.83 mM H_2O_2, 30 min). HEC were freshly prepared from suction blisters ($n = 7$ volunteers) taken from human skin sites pretreated *in vivo* by topical application of 0.05% AGR containing cream or placebo (cream). AGR and the placebo preparation were applied topically twice a day for 7 days. Data are normalized on correspondent unstressed HEC from placebo-treated skin sites (C1 = 100% C2, unstressed HEC from AGR-treated skin sites).

digestion technique.[23] The *ex vivo/in vitro* analysis of the efficacy of topically applied antioxidant preparations is carried out applying the same *in vitro* parameters and methods used in the first step of the screening protocol. However, in contrast to the protocol of step one, cells freshly prepared from the skin sites are not reincubated with antioxidants *in vitro*, but only stressed *in vitro* as described in step one of the screening protocol.

Ex Vivo/in Vitro Results

Shown for the flavonoid AGR, positive results from the *in vitro* pre-screening are confirmed in *ex vivo* studies. In human epidermis cells isolated from skin sites treated with verum preparation (AGR, 0.05%), the level of intracellular thiols was protected significantly against oxidative depletion compared with vehicle or untreated skin sites (Fig. 6), whereas protection of the mitochondrial membrane potential by AGR appears to be rather poor (data not shown). In contrast to the efficacy profile of α-glucosyl rutin, the results of topically applied ubiquinone 10 (0.01%) indicate a significant efficacy in the protection of the mitochondrial membrane potential, but

[23] S. Stenn, R. L. Moellmann, J. Madri, and E. Kuklinska, *J. Invest. Dermatol.* **93**, 287 (1989).

only a low efficacy potential in the protection of the intracellular thiol level against oxidative depletion (data not shown).

In Vivo Screening of Antioxidants

The topical application of antioxidant preparations for prophylaxis and treatment of sun damage and skin aging is a standard approach in the skin care industry, but data for claim supports derive predominantly from *in vitro* studies on cell lines or animal tests.[1,17] For proof of an *in vivo* efficacy of topically applied antioxidants on human skin, however, only a few *in vivo* methods have been applied and discussed in the literature.[15–14,24] Some of these *in vivo* methods focus on very specific, single aspects of oxidative damage in skin, e.g., peroxidation of stratum corneum lipids and proteins or DNA damage, but these special parameters can hardly reflect the overall damage and oxidative stress response in skin. In this context, it appears to be more reliable to use *in vivo* techniques for the evaluation of an overall effect of topically applied antioxidants. Therefore, we have established the ultraweak photon emission (UPE) technique for noninvasive quantitative investigations of the antioxidant status in human skin.[25] UPE is defined as an emission of electromagnetic radiation in the UV/VIS/IR spectral range from excited molecules and atoms relaxing from electronically excited states to lower energy states.[2]

In Vivo Methods

A detailed description of the UPE method used and the technical setup has been published.[25] Briefly, the principal components of the UPE device are a Peltier-cooled photomultiplier, a signal processor, a data recorder, and a bifurcated light guide connected with light sources and two magnetic shutters. The shutters are used for the timing of the light excitation and for the exclusion of surrounding light and of phosphorescence from skin. For light excitation we use UVA light (50 mJ/cm^2) *in vivo*. The maximum emission range measured has been between 400 and 580 nm. For chemical induction of UPE, the topical application of a benzoyl peroxide formulation (e.g., 5% benzoyl peroxide) is suitable. The UPE measurement on skin proceeds in three steps. After baseline UPE has been recorded on skin, oxidative stress can be induced, e.g., by UVA irradiation and after the excitation light has been shut off, the decay of UVA-induced UPE can be measured. In addition, in clinical trials we have tested the efficacy of

[24] Th. Herrling, L. Zastrow, and N. Groth, *SÖFW J.* **124**, 282 (1998).
[25] G. Sauermann, W. P. Mei, U. Hoppe, and F. Stäb, *Methods Enzymol.* **300**, 419 (1999).

antioxidants *in vivo* in the prophylaxis of polymorphous light eruption (PLE). PLE is supposed to be induced by UVA-dependent oxidative stress.[14]

In Vivo Results

We have used the UPE technique routinely in years for the *in vivo* screening of product efficacy and for the characterization and differentiation of volunteers as to skin types or aging phenomena. In previous investigations we found that UVA-induced UPE from young (18–25 age) versus elderly (55–70 age) skin, as well as UPE from uninvolved atopic versus normal skin, has been enhanced significantly.[25] Topically applied α-glucosyl rutin preparations, selected by our *in vitro* and *ex vivo/in vitro* screening systems, diminished oxidative stress-induced UPE *in vivo* significantly, even at very low doses ($\geq 0.1\%$ AGR), whereas the topical application of 3% vitamin E or 3% vitamin E acetate (Fig. 7) as well as of synthetic antioxidants, i.e., 3% magnesium ascorbyl phosphate, 1% propyl gallate, or 1% butyl hydroxytoluene (BHT), very efficient stabilizers of oxidatively sensitive ingredients in cosmetics, did not show any significant antioxidant effects *in vivo*.[22] The fair *in vivo* efficacy of ingredients such as vitamin E acetate, if applied topically at a higher concentration ($\geq 5\%$) in UV protection, is derived predominantly from its UV absorption capacity measured by reflection spectrometry (data not shown). The *in vivo* inefficacy of most of the topically applied synthetic antioxidants may be caused by an insufficient

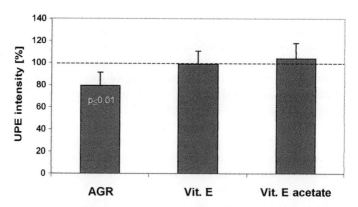

FIG. 7. UVA-induced ultraweak photon emission (UPE) *in vivo* from AGR (0.25%), vitamin E (3%), or vitamin E acetate (3%) versus vehicle-treated skin sites ($n = 20$ volunteers). Antioxidant and vehicle preparation were applied twice a day for 7 days. UPE was measured 14 hr after the last topical application. Data are normalized on UPE from vehicle-treated skin sites.

penetration or bioavailability and/or a lack of communication with the endogenous antioxidant system of human skin.

Finally, based on *in vitro* and *in vivo* data from healthy normal skin and from primary skin cells, we tested α-glucosyl rutin containing preparations for prophylaxis against PLE in a randomized, double-blind, placebo-controlled clinical trial, including 30 volunteers with a history of PLE and 28 matched pair volunteers with healthy, normal skin.[14] The findings of these clinical studies correlate significantly with *in vitro* findings and with data of parallel conducted UPE measurements on the skin of the PLE group versus a matched pair group of volunteers with normal skin, showing a highly significant reduction of UPE in both panels at verum-treated sites compared with vehicle-treated and untreated skin sites. That the *in vivo* efficacy of α-glucosyl rutin is based mainly on antioxidant, but not on UVA filter activity is supported by *in vitro* and *in vivo* efficacy data mentioned previously as well as by the results of *in vivo* reflectance spectrometry (RFS) measurements.[14,26] In these RFS studies *in vivo,* we found a very low UVA filter capacity of α-glucosyl rutin preparations used in comparison to equimolar concentrations to known UVA filters (i.e., Parsol 1789). Therefore, the UVA filter capacity of the α-glucosyl rutin concentrations used for topical treatment appears to be negligible related to the overall UVA protective activity of α-glucosyl rutin, especially in reference to the very high UVA doses ($60–100$ J/cm^2) applied topically in the PLE studies mentioned earlier.

Discussion

In comparison with other techniques applied to the analysis of the *in vitro* parameters used, e.g., HPLC analysis for the evaluation of the intracellular thiol content or Western blotting for the quantitation of intracellular phosphotyrosines, the selected *in vitro* methods appear to be rapid and convenient in handling as well as specific and sensitive enough to be suitable for high-capacity *in vitro* screening systems. The parameters selected in our protocol for *in vitro* screening and for *ex vivo/in vitro* studies seem to be sufficiently predictive for assessing the putative *in vivo* efficacy of antioxidants.

The most frequently used *in vivo* method to measure more general effects of antioxidants in skin is the UV-induced erythema.[14,19] However, this approach does not cover subclinical phenomena supposed to have chronic consequences in skin, but mainly focuses on acute clinical symptoms of UVB-induced damage. UVB-induced skin irritation is not triggered

[26] G. Sauermann, U. Hoppe, and F. Stäb, *Allergologie* **16,** 176 (1993).

predominantly by oxidative stress and does not allow a clear discrimination among the antioxidant, anti-inflammatory, or UVB filter activity of an ingredient. Another approach uses an electron spin resonance (ESR) technique to estimate the antioxidant status of human skin *in vivo* or to claim a so-called radical protection factor (RPF) for determination of the radical-scavenging activity of an antioxidant in preparations by using a test radical as a spin trap.[24] For *in vivo* application of this ESR technique, published results appear to be strongly dependent on the solubility of the test radical, its penetration depth, its intracellular and extracellular localization, and bioavailability, as well as its specifity in interaction with the diverse ROS generated by oxidative stress in different intra- and intercellular compartments of the skin.[24]

Similar considerations and objections apply for the use of this RPF in classifying antioxidant efficacy in cosmetics or in food products, especially if RPF data derived from these types of *in vitro* studies are intended to be used for the prediction and claim of antioxidant efficacy *in vivo*.[16]

To circumvent this type of pitfall for the *in vivo* relevant measurement of antioxidant efficacy, we transferred the principles of the UPE technique, well known for the *in vitro* determination of the rancidity of oils[2] or for the activity of singlet oxygen in biological systems,[27] into an *in vivo* readout system to measure oxidative stress-induced UPE in human skin.[25]

Conclusions

Shown for AGR selected by our *in vitro, ex vivo,* and *in vivo* screening protocol, some topically applied antioxidants can diminish oxidative stress in skin and can protect skin significantly against UV-induced damage and UV-induced immune dysfunctions, such as polymorphous light eruption. The findings of our comparative clinical studies on volunteers with normal and PLE skin correlate significantly with our *in vitro* and *ex vivo/in vitro* efficacy data as well as with the results of ultraweak photon emission measurements *in vivo,* demonstrating the validity and practical suitability of the protocol used for a successful search and selection of *in vivo* functional topical antioxidants.

[27] Cadenas and H. Sies, *Methods Enzymol.* **105,** 221 (1984).

[42] Erythropoietic Protoporphyria: Treatment with Antioxidants and Potential Cure with Gene Therapy

By MICHELINE M. MATHEWS-ROTH

Introduction

Erythropoietic protoporphyria (EPP) is a rare genetic disease in which singlet oxygen is involved in the production of its clinical manifestations and singlet oxygen quenchers are utilized in its treatment. In EPP, the enzyme ferrochelatase, which inserts iron into protoporphyrin to form heme, is defective.[1] This defect allows protoporphyrin to accumulate in the blood, from which it makes its way to the skin, as well as to other organs.

Protoporphyrin is known to be a powerful photosensitizer, which in the presence of light and oxygen will form free radicals and singlet oxygen. These excited species will attack cellular components, which will eventually lead, via a series of reactions that are unelucidated at this time, to the symptoms of pain, burning, or itching of the light-exposed skin experienced by patients suffering from EPP.

This article describes a method that can be used to determine the involvement of singlet oxygen in photosensitized skin and also discuss the present treatment for EPP and the possibilities for curing this disease by means of gene therapy.

An Animal Model to Demonstrate Singlet Oxygen Involvement in Photosensitized Epidermis

We developed an animal model to study the mechanisms of carotenoid photoprotection and protoporphyrin photosensitization in skin.[2] We describe this model here because the method used to detect excited species formation in skin could be used in any study of singlet oxygen photosensitization and prevention of this reaction, as the probe used is relatively specific for singlet oxygen reactions.[3]

Hairless mice (Skh-hairless-1; Temple University; now available from Charles River Laboratories) are divided into three groups: the first group

---

[1] R. J. Desnick, *in* "Harrison's Principles of Internal Medicine" (K. J. Isselbacher, E. Braunwald, J. D. Wilson, J. B. Martin, A. S. Fauci, and D. L. Kasper, eds.), p. 2073. McGraw-Hill, New York, 1994.

[2] M. M. Mathews-Roth, *Photochem. Photobiol.* **40,** 63 (1984).

[3] L. Dubertret, R. Santus, M. Bazin, and T. De Sa e Melo, *Protochem. Photobiol.* **35,** 103 (1982).

receives Roche 10% water-miscible β-carotene beadlets (Hoffmann-LaRoche, Inc., Nutley, NJ) mixed in Purina rodent chow to give a final carotenoid concentration of 3.3%; the second group receives Roche beadlets containing the carotenoid canthaxanthin also to a final concentration of 3.3%; and the third group receives the corresponding amount of placebo beadlets. Protoporphyria is induced in all mice in the three groups by the addition of 0.5%(w/w) collidine (diethyl-1,4-dihydro-2,4,6-trimethyl pyridine-3,5-dicarboxylate, Eastman Kodak Co., No. 2072) to the chow, omitting collidine from the diet every fifth day.[4] To study photochemical reactions in the skin, the skin of the back of all mice in the three groups is excised and the epidermis is separated from the dermis by immersing the piece of skin in a 60° water bath for 30 sec. The epidermal sheets are soaked for 1 hr in a solution containing 5 ml of a saturated solution of 1,3-diphenylisobenzofuran (DPBF, Sigma) in ethanol, 15 ml of dimethyl sulfoxide (DMSO, Sigma), and 30 ml of Dulbecco's phosphate-buffered saline (D-PBS). After the soaking period, the skin is rinsed in D-PBS alone. Preliminary experiments had shown that the concentrations of solutions given here work better for hairless mouse skin than those given by Dubertret et al.[3] The treated epidermis is then placed behind a 5-mm circular diaphragm, which is placed between two pieces of glass. The preparation is exposed to the light of a Sylvania Sun-Gun lamp (Type SG-1, DXN bulb), which is filtered with an infrared filter and a Corning CS 3-68 filter, which cuts off light below 520 nm, as had been done by Dubertret et al.[3] The intensity is 60 W/m^2 and the duration of exposure is 20 min. The absorption spectrum of the epidermal sheet is then read in a sensitive spectrophotometer from 300 to 700 nm. With photooxidation, the absorption of DPBF decreases as compared to that of the dark control. In mice receiving placebo and collidine, the maximum decreased 79%, whereas in mice receiving β-carotene and collidine, it decreased only 47% and in mice receiving canthaxanthin and collidine, it decreased 44%, showing that carotenoids prevented some of the reactions mediated by singlet oxygen.[2]

Treatment of EPP with Antioxidants

Sistrom et al. had demonstrated that one of the main functions of carotenoids in photosynthetic bacteria is to protect cells against lethal photosensitization by their own chlorophyll.[5] It was also shown that carotenoids could

[4] H. Baart de la Faille, R. A. Woutersen, H. van Weelden, and E. H. Baart de la Faille-Kuyper, in "Inflammation: Mechanisms and Treatment" (D. A. Willoughby and J. P. Giroud, eds.), p. 603. University Park Press, Baltimore, MD, 1980.

[5] W. R. Sistrom, M. Griffiths, and R. Y. Stanier, J. Cell. Comp. Physiol. **48,** 459 (1956).

TABLE I
DOSAGE SCHEDULE OF β-CAROTENE

Age (years)	30-mg capsules/day
1–4	2–3
4–8	3–4
8–12	4–5
12–16	5–6
16 and older	6–10

protect against lethal photosensitization in nonphotosynthetic bacteria by either endogenous[6] or exogenous[7] photosensitizers. As a result of these findings, the author hypothesized that carotenoids could protect against lethal photosensitization by protoporphyrin, and this hypothesis was confirmed in an animal model.[8] Because of these results, the author organized a clinical study to determine if the oral administration of β-carotene could prevent or lessen the photosensitivity in EPP.[9] β-Carotene treatment was found to be effective in over 80% of the patients treated, and the effectiveness of β-carotene treatment in EPP has been confirmed by others.[10] In 1975, the U.S. FDA approved the use of β-carotene for this indication. To our knowledge, this was the first successful use of an antioxidant in the treatment of a disease.

The dosage schedule of β-carotene is recommended in Table I.

It is important for EPP patients to ingest the proper formulation of β-carotene to obtain its greatest beneficial effect. It is also important to make sure the preparation is made with pharmaceutical grade β-carotene. The pharmaceutical grade formulation having the highest effective absorption is "dry β-carotene beadlets, 10%" manufactured by Hoffmann-LaRoche. At present, Lumitene (Tishcon Corp.) is the only β-carotene preparation containing Roche beadlets exclusively. Preparations using β-carotene crystals dissolved in vegetable oil, β-carotene-containing algae preparations, or dry beadlets made by other manufacturers are not suitable for use in treating EPP because these preparations are absorbed erratically by the body and, in the case of algae preparations, may contain algal components, which may act as photosensitizers. Most brands of over-the-counter β-

[6] M. M. Mathews and W. R. Sistrom, *Arch. Mikrobiol.* **35,** 139 (1960).

[7] R. Kunisawa and R. Y. Stanier, *Arch. Mikrobiol.* **31,** 146 (1958).

[8] M. M. Mathews, *Nature* **203,** 1092 (1964).

[9] M. M. Mathews-Roth, M. A. Pathak, T. B. Fitzpatrick, L. H. Harber, and E. H. Kass, *J. Am. Med. Assoc.* **228,** 1231 (1970).

[10] M. M. Mathews-Roth, *Ann. N.Y. Acad. Sci.* **691,** 127 (1993).

carotene use these less-effective preparations or may mix Roche beadlets with preparations that are not as well absorbed. These other preparations may work well for vitamin supplementation or other uses, but they do not deliver the high absorption, and thus the high blood levels, needed for the effective treatment of EPP.

Additionally, EPP patients should have an annual erythrocyte protoporphyrin test, a complete blood count, and a blood chemistry panel, including liver chemistries such as bilirubin, serum glutamic oxaloacetic transaminase (SGOT), and serum glutamic pyruvate transaminase (SGPT). In an occasional patient, protoporphyrin causes liver disease, so monitoring liver function is important. EPP patients should also not use any drug or anesthetic that causes cholestasis and should also avoid alcohol and estrogen (e.g., birth control pills, hormone replacement therapy), as these substances can have deleterious effects on an EPP patient's liver.

We have also found that the amino acid L-cysteine, which also has some excited species-quenching ability, may alleviate the symptoms of EPP.[11] We are presently conducting phase III studies. Because these studies are not complete, and FDA approval has not yet been obtained, we cannot at this time suggest the use of cysteine as a photoprotectant in EPP. In view of our results with β-carotene and cysteine, it is possible that any compound with singlet oxygen-quenching capabilities, which has no human toxicity at high doses, may be worth investigating as a potential therapeutic agent for the prevention or lessening of the photosensitivity associated with EPP.

Curing EPP: An Animal Model for Gene Therapy of EPP

Although offering highly effective photoprotective therapy, β-carotene and cysteine have no effect on the genetic defect in ferrochelatase and thus cannot protect the EPP patient against the deleterious effects of protoporphyrin accumulation. The only way that protoporphyrin accumulation can be stopped is by giving patients additional copies of the normal ferrochelatase gene, i.e., by gene therapy targeting the bone marrow. The bone marrow is the target of choice, as it has been shown that the marrow is the source of all abnormally accumulating protoporphyrin in the blood.[12] Although all cells produce abnormal levels of protoporphyrin in EPP, only the hematopoietic system secretes it into the circulation in large amounts.[12] Thus, we are pursuing preclinical studies, which must be done before gene therapy in a patient can be attempted.

[11] M. M. Mathews-Roth, B. Rosner, K. Benfell, and J. E. Roberts, *Photodermatol. Photoimmunol. Photomed.* **10**, 244 (1994).
[12] S. Piomelli, A. A. Lamola, M. B. Poh-Fitzpatrick, C. Seaman, and L. C. Harber, *J. Clin. Invest.* **56**, 1519 (1973).

We have demonstrated that transfecting fibroblasts from EPP patients with the cDNA for normal ferrochelatase via a retroviral vector corrects the metabolic defect,[13] which has been confirmed by others.[14,15] Additionally, we have shown that burst-forming unit erythroid (BFUe) colonies from EPP patients fluoresce when exposed to 400- to 410-nm light, whereas BFUe from normal people do not.[16] The absence of fluorescence in BFUes derived from EPP patients' peripheral blood stem/progenitor cells transduced with normal ferrochelatase will serve as an indicator that the transduced gene functioned effectively to correct the metabolic defect.

We were also able to infect peripheral blood stem/progenitor cells from EPP patients with the retroviral vector LXSN containing the normal human ferrochelatase cDNA, which also contains the NEO gene, conferring resistance to the antibiotic, G-418. The rate of infection was low (approximately 10%), as assessed by G418 resistance in methylcellulose semisolid medium (Mathews-Roth and Wise, unpublished observations). We found that all G-418-resistant BFUe colonies did not fluoresce under 410-nm light, thus demonstrating a "cure" of the EPP disease phenotype in this *in vitro* model.

A mouse model of EPP was developed by workers in France during mutagenesis studies with ethylnitrosourea.[17] These mice are available from Jackson Laboratories (Balb/C-Fech^{m1Pas} strain). We report here data on gene therapy of these animals.[18,19]

We first transplanted female EPP mice that had received 850 cGy of ^{137}Cs radiation with 5×10^6 syngenic bone marrow cells from normal male Balb/C donors. Normal Balb/C mice, transplanted EPP mice, and nontransplanted EPP mice were tested for photosensitivity and for protoporphyrin blood levels at 4 and 7 months. For photosensitivity testing, we expose a 2×3-cm area of the depilated back skin of the mouse to 20 min of exposure to a Spectroline mercury vapor lamp filtered with window glass to eliminate UVB radiation. Blood protoporphyrin is assayed by the method of Piomelli *et al.*[20] We found that nontransplanted EPP mice developed many erythrmatous ulcerative skin lesions in exposed skin, but that normal

[13] M. M. Mathews-Roth, J. L. Michel, and R. J. Wise, *J. Invest. Dermatol.* **104,** 497 (1995).

[14] H. deVermeil, C. Ged, S. Boulechfar, and F. Moreau-Gaudry, *J. Bioenerg. Biomembr.* **27,** 239 (1995).

[15] S. T. Magness and D. A. Brenner, *Hum. Gene. Ther.* **6,** 1285 (1995).

[16] M. M. Mathews-Roth, R. J. Wise, and B. A. Miller, *Blood* **87,** 4480 (1996).

[17] S. Tutois, X. Montagutelli, V. DaSilva, H. Joualt, P. Pouyet-Fesard, K. Leroy-Viard, J. L. Guénet, Y. Nordmann, Y. Beuzard, and J. Ch. Beybach, *J. Clin. Invest.* **88,** 1730 (1991).

[18] R. Pawliuk, M. Mathews-Roth, and P. Leboulch, *Blood* **92**(Suppl. 1), 297a (1998). [Abstract 1217]

[19] R. Pawliuk, T. Bachelot, R. J. Wise, M. M. Mathews-Roth, and P. Leboulch, *Nature Medicine* **5,** 768 (1999).

[20] S. Piomelli, *Clin. Chem.* **23,** 264 (1977).

Balb/C mice and transplanted mice did not. Porphyrin levels also decreased markedly. At 7 months posttransplant, transplanted EPP mice had an average blood protoporphyrin level of 2728 nM, whereas non-EPP Balb/C mice had an average level of 1919 nM and nontransplanted EPP mice had an average level of 17,133 nM.

We then performed a gene transfer experiment. We constructed a MSCV-derived retroviral vector that expresses both the human ferrochelatase cDNA and a gene encoding the green fluorescent protein (EGFP) as a marker. A vector expressing only the EGFP gene was also constructed as a control. BOSC 23 cells were infected with vectors using $CaPO_4$ precipitation, viral supernates were collected at 48 hr and used to infect AM12 cells, and viral supernates from AM12 cells were collected and used to infect the stable ecotropic packaging cell line GPE86. Supernatants from this last infection were collected and filtered through a 0.45-μm filter. Bone marrow cells were harvested from the femurs of male EPP mice that had received 150 mg/kg 5-fluorouracil 4 days previously. Bone marrow cells were then cultured for 48 hr in α minimal essential medium (αMEM) containing 15% fetal calf serum, 10 mg/ml human interleukin (IL)-6, 6 ng/ml murine IL-3, and 100 ng/ml murine Steel factor. After this period, cells were exposed to filtered viral supernatants in Falcon nontissue culture-treated culture dishes coated with fibronectin. The viral supernatants were changed once over a 48-hr period, and the bone marrow cells were then grown for an additional 48 hr in αMEM and the cytokines described earlier to allow expression of the EGFP. EGFP+ cells were then identified, selected, and injected into recipient female EPP mice that had received 850 cGY of whole body irradiation. We determined photosensitivity and protoporphyrin levels at 1, 3, and 4 months postgene therapy. As in the bone marrow transplant experiment, mice that received cells infected with the vector containing the ferrochelatase cDNA were no longer photosensitive. Their protoporphyrin levels decreased to an average of 3787 nM at 7 months, which is within the normal range.[17] Mice that received cells infected with the vector containing only the EGFP were as photosensitive as untreated EPP mice and their protoporphyrin levels remained elevated.

Thus, our preliminary experiments suggest that EPP might be amenable to cure by gene therapy. However, many problems need to be resolved before gene therapy can be tried in an EPP patient. For example, infection rates of peripheral blood stem cells must be maximized, procedures for producing high-titer vectors that do not form recombinant virus must be perfected, and tests to demonstrate the persistance of transduced cells must be performed. We cannot predict when gene therapy for EPP will be available, but we are working to make it available as soon as possible. Until then, β-carotene treatment offers the best relief for the symptoms of EPP.

[43] Porphyrias: Photosensitivity and Phototherapy

By MAUREEN B. POH-FITZPATRICK

Introduction

Porphyrias are a heterogeneous group of inherited or acquired metabolic disorders, all of which result from partial deficiencies in the activity of enzymes regulating heme biosynthesis (Table I). Individuals with some of the several porphyric syndromes may exhibit cutaneous sensitivity to visible light exposure as a major clinical manifestation. Light-induced skin lesions of patients with porphyrias result from oxygen-mediated photochemical reactions. These reactions are initiated by the photoexcitation of porphyrin molecules that are produced in large quantities during faulty operation of the heme synthetic pathway. Medical management of porphyrin-sensitized phototoxicity in patients with various forms of porphyrias usually begins with recommendations for avoidance of sunlight by lifestyle changes, by physical means of light attenuation such as hats, gloves, and long-sleeved clothing or plastic filtering films applied to windows, and by application to the skin of topical sunblock formulations containing agents that reflect visible light. In many cases, however, such relatively risk-free measures are impractical and/or insufficient, stimulating ongoing searches for additional means of increasing light tolerance. Ingestion of energy-quenching agents such as β-carotene or cysteine has improved the light tolerance of some patients with some porphyrias with photocutaneous manifestations, but the degree of protection gained has not been equal in all porphyrias or in all patients with any porphyria. Undesirable side effects of such systemic photoprotective agents may also discourage their use in some cases. These considerations have led to the investigation of the rather paradoxical application of phototherapies in porphyrias. Contemporary methods of phototherapy for skin diseases involve deliberate exposure of skin to selected wavelengths of ultraviolet radiation, including longer ultraviolet wavelengths near the visible light spectrum in which the action spectrum for phototoxicity in patients with porphyrias is maximal. This article reviews the pathophysiology of photosensitivity in porphyrias and the theory and practice of application of ultraviolet phototherapy as its treatment.

Photobiology and Photochemistry of Porphyrins

The predominant sites of human heme formation are maturing bone marrow erythrocyte precursor cells and hepatocytes. In juvenile erythroid

TABLE I

ENZYMES OF HEME BIOSYNTHESIS ASSOCIATED WITH PORPHYRIAS

Defective enzyme	Porphyria
Aminolevulinic acid dehydratase	Aminolevulinic acid dehydratase porphyria
Porphobilinogen deaminase	Acute intermittent porphyria
Uroporphyrinogen III synthase	Congenital erythropoietic porphyria
Uroporphyrinogen decarboxylase	Porphyria cutanea tarda
Coproporphyrinogen oxidase	Hereditary coproporphyria
Protoporphyrinogen oxidase	Variegate porphyria
Ferrochelatase	Erythropoietic protoporphyria

cells, the heme produced is chiefly used to form hemoglobin, whereas in hepatocytes, it is incorporated into hemoprotein enzymes (cytochromes, catalases, peroxidases). In both cell lines, heme is the end product of a series of enzymatically regulated biochemical conversions of porphyrin precursor and porphyrinogen intermediaries that yield the final intermediary protoporphyrin. Heme is then formed by insertion of ferrous iron into the protoporphyrin molecular nucleus, a process mediated by the inner mitochondrial membrane enzyme ferrochelatase. Normal heme synthesis is efficient, with very few intermediary substrates escaping conversion. When genetically determined or acquired factors interfere with the normal function of different enzymes in the pathway, however, various intermediaries and by-products of heme synthesis accumulate. Excessively accumulated porphyrin by-products and the intermediary protoporphyrin, which are all photoactive molecules, can interact with long ultraviolet and visible light rays to cause cutaneous photosensitivity. Excess accumulations of nonphotoactive porphyrin precursors are associated with a diverse array of systemic neurovisceral and neuropsychiatric disorders that can reach life-threatening severity but are not the focus of this article and will not be discussed further. Either photocutaneous or neurological signs and symptoms, or both, clinically characterize the several porphyric syndromes (Table II).

Diagnosis of each porphyria is usually made by laboratory demonstration of its characteristic biochemical profile of excess porphyrins and/or precursors in blood, urine, and fecal specimens. The unique profile that characterizes each porphyria is determined by the specific enzyme defect with which the syndrome is linked. With the exception of acquired forms of porphyria cutanea tarda, each porphyric syndrome is caused by mutations in the gene encoding the protein structure of its associated defective enzyme (Table I). Activity levels of these defective enzyme proteins are reduced to various fractions of normal and are virtually abrogated in some cases. In some inheritable porphyrias, in addition to a mutant genotype, the

TABLE II
CLINICAL MANIFESTATIONS OF PORPHYRIAS

Cutaneous photosensitivity
   Erythropoietic protoporphyria
   Congenital erythropoietic porphyria
   Porphyria cutanea tarda
   Variegate porphyria
   Hereditary coproporphyria
Neurological dysfunction
   Aminolevulinic acid dehydratase porphyria
   Acute intermittent porphyria
   Variegate porphyria
   Hereditary coproporphyria

adverse influence of endogenous or exogenous inducing factors such as hormonal fluctuations, infections, or drug use is often required for overt expression of the disease phenotype.

The biochemical profiles of porphyrias associated with cutaneous photosensitivity are characterized by high levels of porphyrins, which are oxidation products of their corresponding porphyrinogens. The electronic configuration of these oxidized tetrapyrrole macrocycles enables their maximal absorption of light energy in the violet visible light region of the electromagnetic spectrum (400–410 nm), with lesser absorption peaks at intervals extending into the red region (600–650 nm). Once promoted to singlet excited-state levels by the absorbed energy, porphyrin molecules can return to their stable ground states by emission of energy as light (fluorescence) and by transference of energy in chemical reactions. Singlet-excited porphyrins are short-lived and have a limited probability of reacting directly with biological targets, but readily undergo transition to triplet-excited states, which are longer-lived and can transfer energy to oxygen. The resultant singlet-excited oxygen and other reactive oxygen species (peroxides, superoxide anion, hydroxyl radicals) can all participate in porphyrin-sensitized photochemistry. Reactions causing the cutaneous photosensitivity of porphyrias are predominantly oxygen mediated or "photodynamic" in nature. Porphyrin-sensitized photodynamic reactions of biologic importance include peroxidation of cell membrane lipids and cross-linking of cell membrane and intracellular proteins, leading to loss of membrane integrity and function and disruption of intracellular organelles. A plethora of inflammation-related events in the skin appear to play interconnected roles in the production of porphyric phototoxicity. These include the disruption of dermal capillary endothelial cell membranes with the subsequent leakage of intravascular contents into the dermis, complement activation, mast cell degranulation with release of inflammatory mediators and proteases, neutrophil recruitment, lysosomal disruption with release of hydrolases and other enzymes, and the effects of released hydrolases and proteases on structural proteins such as elements of the basement membrane zone or on functional proteins such as enzymes.

Although all clinically relevant porphyrins are efficient photosensitizers, phototoxicity in porphyrias is manifested in two patterns: immediate and delayed. The immediate phototoxic reaction begins during sunlight exposure as stinging or burning pain in exposed skin. The affected skin can become massively swollen and purpuric over the next several hours, but this impressive reaction resolves after a few days of sun avoidance, leaving relatively little permanent visible scarring. This immediate pattern is most often displayed by individuals with erythropoietic protoporphyria. The delayed pattern is common to all other porphyrias with photocutaneous features; mechanical fragility and subepidermal blistering of exposed skin appearing many hours to days after sun exposure are its initial manifestations. Later sequelae developing in porphyric individuals who manifest delayed phototoxicity can include atrophic and hypertrophic scarring that in concert with destruction of cartilage and bone can cause dysmorphic changes of head, neck, and hands of mutilating severity. Other chronic changes reflecting cumulative photodamage of the delayed pattern are destruction of hair follicles of the scalp, eyebrows, and lashes; excessive growth of hair on face and extremities; mottled brownish hyperpigmentation; waxy or yellow thickened plaques on the head, neck, chest, and back that may become calcified; and detachment of nail plates from their beds (onycholysis) that may lead to nail shedding. Rationales for there being these two recognizably different patterns have been largely based on differences in solubility properties among porphyrins: lipophilic porphyrins more efficiently sensitizing photodynamic damage to lipoprotein structures (i.e., cellular or subcellular membranes), whereas hydrophilic porphyrins more efficiently sensitize damage to aqueous cytosolic components or other water-soluble targets.[1] In erythropoietic protoporphyria, in which sunlight commonly elicits skin pain acutely, the dicarboxylic protoporphyrin molecule, which is the major accumulated metabolite, is highly hydrophobic and lipophilic. The rationale for this acute phototoxicity postulates that protoporphyrin is more likely to be in close proximity to lipid-rich structures, the photodamage of which engenders a rapidly evolving cascade of inflammatory events. In porphyrias typified by delayed-onset phototoxicity, the porphyrins accumulated are predominantly of more highly polycarboxylated (and therefore less lipophilic) moieties that are more likely to be

[1] S. Sandberg and I. Romslo, *Clin. Chim. Acta* **109,** 193 (1981).

distributed in or near cytosolic or other aqueous sites of skin biomolecular elements, leading to the idea that a different panoply and evolution of events may occur after photoexcitation. Although immediate and delayed phototoxicity patterns are clinically distinctive, features of each accompany a predominant expression of the other often enough to support the concept that both result from similar photodynamic insults to an array of critical biomolecular structures that can be mediated by all of the clinically relevant porphyrins, but with varying degrees of efficiency yielding, in most in stances, relatively polar displays of a spectrum of manifestations.

Phototherapy of Porphyrias

The most uniformly beneficial therapy for porphyric photosensitivity is sunlight avoidance. For mildly affected individuals who can adjust their lifestyles to eliminate exposure to the most intense light radiation, reduction of symptoms may be sufficient without any additional therapy. Most patients, however, want and need more relief. Use of sun-exclusive clothing, window and lamp filters, and topical ultraviolet A (UVA) and visible light "sunblock" formulations may further reduce the light reaching the skin, but infrequently provide total relief. Systemically administered agents aimed at interfering with porphyrin-sensitized photochemical reactions by oxyradical-scavenging mechanisms appear to be quite protective for at least some patients, particularly those with the immediate phototoxicity of erythropoietic protoporphyria, but are not uniformly effective in all patients with this disorder and have not been as effective in patients with porphyrias characteristically exhibiting the delayed-type photoxicity pattern. Melanin is a brownish polyquinone pigment that is naturally present in the epidermal layer of the skin and serves as an endogenous sunscreen with a broad absorption spectrum extending throughout the ultraviolet and visible light regions. The melanin content of the skin can be enhanced to varying degrees in different human skin types by sun exposure. Augmentation of constitutive melanin levels by controlled ultraviolet irradiation using artificial sources has therefore been proposed as a rational therapy for patients with several photosensitivity disorders, including porphyrias. Ultraviolet phototherapy is most often performed using sources emitting radiation predominantly of shorter wavelengths conventionally referred to as ultraviolet B (290–320 nm) or ultraviolet A (320–400 nm).

Ultraviolet B radiation is more efficient than ultraviolet A at producing the skin reaction commonly called "sunburn" and at inducing facultative melanization. UVB is more distant than UVA from the visible light wavelengths that excite porphyrins most effectively. While most of the UVB radiation incident on human skin is absorbed by elements of the epidermis,

UVA penetrates well into the dermis and can affect dermal cells and connective tissue profoundly. When used in combination with certain photosensitizing furocoumarin compounds called "psoralens," UVA becomes a far more efficient modality for stimulating melanogenesis. Psoralen–UVA phototherapy is commonly referred to as "PUVA." Both UVB and PUVA therapies have other skin-related and systemic effects that pose important risks that must be considered before recommending the use of either in any given patient with any skin disorder.

Ultraviolet B Phototherapy

Ultraviolet B phototherapy can be delivered by several types of sources; cabinets fitted with arrays of long fluorescent tube lamps that can irradiate the entire body are used most frequently in current clinical practice. Conventional UVB phototherapy lamps emit a relatively broad band of energy across the UVB region. Narrow band UVB-emitting lamps constraining the bulk of energy output to between 311 and 313 nm have been employed to treat a small number of patients with porphyrias, with apparent benefit.[2,3] In addition to increasing the number of melanocytes and size and number of melanosomes, other UVB effects that may contribute to reducing cutaneous photosensitivity include thickening of the stratum corneum (the "horny" outermost layer of the skin), as well as of the underlying epidermis. UVB also alters production of cytokines by cells of the epidermis and dermis and produces an array of other effects influencing the many immunological functions of the skin.

Short-term risks of UVB include excessive exposure with resultant painful erythema and blistering ("sunburns") and inadvertent burns of the eyes, which must be shielded assiduously during therapy. Development of skin cancers years after UVB treatment is a long-term risk that increases with chronicity and intensity of exposure, as does the premature skin aging noted in individuals treated with this modality repetitively.

Protocols for courses of UVB therapy for patients with porphyrias have been described,[2–4] but the total numbers of patients treated by each are small, and whether benefits reported would be obtained in larger studies remains to be confirmed. In these studies, irradiations three to five times per week for several weeks with UVB lamps appeared to confer increased sunlight tolerance after courses of up to 30 treatments given in the spring. The resulting total doses for the entire course of treatments of ~1,000 to

[2] L. J. Warren and S. George, *Australas. J. Dermatol.* **39,** 179 (1998).
[3] P. Collins and J. Ferguson, *Br. J. Dermatol.* **132,** 956 (1995).
[4] R. Roelandts, *Dermatology* **190,** 330 (1995).

~16,000 mJ/cm^2 are, in general, smaller than cumulative doses common among patients with disorders such as psoriasis, who often require intermittent courses more frequently than once a year and continuous maintenance treatments over years to achieve acceptable levels of lesion control. If a course of UVB irradiation must be repeated each spring to increase summer sunlight tolerance for individuals with porphyria over many years, particularly if the practice begins in childhood, cumulative damage and long-term risks will become more significant concerns.

Psoralen–UVA Photochemotherapy

PUVA therapy has been used in a few cases of erythropoietic protoporphyria with reported benefits.[4,5] The artificial sources of UVA radiation most often used in clinical practice are arrays of fluorescent tube lamps engineered to emit broad-band UVA energy peaking at ~350 nm with only a small fraction of the total output beyond 380 nm. PUVA is a more cumbersome therapy than UVB, as patients must ingest a drug (usually 8-methoxypsoralen or trimethylpsoralen) 2 hr prior to exposure that sensitizes the skin to UVA not only during the controlled irradiation, but for 8–12 hr afterward. During the entire day of treatment, therefore, patients are at risk for inadvertent exposure to ambient sunlight with resultant phototoxicity manifested as painful erythema and florid blistering of accidentally exposed skin. Individuals with cardiac, hepatic, or renal dysfunctions or with photosensitivity disorders are not considered good candidates for PUVA treatment, so pretreatment clinical and laboratory evaluations to exclude these contraindications are required. Baseline ophthalmological evaluations are recommended to search for existing cataracts in the ocular lenses of patients for whom PUVA therapy is contemplated because cataract formation is a potential side effect. UVA-impermeable wrap-around eyeglasses or shields must be worn for the entire day of treatment once the psoralen is ingested to limit this risk, and repeated eye examinations are recommended if therapy continues over time. Many commonly used drugs are potential photosensitizers that can be activated by UVA, so candidates for PUVA therapy must avoid their use.

The action spectrum for PUVA-induced erythema in human skin is maximal between 320 and 335 nm. The erythematous inflammation produced by PUVA does not peak until 48–72 hr after exposure, compared to that of UVB "sunburn," which peaks between 12 and 24 hr. Therefore, to avoid exceeding therapeutic thresholds for the production of additive excessive redness and discomfort, PUVA treatments are usually delivered

[5] A. M. Ros, *Photodermatology* **5,** 148 (1988).

no more frequently than every other day, whereas UVB treatments can be delivered safely on a daily schedule. PUVA photochemistry encompasses both oxygen-independent (type I) formation of monofunctional and bifunctional psoralen adducts in DNA and oxygen-dependent (type II) reactions involving the transfer of energy to molecular oxygen. Precisely how PUVA exerts its beneficial effects is still not entirely clear. The beneficial effect of PUVA in the hyperproliferative skin disease psoriasis may reflect dampening of accelerated DNA synthesis as a result of UVA-induced psoralen–DNA cross-linking. PUVA-induced alterations of the numbers and functions of immunocompetent cells resident in the skin or able to traffick to and from the skin may also be critically important in the overall therapeutic effect observed clinically in psoriasis and may be of paramount importance in other disorders for which PUVA is effective, such as cutaneous T-cell lymphoma and atopic dermatitis. There is no information available to suggest that such mechanisms are responsible in any way for the benefits noted in cases of porphyria.

Protocols for courses of PUVA therapy for porphyria with increasing doses over several weeks have been described.[4,5] As for UVB, the mechanisms by which PUVA affects reduction in porphyrin-induced photosensitivity can be speculated to be several, as psoralen-mediated photochemistry results in numerous responsive structural and functional biological changes in the skin. As with UVB, however, it seems most likely that the most important of these is the pronounced enhancement of melanin pigmentation PUVA therapy can produce.

PUVA has a risk profile[6] that generally exceeds that of UVB for both short- and long-term adverse effects. While UVB "sunburn" is painful, the skin pain of an acute PUVA phototoxic reaction is quite severe; the potential for cataract formation is only associated with PUVA. While shielding of male patients' genitalia to reduce acute burns and long-term risk of skin cancer formation and wearing of ultraviolet-impermeable goggles are important precautions during both UVB and PUVA treatments, post-treatment protective eyewear is required only for PUVA. Ingestion of 8-methoxypsoralen often causes nausea, requiring the concomitant use of antiemetics. Bothersome itching of PUVA-treated skin is a common complaint; sensations of pain in treated skin that are not due to an acute reaction occur occasionally, as does excessive hair growth. Chronic UVB radiation is carcinogenic, but increased frequencies of squamous cell, basal cell, and melanoma skin cancers after PUVA have been recognized as significant

[6] W. L. Morison, R. D. Baughman, R. M. Day, D. Forbes, H. Hönigsmann, G. G. Krueger, M. Lebwohl, R. Lew, L. Naldi, J. A. Parrish, M. Piepkorn, R. S. Stern, G. D. Weinstein, and E. Whitmore, *Arch. Dermatol.* **134,** 595 (1998).

long-term risks that are still being quantified. Accelerated aging of skin with severely degenerated dermal elastic tissue was noted early in PUVA clinical trials. Persistent large and irregularly shaped flat brown skin lesions called "PUVA lentigenes" or "PUVA freckles" that contain increased numbers of melanocytes and morphologically abnormal melanosomes are PUVA sequelae that occur often and may signal an increased risk of melanoma. The many and profound effects of both UVB and PUVA on immunological functions of the skin are still being elucidated, but it is clear that these play important roles in carcinogenesis. Due to concern about its long-term potential for adverse effects, PUVA therapy is used infrequently for children with disorders in which UVB is also effective.

Both UVB and UVA induce matrix metalloproteinases, a family of enzymes responsible for the degradation of connective tissue proteins. Studies of UVA and light (340–450 nm) radiation on the synthesis of matrix metalloproteinases in human dermal fibroblasts showed that interstitial collagenase, 72-kDa type VI collagenase, and stromelysin were all induced after radiation, whereas synthesis of their major inhibitory enzyme was not affected.[7] Photosensitization by uroporphyrin enhanced metalloproteinase induction after radiation, further perturbing the normal balance of these enzymes. This finding led to the conclusion that photocutaneous lesions in porphyria cutanea tarda patients, who accumulate uroporphyrin, may reflect the destruction of dermal and basement zone proteins as a result of the unbalanced synthesis of matrix metalloproteinases. Another study showed that plasmid DNA incubated with protoporphyrin-rich serum of patients with erythropoietic protoporphyria was damaged photodynamically after UVA irradiation.[8] Such observations, while *in vitro,* add to concern that damage sustained by patients with porphyrias who might well be subjected to several treatments with UVA every year for many years may not be entirely inconsequential. The relative risks of this therapy compared to its benefits in this patient population remain largely unknown, however, so that little firm scientific guidance can be offered to clinical practitioners. In the absence of data to the contrary from comparative clinical trials, it seems reasonable to infer that in most instances, especially in porphyric children who warrant phototherapy, UVB may be preferable to PUVA, as the benefit may be gained with a lesser burden of risk.

[7] G. Herrmann, M. Wlaschek, K. Bolsen, K. Prenzel, G. Goerz, and K. Scharffetter-Kochanek, *J. Invest. Dermatol.* **107,** 398 (1996).
[8] H. Hashizume, Y. Tokura, T. Oku, Y. Iwamoto, and M. Takigawa, *Arch. Dermatol. Res.* **287,** 586 (1995).

[44] Carotenoids in Human Skin: Noninvasive Measurement and Identification of Dermal Carotenoids and Carotenol Esters

By WILHELM STAHL, ULRIKE HEINRICH, HOLGER JUNGMANN, HAGEN TRONNIER, and HELMUT SIES

Introduction

Skin is the major light-exposed tissue and thus is susceptible to photooxidative stress. *In vitro* studies demonstrated that carotenoids interact with light of different wavelengths and are capable of scavenging reactive oxygen species, especially those formed in photooxidative reactions.[1,2] Carotenoids efficiently scavenge singlet molecular oxygen and excited triplet carbonyls, which are formed on photoexcitation of sensitizer molecules. Protective effects of carotenoids in the skin were shown in the treatment of a genetic disorder, erythropoietic protoporphyria.[3] In this disease, exceptionally high amounts of the photosensitizer protoporphyrin generate reactive intermediates in photooxidative reactions. It was suggested that the excited sensitizers and/or subsequently formed reactive triplet carbon species or singlet molecular oxygen are the reactive agents responsible for damaging reactions with biological macromolecules. The symptoms of erythropoietic protoporphyria are alleviated on treatment with high doses of β-carotene.[4]

Dermal accumulation of β-carotene attracted attention, as it has been speculated that carotenoids contribute to protection against acute and chronic exposure to ultraviolet (UV) light.[5] Although the protective effects of carotenoids toward skin lesions are still under investigation,[5-7] β-carotene supplements are in use as oral sun protectants. Little is known about the

[1] W. Stahl, A. R. Sundquist, and H. Sies, *in* "Vitamin A in Health and Disease" (R. Blomhoff, ed.), p. 275. Dekker, New York, 1994.

[2] C. A. Rice-Evans, J. Sampson, P. M. Bramley, and D. E. Holloway, *Free Radic. Res.* **26,** 381 (1997).

[3] M. M. Mathews-Roth, *Ann. N.Y. Acad. Sci.* **691,** 127 (1993).

[4] J. von Laar, W. Stahl, K. Bolsen, G. Goerz, and H. Sies, *J. Photochem. Photobiol. B Biol.* **33,** 157 (1996).

[5] H. K. Biesalski, C. Hemmes, W. Hopfenmüller, C. Schmid, and H. P. M. Gollnick, *Free Radic. Res.* **24,** 215 (1996).

[6] M. Garmyn, J. D. Ribaya-Mercado, R. M. Russell, J. Bhawan, and B. A. Gilchrest, *Exp. Dermatol.* **4,** 104 (1995).

[7] J. D. Ribaya-Mercado, M. Garmyn, B. A. Gilchrest, and R. M. Russell, *J. Nutr.* **125,** 1854 (1995).

accumulation of carotenoids in skin and their distribution in various skin areas. The dermal levels of carotenoids can be measured noninvasively by reflection photometry, which was used to follow β-carotene levels in skin after supplementation.[8]

Xanthophylls carrying a hydroxyl group can occur as carotenol fatty acid esters. Lutein, zeaxanthin, and cryptoxanthin, as well as capsanthin or violaxanthin, have been identified as parent carotenoids in plants.[9–11] After ingestion of tangerine juice concentrate[12] rich in β-cryptoxanthin esters, no carotenol fatty acid esters were detectable in human chylomicrons or serum. These data lend experimental support to previous suggestions that carotenol fatty acid esters are hydrolyzed prior to release into the systemic circulation.[13] Small amounts of carotenol fatty acid esters have been separated and assigned in human skin by means of high-performance liquid chromatography (HPLC).[14]

Skin Levels of Carotenoids Measured by Reflection Photometry after Supplementation[8]

In order to determine the accumulation of carotenoids in skin, a study on long-term supplementation with 24 mg β-carotene for 12 weeks was performed. Twelve female healthy adults took part in the study. Betatene, an extract of the alga *Dunaliella salina* (Betatene Ltd., Australia), was used as a carotenoid source. On day 0 and after 4, 8, and 12 weeks of treatment, blood samples were collected; an additional blood sample was obtained 2 weeks after the cessation of treatment. Analyses of β-carotene in serum were performed by HPLC as described previously.[15] Carotenoid levels in skin were analysed by noninvasive reflection spectroscopy at the time points of blood collection at different areas of the skin.[8,16]

Reflection Photometry

Reflection spectra were collected between 350 and 850 nm with a Multiscan OS 20 spectrophotometer (MBR GmbH, Herdecke, Germany)

[8] W. Stahl, U. Heinrich, H. Jungmann, J. von Laar, M. Schietzel, H. Sies, and H. Tronnier, *J. Nutr.* **128,** 903 (1998).
[9] F. Khachik, G. R. Beecher, and W. R. Lusby, *J. Agric. Food Chem.* **37,** 1465 (1989).
[10] S. D. Lin and A. O. Chen, *J. Food Biochem.* **18,** 273 (1995).
[11] T. Wingerath, W. Stahl, D. Kirsch, R. Kaufmann, and H. Sies, *J. Agric. Food Chem.* **44,** 2006 (1996).
[12] T. Wingerath, W. Stahl, and H. Sies, *Arch. Biochem. Biophys.* **324,** 385 (1995).
[13] J. A. Olson, *Pure Appl. Chem.* **66,** 1011 (1994).
[14] T. Wingerath, H. Sies, and W. Stahl, *Arch. Biochem. Biophys.* **355,** 271 (1998).
[15] W. Stahl, W. Schwarz, A. R. Sundquist, and H. Sies, *Arch. Biochem. Biophys.* **294,** 173 (1992).
[16] H. Jungmann, U. Heinrich, M. Wiebusch, and H. Tronnier, *Kosmet. Med.* **1,** 50 (1996).

coupled to an all-silica fiberoptic reflectance bundle (Top Sensor Systems, Eerbeek, NL). The spectral resolution was about 1.2 nm; titanium oxide was used as a white reference standard. A 5-W (5 J sec^{-1}) halogen lamp (MBR GmbH, Herdecke, Germany) was used for tissue illumination.

Data obtained by spectroscopy are transformed to a log $(1/R)$ scale, meancentered and normalized. The law of Lambert–Beer is not suitable for the calculation of carotenoid levels in tissues based on data obtained with reflection photometry because of (i) heterogeneous distribution of carotenoids in tissues, (ii) other substances present in the tissue that contribute to reflection spectra, and (iii) the unknown pathlength of the reflected light in the tissue. Thus, further data analyses were performed using software developed for this purpose.[16,17]

The distribution of the chromophore was determined as follows. If S (λ) denotes the measured reflection spectrum of the skin and E (λ) the absorption spectrum of a β-carotene solution in a cuvette, then a mapping M exists with $E(\lambda) = M[S(\lambda)]$, where M is a nonlinear one-to-one mapping. This mapping M depends on the degree of inhomogeneity and not on the type of inhomogeneity[18] and is scale invariant. Therefore, the degree of inhomogeneity can be calculated and an inhomogeneous spectrum can be corrected. All spectra were processed this way.

Correction of the Influence of Other Components on Reflection Spectra

In multivariate analyses, information about every single compound is not required.[19] Because statistical methods are sensitive to changes of measuring conditions, it is possible to assign the deviation to the compound of interest.[20] Applying partial component regression and partial least-square multivariate algorithm, the deviation due to the known β-carotene spectrum can be assessed, if the nonlinear distortion of this spectrum has been corrected previously.[21] Because an inverse mapping of M exists, it is possible to alter M' in a way that the estimated square of the deviation approaches a minimum and thus a corrected spectrum is obtained.[22] For each corrected spectrum, multivariate analytical methods were applied.[23]

[17] E. Postaire, H. Jungmann, M. Bejot, U. Heinrich, and H. Tronnier, *Biochem. Mol. Biol. Interact.* **42,** 1023 (1997).

[18] R. Wodick and D. W. Lübbers, *Pflüg. Arch.* **342,** 29 (1973).

[19] H. Martens and T. Naes, "Multivariate Calibration." Wiley, Chichester, 1991.

[20] C. Jochum, P. Jochum, and B. R. Kowalski, *Anal. Chem.* **53,** 85 (1981).

[21] R. Marbach, "Messverfahren zur IR-spektroskopischen Blutglukosebestimmung." VDI Verlag, Düsseldorf, 1993.

[22] D. W. Lübbers and J. Hoffmann, *Adv. Physiol. Sci.* **8,** 353 (1980).

[23] W. R. Hruschka and K. Norris, *Appl. Spectrosc.* **36,** 261 (1982).

Results

Carotenoid levels in specific parts of the skin and serum are given in Table I. Basal levels in skin are distinctively different within different regions. In the skin of the forehead, palm of the hand, and dorsal skin, relatively high basal levels were found; lower levels were detected in skin of the arm and back of the hand.

Carotenoid levels increased in all parts of the skin on treatment with Betatene. Maximum values were reached after 12 weeks of treatment in all areas. Based on the starting levels, the increase in skin carotenoids was 3.4-fold in forehead, 1.7-fold in dorsal skin, 3.2-fold in palm of the hand, 18-fold in back of the hand, and 2.7-fold at the inside of the arm after 12 weeks of supplementation. After cessation of supplementation the carotenoid levels in skin decreased in all dermal areas.

Serum levels of β-carotene also increased on supplementation. Starting from a mean base value of 0.44 nmol/ml, mean peak levels were detected after 12 weeks of treatment with 1.8 nmol/ml representing a 4.1-fold increase. Two weeks after the end of treatment, mean serum levels were decreased again.

Carotenoid levels determined in the present study are similar to those reported in the literature. In facial skin, mean β-carotene values of about

TABLE I

CAROTENOID CONTENT IN DIFFERENT AREAS OF SKIN AS DETERMINED BY REFLECTION PHOTOMETRY AND SERUM CONCENTRATIONS[a] OF β-CAROTENE AFTER INGESTION OF CAROTENOIDS FROM BETATENE[b]

| Skin | Time (weeks) | | | | | Coefficient (r) of correlation[c] |
| | 0 | 4 | 8 | 12 | 14 | |
	Carotenoid content (nmol/g)					
Arm (inside)	0.07 ± 0.05	0.11 ± 0.08	0.13 ± 0.06	0.19 ± 0.08	0.10 ± 0.07	0.767
Forehead	0.40 ± 0.09	0.85 ± 0.19	1.14 ± 0.20	1.36 ± 0.23	0.60 ± 0.12	0.891
Dorsal	0.22 ± 0.13	0.28 ± 0.10	0.30 ± 0.16	0.37 ± 0.14	0.32 ± 0.10	0.710
Hand (back)	0.03 ± 0.04	0.11 ± 0.09	0.50 ± 0.32	0.54 ± 0.52	0.35 ± 0.25	0.701
Hand (palm)	0.32 ± 0.08	0.82 ± 0.09	0.82 ± 0.08	1.03 ± 0.12	0.71 ± 0.09	0.936
	Carotenoid concentration (nmol/ml)					
Serum	0.44 ± 0.22	1.35 ± 0.51	1.68 ± 0.75	1.80 ± 0.94	1.15 ± 0.67	

[a] Skin levels were analyzed by reflection spectrophotometry; serum levels by HPLC. Data are given as means ± SD ($n = 12$).
[b] Modified from Stahl *et al.*, *J. Nutr.* **128,** 9903 (1998).
[c] Coefficients of correlation: serum vs skin levels of β-carotene.

0.1–0.2 nmol/g wet tissue have been measured by means of HPLC.[24] This is rather close to the basal level determined here, varying from 0.03 to 0.4 nmol/g depending on the skin area. Higher levels at about 1.5 nmol/g wet tissue were found when subcutaneous fat was included in the samples.[6]

The energy of the halogen lamp (5 J sec^{-1}) is sufficient to penetrate skin to a depth of about 0.1–0.85 mm. Thus, the content of carotenoids from subcutaneous fat does not contribute significantly to the measurement using reflection spectroscopy under the present conditions.

Data show that reflection spectroscopy affords spectroscopic data from surfaces of tissues and is a suitable and convenient method to determine carotenoid levels noninvasively in accessible tissues such as skin.

Separation and Identification of Carotenoid Esters in Skin[14]

Sample Preparation

Skin samples are obtained from five adult women who underwent plastic surgery. Visible adipose tissue is removed and the samples are stored at −80° until analysis. For homogenization and extraction of samples the tissues are frozen in liquid nitrogen and broken into small pieces. Collagenase (type IV) from *Clostridium histolyticum* and lipase (type VII) from *Candida rugosa* are added for enzymatic digestion; samples are incubated at 37° for 2 hr.

Thereafter, the mixture is homogenized on ice with an Ultra Turrax, and protease solution (pronase E) from *Streptomyces griseus* is added to the homogenate; incubation is at 37° for 1 hr. After incubation, 1 ml of ethanol is added, and the carotenoids are extracted with 6 ml *n*-hexane/dichloromethane (5/1) containing 1.1 mM butylated hydroxytoluene. The organic solvent is evaporated under a gentle stream of nitrogen, and the dry residues are stored at −80°. For saponification of fractions isolated by HPLC, methanolic potassium hydroxide (10%) is used.

HPLC Analyses

High-performance liquid chromatography is carried out with a Merck/Hitachi Model 655 A-12 ternary solvent delivery system equipped with a Merck/Hitachi Model L-4200 UV/visible detector and a Merck/Hitachi Model L-5000 LC controller. For analytical separations, a 5-μm Suplex pKb 100 column (250 mm length × 4.6 mm i.d.) (Supelco, Bellefonte, PA) with a 20-mm guard column is used. Two sets of HPLC conditions are employed using eluents A and B.

[24] Y.-M. Peng, Y.-S. Peng, and Y. Lin, *Cancer Epidemiol. Biomark. Prev.* **2**, 139 (1993).

Eluent A is used for the separation of carotenol fatty acid esters and contains methanol/acetonitrile/dichloromethane/n-hexane (10/85/2.5/2.5, v/v/v/v). Gradient: isocratically for 10 min; from 10 to 40 min a linear gradient is applied to methanol/acetonitrile/dichloromethane/n-hexane (10/45/22.5/22.5, v/v/v/v). The final composition is held for another 10 min. The flow rate is 0.7 ml/min.

Eluent B is applied for the separation of saponified carotenol fatty acid extracts. It consists of 64% methanol/acetonitrile/2-propanol (54/44/2, v/v/v) and 36% H_2O. Gradient: isocratically for 5 min; from 5 to 15 min a linear gradient is applied to methanol/acetonitrile/2-propanol (54/44/2, v/v/v) with a flow rate of 1 ml/min. The final composition is held for another 10 min. Carotenol fatty acid esters are detected at 450 nm.

Reference carotenoids lutein, zeaxanthin, and β-cryptoxanthin were from Hoffmann-La Roche (Basel, Switzerland). α-Cryptoxanthin is isolated from tangerine juice concentrate.[11] 2′,3′-Anhydrolutein is prepared as described in the literature.[25] Straight, long-chain fatty acid esters are prepared from the parent carotenoids and the appropriate fatty acid chlorides.[11] Structural elucidation of reference material is performed by matrix-assisted laser desorption/ionization postsource decay mass spectrometry (MALDI/PSD/MS).[26,27]

Results

Figure 1 shows a typical chromatogram of the carotenoid pattern in a human skin sample. Lutein, zeaxanthin, and lycopene are partially coeluting under the present conditions and are detectable at retention times between 12 and 20 min. The dominating carotenoid in skin is β-carotene, present mainly in its *all-trans* configuration. Small amounts of 9-*cis* and 13-*cis* β-carotene, as well as α-carotene, are also detectable. Compounds eluting after β-carotene are fatty acid esters of several xanthophylls and have been assigned according to the order of elution with peak numbers 1–18. The whole fraction, including peaks 1–18, was collected and saponified with methanolic potassium hydroxide. Figure 2 shows the HPLC trace (eluent B) of the hydrolyzed sample. Identification of carotenoids was by coelution with synthetic reference material and UV/VIS spectroscopy. The major

[25] F. Khachik, C. J. Spangler, J. C. Smith, L. M. Canfield, A. Steck, and H. Pfander, *Anal. Chem.* **69,** 1873 (1997).

[26] R. Kaufmann, T. Wingerath, D. Kirsch, W. Stahl, and H. Sies, H., *Anal. Biochem.* **238,** 117 (1996).

[27] T. Wingerath, D. Kirsch, R. Kaufmann, W. Stahl, and H. Sies, *Methods Enzymol.* **299,** 390 (1999).

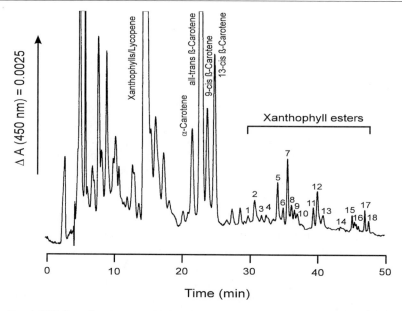

Fɪɢ. 1. HPLC profile of carotenoids in human skin. For identification of xanthophyll esters (peak 1–18), see Table II (from Wingerath *et al.*[14]).

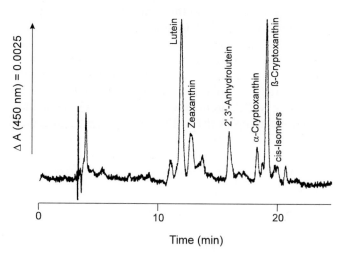

Fɪɢ. 2. HPLC profile of carotenoids present in the xanthophyll ester fraction after saponification. Peaks 1–18 (see Fig. 1) were collected and saponified. Parent carotenoids were separated by HPLC (from Wingerath *et al.*[14]).

parent xanthophylls are lutein, zeaxanthin, 2',3'-anhydrolutein, α-crypto-xanthin, and β-cryptoxanthin.

For further identification, single peaks (1–18) were isolated and saponi-fied. The resulting samples were analyzed by HPLC (eluent B) and identi-fied by coelution with synthetic reference compounds. The concentrations were too low for identification by UV/VIS spectroscopy. The parent carot-enoids detected in peaks 1–18 are listed in Table II. Spiking experiments with several synthetic fatty acid esters of lutein, zeaxanthin, cryptoxanthin, and 2',3'-anhydrolutein were performed in order to assign tentatively some of the xanthophyll esters; data are given in Table II.

The pattern of carotenol esters in different individuals ($n = 5$) was very similar, but absolute levels differed between individuals. In comparison to other carotenoids, xanthophyll ester levels in skin are very low and at the detection limit of the methods applied here. The amount of zeaxanthin monopalmitate (peak 7) and β-cryptoxanthin palmitate (peak 13) was calcu-

TABLE II
Carotenol Fatty Acid Esters in Human Skin Separated by HPLC[a]

Peak[b]	Identified parent xanthophylls[c]	Carotenol ester
1	Lutein	Lutein monoester
2	Lutein	Lutein monolinoleate[d]
3	Lutein/zeaxanthin*	Zeaxanthin monolinoleate[d]
4	Lutein/zeaxanthin/2',3'-anhydrolutein*/ α-cryptoxanthin	2',3'-Anhydrolutein linoleate[d]
5	Lutein*/zeaxanthin/2',3'-anhydrolutein/ α-, β-cryptoxanthin	Lutein monooleate[d]
6	Lutein*/zeaxanthin	Lutein monopalmitate[d]
7	Lutein/zeaxanthin*	Zeaxanthin monopalmitate[d]
8	Lutein/zeaxanthin/2',3'-anhydrolutein*	2',3'-Anhydrolutein oleate[d]
9	2',3'-Anhydrolutein/α-β-cryptoxanthin*	β-Cryptoxanthin linoleate[d]
10	2',3'-Anhydrolutein/α-cryptoxanthin/ β-cryptoxanthin*	β-Cryptoxanthin myristate[d] 2',3'-Anhydrolutein palmitate[d]
11	α-Cryptoxanthin	α-Cryptoxanthin oleate[d]
12	α-Cryptoxanthin/β-cryptoxanthin*	β-Cryptoxanthin oleate[d]
13	β-Cryptoxanthin	β-Cryptoxanthin palmitate[d]
14	Lutein/β-cryptoxanthin*	β-Cryptoxanthin stearate[d]
15	Lutein*/zeaxanthin	Lutein/zeaxanthin diester
16	Lutein*/zeaxanthin	Lutein/zeaxanthin diester
17	Lutein*/zeaxanthin	Lutein/zeaxanthin diester
18	Lutein*/zeaxanthin	Lutein/zeaxanthin diester

[a] From Wingerath et al., Arch. Biochem. Biophys. 355, 271 (1998).
[b] Peak numbers as indicated in Fig. 1.
[c] Major xanthophyll indicated with an asterisk.
[d] Assigned by spiking experiments.

lated using synthetic zeaxanthin monopalmitate and β-cryptoxanthin palmitate as external standards. The level of zeaxanthin monopalmitate and β-cryptoxanthin palmitate was ~0.8 and 0.3 pmol/g skin, respectively β-Carotene levels in human skin are in the range of 30–400 pmol/g.[8,24]

Xanthophyll esters in human skin may originate from different biochemical processes. Esterification of free carotenols can be mediated by enzymatic activity, but alternatively, small amounts of xanthophyll esters may circulate in the blood and subsequently accumulate in tissues.

Acknowledgments

Our research was supported by the Henkel Nutrition and Health Group (Henkel KGaA, Düsseldorf, Germany) and by the Institut Danone für Ernährung e.V. (Munich, Germany).

Section III

Ozone

[45] Reactive Absorption of Ozone: An Assay for Reaction Rates of Ozone with Sulfhydryl Compounds and with Other Biological Molecules

By Jeffrey R. Kanofsky and Paul D. Sima

Introduction

Ozone is toxic to both plants and animals. Because the reactions of ozone with many types of biological molecules are extremely fast, ozone reacts very close to the surface of tissues that are in direct contact with the atmosphere.[1-3] In animals, ozone reacts with biomolecules present in the fluid lining the air passages of lungs.[3] In plants, ozone reacts with biomolecules within the cell walls lining the air passages of leaves.[4]

As ozone is consumed within tissues through reactions with biological molecules, more ozone is drawn into the tissues in a process called reactive absorption.[5-9] Mathematical models of reactive absorption are well established and provide a means of calculating reaction rate constants for gases that react with nonvolatile solutes in liquids.[5-7,9] While most of the development of these mathematical models has occurred within the chemical engineering literature, the measurement of reactive absorption can provide some unique insights about ozone reactions with biological molecules close to the surface of a solution. Using a very simple apparatus, it is possible to obtain values for even very large reaction rate constants.[5-7,9]

[1] D. Giamalva, D. F. Church, and W. A. Pryor, *Biochem. Biophys. Res. Commun.* **133,** 773 (1985).

[2] W. A. Pryor, D. H. Giamalva, and D. F. Church, *J. Am. Chem. Soc.* **106,** 7094 (1984).

[3] W. A. Pryor, *Free Radic. Biol. Med.* **12,** 83 (1992).

[4] W. L. Chameides, *Environ. Sci. Technol.* **23,** 595 (1989).

[5] P. V. Danckwerts, "Gas-Liquid Reactions." McGraw-Hill, New York, 1970.

[6] S. E. Schwartz and W. H. White, *Adv. Environ. Sci. Tech.* **12,** 1 (1983).

[7] S. E. Schwartz, *in* "Precipitation Scavenging and Atmosphere-Surface Exchange" (S. E. Schwartz and W. G. N. Slinn, Coords.), Vol. 2, p. 789. Hemisphere Publishing, Washington, 1992.

[8] E. M. Postlethwaith, S. D. Langford, and A. Bidani, *Toxicol. Appl. Pharmacol.* **125,** 77 (1994).

[9] J. R. Kanofsky and P. D. Sima, *Arch. Biochem. Biophys.* **316,** 52 (1995).

METHODS IN ENZYMOLOGY, VOL. 319

Method

Apparatus

Figure 1 shows an apparatus suitable for studying the reactive absorption of ozone by aqueous solutions containing various biological molecules. A carrier gas containing a small concentration of ozone, typically 30 ppm, flows over a 0.5-ml volume of solution containing the biological molecules being studied. The ozone enters the reactor via a 17-gauge stainless steel tube covered with a fluorocarbon tube. The needle is 1.2 cm above the surface of the solution. The concentration of ozone in the outflow of the reactor is then measured and compared with the concentration of ozone exiting an empty reactor. The difference in concentrations is used to calculate the rate at which ozone is consumed within the solution.

For the flow reactor, it is important to use a material that is not wet by water. The rate at which ozone enters the aqueous solution is related to its surface area. In glass, or other materials wet by water, a thin film of aqueous solution can form along the sides of the flow reactor, greatly increasing the liquid surface area in a nonreproducible manner. It is also important that the materials used in the flow reactor do not react with

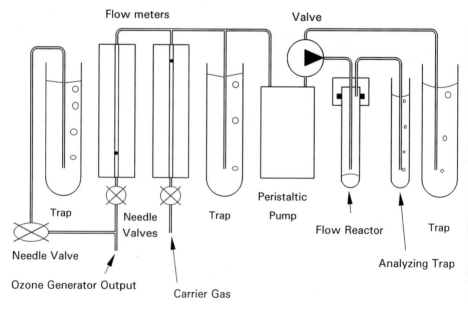

FIG. 1. Apparatus for studying the reactive absorption of ozone by aqueous solutions of biomolecules.

ozone. We have found that 12-mm-diameter polystyrene test tubes work well as flow reactors. Aqueous solutions have a relatively flat meniscus. Further, there is no detectable decrease in ozone concentration in gas that had flowed through the polystyrene tubes.[9]

Ozone is produced using an electrodeless discharge (e.g., Model O3V5-O ozone generator, Ozone Research and Equipment Corp., Phoenix, AZ). About 1% of the oxygen flowing through the discharge is converted into ozone. Using the series of needle valves and flow meters shown in Fig. 1, the ozone is diluted with carrier gas, most commonly oxygen, nitrogen, or air. A peristaltic pump, using silicon rubber tubing, takes a fraction of the diluted ozone gas and provides a constant flow of gas into the flow reactor. A valve diverts the output from the peristaltic pump into either a waste trap or the flow reactor. Except for the peristaltic pump tubing and the polystyrene flow-reactor tube, all other materials in contact with the ozone flow were made of fluorocarbon, stainless steel, or glass.

The outflow from the flow reactor is bubbled through a trap containing 1% (w/v) potassium iodide, 0.1 M monobasic potassium phosphate, and 0.1 M dibasic sodium phosphate. The amount of ozone entering the trap in a given time is assayed using an extinction coefficient of 24,500 cm^{-1} M^{-1} at 352 nm.[10]

The entire apparatus is most safely operated within a chemical fume hood. All of the ozone exiting the apparatus should be discharged through traps containing 10% potassium iodide and 200 mM phosphate buffer at pH 7.0. It is also prudent to monitor the room air for excessive ozone levels.

Data Analysis

Danckwerts and others have developed mathematical models for two-phase systems in which a reactive gas is in contact with a liquid.[5-7] The liquid contains a solute that reacts with the gas. These mathematical models allow one to calculate the gas-solute reaction rate constant from the rate of reactive absorption and from other parameters. For fast ozone-biomolecule reactions that are pseudo-first order in ozone, Eq. (1) is valid.

$$\frac{[O_3]_0}{F} = K_G + \frac{H}{AD^{1/2}k^{1/2}[B]^{1/2}} \tag{1}$$

where F is the net flux of ozone into the liquid, $[O_3]_0$ is the concentration of ozone entering the reactor, K_G is the mass transfer coefficient for the gas phase, H is the Henry's law constant, A is the effective area of the gas-liquid interface, D is the diffusion coefficient for ozone in the aqueous

[10] B. E. Saltzman and N. Gilbert, *Anal. Chem.* **31**, 1914 (1959).

solution, k is the ozone-biomolecule reaction rate constant, and $[B]$ is the biomolecule concentration. Values for H and for D are available in the literature.[11,12] Experimentally, the rate of reactive absorption is calculated from the decrease in ozone concentration caused by passage through the reactor [Eq. (2)]:

$$F = f([O_3]_0 - [O_3]_x) \qquad (2)$$

where f is the flow rate of the carrier gas and $[O_3]_x$ is the ozone concentration in the gas exiting the reactor. Combining Eq. (1) and (2) gives Eq. (3):

$$\frac{[O_3]_0}{[O_3] - [O_3]_x} = fK_G + \frac{fH}{AD^{1/2}k^{1/2}[B]^{1/2}} \qquad (3)$$

From Eq. (3) it is apparent that a plot of $[O_3]_0/([O_3]_0 - [O_3]_x)$ against the reciprocal of the square root of the biomolecule concentration should be linear. It is also apparent that the square root of the ozone-biomolecule reaction rate constant is inversely proportional to the slope of this plot.

Equation (3) is generally used to calculate relative rate constants. The value of A, the effective surface area of the liquid, may be somewhat different from the geometric area calculated, assuming that the liquid is a flat surface. However, for a given flow reactor design, the value of A should be independent of the reaction being studied. Thus, the value of A can be calculated from Eq. (3) using an ozone-biomolecule reaction whose reaction rate constant is known.

Equations (1) and (3) are valid only when three sets of conditions are met. First, the biomolecule concentration must be large compared to the ozone concentration in order to prevent a significant reduction of the biomolecule concentration at the surface of the liquid. Equation (3) will be valid to within 10% when

$$[B] > \frac{\pi k z^2 D [O_3]_i^2 \Delta t}{D_B H^2} \qquad (4)$$

where z is the number of moles of biomolecule B reacting with each mole of ozone, Δt is the time period over which the reaction is studied, and D_B is the diffusion coefficient for biomolecule B.[5] Second, relatively long reaction times must be studied. Eq. (3) will be correct to within 5% if

$$k[B] \Delta t > 10 \qquad (5)$$

[11] J. L. Sotelo, F. J. Beltrán, F. J. Benitez, and J. Beltrán-Heredia, *Water Res.* **23**, 1239 (1989).
[12] V. I. Matrozov, S. A. Kashtanov, A. M. Stepanov, and B. A. Tregubov, *Zh. Prikl. Khim.* **49**, 1070 (1976).

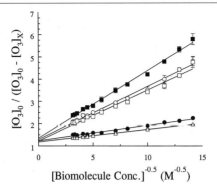

FIG. 2. Effect of biomolecule concentration on the rate of reactive absorption of ozone. Data for thiosulfate, ascorbate, cysteine, methionine, and glutathione are shown by △, ●, □, ○, and ■, respectively. Conditions were 100 mM sodium phosphate, pH 7.0, and a 1.25 ± 0.01-ml sec^{-1} carrier gas flow rate. The ozone concentration in the carrier gas was 20–41 ppm for thiosulfate, 21–29 ppm for ascorbate, 26–37 ppm for cysteine, 20–21 ppm for methionine, and 27–30 ppm for glutathione. From J. R. Kanofsky and P. D. Sima, *Arch. Biochem. Biophys.* **316**, 52 (1995), with permission.

The third requirement is that the reaction must be fast enough to consume all of the ozone within a thin layer of solution near the surface. The thickness of this layer must be small relative to the depth of solution in the reactor. The ozone concentration as a function of distance below the liquid surface is given by Eq. (6):

$$\frac{H[O_3]_y}{[O_3]_i} = \exp\left(-\frac{yk^{1/2}[B]^{1/2}}{D^{1/2}}\right) \tag{6}$$

where $[O_3]_y$ is the ozone concentration at depth y and $[O_3]_i$ is the ozone concentration in the gas phase in contact with the solution.[13] This condition is generally very easy to meet. For conditions we used to study ozone-biomolecule reactions, the ozone concentration will drop by a factor of 1000 at a depth of less than 5×10^{-4} cm.[9]

Results and Discussion

Figure 2 shows the effect of biomolecule concentration on the rate of reactive absorption of ozone. Note that the plots of $[O_3]_0/([O_3]_0 - [O_3]_x)$ against the reciprocal of the square root of the biomolecule concentration are linear for each of the biomolecules studied. Further, note that all of

[13] J. Crank, "The Mathematics of Diffusion," p. 335. Oxford Univ. Press, Oxford, 1975.

TABLE I
RELATIVE OZONE-BIOMOLECULE RATE CONSTANTS
OBTAINED BY REACTIVE ABSORPTION MEASUREMENTS[a]

Biomolecule	No detergent[b]	Detergent[c]
Methionine, pH 7.0	1.00 ± 0.06	1.00 ± 0.08
Ascorbate, pH 2.0	0.14 ± 0.01	
Ascorbate, pH 3.0	0.89 ± 0.04	
Ascorbate, pH 4.0	4.8 ± 0.2	
Ascorbate, pH 5.0	9.7 ± 0.5	
Ascorbate, pH 6.0	12 ± 1	
Ascorbate, pH 7.0	12 ± 1	
Cysteine, pH 7.0	1.1 ± 0.1	
Glutathione, pH 7.0	0.62 ± 0.03	0.39 ± 0.02
Thiosulfate, pH 7.0	18 ± 2	
Thiosulfate, pH 7.0, 3°	12 ± 1	

[a] From J. R. Kanofsky and P. D. Sima, *Arch. Biochem. Biophys.* **316,** 52 (1995), with permission.
[b] 100 mM sodium phosphate.
[c] 100 mM sodium phosphate and 70 mM sodium lauryl sulfate.

the intercepts of the plots are approximately the same, as predicted by Eq. (3), as the intercepts should depend only on the flow rate and the gas-phase mass transfer coefficient. Relative rate constants obtained from reactive absorption measurements are summarized in Table I.

Figure 3 shows a correlation between the relative rate constants obtained from reactive-absorption experiments and literature values for absolute rate constants. Note the excellent correlation for ascorbic acid, studied between pH 2 and pH 7, as well as for methionine. Studies with these reactions were used to calibrate the flow reactor, obtaining a value for the effective surface area.

Compared to ascorbic acid and methionine, the rates of reactive absorption for cysteine and for glutathione were much slower than predicted by ozone reaction rate constants obtained from stopped-flow studies.[2] Reactive absorption measurements give rate constants of 4.4 and 2.5 × 10⁶ M^{-1} sec⁻¹ for cysteine and glutathione, respectively. Previously, a number of explanations have been put forward to explain the inconsistencies in the apparent rate of the ozone-glutathione reaction.[9]

Reduced glutathione is present in the fluid lining the air passages of lungs at a concentration on the order of 100 μM.[14,15] However, the slow

[14] A. M. Cantin, S. L. North, R. C. Hubbard, and R. G. Crystal, *J. Appl. Physiol.* **63,** 152 (1987).
[15] C. E. Cross, A. van der Vliet, C. A. O'Neill, S. Louie, and B. Halliwell, *Environ. Health Perspect.* **102**(Suppl. 10), 185 (1994).

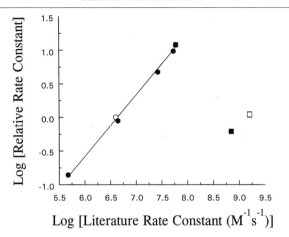

FIG. 3. Rate constants for the reaction of ozone with biomolecules: correlation between reactive absorption measurements in this study and data from the literature (Pryor et al.[2] and Giamalva et al.[1]). The ● data points, in the order shown on the horizontal axis, are for ascorbic acid, pH 2.0, ascorbic acid, pH 3.0, ascorbic acid, pH 4.0, ascorbic acid, pH 5.0, and ascorbic acid, pH 6.0. The data point for ascorbic acid, pH 7.0, overlies the data point for ascorbic acid, pH 6.0. The ○, ■, and □ data points are for methionine, pH 7.0, glutathione, pH 7.0, and cysteine, pH 7.0, respectively. From J. R. Kanofsky and P. D. Sima, Arch. Biochem. Biophys. **316,** 52 (1995), with permission.

rate of reactive absorption of glutathione relative to other antioxidants, such as ascorbate, may lessen the importance of glutathione as a protector against ozone-induced lung damage. Indeed, Langford et al.[16] found that the depletion of ascorbate from rat bronchoalveolar lavage fluid reduced the reactive absorption of ozone much more than the depletion of sulfhydryl compounds.

The concept of reactive absorption also explains some of the difficulties involved in the use of exogenous antioxidants to lessen oxidative damage from ozone. Using a simple two-phase system composed of a gas flowing over a sample of human plasma, Van der Vliet et al.[17] studied the effects of added glutathione and dihydrolipoic acid. They noted that the addition of these antioxidants to plasma increased the rate of reactive absorption of ozone, but provided only minor decreases in various measures of oxidation. In most two-phase systems, increased reactive absorption of ozone will limit the protection provided by an added antioxidant to some degree.

However, the intact respiratory system reactively absorbs all ozone in the absence of added antioxidants. Thus, antioxidants added to the fluid

[16] S. D. Langford, A. Bidani, and E. M. Postlethwait, Toxicol. Appl. Pharmacol. **132,** 122 (1995).
[17] A. Van der Vliet, C. A. O'Neill, J. P. Eiserich, and C. E. Cross, Arch. Biochem. Biophys. **321,** 43 (1995).

lining the air passages cannot increase the ozone absorption and may be more effective in protecting endogenous target molecules than studies with simple two-phase systems suggest.[17] Further, as suggested by Van der Vliet et al.,[17] added antioxidants may shift the location in the bronchial tree where ozone is absorbed to more proximal sites.

In summary, the measurement of reactive absorption of ozone by solutions containing various biomolecules can be easily done using relatively simple equipment. Analysis of the rates of reactive absorption with the theoretical models of Danckwerts and others can provide values for ozone-biomolecule rate constants.[5–7,9] In addition, the concept of reactive absorption provides some unique insights into the limitations of exogenous antioxidants as protectors against ozone.

Acknowledgments

This material is based on work supported by the Office of Research and Development, Medical Research Service, Department of Veterans Affairs. The opinions expressed are those of the authors and not necessarily the opinions of the Department of Veterans Affairs or of the United States Government.

[46] Assay for Singlet Oxygen Generation by Plant Leaves Exposed to Ozone

By Jeffrey R. Kanofsky and Paul D. Sima

Introduction

Ozone is a well-established atmospheric toxin for plants. It is known to diffuse through the open stomata of leaves into their inner air spaces.[1] Chameides[2] has carried out a mathematical analysis of the interaction of ozone with plants and concluded that most of the ozone entering a leaf will react with the ascorbic acid present within the cell walls of the plant cells lining the leaf air spaces. Ascorbic acid concentrations on the order of 300 μM within the cell wall, combined with the large rate constant for the reaction of ozone with ascorbic acid ($k = 6 \times 10^7 \ M^{-1} \ sec^{-1}$ at pH 7 in aqueous solution[3]), ensure that no ozone will reach the plasmalemmas

---

[1] G. Kerstiens and K. J. Lendzian, New Phytol. 112, 13 (1989).

[2] W. L. Chameides, Environ. Sci. Technol. 23, 595 (1989).

[3] D. Giamalva, D. F. Church, and W. A. Pryor, Biochem. Biophys. Res. Commun. 133, 773 (1985).

of the plant cells.[2] Because the reaction of ascorbic acid with ozone produces singlet oxygen in high yield, plants exposed to an ozone-containing atmosphere should generate singlet oxygen.[4,5] The singlet oxygen generated within the cell wall should also not reach the plant cell plasmalemma, as singlet oxygen reacts rapidly with ascorbic acid ($k = 1.6 \times 10^8\ M^{-1}\ \mathrm{sec}^{-1}$ in aqueous solution[6]) and, in addition, is quenched efficiently by water (3.1 μsec lifetime[7]). However, because most of the singlet oxygen is generated close to the outer surface of the cell wall, some singlet oxygen should diffuse back into the air spaces of the leaf.

The measurement of singlet oxygen chemiluminescence at 1270 nm is one of the most specific tests for singlet oxygen generation, but is somewhat limited in sensitivity.[8] The detection of 1270-nm emission from intact tissues normally presents a difficult technical challenge because the intensity of the singlet oxygen emission from intact tissues is usually very small. The lifetime of singlet oxygen within tissues is very short due to the high concentration of biological molecules.[9–11] This short lifetime normally causes the steady-state concentration of singlet oxygen to be very small and consequently limits the intensity of the 1270-nm emission. However, the plant leaf, with its network of air spaces, represents a special situation. Singlet oxygen chemiluminescence at air–tissue interfaces is enhanced significantly due to the diffusion of singlet oxygen into air where the lifetime of singlet oxygen is long and the chemiluminescence is relatively intense.[5,12,13] This effect makes the measurement of singlet oxygen emission from plants exposed to ozone feasible.

Method

The chemiluminescent spectrometer used for these measurements has been described previously.[8] The cuvette used for studies of singlet oxygen

[4] J. R. Kanofsky and P. Sima, *J. Biol. Chem.* **266,** 9039 (1991).

[5] J. R. Kanofsky and P. D. Sima, *Photochem. Photobiol.* **58,** 335 (1993).

[6] M. Rougée and R. V. Bensasson, *C.R. Acad. Sci. Paris* **302** (Series II), 1223 (1986).

[7] S. Y. Egorov, V. F. Kamalov, N. I. Koroteev, A. A. Krasnovsky, Jr., B. N. Toleutaev, and S. V. Zinukov, *Chem. Phys. Lett.* **163,** 421 (1989).

[8] J. R. Kanofsky, *Methods Enzymol.* **319,** 59 (2000).

[9] A. A. Krasnovsky, Jr., *in* "Molecular Mechanisms of Biological Action of Optical Radiation" (A. B. Rubin, ed.), p. 23. Nauka, Moscow, 1988.

[10] J. Moan and K. Berg, *Photochem. Photobiol.* **53,** 549 (1991).

[11] A. Baker and J. R. Kanofsky, *Photochem. Photobiol.* **55,** 523 (1992).

[12] J. R. Kanofsky and P. D. Sima, *Arch. Biochem. Biophys.* **312,** 244 (1994).

[13] J. R. Kanofsky and P. D. Sima, *J. Biol. Chem.* **270,** 7850 (1995).

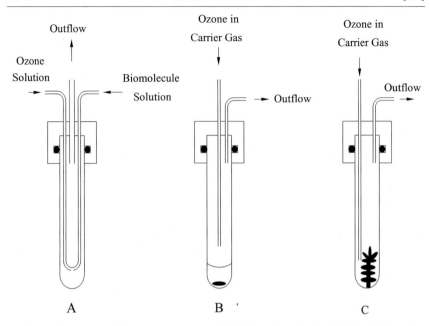

FIG. 1. Cuvettes used for measurements of singlet oxygen emission. (A) Cuvette for one-phase system using continuous flow of reagents. The solutions mix in a small hole at the bottom of the loop of Teflon tubing. (B) Cuvette used for two-phase system. A Teflon-coated magnetic bar mixes the solution rapidly. (C) Cuvette used for plant cuttings.

emission from plants is shown in Fig. 1. A small cutting from a plant is placed in the cuvette. The cuvette is a component of a sealed flow system. A 17-gauge stainless-steel tube carries ozone, diluted in carrier gas, to the bottom of the cuvette. An exit tube at the top of the cuvette collects the ozone that has not reacted with the plant leaves and directs the ozone to an assay tube containing 1% (w/v) potassium iodide, 0.1 M monobasic potassium phosphate, and 0.1 M dibasic potassium phosphate. The quantity of ozone entering the assay tube in a given time is measured using an extinction coefficient of 24,500 cm^{-1} M^{-1} at 352 nm.[14] When the system is first set up, the validity of the iodometric assay should be checked using the indigo method.[15]

A more direct analysis of singlet oxygen generation from the fluid within cell walls is useful to further characterize the mechanism of singlet oxygen generation from plant leaves. In addition to ascorbic acid, a number of other biological molecules, such as glutathione, methionine, and uric acid,

[14] B. E. Saltzman and N. Gilbert, *Anal. Chem.* **31**, 1914 (1959).
[15] H. Bader and J. Hoigné, *Water Res.* **15**, 449 (1981).

have been shown to react with ozone to generate singlet oxygen.[4,5] Thus, the observation of singlet oxygen generation does not mean that ascorbic acid is the main reactant.

A diluted extract of intercellular fluid can be obtained using the vacuum infiltration technique.[16] Leaves are washed with distilled water, after which the leaves are vacuum infiltrated three times. The infiltration fluid contains 40 mM sodium acetate, pH 4.5, and 25 μM ethylenediaminetetraacetic acid (EDTA). EDTA is needed to limit the metal-catalyzed oxidation of ascorbic acid. The pressure in the container holding the leaves and infiltration fluid is reduced to 35 kPa for a few minutes. Then, the intercellular fluid extract is recovered by centrifuging the leaves at 1500g for 10 min. The fluid is passed through a 0.2-μm filter before use.

Singlet oxygen generation from ozone-exposed intercellular extracts can be measured using either a one-phase aqueous system or a two-phase system. In the one-phase system, an ozone-containing aqueous solution is mixed with the intercellular extract using a continuous flow system. The cuvette used for these studies is shown in Fig. 1. There must be no pockets of air within the system and the flow rate must be adjusted so that the reaction is complete while the flow is still within the view of the spectrometer light detector.[5] The one-phase system has the advantage of providing a quantitative value for the amount of singlet oxygen generated per mole of ozone consumed using simple data reduction techniques.[5,8] The spectrometer is conveniently calibrated using the reaction of hydrogen peroxide with hypochlorous acid.[8]

The simplest two-phase system involves the flow of an ozone-containing carrier gas over a small volume of a rapidly stirred aqueous solution.[5,12] The cuvette used to study two-phase systems is shown in Fig. 1.

Two-phase systems containing more than one type of biological molecule sometimes behave differently than one-phase systems.[17] Excluding secondary reactions, the relative consumption of various biological molecules in a one-phase system should depend only on ozone-biomolecule rate constants. In a two-phase system, where ozone-biomolecule reactions occur close to the surface of the liquid, the relative rates of diffusion of the various molecules to the surface are also a factor. Data reduction to obtain quantitative measurements in a two-phase system is much more complex, but quantitative values for singlet oxygen yields can still be obtained from measurements of the intensity of the singlet oxygen emission.[12]

Under most experimental conditions, almost all of the emission comes

[16] F. J. Castillo and H. Greppin, *Environ. Exp. Bot.* **28,** 231 (1988).
[17] J. R. Kanofsky and P. D. Sima, *Arch. Biochem. Biophys.* **316,** 52 (1995).

from singlet oxygen that has diffused back into the gas phase.[12] The intensity of the gas-phase emission, I_{gas}, is given by Eq. (1):

$$I_{gas} = \frac{K_{gas}HYF\alpha_1}{\alpha_3(\alpha_1 + \alpha_2)(HD_3\alpha_3 + D_2\alpha_2)} \tag{1}$$

where $\alpha_n = (D_n\tau_n)^{-\frac{1}{2}}$; D_1, D_2, and D_3 are the diffusion coefficients for ozone in the liquid, for singlet oxygen in the liquid, and for singlet oxygen in the gas, respectively; τ_1, τ_2, and τ_3 are the lifetimes for ozone in the liquid, for singlet oxygen in the liquid, and for singlet oxygen in the gas, respectively; H is Henry's law constant for singlet oxygen; F is the rate of ozone consumption by the biomolecule solution; Y is the singlet oxygen yield from the ozone biomolecule solution; and K_{gas} is a spectrometer calibration constant. Reaction rate constants for ozone-biomolecule reactions being studied can be used to calculate τ_1. Values for τ_2 and τ_3 can be calculated from the singlet oxygen-quenching constants of the various system components. The reaction of ozone with ascorbic acid is a convenient standard to use to obtain K_{gas}.[12] Values for the various parameters necessary for these calculations have been summarized by Kanofsky and Sima.[12] Measured values needed to calculate K_{gas} are the intensity of the emission and the rate of ozone consumption by the biomolecule solution.

Interpretation of the intensity of singlet oxygen emission from an intact, ozone-exposed leaf requires a different mathematical formulation. Equation (1) was derived assuming that the depth of the gas space above the liquid was large compared to the mean diffusion distance for singlet oxygen.[12] The air passages in leaves are small relative to the mean diffusion distance for singlet oxygen in air (about 1 mm[18]). Thus, we can assume that the singlet oxygen concentration in each cross section of the air passages is constant. The air spaces in general have irregular shapes, but as long as the radius of curvature of the passages is large compared to the depth of ozone penetration into the cell wall (about 400 nm for a 300 μM ascorbate concentration), Eqs. (2) and (3) are valid:

$$I_{gas} = \frac{K_{gas}HYF\alpha_1\tau_3V'}{(\alpha_1 + \alpha_2)(D_2\alpha_2 + HV')} \tag{2}$$

where V' is the volume of the air passage per unit surface area. Emission from the plant cell walls is given by Eq. (3)

$$I_{cell} = \frac{K_{cell}YF\tau_2}{1 - D_2\tau_2\alpha_1^2}\left(1 - \frac{\alpha_1V'H + D_2\alpha_1^2\tau_3}{\alpha_2V'H + D_2\alpha_2^2\tau_3}\right) \tag{3}$$

[18] W. R. Midden and S. Y. Wang, J. Am. Chem. Soc. 105, 4129 (1983).

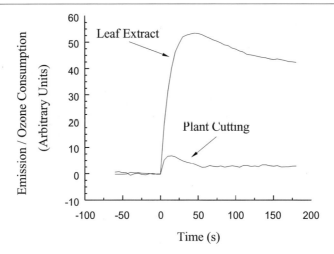

FIG. 2. Time course of 1270-nm emission from the reaction of ozone with *S. album* L. leaf extract and with *S. album* L. plant cuttings. The ascorbic acid concentration in the leaf extract was 324 μM. The ozone concentration was 21 ppm. The curve for the plant cuttings was an average of 10 experiments. The *S. album* L. plant cuttings weighed 64 ± 7 mg. The ozone concentration was 22 ± 5 ppm. The flow rate of the nitrogen carrier gas was 1.25 ml sec^{-1} for both curves. Adapted from J. R. Kanofsky and P. D. Sima, *J. Biol. Chem.* **270**, 7850 (1995), with permission.

where K_{cell} is a spectrometer calibration constant for emission from an aqueous solution.

Results and Discussion

Figure 2 shows the time course 1270-nm emission from an extract of intercellular fluid from *Sedum album* L. leaves exposed to ozone. The assignment of the 1270-nm emission to singlet oxygen is strengthened by the spectral analysis in Fig. 3, showing the expected maximum near 1270 nm. Further evidence that the emission is due to singlet oxygen comes from studies with different carrier gases for the ozone. Because most of the light emission comes from the gas phase, changing the lifetime of singlet oxygen in the gas phase should change the intensity of the emission. At 25°, the lifetime of singlet oxygen in water-saturated oxygen gas containing 22 ppm ozone is 22 msec.[12] Changing the carrier gas from oxygen to nitrogen ($k = 78\ M^{-1}\ sec^{-1}$ for nitrogen[19]) increases the singlet oxygen lifetime to 95 msec.[12] Using Eq. (1), the intensity of the singlet oxygen emission with

[19] W. R. Midden and T. A. Dahl, *Biochim. Biophys. Acta* **1117**, 216 (1992).

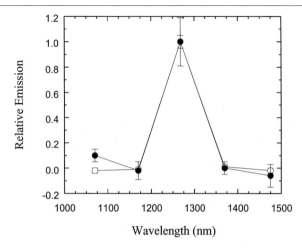

Wavelength (nm)

FIG. 3. Spectral analysis of near-infrared emission from the reaction of ozone with *S. album* L. plant cuttings and from the reaction of ozone with *S. album* L. leaf extract. ○, leaf extract; ●, plant cuttings. The larger set of error bars refers to plant cuttings. Prepared using tabular data from J. R. Kanofsky and P. D. Sima, *J. Biol. Chem.* **270**, 7850 (1995), with permission.

oxygen carrier gas should be 0.237 of the emission with nitrogen carrier gas, in good agreement with the experimental results shown in Table I.

Leaf extracts are useful for determining the specific biological molecules responsible for generating the singlet oxygen. After the *S. album* L. leaf

TABLE I
CARRIER GAS AFFECTS THE INTENSITY OF SINGLE OXYGEN EMISSION[a]

Sample	Carrier gas	Emission/ozone consumed
300 μM ascorbate[b]	Nitrogen	1.00 ± 0.17[c]
300 μM ascorbate[b]	Oxygen	0.19 ± 0.03
Leaf extract[b]	Nitrogen	1.00 ± 0.27
Leaf extract[b]	Oxygen	0.19 ± 0.02
Plant cutting[d]	Nitrogen	0.09 ± 0.02
Plant cutting[e]	Oxygen	0.01 ± 0.01

[a] Adapted from J. R. Kanofsky and P. D. Sima, *J. Biol. Chem.* **270**, 7850 (1995), with permission. Mean ozone concentrations varied from 19 to 25 ppm.

[b] The buffer contained 90 mM potassium chloride, 36 mM sodium acetate, 14.6 mM sodium phosphate, pH 7.0, and 23 μM DTAC.

[c] The ratio of emission to ozone consumed for a 300 μM solution of ascorbic acid at pH 7.0 was assigned a value of 1.00.

[d] The weight of the plant cuttings was 64 ± 7 mg.

[e] The weight of the plant cuttings was 61 ± 1 mg.

extract was treated with ascorbate oxidase, the intensity of the singlet oxygen emission was only 0.19 ± 0.07 of the emission before treatment.[13] The ozone consumed by the extract decreased from 0.29 ± 0.03 to 0.16 ± 0.02 nmol sec^{-1}.[13] This result is consistent with the theoretical work of Chameides showing that ascorbic acid will be a major sacrificial antioxidant in plants exposed to ozone.

Figure 2 also shows the time course of the 1270-nm emission from *S. album* L. cuttings exposed to ozone. Figure 3 shows the expected emission peak near 1270 nm. From Fig. 2 as well as Table I, it is apparent that the intensity of the emission per amount of ozone consumed is much lower from the cuttings than from the intercellular extract. One cause of this is internal absorption and scattering of light by the plant cuttings.[13]

A second cause of the decreased intensity is shown in Fig. 4. Figure 4, obtained using Eqs. (2) and (3), shows that the intensity of the singlet oxygen emission should decrease as the size of the air passages decrease. At very small values of V', the behavior of the system approaches that of the one-phase liquid system. The emission intensity is very small and most of the emission comes from the liquid phase. Further, changing the carrier gas from nitrogen to oxygen has no effect. In contrast, when V' is large, the behavior of the system approaches that of the simple two-phase system. Almost all of the emission comes from the gas phase. The emission intensity is large and begins to approach the emission from the simple two-phase

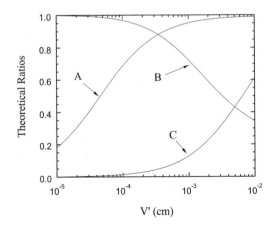

FIG. 4. Gas volume per unit surface area (V') in air passages of a leaf affects singlet oxygen emission. The cell wall ascorbate concentration is assumed to be 300 μM. (A) The fraction of emission coming from the gas phase, (B) the ratio of emission using oxygen carrier gas to emission using nitrogen carrier gas, and (C) the intensity of singlet oxygen emission. The intensity of singlet oxygen emission is scaled so that the emission expected from the simple two-phase system described by Eq. (1) is given a value of 1.00.

system. The ratio of emission with oxygen carrier gas to emission with nitrogen carrier gas decreases, approaching the value of 0.237 found for the simple two-phase system. In interpreting the meaning of Fig. 4, it is useful to think of leaf air passages as simple cylinders. The diameter of the cylinder will be four times the value of V'.

As shown in Table I, replacing the nitrogen carrier gas with oxygen decreases the intensity of emission from the plant cuttings. Unfortunately, the emission from the plant cutting was too weak to obtain an accurate value for the amount of reduction. As shown in Fig. 4, the substantial reduction in emission caused by oxygen carrier gas suggests that V' for *S. album* L. leaves is relatively large. This would imply that the emission should also be large. Consequently, internal absorption and scattering of light must be responsible, in part, for the small emission from the leaves.[13]

In summary, the measurement of singlet oxygen emission from plants exposed to ozone represents a novel technique for studying the chemistry of ozone at plant surfaces. Singlet oxygen is clearly a major intermediate formed in ozone-exposed plants.

Acknowledgments

This material is based on work supported by the Office of Research and Development, Medical Research Service, Department of Veterans Affairs. The opinions expressed are those of the authors and not necessarily the opinions of the Department of Veterans Affairs or of the United States Government.

[47] Ozone Effects on Plant Defense

By CHRISTIAN LANGEBARTELS, DIETER ERNST, JAAKKO KANGASJÄRVI, and HEINRICH SANDERMANN, JR.

Soon after the discovery of visible ozone symptoms on tobacco leaves and *Ponderosa* pine needles, the air pollutant ozone was recognized to generally inhibit plant photosynthesis and growth[1,2] and to influence the likelihood of biotic plant disease.[3] Tree decline in California was attributed

[1] R. Guderian (ed.), "Air Pollution by Photochemical Oxidants: Formation, Transport, Control and Effects on Plants." Ecological Studies 52, Springer, Berlin, 1985.
[2] M. Treshow and F. K. Anderson, "Plant Stress from Air Pollutants." Wiley, Chichester, 1989.
[3] A. S. Heagle, *Annu. Rev. Phytopathol.* **11**, 365 (1973).

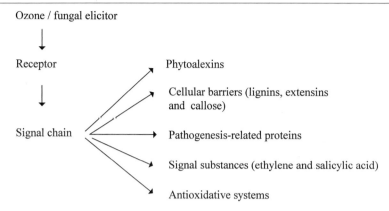

Ozone / fungal elicitor

Receptor

Signal chain

Phytoalexins

Cellular barriers (lignins, extensins and callose)

Pathogenesis-related proteins

Signal substances (ethylene and salicylic acid)

Antioxidative systems

FIG. 1. Ozone as an elicitor of plant defense reactions (modified from Sandermann *et al.*[10]).

to an ozone-enhanced attack by root rot fungi and bark beetles.[1,2,4] Extensive crop loss programs have defined yield reductions by near-ambient ozone,[5,6] but interactions between ozone and pathogens were suppressed in these studies by the use of pesticides.

Ozone is taken up through leaf or needle stomata and is apparently destroyed rapidly in the apoplast compartment.[7] The dose-dependent ozone effects on chloroplast functions and on nuclear gene expression are thought to proceed through signal chains. Ethylene and salicylic acid have been identified as signal substances.[8–10] The detrimental effects of ozone on chloroplast functions and on allocation of photosynthates influence the plant defense status by lowering growth rates and competitiveness. With regard to differential gene expression, ozone represses certain genes for chloroplast proteins and induces a broad range of antioxidant and antimicrobial defensive genes in dependence on plant species and developmental status.[8–10] In analogy to fungal elicitors, as well as microbial and viral pathogens, five classes of transcripts are known to be induced (Fig. 1). In

[4] P. R. Miller and J. R. McBride (eds.), "Oxidant Air Pollution Impacts in the Montane Forests of Southern California," Vol. 134. Springer, Berlin, 1998.

[5] W. W. Heck, O. C. Taylor, and D. T. Tingey, "Assessment of Crop Loss from Air Pollutants." Elsevier, London, 1988.

[6] H. J. Jäger, M. Unsworth, P. De Timmerman, and P. Mathy, "Effects of Air Pollution on Agricultural Crops in Europe." E. Guyot, SA, Brussels, 1992.

[7] R. L. Heath and G. E. Taylor, in "Forest Decline and Ozone: A Comparison of Controlled Chamber and Field Experiments" (H. Sandermann, A. R. Wellburn, and R. L. Heath, eds.), Ecol. Studies, Vol. 127, p. 317. Springer, Berlin, 1997.

[8] H. Sandermann, *Annu. Rev. Phytopathol.* **34**, 347 (1996).

[9] E. J. Pell, C. D. Schlagnhaufer, and R. N. Arteca, *Physiol. Plant.* **100**, 264 (1997).

[10] H. Sandermann, D. Ernst, W. Heller, and C. Langebartels, *Trends Plant Sci.* **3**, 47 (1998).

a number of cases, ozone induction has also been demonstrated at protein and metabolite levels. Several of the induced defensive systems should *a priori* be beneficial for the plant. In fact, ozone is known to protect plants against certain pathogens.[3,11] However, in many other cases, pathogen attack is enhanced, and ozone-treated plants are also more susceptible to other abiotic stressors.[8] In addition, ozone can induce an oxidative burst and subsequent necrotic lesions, as documented in detail for tobacco[12] and birch.[13] There is a need to elucidate the rules that govern the conversion of the molecular ozone effects to either increased resistance or to lesion development and higher susceptibility for abiotic or biotic stressors. A molecular epidemiology approach has been proposed.[14] This article describes the basic techniques that are necessary for this research task. Exposure techniques as well as biochemical, genetic, and phytopathological methods will be covered. It should be noted that at this time there is a development to apply isolated cDNAs or expressed sequence-tagged (EST) clones to filters or slides at high density, followed by reverse Northern analysis. This expression profiling using microarray or chip technology will, in future years, allow the parallel transcript analysis of thousands of genes in the plants' response toward ozone and correlation with known genes of pathogen defense.

Exposure Techniques

The phytotoxic effects of ozone have been studied in indoor systems that range from small cuvettes to highly sophisticated walk-in exposure chambers. Early constructions of indoor exposure systems have been reviewed.[15] The phytotron of the GSF-Research Centre of Environment and Health (Munich) consists of four walk-in chambers, each of which can be subdivided into four separate smaller chambers. Ozone dosage and solar light simulation are of a high technical standard as described previously.[16]

---

[11] W. J. Manning and A. von Tiedemann, *Environ. Pollut.* **88,** 219 (1995).

[12] M. Schraudner, W. Moeder, C. Wiese, W. Van Camp, D. Inzé, C. Langebartels, and H. Sandermann, *Plant J.* **16,** 235 (1998).

[13] R. Pellinen, T. Palva, and J. Kangasjärvi, *Plant J.* **20,** 349 (1999).

[14] H. Sandermann, *Environ. Pollut.,* in press (2000).

[15] R. C. Musselman and B. A. Hale, *in* "Forest Decline and Ozone: A Comparison of Controlled Chamber and Field Experiments" (H. Sandermann, A. R. Wellburn, and R. L. Heath, eds.), Ecol. Studies, Vol. 127, p. 277. Springer, Berlin, 1997.

[16] C. Langebartels, W. Heller, D. Ernst, M. Lippert, C. Lütz, H.-D. Payer, and H. Sandermann, *in* "Forest Decline and Ozone: A Comparison of Controlled Chamber and Field Experiments" (H. Sandermann, A. R. Wellburn, and R. L. Heath, eds.), Ecol. Studies, Vol. 127, p. 163. Springer, Berlin, 1997.

The GSF phytotron chamber is shown schematically in Fig. 2. Studies under seminatural conditions are performed in open-top chambers (OTCs) that have only limited control over light, humidity, and temperature. OTCs have been widely applied in crop loss programs.[5,6] Several types of open- and

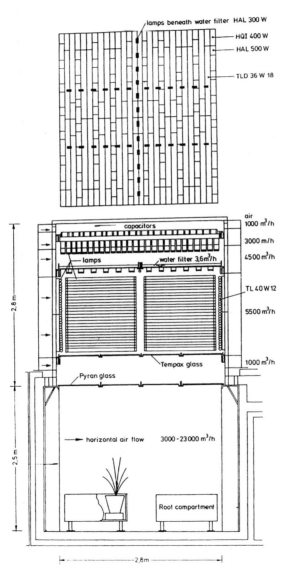

FIG. 2. Schematic view of the phytotron exposure facility at GSF (from Langebartels *et al.*[16]).

closed-top field exposure chambers have been described.[15] These authors also present a useful discussion of possible experimental pitfalls. British researchers have developed solar dome chambers and a pioneering open-air fumigation system (Liphook field exposure experiment). These techniques have also been summarized.[17]

Ozone Exposure of Ozone-Sensitive Tobacco Bel W3 and Other Plant Species

Four plexiglass cuvettes (108 × 71 × 95 cm) placed in a growth cabinet (10 m^2) are used to simultaneously expose 4 × 6 tobacco plants.[18] The cuvettes draw conditioned air from the large cabinet and use vertical, single pass through air circulation. Ozone-free air is applied to all plants during pre- and postcultivation and to control plants during the entire experimental period. To obtain pollutant-free air, ambient air is passed through a series of filters, including particle filters, activated charcoal, KMnO$_4$-coated alumina granules (Purafil II), and charcoal, and through a second set of particle filters. This system effectively removes ozone, SO$_2$, and nitrogen oxides to more than 99%.[16]

For pollutant treatments, ozone is injected by computer-controlled mass flow meters (MKS Company, Eching, Germany) into the incoming stream of conditioned and filtered air. Ozone is generated by silent electrical discharge in dry oxygen, using, for example, the ozone generator Model 500 (Fischer, Meckenheim, Germany). Ozone produced from air should not be used because it contains varying amounts of interfering N compounds. Air from the chambers is sampled using Teflon lines, and ozone concentrations are measured at the plant canopy level with an ultraviolet ozone analyzer (CSI 3100, Columbia Scientific Instruments). Calibration is performed regularly.

A switching system is used to monitor air samples from the four cuvettes (5 min each) and ozone concentrations are recorded automatically. Two circulating fans per cuvette, providing a wind speed of 0.3 to 1.3 m/sec, are installed to ensure uniform exposure of the plants and to overcome boundary layer resistance of the leaves.

Tobacco plants, grown in 14-cm pots in Fruhstorfer standard substrate (type T) for 8 weeks, from 20 to 30 cm high, are used for experimentation. Seeds of the ozone-sensitive cv. Bel W3 and the tolerant Bel B may be ob-

---

[17] A. R. Wellburn, J. D. Barnes, P. W. Lucas, A. R. McLeod, and T. A. Mansfield, in "Forest Decline and Ozone: A Comparison of Controlled Chamber and Field Experiments" (H. Sandermann, A. R. Wellburn, and R L. Heath, eds.), Ecol. Studies, Vol. 127, p. 200. Springer, Berlin, 1997.

[18] C. Langebartels, K. Kerner, S. Leonardi, M. Schraudner, M. Trost, W. Heller, and H. Sandermann, *Plant Physiol.* **95**, 882 (1991).

tained from one of the authors (CL). They are germinated with a 1-mm cover of the substrate and grown in 20 × 5-cm plastic boxes for 4 weeks before being potted into individual containers. Key climate parameters during cultivation and exposure are 25/20° day/night temperatures, a 14-hr photoperiod (6 to 22 hr, 100 to 200 μmol m^{-2} sec^{-1}), and 70% relative air humidity.

Tobacco and tomato plants are exposed to a single pulse or to daily pulses of 150 or 200 nl/liter ozone, respectively, for 5 hr (between 9 and 14 hr). A single pulse will lead to approximately 20 to 30% leaf injury of middle-aged leaves of tobacco cv. Bel W3. Alternatively, ozone field values can be simulated as 2 hr means as ambient (1×) and proportionally to ambient increased (e.g., 1.5×, 2×) or decreased (e.g., 0.2×) levels.[16,19] Leaf injury is determined after 24 or 48 hr on middle-aged leaves 3 to 5. Leaf 1 is defined as the first leaf from the apex that is longer than 8 cm. These leaves show a uniform distribution of lesions on the entire surface of cv. Bel W3, whereas no injury is found on leaves of cv. Bel B.[12] Injury is determined visually as the percentage of leaf area. Data are calibrated with a planimeter (LI-3000A, LI-COR, Nebraska).

Oxidative Burst

In analogy to plant pathogen attack, ozone exposure triggers an oxidative burst in tobacco cv Bel W3[12] as well as in other plant species.[20] Histochemical detection of H_2O_2 and superoxide radicals is performed in tobacco leaf discs (2 cm diameter) or leaf halves infiltrated with 0.1% (w/v) 3,3'-diaminobenzidine × HCl, 10 mM MES buffer (pH 6.5), or 0.1% (w/v) nitroblue tetrazolium (NBT), 10 mM sodium azide, 50 mM potassium phosphate (pH 6.4), respectively.[12] Leaves are infiltrated *in vacuo* (200 mbar, 2 × 30 sec), incubated in the dark for 30 min (20 min for *A. thaliana*), and immersed in a mixture of lactic acid, phenol, and water (81 : 1 : 1, v/v/v) for 2 days in the dark. Alternatively, destaining can be performed in 95% (v/v) ethanol.

Stress Metabolites

Ethylene emission from ozone-treated plants is one of the best early indicators of ozone sensitivity[21] and exhibits maximum values after 1–2 hr (tobacco, tomato, *Arabidopsis*) to 8 hr (European beech). For measurement, tobacco leaf segments (2 × 8 cm) are placed adaxially on filter paper moistened with water or buffer.[18] For experiments with other plant species,

[19] W. Van Camp, H. Willekens, C. Bowler, M. Van Montagu, D. Inzé, P. Reupold-Popp, H. Sandermann, and C. Langebartels, *Bio/Technology* **12**, 165 (1994).
[20] K. Overmyer, J. Kangasjärvi, T. Kuittinen, and M. Saarma, *in* "Responses of Plant Metabolism to Air Pollution and Global Change" (L. De Kok, and I. Stulen, eds.), p. 403. Backhuys Publishers, Leiden, 1998.
[21] F. A. M. Wellburn and A. R. Wellburn, *Plant Cell Environ.* **19**, 754 (1996).

one individual leaflet is routinely used with tomato, two complete leaves with European beech (*Fagus sylvatica* L.), and two leaf rosettes with *Arabidopsis thaliana.* The filter paper is rolled up cylindrically and placed into 7-ml glass tubes, which are then sealed with silicon septa (Verneret, France) and incubated for exactly 1 hr at 20° in the dark. One-milliliter gas samples are withdrawn with a disposable syringe. Ethylene is measured by gas chromatography (AutoSystem XL, Perkin-Elmer, Überlingen, Germany) equipped with a Porapak Q (80–100 mesh, 0.6 m × 2 mm) column and a flame ionization detector. Column, injector, and detector temperatures are 40, 150, and 200°, respectively. N_2 is used as the carrier gas at 30 ml/min. Under these conditions, retention times for ethylene and ethane are 0.75 and 1.0 min, respectively.[18,22]

Levels of l-aminocyclopropane-l-carboxylate (ACC), the precursor of ethylene, correlate with ethylene emission and can conveniently be used to estimate ethylene induction in deep-frozen material.[18,22] Contents of the diamine, putrescine, are inversely correlated with ethylene levels and are determined after dansyl or FMOC derivatization and HPLC (high-performance liquid chromatography) separation.[18,22]

For analyses of phenolic stress metabolites (in particular, phytoalexins), modifications of a published method[23] have proved to work with numerous plant species. About 100 mg leaf or needle material that had been shock-frozen and stored in liquid nitrogen is mixed thoroughly under liquid nitrogen cooling with 100 mg Celite (diatomaceous earth, Sigma) by means of a Microdismembrator II (Braun-Melsungen, Germany; 2.5 min, medium speed). Samples are divided equally into two safe-lock Eppendorf vials. Methanol (500 μl/vial) is added, the sample is dispersed using a vortex mixer, and extraction is allowed to proceed at room temperature in the dark for 2 hr with vortex treatment every 15 min. The samples are cleared by centrifugation (Eppendorf centrifuge, 11,000 rpm, 10 min) and the supernatant is stored at −80°. Prior to HPLC analysis, 25% (v/v) water is added, the sample is again centrifuged, and aliquots (e.g., 10 μl) are injected for HPLC analysis. Several HPLC methods for different classes of phenolic compounds have been described in detail.[23,24]

Ozone-Induced Transcripts

Ozone-induced transcripts can be assigned to five major metabolic areas with known phytopathological relevance[8–10] (Fig. 1). Several ozone-induced

[22] J. Tuomainen, C. Betz, J. Kangasjärvi, D. Ernst, Z.-H. Yin, C. Langebartels, and H. Sandermann, *Plant J.* **12,** 1151 (1997).

[23] D. Rosemann, W. Heller, and H. Sandermann, *Plant Physiol.* **97,** 1280 (1991).

[24] L. Packer (ed.), *Methods Enzymol.* **299,** Part A, Chapters 9–18 (1999).

cDNA libraries have been constructed from sources that include parsley,[25,26] potato,[27] *A. thaliana*,[28] Scots pine,[29,30] birch,[31] saltbush,[32] and spruce.[33] The construction of these cDNA libraries, either by differential screening or differential display, involves commercially available kits, as recommended by the individual manufacturer.

Isolation of RNA from Herbaceous Plant Species

Most standard procedures for the isolation of total RNA from herbaceous plant species described in plant molecular biological manuals will result in functional RNA. In addition, plant RNA extraction kits are available commercially for different herbaceous species, including *A. thaliana*, potato, tobacco, and tomato, which are often used in ozone research. For the isolation of *A. thaliana* leaf RNA, two commercially available kits [RNeasy Plant Mini RNA kit (Qiagen, Hilden); TRIzol Reagent (GIBCA/BRL, Karlsruhe)] can be used successfully. RNA can be purified from leaf samples, within 1 hr per day, and up to 30 samples can be processed per day. For leaves of tobacco, tomato, and potato, the following RNA isolation method has been used successfully.[22,34,35]

> Extraction buffer: 8 M guanidine hydrochloride, 20 mM MES, 20 mM EDTA, 50 mM 2-mercaptoethanol, pH 7.0
> SET (10 mM Tris, pH 7.4, 5 mM EDTA, 1% SDS) saturated phenol/chloroform/isoamyl alcohol (25:24:1, v/v)
> Chloroform/isoamyl alcohol (24:1, v/v)
> 1 M acetic acid
> 3 M potassium acetate, pH 5.2

[25] H. Eckey-Kaltenbach, E. Großkopf, H. Sandermann, and D. Ernst, *Proc. Royal Soc. Edinburgh* **102B**, 63 (1994).
[26] H. Eckey-Kaltenbach, E. Kiefer, E. Grosskopf, D. Ernst, and H. Sandermann, *Plant Mol. Biol.* **33**, 343 (1997).
[27] C. D. Schlagnhaufer, R. E. Glick, R. N. Arteca, and E. J. Pell, *Plant Mol. Biol.* **28**, 93 (1995).
[28] Y. K. Sharma and K. R. Davis, *Free Radic. Biol. Med.* **23**, 480 (1997).
[29] A. Wegener, W. Gimbel, T. Werner, J. Hani, D. Ernst, and H. Sandermann, *Can. J. For. Res.* **27**, 945 (1997).
[30] A. Wegener, W. Gimbel, T. Werner, J. Hani, D. Ernst, and H. Sandermann, *Biochim. Biophys. Acta* **1350**, 247 (1997).
[31] M. Kiiskinen, M. Korhonen, and J. Kangasjärvi, *Plant Mol. Biol.* **35**, 271 (1997).
[32] E. G. No, R. B. Flagler, M. A. Swize, J. Cairney, and R. J. Newton, *Physiol. Plant.* **100**, 137 (1997).
[33] K. Buschmann, M. Etscheid, D. Riesner, and F. Scholz, *Eur. J. Forest Pathol.* **28**, 307 (1998).
[34] D. Ernst, M. Schraudner, C. Langebartels, and H. Sandermann, *Plant Mol. Biol.* **20**, 673 (1992).
[35] H. Willekens, W. Van Camp, M. Van Montagu, D. Inzé, C. Langebartels, and H. Sandermann, *Plant Physiol.* **106**, 1007 (1994).

1. Using a mortar and pestle, grind 0.3 g frozen leaf tissue under liquid nitrogen to a fine powder.

2. Allow to thaw in 2 volumes of extraction buffer, mix well, and extract with 1 volume phenol/chloroform/isoamyl alcohol (10–15 min, Vortex-Vibrator) and separate phases by centrifugation (10,000g for 15 min).

3. Reextract the aqueous phase with 1 volume chloroform/isoamyl alcohol and separate by centrifugation (10,000g for 15 min).

4. Add to the final aqueous phase 1/20 volume 1 M acetic acid and 0.7 volumes of cold ethanol, and precipitate RNA at $-20°$ overnight or at $-70°$ for 30 min.

5. Collect precipitate by centrifugation at 10,000g for 10 min at 4° and wash twice with 1 ml 3 M potassium acetate and once with 70% ethanol.

6. Air dry the pellet and dissolve the dried sample in 50–200 μl sterile water (if necessary, heat up to 60°) and store at $-80°$.

Isolation of total RNA from Tree Species for Northern Analysis and RT-PCR

The isolation of high-quality RNA from tree tissues, especially from gymnosperms, is often hampered by the presence of interfering secondary metabolites. As ozone enters the leaf via stomata, only RNA extraction methods for leaf and needle material are considered here. The method of choice for isolation of total RNA from needles, suitable for Northern blot and reverse-transcriptase polymerase chain reaction (RT-PCR) analysis has been the protocol according to Chang et al.[36] The isolation procedure requires only a few steps, and RNA precipitation can be carried out overnight. Other methods described, which result in similar yields of about 120–150 μg/g fresh weight (A_{260}) and purities of 1.7–2.0 (A_{260}/A_{280}), are more complicated and usually include an aqueous two-phase system or CsCl centrifugation, as well as a phenol extraction step.

The following RNA isolation procedure[36] has been used successfully for the isolation of total RNA from needles of Scots pine seedlings and leaves of grapevine.[29,37]

Solutions and equipment for total needle RNA isolation: Extraction buffer, 2% (w/v) hexadecyltrimethylammonium bromide (CTAB), 2% (w/v) polyvinylpyrrolidone 25 (PVP),* 100 mM Tris–HCl, pH 8.0, 25 mM EDTA, 2 M NaCl, 0.5 g l^{-1} spermidine, 2% (v/v) 2-mercaptoethanol (to be added just before use)

[36] S. Chang, J. Puryear, and J. Cairney, *Plant Mol. Biol. Rep.* **11**, 113 (1993).

[37] R. Schubert, R. Fischer, R. Hain, P. H. Schreier, G. Bahnweg, D. Ernst, and H. Sandermann, *Plant Mol. Biol.* **34**, 417 (1997).

* Alternatively, polyvinylpolypyrrolidone (PVPP) can be used (2:1; w/w, with respect to lyophilized tissue).

Chloroform/isoamyl alcohol (24:1; v/v)
10 M LiCl
SSTE buffer: 10 mM Tris–HCl, pH 8.0, 1 mM EDTA, 1 M NaCl, 0.5% (w/v) SDS

1. Grind frozen needles (0.5 g) together with Celite (tip of a spatula) under liquid nitrogen to a fine powder using a Microdismembrator (Braun-Melsungen) at maximum output for 2 min. Do not allow the tissue to thaw.
2. Add 3 ml of prewarmed extraction buffer (65°) and incubate for at least 15 min at 65° under rigorous shaking (vortex shaker).
3. Extract two times with 1 volume of chloroform/isoamyl alcohol. The phases are separated by centrifugation at 12,000g for 15 min at 4°.
4. Add one-fourth volume 10 M LiCl to the aqueous supernatant and mix. Precipitate the RNA overnight at 4° and centrifuge at 12,000g for 15 min at 4°.
5. Dissolve the pellet in 500 μl SSTE buffer and extract with 1 volume chloroform/isoamyl alcohol. Separate the phases at 12,000g for 5 min at 4°.
6. Add 2 volumes of cold ethanol, mix, and incubate at −80° for 1 hr. Recover the precipitate by centrifugation at 13,000g for 30 min at 4° and remove the supernatant.
7. Dry the pellet under air and resuspend in 32 μl of diethyl pyrocarbonate (DEPC)-treated water.
8. The aqueous solution of RNA can be stored at −80° for a short period of time. For long-term storage, precipitated RNA should be kept under ethanol at −20°.

Isolation of Poly(A)⁺ RNA from Tree Species for the Construction of cDNA Libraries[38]

Intact poly(A)$^+$ RNA has been isolated from needles of Scots pine for the analysis of ozone-induced, *in vitro*-translated proteins,[38] as well as for the construction of an ozone-induced cDNA library, resulting in full-length cDNA clones.[29,30]

Extraction buffer: 4.2 M guanidine thiocyanate, 130 mM diethyldithiocarbamic acid, 5 mM EDTA (pH 8.0), 16 mM lauryl sarcosine, 26 mM sodium citrate (pH 7.0), 2% (w/v) PVP 10,000 (GTC buffer). Immediately before use, add 400 μl 2-mercaptoethanol to 16 ml GTC buffer
CsCl cushion: 5.7 M CsCl, 5 mM EDTA (pH 8.0)
Tris (0.1 M, pH 8.8) saturated phenol/chloroform/isoamyl alcohol (25:24:1; v/v/v)

[38] E. Großkopf, A. Wegener-Strake, H. Sandermann, and D. Ernst, *Can. J. For. Res.* **24**, 2030 (1994).

Chloroform/isoamyl alcohol (24:1; v/v)

3 M sodium acetate, pH 5.2

1. Grind frozen needles (2 g) together with Celite (tip of a spatula) under liquid nitrogen to a fine powder using a Microdismembrator (Braun-Melsungen) at maximum output for 2 min.

2. Immediately transfer the powder into a mixture of 16 ml GTC buffer and gently shake the suspended material at room temperature for 30 min.

3. Centrifuge at 7800g for 20 min at 18°.

4. Add 0.4 g CsCl/ml supernatant and layer the solution on a 2.5-ml CsCl cushion and centrifuge at 150,000g for 16 hr at 4° in a swing-out rotor.

5. Carefully remove the supernatant, resuspend the RNA pellet in 500 μl DEPC-treated water, and transfer to an Eppendorf tube.

6. Extract two times with phenol/chloroform/isoamyl alcohol and once with chloroform/isoamylalcohol.

7. Precipitate RNA with 2.5 volumes of cold ethanol and 0.1 volumes of 3 M sodium acetate overnight at −20°.

8. Centrifuge at 13,000g for 30 min at 4° and remove the supernatant.

9. Dry the pellet under air and dissolve in 105 μl DEPC-treated water.

This method has been used successfully for the isolation of total RNA (120 μg/g fresh weight) from seedlings of Scots pine and older needle years of Scots pine, Mugho pine, and Norway spruce in Northern blot analysis.[29,30,38,39] Exemplary cDNA probes for each of the pathways of Fig. 1 are listed in Table I.

Isolation of Poly(A)$^+$ RNA[36]

For the isolation of poly(A)$^+$ RNA, commercially available kits using poly(U)-Sepharose, oligo(dT)-cellulose, and biotinylated oligo(dT)-primer/streptavidin-bound magnetic particles have been tested. The polyA-Tract system (Promega, Heidelberg, Germany) resulted in reproducible good results with yields of about 1.0–1.5% and purities of 2.0 ($A_{260/280}$) and <2.5 ($A_{260/230}$). Agarose gel electrophoresis showed no discrete ribosomal bands, and cDNA first-strand synthesis resulted in yields of more then 15%.

The procedure is as described by the supplier (protocol for small-scale mRNA isolation, using up to 1 mg of total RNA) with the following modifications:

1. Add 50 μl sample buffer (10 mM Tris, pH 7.5, 1 mM EDTA, 5 M NaCl) to 450 μl dissolved RNA.

2. Wash particle-bound RNAs twice with high salt buffer (10 mM Tris, pH 7.5, 1 mM EDTA, 0.5 M NaCl).

---

[39] C. Zinser, D. Ernst, and H. Sandermann, *Planta* **204,** 169 (1998).

TABLE I
OZONE-INDUCED TRANSCRIPTS[a]

	Gene	Plant species	Plasmid used	Accesion number	Reference
Phytoalexins	STS	Scots pine	pSP-54	S50350	[b]
Cellular barriers	CAD	Norway spruce	pSCAD15	X72675	[c]
		Scots pine	pSCAD15	X72675	[b]
PR proteins	Basic β-1,3-glucanase	Tobacco	PGL 43	M20620	[d]
Signal substances	ACC synthase 2	Tomato	pBTAS1	M83320	[e]
Antioxidative systems	Catalase 2	Tobacco	pCat2A	Z36976	[f]
	GST-1	Arabidopsis	cDNA clone		[g]

[a] Exemplary cDNA probes for each of the ozone-responsive defense pathways of Fig. 1 are given.
[b] C. Zinser, D. Ernst, and H. Sandermann, *Planta* **204,** 169 (1998).
[c] H. Galliano, M. Cabané, C. Eckerskorn, F. Lottspeich, H. Sandermann, and D. Ernst, *Plant Mol. Biol.* **23,** 145 (1993).
[d] D. Ernst, M. Schraudner, C. Langebartels, and H. Sandermann, *Plant Mol. Biol.* **20,** 673 (1992).
[e] J. Tuomainen, C. Betz, J. Kangasjärvi, D. Ernst, Z.-H. Yin, C. Langebartels, and H. Sandermann, *Plant J.* **12,** 1151 (1997).
[f] H. Willekens, W. Van Camp, M. Van Montagu, D. Inze, C. Langebartels, and H. Sandermann, *Plant Physiol.* **106,** 1007 (1994).
[g] Y. K. Sharma and K. R. Davis, *Plant Physiol.* **105,** 1089 (1994).

3. Wash twice with low salt buffer (10 mM Tris, pH 7.5, 1 mM EDTA, 0.1 M NaCl).

4. Elute poly(A)$^+$ RNA with 500 μl DEPC-treated water at 37°.

Production of Ozone-Sensitive Mutants of *Arabidopsis*

The small crucifer *A. thaliana,* a popular model system in genetic studies in plant biology,[40] has several advantages that make it a suitable system in plant/ozone/pathogen interaction studies. There are several defined mutants available from the two *Arabidopsis* stock centers,[†] physical and genetic maps of the chromosomes exist, EST (expressed-sequence-tag) projects identify expressed genes, and the complete genome is expected to be sequenced by the end of year 2000. Information is available from the *Arabidopsis* web site

[40] D. W. Meinke, J. M. Cherry, C. Dean, S. D. Rounsley, and M. Koornneef, *Science* **282,** 662 (1998).

† Arabidopsis Biological Resource Center (ABRC) at the Ohio State University and The Nottingham Arabidopsis Stock Centre (NASC) in the Department of Life Science at the Ohio State University (http://www.aims.cps.msu.edu/aims/) and The University of Nottingham, UK (http://nasc.nott.ac.uk/).

(http://www.arabidopsis.org). Various seed stocks and supplies are available from a commercial supplier, Lehle Seeds (http://www.arabidopsis.com). Detailed information is available on the basic biology and development, as well as growing of *Arabidopsis,* generation, and mapping of mutations and isolating the gene responsible for the mutation.[41–43]

Defense gene expression in most widely used *Arabidopsis* "laboratory" ecotypes, such as Col-0, is induced by ozone similar to other species.[20,44–46] However, if visible lesions are used as a marker of sensitivity, Col-0, for example, is quite resistant to ozone; it can tolerate 300 nl/liter ozone without visible symptoms.[47] Ozone-sensitive ecotypes, such as Cvi-0, have also been identified.[28] The major advantage of *Arabidopsis* is to generate ozone-sensitive mutants and to perform genetic analysis on them. This can ultimately yield resistant/sensitive pairs where a mutation in one gene is the only difference between the strains. Thus, stress reactions induced by ozone, and their importance for pathogen susceptibility, can be studied specifically.

Arabidopsis can be grown in small environmentally controlled cuvettes. In screenings for ozone-sensitive ethylmethane sulfonate (EMS) mutants,[20,48] 15,000–20,000 M_2 plants have been used. In the screening by Conklin *et al.,*[48] EMS-mutagenized Col-0 were exposed to 250 nl/liter ozone for 8 hr 14 days after sowing, using, a density of 40/cm^2, and mutants were selected based on visible tissue damage. In the screening reported by Overmyer *et al.,*[20] 3-week-old EMS-mutagenized Col-0 plants grown at a density of 3000/m^2 were exposed to 250 nl/liter ozone (8 hr) for three consecutive days and individuals displaying visible lesions were selected. An alternative approach for finding ozone sensitive strains in *Arabidopsis* was applied by Sharma and Davis,[28] who screened 36 naturally occurring ecotypes available at the ABRC for ozone sensitivity. They used 300 nl/liter of O_3, 6 hr a day for 3 days, and identified ecotypes that were significantly more sensitive to O_3 exposure.

The seed population mutagenized is referred to as the M_1 generation. Progeny derived from M_1 plants by self-fertilization, the M_2 generation, is the first generation where recessive mutations can be detected and, for this

[41] C. Koncz, N.-H. Chua, and J. Schell (eds.), "Methods in *Arabidopsis* Research." World Scientific Publishing, Singapore, 1992.

[42] E. M. Meyerowitz and C. R. Somerville (eds.), "*Arabidopsis.*" Cold Spring Harbor Laboratory Press, Cold Spring Harbor, NY, 1994.

[43] J. M. Martinez-Zapater and J. Salinas (eds.), "Methods in Molecular Biology: *Arabidopsis* Protocols," Vol. 82. Humana Press, Totowa, NJ, 1998.

[44] Y. Sharma and K. Davis, *Plant Physiol.* **105,** 1089 (1994).

[45] P. L. Conklin and R. L. Last, *Plant Physiol.* **109,** 203 (1995).

[46] Y. K. Sharma, J. León, I. Raskin, and K. R. Davis, *Proc. Natl. Acad. Sci. U.S.A.* **93,** 5099 (1996).

[47] J. Vahala, C. D. Schlagnhaufer, and E. J. Pell, *Physiol. Plant* **103,** 45 (1998).

[48] P. L. Conklin, E. H. Williams, and R. L. Last, *Proc. Natl. Acad. Sci. U.S.A.* **93,** 9970 (1996).

reason, is used for screening. Chemically (EMS) and physically (gamma or fast neutron irradiation) mutagenized M_2 generation seeds are available from Lehle Seeds in Col-0, Ler, and Ws ecotypes. For chemical or irradiation mutagenesis, any other ecotype can be used if the mutagenization is to be performed by the researcher. Detailed information on the mutagenization procedures can be found in books describing general protocols of *Arabidopsis* research.[42,43] Ten thousand to 20,000 M_2 generation seeds are grown at a density of 2500 to 5000 individuals/m^2 to the desired developmental stage. Screening should be performed before the plants start bolting, preferably at 2–3 weeks age. On subsequent ozone exposure, plants showing visible damage are selected and replanted individually. M_3 generation seed is collected. M_3 plants derived from each selected M_2 plant are grown in both high density (same as in screening) and low density (to be used in subsequent work), and the true sensitivity under both growing conditions is verified by exposing them to ozone. If M_3 generation plants show segregation of the phenotype at the 3 : 1 (sensitive : resistant) or 1 : 2 : 1 (sensitive : intermediate : resistant) ratio, sensitive individuals should be selected, M_4 generation seed collected, and the sensitivity of the M_4 plants analyzed. M_3 plants whose M_4 progeny does not segregate for sensitivity should be selected for further work.

The first step of genetic analysis is to cross the mutant line (M_3) with the wild type used. If the sensitivity is a result of a mutation in one Mendelian locus, exposing the F_1 to ozone will reveal whether the mutation is recessive (F_1 generation behaves as wild type), codominant (F_1 intermediate), or dominant (the F_1 generation shows the sensitive phenotype). The F_2 generation (obtained by harvesting seed from the F_1 plants) is used to determine how many loci are responsible for the phenotype by following the Mendelian segregation of sensitivity in the F_2 generation. Seed from selected F_2 generation individuals with a mutant phenotype(s) is used to follow segregation at the F_3 generation to verify the inheritance of the mutation. EMS mutagenization results in individual plants containing several independent mutations in the genome. Ozone sensitivity, backcrossing, exposing the progeny to ozone, and reselection for sensitivity should be repeated for several generations.[43] According to rules for the *Arabidopsis* community,[49] all new mutations should be mapped and crossed with existing mutants mapping to the same location that have a similar phenotype to test complementation.

Biochemical and Gene Expression Analysis of Mutants

The chemical and biochemical analysis described earlier can be used in studies with the mutants. Such an approach with the *soz1* mutant,[48] later renamed *vtc1*, which was isolated by its ozone-sensitive phenotype, led to

[49] D. Meinke and M. Kornneef, *Plant J.* **12**, 247 (1997).

the identification of a defective ascorbic acid biosynthetic pathway.[50] Another ozone-sensitive mutant, termed czn 1, has lower levels of CuZn-SOD protein.[51] In addition to Northern hybridizations, use of the *Arabidopsis* expressed sequence tag (EST) collection (http://www.ncbi.nlm.nih.gov/dbEST/index.html) can be utilized as quotes for transcript detection/profiling. *Arabidopsis* ESTs are available from the Ohio State *Arabidopsis* Stock Center. Contigs of the *Arabidopsis* EST sequences assembled by The Institute for Genomic Research (TIGR) are available at their *Arabidopsis* www pages (http://www.tigr.org/tdb/at/at./html). To identify EST clones of genes of interest, these data bases can be searched with a nucleotide or protein sequence or with a gene product name.

The reverse Northern hybridization technique can be applied to ozone-sensitive mutants with EST clones. Information can be obtained on the downstream effects that a defect in one defined gene responsible for newly gained sensitivity has in antioxidant and pathogen defense gene expression, and also in other processes. In reverse Northern hybridization, cDNA/EST clones are spotted on the membrane and probed with labeled cDNA made from isolated mRNA. PCR-amplified inserts from each EST clone are spotted in equal amounts (10 to 100 ng) to hybridization membranes using a 96-well (or similar) dot-blot manifold. The number of identical membranes required is the same as the number of probes to be used. Isolated mRNA from control and ozone-exposed tissues is used as a template for probes by labeling the first strand of cDNA. Hybridizations and detection are also performed using standard procedures.[52]

Isolation of the Gene Explaining the Phenotype

The ultimate goal in generating and selecting ozone-sensitive mutants is to find and isolate the gene that explains ozone sensitivity. When the mutation has been introduced by chemical mutagenesis, location of the mutation on the genetic map of *Arabidopsis* has to be determined first. For this the mutant has to be crossed with another ecotype polymorphic for the markers to be used, and genetic linkage of the phenotype with markers can be established. Molecular markers (RFLP, SSLP, SSCP, RAPD, AFLP, CAPS) are used for mapping and the procedures are described in detail.[43] BAC und YAC (bacterial and yeast artificial chromosomes) clones covering this region can be searched for at the Arabidopsis Genomic View (http://genome-www3.stanford.edu/Arabidopsis/

[50] P. L. Conklin, *Trends Plant Sci.* **3**, 329 (1998).
[51] D. J. Kliebenstein, P. L. Conklin, A. C. Martin, and R. L. Last, Plant Protein Club 1998 Annual Symposium, Abstract Book (D. Bowles, ed.), pp. 13–14. York, Great Britain (1998).
[52] J. Sambrook, T. Maniatis, and E. F. Fritsch, "Molecular Cloning," 2nd Ed. Cold Spring Harbor Laboratory Press, Cold Spring Harbor, NY, 1989.

chromosomes/), at Cold Spring Harbor Laboratories (http://nucleus.cshl. org/arabmaps/), and at the Max-Planck-Institute Golm, Germany (http:// www.mpimp-golm.mpg.de/101/mpi_mp_map/bac.html). Clones can be obtained from ABRC or the RZPD Berlin, Germany (http:///www.rzpd.de/) (BAC clones). With completion of the *Arabidopsis* physical map, chromosome walking experiments are no longer necessary. BAC or YAC sequence information can be used to simply generate new markers[53] to fine-map the locus of interest. BAC or YAC inserts may then be subcloned into smaller pieces to DNA from another ecotype as needed; the corresponding genomic region has to be cloned first from the respective ecotype.

Pathogen Treatment

Laboratory and field studies with crop plant and tree species and numerous viral, bacterial, and fungal pathogens have shown that ozone can dispose the host plant to either higher tolerance or higher susceptibility. The considerable experimental data base and the current lack of predictability have been well reviewed.[11] However, laboratory studies with model plant species have documented the enhanced pathogen tolerance expected from the induction of plant defense systems by ozone. Only methods from two case studies will be described briefly here.[46,54]

Tobacco/Tobacco Mosaic Virus[54]

Nicotiana tabacum L. cv. Xanthi-nc was employed after UVB or ozone treatment, or without special treatment (control). Plants were inoculated by gently rubbing leaves with the U1 strain of TMV ($5 \mu g \cdot leaf^{-1}$) in the presence of carborundum. Necrotic lesions developed 40–48 hr later. The mean diameter of at least 30 lesions per leaf, with five leaves per treatment, was measured with a stereomicroscope 7 days after inoculation. For example, ozone reduced the control lesion diameter of 1.54 ± 0.01 to 0.53 ± 0.04 mm.

Arabidopsis thaliana/Pseudomonas syringae[46]

Arabidopsis thaliana accession Col-O and several mutant lines were employed with or without prior ozone treatment. *Pseudomonas syringae* pv. *maculicola* (PSM) KD 4326 and a derivative containing the avirulence gene *avrB* are grown in King's B medium containing antibiotics. Leaves are inoculated using either syringe infiltration or immersion methods. *In planta* multiplication rates are determined by plating serial dilutions of leaf extracts on selective King's B medium. Ozone pretreatment reduced *in planta* growth of PSM 4326 some 10-fold. *In planta* growth of the avirulent control strain was always much lower.

[53] M. T. Hauser, F. Adhami, M. Dorner, E. Fuchs, and J. Glössl, *Plant J.* **16,** 117 (1998).
[54] N. Yalpani, A. J. Enyedi, J. Leon, and I. Raskin, *Planta* **193,** 372 (1994).

[48] High-Pressure Liquid Chromatography Analysis of Ozone-Induced Depletion of Hydrophilic and Lipophilic Antioxidants in Murine Skin

By Stefan U. Weber, Sumana Jothi, and Jens J. Thiele

Introduction

Ozone (O_3) is best known for its role in the stratosphere where it filters out much of the short-wave spectrum of solar ultraviolet (UV) radiation that would otherwise penetrate the earth. However, it is also well known that this protective "ozone layer," if placed at ground level (troposphere), would be highly toxic and would react with various molecules susceptible to oxidative attack. Due to the breakdown of the stratospheric ozone layer in conjunction with industrial emission, tropospheric ozone has become a major environmental pollutant. Our increased exposure to this powerful oxidant has various health implications currently under investigation. Ozone is well known to exert profound effects on the respiratory tract. Moreover, studies suggest that ozone also induces biochemical changes in murine skin.[1]

Due to its barrier structure and function, the skin is highly susceptible to external oxidative agents. More specifically, the outermost layer of the skin, the *stratum corneum* (SC), has been identified as the primary target for ozone-induced damage. It is composed of enucleated corneocytes consisting of core and envelope proteins. The envelope proteins are linked to lipids that interact with the lipid matrix surrounding the corneocytes.[2] The integrity of the SC is required for the regulation of the transepidermal water loss, and SC dysfunctions are involved in several skin disorders.[3] The SC houses a protective antioxidant network that is depleted after exposure to ozone. Our group has studied extensively the effects of ozone on hydrophilic and lipophilic antioxidants as well as on lipid oxidation products in murine skin.[4] The SC consistently proved to be the most affected by ozone exposure in comparison to epidermis or whole skin. A single exposure of 1 ppm for 2 hr caused a significant depletion of vitamin E in murine SC,

---

[1] J. J. Thiele, M. Podda, and L. Packer, *Biol. Chem.* **378**, 1299 (1997).

[2] P. M. Elias and K. R. Feingold, *Semin. Dermatol.* **11**, 176 (1992).

[3] M. Mao-Qiang, K. R. Feingold, C. R. Thornfeldt, and P. M. Elias, *J. Invest. Dermatol.* **106**, 1096 (1996).

[4] J. J. Thiele, M. G. Traber, K. Tsang, C. E. Cross, and L. Packer, *Free Radic. Bio. Med.* **23**, 385 (1997).

and repeated exposure to this dosage also resulted in increased malondialdehyde (MDA) formation, an indicator of lipid oxidation.[5]

This article reports a highly sensitive method for detecting antioxidants in murine SC before and after *in vivo* ozone exposure. We have optimized our noninvasive tape-stripping technique and extraction followed by high-pressure liquid chromatography (HPLC) detection and analysis of uric acid, vitamin C, glutathione, and vitamin E.

Ozone Exposure

For ozone exposure, a Sander ozonizer Model IV, Eltze, Germany, and a Dasibi ozone detector Model 1003-AH, Glendale, California, are employed. The ozone is produced from oxygen by electrical discharge, mixed with filtered, ozone-free ambient air, and fed into a stainless-steel exposure chamber at a constant flow rate of 200 liter/min. Ozone levels can be adjusted to values of 0 to 10 ppm and monitored with the ozone detector.

All animal research must be approved by a local animal research committee. Animals are anesthetized by intraperitoneal injection of sodium pentobarbital (50 mg/kg body weight, Nembutal, Abbott Laboratories, Chicago, IL). Animals remain anesthetized throughout the exposure period. Mice are exposed for desired time periods at desired concentrations of ozone. After ozone exposure, mice are given a second dose of pentobarbital and are allowed to breathe normal air for 20 min before tape stripping is performed.

Stratum Corneum Isolation: Tape Stripping

Mice should remain under anesthesia during this procedure. Using optimal adhesive material is important to avoid undesirable interference with HPLC analysis. After testing several different types of adhesive tape, we found D-Squame adhesive tape disks (22 mm diameter, CuDerm Corporation, Dallas, TX) to be optimal. These precut disks are made specifically for SC removal, are suitable for the thin murine SC, and do not interfere with electrochemical detection of urate, ascorbate, glutathione, or α-tocopherol.

On average, 10 tape strippings result in the removal of the largest portion of murine SC. Ten consecutive tape strips remove an SC layer of approximately 3.5 μm, assuming a uniform density of 1 g/cm^3. We found the average weight of one tape strip (surface area ~4.0 cm^2) to be 30 ± 9 μg/cm^2. After tape stripping, pooled disks (six) should be weighed on a

[5] J. J. Thiele, M. G. Traber, T. G. Polefka, C. E. Cross, and L. Packer, *J. Invest. Dermatol.* **108**, 753 (1997).

sensitive scale able to measure 0.5 to 1.5 mg. We recommend weighing the tapes on aluminum foil to avoid static charge interference.

Step 1: Before Ozone Exposure

The tape-stripping procedure must be uniform in order to maximize reproducibility. The first D-Squame disk is adhered smoothly to one side of the dorsal surface of the mouse. (*Note:* Using a marker, indicate where the first tape is placed down so that subsequent tapes may be placed in the same spot.) Tape is flattened equally over skin three times and then removed gently with moderate and even traction. The first disk is discarded and the procedure is repeated with a second disk in the same location. (These first two tape strips pick up surface lipids and environmental contaminants and are therefore discarded when detecting for antioxidants.) The described tape-stripping procedure is repeated on the six subsequent SC layers and all six tapes are pooled for each mouse. Tapes should be stored at $-80°$ in the dark for no more than 30 min until analyzed to protect the activity of the antioxidants.

Step 2: After Ozone Exposure

Twenty minutes after the second injection of anesthetic, tape stripping is perform as just described, on the *contralateral* side of the dorsum.

Antioxidant Measurements

Various antioxidants can be extracted from the collected SC and can then be measured by HPLC. This section describes the extraction methods and HPLC analysis for hydrophilic antioxidants (ascorbate, urate, and gluta-thione) and for the major lipophilic antioxidant (vitamin E from SC). The underlying methods for ascorbate and urate are publications[6,7] for glutathione[8] and for vitamin E.[9–11] When designing an experiment, it is important to note that ascorbate, urate, and glutathione can be analyzed from the same tape. In the same extract, the protein concentration can be determined as well for normalization. For lipophilic extraction, a separate tape is needed, as described in the flow chart (Fig. 1).

[6] K. R. Dhariwal, W. O. Hartzell, and M. Levine, *Am. J. Clin. Nutr.* **54,** 712 (1991).

[7] J. J. Thiele, J. K. Lodge, J. H. Choi, and L. Packer, *in* "Springer Lab Manual" (H. Sternberg and P. S. Timiras, eds.). Springer-Verlag, Heidelberg, in press.

[8] D. L. Rabenstein and R. Saetre, *Anal. Chem.* **49,** 1036 (1977).

[9] J. K. Lang and L. Packer, *J. Chromatogr.* **385,** 109 (1987).

[10] J. K. Lang, K. Gohil, and L. Packer, *Anal. Biochem.* **157,** 106 (1986).

[11] J. J. Thiele and L. Packer, *Methods Enzymol.* 413 (1999).

Ascorbate and Urate

Reagents

Standards: L-ascorbic acid (Sigma) and uric acid (Sigma) #685-1. For the mobile phase: Q12 ion-pair cocktail (0.5 M dodecyltriethylammonium phosphate; Regis Technologies, Inc., Martin Grove, IL) USA. For sample extraction: 6-ml capped polypropylene round-bottom tubes (Becton-Dickinson, Lincoln Park, NJ); phosphate-buffered saline (PBS) with 1 mM ethylenediaminetetraacetic acid (EDTA), filtered through chelex resin, and 10 mg/ml butylated hydroxytoluene.

Equipment

Pump: LC-10AD (Shimadzu, Kyoto, Japan). Column: Microsorb C18, 12 cm, 3-μm particle size (Rainin Instruments, Inc., Woburn, MA); All-Guard precolumn system (Alltech, Deerfield, IL); 50-μl injection loop. Detection system: LC-4B amperometric electrochemical detector with a glassy carbon electrode (Bioanalytical Systems, West Lafayette, IN). Mobile phase: A fresh mobile phase should be made each day of analysis from a stock solution of 800 mM sodium acetate buffer, pH adjusted to 4.75 with glacial acetic acid. This stock may be stored at 4° for 1 month. Take 50 ml of stock and dilute this to a final volume of 900 ml with HPLC-grade water. Add 0.2 g of disodium EDTA and dissolve by mixing. Add 50 ml of HPLC-grade methanol and 3 ml of Q12 ion-pair cocktail. Bring volume up to 1 liter with HPLC-grade water, filter, and sonicate to deaerate.

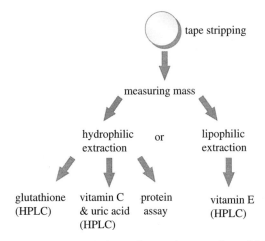

Fig. 1. Flow chart of tape stripping and extraction procedure of SC samples.

Operation

An isocratic solvent delivery system is used with a flow of 1 ml/min. The electrochemical detector is operated at a 0.5-V potential. The column should be equilibrated in the mobile phase, preferably overnight, with a low flow rate, e.g., 0.1 ml/min. After the electrode is turned on, it may take from 2 to 4 hr for the electrode to stabilize. The retention time may still shift even after equilibration.

Standard Preparation

For optimal results, standards should be prepared in the mobile phase each day of analysis from a concentrated stock solution or from the purified product in powder form and stored on ice in covered tubes to avoid photooxidation. Standards covering the full sample range should be analyzed before and after the samples.

Sample Preparation

Using clean forceps, pooled tapes (groups of three) are placed into polypropylene tubes with the adhesive side facing inward; 960 μl of PBS-EDTA and 60 μl of butylated hydroxytoluene are aliquoted into each tube; and tubes are vortexed at high speed for 2 min. Samples are centrifuged at 500g for 3 minutes and the tapes are removed. This procedure is repeated with more tapes as needed. After the last extraction the supernate is then filtered and ready for immediate injection. *Note:* Even storage at −80° or in liquid nitrogen with subsequent thawing decreases the recovery of the extracted ascorbate/urate. If multiple determinations are necessary, make aliquots and place them immediately at −80° or in liquid nitrogen to minimize freeze/thaw cycles.

Retention times: Urate, 2.6 min; ascorbate, 4.1 min (Fig. 2)

Sensitivity limit: Concentrations of 0.05 μM ascorbate and urate are detectable

Recovery (estimated): Urate, 80–85%; ascorbate, 82–87%

Glutathione

Reagents

Standards: glutathione (reduced) (Sigma). For mobile phase: monochloroacetic acid, For electrode: triple distilled mercury (Aldrich). For sample extraction: identical to extraction of vitamin C

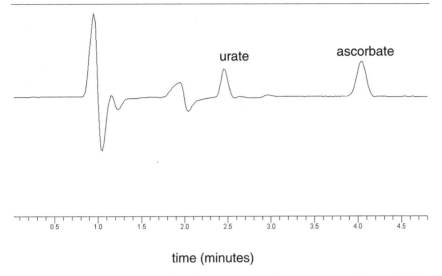

FIG. 2. Typical chromatogram for simultaneous urate and ascorbate analysis in pooled SC.

Equipment

Pump: LC-10AD (Shimadzu). Column: Altima C-18, 250 × 4.6 mm (Alltech), 5-μm particle size; All-Guard precolumn system (Alltech): 50-μl injection loop. Detection system: LC-4B amperometric electrochemical detector with a gold/mercury electrode (Bioanalytical Systems). Mobile phase: Dissolve 9.4 g monochloroacetic acid into 950 ml of HPLC-grade water; adjust pH to 3–3.5 with sodium hydroxide pellets (approximately 25–30). Then add 40 ml HPLC-grade methanol and bring volume up to 1 liter with HPLC-grade water. Filter and degas with helium.

Operation

The system is run at a flow rate of 1 ml/min at an applied voltage of 0.154. The electrode should be polished and plated with mercury the day before analysis according to the instructions provided. Also, the mobile phase must be continually degassed throughout the operation to eliminate interference from oxygen.

Standard Preparation

Standards should be prepared in an identical manner as for ascorbate and urate.

Sample Preparation

The procedure for glutathione extraction is identical to that for ascorbate and urate with the exception that before injecting samples, 50 μl of the supernate is diluted with the mobile phase. This step improves peak shape, as methanol interferes with glutathione analysis. If glutathione, ascorbate, and urate are to be analyzed in the same sample, take separate aliquots from the same extraction. Analyze ascorbate and urate immediately and store aliquots for glutathione analysis at −80° or in liquid nitrogen.

 Retention time: 4.9 min (Fig. 3)
 Sensitivity limit: 0.05 μM glutathione is detectable
 Recovery (estimated): 80–86%

Vitamin E

Reagents

Standards: α-tocopherol (Sigma). For mobile phase: lithium perchlorate. For sample extraction: 50-ml capped polypropylene centrifuge tubes (Corning Costar Corporation, Cambridge, MA); chelexed PBS containing 1 mM EDTA; 10 mg/ml butylated hydroxytoluene (BHT); 0.1 M sodium dodecyl sulfate (SDS); glass conical centrifuge tubes with caps (Kimble).

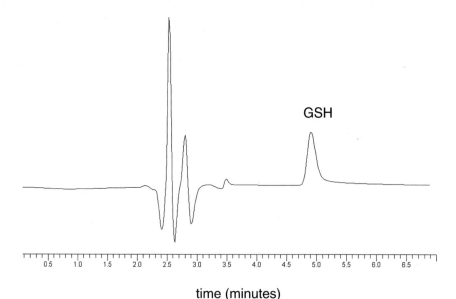

time (minutes)

FIG. 3. Typical chromatogram for glutathione analysis from pooled SC.

Equipment

Pump: LC-10AD (Shimadzu). Column: Ultrasphere ODS C-18, 4.6 mm i.d., 25 cm, 5-μm particle size column (Beckman, Fullerton, CA); All-Guard precolumn system (Alltech); SIL-10A autoinjector with a sample cooler operated by a SCL-10A system controller (both from Shimadzu). Detection system: LC-4B amperometric electrochemical detector with a glassy carbon electrode (Bioanalytical Systems). Mobile phase: HPLC-grade methanol : ethanol (1 : 1.95, v : v) with 20 mM lithium perchlorate. The mobile phase should be filtered and degassed by bubbling with nitrogen or by sonicating before use. It may be stored at 4° for up to 2 weeks but should be filtered and degassed again before use.

Operation

An isocratic solvent delivery system is used with a flow of 1.2 ml/min. The electrochemical detector is operated at a 0.5-V potential, with a full recorder scale at 50 nA. The column should be allowed to equilibrate for at least 2 hr.

Data Analysis

Data can be collected using a Perkin Elmer Interface and analyzed by Turbochrom software (Perkin Elmer Nelson, Cupertino, CA).

Standard Preparation

α-Tocopherol should be dissolved in HPLC-grade ethanol to a final stock concentration of approximately 50–100 μM. An accurate concentration can be determined spectrophotometrically using the molar extinction coefficient, $\varepsilon = 3236 \ M^{-1} \ cm^{-1}$.

For optimal results, standards should be prepared in the mobile phase each day of analysis from concentrated stock solutions and stored on ice in covered tubes to avoid photooxidation.

Sample Preparation

After weighing, and with clean forceps, pooled tapes (groups of six) are placed into 50-ml polypropylene tubes. Two milliliters of PBS-EDTA, 50 μl BHT, 1 ml 0.1 M SDS, and 4 ml HPLC-grade ethanol are aliquoted into each tube. Tubes are vortexed at high speed for 1 min at room temperature; care should be taken to avoid clumping of the disks in the bottom of the tube. Four milliliters of HPLC-grade hexane is added to each tube and then tubes are vortexed vigorously for 1 min. The entire contents of the

50-ml tubes, except the disks, are transferred into 10-ml glass conical tubes, which are centrifuged for 3 min at 300g.

After centrifugation, 3 ml of the top hexane layer is transferred carefully into fresh 10-ml conical tubes. We recommend the use of a positive displacement pipette for the transfer of hexane. The hexane is brought to dryness under nitrogen. (Drying needles must be cleaned thoroughly with ethanol before reapplying to samples). The residue is resuspended in 500 μl ethanol:methanol (1:1, v:v) by vortexing for 15 sec. Contents are transferred into Eppendorf microcentrifuge tubes and centrifuged for 1 min at maximum speed. A known amount of the supernate is analyzed by HPLC as soon as possible to avoid oxidation of antioxidants. It is important to include dilution factors in the final calculations.

Retention time: 4.1 min (Fig. 4)
Sensitivity limit: 0.1 μM α-tocopherol is detectable
Recovery: 70–75%
Variance: <10%

Identification and Quantification of Peaks

A sample of extracted tape without SC needs to be included in the analysis to monitor for any interfering peaks. The identity of peaks from hydrophilic and lipophilic extracts is checked by spiking the sample with a known amount of the antioxidant. A 100% coelution with the sample is

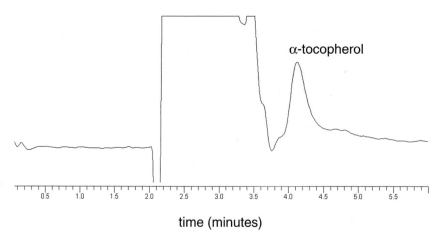

time (minutes)

Fig. 4. Typical chromatogram for α-tocopherol analysis extracted from tape strips. A massive conglomerate of several peaks from organic tape compounds precedes the vitamin E peak.

a good indicator for the authenticity of the peak. For quantification, samples are compared to standard curves. Linearity of the standard curve with the described setup is given in a range from 0 to 100 pmol for ascorbate, urate, and glutathione and a range from 0 to 50 pmol for vitamin E.

Troubleshooting

The sensitivity of the HPLC system in each case depends on the condition of the electrode. If a decrease in sensitivity occurs, polishing the electrode will help (A-1302 Electrode Polishing and Care System, Bioanalytical Systems).

Ascorbate/Urate

Equilibration of the system may take up to 4 hr. If a shift in retention time is observed, there is an inadequate association of the ion pair in the mobile phase with ascorbate/urate. If the retention time is less than 3 min for ascorbate, the concentration of the Q12 ion-pair cocktail can be increased. After use of the ion-pair cocktail, the HPLC system should be cleaned with HPLC-grade water overnight. Any metaphosphoric acid in the extract interferes with the specific peaks in the chromatogram.

Glutathione

If decreased sensitivity or increased noise occurs, polish and replate the mercury electrode.

Vitamin E

When three-fourths of the hexane has dried down, a film of glue may form, preventing the rest of the hexane from evaporating. When this occurs, the film is punctured with either a clean pipette tip or the drying needle. The HPLC system described for vitamin E analysis is generally very stable. If the baseline is noisy, it may be due to air bubbles trapped at the reference electrode, which can be removed by allowing the reference chamber to fill with the mobile phase. Butylated hydroxytoluene should not interfere, as it retains less than α-tocopherol. The use of different tapes may be problematic, as each type of tape, sometimes even each batch, contains a different composition of organic compounds that may interfere with the detection of vitamin E.

Discussion

The methods described allow detection of hydrophilic and lipophilic antioxidants from minute samples of SC sticking to cellophane tapes. The major advantage of these methods is the fact that they are essentially noninvasive, as the SC, as a dead layer, is free of vasculature and nerve endings. Mice do not necessarily have to be sacrificed after the experiment. Mouse SC regenerates within 2 weeks. The methods can be applied to analyze the biochemical alterations in response to ozone, but also to other stressors, such as ultraviolet light or chemicals. Moreover, the penetration of topically applied antioxidants can be monitored without difficulty. Our methods are equally suitable for application to human skin. From other studies it is known that the sampling of SC from humans is generally well tolerated.[12]

In summary, the powerful methods described in this article have proved to be valuable tools for determining the effects of ozone on the outermost layer of the skin, the SC. These methods may therefore assist in evaluating the risk of environmental stressors to human health.

Acknowledgments

The authors thank Wesley Sangsoo Park and Chate Luu for excellent technical assistance and Drs. John Lodge (London), Fabio Virgili (Roma), Amir Tavakkol, and Thomas Polefka for helpful discussion. This research was in part supported by Colgate-Palmolive.

[12] J. J. Thiele, M. G. Traber, and L. Packer, *J. Invest. Dermatol.* **110,** 756 (1998).

[49] Reactions of Vitamin E with Ozone

By Daniel C. Liebler

Vitamin E (α-tocopherol, **1**) protects tissues against oxidative damage by ozone, which is a major component of smog and a powerful oxidant.[1–3] In the lung, α-tocopherol is present both in pulmonary epithelial cell membranes and, together with water-soluble antioxidants, in the lung-lining fluid layer.[4] Ozone produces deleterious effects both by direct oxidation of biomolecules and by reactions with lipids that yield hydrogen peroxide,

[1] W. A. Pryor, *Am. J. Clin. Nutr.* **53,** 702 (1991).
[2] U.S. Environmental Protection Agency, *in* "Air Quality Criteria for Ozone and Other Photochemical Oxidants." EPA Report No. EPA-600/8-79-004. Environmental Criteria and Assessment Office, U.S. EPA, Research Triangle Park, NC, 1978.
[3] M. G. Mustafa, *Free Radic. Biol. Med.* **9,** 245 (1990).
[4] U. P. Kodavanti, D. L. Costa, J. Richards, K. M. Crissman, R. Slade, and G. E. Hatch, *Exp. Lung Res.* **22,** 435 (1996).

aldehydes, and free radicals.[5-8] α-Tocopherol may provide antioxidant protection by reactions with ozone itself and with radicals formed subsequent to ozonation of lipids. Much is known about the fate of α-tocopherol in reactions with peroxyl radicals.[9-15] Current understanding of the reactions of α-tocopherol with ozone is considered in this article.

Ozone reacts rapidly with α-tocopherol in aqueous micelles and in organic solvents. Detection of the tocopheroxyl radical **2** by electron spin resonance (ESR) spectroscopy during the ozonation of α-tocopherol in carbon tetrachloride suggested that the reaction proceeds in part by electron transfer.[16] α-Tocopherolquinone **3** was identified as a stable end product of the oxidation.

Ozonation of α-tocopherol in acetonitrile followed by reversed-phase high performance liquid chromatography (HPLC) analysis of the products (see later) indicated that α-tocopherol was oxidized to several products, including α-tocopherolquinone **3** and its immediate precursor 8a-hydroxytocopherone **4**.[17] Formation of 8a-hydroxytocopherone **4** may involve either sequential electron transfers from α-tocopherol and hydrolysis of the tocopherone cation or ozone addition at C-8a, elimination of oxygen and rearrangement. This latter mechanism may involve the release of singlet oxygen as reported previously.[18]

[5] R. M. Kafoury, W. A. Pryor, G. L. Squadrito, M. G. Salgo, X. Zou, and M. Friedman, *Toxicol. Appl. Pharmacol.* **150,** 338 (1998).

[6] E. M. Postlethwait, R. Cueto, L. W. Velsor, and W. A. Pryor, *Am. J. Physiol.* **274,** L1006 (1998).

[7] W. A. Pryor and D. F. Church, *Free Radic. Biol. Med.* **11,** 41 (1991).

[8] W. A. Pryor, M. Miki, B. Das, and D. F. Church, *Chem. Biol. Interact.* **79,** 41 (1991).

[9] D. C. Liebler, P. F. Baker, and K. L. Kaysen, *J. Am. Chem. Soc.* **112,** 6995 (1990).

[10] D. C. Liebler and J. A. Burr, *Biochemistry* **31,** 8278 (1992).

[11] D. C. Liebler, J. A. Burr, S. Matsumoto, and M. Matsuo, *Chem. Res. Toxicol.* **6,** 351 (1993).

[12] D. C. Liebler and J. A. Burr, *Lipids* **30,** 789 (1995).

[13] M. Matsuo, S. Matsumoto, Y. Iitaka, and E. Niki, *J. Am. Chem. Soc.* **111,** 7179 (1989).

[14] R. Yamauchi, T. Matsui, K. Kato, and Y. Ueno, *Lipids* **25,** 152 (1990).

[15] R. Yamauchi, Y. Yagi, and K. Kato, *Biochim. Biophys. Acta* **1212,** 43 (1994).

[16] D. H. Giamvala, D. F. Church, and W. A. Pryor, *J. Am. Chem. Soc.* **108,** 6646 (1986).

[17] D. C. Liebler, S. Matsumoto, Y. Iitaka, and M. Matsuo, *Chem. Res. Toxicol.* **6,** 69 (1993).

[18] J. R. Kanofsky and P. Sima, *J. Biol. Chem.* **266,** 9039 (1991).

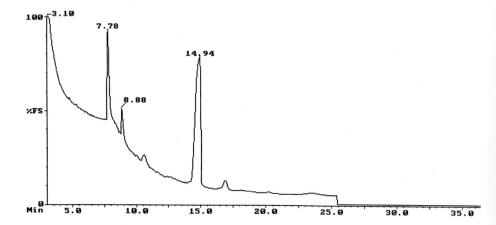

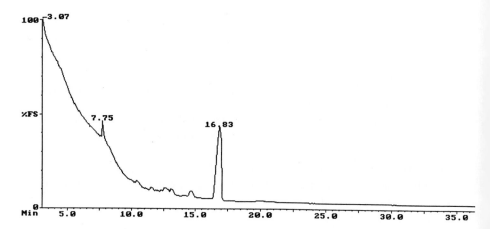

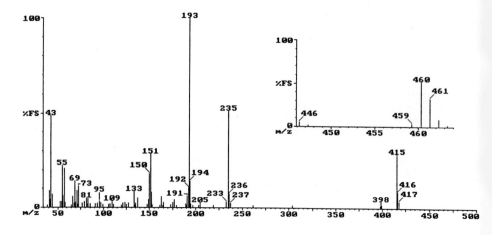

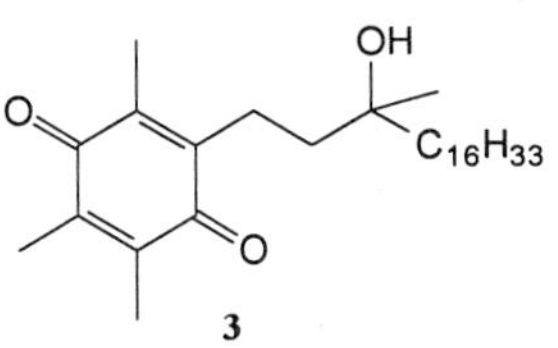

3 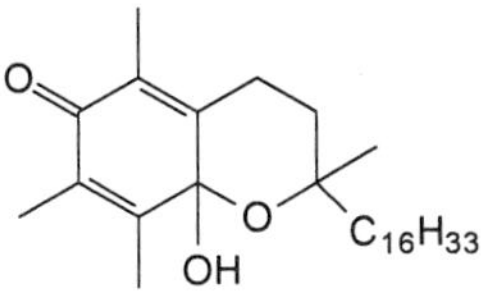   4

The other principal products of α-tocopherol ozonation in acetonitrile were identified as two epimers of a novel spiro product, 10-acetyl-7-(4′,8′,12′-trimethyltridecyl)-3,4,7-trimethyl-2-oxo-1,6-dioxaspiro[4.5]deca-3,9-diene 5, which differ in absolute configuration at the spiro carbon.[17] The X-ray crystal structure of the analogous spiro product derived from the ozonation of the vitamin E model compound 2,2,5,7,8-pentamethylchroman-6-ol was determined. These spiro products display ultraviolet (UV) absorbance maxima at a surprisingly low wavelength (219 nm, $\varepsilon = 14{,}500$ cm^{-1}), which may reflect an unusual electronic effect of the spiro structure on the absorbance of the conjugated carbonyl chromophores.

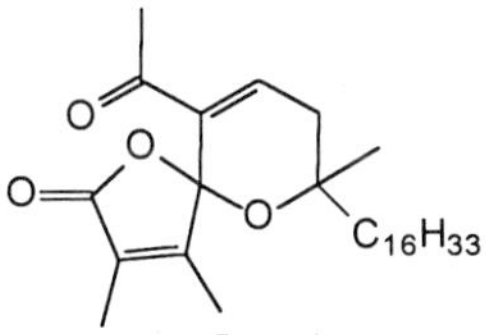

5

Whereas α-tocopherolquinone 3 and 8a-hydroxytocopherone 4 are formed by oxidants other than ozone, spiro products 5 appear to be uniquely formed by the ozonation of α-tocopherol.[17] These products thus could serve as markers for the exposure of membranes containing α-tocopherol to ozone, although this hypothesis remains untested. Methods to produce spiro products 5 by ozonation of α-tocopherol and to analyze the spiro products by gas chromatography–mass spectrometry (GC-MS) are described.

Ozonation of α-Tocopherol

(*Caution:* Ozone is a very strong oxidant and is highly toxic! All procedures involving ozone should be done in a well-ventilated hood equipped

---

Fig. 1. GC-MS analysis of spiro products 5. (Top) Analysis of the 5S,7R epimer (eluting at 14.9 min). (Center) Analysis of the 5R,7R epimer (eluting at 16.8 min). (Bottom) Electron ionization mass spectrum of the 5S,7R epimer at 70 eV. Mass spectra of both products were essentially identical (not shown).

with a safety shield. Excess, unreacted ozone should be destroyed by passage through a solution of potassium iodide.) Ozone may be prepared by passing a stream of air through a silent arc discharge type ozone generator at a flow rate of approximately 20 ml min^{-1}. Although research-grade arc discharge ozone generators can be quite expensive, ozone may also be produced by a similar, relatively inexpensive apparatus sold in pet shops for the purification of water in fish tanks (Sander Aquarientechnik, Am Osterberg, Germany). Although it is not necessary to calibrate the ozone output of the apparatus for purposes of preparative oxidation of α-tocopherol, procedures for doing so have been described.[16] The air-ozone stream is bubbled through a 4 mM solution of α-tocopherol in acetonitrile (70 ml) at ambient temperature. Samples of the mixture may be sampled for HPLC analysis on a Spherisorb ODS-2, 5-μm, 4.6 × 250-mm column eluted with methanol:water (95:5, v/v) at 1 ml min^{-1}. Product elution is monitored at 220 nm for spiro products **5** and at 270 nm for α-tocopherolquinone **3**. Two epimers of spiro product **5** are resolved by reversed-phase HPLC and elute at approximately 3–5 min, whereas α-tocopherolquinone elutes at approximately 7 min and α-tocopherol at 9 min. Based on nuclear magnetic resonance analyses, the earlier-eluting spiro product **5** was tentatively assigned as the 5S,7R epimer, whereas the latter eluting product was assigned as the 5R,7R epimer.

GC-MS Analysis of Spiro Products 5

Spiro products **5** may be analyzed by GC-MS, which permits highly sensitive and specific detection by selected ion monitoring. GC-MS analyses were done on a Fisons MD 800 instrument (Beverly, MA) equipped with a Carlo Erba 8035 gas chromatograph, on column injector, and a Fisons A200S autosampler. The product may be extracted from a liposomal or membrane sample according to previously published procedures.[19,20] The final hexane or heptane extract may be evaporated under nitrogen, and the product-containing residue is dissolved in a minimum amount (ca. 100 μl) for analysis. (Unlike α-tocopherol and its oxidation products, spiro product **5** contains no free hydroxyl groups and thus does not require derivatiziation prior to analysis.) Spiro product **5** may be resolved on a J&W DB-5ms 30 m, 0.25 mm column (J&W Scientific, Folsom, CA). The column was held isothermally at 150° for 1 min, then temperature programmed at 25° min^{-1} to 280°. Products were analyzed by electron ionization MS at 70 eV.

[19] D. C. Liebler, J. A. Burr, L. Philips, and A. J. L. Ham, *Anal. Biochem.* **236,** 27 (1996).
[20] G. W. Burton, A. C. Webb, and K. U. Ingold, *Lipids* **20,** 29 (1985).

The two diastereomeric spiro products **5** were well resolved, with the 5S,7R epimer eluting at 14.9 min (Fig. 1, top) and the 5R,7R epimer eluting at 16.8 min (Fig. 1, center). Both yielded identical mass spectra, with a weak molecular ion at m/z 460 (Fig. 1, bottom, inset). More intense fragment ions result from loss of the phytyl side chain (m/z 235) and retro Diels–Alder cleavage of the six-membered ring (m/z 193). The latter is the base peak and would be an appropriate ion for selected ion monitoring detection of either epimer of **5**.

Acknowledgments

Work in the author's laboratory was supported in part by NIH Grants CA47943, CA59585, and ES06694. The author gratefully acknowledges the expert technical assistance of Jeanne A. Burr.

[50] Induction of Nuclear Factor-κB by Exposure to Ozone and Inhibition by Glucocorticoids

By KIAN FAN CHUNG and IAN M. ADCOCK

Effect of Environmental Ozone on Lungs

Ozone is an important component of the photochemical oxidation product of air pollution involving substrates emitted from automobile engines. It is mainly generated from hydrocarbons and nitrogen dioxide in the presence of ultraviolet radiation. Average levels are usually lowest in the winter and highest in the summer, with concentrations usually peaking in mid-afternoon. Annual average levels of ozone in industrialized countries range from 20 to 40 parts per billion, but concentrations of 100 ppb or more are often reached. Attention has been drawn to the potential adverse effects on respiratory health because of its potential toxic effects related to its oxidative properties.

Epidemiological studies have consistently shown the occurrence of symptoms and a decline in lung function even after short-term exposure to ozone.[1–3] There is an association between short-term levels of ozone

[1] P. L. Kinney, J. H. Ware, and J. D. Spengler, *Arch. Environ. Health* **43**, 168 (1988).
[2] M. Castillejos, D. R. Gold, D. Dockery, T. Tosteson, T. Baum, and F. E. Speizer, *Am. Rev. Respir. Dis.* **145**, 276 (1992).
[3] C. Braun Fahrlander, N. Kunzli, G. Domenighetti, C. F. Carell, and U. Ackermann Liebrich, *Pediatr. Pulmonol.* **17**, 169 (1994).

levels and hospital admissions for respiratory disorders in subjects with preexisting lung disease.[4,5] Long-term exposure to ozone is associated with symptoms of chronic respiratory disease and with an annual rate of decline of lung function. With acute exposure studies, ozone causes a transient reduction in lung function.[6] Patients with asthma show a greater response than normal subjects.[7–9] Ozone also causes an increase in bronchial responsiveness to methacholine.[7,9,10]

Inflammatory Effects of Ozone in Lungs

Because of its relatively low solubility in water, ozone can penetrate into the lung periphery to cause lipid peroxidation. It induces cellular and various biochemical changes that can be detected in bronchoalveolar lavage fluid.[8,10–14] A neutrophil influx is a prominent feature, appearing in the upper as well as in the lower airways. Elevated levels of prostaglandins $F_{2\alpha}$ and E_2 and thromboxane B_2 have been recorded in bronchoalveolar lavage fluid. Levels of cytokines such as interleukin (IL)-6, IL-8, and GM-CSF are also reported. Ozone also causes cellular damage and an increase in epithelial permeability, as reflected by increased levels of lactate dehydrogenase, total protein, fibronectin, fibrinogen, IgG, and albumin.[11,12,14–16]

Studies in animals show that exposure to ozone induces airway and alveolar cell necrosis, together with an inflammatory cellular response characterized by an influx of neutrophils.[17–19] There is also increased expression

[4] R. P. Cody, C. P. Weisel, G. Birnbaum, and P. J. Lioy, *Environ. Res.* **58,** 184 (1992).

[5] R. T. Burnett, R. E. Dales, M. E. Raizenne *et al., Environ. Res.* **65,** 172 (1994).

[6] M. J. Hazucha, D. V. Bates, and P. A. Bromberg, *J. Appl. Physiol.* **67,** 1535 (1989).

[7] J. W. Kreit, K. B. Gross, T. B. Moore, T. J. Lorenzen, J. D'Arch, and W. L. Eschenbacher, *J. Appl. Physiol.* **66,** 217 (1989).

[8] M. A. Basha, K. B. Gross, C. J. Gwizdala, A. H. Haidar, and J. Popovich, Jr., *Chest* **106,** 1757 (1994).

[9] R. Jorres, D. Nowak, and H. Magnussen, *Am. J. Respir. Crit. Care Med.* **153,** 56 (1996).

[10] J. Seltzer, B. G. Bigby, M. Stulborg, M. J. Holtzman, and J. A. Nadel, *J. Appl. Physiol.* **60,** 1321 (1986).

[11] H. S. Koren, R. B. Devlin, and D. E. Graham, *Am. Rev. Respir. Dis.* **139,** 407 (1989).

[12] R. B. Devlin, W. F. McDonnell, R. Mann *et al., Am. J. Respir. Cell Mol. Biol.* **4,** 72 (1991).

[13] R. M. Aris, D. Christian, P. Q. Hearne, K. Kerr, W. E. Finkbeiner, and J. R. Balmes, *Am. Rev. Respir. Dis.* **148,** 1363 (1993).

[14] G. G. Weinmann, M. C. Liu, D. Proud, M. Weidenbach Gerbase, W. Hubbard, and R. Frank, *Am. J. Respir. Crit. Care Med.* **152,** 1175 (1995).

[15] J. R. Balmes, L. L. Chen, C. Scannell *et al., Am. J. Respir. Crit. Care Med.* **153,** 904 (1996).

[16] C. Scannell, L. Chen, R. M. Aris *et al., Am. J. Respir. Crit. Care Med.* **154,** 24 (1996).

[17] D. M. Hyde, W. C. Hubbard, V. Wong, R. Wu, K. Pinkerton, and C. G. Plopper, *Am. J. Respir. Cell Mol. Biol.* **6,** 481 (1992).

[18] M. Sagai, K. Arakawa, T. Ichinose, and N. Shimojo, *Toxicology* **46,** 251 (1987).

[19] E.-B. Haddad, M. Salmon, J. Sun *et al., FEBS Lett.* **363,** 285 (1995).

of several proinflammatory cytokines and enzymes, such as IL-1β, TNF-α, cytokine-induced neutrophil chemoattractant (CINC), macrophage inflammatory protein-2, and inducible nitric oxide synthase.[19–22] Ozone induces epithelial cell necrosis with a consequent increase in epithelial cell proliferation, which has been observed either through incorporation of 5-bromodeoxyuridine or tritiated thymidine labeling, which is probably a reflection of the epithelial cell repair process.[18,23–25]

Studies in rats have indicated that the acute epithelial injury induced by ozone is likely to be a direct effect of ozone and not an indirect effect of neutrophils.[26] The proliferative effect on airways epithelium could be inhibited by an antioxidant and also by the antiinflammatory agent, corticosteroids.[27]

Mechanisms of Ozone-Induced Airway Inflammation

The mechanisms by which ozone can induce upregulation of cytokine and enzyme expression in the airways are not clear. Direct oxidative mechanisms may play a role. Oxidizing agents such as hydrogen peroxide can induce an increase in the binding activity of the transcription factor, nuclear factor κB (NF-κB), in certain cell lines *in vitro.*[28–30] However, proinflammatory cytokines released during ozone exposure, such as IL-1β and TNF-α, may induce NF-κB.[31–34] NF-κB is a member of a family of transcription

[20] E.-B. Haddad, M. Salmon, H. Koto, P. J. Barnes, I. Adcock, and K. F. Chung, *FEBS Lett.* **379,** 265 (1996).

[21] K. J. Pendino, R. L. Shuler, J. D. Laskin, and D. L. Laskin, *Am. J. Respir. Cell Mol. Biol.* **11,** 279 (1994).

[22] E.-B. Haddad, S. Liu, M. Salmon, H. Koto, P. J. Barnes, and K. F. Chung, *Environ. Toxicol. Pharmacol.* **293,** 287 (1995).

[23] P. Rajini and H. Witschi, *Toxicol. Appl. Pharmacol.* **130,** 32 (1995).

[24] P. Rajini, T. R. Gelzleichter, J. A. Last, and H. Witschi, *Toxicology* **83,** 159 (1993).

[25] M. Salmon, H. Koto, O. T. Lynch *et al., Am. J. Respir. Crit. Care Med.* **157,** 970 (1998).

[26] K. Donaldson, G. M. Brown, D. M. Brown, J. Slight, W. M. Maclaren, and J. M. Davis, *Res. Rep. Health Eff. Inst.* **1** (1991).

[27] M. Salmon, H. Koto, O. T. Lynch *et al., Am. J. Respir. Crit. Care Med.* **157,** 970 (1998).

[28] R. Schreck, K. Albermann, and P. A. Baeuerle, *Free Radic. Res. Commun.* **17,** 221 (1992).

[29] M. B. Toledano and W. J. Leonard, *Proc. Natl. Acad. Sci. U.S.A.* **88,** 4328 (1991).

[30] I. M. Adcock, C. R. Brown, O. Kwon, and P. J. Barnes, *Biochem. Biophys Res. Commun.* **199,** 1518 (1994).

[31] H. P. Hohmann, M. Brockhaus, P. A. Baeuerle, R. Remy, R. Kolbeck, and A. P. van Loon, *J. Biol. Chem.* **265,** 22409 (1990).

[32] G. Messer, E. H. Weiss, and P. A. Baeuerle, *Cytokine.* **2,** 389 (1990).

[33] L. Osborn, S. Kunkel, and G. J. Nabel, *Proc. Natl. Acad. Sci. U.S.A.* **86,** 2336 (1989).

[34] I. M. Adcock, H. Shirasaki, C. M. Gelder, M. J. Peters, C. R. Brown, and P. J. Barnes, *Life Sci.* **55,** 1147 (1994).

regulatory factors showing a common structural motif for DNA-binding and dimerization. It regulates the transcription of many genes, particularly those of cytokines; NF-κB-binding motifs have been described in the proximal promoter regions of cytokine genes such as MIP-2, TNF-α, IL-8, granulocyte–macrophage–colony stimulating factor (GM-CSF), MGSA/GRO, MIPO-1α, and IL-6 genes and appear to be important in regulating their transcription.[35–40] In unstimulated cells, NF-κB is bound to an inhibitor l-κB, which keeps the complex localized to the cytoplasm. After stimulation, NF-κB dissociates from l-κB and translocates to the nucleus, where it can bind to specific sequences in the promoter and induce transcription.[39]

Chemokines in Ozone-Induced Airway Neutrophilia

Because ozone exposure induces neutrophilia in the lungs, we determined the importance of chemoattractants for neutrophils in this process. Macrophage inflammatory protein-2 is a CXC chemokine that is expressed in the lungs of rats and mice exposed to ozone.[19,41] CINC is a rat cytokine chemoattractant that was first purified from medium conditioned with a cytokine-induced epithelial clone of normal rat kidney.[42] It is within the family of the CXC chemokines, which in the rat includes MIP-2. CINC shares the greatest sequence homology (69%) with the human CXC chemokine melanoma growth stimulating activity/GRO (MGSA/GRO) and is less homologous with IL-8.[43] CINC is a potent inducer of neutrophil chemotaxis and infiltration *in vivo* and *in vitro*.[44,45] We showed that CINC is an important chemokine, causing ozone-induced neutrophil chemoattrac-

[35] M. Grove and M. Plumb, *Mol. Cell. Biol.* **13,** 5276 (1993).

[36] T. M. Danoff, P. A. Lalley, Y. S. Chang, P. S. Heeger, and E. G. Neilson, *J. Immunol.* **152,** 1182 (1994).

[37] J. Hiscott, J. Marois, J. Garoufalis *et al., Mol. Cell. Biol.* **13,** 6231 (1993).

[38] M. Grilli, A. Chen-Tran, and M. J. Lenardo, *Mol. Immunol.* **30,** 1287 (1993).

[39] P. A. Bauerle, *Biochim. Biophys. Acta* **1072,** 63 (1991).

[40] U. Widmer, K. R. Manogue, A. Cerami, and B. Sherry, *J. Immunol.* **150,** 4996 (1993).

[41] K. E. Driscoll, L. Simpson, J. Carter, D. Hassenbein, and G. D. Leikauf, *Toxico. Appl. Pharmacol.* **119,** 306 (1993).

[42] K. Watanabe, S. Kinoshita, and H. Nakagawa, *Biochem. Biophys. Res. Commun.* **161,** 1093 (1989).

[43] J. J. Oppenheim, C. O. C. Zacharide, N. Mukaida, and K. Matsushima, *Annu. Rev. Immunol.* **9,** 617 (1991).

[44] K. Watanabe, F. Koizumi, Y. Kurashige, S. Tsurufuji, and H. Nakagawa, *Exp. Mol. Pathol.* **55,** 30 (1991).

[45] M. Iida, K. Watanabe, M. Tsurufuji, K. Takaishi, Y. Iizuka, and S. Tsurufuji, *Infect. Immun.* **60,** 1268 (1992).

tion.[46] With the identification and cloning of the CINC gene, an NF-κB motif has been found in the upstream region, and no other *cis*-acting regulatory elements have been identified on up to 1 κb upstream region that has been cloned.[47] This suggests that NF-κB is likely to be involved in CINC gene transcription.

We determined whether ozone induces NF-κB activation in the rat following exposure to ozone and its time course in relation to expression of the rat chemokine CINC mRNA and to neutrophilic inflammation. We found that the oxidative effects of ozone lead to an increased DNA binding of NF-κB in the lungs associated with an increased transient expression of CINC mRNA. Glucocorticoids are used to inhibit NF-κB binding,[48] possibly through an induction of I-κB[49]; these antiinflammatory drugs inhibit CINC mRNA expression together with neutrophilic inflammation.[20] The CINC promoter does not contain glucocorticoid response elements,[47] suggesting that this inhibitory effect of corticosteroids on CINC expression is mediated via the inhibition of transcription factors such as NF-κB. Direct interactions between the activated glucocorticoid receptor and NF-κB have been well described and appear to be one of the major actions of corticosteroids in suppressing inflammation in many cell types.[48,50,51]

Methods for Studying Effects of Ozone in Rats

Ozone is generated by passing laboratory air through a Sander ozoniser (Model IV; Erwin Sander GmbH, Uetze, Germany) at a rate of 0.5 liter/min and mixed with compressed air at a rate of 8 liters/min. The ozone concentration is measured at regular time points using specific sampling tubes (Drager Ltd., Blyth, UK) introduced via an outlet port to the chamber. Four groups of pathogen-free brown Norway rats (250–350 g) are exposed either to ozone (OZ, 3 ppm for 6 hr) or to filtered air. Synthetic glucocorticoid dexamethasone (DEX) is administered (3 mg/kg given ip) 1 hr prior to filtered air (DEX) or ozone (DEX/OZ) exposures. After 2, 8, or 24 hr following exposure, rats are sacrificed, and bronchoalveolar ravage is performed, followed by the removal of lungs.

The lungs are lavaged 10 times with 2-ml aliquots of saline solution

[46] H. Koto, M. Salmon, B. Haddad el, T. J. Huang, J. Zagorski, and K. F. Chung, *Am. J. Respir. Crit. Care Med.* **156,** 234 (1997).

[47] N. Mukaida and K. Matsushima, *Cytokine* **4,** 41 (1992).

[48] A. Ray and K. E. Prefontaine, *Proc. Natl. Acad. Sci. U.S.A.* **91,** 752 (1994).

[49] N. Auphan, J. A. DiDonato, C. Rosette, A. Helmberg, and M. Karin, *Science* **270,** 286 (1995).

[50] I. M. Adcock, C. R. Brown, H. Shirasaki, and P. J. Barnes, *Eur. Respir. J.* **7,** 2117 (1994).

[51] R. I. Scheinman, A. Gualberto, C. M. Jewell, J. A. Cidlowski, and A. S. Baldwin, Jr., *Mol. Cell. Biol.* **15,** 943 (1995).

through the tracheostomy. The lavage fluid is centrifuged ($500g$ for 10 min at 4°), and the cell pellet is resuspended in Hanks' balanced salt solution. Total cell counts are made by counting 10 μl of the cell suspension added to 90 μl of Kimura stain, with cell counting in a Neubauer chamber under a light microscope. Differential cell counts are made on cytospin preparations, which are provided by centrifuging at 300 rpm for 5 min and staining with May–Grunwald Giemsa stain. Cells are identified by standard morphology as macrophages, neutrophils, eosinophils, lymphocytes, basophils, and epithelial cells.

After lavage, the lungs are removed and then used for mRNA extraction and electrophoretic mobility shift assays.

Reverse Transcription–Polymerase Chain Reaction (RT-PCR) and Sequencing of cDNA

For the preparation of CINC cDNA, a brown Norway rat is administered lipopolysaccharide (LPS) (*Escherichia coli,* 100 μg intrathecally). Total cellular RNA from lung recovered from a LPS-stimulated rat is isolated according to the method of Chomzynski and Sacchi[52] and reversely transcribed at 42° for 60 min. PCR is performed as described previously.[19] The primers are designed from the published sequences for CINC[53] and GAPDH cDNAs.[54] PCR products are electrophoresed in 2% agarose gels to visualize the CINC and GAPDH bands. The sizes of the PCR products generated are 205 for CINC and 309 bp for GAPDH. The PCR product is excised and purified using Geneclean II (Stratech, Luton, UK). The CINC PCR product is analyzed by sequencing. The cycle sequence is performed on 100 ng of PCR product and 1 pmol of each primer. Cycling conditions are 30 sec at 95°, 30 sec at 55°, and 30 sec at 72° using exo pfu (Stratagene, Cambridge, UK). Purified products show a 100% homology with the published sequence of rat CINC cDNA[53] and are then used as cDNA probes for Northern blot analysis.

Method for Northern Blot Analysis

The 205-bp CINC PCR product and the 1272-bp *Pst*I fragment from rat GAPDH cDNA are labeled by random priming using [α-^{32}P] dCTP (3000 Ci mmol^{-1}; Amersham, UK). Total cellular or messenger RNAs from rat lung or bronchoalveolar ravage cells are subjected to electrophoresis

[52] P. Chomczynski and N. Sacchi, *Anal. Biochem.* **162,** 156 (1987).
[53] S. Huang, J. D. Paulauskis, J. J. Godleski, and L. Kobzik, *Am. J. Pathol.* **141,** 981 (1992).
[54] P. Fort, L. Marty, M. Piechaczyk *et al., Nucleic Acids Res.* **13,** 1431 (1985).

on a 1% agarose/formaldehyde gel and blotted onto Hybond-N membranes (Amersham, UK). Prehybridization and hybridization are carried out at 42° as described previously.[19] Each blot is washed to a stringency of $0.1\times$ SSC/0.1% SDS for 30 min at 60° and exposed at −80° for 1–7 days to Kodak X-OMAT S film. Autoradiographic bands are quantified by densitometric scanning (Quantity One software, PDI, New York, NY).

Method for Electrophoretic Mobility Shift Assays (EMSA)

Nuclear proteins are extracted from lung tissues as described later in detail.[55] Double-stranded oligonucleotides encoding the consensus target sequence of NF-κB (and flanking regions) present in the CINC promoter (5'-CTCCGGGAATTTCCCTGGC-3') (R&D systems, Abingdon, UK) are end labeled using [α-^{32}P]ATP and T4 polynucleotide kinase. Ten micrograms of nuclear protein from each sample is incubated with 50,000 cpm of labeled oligonucleotide in 25 μl incubation buffer [4% glycerol, 100 mM NaCl, 1 mM EDTA, 1 mM dithiothreitol (DTT), 10 mM Tris–HCl, pH 7.5, 0.8 mg/ml sonicated salmon sperm DNA] for 20 min at 25°. Protein–DNA complexes are separated on a 6% polyacrylamide gel using $0.25\times$ Tris–borate–EDTA running buffer. The retarded band is detected by autoradiography and quantified by laser densitometry. Specificity is determined by the addition of excess unlabeled double-stranded oligonucleotides and by using a nonspecific competitor consisting of the transcription factor activator protein-1 (AP-1).

Details of EMSA Preparation

Preparation of Nuclear Proteins

1. Cells are washed in cold media following incubation and pelleted by microcentrifugation (tissue samples are similarly washed).

2. Resuspend cell pellets in 100 μl buffer A per 10^6 cells and incubate for 10 min at 4°. Tissue samples are finely chopped in buffer A before incubation.

3. Collect nuclei by a brief centrifugation in a microfuge. Remove supernatant (cytosol fraction) and store at −70°.

4. Resuspend nuclear pellet in 15 μl buffer B per 10^6 nuclei and incubate for 15 min on ice.

[55] I. M. Adcock, C. R. Brown, C. M. Gelder, H. Shirasaki, M. J. Peters, and P. J. Barnes, *Am. J. Physiol* **37,** C331 (1995).

5. Microfuge (14,000g) sample for 10 min at 4° and collect supernatant (nuclear fraction).

6. Dilute in buffer C to 75 μl per 10^6 cells and store at $-70°$.

Buffer A: 10 mM HEPES (pH 7.9), 1.5 mM MgCl$_2$, 10 mM KCl, 0.5 mM DTT, and 0.1% NP-40

Buffer B: 20 mM HEPES (pH 7.9), 1.5 mM MgCl$_2$, 0.42 M NaCl, 0.5 mM DTT, 0.2 mM EDTA, 25% glycerol, and 0.5 mM phenylmethyl-sulfonyl fluoride (PMSF)

Buffer C: 20 mM HEPES (pH 7.9), 50 mM KCl, 0.5 mM DTT, 0.2 mM EDTA, and 0.5 mM PMSF

PMSF is dissolved to 100 mM in isopropanol.

T_4 Polynucleotide Kinase (T_4 PNK) End Labeling of Oligonucleotide

Set up the following reaction in a tube: 1 μl (2 pmol) double-stranded oligonucleotide, 1 μl (10 units) of T_4 PNK, 20 μCi [γ-^{32}P]ATP, 1 μl 10× T_4 PNK buffer, and H$_2$O to 10 μl. Incubate for 30 min at 37°.

Add 200 μl of STE buffer (10 mM Tris, pH 8.0, 100 mM NaCl, 1 mM EDTA) and separate the labeled oligonucleotide from the unincorporated label by centrifugation through a 1-ml Sephadex G-25 column. The probe passes through the column whereas unincorporated label remains in the column matrix.

Electrophoretic Band Shift Assay

1. Set up in separate microfuge tubes.

	Binding assay	Competition assay
Sample protein	5–10 μg	5–10 μg
5× incubation buffer	5 μl	5 μl
Competitor oligonucleotide	0 μl	0.5–2.0 μl
H$_2$O to final volume 24 μl		

Competition assays may be performed in parallel with binding assays using 2.0, 1.0, and 0.5 μl of the specific competitor oligonucleotide.

2. Preincubate at 4° for 20 min.

3. Add 5–10 × 10^5 cpm ^{32}P-labeled oligo nucleotide in 1 μl.

4. Incubate at 4° for 40 min.

5. Stop the reaction by adding bromphenol blue gel loading buffer and separate by electrophoresis in a 6% nondenaturing PAGE with 0.25× TBE as buffer.

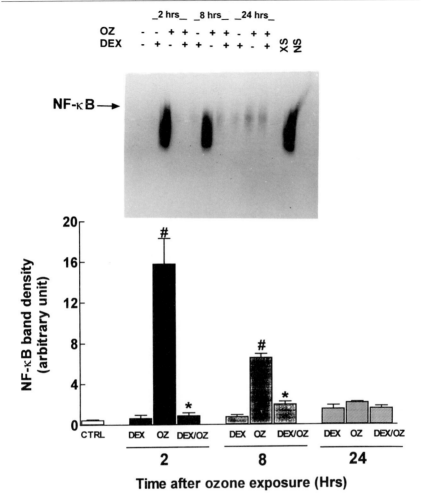

FIG. 1. Effect of ozone and dexamethasone on NF-κB-binding activity. (Top) An electro-phoretic gel mobility shift assay showing NF-κB-binding activity (arrowed) in 10 μg rat lung cell nuclear proteins from individual animals exposed to filtered air (CTRL) or to ozone (3 ppm for 6 hr) in the presence or absence of dexamethasone (DEX) pretreatment for various time points after exposure. (Bottom) Densitometric scanning of band shift data. Ozone (2 and 8 hr postexposure) produces a marked increase in NF-κB-binding activity. This increase is suppressed by pretreatment with dexamethasone. The specificity of binding is confirmed by the addition of 100-fold excess unlabeled NFκB oligonucleotide (XS) and by using a nonspecific (NS) competitor (activator protein-1 or AP-1). Values are mean ± SEM of six to eight animals. #$p < 0.05$ compared to control (CTRL). [Reproduced with permission from E.-B. Haddad, M. Salmon, H. Koto, P. J. Barnes, I. Adcock, and K. F. Chung, *FEBS Lett.* **379,** 265 (1996).]

5× incubation buffer: 50 m*M* Tris, pH 7.5, 500 m*M* NaCl, 5 m*M* DTT, 5 m*M* EDTA, 20% (v/v) glycerol, and 0.4 mg/ml sonicated salmon sperm DNA.

Ozone Induction of NF-κB-Binding Activity and Its Inhibition by Dexamethasone

Electrophoretic mobility shift assays on nuclear extracts show a marked increase in NF-κB DNA-binding activity following ozone exposure (Fig. 1). This activity is maximal at 2 hr and declines thereafter to control levels after 24 hr postozone exposure. A significant level of NF-κB-binding activity is still seen at 8 hr after the cessation of ozone exposure. Dexamethasone alone does not have any effect on baseline NF-κB DNA-binding activity, but attenuates this activity in ozone-exposed rats toward control levels at 2 and 8 hr after exposure to ozone.

Ozone Induces CINC mRNA Expression in Rat Lung and Inhibition by Dexamethasone

Ozone-exposed rats (3 ppm for 6 hr) demonstrate a significant increase in neutrophil counts in bronchoalveolar lavage fluid at 2, 8, and 24 hr

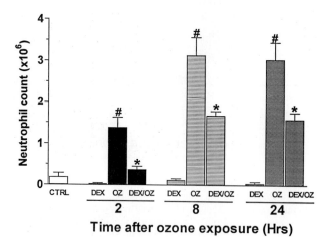

FIG. 2. Effect of ozone and/or dexamethasone on neutrophil accumulation in bronchoalveolar lavage fluid and time course of neutrophil accumulation of rats exposed to ozone. There is an increase in neutrophils recovered from 2 to 24 hr, which can be prevented by dexamethasone. Data shown as mean ± SEM of six to nine rats in each group. #$p < 0.05$ compared to control (CTRL); *$p < 0.05$ compared to ozone (OZ). [Reproduced with permission from E.-B. Haddad, M. Salmon, H. Koto, P. J. Barnes, I. Adcock, and K. F. Chung, *FEBS Lett.* **379**, 265 (1996).]

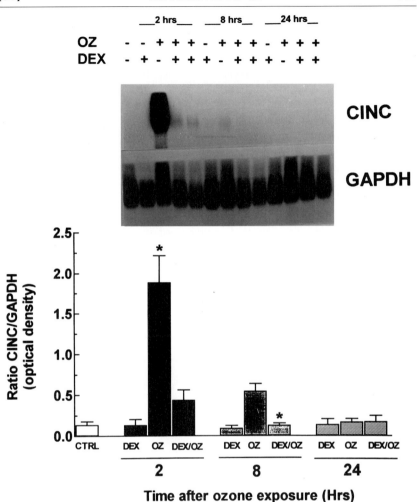

FIG. 3. Effect of dexamethasone on ozone-induced CINC mRNA expression. Northern blot of CINC mRNA expression in mRNA purified from rat lung after exposure to ozone (OZ) or to filtered air (CTRL). One hour prior to ozone (DEX/OZ) or to filtered air (DEX), dexamethasone is administered; the lungs are then harvested at various time points after ozone and assayed for CINC mRNA expression. (Top) CINC mRNA from rats in each experimental group, together with GAPDH mRNA expression. (Bottom) Mean ± SEM of densitometric measurements of autoradiograms obtained in at least six animals in each group. CINC mRNA signals are expressed as a ratio to GAPDH mRNA to account for differences in loading or transfer of mRNA. Ozone induces a significant increase in CINC mRNA expression 2 hr postexposure (*$p < 0.05$ compared to control), an effect suppressed by dexamethasone. Eight hours after exposure, there is a significant reduction in CINC mRNA expression by dexamethasone (*$p < 0.05$ compared to ozone at 8 hr). [Reproduced with permission from E.-B. Haddad, M. Salmon, H. Koto, P. J. Barnes, I. Adcock, and K. F. Chung, *FEBS Lett.* **379,** 265 (1996).]

postozone exposure (Fig. 2). This increase was maximal at 8 hr and remained higher up to 24 hr postexposure. There is no significant difference in total cell counts, macrophages, eosinophils, or lymphocytes recovered in BAL fluid in animals exposed to ozone or filtered air. Ozone exposure results in a marked increase in CINC mRNA 2 hr after exposure and declines rapidly thereafter (Fig. 3). The CINC mRNA signal is barely detectable in RNA obtained from animals exposed to filtered air only. A similar profile of CINC mRNA expression is found in Northern analysis of RNA extracted from bronchoalveolar lavage cells.

Dexamethasone alone has no effect on the level of CINC mRNA expression over the time course investigated. However, it almost completely inhibits (87–92% reduction) ozone-induced CINC-mRNA expression to control levels 2 and 8 hr after ozone exposure (Fig. 3). The inhibitory effect of dexamethasone is also observed in terms of the number of neutrophils recovered in BAL fluid. Dexamethasone significantly reduced ozone-induced neutrophilia in bronchoalveolar lavage fluid up to 24 hr after exposure without any effect on total cell number, eosinophils, lymphocytes, or macrophages (Fig. 2).

[51] Detection of 4-Hydroxy-2-nonenol Adducts Following Lipid Peroxidation from Ozone Exposure

By Luke I. Szweda, Pamela A. Szweda, and Andrij Holian

Introduction

Peroxidation of membrane lipids results in free radical-mediated fragmentation of polyunsaturated fatty acids, giving rise to various aldehydes, alkenals, and hydroxyalkenals.[1–4] Many of these products are cytotoxic and their effects are proposed to be mediated by reactivity toward specific proteins.[4] 4-Hydroxy-2-nonenal (HNE), an α,β-unsaturated aldehyde, is a major and the most cytotoxic product of lipid peroxidation.[2–4] As such, HNE is likely an important mediator of free radical damage to cells. Incubation

[1] A. Bindoli, *Free Radic. Biol. Med.* **5,** 247 (1988).

[2] A. Benedetti, M. Comporti, and H. Esterbauer, *Biochim. Biophys. Acta* **620,** 281 (1980).

[3] H. Esterbauer, A. Benedetti, J. Lang, R. Fulceri, G. Fauler, and M. Comporti, *Biochim. Biophys. Acta* **876,** 154 (1986).

[4] H. Esterbauer, R. J. Schaur, and H. Zollner, *Free Radic. Biol. Med.* **11,** 81 (1991).

METHODS IN ENZYMOLOGY, VOL. 319

of proteins with HNE results in enzyme inactivation[4-8] followed by the formation of inter- and intramolecular protein cross-links.[6,9-12] The HNE cross-linked protein exhibits fluorescence with spectral properties similar to pigments that accumulate during the progression of certain degenerative diseases (ceroid) and aging (lipofuscin).[10,13] Furthermore, the HNE cross-linked protein is resistant to proteolysis and acts as a potent noncompetitive inhibitor of the multicatalytic proteosome.[10-12,14-16]

The ε-amino group of lysine, imidazole nitrogens of histidine, and the sulfhydryl moiety of cysteine can react at the double bond (C3) of HNE to form 1:1 Michael adducts.[4-6,9,17-19] These reactions are responsible for protein inactivation and occur by a bimolecular collision-based mechanism.[4-7] As determined in model studies with amino acids or amino acid analogs, the relative reactivities for 1,4 (Michael) addition to HNE are cysteine > histidine > lysine.[4] Nevertheless, the susceptibility of protein to HNE inactivation appears primarily due to the relative accessibilities of reactive residues on protein to HNE and the importance of these residues to protein structure and function.[5] The ε-amino group of lysine can react further with the carbonyl moiety of Michael adducts to generate 2:1 amino acid–HNE Schiff base adducts responsible for protein cross-links.[9,10,12] In addition, the oxidative rearrangement of the 2:1 lysine–HNE Schiff base cross-link results in the formation of a fluorescent 2-hydroxy-3-imino-1,2-dihydropyrrol derivative.[13,20,21] This information has allowed us to develop

[5] L. I. Szweda, K. Uchida, L. Tsai, and E. R. Stadtman, *J. Biol. Chem.* **268,** 3342 (1993).

[6] K. Uchida and E. R. Stadtman, *J. Biol. Chem.* **268,** 6388 (1993).

[7] J. J. Chen, H. Bertrand, and B. P. Yu, *Free Radic. Biol. Med.* **19,** 583 (1995).

[8] O. Ullrich, W. G. Siems, K. Lehmann, H. Huser, W. Ehrlich, and T. Grune, *Biochem. J.* **315,** 705 (1996).

[9] J. A. Cohn, L. Tsai, B. Friguet, and L. I. Szweda, *Arch. Biochem. Biophys.* **328,** 158 (1996).

[10] B. Friguet, E. R. Stadtman, and L. I. Szweda, *J. Biol. Chem.* **269,** 21639 (1994).

[11] B. Friguet, L. I. Szweda, and E. R. Stadtman, *Arch. Biochem. Biophys.* **311,** 168 (1994).

[12] B. Friguet and L. I. Szweda, *FEBS Lett.* **405,** 21 (1997).

[13] L. Tsai, P. A. Szweda, O. Vinogradova, and L. I. Szweda, *Proc. Natl. Acad. Sci. U.S.A.* **95,** 7975 (1998).

[14] R. E. Pacifici, Y. Kono, and K. J. Davies, *J. Biol. Chem.* **268,** 15405 (1993).

[15] A. J. Rivett, *J. Biol. Chem.* **260,** 12600 (1985).

[16] D. C. Salo, R. E. Pacifici, S. W. Lin, C. Giulivi, and K. J. Davies, *J. Biol. Chem.* **265,** 11919 (1990).

[17] K. Uchida, K. Itakura, S. Kawakishi, H. Hiai, S. Toyokuni, and E. R. Stadtman, *Arch. Biochem. Biophys.* **324,** 241 (1995).

[18] L. Tsai and E. A. Sokoloski, *Free Radic. Biol. Med.* **19,** 39 (1995).

[19] D. V. Nadkarni and L. M. Sayre, *Chem. Res. Toxicol.* **8,** 284 (1995).

[20] G. Xu and L. M. Sayre, *Chem. Res. Toxicol.* **11,** 247 (1998).

[21] K. Itakura, T. Osawa, and K. Uchida, *J. Organ. Chem.* **63,** 185 (1998).

polyclonal antibodies for the detection of protein–HNE adducts in complex biological samples.[9,13,17,22] It should be noted that while HNE is not the only reactive product of lipid peroxidation, HNE shares many chemical and structural features with other lipid peroxidation products.[2–4] Thus, detection of the HNE-modified protein serves as an important index of free radical/lipid peroxidation-derived damage, and characterization of specific protein targets of HNE modification may identify proteins highly susceptible to oxidative modification(s).

Although ozone exposure has long been known to cause respiratory tract injury and inflammation, the mechanism(s) of its cytotoxicity remains unknown. Evidence suggests that ozone does not induce cytotoxicity via direct epithelial interactions and that aldehydes are formed in the lung lining fluid.[23–28] Furthermore, studies in both murine and human ozone exposures (0.25–0.4 ppm for 1–3 hr) have shown that HNE–protein adducts were formed *in vivo* and that exogenous HNE could mimic many of the cellular effects of ozone exposure.[29–31] Consequently, these findings support the notion that ozone, and likely other environmental agents that may stimulate lipid peroxidation, causes tissue injury through the formation of reactive aldehydes such as HNE.

Preparation and Characterization of Anti-HNE Polyclonal Antibodies

Principle

We have developed a series of antibody preparations specific to distinct protein–HNE modifications or their reductively stabilized deriva-

[22] K. Uchida, L. I. Szweda, H. Z. Chae, and E. R. Stadtman, *Proc. Natl. Acad. Sci. U.S.A.* **90,** 8742 (1993).

[23] E. M. Postlethwait, R. Cueto, L. W. Velsor, and W. A. Pryor, *Am. J. Physiol. Lung Cell Mol. Physiol.* **274,** L1006 (1998).

[24] S.-C. Hu, A. Ben-Jebria, and J. S. Ultman, *J. Appl. Physiol.* **73,** 1655 (1992).

[25] S. D. Langford, A. Bidani, and E. M. Postlethwait, *Toxicol. Appl. Pharmacol.* **132,** 122 (1995).

[26] W. A. Pryor, *Free Radic. Biol. Med.* **12,** 83 (1992).

[27] J. S. Ultman, *in* "Air Pollution, the Automobile, and Public Health" (A. Y. Watson, R. R. Bates, and D. Kennedy, eds.), p. 323. National Academy Press, Washington, DC, 1988.

[28] R. M. Uppu, R. Cueto, G. L. Squadrito, and W. A. Pryor, *Arch. Biochem. Biophys.* **319,** 257 (1995).

[29] R. F. Hamilton Jr., M. E. Hazbun, C. A. Jumper, W. L. Eschenbacher, and A. Holian, *Am. J. Respir. Cell Mol. Biol.* **15,** 275 (1996).

[30] R. F. Hamilton, Jr., L. Li, W. L. Eschenbacher, L. Szweda, and A. Holian, *Am. J. Physiol.* **274,** L8 (1998).

[31] A. Kirichenko, L. Li, M. T. Morandi, and A. Holian, *Toxicol. Appl. Pharmacol.* **141,** 416 (1996).

tives.[9,13,17,22] This allows for the development of flexible analytical strategies for the detection and immunopurification of HNE-modified protein(s) in biological samples and for the quantitation of the level of HNE modification. In the case of ozone exposure, ozone initiates lipid peroxidation in the lung lining fluid generating HNE, among other aldehydes. HNE reacts biomolecularly with accessible Cys, His, or Lys residues on proteins, which can be detected in either the native complex or the reduced stable complex with one or more of the antibodies described later. These HNE–protein adducts can be detected and/or quantitated using Western blotting, ELISA, or a variety of histochemical approaches.

Preparation and Characterization of Polyclonal Antibody Specific to Amino Acid–HNE Adducts

As described previously,[17,22] rabbits were immunized with HNE-treated keyhole limpet hemocyanin (KLH). After immunopurification, the antibody specificity was tested by competitive ELISA and Western blot analyses and found to recognize 1:1 lysine–, histidine–, and cysteine–HNE Michael adducts and the 2:1 lysine–HNE Schiff base cross-link.[17,22] The antibody did not recognize protein modified by malondialdehyde, another major product of lipid peroxidation, or Michael adducts formed between N_α-acetylcysteine and acrolein, t-2-pentenal, or t-2-nonenal. Antibody binding requires the presence of the 4-hydroxyl group and is sensitive to the chain length of the modifying 4-hydroxy-2-alkenal.[17] Therefore, this antibody preparation is highly specific to HNE-derived protein modifications and is particularly useful for immunohistochemical studies. It can also be used for Western blot analysis and ELISA as long as precautions are taken to take into account possible adduct reversal. This antibody is available from Calbiochem.

Preparation and Characterization of Antibody to Reductively-Stabilized Amino Acid–HNE Adducts

It has been determined that HNE modifications to protein are stable and do not undergo further reactions at 4°. In addition, the inclusion of 2-mercaptoethanol in samples for Western blot analysis prevents further reactions of HNE with protein. However, both 1:1 and 2:1 amino acid–HNE adducts can dissociate at low pH (≤ 2) and can undergo further reactions at elevated temperatures ($\geq 10°$), chemical properties that affect the accuracy with which these modifications can be detected and quantified.[4,5,10] Unlike immunohistochemical and Western blot analysis, certain experiments require exposure of samples to harsh conditions prior to analysis (acid hydrolysis of protein samples). Amino acid–HNE adducts can be effectively trapped in stable forms by treatment of protein samples with a suitable reductant prior to certain analyses.[5,10] Chemical reduction preserves HNE-derived modifications to protein and serves only to stabilize HNE adducts to harsher conditions of sample preparation. We therefore prepared polyclonal antibody specific to reductively stabilized amino acid–HNE adducts.[9]

We have previously generated an antibody to the NaBH₄-reduced form of glucose-6-phosphate dehydrogenase treated with HNE.[9] In addition, antisera have been raised to the NaBH₄-reduced form of HNE-treated KLH. Polyclonal antibodies were then affinity purified to reduced amino acid–HNE adducts. As judged by competitive ELISA and Western blot analyses, purified antibodies were highly specific to reduced HNE-derived modifications, exhibiting no cross-reactivity toward reduced forms of adducts arising from the lipid peroxidation product malonaldehyde or structurally related products, such as 4-hydroxy-2-pentenal or t-2-nonenal. Reduced lysine–, histidine–, and cysteine–HNE (1:1) Michael adducts and the reduced 2:1 lysine–HNE cross-link all compete for antibody binding, indicating that the epitope recognized by the antibodies is the reduced HNE-derived portion of the adducts.[9] Therefore, selective reduction of HNE adducts in combination with antibodies to reduced HNE adducts can be used to distinguish between 1:1 amino acid–HNE Michael adducts and 2:1 lysine–HNE Schiff base cross-links. NaBH₄ reduces both imine (Schiff base) and carbonyl groups, whereas NaCNBH₃ reduces only imines. Therefore, antibody binding following treatment of samples with NaBH₄ reflects the presence of both Michael adducts and Schiff base cross-links, whereas antibody binding following reduction with NaCNBH₃ reflects solely the presence of Schiff base cross-links. Although this antibody is useful for Western blots and ELISA, it is particularly useful for detecting HNE-adducted proteins treated under strongly denaturing conditions or analysis of acid-hydrolyzed peptides to identify sites of modification.

Preparation and Characterization of Polyclonal Antibody to the Fluorescent Lysine–HNE Cross-Link

Authentic N_α-acetyllysine–HNE fluorophore, prepared as described previously,[13] was conjugated to KLH using 1-ethyl-3-(3-dimethylaminopropyl) carbodiimide (EDC). Rabbits were immunized with fluorophore-linked KLH according to standard immunization protocols, antisera obtained, and polyclonal antibody immunopurified.[13] As judged by competitive ELISA and Western blot analyses, the antibody is highly specific to the HNE-derived portion of the fluorophore. No cross-reactivity toward lysine–, cysteine–, or histidine–HNE Michael adducts or the nonfluorescent 2:1

lysine–HNE Schiff base cross-link was observed. Antibody binding could, however, be competed with analogous fluorophores produced by the reaction of other primary amines (ethylamine and γ-aminobutyric acid) with HNE, thereby confirming that the epitope recognized by the antibody is the HNE-derived chromophore.[13] This antibody is useful for immunohistochemistry and Western blotting to detect long-term buildup of lipid peroxidation-derived damage and is available from Calbiochem.

Detection of 4-Hydroxy-2-nonenal-Modified Protein

Reduction Protocols

Immuno(histo)chemical analyses utilizing antibodies specific to chemically reduced amino acid–HNE adducts require prior treatment of protein/tissue samples with a suitable reductant.[9] Sodium borohydride reduces carbonyl and imine (Schiff base) functional groups and can therefore be used to reduce 1:1 amino acid–HNE Michael adducts and 2:1 amino acid–HNE Schiff base cross-links.[5,10] Under certain conditions, sodium cyanoborohydride selectively reduces imine functional groups, and antibody binding detected following treatment of samples with this reductant corresponds solely to the presence of Schiff base cross-links.[9] Thus, HNE-derived Michael adducts and Schiff base cross-links can be distinguished immunochemically by using different reduction protocols prior to analysis. It should be noted that chemical reduction also enables the introduction of radioisotopic labels to amino acid–HNE adducts, and levels of specific Michael adducts and/or Schiff base cross-links can therefore be quantitated by HPLC or other methodologies.[9,10]

Sodium Borohydride Reduction[5,9,10]

For immunoblot analysis or radioisotopic labeling, a solution of 100–150 mM NaBH$_4$ or NaB^{3}H$_4$ in aqueous NaOH is diluted 1/10 into biological homogenate (0.5 mg/ml protein) or protein solution (0.25 mg/ml). Samples are typically suspended in an appropriate buffer (0.1 M Na$_2$HPO$_4$, pH 7.2). Depending on the buffer selected, the concentration of NaOH is empirically adjusted such that the pH of the reaction mixture remains below 8.0 throughout the reduction reaction (the final NaOH concentration is typically 0.1–10 mM). After an appropriate period of time (10–30 min) at 25°, excess sodium borohydride is scavenged by the addition of an aliquot of aqueous HCl, to a final pH of approximately 4–5. Protein samples are then prepared for analysis.

For immunohistochemical analysis, tissue sections or cells (on slides, coverslips, or grids, as appropriate) are treated with $NaBH_4$ in an appropriate buffer (PBS). Conditions should be optimized, depending on the tissue. In certain cases, it may be necessary to treat the tissue sections/cells with Triton X-100 (0.02–2%, depending on the tissue/cell source) prior to reduction to ensure the accessibility of cellular components to the reductant. Following reduction, sections/cells are rinsed several times with an appropriate buffer (typically PBS).

Sodium Cyanoborohydride Reduction[9]

For immunoblot analysis or radioistopic labeling, a solution of 1 M $NaCNBH_3$ (or tritiated derivative) in water or an appropriate buffer is diluted 1/10 into biological homogenate (0.5 mg/ml) or protein solution (0.25 mg/ml). In the reduction mixture, the final pH should be approximately 6.0. After 10–30 min at 25°, the reduction reaction is quenched by the addition of an aliquot of aqueous HCl, to a final pH of approximately 4–5. For immunohistochemical analysis, tissue sections or cells (on slides, coverslips, or grids, as appropriate) are treated with $NaCNBH_3$ in buffer or water (pH ~6.0). As with sodium borohydride reduction (see earlier discussion) conditions should be optimized, depending on tissue/cell source. Sections/cells are then rinsed several times with an appropriate buffer (typically PBS).

Western Blot Analysis

Samples to be probed using antibody to reduced forms of amino acid–HNE adducts are treated with an appropriate reductant prior to analysis as described earlier[9] Proteins are resolved by SDS–PAGE (25 mM Tris base, 0.15 M glycine, 0.1% SDS) and then electroblotted (25 mM Tris base, 0.15 M glycine, 10% methanol) onto an appropriate membrane. After blocking (0.2% I-block from Tropix in PBS, 0.1% Tween 20), membrane-immobilized proteins are incubated with primary antibody, as antisera or in purified form, diluted 1/1000 to 1/5000 in blocking buffer. If the blot is exposed to primary antibody overnight at room temperature, a prior blocking step is not required. The blot is then rinsed (PBS, 0.1% Tween 20) and exposed to secondary antibody (goat antirabbit IgG conjugated to alkaline phosphatase). After several rinses with wash buffer (PBS, 0.1% Tween 20) and substrate buffer (20 mM Tris base, 1 mM $MgCl_2$, pH 9.8), primary antibody binding is visualized on incubation of the blot with appropriate chemiluminescent substrates/enhancers (CSPD and Nitroblock from Tropix) and autoradiography.[9,22] Relative levels of HNE modification can be estimated by including a standard curve constructed using protein containing a known quantity of HNE adduct(s) and densitometric analyses. For competitive Western blot analysis, incubation of membrane-immobilized protein with primary antibody is performed in the presence of known concentrations of specific competitors. Relative changes in the degree of antibody binding are assessed by densitometry.

ELISA

Protein samples (typically up to 400 ng/well) are adsorbed to microtiter plates. After a blocking step (0.2% I-block from Tropix in PBS, 0.05% Tween 20), samples are incubated with primary antibody diluted into blocking buffer. As noted earlier, the use of antibodies to reduced amino acid–HNE adducts requires prior treatment of samples with an appropriate reductant. Anti-HNE sera of purified antibody is typically diluted 1/500–1/1000; however, a linear range should be established experimentally.[13,22] The wells are then rinsed several times and exposed to an appropriate secondary antibody (goat antirabbit IgG conjugated to alkaline phosphatase). After rinsing with wash buffer (PBS, 0.05% Tween 20) and substrate buffer (20 mM Tris base, 1 mM MgCl$_2$, pH 9.8), the immobilized protein is treated with an appropriate colorimetric substrate. Primary antibody binding is visualized either as the increase in absorbance (for the alkaline phosphatase substrate p-nitrophenol phosphate, at 405 nm) or the rate of increase in absorbance relative to a negative control. For both cases, the values must be within the linear range. Relative levels of HNE modification can be estimated by including a standard curve constructed using pure protein containing a known quantity of the appropriate HNE adduct. For competitive ELISA, incubation of samples with a primary antibody is performed in the presence of known concentrations of specific competitors.

Immunohistochemical Analysis

Antibodies can be used in combination with light, fluorescence, confocal, and electron microscopy for immunohistochemical localization of the HNE-modified protein. In addition, immunogold electron microscopy is useful for quantitating relative levels of HNE modification(s). Tissue samples or cells are fixed, embedded, and sectioned as required, according to standard protocols appropriate for the microscopic technique to be utilized.[32–35]

[32] P. Klivenyl, D. St. Clair, M. Wermer, H.-C. Yen, T. Oberley, L. Yang, and M. F. Beal, *Neurobiol. Dis.* **5,** 253 (1998).
[33] K. E. Muse, T. D. Oberley, J. M. Sempf, and L. W. Oberley, *Histochem. J.* **26,** 734 (1994).
[34] H. C. Mutasa, *Histochem. J.* **21,** 249 (1989).
[35] K. Uchida, M. Shiraishi, Y. Naito, Y. Torii, Y. Nakamura, and T. Osawa, *J. Biol. Chem.* **274,** 2234 (1999).

Immersion of fresh tissue (5 × 2 × 2 mm for light, fluorescence, and confocal microscopy; 1 mm^3 for electron microscopy) in a relatively large volume of 10% neutral-buffered formalin (3 hr at 25°; 24 hr at 4°) provides excellent fixation.[33] For light, fluorescence, and confocal microscopy, the tissue is then washed extensively (PBS) and serially saturated with 10% and then 20% sucrose in PBS. Tissue is embedded in OCT TissueTek medium, sectioned (typically 10 μm), and picked up on gelatin-subbed glass slides. For immunogold electron microscopy, fixed tissue is embedded in LR White resin, sectioned (70–80 nm), and placed on nickel grids as described previously.[32–34] Samples to be analyzed using antibodies to reduced forms of the amino acid–HNE adduct are treated with reductant (3 × 5 min 1 mg/ml NaBH$_4$ in PBS, 2.0% Triton X-100). Immunohistochemical staining is performed using primary antibody dilutions of 1/100–1/1000 in blocking buffer. After several washes, samples are incubated with secondary antibody (goat antirabbit IgG conjugated to an appropriate detecting system for light, fluorescence, and confocal microscopy; gold-conjugated protein A for immunogold electron microscopy). Primary antibody binding is visualized after the appropriate workup.[32–34,36,37]

[36] A Yoritaka, N. Hattori, K. Uchida, M. Tanaka, E. R. Stadtman, and Y. Mizuno, *Proc. Natl. Acad. Sci. U.S.A.* **93**, 2696 (1996).
[37] Y. Ando, T. Brannstrom, K. Uchida, N. Nyhlin, B. Nasman, O. Suhr, T. Yamashita, T. Olsson, M. El Salhy, M. Uchino, and M. Ando, *J. Neurol. Sci.* **156**, 172 (1998).

[52] Synthesis of Inflammatory Signal Transduction Species Formed during Ozonation and/or Peroxidation of Tissue Lipids

By Giuseppe L. Squadrito, Maria G. Salgo, Frank R. Fronczek, and William A. Pryor

Lipid peroxidation and ozonation produce oxidized lipids that are bioactive and can act as signal transducers. The synthesis and isolation of these oxidized lipid products in pure form will allow the testing of hypotheses regarding their role as potential relay molecules. Of special interest are oxidized lipids derived from 1,2-diacyl-*sn*-glycero-3-phosphocholine that feature a shortened acyl chain on the second position ending with an aldehyde function. These platelet-activating factor analogs are not only produced during ozonation but also during the peroxidation of 1,2-diacyl-*sn*-glycero-3-phosphocholines. Another class of biologically active lipids are the 1-hydroxy-1-hydroperoxyalkanes formed *in vivo* from the addition of water to the carbonyl oxides that are intermediates in the ozonation of unsaturated lipids in aqueous media. 1-Hydroxy-1-hydroperoxyalkanes are conveniently generated *in situ* from the hydrolysis of bis(1-hydroxyalkyl) peroxides. Bis(1-hydroxyalkyl)peroxides are not themselves direct lipid ozonation products, but are stable crystalline compounds that slowly hydrolyze to the corresponding 1-hydroxy-1-hydroperoxyalkanes and aldehydes.

This chapter reports the syntheses of 1-palmitoyl-2-[3-(8-octanoyl)-5-(1-octyl)-1,2,4-trioxolane]-*sn*-glycero-3-phosphocholine (the Criegee ozonide of 1-palmitoyl-2-oleoyl-*sn*-glycero-3-phosphocholine), 1-palmitoyl-2-(9-oxononanoyl)-*sn*-glycero-3-phosphocholine, 1-palmitoyl-2-(1-hydroxy-1-hydroperoxynonanoyl)-*sn*-glycero-3-phosphocholine, bis(1-hydroxyheptyl)peroxide, and bis(1-hydroxynonyl)peroxide (the percursors of the corresponding 1-hydroxy-1-hydroperoxyalkanes).

Introduction

Ozone is present in smoggy air[1–4] and reacts with lipids at the air/tissue boundary in the lung,[5–8] producing bioactive oxidized lipids when polluted air is breathed.[9–11] Ozone is so reactive toward unsaturated lipids that it reacts as it traverses the lung lining fluid layer or, where it is thin or absent, within lung epithelial cell membranes.[5,6] Unsaturations in membrane lipids are present in unsaturated fatty acids (UFA) of the glycerophospholipids that form the membrane. Lipid-derived ozonation products can then penetrate further into the underlying tissue, either themselves causing pathology

[1] I. Jaspers, L. C. Chen, and E. Flescher, *J. Cell. Physiol.* **177**, 313 (1998).
[2] M. C. Madden, M. Friedman, L. A. Dailey, and J. M. Samet, *Inhal. Toxicol.* **10**, 795 (1998).
[3] W. J. McKinney, R. H. Jaskot, J. H. Richards, D. L. Costa, and K. L. Dreher, *Am. J. Respir. Cell Mol. Biol.* **18**, 696 (1998).
[4] Q. Zhao, L. G. Simpson, K. E. Driscoll, and G. D. Leikauf, *Am. J. Physiol.* (*Lung Cell. Mol. Physiol.*) **274**, L39 (1998).
[5] W. A. Pryor, *Free Radic. Biol. Med.* **12**, 83 (1992).
[6] R. M. Uppu, R. Cueto, G. L. Squadrito, and W. A. Pryor, *Arch. Biochem. Biophys.* **319**, 257 (1995).
[7] W. A. Pryor, G. L. Squadrito, and M. Friedman, *Free Radic. Biol. Med.* **19**(6), 935 (1995).
[8] W. A. Pryor, G. L. Squadrito, and M. Friedman, *Toxicol. Lett.* **82/83**, 287 (1995).
[9] W. A. Pryor, E. Bermúdez, R. Cueto, and G. L. Squadrito, *Fundam. Appl. Toxicol.* **34**, 148 (1996).
[10] R. Cueto, G. L. Squadrito, and W. A. Pryor, *Methods Enzymol.* **233**, 174 (1994).
[11] R. Cueto, G. L. Squadrito, E. Bermúdez, and W. A. Pryor, *Biochem. Biophys. Res. Commun.* **188**, 129 (1992).

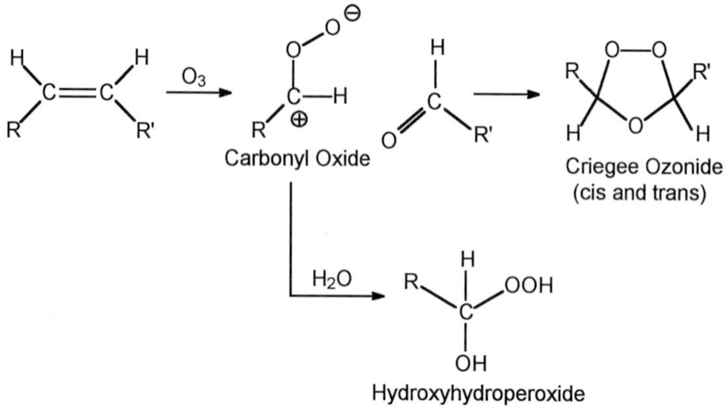

FIG. 1. Simplified mechanism of Criegee ozonation of an olefin showing the intermediate carbonyl oxide that can react with the aldehyde fragment and form a Criegee ozonide or be trapped by water and form a hydroxyhydroperoxide.

or triggering the release of endogenous proinflammatory mediators.[7,8] Thus, the toxicity of ozone must be conceived as a cascade of chemical and biochemical reactions responsible for the effects of ozone.

Ozonation of UFA-containing lipids yields products that can initiate lipid peroxidation, and lipid peroxidation yields oxidatively fragmented lipids that can act as mediator molecules. Thus, ozone can lead to both ozonation and peroxidation, and both processes yield oxidized lipids that can act as mediator molecules. Each process yields a unique spectrum of products that depends on the degree and position of unsaturation(s) in the lipid.[9–11]

Ozone reacts rapidly with carbon–carbon double bonds according to the Criegee mechanism.[12] In aqueous medium, water can trap the intermediate carbonyl oxides before they can recombine to form the Criegee ozonide, as shown in Fig. 1. 1,2-Diacyl-*sn*-glycero-3-phosphocholines (commonly called phosphatidylcholines or lecithins) usually contain an UFA at the *sn*-2 position and are important cell membrane constituents that represent the lipid class from which the majority of oxidized lipids arise.[6,9] Figure 2 depicts the ozonation of a monounsaturated phospholipid in an aqueous medium. When a polyunsaturated fatty acid (PUFA) is at the *sn*-2 position, it gives rise to product multiplicity as each double bond becomes nearly equally accessible to ozone, and more products are possible. Lipid fragments are produced, consisting of aldehydes and 1-hydroxy-1-hydroperoxides, together with a small yield of Criegee ozonides.[12]

[12] G. L. Squadrito, R. M. Uppu, R. Cueto, and W. A. Pryor, *Lipids* **27**, 955 (1992).

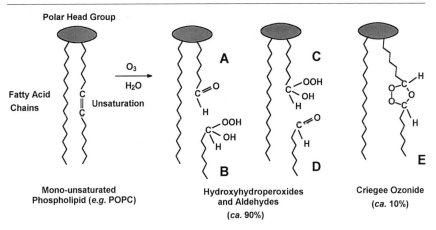

FIG. 2. Ozonation of a monounsaturated diacylglycerophospholipid in an aqueous medium.

Lipid peroxidation also gives rise to oxidatively fragmented lipid aldehydes by a process in which an alcoxyl radical intermediate undergoes β scission (Fig. 3). PUFA undergo peroxidation more easily than monounsaturated fatty acids (MUFA) and give multiple products. Alcoxyl radicals can fragment by scission in two ways, as shown in Fig. 3. The position of the alcoxyl radical center in the fatty acid can also vary in PUFA residues. Nevertheless, the total number of possible lipid aldehydes from lipid peroxidation is rather small because of the limited number of common PUFA.

Large and small fragments are formed when a 1,2-diacyl-*sn*-glycero-3-phosphocholine forms aldehydic products by either ozonation or peroxida-

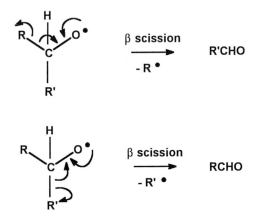

FIG. 3. β scission pathways of an alcoxyl radical that lead to aldehydes.

tion. Large fragments contain the glycerol backbone, whereas small fragments are simple linear aldehydes. The formation of lipid fragments produced by ozonation of lipids is shown in Fig. 2, and fragments generated by β scission of alcoxyl radicals produced during lipid peroxidation are shown in Fig. 3, where one of the residues (R or R') contains the glycerol backbone of the 1,2-diacyl-sn-glycero-3-phosphocholine. The synthesis and isolation of oxidized lipid products in pure form will allow the testing of hypotheses regarding their role as potential relay molecules.

Synthesis Strategies

Glycerophospholipid Aldehydes and
Glycerophospholipid Hydroxyhydroperoxides

We make use of the ozonation of glycerophospholipids under controlled conditions to obtain Criegee ozonides. We then reduce the Criegee ozonides to aldehydes and synthesize hydroxyhydroperoxides by the nucleophilic addition of hydrogen peroxide to aldehydes (Fig. 4). The lipid fragment sizes can be selected by choosing the appropriate position of the carbon–carbon double bonds on the fatty acid at the sn-2 position of the glycerophospholipid that undergoes ozonation. When ozonation is conducted in anhydrous dichloromethane (DCM), chilled to $-78°$ in a dry ice/acetone mixture, hydrolysis of the carbonyl oxide does not take place and the Criegee ozonide is the main product (Fig. 1). The Criegee ozonides can then be converted easily to aldehydes by reaction with zinc in acetic acid. Hydroxyhydroperoxides can be obtained by allowing the glycerophospholipid aldehydes to react with H_2O_2 dissolved in DCM [which can be obtained directly by

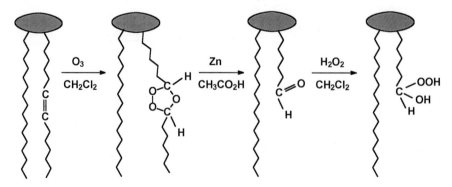

Fig. 4. Synthesis strategy to glycerophospholipid Criegee ozonides, aldehydes, and hydroxyhydroperoxides.

extracting H_2O_2 (30%) with DCM]. The solvent is then evaporated under a stream of dry nitrogen and any excess H_2O_2 is evaporated in a vacuum line. Unlike with linear aldehydes (see later), the formation of bisperoxides is minimal with glycerophospholipid aldehydes due to the bulkier residues in these aldehydes.

1-Hydroxy-1-hydroperoxyalkanes

1-Hydroxy-1-hydroperoxides are important products of lipid ozonation in liposomal suspensions, a system that resembles cellular membranes.[12] 1-Hydroxy-1-hydroperoxyalkanes are difficult to isolate and handle because they exist in equilibrium with the aldehyde and hydrogen peroxide, according to Eq. (1).

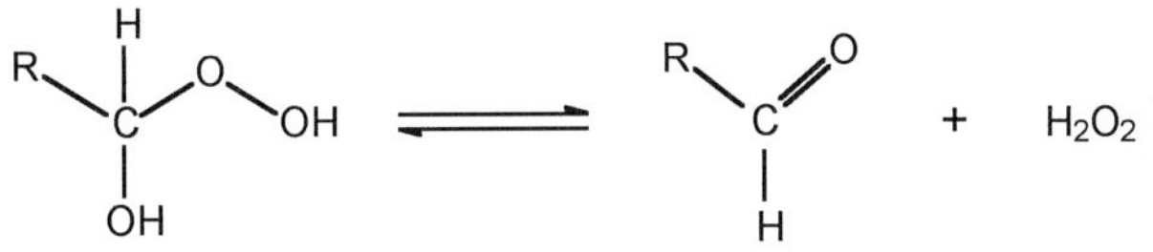

$$\text{(1)}$$

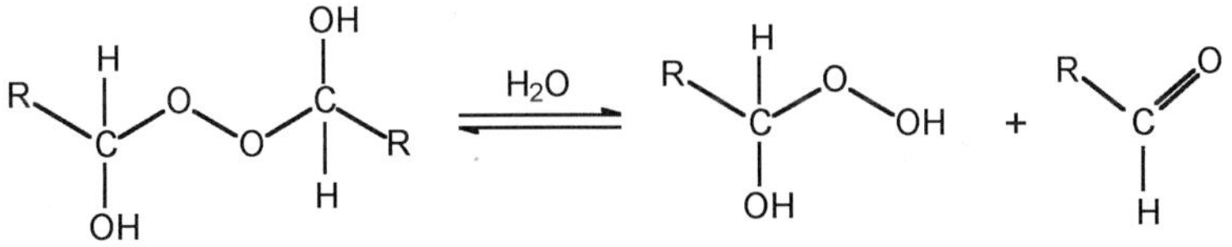

$$\text{(2)}$$

In practice it is convenient to make use of the synthesis strategy shown in Fig. 5. The strategy is bases in using the reverse reaction in Eq. (1) rather than of ozonation of an olefin to arrive to the 1-hydroxy-1-hydroperoxides. The reaction can be accomplished by stirring a solution of the aldehyde in DCM with H_2O_2 (30%). Concentration of the organic layer affords the corresponding bis(1-hydroxyalkyl)peroxide, the condensation product of two molecules of aldehyde, and one molecule of H_2O_2 [the reverse Eq. (2)], which can be purified further by recrystallization when using simple aldehydes with a carbon backbone longer than six carbon atoms. We find bisperoxides to be stable compounds that can be kept at $-20°$ for months

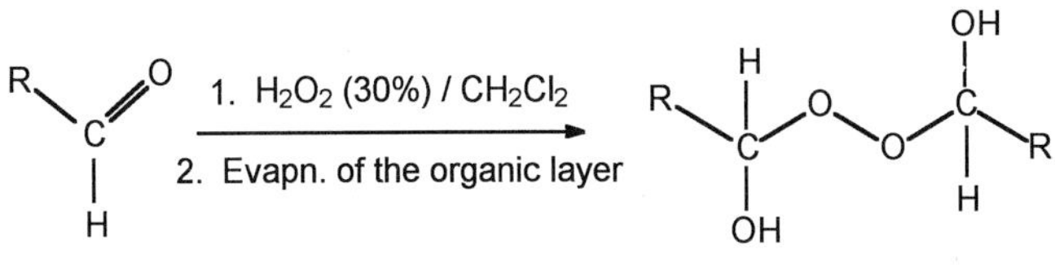

Fig. 5. Synthesis strategy to bis(1-hydroxyalkyl)peroxides.

without decomposition. Bisperoxides can be used as precursors of the 1-hydroxy-1-hydroperoxides to which they spontaneously hydrolyze.

Following the strategies described earlier, we describe the synthesis of 1-palmitoyl-2-[3-(8-octanoyl)-5-(1-octyl)-1,2,4-trioxolane]-*sn*-glycero-3-phosphocholine (**1**) [the Criegee ozonide of 1-palmitoyl-2-oleoyl-*sn*-glycero-3-phosphocholine (POPC)], 1-palmitoyl-2-(9-oxononanoyl)-*sn*-glycero-3-phosphocholine (**2**), 1-palmitoyl-2-(1-hydroxy-1-hydroperoxynonanoyl)-*sn*-glycero-3-phosphocholine, bis(1-hydroxyheptyl)peroxide, and bis(1-hydroxynonyl)peroxide. Compounds (**2**) and (**3**) can arise *in vivo* from the ozonation of any monounsaturated 1-palmitoyl-2-acyl-*sn*-glycero-3-phosphocholines with an unsaturation at C9–C10 at the *sn*-2 position of the glycerol backbone or from the ozonation of any 1-palmitoyl-2-acyl-*sn*-glycero-3-phosphocholine with an unsaturation at C9–C10 and other unsaturations between C atoms of higher number but no other unsaturations between C atoms of lower number. Compound (**2**) can also arise from the lipid peroxidation of any 1-palmitoyl-2-acyl-*sn*-glycero-3-phosphocholine with an unsaturation at C9–C10 and other unsaturations between C atoms of higher number but no other unsaturations between C atoms of lower number. Compounds **4** and **5** are precursors of the corresponding 1-hydroxy-1-hydroperoxyalkanes and aldehydes to which they hydrolyze. 1-Hydroxy-1-hydroperoxyheptane would be produced from the ozonation of palmitoleate-containing lipids or any other lipids containing n-7 unsaturated fatty acids, whereas 1-hydroxy-1-hydroperoxynonane would be produced from the ozonation of oleate-containing lipids or any other lipids containing n-9 unsaturated fatty acids.

Materials

1-Palmitoyl-2-oleoyl-*sn*-glycero-3-phosphocholine (Avanti Polar Lipids), heptanal (Aldrich), nonanal (Alfa), and hydrogen peroxide (30%) (Mallinckrodt) are used as received. Ozone is generated using a Sander 200 ozonizer from Erwin Sander (Uetze-Eltze, Germany). Diol solid-phase extraction columns (3 ml) are purchased from Varian. Proton (^{1}H) and carbon (^{13}C) nuclear magnetic resonance (NMR) spectra are obtained using solutions of CDCl$_3$ with tetramethylsilane as an internal standard on Bruker 200 or 400 AM spectrometers.

Caution: Ozone is an extremely strong oxidizing agent and is highly cytotoxic. Excess ozone must be destroyed in a solution of potassium iodide and the reactions must be conducted in a good fume hood. A safety shield should be used because peroxides can be explosive.

Synthesis of 1-Palmitoyl-2-[3-(8-octanoyl)-5-(1-octyl)-1,2,4-
   trioxolane]-*sn*-glycero-3-phosphocholine (POPC Criegeee
   Ozonide) (**1**)

    POPC Criegee ozonide is synthesized using a method similar to one
described previously.[12] Briefly, POPC (200 mg) is dissolved in DCM (50
ml) and chilled to −78° in a dry ice/acetone mixture. A stream of 1% ozone
in oxygen is bubbled through the solution at a flow rate of 50 ml/min
until the solution turns slightly blue, indicating that POPC has reacted
completely. Excess ozone is flushed with nitrogen, and the reaction mixture
is brought to room temperature and the solvent evaporated in a rotary evaporator.

Synthesis of 1-Palmitoyl-2-(9-oxononanoyl)-*sn*-glycero-3-
   phosphocholine (**2**)

    The mixture of *cis*- and *trans*-POPC Criegee ozonides produced as
described earlier is stirred with 130 mg Zn powder and 1.0 ml water/20 ml
acetic acid for 2 hr at room temperature. The mixture is diluted with 50
ml of chloroform, washed with 3 × 25 ml of water, and dried over sodium
sulfate. Traces of small molecular weight by-products are removed using
a diol solid-phase extraction column; the chloroform containing the reduced
compound is applied to the column followed by elution with methanol.
The eluent is collected, evaporated, and dried under vacuum. The two-step
synthesis of **2** described here is prefereable to the ozonation of POPC in

water,[13] where the overoxidation of the aldehydic product lowers yields and requires laborius separation methods. Contrastingly, ozonation of POPC in DCM stops at the Criegee ozonide, which can be selectively reduced to **2** with Zn in acetic acid with good yields and purified from smaller molecular weight fragments by a simple solid-phase extraction method. ^{1}H NMR: 9.74 (1H, t), 5.17 (1H, m), 4.36 (1H, dd), 4.27 (2H, m), 4.10 (1H, dd), 3.91 (2H, m), 3.74 (2H, m), 3.31 (9H, br s), 2.40 (2H, td), 2.30-2.22 (4H, m), 1.55 (6H, m), 1.30-1.20 (30H, m), 0.86 (3H, t). ^{13}C NMR: 202.68, 173.48, 173.03, 70.46, 66.25, 63.43, 62.87, 59.31, 54.27, 43.78, 34.14, 34.07, 29.66-28.29 (13C), 31.87, 24.84, 24.73 (2C), 22.63, 21.91, 14.06. The FAB-MS spectrum of **2** affords an M + 1 peak at 650.3 m/e (calc. 650.9 m/e).

Synthesis of 1-Palmitoyl-2-(1-hydroxy,1-hydroperoxynonanoyl)-*sn*-glycero-3-phosphocholine (3)

Fifteen mg of 1-palmitoyl-2-(9-oxononanoyl)-*sn*-glycero-3-phosphocholine is dissolved in 1 ml of DCM containing H_2O_2. The solution of H_2O_2 in DCM is prepared by extracting H_2O_2 (30%; 30 ml) with DCM (20 ml). The solution is vortexed gently and allowed to react for 15 min. After this period, the solvent is evaporated under a gentle stream of nitrogen, affording 1-palmitoyl-2-(1-hydroxy,1-hydroperoxynonanoyl)-*sn*-glycero-3-phosphocholine.

Synthesis of Bis(1-hydroxyalkyl)peroxides: Bis(1-hydroxy-heptyl)peroxide (4) and Bis(1-hydroxynonyl)peroxide (5)

Bis(1-hydroxyheptyl)peroxide (4)

[13] J. Santrock, R. A. Gorski, and J. F. O'Gara, *Chem. Res. Toxicol.* **5**, 134 (1992).

Bis(1-hydroxynonyl)peroxide (5)

Attempts to isolate the 1-hydroxy-1-hydroperoxyalkanes resulted in conversion to the more stable corresponding *meso* and *d,l* forms of the bis(1-hydroxyalkyl)peroxides in about equal amounts. Hydrogen peroxide (30%, 20 ml) is added at once with stirring to a solution of the aldehyde in DCM (0.80 *M*, 10 ml) at room temperature and stirred for 1 hr. The organic layer is then separated, and the aqueous layer is further extracted twice with 20 ml of DCM. The organic fractions are combined and dried over anhydrous sodium sulfate, the solvent evaporated, and the residue recrystallized twice from a 1:1 DCM:petroleum ether (30–60°) mixture (70% yield). Examination of recrystallized material produces crystals of suitable quality for X-ray crystallographic determination. ^{1}H NMR: *meso*-**4**: 5.23 (1H, br t), 3.87 (1H, s, O-H), 1.70-1.50 (2H, m), 1.45-1.35 (2H, m), 1.35-1.25 (6H, m), 0.85 (3H, t). ^{13}C NMR: *meso*-**4**: 100.99, 33.24; *d,l*-**4**: 100.67, 32.95; and (both compounds) 31.63, 29.00, 24.51, 14.01. ^{1}H NMR: mixture of *meso*- and *d,l*-**5**: 5.22 (1H, br t), 1.70-1.50 (2H, m), 1.45-1.35 (2H, m), 1.35-1.10 (10H, m), 0.85 (3H, t). ^{13}C NMR: *meso*-**5**: 100.93, 32.85; *d,l*-**5**: 100.60, 33.16; and (both compounds) 31.76, 29.3-29.0 (3C), 24.48, 22.56, 13.99.

Discussion on the Synthesis of Bis(1-hydroxyalkyl)peroxides

The chemical literature seldom makes reference to bis(1-hydroxyalkyl)peroxides, and the following information will help the characterization of reaction mixtures that contain isomeric bis(1-hydroxyalkyl)peroxides and 1-hydroxy-1-hydroperoxyalkanes. [C-1 in 1-hydroxy-1-hydroperoxyalkanes and in bis(1-hydroxyalkyl)peroxides is a chiral center, therefore 1-hydroxy-1-hydroperoxyalkanes exist as a pair of enantiomers and bis(1-hydroxyalkyl)peroxides exist as a pair of enantiomers and a *meso* form.] Analysis of reaction mixtures of heptanal or nonanal with hydrogen peroxide by ^{13}C NMR revealed the presence of three resonances at about 101.6-101.5 (the minor component), 101.0-100.9, and 100.7-100.6 ppm. Recrystallization of the crude reaction mixtures resulted in the disappearance of resonances around 101.6-101.5 ppm (corresponding to the C-1 of 1-hydroxy-1-hydroperoxyalkanes), leaving two resonances of similar intensities at 101.0-100.9 and 100.7-100.6 ppm. Resonances at 101.6-101.5 ppm were regenerated (together with an equimolar amount of the corresponding aldehyde) on the addition of a small amount of water. NMR analysis of the partial hydrolysis of *meso*-**4** was accomplished after the addition of a small amount

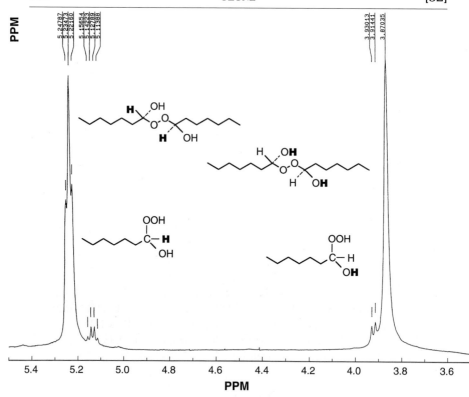

FIG. 6. Section of the ^{1}H NMR spectrum of partially hydrolyzed *meso*-bis(1-hydroxyheptyl)peroxide. Resonances correspond to protons in boldface.

of water (up to 10 μl) to the NMR tube containing *meso*-**4** dissolved in 1 ml of CDCl$_3$. Sections of ^{1}H NMR and ^{13}C NMR spectra of partially hydrolyzed *meso*-**4** are shown in Fig. 6. The partial hydrolysis produced a doublet of triplets at 5.13 ppm (corresponding to the methine proton of the 1-hydroxy-1-hydroperoxyheptane), a singlet at 9.19 ppm (corresponding to the hydroperoxy proton), and split the hydroxylic proton singlet into a doublet in the ^{1}H NMR spectrum, reflecting equilibria among bis(1-hydroxyalkyl)peroxides, 1-hydroxy-1-hydroperoxyalkanes, aldehydes, and hydrogen peroxide depicted in Eqs. (1) and (2).

The assignment of the $-$OH and $-$OOH protons was confirmed by an exchange experiment with D$_2$O that resulted in their disappearance. Mixtures of isomeric bis(1-hydroxyalkyl)peroxides and 1-hydroxy-1-hydroperoxyalkanes appear to have been identified incorrectly as containing just the latter,[14] despite the presence of more NMR resonances that could be explained as arising from a single pure compound. Under conditions of high dilution, as occurs in biological systems exposed to environmental ozone levels, only aldehydes, 1-hydroxy-1-hydroperoxyalkanes, and hydrogen peroxide would be expected to be present. Bis(1-hydroxyalkyl)peroxides, although not themselves direct products of lipid ozonation, are synthesized easily and are convenient stable precursors of the difficult to isolate 1-hydroxy-1-hydroperoxyalkanes; they can be refrigerated and stored for several months without appreciable decomposition and are suitable for mechanistic studies.

X-Ray Crystallographic Analysis

Diffraction data for *meso*-**4** and -**5** were collected on an Enraf-Nonius CAD4 diffractometer equipped with CuK_α radiation ($\lambda = 1.54184$ Å) and a graphite monochromator. Intensity data were measured by ω-2θ scans of variable rate. For each compound, a quadrant of data was collected within the limits $2 < \theta < 75°$. Data reduction included corrections for background, absorption, Lorentz, and polarization effects. Absorption corrections were based on ψ scans. The structures were solved using direct methods and refined by full-matrix least squares, treating nonhydrogen atoms anisotropically, using Enraf–Nonius MolEN programs.[15] Hydrogen atoms were refined isotropically for *meso*-**4** and placed in calculated positions for *meso*-**5**. The OH hydrogen atom of *meso*-**5** appears to be disordered and was modeled by two half-populated positions. Crystal data are *meso*-**4**: $C_{14}H_{30}O_4$, M_r 262.4, monoclinic space group $P2_1/c$, a = 18.569(4), b = 4.3840(5), c = 9.941(2)Å, $\beta = 94.57(2)°$, V = 806.7(5)Å^3, Z = 2, $d_c = 1.080$ g cm^{-3}, T = 23° $\mu(CuK_\alpha) = 5.9$ cm^{-1}. Of 1647 unique data, 1344 had I > $3\sigma(1)$ and were used in the refinement. Convergence was achieved with $R = 0.040$ and $R_w = 0.052$. *meso*-**5**: $C_{18}H_{38}O_4$, M_r 318.5, monoclinic space group $P2_1/c$, a = 22.940(5), b = 4.3639(7), c = 9.9400(11)Å, $\beta = 92.71(2)°$, V = 993.9(5)Å^3, Z = 2, $d_c = 1.064$ g cm^{-3}, T = 23°, $\mu(CuK_\alpha) = 5.5$ cm^{-1}. Of 1995 unique data, 1472 had I > $3\sigma(I)$ and were used in the refinement. Convergence was achieved with $R = 0.091$ and $R_w = 0.114$.

Conclusions

Studies directed to better understand the mechanisms involved in the toxicologies of lipid ozonation and peroxidation demand the synthesis of

---

[14] G. D. Leikauf, Q. Zhao, S. Zhou, and J. Santrock, *Am. J. Respir. Cell Mol. Biol.* **9**, 594 (1993).

[15] C. K. Fair, "MolEN: An Interactive System For Crystal Structure Analysis." Enraf-Nonius, Delft, The Netherlands, 1990.

products derived from these processes. These lipid-derived ozonation and peroxidation products have distinct effects as inflammatory mediators in human diseases[16,17] by activating eicosanoid metabolism[18] and phospholipases[19] and by inducing the release of platelet activating factor and interleukins 6 and 8.[20] Syntheses described in this article can be extended to other homologous compounds and are important to allow the testing of hypotheses regarding relay molecules in the toxicologies of lipid ozonation and peroxidation.

Acknowledgment

This work was supported by a grant from the National Institutes of Health.

[16] G. A. Zimmerman, S. M. Prescott, and T. M. McIntyre, *J. Nutr.* **125,** 1661 (1995).

[17] T.-A. Imaizumi, D. M. Stafforini, Y. Yamada, T. M. McIntyre, S. M. Prescott, and G. A. Zimmerman, *J. Int. Med.* **238,** 5 (1995).

[18] G. D. Leikauf, Q. Zhao, S. Zhou, and J. Santrock, *Am. J. Respir. Cell Mol. Biol.* **9,** 594 (1993).

[19] R. M. Kafoury, W. A. Pryor, G. L. Squadrito, M. G. Salgo, X. Zou, and M. Friedman, *Toxicol. Appl. Pharmacol.* **150,** 338 (1998).

[20] R. M. Kafoury, W. A. Pryor, G. L. Squadrito, M. G. Salgo, X. Zou, and M. Friedman, *Am. J. Respir. Crit. Care Med.* **160,** 1934 (1999).

Section IV

General Methods

[53] Assay for Redox-Sensitive Transcription Factor

By Madan M. Chaturvedi, Asok Mukhopadhyay,
and Bharat B. Aggarwal

Introduction

The response of mammalian cells to external stimuli involves the transduction of signals back and forth across the plasma membrane between the cytoplasm and the nucleus. Reactive oxygen species (ROS) or oxidants such as hydrogen peroxide (H_2O_2), superoxide anions, hydroxyl radicals, nitric oxide (NO), or peroxynitrite also modulate cell-signaling pathways in mammalian cells.[1] Oxidative stress involving these molecules has been implicated in the pathophysiology of various diseases and in tumor promotion and cell growth.[2,3] Further, even sublethal levels of ROS can induce signal transduction and bypass normal second messengers to activate the cell and induce cellular adoptive responses to oxidative stress. In a feedback loop, the product of genes that are newly induced in response to oxidative stress may confer protection against subsequent challenge, repair ROS-mediated damage of cellular components, or serve to signal stress to neighbouring cells or cells in other part of the organism.[1]

In addition to regulating transmembrane signaling, oxidative stress influences gene expression by modulating such transcription factors as activator protein-1 (AP-1), nuclear factor-κB (NF-κB), and somatotropin-releasing factor/tissue-coding factor (SRF/TCF).[4–6] The two best-characterized redox-sensitive transcription factors are NF-κB and AP-1. NF-κB is a nuclear transcription factor that resides in its inactive states in the cytoplasm as a heterotrimer consisting of p50, p65, and IκBα subunits. On activation of the complex, IκBα sequentially undergoes phosphorylation, ubiquitination, and degradation, thus releasing the p50-p65 heterodimer

---

[1] H. J. Forman and E. Cadenas (eds.), "Oxidative Stress and Signal Transduction." Chapman & Hall, New York, 1997.

[2] P. A. Cerutti, *Science* **227,** 375 (1985).

[3] M. Meyer, R. Schreck, J. M. Muller, and P. A. Baeuerle, *in* "Oxidative Stress, Cell Activation and Viral Infection" (C. Pasquier, C. Auclair, R. Y. Oliver, and L. Packer, eds.), p. 217. Birkhauser Verlag, Switzerland, 1994.

[4] R. Schreck, *Trends Cell Biol.* **1,** 39 (1991).

[5] J. M. Muller, R. A. Rupee, and P. A. Baeuerle, *Methods* **11,** 301 (1997).

[6] K. Schulze-Osthoff, M. Bauer, M. Vogt, S. Wesselborg, and P. A. Baeuerle, *in* "Oxidative Stress and Signal Transduction" (H. J. Forman and E. Cadenas, eds.), p. 239. Chapman & Hall, New York, 1997.

for translocation to the nucleus.[7,8] AP-1 is a heterodimer of c-Jun and c-fos. The activation of c-Jun requires its phosphorylation by c-Jun N-terminal kinase (JNK), which belongs to the MAP kinase family. JNK is particularly activated by various stress conditions, including oxidative stress, and thus have acquired another name, stress-activated protein kinases (SAPK).[8,9]

To delineate the mechanism of response to oxidative stress, an assay is required for transcription activation. This article describes an assay, the electrophoretic mobility shift assay (EMSA), used commonly to assay for redox-sensitive transcription factors.[10] The method described is very suitable for measuring the activation of transcription factors such as NF-κB in response to a variety of prooxidant stimuli, including tumor necrosis factor (TNF), interleukin-1, H_2O_2, ceramide, okadaic acid, lipopolysaccharides, and phorbol myristate acetate (PMA). The method is extremely sensitive, as the activation of NF-κB can be detected within minutes after treatment of cells with as little as 1 pM TNF.[11] We have also used the protocol to detect the activation of AP-1 by a variety of stimuli.[12–17] The involvement of ROS in NF-kB activation is demonstrated by its inhibition by pyrrolidine dithiocarbamate.[17,18] The nuclear extract prepared by the method described here[11,19] has also been used to assay other constitutively active transcription factors such as Oct-1, SP-1, and TF-IID.[13,14,19–21]

[7] A. Ashkenazi and V. M. Dixit, *Science* **281,** 1305 (1998).

[8] M. Karin and M. Delhase, *Proc. Natl. Acad. Sci. U.S.A.* **95,** 9067 (1998).

[9] T. Yuasa, S. Ohno, J. H. Kehrl, and J. M. Kyriakis, *J. Biol. Chem.* **273,** 22681 (1998).

[10] F. M. Ausubel, R. Brent, R. E. Kingston, D. D. Moore, J. G. Seidman, J. A. Smith, and K. Struhl, "Current Protocols in Molecular Biology," Vol. 2, Chapter 12. Wiley, New York, 1995.

[11] M. M. Chaturvedi, R. LaPushin, and B. B. Aggarwal, *J. Biol. Chem.* **269,** 14575 (1994).

[12] M. M. Chaturvedi, A. Kumar, B. G. Darnay, G. B. N. Chainy, S. Agarwal, and B. B. Aggarwal, *J. Biol. Chem.* **272,** 30129 (1997).

[13] S. Singh and B. B. Aggarwal, *J. Biol. Chem.* **270,** 24995 (1995).

[14] K. Natarajan, S. Singh, T. R. Burke, Jr., D. Grunberger, and B. B. Aggarwal, *Proc. Natl. Acad. Sci. U.S.A.* **93,** 9090 (1996).

[15] S. K. Manna, H. J. Zhang, T. Yan, L. W. Oberley, and B. B. Aggarwal, *J. Biol. Chem.* **273,** 13245 (1998).

[16] K. Natarajan, S. K. Manna, M. M. Chaturvedi, and B. B. Aggarwal, *Arch. Biochem. Biophys.* **352,** 59 (1998).

[17] D. K. Giri and B. B. Aggarwal, *J. Biol. Chem.* **273,** 14008 (1998).

[18] R. Schreck, B. Meier, D. N. Mannel, W. Droge, and P. A. Baeuerle, *J. Exp. Med.* **175,** 1181 (1992).

[19] E. Schreiber, P. Matthias, M. M. Muller, and W. Schaffner, *Nucleic Acids Res.* **17,** 6419 (1989).

[20] A. Kumar, S. Dhawan, N. J. Hardegen, and B. B. Aggarwal, *Biochem. Pharmacol.* **55,** 775 (1998).

[21] S. Singh and B. B. Aggarwal, *J. Biol. Chem.* **270,** 10631 (1995).

Materials

Penicillin, streptomycin, RPMI 1640, DMEM, and fetal bovine serum (FBS) are obtained from GIBCO-BRL (Life Technology, Inc., Grand Island, NY). ^{32}P-labeled [γ-32p]ATP with a specific activity of 7000 Ci/mmol is obtained from ICN pharmaceutical, Inc. (Costa Mesa, CA). Bacteria-derived recombinant human TNF, purified to homogeneity with a specific activity of 5×10^7 units/mg, is kindly provided by Genentech, Inc. (South San Francisco, CA). Bovine serum albumin (BSA) is purchased from United States Biochemical Corp. (Cleveland, OH) and poly(dI:dC) from Pharmacia Biotech (Alameda, CA). Forty five-mer oligonucleotides having the NF-κB-binding sequence are custom synthesized and supplied by GIBCO-BRL (Life Technology, Inc.). U937 cells (a human histiocytic lymphoma) are obtained from the American Type Culture Collection (Rockville, MD), and the human myelomonoblastic leukemia cell line ML-1a is kindly provided by Dr. Ken Takeda of Showa University, Japan. The cells are checked periodically for mycoplasma contamination using the DNA-based assay kit purchased from Gene Probes (San Diego, CA). The polyclonal antibodies used are as follows: anti-p65, against the epitope corresponding to amino acids mapping within the amino-terminal domain of human NF-κBp65; anti-p50, against a peptide 15 amino acids long mapping at the NLS region of NF-κBp50; anti-IκBα, against amino acids 297–317 mapping at the carboxy terminus of IκBα/MAD-3; and anti-cyclin D1, against amino acids 1–295, which represents full-length cyclin D1 of human origin. All these antibodies are procured from Santa Cruz Biotechnology, Inc. (Santa Cruz, CA). All other biochemicals are purchased from Sigma Chemical Co. (St. Louis, MO). Plasticware is obtained from Becton-Dickinson Labware (Lincoln Park, NJ).

Reagents: Preparation of Stock Solutions

1 M HEPES, pH 7.9. Dissolve 23.83 g HEPES (MW 238.3) in 70–80 ml deionized H$_2$O (dH$_2$O), adjust pH to 7.9 with 10 M NaOH ($\sim$6 ml), and make up the volume to 100 ml. Sterilize by autoclaving and store at room temperature.

2 M KCl. Dissolve 14.91 g KCl (MW 74.55) in 50–60 ml dH$_2$O and make up the volume to 100 ml. Sterilize by autoclaving and store at room temperature.

5 M NaCl. Dissolve 29.22 g NaCl (MW 58.44) in 80 ml dH$_2$O and make up the volume to 100 ml. Sterilize by autoclaving and store at room temperature.

0.1 M EGTA, pH 7. Suspend 1.902 g EGTA (MW 380.4) in 30–40 ml of dH$_2$O. Measure pH while stirring and adjust with 10 *M* NaOH. It will dissolve completely when pH approaches around 7. Adjust the volume to 50 ml, sterilize by autoclaving, and store at room temperature.

0.5 M EDTA, pH 8. Suspend 93.05 g EDTA-disodium salt (MW 372.2) and 10 g NaOH in 450 ml dH$_2$O. Measure pH while stirring and adjust with 10 *M* NaOH. It will dissolve completely when pH approaches around 8. Adjust the volume to 500 ml, sterilize by autoclaving, and store at room temperature.

1 M Dithiothreitol (DTT). Dissolve 0.7725 g DTT (MW 154.5) in 4 ml dH$_2$O. Make up the volume to 5 ml. Sterilize by filtration through a 0.45-micron filter. Dispense in 0.5-ml aliquots and store frozen at −20°.

10% NP-40. Place 1 ml 100% NP-40 in a 15-ml Falcon tube and add 9 ml sterile dH$_2$O. Mix gently by inverting the tube. Avoid vortexing, which may cause frothing. Dispense in 1-ml aliquots and store at room temperature.

1 mg/ml Leupeptin. Prepare 10 mg/ml leupeptin by dissolving in sterile dH$_2$O. Dilute 10-fold to prepare the working stock of 1 mg/ml. Dispense into smaller aliquots and store frozen at −20°.

1 mg/ml Aprotinin. Dilute commercially available 2 mg/ml aprotinin twofold in sterile dH$_2$O and store at 4°.

250 mg/ml Benzamidine. Dissolve 1.25 g benzamidine in 4 ml sterile dH$_2$O in a 15-ml Falcon tube. Make up the volume to 5 ml. Dispense into 1-ml aliquots and store at −20°. For best results, use freshly aliquoted solutions of protease inhibitors (particularly aprotinin, leupeptin, and benzamidine). Aliquot inhibitors in small volumes so that they are not subjected to more than three or four freeze–thaw cycles.

100 mM Phenylmethylsulfonyl Fluoride (PMSF). Dissolve 0.0871 g PMSF (MW 174.2) in 5 ml isopropanol and store at room temperature in a brown bottle.

1 mg/ml Poly(dI:dC). Dissolve poly(dI:dC) to a concentration of 1 mg/ml according to manufacturer's recommendation. To the stock of 1.25 mg poly(dI:dC), add 1.25 ml sterile dH$_2$O, transfer to a microfuge tube, incubate at 45° for 5 min, transfer on ice, aliquot 50 μl into each tube, and store frozen at −20°.

6× DNA Loading Dye. Dissolve 25 mg each of xylene cyanol FF and bromophenol blue in 7 ml sterile dH$_2$O in a 15-ml Falcon tube. Add 3 ml sterile 100% glycerol, mix, dispense into 1-ml aliquots, and store at 4°.

5× EMSA Buffer. Dissolve 151.43 g Tris (MW 121.14) and 750 g glycine (MW 75) in 4.5 liters of dH_2O. Add 100 ml 0.5 M EDTA (pH 8). Make up the volume to 5 liters. Make sure the pH is 8.5. Sterilize by autoclaving and store at room temperature.

Sephadex G-50 (Fine). Add 10 g Sephadex to ~200 ml of dH_2O. Allow it to swell at room temperature for 3 to 4 hr. Decant fine particles, sterilize by autoclaving, and store at 4°.

30% Acrylamide Solution. Weigh out 29.2 g acrylamide and 0.8 g bisacrylamide and dissolve by adding 20–30 ml dH_2O. Make up the volume to 100 ml and store at 4° in a brown bottle. Take precautions against spilling while handling acrylamide and bisacrylamide as they are neurotoxins.

10% Ammonium Persulfate. Dissolve 1 g ammonium persulfate in 10 ml dH_2O and store at 4°. Use within a month.

Cell Lysis Buffer. To prepare this buffer, mix 1 ml HEPES (1 M), 0.5 ml NaCl (2 M), 0.02 ml EDTA (0.5 M), 0.1 ml EGTA (0.1 M), and 96.28 ml sterile dH_2O. Lysis buffer is stored at 4°. Just before use add 0.1 ml DTT (0.1 M), 0.5 ml PMSF (100 mM), 0.2 ml leupeptin (1 mg/ml), 0.2 ml aprotinin (1 mg/ml), and 0.2 ml benzamidine (250 mg/ml) to 9.79 ml of the lysis buffer.

Nuclear Extraction Buffer. To prepare this buffer, mix 1 ml HEPES (1 M), 4 ml NaCl (5 M), 0.1 ml EDTA (0.5 M), 0.5 ml EGTA (0.1 M), and 43.1 ml sterile dH_2O. This buffer is stored at 4°. Just before use, add 0.01 ml DTT (0.1 M), 0.01 ml PMSF (100 mM), 0.002 ml leupeptin (1 mg/ml), 0.002 ml aprotinin (1 mg/ml), and 0.002 ml benzamidine (250 mg/ml) to 0.974 ml of the buffer.

10× Binding Buffer. 200 mM HEPES, pH 7.9, 4 mM EDTA, pH 8, 4 mM DTT, and 50% glycerol are mixed in a sterile tube using stock solutions and sterile 100% glycerol. Dispense in 1-ml aliquots and store at −20°.

5' End Labeling of NF-κB Oligonucleotides. The wild-type oligoncleotide used for EMSA in our laboratory is a 45-mer sequence derived from HIV-LTR that contains two NF-κB-binding sites.[22] The sequences of the two complementary strands are as follows: NF-5' (5' TTG TTA CAA GGGACT TTCC GC TG GGGACTT TCC AGG GAG GCG TGG 3') NF-3' (5' CCA CGC CTC CCT GGAAAG TCC CC A GC GGAAAGTCCC TTG TAA CAA 3'). They are dissolved at a concentration of 1000 pmols/μl and stored at −20°. The working stocks are prepared by further serially diluting to 100, 10, and 1 pmol/μl. The following oligonucleotides with mutated NF-κB-binding sites where the conserved GGG are changed to CTC[22] are also custom synthesized: MUT-5' (5'TTG TTA CAA CTC ACT TTC C GC TGC TCA CTT TCC AGG GAG GCG TGG 3')

---

[22] G. Nabel and D. Baltimore, *Nature* **326,** 711 (1987).

MUT-3' (5'CCA CGC CTC CCT<u>GGA AAG TGA G</u> CA GCG <u>GAA AGT GAG</u> TTG TAACAA 3'). The mutant oligoncleotides are used to check the specificity of NF-κB.

Labeling Reaction. Because these oligonucleotides have a free OH at their 5' ends, they can be labeled easily using [γ-^{32}P]ATP and T4 polynucleotide kinase. Usually 2 pmol of one of the strands is labeled and annealed with the excess of unlabeled complementary oligonucleotide to make it double-stranded for EMSA. Perform the following steps in the room/area designated for use of radioactive materials. Also, while handling radioactive material, use the safety guidelines recommended by the institution.

1. Thaw out the oligonucleotides, [γ-^{32}P]ATP and T4 polynucleotide kinase buffer on ice. Mix the following components in a microfuge tube on ice:

Chemical	Volume (μl)
NF-5' or NF-3' (1 pmol/μl)	2
10× T4 polynucleotide kinase buffer	5
Sterile dH$_2$O	37
[γ-^{32}P]ATP (100 μCi/μl)	2
T4 polynucleotide kinase	4
Total volume	50

2. Mix by tapping, pulse spin, and incubate at 37° for 30 min.
3. Stop the reaction by adding 2 μl of 0.5 *M* EDTA, pH 8, and chill on ice.
4. Prepare a spun column using a 5-ml Bio-Rad disposable column. Put a column (sterilized by autoclaving) in a 50-ml Falcon tube in which a hole has been drilled to make a stand for the column. Pack approximately 5 ml of Sephadex G-50 slurry (i.e., up to the neck of the column) into the column.
5. To pack the column, keep the column in the improvised stand and centrifuge it in a Beckman GPR centrifuge using a GH 3.7 rotor at 2000 rpm at room temperature. Once the speed has reached 2000 rpm, centrifuge for 4 min.
6. Stop the centrifuge manually after 4 min. Remove the column and transfer it to another stand, which now contains a microfuge tube for the collection of eluate. The column should be placed in such a way that the nozzle of the column is placed in the collection tube.
7. Apply the labeling reaction on top of the column.
8. For elution, centrifuge exactly as done in step 5 and 6. After the end of the run, the collection tube should contain exactly the same volume as applied on the column. The eluate contains the labeled oligonucleotide. All the unincorporated [γ-^{32}P]ATP is trapped in

the column. Hence, discard the column carefully in a radioactive waste disposal container. A PD-10 column supplied by Pharmacia Fine Chemicals can be substituted for the improvised spun column.

9. Count the eluate using a Packard β liquid scintillation counter. A good labeling should give counts between 7 and 9×10^6 cpm; thus the specific activity of the oligonucleotide will be 1.6–2 μCi/pmol. In using a $[\gamma\text{-}^{32}\text{P}]\text{ATP}$ that has a specific activity of 7000 Ci/mmol (or 7 μCi/pmol), a 100% labeling would correspond to a specific activity of $\sim 7 \times 10^6$ cpm/pmol of oligonucleotide. This, however, assumes a counting efficiency of the counter as 100%.

10. To prepare the double-stranded oligonucleotide, add 1 μl of corresponding complementary oligoncleotide to the labeled oligonucleotide eluate. The concentration of complementary oligonucleotide should be 100 pmol/μl. If NF-5' is used for labeling, then use NF-3' for annealing. If Mut-5' is labeled, then use Mut-3' for annealing. A 20- to 50-fold molar excess of unlabeled complementary oligonucleotide is used to ensure the complete conversion of labeled single-stranded oligonucleotide into double-stranded oligonucleotide.

11. Keep the mixture of oligonucleotides in boiling water for 3 min for denaturation and then pulse spin. Then leave at room temperature for 15–30 min for annealing.

12. Transfer on ice and dilute to a concentration of 4 fmol/μl using sterile dH$_2$O. If 2 pmol is used for labeling and annealing, and if it is diluted to 500 μl, it would give a concentration of 4 fmol/μl. Store the labeled oligonucleotide at $-20°$. The excess of unlabeled complementary oligonucleotide does not interfere with the specific binding of transcription factors.

13. One may alternatively use commercially available consensus double-stranded oligonucleotides for NF-κB. Use 1 pmol of double-stranded oligonucleotide (equivalent to the 2 pmol of 5' ends) for labeling. Perform steps 1 to 9. Steps 10 to 12 are not required when double-stranded oligonucleotides are used. Also, as each double-stranded oligonucleotide would have two 5' ends, their specific activity would be twice as high as single-stranded oligonucleotides.

Treatment of Cells and Preparation of Extracts

The following protocol is designed to treat cells with different oxidants such as H$_2$O$_2$ and prooxidants such as TNF. The optimal dose and effect of a modulator for the activation of transcription factors are to be determined by performing pilot experiments.

Treatment of Cells Grown in Suspension

Count the cells and collect the cell pellet by centrifugation and suspend it in fresh medium (complete) at a concentration of 2×10^6/ml. Transfer 1 ml suspension to a 24-well plate for treatment with TNF or other modulators. Mix the cells with a P-1000 micropipette after the addition of TNF or other modulators. The cells must be grown following a regular splitting schedule. For example, if a cell is split every third day, one should not force the cells to split on a 1- or 2-day basis, which could cause stress and change the cell physiology, thus affecting the reproducibility of the results. If one needs cells every day to perform experiments, one may design a cell growth schedule so that cells seeded 3 days earlier are available every day. In addition, at no point should one perform experiments with overgrown cells (i.e., cells from the stationary phase of growth). Such cells as U937 or ML-la grown to a density of $0.8–1.0 \times 10^6$/ml (i.e., when the cells are in late log phase of growth) would be good for most purposes.

Adherent cells such as L-929 or HeLa should be grown in 6-well plates until they become semiconfluent. If a longer treatment with TNF or other activator is required, suspend cells at a lower density. It would be desirable to perform a pilot experiment to find out whether the modulator is cytostatic or cytotoxic. For example, if the cells are to be treated for 24 hr with a drug that is cytostatic but not cytotoxic, it is advisable to suspend the cells at a density of 0.5×10^6/ml (a 4-ml volume may be taken to get 2×10^6 cells) for treatment. Accordingly, if the treatment is to be done for a period longer than 24 hr and with a drug that is not cytostatic, the cells may be suspended at lower density. A modulator is typically added in a volume of 10–20 μl to the cells suspended in a 1-ml medium. For example, if a 10 pM final concentration of TNF is to be achieved, 10 μl of 1 nM TNF should be added in 1-ml cell suspension. If a modulator is not soluble in medium but soluble in solvents such as DMSO, then keep the volume added to the cell suspension low (5–10 μl).

Incubate the cell suspension at 37° for the desired time. While the cells are being incubated, label microfuge tubes and keep them on ice (prechilled microfuge tubes). Once the incubation is over, transfer the cells to prechilled microfuge tubes with care to remove the cells completely. To achieve this, the cells may first be mixed with a P-1000 and transferred in two steps using a volume of 0.5 ml every time. The chilling step is necessary to stop the incubation, particularly when the treatment is being done for a shorter period. Once chilled, cells may be left on ice for 30–60 min if many samples are being handled at a time. Otherwise, it is preferable to start the preparation of nuclear extracts immediately.

Treatment of Adherent Cells

Trypsinize, count, and centrifuge the cells to collect a cell pellet. For cells such as L929, HepG2, and HeLa, suspend in fresh medium (complete) at a concentration of 1×10^6 cells/ml and transfer 2 ml (equivalent to 2×10^6 cells) to a 6-well plate. For bigger cells, such as human diploid FS4 cells,[23] suspend at a concentration of $0.2–0.5 \times 10^6$/ml. Transfer 4–10 ml in either 60-mm (for 4 ml) or 100-mm (10 ml) petri dishes. Incubate at 37° for 4 hr to overnight so that cells adhere and flatten out. The time required may vary from cell type to cell type; observation under a phase-contrast microscope is necessary to determine this time. Add TNF or the modulators in the desired concentration. Mix by swirling the medium and incubate at 37°. To terminate the incubation, remove medium and replace with ice-cold PBS. Trypsinize or scrape the cells and transfer to 15-ml Falcon centrifuge tubes. Centrifuge, remove the supernatant, suspend the pellet in 1 ml ice-cold PBS, and transfer the suspension to prechilled microfuge tubes.

Preparation of Cytoplasmic and Nuclear Extracts

All subsequent steps are performed between 0 and 4° unless otherwise indicated. The success of preparing a good nuclear extract depends on keeping the temperature around 0°. If there is a step that needs to be done at room temperature, do it as quickly as possible and return the tube to ice immediately.

1. Centrifuge the tubes containing the cell suspension in a refrigerated microfuge at full speed (14,000g) for 1 min. Use a fine needle to aspirate the supernatant, being careful not to remove the cell pellet. Novices should practice with a Pasteur pipette with the end drawn as a thin jet and bent a little. While removing the supernatant, the tip of the bent Pasteur pipette is kept away from the pellet. This helps keep the flow of suction away from the pellet and hence reduces the chance of losing the cell pellet.
2. Add 1 ml ice-cold PBS and suspend the cells with a P-200 micropipette.
3. Centrifuge and aspirate the supernatant as in step 1. The PBS should be removed completely for proper lysis of the cells in subsequent steps.

[23] B. B. Aggarwal, K. Totpal, R. LaPushin, M. M. Chaturvedi, O. M. Pereira-Smith, and J. R. Smith, *Exp. Cell Res.* **218,** 381 (1995).

4. Add 400 μl ice-cold lysis buffer and suspend the cell pellet using P-200. When using cells such as FS-4, suspend in 800 μl of lysis buffer. If the cytoplasmic extract is to be saved at a higher protein concentration for Western blotting and other analysis, suspend the pellet only in 100 μl of lysis buffer.

5. Incubate the cell suspension on ice for 15 min. Because the lysis buffer is hypotonic, cells will swell in lysis buffer.

6. Add 12.5 μl of 10% NP-40 for every 400 μl of cell suspension from step 4. If the volume of lysis buffer used at step 4 is reduced or increased, adjust the volume of 10% NP-40 accordingly. NP-40 is a nonionic detergent that is used to solubilize the plasma membrane to release nuclei. The nuclear pellet from properly lysed cells (i.e., after step 8) is difficult to visualize, but the pellet becomes visible once the cytoplasmic extract is removed. If a yellowish pellet is visible before removing the cytoplasmic extract, the lysis conditions should be modified, either by increasing the amount NP-40 or by including a washing step with lysis buffer containing NP-40. The final concentration of NP-40 can be kept anywhere between 0.25 and 0.5%. NP-40 up to 0.5% does no harm to the nuclei. However, it does solubilize the outer nuclear membrane. Moreover, if the cells are left longer in NP-40, the inner nuclear membrane may be dissolved, releasing the chromatin material, which is not desirable.

7. Mix the cell suspension on a vortex machine vigorously for 10 sec to lyse the cells.

8. Centrifuge for 1 min and remove the supernatant (cytoplasmic extracts) by using P-200.

9. If the cytoplasmic extract is to be saved (for detection of IκBα, p50, or p65 by Western blot analysis), transfer the supernatant to a prechilled microfuge tube. Store the cytoplasmic extract frozen at $-70°$.

10. Make sure that the supernatant (cytoplasmic extract) is completely removed. If not, use either P-200 or a pasteur pipette whose end is drawn into a thin jet.

11. To the pellet add 25 μl of ice-cold nuclear extraction buffer. Use 50 μl for FS-4 cells.[23]

12. Incubate on ice for 30 min with intermittent vortexing, centrifuge at 14,000g for 5 min, and transfer the nuclear extract to a prechilled microfuge tube. The 0.4 M NaCl in the nuclear-extraction buffer extracts the loosely bound nonhistone chromosomal proteins (which includes all transcription factors and enzymes needed for transcription and replication). When the nuclear pellet is extracted in such a buffer, the remaining pellet contains DNA and histones, which

make the pellet gluey. Consequently, the pellet may stick on the wall of the tube during extraction. Hence, care should be taken to ensure that the pellet remains submerged in buffer during extraction. While removing the nuclear extract, sometimes the pellet may also be sucked. If that happens, eject the nuclear extract back into the tube and begin again. The pellet usually sticks to the wall of the tube, but sometime it sticks to the pipette tip, at which point it can be discarded. In this case, use a fresh tip to remove the clear nuclear extract from the tube. Remove a small aliquot (2–4 μl) from the nuclear extract for protein estimation by the Bradford reagent.[24] If not used immediately, quick-freeze the nuclear extracts in liquid nitrogen and store at $-70°$.

Electrophoretic Mobility Shift Assay

Preparation of Polyacrylamide Gels

Normally, a 7.5% gel in Tris–glycine–EDTA buffer[25] gives better resolution for the NF-κB–DNA complex. However, if a supershift assay is to be performed, the gel percentage may be reduced to 5%. The NF-κB–DNA complex may also be resolved using 0.5$\times$ Tris–borate–EDTA buffer instead of Tris–glycine–EDTA buffer.

1. Before starting the binding reaction, cast a 7.5% polyacrylamide gel.
2. Clean the glass plates. Setup the gel mould (16 $\times$ 18 $\times$ 0.15 cm) and prepare the gel by mixing the following solutions. A thick gel ($\sim$1.5 mm) gives a better resolution for 24 μl sample loading of the EMSA reaction mix than 1-mm-thick gels. Unlike SDS–PAGE (where due to a difference in pH and ionic strength stacking of protein samples may result in sharper bands), EMSA gels are native gels with a continuous buffer system, and stacking of a loaded sample does not occur. Hence, increasing the gel thickness enhances the resolution of bands.

Stock solution	Volume (ml)
Acrylamide and bisacrylamide	12.5
5$\times$ EMSA buffer	10
dH$_2$O	27.06
10% ammonium persulfate	0.40
TEMED	0.04
Total volume	50

[24] M. M. Bradford, *Anal. Biochem.* **72**, 248, (1976).
[25] H. Singh, J. H. Lebowitz, A. S. Baldwin, Jr., and P. A. Sharp, *Cell* **52**, 415 (1988).

3. Pour the gel and let it polymerize for 30 to 45 min or overnight.
4. Once the gel is polymerized, remove the bottom spacer and the comb, rinse the wells with water, and fix the gel to the chamber. Add 1× EMSA buffer. Flush the wells with buffer using a 10-ml hypodermic syringe and needle. Also, if there is an air bubble at the bottom of the gel, remove it by flushing with a syringe.
5. Prerun the gel in 1× EMSA buffer at 150 V (~40 mA) at room temperature for at least 30 min. A prerun is essential to let the unpolymerized excess acrylamide, bisacrylamide, ammonium persulfate, and TEMED move ahead of the sample.

Binding Reaction

1. Remove the nuclear extracts from −70° and let them thaw on ice. In the meantime, prepare the binding reaction mixture.
2. A typical binding reaction mixture for each sample comprises the following components, mixed on ice in a microfuge tube[26]:

Stock solution	Volume (μl)	Sequence of addition
dH$_2$O	6	1
10× binding buffer	2	2
Poly(dI:dC)	2	3
^{32}P-labeled ds NF-κB oligonucleotide	4	4
10% NP-40	2	5
Total volume	16	

If competitor oligonucleotides or antibodies are also to be added, the volume of H$_2$O should be adjusted accordingly. Usually, the volume of competitor oligonucleotide and antibodies are kept within 2 μl. While preparing the binding reaction mixture, mix the contents by pipetting in and out with the same pipette. NP-40, used to enhance the binding of transcription factors to DNA,[27] should be added last to avoid frothing. Also, do not use a vortex to mix the contents. Rather, gently invert or tap the tube two or three times and pulse spin to drive the contents to the bottom of the tube. It is convenient to prepare a common binding mix in one tube and dispense equal volumes to each tube containing nuclear extracts. This would minimize the error due to the addition of many components in a small volume of binding reaction.

[26] M. A. Collart, P. A. Baeuerle, and P. Vassalli, *Mol. Cell. Biol.* **10**, 1498 (1990).
[27] H. H. Hassanain, W. Dai, and S. L. Gupta, *Anal. Biochem.* **213**, 162 (1993).

3. To 4 μl of nuclear extracts in each tube, add 16 μl of the just-described mixture and mix with the same pipette. Usually a nuclear extract prepared from 2 × 10⁶ cells such as U-937 should give a protein concentration of 0.8–1.2 μg/μl. Hence, if 4 μl is taken in a binding reaction, it gives a protein concentration of 3.2–4.8 μg/reaction. If the variation in protein concentrations is within 5% among different samples, there is no need to adjust the protein concentration. However, if the variation is large, one must adjust the concentration so that each binding reaction receives a roughly equal amount of protein. The volume of nuclear extract in a binding reaction mixture should usually be kept at 4 μl, which gives a final 80 mM concentration of NaCl. Because the nuclear extracts prepared by the method are not dialyzed to remove the salt, the volume of nuclear extract in a binding reaction should not exceed 6 μl. The final salt concentration in no case should exceed 125 mM.

4. Pulse spin in a refrigerated microfuge, incubate in a 37° water bath for 15 min, and transfer the tubes on ice.

5. Add 4 μl of DNA-loading dye. Mix by tapping, pulse spin to collect everything at the bottom of the tube, and transfer the tubes on ice.

6. Stop the prerun, flush the wells again with EMSA buffer, and load the sample using P-100. As the sample contains 1% NP-40, avoid sucking air bubbles before loading. These air bubbles are very tiny and may cause an upward flow of sample in the well while loading, resulting in a loss of sample and an unexpected variation in results.

7. Electrophorese at 150 V (~40 mA) for 3 hr until the bromphenol blue migrates to 1–2 cm from the bottom of the gel. Disconnect the power supply and remove the gel from the chamber.

8. Insert a thin wedge-shaped article between the glass plates from one corner of the gel mould. Apply a twisting pressure to lift one of the glass plates (usually the top smaller one) carefully without disturbing the gel. Mark the orientation of the gel by trimming the right-hand top corner of the gel. Place a piece of filter paper on the top to cover the gel. Press the filter paper uniformly but gently. From one corner, carefully lift the filter paper on which the gel is firmly stuck. While removing the glass plates, it is better to remove the notched glass plate. If the gel is lifted from a larger plate on a filter paper, the orientation of the gel remains the same as loading the sample from left to right. Sometimes it is advisable to load the samples asymmetrically so that even if the orientation gets reversed while lifting the gel on filter paper, one can always figure out the sequence of loading. In EMSA, the asymmetry can be defined easily by always loading the free oligonucleotide on one side of the gel.

9. Cover the gel with Saran wrap and dry it in a gel dryer at 80° for 1 hr.

10. Once the gel has dried down, remove from the gel dryer, tape the corners, and expose on a PhosphorImager screen (Molecular Dynamics). A 10- to 12-hr exposure using a Molecular Dynamics screen gives a well-saturated image if the oligonucleotide is freshly labeled. One may adjust the contrast and brightness to produce a publication-quality image. The contrast should be adjusted in such a way that it produces an image that matches with the result of quantification. For quick information, even a 2-hr exposure would be alright. If a PhosphorImager is not available, one can expose the dried gel on a X-ray film. In that case the length of the exposure should be optimized.

Supershift and Cold Competition

For supershift, add 1 μl antibodies (these are prepared at 10× higher concentration compared to the antibodies for Western blotting) in a binding reaction without the ^{32}P-labeled oligonucleotide. Add 4 μl of nuclear extract and incubate at room temperature for 5 min. Transfer on ice, add 4 μl of the ^{32}P-labeled oligonucleotide, and proceed as for a regular binding reaction.

For a cold competition assay, prepare a double-stranded NF-κB wild-type oligonucleotide as follows: Take 10 μl each of NF-5′ (10 pmol/μl) and NF-3′ (10 pmol/μl). Denature by keeping in a boiling water bath for 3 min. Pulse spin and leave overnight at room temperature for annealing. The next day, dilute by adding 230 μl of sterile dH$_2$O. This gives a final concentration of 400 fmol/μl of double-stranded NF-κB oligonucleotides. Aliquot in 25 μl volume and store frozen at $-20°$. Use 2 μl of 400 fmol/μl double-stranded oligonucleotides in a manner similar to that described for supershift. This gives 50× molar excess if 16 fmol of labeled oligonucleotides is used in a binding reaction.

Western Blotting for IκBα. For IκBα, p50, and p65 Western blot assays, use 25–30 μg of cytoplasmic extracts. Following sodium dodecyl sulfate (SDS)–polyacrylamide gel (10%) electrophoresis (PAGE), transfer the proteins to nitrocellulose membranes electrophoretically. Block the membranes with phosphate-buffered saline with 0.5% Tween 20 (PBST) containing 5% fat-free milk and then expose them to IκBα, p50, or p65 antibodies at 1 to 3000 dilution. Wash the membranes with PBST, and treat them with secondary antibody conjugated to horseradish peroxidase. Visualize the antigen-antibody reaction on film using an enhanced chemiluminescence (ECL) assay and Amersham ECL reagent.

Results

Typical results obtained from EMSA of NF-κB are shown in Fig. 1. In this experiment, the dose response (Fig. 1A) and the kinetics (Fig. 1B) of activation of NF-κB by TNF have been studied. Also, the specificity of the NF-κB complex was established by competition and supershift assays (Fig. 1C). Briefly, 2×10^6 ML-1a cells in 1 ml volume were treated with different concentrations of TNF (1 to 10,000 pM) for 30 min. Thereafter, the nuclear extracts were prepared and assayed for NF-κB activation by EMSA as described earlier. Results (Fig. 1A) indicate that TNF activated NF-κB at a concentration as low as 1 pM. However, a strong activation was obtained only at the 10 pM TNF concentration, and beyond 100 pM, the activation was saturated.

Results of supershift and cold competition are shown in Fig. 1C. The NF-κB was supershifted only when antibodies against p50 and p65 were used, but not when unrelated antibodies, such as anti-c-Rel or anti-cyclin D1, were used. When a 25-fold molar excess of unlabeled NF-κB oligonucleotide was added to the binding reaction mixture, it competed for binding sites and so no binding with the labeled oligonucleotide was observed, suggesting that the complex formed was specific. To ascertain the specificity further, binding was carried out with a labeled mutant oligonucleotide in which the conserved GGG required for NF-κB binding was mutated to CTC. Again no band for NF-κB was observed, suggesting that the complex formed required the conserved GGG for binding.

The kinetics of activation is shown in Fig. 1B. TNF at 10 pM concentration activated NF-κB within 10 min of treatment. The activation peaked around 15 min and then declined. This is a typical transient activation of NF-κB by TNF. The activation of NF-κB correlated with the degradation of the NF-κB inhibitor (IκBα) that was degraded initially, but later was resynthesized in response to NF-κB (see Fig. 1D) because the IκB promoter is under the control of NF-κB. NF-κB is extremely sensitive because TNF activates it when only 2–3% of TNF receptors are occupied by the ligand.[28]

Troubleshooting of Common Problems Encountered in the Assay of NF-κB by EMSA

1. EMSA Bands Are Not Sharp and the Pattern is Smeared

Explanations. This is one of the most commonly observed problem and may be due to the degradation of proteins in the crude nuclear extracts.

---

[28] H. Chen and B. B. Aggarwal, *J. Biol. Chem.* **269**, 31424 (1995).

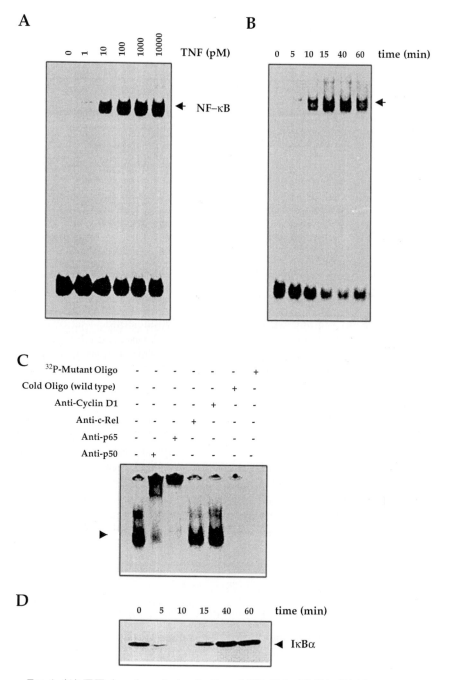

FIG. 1. (A) TNF dose-dependent activation of NF-κB by EMSA. (B) Time course of TNF-dependent NF-κB activation by EMSA. (C) Supershift and cold competition of NF-κB activation. (D) Time course of TNF-dependent degradation of IκBα.

In a typical EMSA for NF-κB, two bands are observed corresponding to p50/p50 homodimer and p50/p65 heterodimer. The homodimer band migrates slightly slower than the heterodimer band. Further, because NF-κB belongs to the rel family of transcription factors, some cells have other isoforms of NF-κB. The composition and specificity of each band should be verified by supershift (using specific antibodies) and cold competition (using excess unlabeled oligonucleotide). The heterodimer is induced by oxidants and prooxidant cytokines. However, sometimes due to the degradation of proteins in the nuclear extracts, the inducible heterodimer band either comigrates or migrates slower than the homodimer band. Sometime even a very broad band may be seen. The problem of degradation is typical for cells of monocytic origin, such as U937 and ML-1a, which have high concentrations of intracellular proteases. This problem is usually not observed with cell such as HeLa, FS4, Jurkat. Also, as mentioned in the section on "preparation of stock solutions," such a problem may be overcome by using fresh stocks of protease inhibitors. The protease inhibitors aprotinin and leupeptin lose their activity after repeated freezing and thawing. In the cocktail of protease inhibitors, one may even include more inhibitors. However, the use of protease inhibitors such as TPCK is not recommended, as TPCK inhibits the activation of NF-κB by binding covalently to the p50 subunit.[29]

2. Highly Variable Protein Contents in Nuclear and Cytoplasmic Extracts

Explanations. The protocol is designed such that one can work with a minimal number of cells (i.e., 2×10^6) and can handle a large number of experimental sets simultaneously. Up to 48 samples have been processed simultaneously in our laboratory. Hence, if extreme care is not taken during the preparation of nuclear extracts, one often encounters unexpected variations. Typical problems are loss of cells and nuclear pellet while removing supernatants and improper extraction of the nuclear pellet.

3. High Background and Radioactivity Sticks in the Well

Explanations. These problems are rarely seen and are often associated with the use of old [γ-^{32}P]ATP (more than a month old) or water that has not been deionized properly. We always recommend using Millipore-deionized water, as some contaminating trace amounts of divalent cations may interfere with denaturation and annealing while preparing the probe.[30] If a commercially supplied double-stranded oligonucleotide is used, then

[29] T. S. Finco, A. A. Beg, and A. S. Baldwin, *Proc. Natl. Acad. Sci. U.S.A.* **91,** 11884 (1994).
[30] W. F. Dove and N. Davidson, *J. Mol. Biol.* **5,** 467 (1962).

this problem is less likely to occur. However, we have adopted the process of annealing to convert an end-labeled single-stranded oligonucleotide into a double-stranded oligonucleotide. A 50- to 100-fold molar excess of unlabeled complementary oligonucleotide is used in a very stringent condition for annealing. This, indirectly, becomes advantageous in producing clean bands because some single-stranded DNA-binding proteins present in crude nuclear extract are titrated out by the excess of single-stranded complementary oligonucleotide. A smeared background may also be generated if the spin column is not made properly, and the probe may become contaminated with unincorporated $[\gamma\text{-}^{32}P]ATP$.

4. Fold Activation Varies Greatly among Experimental Replicates

Explanations. Although the NF-κB p50/65 heterodimer is an inducible transcription factor, some low basal activation is always found. Therefore, the fold induction is to be calculated as a ratio of induced activation over uninduced activation. However, because NF-κB is also induced by a variety of other stress conditions and if the cells for replicate experiments are not grown under identical conditions, the basal level of NF-κB may vary to a great extent. This causes a great degree of variation in the fold activation among experimental replicates.

5. Visual Inspection of EMSA Bands Does Not Match with the Quantification of Radioactivity

Explanations. This problem is observed if one has generated an autoradiogram using a short exposure on X-ray film or obtained an image using a PhosphorImager. When a PhosphorImager is used to generate an image, the brightness and contrast should be adjusted strictly in accordance with the values from quantification. An artifact can be generated easily where the low basal activity of NF-κB is not seen, and the induced activity is seen as a dark band. In such a situation, visually, the fold of activation can be interpreted as very high. The same problem may be observed if the exposure is done on X-ray film for only a short period of time. X-ray films do not have a linear sensitivity to detect low radioactivity usually present in the basal band of NF-κB. Hence, it is recommended that while using X-ray films, an autoradiogram with both a short- and a long-term exposure be generated. The fold activation should then be evaluated carefully if the autoradiograms are scanned and the radioactivity in the gel is not counted directly.

Acknowledgment

This research was supported by The Clayton Foundation.

[54] Fluorescent Fatty Acid to Monitor Reactive Oxygen in Single Cells

By E. H. W. Pap, G. P. C. Drummen, J. A. Post, P. J. Rijken, and K. W. A. Wirtz

Introduction

Reactive oxygen species (ROS) are formed during normal cellular metabolism and under the influence of environmental factors such as sunlight and ozone. Scavenging of ROS by antioxidants and repair of oxidative damage are elementary processes required for cell survival. Any imbalance between oxidant and antioxidant activities affects ROS production and could potentially lead to pathogenesis or the progression of disease states.[1,2] Despite enormous interest in the effects of ROS on cell function and human health, their detection is still a major challenge. Because the current assays are cumbersome, their application is often the limiting factor in the analysis of oxidant activities and antioxidant protection. The ideal assay should be simply to provide quantitative information on ROS activities, preferably with spatiotemporal resolution. It should report on oxidative damage in cells and allow for the screening of antioxidants. Assays based on the detection of secondary reaction products (e.g., aldehydes, isoprostanes) provide a wealth of information about the oxidation processes involved, but are indirect and elaborate and do not provide (sub)cellular resolution.

Intracellular ROS measurements with oxidation-sensitive fluorophores, such as dichlorofluorescin, lack accuracy due to variations in dye uptake and compartmentalization. In Ca^{2+} and pH imaging, these inaccuracies have been circumvented by the use of parameter-sensitive ratio dyes.[3] These dyes have in common that the fraction of dye that is complexed to one of these ions can be deduced from the fluorescence spectrum. From the ratio of two fluorescence images recorded at wavelengths specific for complexed and noncomplexed dye, the ion concentration can be calculated in each cellular compartment. This article describes the oxidative sensitivity of the fluorophore BODIPY[581/591] and its application in the detection of intracellular ROS activities. On oxidation, fluorescence excitation and emission of this probe is shifted from red to green. The ratio of green (oxidized) to total (green + red) fluorescence eliminates variations caused by heterogeneous

[1] C. Rice Evans and R. Burdon, *Prog. Lipid Res.* **32,** 71 (1993).
[2] J. M. Gutteridge, *Med. Lab. Sci.* **49,** 313 (1992).
[3] G. R. Bright, G. W. Fisher, J. Rogowska, and D. L. Taylor, *J. Cell. Biol.* **104,** 1019 (1987).

METHODS IN ENZYMOLOGY, VOL. 319

probe uptake and distribution. This property is unique among oxidation-sensitive probes and facilitates the visualization of ROS activities on a subcellular level. In addition to microscopic applications, the probe provides sufficient sensitivity to determine on-line the protection of cellular and vesicle membranes by antioxidants in multiwell plates.

The multiwell approach is of particular interest in the dissection of antioxidant networks. In this concept, antioxidants do not function singly, but are interconnected in a redox-based antioxidant cascade. That is to say, electrons accepted by a particular antioxidant can be donated to the next redox partner in the cascade, thereby regenerating the antioxidant. It has been demonstrated repeatedly that such an interaction exists between tocopherol and ascorbate.[4-6] It will be shown that the interaction between tocopherol and ascorbate in phospholipid vesicles leads to a reduced oxidation rate of BODIPY[581/591] as compared to controls in which either one of these antioxidants is present. Further properties and limitations of the technology will be discussed as illustrated by the production of ROS during cell death.

Spectra and Properties of C11-BODIPY[581/591]

C11-BODIPY[581/591] (Molecular Probes, Junction City, OR) can be used as a fluorescent indicator of reactive oxygen activities.[6a] The susceptibility of the probe toward reactive oxygen is comparable to that of endogenous fatty acids. It is lipophilic and can be incorporated readily into vesicles or cellular membranes. Oxidation of C11-BODIPY[581/591] can be initiated by a variety of oxy-, peroxy-, or hydroxyl radicals, but not by superoxide, nitric oxide, or hydroperoxides. The probe is also not sensitive for singlet oxygen released either by naphthalene-derived endoperoxides or by eosin. The probe may be used in a similar manner as other oxidation-sensitive probes (e.g., dichlorofluorescin), but offers the advantage that the excitation and emission maximums of the oxidized form are different from those of the nonoxidized form. The fluorescence spectrum of C11-BODIPY[581/591] (0.1 mol%) in egg phosphatidylcholine (PC) vesicles (200 μM) is shown in Fig. 1A. The spectrum is a composite of the fluorescence of the nonoxidized form (peak at 581/591 nm) and of the oxidized form (peak at 500/510 nm). Fluorescence measurements were carried out at ambient temperature, 30

[4] J. J. van den Berg, F. A. Kuypers, B. Roelofsen, and J. A. Op den Kamp, *Chem. Phys. Lipids* **53,** 309 (1990).
[5] C. E. Thomas, L. R. McLean, R. A. Parker, and D. F. Ohlweiler, *Lipids* **27,** 543 (1992).
[6] H. Sies and W. Stahl, *Am. J. Clin. Nutr.* **62,** 1315s (1995).
[6a] E. H. W. Pap, G. P. Drummen, V. J. Winter, T. W. Kooij, P. Rijhen, K. W. Wirtz, J. A. Op den Kamp, J. W. Hage, and J. A. Post, *FEBS Lett.* **453:3,** 278, (1999); Y. M. A. Nagib, *Anal. Biochem.* **265:2,** 290 (1998).

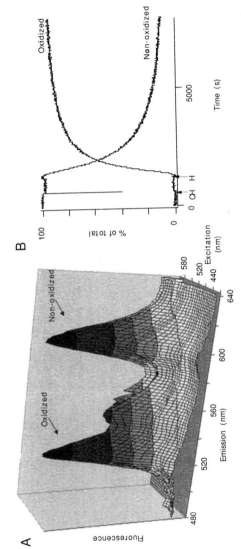

Fig. 1. (A) Oxidation-induced spectral shift of C11-BODIPY[581/591] two-dimensional fluorescence spectrum of C11-BODIPY[581/591] in egg PC vesicles after 30 min of exposure to cumene hydroperoxide (100 μM) and hemin (80 nM). The spectrum was reconstructed from a series of emission spectra that were taken at different excitation wavelengths. (B) The time course of the red (nonoxidized) and green (oxidized) fluorescence after the addition of cumene hydroperoxide (CH) and hemin (H) is plotted.

min after exposure to cumene hydroperoxide (100 μM) and hemin (80 nM) using a Quantum Master spectrofluorometer (PTI, Surbiton, Surrey, UK). Decay of the red and increase of the green fluorescence are shown in Fig. 1B.

Labeling of Cells with C11-BODIPY[581/591]

At least 24 hr prior to the experiment, cultured cells are detached from their culture flask using trypsin/EDTA and transferred to dishes containing coverslips (diameter of 24 mm). A stock solution of C11-BODIPY[581/591] in dimethylformamide (20 mM) is first diluted 100 times in fetal calf serum and then administered to the growth medium (5 vol%), yielding a final probe concentration of 10 μM. A 30-min incubation of the cells is sufficient to give a strong labeling of the cellular membranes. Negative effects on cell viability after prolonged exposure to C11-BODIPY[581/591] (up to 4 days) were not detected.

Confocal Laser Scanning Microscopy and Image Processing

All images were taken with a Leica TCSNT confocal laser scanning system on an inverted microscope DMIRBE (Leica Microsystems, GmbH, Heidelberg, Germany) using an argon–krypton laser as the excitation source. In principle, each fluorescence microscope with separate excitation and detection for red and green fluorescence can be used to monitor the oxidation of C11-BODIPY[581/591] in cells.

Prior to collecting the images, the cells are rinsed with enriched phosphate-buffered saline (PBS$^+$ 137 mM NaCl, 2.7 mM KCl, 0.5 mM MgCl$_2$, 0.9 mM CaCl$_2$, 1.5 mM KH$_2$PO$_4$, 8.1 mM Na$_2$HPO$_4$, and 5 mM glucose, pH 7.4) to remove the nonincorporated dye. Subsequently, the coverslip is placed in the temperature (37°)-controlled holder of the microscope, filled with 0.5 ml PBS$^+$. The green and red fluorescence of C11-BODIPY[581/591] are acquired simultaneously using double wavelength excitation (laser lines 488 and 568 nm) and detection (emission band-pass filters 530/30 and 590/30). The images are processed on a Power Macintosh computer using the program NIH image (Wayne Rasband, NIMH, Bethesda, MD). From each set of green and red images, the fraction of oxidized C11-BODIPY[581/591] is calculated. The intensity of a pixel with coordinates x,y in the fluorescence images of the green ($I^g_{x,y}$) and red ($I^r_{x,y}$) channels correspond to Eqs. (1) and (2):

$$I^r_{x,y} = [B]_{x,y} (1 - F_{x,y}) c_r \qquad (1)$$
$$I^g_{x,y} = [B]_{x,y} F_{x,y} c_g \qquad (2)$$

where $[B]_{x,y}$ corresponds to the total concentration of probe, $F_{x,y}$ to the fraction of oxidized probe, and c_r and c_g to constants that are related to the extinction coefficient of the red and green fluorescent forms of the probe and to wavelength-specific instrument properties. From Eqs. (1) and (2) it can be derived that $F_{x,y}$ corresponds to Eq. (3):

$$F_{x,y} = \frac{I_{x,y}^g}{I_{x,y}^g + I_{x,y}^r \dfrac{c_g}{c_r}} \tag{3}$$

Thus, by dividing the green image intensities by the sum of the green and red image intensities, an image is obtained that expresses the fraction of oxidized probe. In this image the term $[B]_{x,y}$ is eliminated. Therefore, values in this image are independent of heterogeneous dye uptake and uneven distribution both between cells and even between subcellular organelles. The term c_g/c_r is constant for each microscopic configuration (excitation wavelengths and intensities, detection wavelengths and sensitivity). With a constant setup, this term has to be determined only once by dividing the image intensities of C11-BODIPY[581/591]-loaded cells after complete oxidation (green) by the image intensities before oxidation (red).

ROS Production during Cell Death

To illustrate the application of C11-BODIPY[581/591], rat-1 fibroblasts were deprived from serum to induce apoptosis.[7] After 24 hr treatment, the cells were loaded with C11-BODIPY[581/591] and mounted on the confocal laser scanning microscope. Extensive oxidation of the probe was observed in about 15% of the cells (green cells), whereas the probe remained intact in the other cells (red cells) (Fig. 2A, left) (see color insert). At the time of acquisition the cells exclude propidium iodide, indicating that the plasma membrane is still intact. On prolonged exposure, nuclei of only the green cells became stained by propidium iodide, indicating that the plasma membrane integrity of these cells is lost (data not shown). This observation confirms that reactive oxygen is formed during the process of cell death.[8,9] From the individual images of green and red fluorescence, the fraction of oxidized probe was calculated (Fig. 2A, right). The variation in color confirms the large intercellular variance of reactive oxygen production (fraction oxidized varies from 0.17 to 0.89). The right-hand side of Fig. 2 further

[7] J. L. Kummer, P. K. Rao, and K. A. Heidenreich, *J. Biol. Chem.* **272**, 20490 (1997).
[8] K. Banki, E. Hutter, E. Colombo, N. J. Gonchoroff, and A. Perl, *J. Biol. Chem.* **271**, 32994 (1996).
[9] E. Gulbins *et al.*, *Immunology* **89**, 205 (1996).

illustrates that the image processing eliminates intensity fluctuations as a result of heterogeneous probe distribution. As seen in the left-hand side of Fig. 2, organelles around the nucleus contain much higher probe levels than the nucleus itself and the cellular extensions. However, these cellular structures are hardly distinguishable in the right-hand side of Fig. 2. Magnification of the cell outlined in Fig. 2A shows that a large part of the probe is oxidized, particularly at the cellular borders (Fig. 2B). In some worm-like cellular structures, which represent mitochondria (determined by double labeling with rhodamine 123), the probe is hardly oxidized. This indicates that despite their prominent role in reactive oxygen production, a part of the mitochondrial population is well protected against oxidation in this stage of cell death.

Multiwell Application to Determine the Efficacy of Antioxidants

The spectral properties of C11-BODIPY$^{581/591}$ in vesicles and cells are highly suitable for analyzing its protection against oxidation by single antioxidants and by antioxidant mixtures. From the protection of C11-BODIPY$^{581/591}$ in lipid vesicles, the conditions that regulate the coupling between redox partners can be established. By taking these conditions into account, the functional significance and efficiency of the redox couple can be measured in monolayers of living cells cultured in multiwell plates.

This approach is illustrated by an experiment in which the protection of C11-BODIPY$^{581/591}$ in vesicles by various concentrations of tocopherol and ascorbate is measured. Vesicles are made by injection (10 μl) of an ethanolic solution of egg PC (20 mM,) C11-BODIPY$^{581/591}$ (20 μM) and various concentrations (0–40 μM) of tocopherol (an equal molar mixture of α and γ isomers) into 1 ml PBS.[10] The vesicles are dispensed into a 96-well plate whereupon various concentrations of ascorbate are added. Oxidation of C11-BODIPY$^{581/591}$ is initiated by the addition of cumene hydroperoxide (1 mM) and CuSO$_4$ (20 μM). Decay of the red fluorescence is collected by a Tecan multiwell spectrofluorometer (Tecan, Ltd., UK) using excitation and emission band-pass filters of 570/20 and 620/20 nm, respectively. The fluorescence is collected for 2 hr with time intervals of 3 min. During the measurements, the temperature is kept constant at 37°. The background signal of vesicles without C11-BODIPY$^{581/591}$ corresponds to approximately 5% of the signal obtained from labeled vesicles. A selection of the decays of the red C11-BODIPY$^{581/591}$ fluorescence is presented

---

[10] S. Batzri and E. D. Korn, *Biochim. Biophys. Acta* **298** (1973).

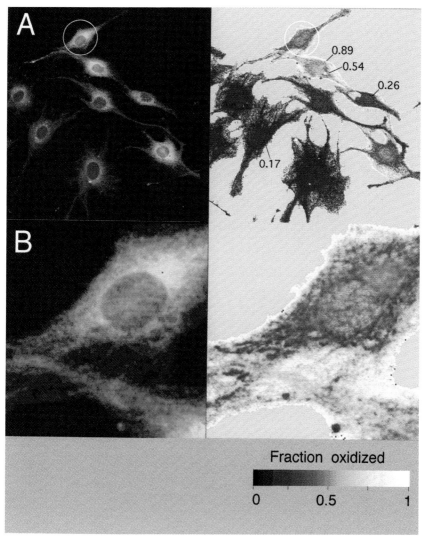

Fig. 2. Images of C11-BODIPY$^{581/59}$-labeled rat-1 fibroblast cultured for 24 hr in serum-deprived medium. (Left) Images of green and red fluorescence are merged. (Right) The fraction of oxidized probe is presented. A magnification of the cell outlined in A is presented in B.

in Fig. 3 (dotted lines). When no antioxidants are present, two-thirds of the C11-BODIPY$^{581/591}$ is oxidized after 2 hr (Fig. 3, panel a1). The decay of C11-BODIPY$^{581/591}$ fluorescence is almost linear, indicating that the production of radicals is constant during the 2-hr period. Little protection of the probe is obtained by ascorbate in the absence of tocopherol (Fig. 3, panels a1–a6). High levels of tocopherol (0.1–0.2 mole%) result in an almost complete protection of the probe in the first hour of the measurement (see Fig. 3, panels d1–e1). However, after 2 hr of incubation, the rate of C11-BODIPY$^{581/591}$ oxidation in these samples is about identical to the rate of oxidation in the absence of antioxidant, indicating that with time the protection by tocopherol is lost. When both ascorbate and tocopherol are present, a strong protection is obtained that lasts during the entire measurement (Fig. 3, see panels d5–e6). This observation is in accordance with previous observations that ascorbate is able to regenerate tocopherol from its oxidized form.[4–6]

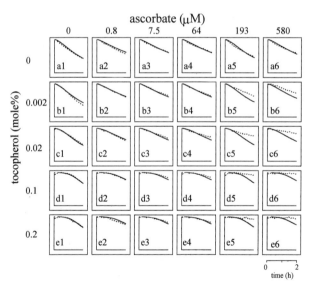

FIG. 3. Normalized fluorescence decays of C11-BODIPY$^{581/591}$ in egg PC vesicles in the presence of various concentrations of tocopherol and ascorbate. Oxidation was initiated by exposure of the vesicles to cumene hydroperoxide (1 mM) and CuSO$_4$ (20 μM). Fluorescence was recorded for 2 hr. Experimental decays (dashed lines) acquired in the presence of a single antioxidant (a1–a6, a1–e1) were fitted numerically (solid lines). Using optimized values from this analysis, decays in the presence of both antioxidants were predicted (solid lines in b2–e6).

Numerical Analysis of Decays

A simple reaction scheme for oxidation of the chromophore and its protection by two antioxidants is presented in Fig. 4. Radicals (R·) are formed from cumene hydroperoxide and metal ions (Me) with a rate constant k_i. The radicals can combine with each other (termination reaction) with rate constant k_i. They can also oxidize the red fluorescent chromophore (Bo) to the green fluorescent form (Bo_{ox}). The rate of decline of red fluorescent chromophores (dBo/dt) will depend on its actual concentration [Bo], the concentration of oxidant ([R·]), and the oxidative sensitivity of the chromophore (k_b). The presence of a radical scavenging antioxidant such as tocopherol [toc] will lower the actual [R·], thereby reducing the rate of Bo oxidation. The tocopherol radical that is generated after scavenging may be eliminated (e.g., by dimerization) or it may be reduced by ascorbate (asc). Numerical analysis of the decays with a set of differential equations renders antioxidant efficiencies (k_{asc} and k_{toc}). When all variables are optimized for the condition where either of the two antioxidants are present (Fig. 3, panels a1–a6 and a1–e1), the decay of C11-BODIPY[581/591] may be predicted for the condition where both antioxidants are present (Fig. 3, panels b2–e6). In the latter case, when the predicted curve (solid line) declines faster than the measured one (dotted line), one may conclude that a combination of the two antioxidants provides a better protection than expected on the basis of their additive effects.

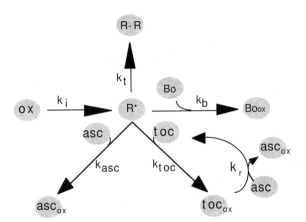

FIG. 4. Reaction scheme of the generation of reactive oxygen and the scavenging by tocopherol and ascorbate.

Pitfalls

When applying this method for the quantification of oxidant activities and the efficacy of antioxidants, one should be aware of the following pitfalls.

1. Probe Leakage. When C11-BODIPY$^{581/591}$ in fetal calf serum is administered to rat-1 fibroblasts, the probe is incorporated into the cell within 30 min. As can be expected from its lipophilicity, the probe resides in cellular membranes after washing labeled cells with PBS buffer or serum-free growth medium. However, subsequent incubation in fresh serum-containing medium results in a gradual decrease of the total intensity of the C11-BODIPY$^{581/591}$ fluorescence due to a back exchange of the fluorophore in the medium.

2. Intracellular Probe Diffusion. A general problem in the detection of oxidation processes on a subcellular level is that many oxidant-sensitive probes have high diffusion rates in the cell. When the probe diffusion throughout a cell is high as compared to its oxidation rate, spatial information about ROS activities in particular organelles is lost. The lipophilicy of C11-BODIPY$^{581/591}$ warrants a relatively long retention of the probe in membrane structures. We observed that a spot of laser-bleached C11-BODIPY$^{581/591}$ in a cell can be discriminated from its surrounding cellular structures for several minutes. This observation confirms that the diffusion of the probe is slow.

3. Quenching and Excimer Formation. At high mole fractions, three different mechanisms might lead to nonlinearities of the fluorophore–fluorescence relationship.

i. SELF-QUENCHING. Fluorophores reabsorb their emitted fluorescence.

ii. RESONANCE ENERGY TRANSFER. The nonoxidized form of C11-BODIPY$^{581/591}$ quenches the fluorescence of the oxidized form by absorbing its excitation energy.

iii. EXCIMER FORMATION. Dimers are formed between excited and non-excited BODIPY monomers.[11,12] The fluorescence emission of these excimers is red shifted with respect to that of the monomers. Thus the red excimer fluorescence of oxidized C11-BODIPY$^{581/591}$ could seriously interfere with data interpretation. At the labeling procedure described here, the average mole fraction of the probe in the cellular membranes is approximately two orders of magnitudes lower than required for excimer formation. Nevertheless, caution should be observed, as nonrandom distribution of C11-BODIPY$^{581/591}$ could result in excimers in local cellular structures. Red

[11] G. M. Makrigiorgos, *J. Biochem. Biophys. Methods* **35**, 23.
[12] R. P. Haugland, "Handbook of Fluorescent Probes and Research Chemicals." Molecular Probes, Eugene, OR, 1996.

excimer fluorescence of oxidized C11-BODIPY$^{581/591}$ depends on excitation in the wavelength range of 480–500 nm and can be distinguished from red fluorescence of the nonoxidized form having an excitation between 560 and 580 nm.

4. Photooxidation of C11-BODIPY$^{581/591}$. High intensity of excitation light (550–590 nm) results in a photoconversion of C11-BODIPY$^{581/591}$ to its oxidized form. One should be aware of this effect, particularly when sequential images are taken from the same samples. After acquisition, the contribution of photooxidation can be quantified by taking an image with lower magnification of the same area. If the extent of oxidation in cells at the border of this latter image is similar to the oxidation in the cells that have been illuminated for different times, the photoinduced oxidation may be neglected.

[55] Noninvasive Techniques for Measuring Oxidation Products on the Surface of Human Skin

By Daniel Maes, Tom Mammone, Mary Ann McKeever, Ed Pelle, Christina Fthenakis, Lieve Declercq, Paolo U. Giacomoni, and Ken Marenus

Introduction

Oxidative stress is defined as the result of an imbalance between prooxidants and antioxidants in living cells. This imbalance leads to molecular damage (peroxidation of lipids, metal-catalyzed carbonylation of proteins, oxidative damage to DNA) as well as to modifications in cellular morphology (zeiosis, modification of chromatin structure) and tissue alterations. One way to illustrate the physiological consequences of oxidative stress imposed on human skin and the relevance of antioxidants is to analyze histology sections before and after an oxidative stress in the presence or absence of antioxidants. Figures 1A and B show the epidermis of a human volunteer before and 3 hr after an exposure to ultraviolet radiation, respectively. Figure 1C shows the epidermis of the same volunteer 3 hr after exposure; the exposure to ultraviolet radiation was performed in the presence of antioxidants. Morphological modifications of the ultraviolet-exposed epidermis are remarkable, particularly the loss of cell-to-cell contact and intercellular spongiosis. Protection afforded by the antioxidants against these morphological changes is nearly complete.

However, histology methods are invasive and impractical for studies involving large cohorts of volunteers. Noninvasive techniques are therefore preferred for the analysis of the molecular consequences of the oxidative stresses induced by UVA, singlet oxygen, or ozone. For example, one might be interested in learning about the formation of oxidation products on the surface or in the different layers of the epidermis exposed to these environmental factors and its inhibition by antioxidants. This article describes some methodologies for investigating the effects of prooxidants and antioxidants on human epidermis.

Materials and Methods

Collection of Ethanol-Soluble Samples from the Skin Surface

A French cup (25 mm diameter) is applied to the skin surface and held firmly in order to avoid leakage. One milliliter of ethanol is introduced and stirred for about 30 sec with a rubber policeman, and then the ethanol is transferred to a polypropylene tube. Add a second milliliter of ethanol, stir, and transfer the ethanol to the same polypropylene tube. Repeat the operation on another site and pool the ethanol samples.

Centrifuge the 4 ml in order to decant squames and other insoluble material. Alternatively, they can be filtered through hydrophobic polypropylene filters (ReZist) with a 0.45-μm pore diameter (or other filters provided that their affinity for ethanol-soluble material is checked carefully) and the ethanol sample is evaporated in a Speed-Vac vacuum evaporator (Savant, Holbrook, NY) for analyses as described later.

Analysis of Epidermal Layers via Tape Stripping

Oxidation products and natural defense factors can be generated ubiquitously, and topically applied antioxidants can diffuse through the stratum corneum. It might be important to be able to describe their distribution across the epidermis. In order to do so, after treatment with prooxidants or the application of antioxidants, the site of generation of oxidation products or the extent of the diffusion of the antioxidants are monitored by tape stripping. Sheets of corneocytes or suprabasal keratinocytes adhere to the tape during the stripping, and the biochemical analysis of material removed with different stripping yields information about the presence of the molecules of interest in the different epidermal layers.

Strips of tape (Scotch Brand, Saint Paul, MN) are applied and pressed on the skin within an outlined template area. The tape is then removed by gently pulling in a downward direction. The procedure is repeated 20 times,

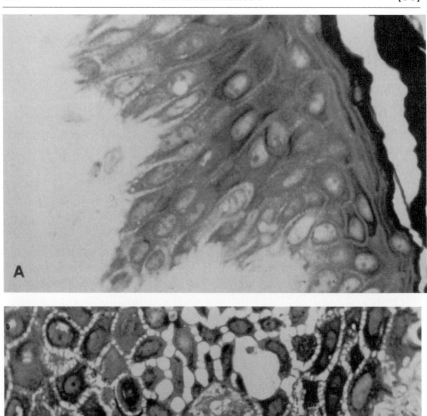

FIG. 1. Histology sections (toluidine blue stain) before and after exposure to ultraviolet radiation from a solar simulator. Experiments performed as described by Giacomoni et al.[6] (A) unexposed control, (B): 3 hr after 3 MED, and (C) 3 hr after exposure to 3 MED in the presence of 1% vitamin E acetate (2 μl per cm^2). Courtesy of Dr. Walter Malorni.

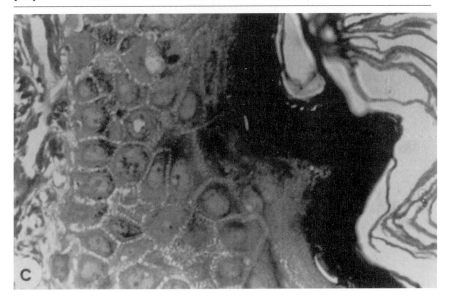

FIG. 1. (*continued*)

and a new site is stripped for every time point. After the stripping, the tape is soaked in an appropriate solvent [e.g., water for ascorbic acid, methanol for retinol and retinoic acid, isopropyl alcohol for vitamin E acetate, and acetonitrile for butylated hydroxytoluene (BHT)]. The sample is then vortexed until the glue separates from the cellophane body of the tape and is centrifuged for 20 min at low speed at room temperature. The supernatant is then subjected to analysis via high-performance liquid chromatography (HPLC) as described later.

Colorimetric Analysis of Skin Lipid Hydroperoxides

This is an adapted version of the method of Ohishi *et al.*[1] that determines the presence of lipid hydroperoxides and was adapted to evaluate UV-induced oxidative reactions in skin model.[2] A kit with the reactants is available commercially through Kamiya Biomedical Company (Thousand Oaks, CA). Samples from skin surface (or from other sources such as reconstructed epidermis) are dried in a Speed-Vac vacuum evaporator and are recovered in 1.1 ml of Good's buffer (100 mM 3-morpholinopropanesul-

[1] N. Ohishi, H. Ohkawa, A. Miike, T. Tatano, and K. Yagi, *Biochem. Int.* **10,** 205 (1985).
[2] E. Pelle, T. Mammone, M. Combatti, K. Marenus, and D. Maes, *J. Invest. Dermatol.* **100,** 595 (1993).

fonic acid titrated to pH 6) containing 14 U/ml ascorbic acid oxidase and 1.3 U/ml lipoprotein lipase. After incubating at 30° for 5 min, 2 ml of Good's buffer containing 67.5 μg/ml of hemoglobin and 0.04 mM 10-N-methylcarbamoyl-3,7-dimethylamino-H-phenothiazine (MCDP) are added to the solution. Hemoglobin catalyzes the reaction of MCDP with hydroperoxides to yield methylene blue. The absorbance of the solution at 675 nm 10 min after the beginning of the incubation at 30° gives a measure of the amount of methylene blue formed. Quantitative results are obtained using a solution of 50 μM cumene hydroperoxides in Good's buffer as standard in the presence of ascorbic acid oxidase and lipoprotein lipase (as described earlier). This technique has been utilized successfully to detect and measure lipid peroxides in stratum corneum on exposure of human epidermis to UVB irradiation.[3]

Analysis of Squalene and Squalene Peroxide by HPLC

Squalene is considered to be the first target lipid on the skin surface to incur oxidative stress by sunlight exposure.[4] The current HPLC method consists of an on-line system for the separation of squalene and squalene peroxide from other lipids on a reversed-phase C_{18} column. Squalene is detected directly after column separation with a UV detector set at 220 nm. Squalene peroxide is detected on-line after a postcolumn reaction with a solution of an iron(II) salt and sulfosalicylic acid, resulting in a colored complex that can be detected with a visible light detector set at 510 nm. Results are expressed as mol% squalene peroxide versus residual squalene.

In order to prepare a squalene standard, dissolve 25 mg of squalene in 10 ml ethanol in a 50-ml gauged vial with vigorous stirring and fill the vial with ethanol to 50 ml. Mix. Dilute this 0.05% solution in ethanol 10 times, aliquot in 1-ml vials, store at −20°, and use as a standard.

The squalene peroxide standard solution is prepared by dissolving in a 100-ml gauged vial 1 g of squalene in 20 ml ethanol with vigorous stirring, adding 10 ml of a 0.5 mM methylene blue solution in ethanol, filling the vial with ethanol up to 100 ml, and mixing. The resulting solution is poured into a beaker with a diameter of 10 cm and is irradiated with a dose of 1.8 J/cm2 of UVB. Methylene blue is removed by solid-phase extraction using silica cartridges (Maxi-clean silica cartridge 600 mg, Alltech) (one cartridge per 50 ml). The squalene peroxide level of this solution is determined by titration using a standardized sodium thiosulfate solution. In an Erlenmeyer

[3] E. Pelle, N. Muizzuddin, T. Mammone, K. Marenus, and D. Maes, *Photoderm. Photoimmun. Photomed.* **15,** 115 (1999).

[4] Y. Kohno, O. Sakamoto, T. Nakamura, and T. Miyazawa, *Yakagaku* **42,** 204 (1993).

flask, mix 5 ml of squalene peroxide with 10 ml of a 60:40 (acetic acid:chloroform) solution. Degas. Add 0.1 ml of saturated KI. Allow to react for at least 30 min at room temperature. Degas. Add 50 ml water. Add dropwise 0.001 N $Na_2S_2O_3$ until a bright yellow color appears. Add 5 drops of starch indicator solution (1% in water), which will generate an indigo color, and continue adding sodium thiosulfate until the indigo color fades away. [The sodium thiosulfate solution can be standardized by a redox titration against a primary standard of 0.001 N KIO_3 freshly prepared or kept in the dark. (Note that for a 1 N solution, 35.66 g of KIO_3 is dissolved in one liter.) 5 ml of 0.001 N KIO_3 is mixed with 10 ml of 1N H_2SO_4 and 0.1 ml of saturated KI. Add the sodium thiosulfate solution to be titrated dropwise until a bright yellow color appears, add starch indicator to yield an indigo color and continue the titration until the indigo fades away.]

Titrated squalene peroxide can be aliquoted in 1-ml vials and used as standard. Squalene peroxide can be kept for several years if stored in the dark at $-20°$. The ethanol-soluble, dry material obtained from the skin surface (see earlier discussion) is recovered with twice 110 μl of ethanol. Ninety microliters is transferred in a vial fitting the autoinjector device of an HPLC chromatography apparatus equipped with a Nucleosil C_{18}, 4.6 × 150-mm column (Macherey-Nagel, Düren, Germany) in a column oven at 50°.

After injecting the sample onto the system, the column is eluted with a mobile phase, starting with 90% ethanol/10% water, then gradually evolving in 5 min to 90% ethanol, 10% acetonitrile, 0.02% acetic acid and running at this condition for 15 min. Afterward the system is reconditioned with the initial mobile phase. The flow rate is 0.8 ml/min. The chromatographic system is equipped with a UV detector set at 220 nm, which allows quantification of the UV-absorbing material. Squalene elutes with a retention time of 17 min.

Immediately after passing through the UV detector, the eluent is mixed with a solution of 0.04% $FeSO4 \cdot 7H_2O$, 1.6% sulfosalicylic acid in methanol/water 90/10 acidified with 0.1% acetic acid via a low dead volume T switch at a rate of 0.2 ml/min. The resulting mixture passes through a postcolumn reactor (5 m × 0.5 mm ID knitted PTFE tube in a column oven at 75°) where ferrous iron is oxidized by the peroxides to yield ferric iron that will react with the sulfosalicylic acid to form a purple complex.[5] This complex elutes around 13 min and can be quantified with a second UV/VIS detector set at 510 nm.

[5] J. Fries and H. Getrost, "Organic Reagents for Trace Analysis," p. 202. Merck, Darmstadt, 1977.

*Analysis by HPLC of Topically Applied Xenobiotics Recovered
   in Tape-Stripped Samples*

The choice of the column and of the mobile phase of HPLC analysis depends on the chemical properties of the molecules to be investigated. A reversed-phase C_{18} 15-cm column (B&J, Muskelgon, MI) is appropriate for retinoic acid, vitamin E, and phenoxyethanol, whereas a C_{18} 30-cm column seems to be more appropriate for BHT and a Prodigy 5 ODS 2 15-cm column (Phenomenex, Torrence, CA) can be used for ascorbic acid and retinol.

So far as the mobile phases are concerned, BHT is eluted in 90:10 (acetonitrile:5% acetic acid in water), retinoic acid is eluted with 70:30 (acetonitrile:5% acetic acid-0.02% triethanolamine in water), ascorbic acid is eluted with 95:05 (0.1% H_3PO_4 with 5 mM Na pentane sulfate:acetonitrile), retinol is eluted with 75:25 (acetonitrile–meythanol), vitamin E acetate is eluted with 5:95 (acetonitrile:methanol), and phenoxyethanol is eluted with 20:80 (acetonitrile:water).

In these conditions, elution times for these chemicals are retinol, 4 min; retinoic acid, 3 min; ascorbic acid, 3 min; penoxyethanol, 4 min; vitamin E acetate, 7 min; and BHT, 5 min; they are detected by UV absorption at the following wavelengths: BHT, 280 nm; vitamin E acetate, 285 nm; phenoxyethanol, 220 nm; retinol, 320 nm; ascorbic acid, 254 nm; and retinoic acid, 350 nm.

The amount of chemical found in a sample from a stripping is quantitated by external standardization.

Antioxidant Mix for Topical Applications

The cream used in some of the following experiments contains 2% α-tocopherol acetate, 0.1% butylated hydroxytoluene, 0.1% Rosemary Natural Antioxidant, 1% magnesium ascorbyl phosphate, 0.5% *N*-acetylcysteine, 0.1% ubiquinone, and 0.1% tocopherol cysteamine.

Results and Discussion

This section describes some experimental results obtained with the methodologies described earlier.

Lipid peroxides can be generated by UVA irradiation. Their amount can be determined quantitatively by following the generation of methylene

---

[6] P. U. Giacomoni, J. F. Nadaud, E. Straface, G. Donelli, M. Heenen, and W. Malorni, *Arch. Dermatol. Res.* **290,** 163 (1998).

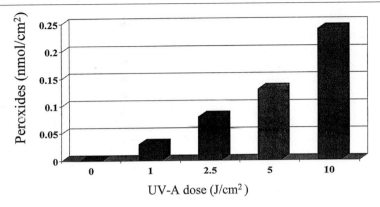

FIG. 2. Generation of lipoperoxides in 1-cm^2 samples of reconstructed epidermis (Advanced Tissue Science, San Diego, CA) exposed to increasing doses of UVA radiation emitted by two FS40 black light lamps (General Electric, Schenectady, NY) (0.69 mW/cm^2). Nanomoles of lipid peroxides generated per cm^2 versus UVA dose (J/cm^2).

blue after completion of a hemoglobin-catalyzed reaction. Figure 2 shows the amount of total lipid peroxides generated in 1-cm^2 samples of reconstructed epidermis subjected to increasing doses of UVA radiation. It can be seen that the generation of peroxides is linear up to 10 J/cm^2.

In order to learn about the effect of environmental factors on skin surface peroxides, we measured the relative amount of squalene peroxide on the forearms of 14 volunteers exposed to the environment at 8 AM in the fall (no sunlight, therefore no ultraviolet) and in the summer (mainly UVA radiation). The results are reported in Table I, where it is clear that the formation of ethanol extractable squalene peroxide can be attributed to environmental UVA radiation in the first hours of the day.

TABLE I

RELATIVE LEVEL OF SQUALENE PEROXIDE ON THE SURFACE OF THE VOLAR ASPECT OF THE FOREARMS OF 14 VOLUNTEERS IN TOURS, FRANCE[a]

Times	Squalene peroxide (expressed as mol% of total squalene)
8 AM July	4.1 ± 0.5
8 AM October	0.8 ± 0.01

[a] The exposure of the forearms to environmental factors started around 7 AM.

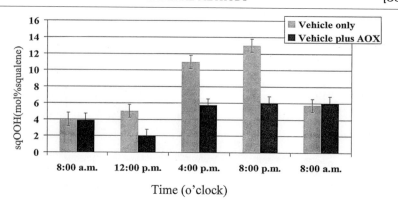

FIG. 3. Time course of accumulation of squalene peroxide in the volar aspect of the forearms of 14 volunteers exposed to environmental conditions on a sunny July day in Tours, France.

We also measured the time course of accumulation of squalene peroxide on the surface of the skin of the volar aspect of the forearms of 14 volunteers exposed to environmental factors during one full summer day and the benefits of a topical application of antioxidants. At the beginning of the

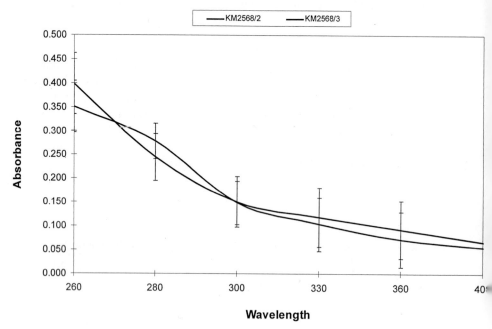

FIG. 4. Absorption spectrum of the vehicle (KM2568/2) and of the vehicle plus the antioxidant cocktail (KM2568/3) used for determining the effect of topical application of antioxidants.

experiment (8 AM), one forearm was treated with a cream containing the antioxidant cocktail described earlier and the other forearm was treated with a cream consisting of the same vehicle, without the antioxidant cocktail. Treatment consisted of application of 2 μl/cm^2. Results are reported in Fig 3. In the vehicle-treated control, the relative amount of squalene peroxide increases from 8 AM to 8 PM and returns to low levels during the night, possibly because of the action of endogenous antioxidants. In antioxidant-treated surfaces, however, the level of squalene peroxide drops by about 50% in the first few hours and then increases with time, although at a much slower rate than in vehicle-treated regions.

In order to make sure that this difference is not the consequence of a UV filter effect of some antioxidant ingredient, absorption spectra of the vehicle alone and of the vehicle containing the antioxidant cocktail have been measured. Ten independent measurements were made by spreading the creams (2 μl/cm^2) on 10 disks of 3M Transpore tape (3M Healthcare, Minneapolis, MD) and inserting the disk in the sample holder of a SPF 90 spectrophotometer (The Optometrics Group, Ayer, MS). No difference was detected among the samples as shown in Fig. 4, in which average optical densities and standard deviations at different wavelengths are reported.

Topically applied chemicals might diffuse through the *stratum corneum.* In order to learn about their rates of penetration, the chemicals in cosmetically suitable vehicles were first applied at different spots on the volar aspect of the forearm of 30 volunteers and then tape stripped at different times. The amount of chemical present in different epidermal layers can

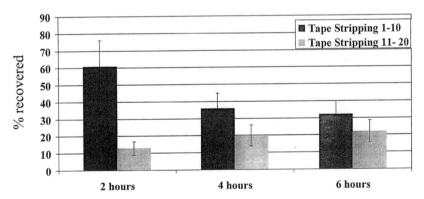

FIG. 5. Relative amount of ascorbic acid (percentage of applied amount) in sheets of the epidermis of 10 volunteers (pool of 1st through 10th and 11th through 20th tape strippings) 2, 4, and 6 hr after application in a silicon-based vehicle currently used for cosmetic purposes. Ascorbic acid was 30 mg/ml in the cream. Each 5-cm^2 zone was treated with 2 mg/cm^2 of cream and thus received 300 μg ascorbic acid.

be analyzed as described earlier. The amount found at different times in the pool of tape stripping 1 to 10 and 11 to 20, for instance, gives an insight into the rate of penetration of the chemical through the upper 10th layer of the *stratum corneum*. In Fig. 5, the relative amount of antioxidants (percentage of totally applied antioxidant) found in the layers removed with strippings 1 through 10 and 11 through 20 is reported versus time.

This kind of measurement allows one to learn about the rate of penetration of topically applied molecules as a function of their chemical properties (when applied in the same vehicle) as well as a function of the physicochemical properties of the vehicle. The determination of peroxides in strippings that remove epidermal layers at different depths is also very helpful to learn about the presence of natural prooxidants and of antioxidants in the epidermis.

Acknowledgment

The authors are indebted to Vicki Giordano for patient help in establishing iconography.

Author Index

Numbers in parentheses are footnote reference numbers and indicate that an author's work is referred to although the name is not cited in the text.

A

Abakerli, R. B., 86
Abe, H., 197(14), 198
Abels, C., 137, 141(33)
Ackermann Liebrich, U., 551
Adam, W., 3, 427, 437
Adcock, I. M., 551, 553, 555, 557
Adhami, F., 534
Afshari, E., 66
Agapakis-Caussé, C., 457, 464(16)
Agarwal, M. L., 329, 345, 346, 348, 349, 352(14), 353(9)
Agarwal, R., 345, 352, 353(8, 23), 391, 395(66, 67)
Agarwal, S., 586
Aggarwal, B. B., 274, 585, 586, 593, 594(23), 599
Agnez, L. F., 3
Agou, F., 274
Ahmad, N., 342, 352, 353, 353(23, 24)
Ahmed, M. U., 429, 431(18)
Ahn, B., 430, 431(30)
Ahrens, C., 17(61), 18, 130, 301, 303, 306(4), 341
Aiba, H., 341, 422
Akaberli, R. B., 71
Akande-Adebakin, S. L., 381
Alam, Z. I., 402
Alapetite, C., 158
Albermann, K., 553
Alberts, A. D., 374
Albro, P. W., 110
Alda, J. O., 396
Alder, 13
Alessi, D. R., 262, 269
Ali, H., 188, 390, 392(61), 398
Allen, C. M., 376
Allgren, R. L., 52

Allouche, D., 121, 122(22), 127(22)
Al-Mehdi, A. B., 428(15), 429, 431(15)
Altmiller, D. H., 152
Ames, A. N., 218
Ames, B. C., 466
Ames, B. N., 30, 36, 36(19), 93, 144, 145, 150(6), 151(20), 158, 219, 224, 401, 402, 431(53), 434
Amici, A., 428, 429(5), 430, 431(30), 436(5)
An, J. Y., 393, 397(81), 398(81)
An, M. T., 350
Ananthaswamy, H. N., 154, 417, 418, 418(2), 460
Andersen, T. N., 49
Anderson, B. R., 70, 74(24)
Anderson, F. K., 520, 521(2)
Anderson, L., 408
Anderson, S., 367
Anderson, V. E., 431(46), 434
Andersson, B., 78
Ando, M., 570
Ando, Y., 570
Andre, J. C., 93
Angel, P., 141, 258
Anne, D. L., 331
Antunez, A. R., 345, 348
Aoyama, H., 193, 194(18)
Appelqvist, L. A., 112
Applegate, L. A., 161, 361
Aragane, Y., 319
Arakawa, K., 552, 553(18)
Aranda, F. J., 107, 108(16)
Ardizzoni, A., 327
Argüllo, G. A., 398
Aris, R. M., 552
Arkin, M. R., 169(18), 170, 171(18), 173, 173(18), 176(18), 184(18)
Armitage, B., 166
Arnfield, M. R., 386

623

Arnheim, N., 369
Aro, E.-M., 78
Aronovitch, J., 428
Arrigo, A. P., 274
Arteca, R. N., 521, 527, 527(9)
Aruoma, O. I., 402, 411(14), 412, 413(60), 414(61), 415, 415(61), 416(60, 61)
Arwert, F., 5, 50, 51(15), 135, 198
Asada, K., 77, 80
Ash, D. V., 377
Ashkenazi, A., 586
Ashworth, A., 262
Athar, M., 121, 391, 395(66, 67)
Auberger, P., 274
Aubry, J. M., 3, 5, 5(11), 6, 7(18), 8(18), 9, 10, 10(18), 11(18, 19), 17, 18, 50, 51(15–18), 54, 54(17), 57(18), 131, 135, 135(20), 137, 138(20), 139(20), 197, 198, 199, 224, 226, 227, 291
Auer, H.-P., 141, 258
Augusto, O., 70
Auletta, M., 291
Auphan, N., 555
Aust, S. D., 409
Austin, L. M., 302
Ausubel, F. M., 279, 586
Axelrod, B., 66, 111
Azaroual, N., 50, 51(17), 54(17), 227

B

Baart de la Faille, H., 480
Baart de la Faille-Kuyper, E. H., 480
Baas, A. S., 140
Baccarani-Contri, M., 467
Bachmann, B., 212, 215(15)
Bachowski, G. J., 87, 88, 89, 90(16, 27), 92(22), 93, 93(22), 94(34), 95, 96(27), 191, 390, 391(59), 392(59)
Baden, H. P., 365, 418
Bader, H., 514
Badylak, J. A., 212
Baeuerle, P. A., 119, 120, 273, 274, 553, 585, 586, 596
Baichwal, V. R., 273
Baird, W. M., 144
Baker, A., 21, 86, 388, 389, 513
Baker, J. C., 30, 36(19), 219
Baker, P. F., 547

Balassy, A., 385
Balasubramanian, D., 429
Baldwin, A. S., 119, 601
Baldwin, A. S., Jr., 555, 595
Bale, A. E., 418
Balmes, J. R., 552
Baltimore, D., 273, 589
Baltschun, D., 230
Banerjee, S. K., 185
Banki, K., 607
Bannwarth, W., 177
Barber, J., 78
Barbosa, M., 120
Bard, J. A., 29, 35(1), 36(1)
Barnes, J. D., 524, 532(16)
Barnes, P. J., 273, 553, 555, 555(20), 557
Barone, A. D., 176, 180(31)
Barrett, T., 111
Barsacchi, R., 69
Barthélémy, C., 6, 7(18), 8(18), 10(18), 11(18), 137, 198
Barton, J. K., 165, 166, 167(11), 168, 168(22), 169(11, 14, 15, 18, 20, 22), 170, 170(14), 171(11, 14, 15, 17–20), 172, 173, 173(12, 13, 18, 20), 175, 176, 176(18), 181, 181(14), 184(11, 14, 18), 185(22), 187, 425, 426(21)
Baserga, R., 282
Basha, M. A., 552
Bashir, S., 405
Bass, D. A., 336
Basu, A. K., 163
Dasu-Modak, S., 18, 130, 154, 155(7), 291, 292(7)
Bates, D. V., 552
Bates, E. E., 274
Battino, R., 106, 107(14)
Batzri, S., 608
Bauer, H. D., 233
Bauer, M., 585
Bauerle, P. A., 554
Baughman, G. L., 50
Baughman, R. D., 492
Baum, T., 551
Baumstark, A. L., 13
Baxevanis, C. N., 467, 478(20)
Bayly, J. G., 66
Baynes, J. W., 429, 431(18)
Bazin, M., 227, 479, 480(3)
Beal, M. F., 431(46), 434, 569, 570(32)

Beatty, E., 403
Beaucage, S. L., 176, 180(31)
Bechara, E. H. J., 70
Bechara, E. J. H., 3
Beckman, B. K., 145, 151(20)
Beckman, J. S., 431(49), 434
Beckman, K. B., 402
Beck-Speier, I., 431(44), 434
Beckwith, A. L. J., 87, 90(20), 390
Beecher, G. R., 495
Beer, J. Z., 311, 312(7), 313, 315(7)
Beg, A. A., 119, 601
Begley, T. P., 185
Beijersbergen van Henegouwen, G. M. J.,
    123, 133
Beilski, 14
Bejot, M., 496
Belaidi, J. P., 457, 464(16)
Bellemare, F., 102, 103, 107(7), 108(7)
Bellevergue, P., 269
Bellus, D., 219
Beltrán, F. J., 508
Beltrán-Heredia, J., 508
Benard, F., 390
Bender, K., 120, 258, 274
Bendich, A., 404
Benedetti, A., 562, 564(2, 3)
Benerjee, D., 349
Benfell, K., 482
Ben-Hur, E., 197, 208, 392
Benitez, F. J., 508
Ben-Jebria, A., 564
Bennett, B. L., 120
Bennett, L. G., 108
Benoit, A., 160(64), 162, 164(64), 362
Bensasson, R. V., 18, 54, 227, 234, 513
Beonders, J., 157
Berberan-Santos, M. N., 107, 108(16)
Berenbaum, M. C., 343, 381
Berg, K., 86, 141, 513
Berg, R. J. W., 162, 365
Berger, M., 14, 15(39), 144, 150, 151, 151(5),
    152, 361, 390, 427, 436
Berger, N. A., 350
Berggren, M., 467, 468(19), 477(19)
Berk, B. C., 140
Berlett, B. S., 428, 429, 429(2), 431(45), 434
Bermudez, E., 194, 196, 571, 572(11)
Bernard, F., 150
Bernardi, P., 111, 329

Berneburg, M., 286, 298, 366, 367, 372(4,
    11), 373(11)
Bertgold, D. S., 403
Bertino, J. R., 349
Bertoloni, G., 198
Bertrand, F., 130
Bertrand, H., 563
Bessho, T., 158
Bessia, C., 274
Betz, C., 526, 527(21)
Beutner, S., 226, 230
Beuzard, Y., 483, 484(17)
Beybach, J. Ch., 483, 484(17)
Bhalla, K., 332
Bhatti, M., 197, 198(9)
Bhawan, J., 494
Bianchini, F., 152, 153, 411, 412(57)
Bickers, D. R., 121, 391, 395(66, 67)
Bidani, A., 505, 511, 564
Bielski, B. H. J., 234
Bienkowski, M. J., 68
Biesalski, H. K., 494
Bigby, B. G., 552
Bihari, I. C., 300
Bilska, M., 30, 36(17)
Bilski, P., 30, 36(17), 41, 110, 129
Bindoli, A., 562
Birch-Machin, M. A., 367, 371(5), 372(5),
    376(5)
Birnbaum, G., 552
Bischoff, F., 191
Bishop, S. M., 78
Bjørås, B., 437
Black, H. S., 190, 191, 193(9)
Blackwell, B., 108
Blanchard, L., 385
Blatt, T., 465
Blattner, C., 257, 258
Blaudschun, R., 129, 130
Blazek, E. R., 14, 155
Bloedorn, B., 226, 237, 238(27)
Bloodworth, A. J., 13
Blount, B. C., 431(53), 434
Blum, A., 389
Blumenberg, M., 286
Bodesheim, M., 37, 387
Boehm-Wilcox, C., 191, 195(12)
Bogdahn-Rai, T., 377, 379(6)
Boger, P., 111
Bogyo, M., 278

Bohm, F., 389
Böhmer, A., 268, 269, 269(27)
Böhmer, F.-D., 255, 258, 259(8), 261(8), 263(8), 265, 268, 269, 269(27)
Böhmer, S.-A., 265
Bohnsack, K., 466, 467(14), 475(14), 476(14), 477(14)
Bohr, V. A., 438
Boiteux, S., 145, 155, 156, 165, 438, 443
Boitreux, S., 111
Bolen, J. B., 111
Bollag, G., 262
Bolsen, K., 130, 137, 141(33), 294, 493, 494
Bolton, P. H., 158
Bonham, A., 309
Bonizzi, G., 129
Bonnett, R., 142, 378, 379(10), 381, 381(10)
Bonnia, F., 466, 475(15)
Borden, A., 185
Bories, M. F., 450
Borner, C., 159, 329, 331
Bors, W., 223, 395
Börsch-Haubold, A. G., 269
Bos, J. L., 261, 262
Bossmann, S. H., 45, 166, 173(13)
Bossy-Wetzel, E., 329
Botsivali, M., 223
Bottigli, U., 69
Boukamp, P., 285, 439
Boulanger, Y., 108
Boulechfar, S., 483
Bourdelande, J. L., 41
Bourgeron, T., 372
Bourre, F., 14
Bours, V., 121, 125, 129, 131
Bouvier, J., 27
Boveris, A., 13, 465
Bowler, C., 525
Bown, S. G., 379
Boye, E., 388
Boyle, R. W., 396
Bradford, M. M., 288, 595
Braiser, A. R., 287
Bramley, P. M., 494
Brannstrom, T., 570
Brasch, R. C., 85
Brash, D. E., 365, 418
Braslavsky, S. E., 37, 40, 43, 47(10), 223
Brasseur, N., 188, 390, 392(61)
Brauer, H. D., 39, 43, 48(23)

Brault, D., 50, 51(18), 57(18)
Braun, A. M., 45
Braun Fahrlander, C., 551
Bredesen, D., 329
Breitkreutz, D., 285, 439
Brendzel, A. M., 70, 74(24)
Brenneisen, P., 17, 18(59), 111, 129, 130, 143, 154, 155(4), 291
Brenner, D. A., 262, 483
Brent, R., 279, 586
Breslin, D. T., 166, 425
Breuer, D. K., 73
Breuer, W., 295
Brewer, D., 102, 105(9), 106(9)
Bridges, A. J., 268
Bridges, B. A., 110, 111
Bright, G. R., 603
Britton, G., 106
Briviba, K., 3, 14, 14(6), 16, 17, 18, 18(57–59), 19(6, 60, 62), 111, 121, 129(21), 130, 131, 135, 135(19, 20), 138(19, 20), 139(20), 141(21), 143, 143(9), 154, 155(4), 198, 199(16), 222, 223, 224, 225, 225(14), 226, 230, 273, 280, 284, 291, 294, 298, 358, 464
Brockhaus, M., 553
Brodsky, M. H., 30, 36(19), 219
Bromberg, P. A., 552
Brown, C. R., 553, 555, 557
Brown, D. M., 155, 553
Brown, G. M., 553
Brown, J. E., 295
Brown, K., 119, 122
Brown, M. D., 370
Brown, N. J., 137, 141(32)
Brown, S. C., 337, 340(24)
Bruder, H. J., 258
Bruder, J. T., 141
Bruijnzeel-Koomen, C. A. F. M., 296, 300(2)
Brummer, J. G., 52
Bruskov, V. I., 362
Bruynzeel-Koomen, C. A. F. M., 300
Bryant, R. G., 386
Buchko, G. W., 144, 150, 150(6), 151, 427
Buckley, N. M., 158
Budnik, A., 298
Bueding, E., 408
Buettner, G. R., 88
Buettner, R., 273, 280, 291, 464
Bun-Hur, E., 16
Burack, L. H., 302

Burdon, R., 603
Burgering, B. M., 261
Burke, T. R., Jr., 586
Burkitt, M. J., 427
Burn, J. L., 137, 141(32)
Burnett, R. T., 552
Burr, J. A., 547, 550
Burrows, C. J., 15, 183
Burton, G. W., 35, 234, 550
Busch, M., 212
Bush, K. M., 431(41), 434
Buss, H., 431(36), 432
Butler, A., 59, 64(7)
Butler, J., 59
Butler, K. W., 108
Buytenhek, M., 67, 73
Bye, S., 405
Byteva, I. M., 66, 223

C

Cabantchik, Z. I., 295
Cacciuttolo, M. A., 401, 411(6)
Cadenas, E., 13, 67, 69, 70(15), 75, 76, 76(37), 223, 227, 388, 478, 585
Cadet, J., 14, 15, 15(39, 40), 111, 143, 144, 150, 150(6), 151, 151(5), 152, 153, 155, 161, 361, 390, 411, 412, 412(57), 427, 436
Cahill, D. S., 161
Cai, H. X., 392
Cai, R., 29, 36(9, 10, 12, 13)
Cairney, J., 527, 528
Cairns, N., 402
Calberg-Bacq, C. M., 14, 121, 129(20)
Camici, P., 69
Campbell, A. K., 465, 475(2)
Candeias, L. P., 144
Canfield, L. M., 499
Canfield, P. J., 191, 195(12)
Cantin, A. M., 510
Cantoni, O., 408
Cao, Z., 120
Capel, P. J. A., 365
Capon, D. J., 419
Capraro, H.-G., 383
Carapeto, F. J., 396
Cardinale, I., 302
Carell, C. F., 551
Carlsson, D. J., 234, 237

Carmical, J. R., 438
Carrell, T., 185
Carter, J., 554
Carthy, C. M., 350
Caruthers, M. H., 176, 180(31)
Casadei de Baptista, R., 70
Casals, M., 43, 45(20)
Castedo, M., 337
Castillejos, M., 551
Castillo, F. J., 515
Catalani, L. H., 3
Cato, A. C. B., 271
Cavelieri, L. F., 404
Cavigelli, M., 259
Cazin, B., 9, 50, 51(16)
Cerami, A., 554
Cerutti, P. A., 585
Cervoise, I., 50, 51(17), 54(17), 227
Chadwick, C. A., 157
Chae, H. Z., 564, 565(22)
Chae, W.-K., 52
Chai, Y.-C., 429, 431(19, 43), 434
Chainy, G. B. N., 586
Chambers, D., 465
Chameides, W. L., 505, 512, 513(2)
Chamulitrat, W., 111
Chan, J. T., 190, 193(9)
Chan, T. P., 431(36), 432
Chance, B., 13, 465
Chang, S., 528
Chang, Y. C., 391
Chang, Y. S., 554
Chanon, M., 393, 395(77), 396, 396(77), 397(77)
Chapman, J. D., 386
Chardon, A., 445, 449, 450, 461(7)
Charlier, M., 185
Charon, D., 269
Châteauneuf, A., 162, 163(65), 362
Chaturvedi, M. M., 585, 586, 593, 594(23)
Chaubo, C., 457, 464(16)
Chavalle, P., 50, 51(18), 57(18)
Chen, A., 338, 340(26)
Chen, A. O., 495
Chen, C., 29
Chen, E. Y., 419
Chen, H., 599
Chen, J., 434
Chen, J. J., 563
Chen, J. Y., 392

Chen, K. T., 392
Chen, L., 552
Chen, L. B., 338, 340(26)
Chen, L. C., 571
Chen, L. L., 552
Chen, M. F., 233
Chen, S. M., 392
Chen, T., 63
Chen, W., 465
Cheng, K. C., 161
Chen-Tran, A., 554
Chernov, M. V., 349, 352(14)
Chernova, O. B., 349, 352(14)
Chernyak, B. V., 329
Cherry, J. M., 531
Chevillard, S., 165
Chevion, M., 428
Chevretton, E. B., 381
Chiba, K., 191, 193, 194(10, 18)
Chien, K. C., 36
Chignell, C. F., 30, 36(17), 41, 86, 110, 111, 129
Chimel, K., 415
Ching, T. Y., 233
Chinnaiyan, A. M., 319
Choi, J. H., 538
Choi, S., 278
Chomczynski, P., 298, 556
Chopp, M., 398
Chou, P. T., 226
Chow, J., 381
Chretien, D., 372
Christensen, R. B., 185
Christian, D., 552
Chua, N.-H., 531
Chung, K. F., 551, 553, 555, 555(20)
Church, D. F., 505, 510(2), 512, 547
Cidlowski, J. A., 555
Cikryt, P., 226
Cilento, G., 3
Civoli, F., 336
Clark, K., 278
Clark, R. M., 29, 36(11)
Clarkson, S. G., 158
Clay, M. E., 345, 348
Claycamp, H. G., 405
Clayton, D. A., 367, 373(6)
Clennan, E. L., 121, 124(19), 133, 233
Clift, R. E., 269
Clifton, C., 233
Climent, I., 430, 431(30)

Clough, R. L., 227
Cochran, A. G., 185
Cody, R. P., 552
Coffer, P. J., 261
Coffing, S. L., 144
Cohen, C. C., 453, 457
Cohen, P., 262, 269, 270
Cohn, J. A., 563, 564(9), 565(9), 566(9), 567(9), 568(9)
Cole, C., 451
Collart, M. A., 596
Collins, A. R., 151, 153, 402, 437
Collins, P., 490
Collis, C. S., 403, 416(25)
Colombo, E., 607
Colon, M., 115, 116(60)
Colussi, V. C., 354
Combatti, M., 615
Comporti, M., 562, 564(2, 3)
Conklin, P. L., 531, 532, 533, 533(46)
Connolly, D., 286
Constam, D. B., 323
Coohill, T. P., 156, 417
Cook, V. M., 144
Coombs, B., 450
Cooper, P. K., 158
Copeland, R. A., 269
Corash, L., 197
Corbett, J. T., 110
Corey, E. J., 278
Corsini, E., 260
Cortopassi, G. A., 369
Cosgrove, T. P., 431(41), 434
Cossariza, A., 327, 467
Costa, D. L., 546, 571
Costabel, U., 430, 431(29, 44), 434
Costantini, P., 329
Coste, H., 269
Cotgreave, I. A., 467, 468(19), 477(19)
Cotter, T. G., 309, 319, 330
Cottin, M., 457
Cotton, M. L., 63
Couet, W. R., 85
Courtois, G., 274
Coutinho, A., 107, 108(16)
Coven, T. R., 302
Cowley, S., 262
Crank, J., 509
Craxton, M., 270
Crews, B. C., 69

Crisp, M., 405, 410
Crissman, H. A., 330
Crissman, K. M., 546
Cross, C. E., 402, 415, 429, 430(22), 434, 510, 511, 512(17), 536, 537
Crowley, J. R., 429, 431(20)
Crystal, R. G., 510
Cuenca, O., 43, 45(20)
Cuenda, A., 269, 270
Cueto, R., 547, 564, 571, 572, 572(6, 9–11), 573, 575(12), 577(12)
Cullis, P. M., 427
Cunniffe, S. M. T., 437
Cunningham, R. P., 158, 437, 438, 439(6)
Cuong, N. K., 50
Currie, A. R., 309
Curtis, J. F., 75, 76(38)
Cuthbert, C., 283
Cutler, R. G., 158
Czapski, G., 428
Czech, W., 297, 298, 299(9), 300(9), 306

D

Dadak, A., 110, 111(6)
Dahl, T. A., 110, 112, 114(40), 198, 516, 517
Dahl, W. R., 113
Dai, W., 596
Daifuku, K., 101
Dailey, L. A., 571
Dales, R. E., 552
Dally, H., 437
Danckweerts, P. V., 505, 508(5)
Dandliker, P. J., 165, 168(22), 169(22), 170, 185(22), 187
Danenberg, K., 349
Danenberg, P., 349
Danoff, T. M., 554
Danpure, H. J., 290
D'Arch, J., 552
Dargusch, R., 465
Darmanyan, A. P., 233
Darnay, B. G., 274, 586
Darzynkiewicz, Z., 330
Das, B., 547
DaSilva, V., 483, 484(17)
Daub, H., 268, 269
Daub, M. E., 110
Daulerio, A. J., 269

David, S. S., 183
Davidson, N., 601
Davies, A. G., 87, 90(20), 390
Davies, K. J., 428, 431(7), 431(56, 57), 434, 563
Davies, M. J., 111, 428, 430
Davies, R. E., 364
d'Avignon, A., 429, 431(20)
Davila, J., 384, 394
Davis, D. A., 431(42), 434
Davis, J., 465
Davis, J. M., 553
Davis, K. R., 527, 531, 532(27), 535(44)
Davis, P. A., 429, 430(22)
Davis, R. J., 111, 131, 141, 270
Davison, I. G. E., 87, 90(20), 390
Day, R. M., 492
Daya-Grosjean, L., 162, 164(68), 365
Deadwyler, G., 113
Dean, C., 531
Dean, P. M., 5
Dean, R. T., 428, 431(11)
Debertret, L., 479, 480(3)
de-Bont, H. J., 283
Deby, C., 283
Deby-Dupont, G., 283
Dechatelet, L. R., 336
Decker, K., 298, 299(11)
Declercq, L., 612
Decreau, R., 393, 395(77), 396, 396(77), 397(77)
Decuyper-Debergh, D., 157, 160(37)
Dedoussis, G. V. Z., 467, 478(20)
Deering, H., 197
Degenhardt, T. P., 429
Degraff, W., 85
de Gruijl, F. R., 156, 158(31), 162, 331, 359, 362, 363, 364, 365, 366
de Laat, A., 156, 158(31), 162, 331, 363
de Laat, J. M. T., 362, 363
Delatour, T., 151, 152, 411, 412(57)
DeLeo, V. A., 364
Delhase, M., 586
Delmelle, M., 78
Delsignore, M. E., 431(56), 434
De Mascio, P., 14
De Mol, N. J., 123, 133
Demple, B., 438
Deneke, C. F., 69, 74(18)
Deng, T., 111

Denny, R. W., 157, 223, 227
de Oliveira, R. C., 3, 160, 161(50)
Derijard, B., 111
De Sa e Melo, T., 479, 480(3)
Deshui, Z., 234
Desnick, R. J., 479
De Timmerman, P., 521, 523(6)
Devary, Y., 258
Devasagayam, T. P., 150, 155, 226, 367, 402
Deveraux, Q., 329
deVermeil, H., 483
Devine, R. M., 208
Devlin, R. B., 552
Devor, E. J., 410
de Vries, A., 162, 365
de Vries-Smits, A. M., 262
DeWeese, T. L., 158
Dewilde, A., 6, 11(19), 17, 135, 197, 199
DeWitt, D. L., 68, 69(4), 73
D'Ham, C., 151
Dhami, S., 78
Dhariwal, K. R., 538
Dhawan, S., 586
Dianne, E. G., 331
Dick, L., 278
DiDonato, J. A., 119, 258, 274, 555
Diepgen, T., 297, 306
Dietz, R., 68
Diffey, B. L., 245, 249, 254, 446, 448, 449, 449(2), 461(7)
Dignam, D., 124
Di Mascio, P., 3, 4, 7, 13(24), 14, 14(24), 16, 71, 112, 135, 137(29), 155, 157, 160, 161(50), 198, 222, 224, 226, 283, 284, 304
Dinh, T.-H., 258
Diplock, A. T., 78, 403, 416(25)
Disdaroglu, M., 401, 411(6)
Dive, C., 325
Dixit, V. M., 319, 586
Dixon, J. E., 266
Dixon, W. J., 110
Dizdaroglu, M., 152, 155, 161, 401, 402, 402 (2, 11), 403, 404, 411(2, 12, 14), 416(12), 438
Djavaheri-Mergny, M., 130
Doba, T., 35
Dobson, V. L., 437
Dockery, D., 551
Doda, J. N., 367, 373(6)
Dodds, D. R., 176, 180(31)

Dohno, C., 168(21), 169(21), 170
Dolphin, D., 381
Domenighetti, G., 551
Domigan, N. M., 431(36), 432
Donaldson, K., 553
Donelli, G., 614(6), 618
Dorion, D., 381
Dorner, M., 534
Dorsey, K., 431(42), 434
Dossou, K. G., 453
Do-Thi, H. P., 14
Dougherty, T. J., 86, 100(6), 121, 343, 344(4), 345(4), 346(4), 353(4), 377, 379(4), 381(4), 386(4), 387
Douki, T., 14, 15, 15(39, 40), 143, 144, 151, 151(5), 152, 153, 361, 411, 412, 412(57), 436
Douli, T., 151
Dove, W. F., 601
Downward, J., 262
Doza, Y. N., 270
Dreher, K. L., 571
Drieg, T., 294
Driggers, W. J., 373
Driscoll, K. E., 554, 571
Drobetsky, E. A., 162, 163(65), 362
Dröge, W., 120, 586
Drumen, G. P. C., 603
Du, X., 429
Dubbelman, T. M. A. R., 393
Dubertret, L., 156, 160(64), 162, 164(64), 362
DuBow, J., 29
Dudley, D. T., 269
Dufraisse, C., 5
Duhamel, L., 269
Dumaz, N., 162, 164(68), 365
Dunford, H. B., 63
Dunlap, W. C., 29, 30, 36, 36(18, 20)
Dunn, D. A., 417
Dunsbach, R., 24, 45
Duprat, F., 9, 50, 51(16)
Durán, N., 70
Duretz, A., 15
Duretz, B., 152
Durner, J., 111
Durocher, G., 385
Durrant, J. R., 78
Dusinka, M., 153
Duthie, S. J., 437
Dutrillaux, B., 165

E

Earl, R. A., 269
Echeveri, F., 319
Eckey-Kaltenbach, H., 527
Eckhardt, R., 113
Egan, R. W., 68, 72, 73(29), 75(29)
Egawa, Y., 36
Eggelte, H. J., 13
Egorov, S. Y., 42, 60, 66(9), 388, 389, 513
Ehman, R. L., 85
Ehrenberg, B., 391
Ehrlich, W., 563
Eide, L., 437
Eiserich, J. P., 434, 511, 512(17)
Eising, R., 115
Elias, P. M., 536
Eling, T. E., 69, 74, 75, 76(38)
Ell, C., 377
Ellerby, L., 329
Ellis, T., 283
Elmets, C. A., 354
El Mohsni, S., 137
El Salhy, M., 570
Elstner, E. F., 77, 78(4)
Enari, M., 332
Eneff, K. L., 155
Engel, P. S., 52
England, T. G., 403, 412, 413(60), 414(61), 415(61), 416(60, 61)
Enninga, I. C., 157, 158(34), 162(34), 361, 363(3)
Enyedi, A. J., 535
Epe, B., 111, 121, 131, 144, 145, 145(1), 151, 153, 154, 155(8), 156, 157, 158, 361, 362, 436, 437, 438, 439(8), 441, 441(8), 442, 443, 444, 444(45)
Epple, R., 185
Epstein, W., 466
Epsztejn, S., 295
Erber, W. N., 212
Erickson, L. C., 441
Ernst, D., 520, 521, 522, 523(15), 524(15), 525(15), 526, 527, 527(10, 21), 528, 528(28, 29), 529, 529(28, 29), 530, 530(28, 29, 36)
Eschenbacher, W. L., 552, 564
Essigmann, J. M., 159, 161, 163
Esterbauer, H., 562, 563(4), 564(2–4), 565(4)
Evans, D. F., 223

Evans, D. L., 325
Evans, H. H., 343, 344(2), 345, 345(2), 346(2), 353(2)
Evans, H. M., 220
Evans, M. D., 405
Everett, R. R., 59, 64(7)
Ewing, R. A. G., 441
Eyers, 270

F

Fabian, J., 237
Fabiano, A.-S., 121, 122(22), 127(22)
Fair, C. K., 582
Fairbanks, G., 88
Fales, I. M., 431(42), 434
Falgueyret, J. P., 68
Faljoni, A., 70
Fang, G., 332
Farahifar, D., 274
Faria Oliveira, O. M. M., 70
Farmillo, A., 237, 391
Farr, A. L., 89, 137
Farr, P., 449, 461(7)
Fauler, G., 562, 564(3)
Favata, M. F., 269
Feelfs, O., 329
Feeser, W. S., 269
Feig, D. I., 161
Feingold, K. R., 536
Feitelson, J., 41, 111
Feix, J. B., 30, 36(16)
Feldman, P., 208
Feng, L., 120
Fenteany, G., 278
Ferenci, K., 302
Ferguson, J., 490
Ferran, P., 166, 170(6)
Ferrario, A., 354, 356, 357
Ferraudi, G., 398
Feyes, D. K., 345, 352, 353(8, 23), 354
Fijishima, A., 29, 36(9, 10, 12, 13)
Filon, A. R., 157, 158(34), 162(34), 361, 363(3)
Finco, T. S., 601
Fingar, V. H., 386, 395(30), 398
Finkbeiner, W. E., 552
Finnegan, M. T. V., 405
Firey, P. A., 86, 389

Fischer, R., 528
Fischer-Nielsen, A., 145
Fisher, A. M., 349, 357
Fisher, E. H., 176, 180(31)
Fisher, G. J., 141
Fisher, G. W., 603
Fisher, R. M., 29, 36(6)
Fitzpatrick, T. B., 481
Flagler, R. B., 527
Flescher, E., 571
Flewig, G., 294
Flood, J., 237
Flores, T., 164
Flower, R. J., 73
Floyd, R. A., 152, 155, 401, 402(3)
Font, J., 41
Fontana, A., 323
Foote, C. S., 36, 48, 66, 71, 85, 86, 121, 124(19),
    133, 144, 149, 150, 157, 166, 170(6), 223,
    227, 233, 234, 385, 391, 418
Forbes, D., 492
Forbes, P. D., 364
Forestier, S., 448
Forman, H. J., 585
Fornace, A. J., Jr., 437
Fort, P., 556
Fossum, T., 248
Foster, T. H., 386
Fourtanier, A., 464
Fragata, M. J., 102, 103, 107(7), 108(7)
Franceschi, C., 327, 467
Frank, R., 552
Frankel, E. N., 32
Franzoso, G., 119, 122, 125
Frederick, J. E., 374
Freeman, B. A., 431(41), 434
Freeman, S. E., 361
Frei, B., 93, 226
Frenk, E., 161, 361
Frenkel, K., 405
Fridovich, I., 221
Friedberg, E. C., 367, 373(6)
Friedman, C. A., 441
Friedman, L. R., 329, 348
Friedman, M., 547, 571, 572(7, 8), 582
Friedman, T., 155
Fries, J., 617
Friguet, B., 563, 564(9), 565(9, 10), 566(9, 10),
    567(9, 10), 568(9)
Frimer, A. A., 3

Fritsch, C., 131, 137, 141, 141(21, 33), 358
Fritsch, E. F., 180, 183(34), 221, 534
Frohn, W., 43
Fronczek, F. R., 570
Fry, D. W., 268
Fthenakis, C., 612
Fu, M. X., 429, 431(18)
Fu, S. L., 428, 431(11)
Fu, Y., 39, 113, 390
Fuchs, E., 534
Fuchs, J., 386, 393, 397(33), 398(33), 465, 466,
    466(1, 5), 467(1), 475(1), 478(16)
Fuerstenberger, G., 364
Fujimori, K., 36, 112, 199, 217, 218(5), 222
Fujimoto, T. T., 391
Fujisawa, K., 168(21), 169(21), 170
Fukuzawa, K., 101, 102(3), 103(3–5), 104(3–
    5), 107, 108, 108(17, 18), 109(5), 112
Fulceri, R., 562, 564(3)
Funk, C. D., 73
Funk, M. O., 93
Furuno, T., 341
Fusenig, N. E., 285, 439
Futterman, G., 113

G

Gabe, E. J., 35
Gaboriau, F., 156
Gaboury, L., 385
Gabrielska, J., 106, 109(13)
Gabrielson, E. W., 393
Gaczynska, M., 278
Gaestel, M., 258, 259(8), 261(8), 263(8)
Gaffney, D. K., 198
Gailani, M., 418
Gailus, V., 111
Gaimvala, D. H., 547
Gajewski, E., 155, 161, 401, 411(6)
Galaris, D., 76
Gale, P. H., 72, 73(29), 75(29)
Gallagher, C. H., 191, 195(12)
Gallagher, T. F., 270
Galli, C. L., 260
Gandin, E., 78, 82
Gange, R. W., 291, 361
Gantchev, T. G., 392, 395, 397, 398(73)
Gao, X., 120
Garavito, R. M., 68, 69(4)

Garcia, G. E., 120
García, N. A., 43, 50
Garland, D., 430, 431(30)
Garmyn, M., 153, 443, 444(45), 494
Garner, M. M., 288
Garoufalis, J., 554
Gasparro, F. P., 319, 320(17)
Gasparutto, D., 151, 427
Gasper, S. M., 169(19), 170
Gattermann, N., 286
Gaullier, J. M., 227
Gaut, J. P., 429, 431(20)
Gay, G., 137
Gazit, A., 268
Geacintov, N. E., 197, 198(11), 400
Gebauer, P., 59, 61(2), 64(2)
Gebel, S., 141, 258, 271
Ged, C., 483
Gedik, C. M., 151, 153, 402
Geiger, P. G., 88, 89, 93, 94, 94(23, 35), 96, 100(36), 113, 192, 390
Geissler, C., 403
Gelder, C. M., 553, 557
Geller, G. G., 233
Gelzleichter, T. R., 553
Genestier, L., 319
Gensch, T., 223
Gentil, A., 158, 161
George, S., 490
Gerguis, J., 191
Gerot, O., 427
Gershoni, J. M., 434
Getrost, H., 617
Gèze, M., 156
Giacomoni, P. U., 408, 612, 614(6), 618
Giamalva, D., 505, 510(2), 512
Giasson, R., 385
Gibbs, J. B., 262
Gibson, P., 249
Gibson, S. L., 386
Giese, B., 168, 169(16)
Gijzeman, O. L. J., 39
Gilaberte, Y., 396
Gilbert, N., 507, 514
Gilbert, W., 418, 420(9), 421(9), 424(9)
Gilchrest, B. A., 494
Gilleaudeau, P., 302
Gimbel, W., 527, 528(28, 29), 529(28, 29), 530(28, 29)
Giorgi, L. B., 78

Girard, D., 385
Giraud, M., 227
Girault, I., 153
Giri, D. K., 586
Girotti, A. W., 85, 86, 87, 88, 89, 90(16, 27), 92(22), 93, 93(22), 94, 94(23, 34, 35), 95, 96, 96(27), 97, 100(36), 111, 113, 188, 189(1), 191, 192, 192(1), 389, 390, 391(59), 392(59), 395
Giulivi, C., 76, 428, 431(7), 563
Glick, R. E., 527
Glössl, J., 534
Godar, D. E., 156, 159(23), 301, 307, 309, 310, 311, 312(7), 313, 314, 314(6), 315(7), 316(4, 5), 317, 317(4, 5), 319(5, 6), 320(5), 321(5), 322(4, 5), 323(4, 5), 327(5), 328(5), 329, 329(5), 363
Godinger, D., 428
Godleski, J. J., 556
Goeddel, D. V., 120, 419
Goedert, M., 270
Goerz, G., 130, 137, 141, 141(33), 294, 493, 494
Goetz, A. E., 137, 141(33)
Gohil, K., 538
Gold, D. R., 551
Goldberg, A. L., 278
Goldberg, L. H., 418
Goldstein, J., 197, 198(11), 400
Gollnick, H. P. M., 494
Gomer, C. J., 121, 343, 344(4), 345(4), 346(4), 349, 353(4), 354, 355, 356, 357, 377, 379(4), 381(4), 386(4), 387
Gomez-Fernandez, J. C., 107, 108(16)
Gonchoroff, N. J., 607
González, M., 45
González-Moreno, R., 41
Goodman, T. M., 250
Goodwin, D. C., 69
Goodwin, G. H., 289
Gordon, J., 434
Gorman, A. A., 63, 227, 319, 388, 390(41), 391(41), 392, 392(41), 395(41)
Gorospe, M., 132, 140, 140(22)
Gorski, R. A., 578, 579(13)
Goto, N., 208
Göttlicher, M., 120, 258, 274
Gottlieb, A. R., 302
Gottlieb, P., 16
Gotto, A. M., 428
Grabbe, S., 297

Graczyk, A., 197
Graf, H., 76
Graham, D. E., 552
Gramm, C., 278
Grandidier, M. H., 457
Grant, C. S., 208
Granville, D. J., 349, 350, 350(16), 357
Green, D. R., 319, 329, 332
Greenberg, M. M., 438
Greenoak, G. E., 191, 195(12)
Gregory, C. D., 325
Grellier, P., 208
Gremaud, E., 152, 411, 412(56)
Greppin, H., 515
Grether-Beck, S., 17, 17(61), 18, 19(60), 121,
    129(21), 130, 143(9), 154, 273, 280, 291,
    298, 301, 303, 306(4), 341, 367, 372(11),
    373(11), 464
Grewe, M., 17, 17(61), 18, 19(60), 121,
    129(21), 130, 143(9), 154, 280, 286, 296,
    297, 298, 299(9, 11), 300, 300(2, 8, 9), 301,
    303, 305, 306(4), 341, 371, 464
Grewer, C., 39
Griffiths, H. R., 402, 403(18), 405
Griffiths, M., 480
Grilli, M., 554
Grist, E., 331, 365
Gritzapis, A. D., 467, 478(20)
Groenendijk, R. T. L., 157, 158(34), 162(34),
    361, 363(3)
Grollman, A. P., 161, 362, 438
Groopman, J. D., 158
Groskopf, E., 527, 529, 530(36)
Gross, E., 391
Gross, K. B., 552
Gross, S., 258, 259(8), 261(8), 263(8)
Grossweiner, L. I., 389
Groth, N., 393, 466, 475, 478(16, 24)
Grove, M., 554
Grunberger, D., 586
Grune, T., 563
Grunwald, D., 337, 340(24)
Gruszecki, W. I., 106, 109(13)
Grzybowski, J., 197
Gu, Z.-W., 428
Gualberto, A., 555
Guan, L., 115
Guderian, R., 520, 521(1)
Gudermann, T., 269
Guénet, J. L., 483, 484(17)

Guiller, A., 152
Guiller, T., 15
Guiraud, H. L., 36
Gulbins, E., 607
Gumulka, J., 189
Gunther, M. R., 69
Guo, A., 68
Gupta, A. K., 385
Gupta, R. C., 405
Gupta, S., 352, 353, 353(24)
Gupta, S. L., 596
Gurinovich, G. P., 66
Gustafson, D. P., 212
Gutteridge, J. M. C., 87, 155, 317, 319(11),
    320(11), 401, 402(1), 408, 409, 603
Guy, A., 161
Guyton, K. Z., 132, 140, 140(22)
Gwizdala, C. J., 552
Gyufko, K., 298, 299(9), 300(8, 9), 371

H

Haberman, H. F., 395
Hacham, H., 361
Haddad, B., 555
Haddad, E.-B., 552, 553, 553(19), 554(19),
    555(20), 556(19), 557(19)
Hadjur, C., 391, 393, 393(69), 397(69)
Hadshiew, I., 466, 467(14), 475(14),
    476(14), 477(14)
Hagen, M. P., 404
Hagen, T. M., 401, 466
Hager, L. P., 59, 61(2), 64(2)
Hah, J. Ch., 115
Haidar, A. H., 552
Haidle, C. W., 16
Hain, R., 528
Hainaut, P., 164
Hair, M. L., 29, 30, 35(3), 36(3, 15)
Haites, N. E., 410
Hajjam, S., 6, 7(18), 8(18), 10(18), 11(18), 17,
    137, 198, 199
Hajra, A. K., 112
Haldane, F., 367, 371(5), 372(5), 376(5)
Hale, B. A., 522, 524(14)
Halko, D. J., 63
Hall, A., 262
Hall, D. B., 166, 167(11), 168, 169(11, 14, 15,
    20), 170, 170(14), 171(11, 14, 15, 20),

173(20), 181, 181(14), 184(11, 14), 425, 426(21)
Hall, R. D., 86, 111, 389
Hall, R. T., 93
Halliwell, B., 87, 152, 317, 319(11), 320(11), 401, 402, 402(1, 11), 403, 404, 405(31), 407(31), 408, 408(31), 409, 409(31), 410(31), 411(14), 412, 413(60), 414(61), 415, 415(16, 61), 416, 416(25, 26, 60, 61), 429, 430(22), 434, 510
Halperin, A. J., 365, 418
Ham, A. J. L., 550
Hamada, S., 112
Hamberg, M., 73, 412, 416(59)
Hamblett, I., 63, 227
Hamilton, C. A., 319
Hamilton, R. F., Jr., 564
Hampton, J. A., 398, 399
Han, D., 466
Hanania, J., 197
Haney, C. A., 196
Hang, B., 401
Hani, H., 527, 528(28, 29), 529(28, 29), 530(28, 29)
Hankovszky, H. O., 85
Hansberg, W., 110, 113, 114, 114(52, 53), 118(52)
Hansen, H. J., 45
Hanson, D. L., 364
Hanson, K. M., 154
Harashima, K., 220
Harber, L. C., 482
Harber, L. H., 481
Harbour, J. R., 29, 30, 35(3), 36(3, 15)
Hardegen, N. J., 586
Harford, J. B., 295
Häring, M., 438
Harm, W., 248
Harman, L. S., 75, 76(38)
Harriman, A., 384, 394
Harris, C. C., 164
Harris, G., 405
Harris, P. L. R., 431(46, 49), 434
Harrison, L., 438
Hartman, P. E., 110, 112, 113, 114(40), 198
Hartman, R. F., 185
Hartwig, A., 437
Hartzell, W. O., 538
Harvey, E. J., 345
Harvey, J. W., 208(12), 212

Harvey, N. S., 319
Hasegawa, Y., 30, 36(18)
Hashimoto, K., 29, 36(9, 10, 12, 13)
Hashizume, H., 493
Haskill, S., 119
Hassan, H. M., 221
Hassanain, H. H., 596
Hassenbein, D., 554
Hatch, G. E., 546
Hattori, N., 570
Haugland, R. P., 611
Haun, M., 70
Haun, R. S., 266
Hauser, M. T., 534
Hauyashi, Y., 333
Hayakawa, A., 347
Hayakawa, M., 274
Hayami, M., 234
Hayashi, Y., 426
Hayishi, O., 72, 73(30), 75(30)
Hazbun, M. E., 564
Hazell, L. J., 428, 431(12)
Hazen, S. L., 428(14), 429, 431(14, 20)
Hazucha, M. J., 552
He, J., 329, 348, 350, 351, 357
He, Y. Y., 393, 397(81), 398(81)
Heagle, A. S., 520, 522(3)
Hearne, P. Q., 552
Heath, R. L., 521
Heck, W. W., 521, 523(5)
Heeger, P. S., 554
Heenen, M., 614(6), 618
Heftler, N. S., 302
Hegler, J., 145, 437, 439(8), 441(8)
Heibel, G. E., 43
Heidenreich, K. A., 607
Heinecke, J. W., 428(14), 429, 431(14, 20, 48), 434
Heinen, G., 18, 130
Heinrich, U., 494, 495, 496, 496(16), 497, 502(8)
Heit, G., 45
Helbock, H. J., 145, 151(20), 402
Held, A. M., 63
Heldin, C. H., 265, 268
Helene, C., 185
Heller, W., 521, 522, 523(15), 524, 524(15), 525(15), 526, 526(17), 527(10)
Hellis, P. F., 185
Helman, W. P., 25, 44, 47(24), 132, 225

Helmberg, A., 555
Helmer, M. E., 68
Hemmes, C., 494
Henderson, B., 197, 198(9)
Henderson, B. W., 86, 100(6), 343, 344(4), 345(4), 346(4), 353(4), 377, 379(4), 381(4), 386, 386(4), 391, 395(30, 65), 398
Henderson, W. A., 238
Hendrich, S., 429, 431(19)
Hendry, G. A. F., 77
Henkel, T., 119, 273
Henninger, H. P., 298, 299(11)
Henzl, H., 437
Herbert, K. E., 402, 403(18), 405
Hernandez, T., 164
Hernandez Blanco, I. A., 239
Herrlich, A., 269
Herrlich, P., 120, 141, 255, 257, 258, 259, 259(8), 261, 261(8, 9), 263(8, 9), 268(9, 14), 269, 269(9), 271, 274
Herrling, T., 393, 466, 475, 478(16), 478(24)
Herrmann, G., 493
Hersh, L. B., 323
Herzberg, S., 5, 50, 51(15), 135, 198
Hetzel, F. W., 398
Hiai, H., 563, 564(17), 565(17)
Hiatt, C. W., 208, 212, 212(8)
Hibi, M., 111
Hideg, É., 77, 78, 79(12), 80, 80(12, 13), 82, 83, 84, 84(23), 110
Hideg, K., 77, 83, 84, 84(23), 85, 110
Hieda, K., 360
Hilfenhaus, J., 208
Hild, M., 233
Hilf, R., 386
Hill, F. L., 87, 89, 91(18, 28), 92(28)
Hiraku, Y., 336, 422, 423(16), 426(16)
Hiratsuka, A., 194
Hirayama, O., 112
Hirobe, M., 199
Hirth, A., 377, 379(6)
Hiscott, J., 554
Hla, T., 73
Hoar, D. I., 410
Hobbs, F., 269
Hober, D., 17, 199
Hochstein, P., 76
Hodes, M. E., 405, 410
Hodis, H. N., 196
Hoebeke, M., 123, 283, 393

Hofer, T., 405, 408(38), 409(38)
Hoffmann, J., 496
Hoffmann, R., 298, 299(11)
Hoffmann, Th., 239
Hogg, A., 130, 294
Hohmann, H. P., 553
Hoigné, J., 514
Holbrook, N. J., 132, 140, 140(22)
Holian, A., 562, 564
Holland, P. V., 207
Hollis, R. C., 50
Holloway, D. E., 494
Hollstein, M., 164
Holmes, M. G., 253
Holmlin, R. E., 165, 166, 167(11), 168(22), 169(11, 22), 170, 171(11), 175, 176, 176(28), 177(28), 184(11), 185(22), 425, 426(21)
Holroyd, C. D., 410
Holtzman, M. J., 552
Holwitt, E., 404
Hölzle, E., 466, 467(14), 475(14), 476(14), 477(14)
Holzmann, B., 271
Hönigsmann, H., 492
Hood, L., 405
Hoogland, H., 59, 64(6), 111
Hoorweg, J. J., 409
Hopf, F. R., 399
Hopfenmüller, W., 494
Hoppe, U., 158, 159(47), 475, 476(25), 477, 478(25)
Hopwood, S. E., 208(13), 212
Hora, K., 120
Hori, K., 112, 222
Horiuchi, K. Y., 269
Hornung, J., 285, 439
Horowitz, B., 16, 197, 198(11), 208, 400
Hosokawa, Y., 301, 308
Hou, C., 48
Houk, K. N., 166, 170(6)
Hourseau, C., 449, 450, 461(7), 464
Hruschka, W. R., 496
Hryhorenko, E. A., 319
Hsi, L. C., 69
Hsu, F. F., 429, 431(20, 48), 434
Hu, M.-L., 428, 429
Hu, S.-C., 564
Hu, Y.-Z., 393, 397(82)
Huang, C., 103

Huang, S., 556
Huang, T. J., 555
Huang, Y., 332
Huang, Z. Q., 434
Hubbard, R. C., 510
Hubbard, W. C., 552
Hudson, B., 173
Hughes, L., 35, 234
Hundall, T., 78
Hunt, D. W. C., 349, 350, 350(16), 357
Hunting, D. J., 397
Hurst, J. K., 63
Huser, H., 563
Hutter, E., 607
Hwang, D., 120, 278
Hyde, D. M., 552

I

Iani, V., 377, 385(9)
IARC, 331, 417
Ibrado, A. M., 332
Ichinose, T., 552, 553(18)
Ihringer, F., 393
Iida, M., 554
Iizuka, Y., 554
Ikebata, W., 107, 108(17, 18)
Ikebe, S., 369
Ikebushi, K., 197(14), 198
Ikesu, S., 112
Iltaka, Y., 547
Imai, N., 29, 30, 36(18)
Imaizumi, T.-A., 582
Imakubo, K., 42
Imbert, V., 274
Inaba, H., 69, 70(20)
Ingold, K. U., 35, 234, 550
Inokami, Y., 101, 103(4, 5), 104(4, 5), 109(5)
Inoue, K., 5, 7(16), 101, 216, 217(4)
Inoue, M., 29, 30, 36, 36(18, 20), 341
Inoue, S., 166, 331, 333(4), 341, 421, 422, 423(14), 425, 426, 426(14), 427
Inzé, D., 522, 525, 525(12), 527
Iordanov, M. S., 258
Ireland, J. C., 29, 36(11)
Ischiropoulos, H., 428(15), 429, 431(15), 434
Isele, U., 383
Ishiguro, K., 217, 218(5)
Ishisaka, R., 341

Israel, A., 274
Issler, S. L., 29, 30, 36(15)
Itakura, K., 563, 564(17), 565(17)
Ito, E., 29, 36(13)
Ito, K., 166, 331, 333(4), 362, 417, 421, 422, 423, 423(5, 14–16), 425, 426, 426(14–16)
Ito, T., 121
Itoh, K., 29, 36(9)
Itoh, S., 36, 112
Iu, K.-K., 41
Iwanatsu, A., 332
Izawa, Y., 220
Izuta, S., 161

J

Jablonski, J.-A., 177
Jacobsen, J. R., 185
Jacovina, A. T., 141
Jadoul, L., 283
Jaeger, C. D., 29, 35(1), 36(1)
Jäger, H. J., 521, 523(6)
Jagger, J., 248
Jahnke, A., 17, 19(60), 121, 129(21), 130, 143(9), 154, 280, 298, 464
Jakobs, A., 393
Jannink, L. E., 393
Jardon, P., 233
Jaruga, P., 438
Jaskot, R. H., 571
Jaspers, I., 571
Jawed, S., 345, 353(8)
Jaworski, K., 189, 190(6)
Jekô, J., 85
Jemmerson, R., 332
Jenkins, A. J., 429, 431(18)
Jenks, W. S., 233
Jenner, A., 401, 402, 412, 413(60), 414(61), 415, 415(16, 61), 416(60, 61)
Jensen, K. G., 145
Jeong, J. K., 160
Jeunet, A., 391, 393, 393(69), 395(77), 396, 396(77), 397(69, 77)
Jewell, C. M., 555
Jiang, Ch.-K., 286
Jiang, H., 350
Jiang, L. J., 393, 397(81, 82), 398(81)
Jit, P., 7
Jochum, C., 496

Jochum, P., 496
Joenje, H., 5, 50, 51(15), 135, 198, 409
Johann, T. W., 175
John, S. W. M., 405
Johns, H. E., 248
Johns, M. B., Jr., 405
Johnson, E. A., 112
Johnson, F. M., 391
Johnson, J. P., 17, 19(60), 121, 129(21), 130, 143(9), 154, 280, 298, 464
Jonat, C., 271
Jones, A. D., 434
Jones, T. W., 86, 389
Jordon, P., 391, 393(69), 397(69)
Jori, G., 86, 198, 343, 344(4), 345(4), 346(4), 353(4), 377, 379(4), 381(4), 383, 386(4), 389, 398
Jorres, R., 552
Joshi, P. C., 151, 427
Jothi, S., 536
Joualt, H., 483, 484(17)
Jovanovic, S. V., 166
Joyce, A., 108
Juedes, M. J., 160
Julliard, M., 393, 395(77), 396, 396(77), 397(77)
Jumper, C. A., 564
Jung, C.-H., 431(43), 434
Jungmann, H., 494, 495, 496, 496(16), 497, 502(8)
Jürgens, O., 43, 45(20)
Jurgensmeier, J. M., 329

K

Kacher, M. L., 233
Kafoury, R. M., 547, 582
Kagan, J., 112
Kaiser, S., 101, 105(1), 112, 224, 226, 283
Kajiwara, 14
Kajiwara, T., 216
Kálai, T., 77, 82, 83, 84, 84(23), 85
Kalashnikova, G., 467
Kalyanaraman, B., 30, 36(16), 428, 431(9)
Kamada, H., 29, 35(2), 36(2)
Kamal-Eldin, A., 112
Kamalov, V. F., 42, 60, 66(9), 388, 389, 513
Kambayashi, Y., 36, 112, 199, 216, 222
Kamiya, H., 163

Kamiya, Y., 35
Kammerijer, A., 450
Kamo, N., 197(14), 198
Kaneko, Y., 112
Kang, S., 141
Kangasjärvi, J., 520, 525, 526, 527, 527(21), 531(19), 532(19)
Kanner, S. B., 111
Kanno, T., 119, 341
Kanofsky, J. R., 21, 39, 50, 59, 63, 64, 64(1, 3, 4, 6, 7), 66, 66(1), 71, 76, 76(27), 85, 86, 111, 113, 216, 226, 388, 389, 390, 505, 507(9), 510(9), 511, 512, 513, 515, 515(4, 5, 8, 12), 516(12), 517(12), 518, 519(13), 547
Kaplan, M. L., 198
Kapp, A., 17(61), 18, 130, 297, 301, 303, 306, 306(4), 341
Karakaja, A., 438
Karin, M., 111, 119, 120, 131, 258, 259, 273, 274, 555, 586
Karmann, K., 17(61), 18, 130, 301, 303, 306(4), 341
Karnovsky, M. J., 3
Kartha, V. B., 66
Kasahara, E., 36
Kasai, H., 152, 161, 163, 333, 426, 427
Kasha, M., 12, 13, 223
Kashiba-Iwatsuki, M., 30, 36(20)
Kashtanov, S. A., 508
Kasibhatla, S., 319
Kass, E. H., 481
Kastan, M. B., 257
Kato, K., 547
Kaufman, F., 39
Kaufman, J., 431(42), 434
Kaufmann, R., 495, 499, 499(11)
Kaur, H., 402
Kawanishi, M., 336
Kawanishi, S., 156, 159(24), 166, 331, 333(4), 336, 341, 362, 417, 421, 422, 423, 423(5, 14–16), 425, 426, 426(14–16), 427, 563, 564(17), 565(17)
Kawasaki, C., 29, 36(12)
Kay, R. J., 274
Kaysen, K. L., 547
Kaziro, Y., 262
Keana, J. F. W., 393
Kearms, D. K., 216
Kearns, D. R., 14, 102, 105(8)

Keck, R. W., 398, 399
Kehrl, J. H., 586
Keiderling, T., 112
Keith, A., 197
Keitt, A. S., 208(12), 212
Kelfkens, G., 364, 366
Keller, G., 197
Keller, R. A., 102, 105(9), 106(9)
Kelley, J. F., 343
Kelley, S. O., 165, 169(20), 170, 171(20), 172, 173(20), 176
Kelly, B., 381
Kelly, F. J., 282
Kelly, K., 125
Kelly, K. F., 410
Kelly, M., 403, 416(25)
Kemler, I., 323
Kennedy, J. C., 381
Kenney, M. E., 348, 354
Kensler, T. W., 140
Kerner, K., 524, 526(17)
Kerr, J. F. R., 309
Kerr, K., 552
Kerstiens, G., 512, 519(1)
Kessar, 7
Kessel, D., 328, 343, 344(4), 345(4), 346(4), 348, 349(11), 353(4), 377, 379(4), 381(4), 386(4)
Keszthelyi, T., 49
Kettle, A. J., 428(15), 429, 431(15)
Keyhani, R., 465
Keyse, S. M., 86, 130, 291, 294
Khachik, F., 495, 499
Khan, A. U., 3, 12, 13, 16, 50, 59, 61(2, 5), 64(2), 71, 76(28), 223, 226
Khoo, J. C., 431(54), 434
Kidd, L. R., 158
Kiefer, E., 527
Kielbassa, C., 144, 145(1), 153, 157, 361, 362, 436, 442, 443, 444, 444(45)
Kiiskinen, M., 527
Kiistala, 467, 478(18)
Kikuchi, T., 302
Kim, C. N., 332
Kim, H.-P., 115
Kim, J. G., 431(55), 434
Kim, S.-T., 185
Kim, T. H., 460
Kimura, T., 91
King, D. S., 185

Kingston, R. E., 279, 586
Kinney, P. L., 551
Kino, K., 183
Kinoshita, S., 554
Kirichenko, A., 564
Kirihara, J. M., 111
Kirilovsky, J., 269
Kiriyama, Y., 208
Kirk, H. E., 274
Kirsch, D., 495, 499, 499(11)
Kirschning, C. J., 120
Kitamura, H., 29, 36(12)
Kitao, T., 237
Klausner, R. D., 295
Klein, C. B., 405
Kleinjans, J. C. S., 408
Klein-Struckmeier, A., 212, 215(15)
Kliebenstein, D. J., 533
Klivenyl, P., 569, 570(32)
Klostermann, P., 29, 36(11)
Klotz, L. O., 3, 14(6), 17(61), 18, 19(6, 62), 111, 130, 131, 135(19, 20), 138(19, 20), 139(20), 141(21), 154, 224, 226, 273, 284, 291, 294, 301, 303, 306(4), 341, 358
Klug, D. R., 78
Knapp, W., 465
Knebel, A., 255, 258, 259, 259(8), 261(8, 9), 263(8, 9), 268(9), 269, 269(9)
Knorr, R., 177
Knox, C. N., 123, 133
Knox, R. S., 386
Knüver-Hopf, J., 212
Kobayashi, K., 220
Kobayasi, Y., 112
Kobzik, L., 556
Kochevar, I. E., 20, 27, 28, 154, 319, 417
Kodavanti, U. P., 546
Kogelnik, A. M., 370
Kohn, K. W., 441
Kohno, Y., 36, 219, 616
Koivuniemi, A., 78
Koizumi, F., 554
Koizumi, H., 220
Kojima, E., 208
Kolbeck, R., 553
Komiyama, T., 36, 199, 217, 218(5)
Konaka, R., 29, 30, 36, 36(18)
Koncz, C., 531
Kondo, M., 191
Kongshaug, M., 141

Kono, Y., 563
Koornneef, M., 531
Kopecek, J., 111, 429
Kopeckova, P., 111, 429
Koppenol, W. H., 59
Korbelik, M., 343, 344(4), 345(4), 346(4), 353(4), 377, 379(4), 381(4), 386(4)
Koren, H. S., 552
Korhonen, M., 527
Korman, N. J., 345, 353(8)
Korn, E. D., 608
Kornneef, M., 533
Koroteev, N. I., 42, 60, 66(9), 388, 389, 513
Kortbeek, H., 5, 135, 198
Kortbeek, U., 50, 51(15)
Korytowski, W., 85, 87, 88, 90(16), 92(22), 93, 93(22), 94, 94(23, 34, 35), 95, 96, 100(36), 113, 191, 192, 390, 391(59), 392(59)
Kosaka, M., 29, 36(12)
Koto, H., 553, 555, 555(20)
Kovalenko, M., 268
Kowall, N., 431(46), 434
Kowalski, B. R., 496
Koyama, M., 108
Kraljic, I., 23, 137, 227
Kransnovskii, A. A., 389
Krasnovsky, A. A., 388
Krasnovsky, A. A., Jr., 42, 60, 66(9), 388, 513
Krause, G. H., 80
Kreit, J. W., 552
Kretz-Remy, C., 274
Kreutzer, D. A., 159
Krieg, T., 18, 130
Krieger, A., 80
Krinsky, N. I., 69, 74(18)
Kripke, M. L., 460
Krishna, M. C., 85
Kristiansen, M., 227
Kroemer, G., 321, 329(20), 332, 337
Kroon, E. D., 363
Krueger, G. G., 492
Krueger, J. G., 302
Kruijer, W., 261
Krutmann, J., 17, 17(61), 18, 19(60), 121, 129(21), 130, 143(9), 154, 273, 280, 286, 291, 296, 297, 298, 299(9), 300(2, 8, 9), 301, 302, 302(16), 303, 305, 306, 306(4), 308, 341, 366, 367, 371, 372(11), 373(11), 464
Kubota, Y., 29, 36(9, 10, 12, 13)

Kuehl, F. A., Jr., 68, 72, 73(29), 75(29)
Kuittinen, T., 525, 531(19), 532(19)
Kukielczak, B. M., 129
Kuklinska, E., 474
Kulig, M. J., 87, 91(18), 188, 390
Kulmacz, R. J., 68, 73
Kulms, D., 319
Kumadaki, I., 107, 108, 108(17)
Kumar, A., 586
Kumar, S., 168, 171(17)
Kummer, J. L., 607
Kunala, S., 418
Kung, H. C., 158
Kunisawa, R., 481
Kunkel, S., 553
Kunkel, T. A., 161
Kunzli, N., 551
Kurashige, Y., 554
Kuroda, M., 113
Kurokawa, Y., 333, 426
Kürten, V., 367, 372(11), 373(11)
Kuwabara, M., 197(14), 198
Kuypers, F. A., 604, 609(4)
Kvam, E., 145, 156, 157(26), 291, 408, 427, 444
Kvan, E., 295
Kwon, O., 553
Kyriakis, J. M., 586

L

Laemmli, U. K., 433
Lafleur, M. V. M., 14, 409
LaFrancois, C. J., 416
Lahiri, D. K., 405
Lalley, P. A., 554
Lambert, C., 28, 63, 227, 319
Lambrecht, B., 212, 215(15)
Lamola, A. A., 482
Lamprecht, J., 113
Lamy, M., 283
Land, E. J., 227
Lander, H. M., 141
Lands, W. E. M., 68
Lane, W. S., 278
Laneuville, O., 73
Lang, J. K., 538, 562, 564(3)
Langebartels, C., 520, 521, 522, 523(15), 524, 524(15), 525, 525(12, 15), 526, 526(17), 527(10, 21)

Langeveld-Wildschut, A. G., 296, 300(2)
Langeveld-Wildschut, E. G., 300
Langford, S. D., 505, 511, 564
Langlois, R., 150, 390, 392(61)
Lanockiand, K., 197
Lanzendörfer, G., 467
Laperrière, A., 385
LaPushin, R., 586, 593, 594(23)
Larkin, H. E., 329, 346, 348, 353(9)
Larkins, L. K., 112
Larkö, O., 446, 449(2)
Larrier, N. A., 158
Larsen, D. L., 5
Laskin, D. L., 553
Laskin, J. D., 553
Last, J. A., 553
Last, R. L., 531, 532, 533, 533(46)
Lautier, D., 161, 361
Laval, J., 155, 438
Lawrence, C. W., 185
Lawson, T. A., 408
Lazarow, P., 112
Leadon, S. A., 158, 163
Lebeau, J., 165
Leboulch, P., 483
Lebovitz, R. M., 124
Lebowitz, J. H., 595
Lebwohl, M., 111, 145, 336, 492
Leckonby, R. A., 52
LeClerc, J. E., 185
Ledbetter, J. A., 111
LeDoux, S. P., 373
Lee, A. S., 354
Lee, F. L., 35
Lee, H.-C., 367, 372(3)
Lee, H. H., 431(53), 434
Lee, J. C., 270
Lee, J.-S., 115
Lee, J. Y., 269
Lee, P. C. C., 394
Lee, S., 112
Lee, W., 314
Lee, Y.-L., 432, 434(34)
Lees, A. J., 402
Leevers, S. J., 262
Leewenburgh, C., 431(48), 434
Leffell, D. J., 418
Legrand-Poels, S., 121, 131, 273
Lehmann, J., 158, 159(47)
Lehmann, K., 563

Leikauf, G. D., 554, 571, 581, 582
Leithauser, M., 68
Lemon, S. M., 208
Lenard, J., 16, 197, 198(12)
Lenardo, M. J., 554
Lendzian, K. J., 512, 519(1)
Lengebartels, C., 527
Lenger, W. A., 191
Lengfelder, E., 395
Lennon, S., 309
Lenny, L., 197, 198(11), 400
Lenz, A., 430, 431(29, 30), 431(44), 434
León, J., 531, 535, 535(44)
Leonard, W. J., 553
Leonardi, S., 524, 526(17)
Le Page, F., 158, 161
Leroy-Viard, K., 483, 484(17)
Leuenberger, H., 383
Levalois, C., 165
Levine, M., 226, 538
Levine, R. L., 428, 429, 429(5), 430, 431, 431(6, 29, 30, 35, 42, 45), 432, 434, 434(31), 436(5, 27, 31)
Levinson, A. D., 419
Levitzki, A., 268
Levy, J. G., 142, 349, 350, 350(16), 356, 357, 381
Lew, R., 492
Lewisch, S. A., 428, 431(6)
Li, J. W., 120
Li, L., 564
Li, N., 120, 273, 274
Li, T., 385
Li, W. W., 259
Li, X., 141
Li, Y.-S., 354
Liebler, D. C., 111, 546, 547, 550
Lii, C.-K., 429, 431(19)
Lim, M., 431, 434(31), 436(31)
Lin, A., 365
Lin, C.-W., 27
Lin, F., 89, 94, 100(36), 113, 192
Lin, J., 141
Lin, M., 338, 340(26)
Lin, P., 141
Lin, S. D., 495
Lin, S. W., 563
Lin, Y., 498, 502(24)
Lin, Y. Y., 196
Linding, B. A., 50, 51(12)

Linhardt, M. S., 405
Lint, T. F., 70, 74(24)
Lion, Y., 78, 82
Lioy, P. J., 552
Lippert, M., 522, 523(15), 524(15), 525(15)
Lippin, A., 208
Litfin, M., 257
Liu, D. J., 381
Liu, L., 332
Liu, M. C., 552
Liu, S., 553, 555(20)
Liu, S. H., 238
Liu, X., 332
Liu, Y., 110, 111(6), 132, 140(22)
Livolsi, A., 274
Llanglois, R., 188
Lledías, F., 110, 113, 114(52, 53), 118(52)
Lloyd, 16
Lloyds, R. S., 438
Lo, W. B., 190
Lock, 7
Lockshin, R. A., 309
Lodge, J. K., 538
Loeb, L. A., 161, 401
Loechler, E. L., 163
Loew, G., 173
Loewen, P. C., 115
Loft, S., 145
Lohman, P. H. M., 156, 158(28), 363
Loll, P. J., 68
Long, E. C., 166, 173(12)
Lopez, M., 14
Lorenzen, T. J., 552
Loriolle, F., 269
Losev, A. P., 66
Lott, M. T., 370
Louie, S., 510
Lovering, G., 227
Lown, J. A. G., 212
Lowry, O. H., 89, 137
Lu, F. D., 392
Lübbers, D. W., 496
Lucas, A. D., 156, 159(23), 317, 363
Lucas, P. W., 524, 532(16)
Luger, T. A., 297, 319
Lumb, A., 7
Luna, M. C., 355, 356
Lunec, J., 402, 403(18), 405
Luo, Y., 328, 348, 349(11)
Lusby, W. R., 495

Lusztyk, J., 397
Lütz, C., 522, 523(15), 524(15), 525(15)
Ly, D. K. Y., 166
Lynch, M., 27, 28
Lynch, O. T., 553
Lyons, T. J., 429, 431(18)

M

Maccoll, A., 87, 90(20), 390
Mackenzie, L. A., 255
Maclaren, W. M., 553
Macpherson, A. N., 78
MacRoberts, A., 197, 198(9), 379
Madden, M. C., 571
Madisen, L., 410
Madri, J., 474
Madzak, C., 14, 157
Maeda, Y., 208
Maes, D., 612, 615, 616
Magness, S. T., 483
Magnussen, H., 552
Magolda, R. L., 269
Magun, B. E., 258
Magun, J. L., 258
Mahboubi, A., 319
Maher, P., 465
Maheswara Rao, N. U., 85
Mahler, H.-C., 444
Maidt, L., 155
Maier, K. L., 431(44), 434
Maiorino, M., 69
Mair, R. D., 93
Makrigiorgos, G. M., 611
Malik, Z., 197
Malone, M. E., 427
Malorni, W., 614(6), 618
Mammone, T., 612, 615, 616
Mancini, J. A., 68
Mandard-Cazin, B., 18, 54
Maniatis, T., 180, 183(34), 221, 534
Mann, B., 120
Mann, K. J., 351
Mann, R., 552
Manna, S. K., 586
Mannel, D. N., 120, 586
Manning, W. J., 522, 535(11)
Manogue, K. R., 554
Manos, E. J., 269

Mansfield, T. A., 524, 532(16)
Mao-Qiang, M., 536
Marbach, R., 496
Marchetti, P., 337
Marcus, S. L., 377, 381
Marenus, K., 612, 615, 616
Margolis-Nunno, H., 16, 197, 198(11), 400
Margot, A., 161
Marinovich, M., 260
Markham, A., 285, 439
Markovic, J., 152
Marks, F., 364
Marks, T. J., 166
Marnett, L. J., 68, 69, 74
Marois, J., 554
Marquis, I., 156
Marron, N. A., 52
Marrot, L., 457, 464(16)
Marsden, C. D., 402
Marshall, C. J., 262
Marston, G., 389
Martens, H., 496
Martí, C., 43, 45, 45(20)
Martin, A. C., 533
Martin, B., 208(13), 212
Martin, H. D., 226, 230
Martin, J. C., 434
Martin, R. L., 22, 102, 232
Martin, S. J., 309, 330
Martínez, C., 45
Martinez-Zapater, J. M., 531, 532(41), 533(41), 534(41)
Martire, D. O., 227
Marty, L., 556
Maruyama, T., 222
Marvel, 7
Mascio, P., 101, 105(1)
Mashima, R., 29, 30, 36(18)
Mason, J. T., 103
Mason, R. P., 69, 75, 76(38), 111
Matheson, M., 191, 195(12)
Mathews, M. M., 481
Mathews-Roth, M. M., 479, 480(2), 481, 482, 483, 494
Mathias, J.-M., 212
Mathy, P., 521, 523(6)
Mathy-Hartert, M., 283
Matray, T. J., 438
Matroule, J.-Y., 119, 129
Matrozov, V. I., 508

Matsui, M. S., 364
Matsui, T., 547
Matsumoto, K., 208
Matsumoto, S., 547
Matsumoto, Y., 347
Matsunaga, T., 360
Matsuo, M., 108, 547
Matsuoka, M., 237
Matsushima, K., 554, 555
Matsuura, K., 101, 102(3), 103(3), 104(3), 112
Matsuura, T., 5, 7(16), 216, 217(4)
Matteucci, M., 176, 180(31)
Matthias, P., 586
Mauthe, R. J., 144
Maxam, A. M., 418, 420(9), 421(9), 424(9)
Maytum, D. J., 361
Mayurama, T., 112
Mazière, J.-C., 130, 208
Mazilier, C., 448
Mazinger, E., 291
McBride, J. R., 521
McBride, L. J., 176, 180(31)
McCaul, S., 248
McCormick, F., 262
McDonnell, W. F., 552
McGarvey, D. J., 39
McGowan, A., 319
McIndoe, R. A., 405
McIntyre, T. M., 582
McKeever, M. A., 612
McKenna, G. J., 365
McKinlay, A. F., 448
McKinney, W. J., 571
McLafferty, F. W., 196
McLean, A. J., 39
McLean, L. R., 604, 609(5)
McLeod, A. R., 524, 532(16)
McManus, B. M., 350
McMaster, J. S., 278
McNamara, M. T., 85
McPhillips, F., 141
Mechin, R., 111
Medeiros, M. H. G., 3, 16
Medford, R. M., 431(55), 434
Meffert, H., 297
Megahed, M., 301
Meggers, E., 168, 169(16)
Meghji, S., 197, 198(9)
Mei, W. P., 475, 476(25), 478(25)
Meier, B., 120, 586

Meier, R., 270
Meinke, D. W., 531, 533
Meister, A., 260
Melnick, C., 212
Melvin, T., 437
Menck, C. F. M., 4, 14, 155, 157, 160, 161(50), 224
Mendenhall, G. D., 237
Menon, I. A., 395
Mercurio, F., 119, 120
Mergny, J. L., 130
Merkel, P. B., 102, 105(8)
Merson-Davis, L. A., 427
Messer, G., 553
Metze, D., 319
Meunier, J. R., 457, 464(16)
Meyer, M., 585
Meyerowitz, E. M., 531, 532(40)
Mia, N., 197
Miasawa, N., 220
Michaeli, A., 41, 111
Michel, C., 223, 395
Michel, J. L., 483
Michel-Beyerle, M. E., 168, 169(16)
Midden, M. R., 294
Midden, W. R., 112, 113, 114(40), 198, 516, 517
Miike, A., 615
Miki, M., 547
Mikkelsen, K. V., 49
Miles-Richardson, G. E., 59, 64(4)
Miller, A. C., 391, 395(65)
Miller, B. A., 483
Miller, E., 431(54), 434
Miller, P. R., 521
Miller, S. A., 310, 311, 312(7), 314, 314(6), 315(7), 319(6)
Milne, L., 427
Milner, A. E., 325
Minnick, D. T., 161
Mira, D., 76
Misawa, N., 112, 222
Mistry, N., 402, 403(18)
Mistry, P., 402, 403(18)
Mitchell, D. L., 360
Mitchell, D. M., 145
Mitchell, J. B., 85
Mittler, R. S., 111
Miyamoto, S., 119
Miyata, N., 155
Miyazawa, T., 92, 93(30), 219, 616

Miyoshi, T., 30, 36(18)
Mizuki, H., 199
Moan, J., 86, 141, 343, 344(4), 345(4), 346(4), 353(4), 377, 379(4), 381(4), 385(9), 386(4), 388, 418, 513
Mochinik, P. A., 218
Moeder, W., 522, 525(12)
Moellmann, R. L., 474
Mohan, R. R., 345, 353(8)
Mohr, H., 16, 197, 198, 198(13), 199(16), 207, 212, 215(15), 216, 224, 225(14)
Mohsni, S. E., 23
Moldeus, P. W., 467, 468(19), 477(19)
Molko, D., 153
Möller, K., 156, 158
Möller, L., 405, 408(38), 409(38)
Möller, W., 416
Mondal, K., 119
Monnier, P., 393
Monroe, B. M., 223, 227, 233, 237, 238, 238(16, 28), 239(28)
Montagutelli, X., 483, 484(17)
Montenegro, L., 466, 475(15)
Montesano, R., 164
Moore, D. D., 586
Moore, J. R., 250
Moore, J. V., 379
Moore, T. B., 552
Morand, O. H., 113
Morandi, M. T., 564
Moreau-Gaudry, F., 483
Moreno, M., 111
Morgan, S. E., 257
Morimoto, K., 333, 408
Morin, B., 14, 15(39), 144, 151, 361, 436
Morison, W. L., 492
Morita, A., 17(61), 18, 130, 297, 301, 302, 303, 305, 306(4), 308, 328, 330(24), 341
Moriya, M., 161
Morlière, P., 129, 156
Morowitz, H. J., 248
Morrice, N., 270
Morris, J. S., 119
Moses, F. G., 238
Mosmann, T., 202, 455
Mosoni, L., 431(45), 434
Motchnik, P. A., 36, 224
Mottola-Hartshorn, C., 338, 340(26)
Mouithys-Mickalad, A., 283
Moureu, 5

Moustacchi, E., 158
Moyal, D., 445, 449, 450, 459, 461(7), 464
Moysan, A., 156
Mrowea, J. J., 237, 238(28), 239(28)
Mruzak, M. H., 87, 90(20)
Mruzek, M. H., 390
Muel, B., 160(64), 162, 164(64), 362
Mueller, D. M., 431(48), 434
Mueller-Dieckmann, C., 274
Muench, H., 202
Muir, J. J., 208
Muizzuddin, N., 616
Mukai, K., 36, 101, 112
Mukaida, N., 554, 555
Mukhopadhyary, T., 418
Mukhopadhyay, A., 585
Mukhtar, H., 121, 342, 345, 346, 352, 353,
    353(7–9, 23, 24), 354, 391, 395(66, 67)
Muller, J. G., 15, 183
Muller, J. M., 585
Müller, K., 50, 51(14)
Muller, M. M., 586
Müller-Breitkreutz, K., 16, 197, 198, 198(13),
    199(16), 216, 224, 225(14)
Mulvihill, J. W., 354
Mundra, K., 7
Munnich, A., 372
Murant, R. S., 386
Murasecco, P. S., 45
Murata, M., 427
Murphy, M. E., 69, 70(16), 112, 224, 226
Murphy, P., 208
Murray, B. W., 120
Murray, R. W., 9, 50
Muse, K. E., 569, 570(33)
Musselman, R. C., 522, 524(14)
Mustafa, M. G., 546
Mutasa, H. C., 569, 570(34)
Myrand, S. P., 269

N

Nabeja, R. T., 302
Nabel, G. J., 553, 589
Nabi, Z., 283
Nackerdien, Z., 401, 411(6)
Nadaud, J. F., 614(6), 618
Nadel, J. A., 552
Nadkarni, D. V., 563

Nadler, S. G., 111
Naes, T., 496
Nagan, N., 112
Nagano, T., 199
Nagaoka, S., 36, 112
Nagata, S., 332
Nagelkerke, J. F., 283
Naito, Y., 569
Nakagawa, H., 554
Nakagawa, M., 220
Nakajima, M., 333, 408
Nakajima, S., 384
Nakamura, K., 112, 220
Nakamura, T., 168(21), 169(21), 170, 219, 616
Nakamura, Y., 569
Nakano, M., 36, 112, 199, 216, 217, 218(5), 222
Nakatani, K., 166, 168(21), 169(21), 170
Nakazumi, H., 237
Naldi, L., 492
Naora, H., 113
Nardello, V., 50, 51(17, 18), 54(17), 57(18),
    227
Nasman, B., 570
Nastainczyk, W., 68, 75, 76(37)
Nataraj, A. J., 154
Natarajan, K., 586
Natoli, G., 274
Nau, M. N., 39
Navathe, S. B., 370
Naylor, G. J., 208(13), 212
Neckers, D. C., 112
Neff, W. E., 32
Neilson, E. G., 554
Neininger, A., 258, 259(8), 261(8), 263(8)
Nelson, W. G., 158
Nesland, J. M., 141, 377, 385(9)
Neuner, A., 45
Neverov, K. V., 388
Newmeyer, D. D., 329
Newton, R. J., 527
Nguyen, K. C., 10
Ni, W., 115
Niedner, R., 297, 306
Nieuwint, A. W. M., 5, 50, 51(15), 135, 198
Nigro, R. G., 14, 160, 161(50)
Nikaido, O., 157, 360
Niki, E., 35, 107, 108(15), 191, 547
Nio, Y., 113
Nishimura, S., 161, 333, 426
Nitza, Y., 197

Niu, S., 166, 173(13)
Nizuno, Y., 570
No, E. G., 527
Noda, H., 29, 35(2), 36(2)
Nohl, H., 110, 111(6)
Nojima, T., 217, 218(5)
Nonell, S., 37, 40, 41, 43, 45, 45(20), 47(10)
Nordheim, A., 261, 268(14)
Nordmann, Y., 483, 484(17)
Norris, K., 496
North, S. L., 510
Nourooz-Zadeh, J., 416
Nouspikel, T., 158
Nowak, D., 552
Nowak, T., 208
Nugteren, D. H., 67
Nukhtar, H., 342
Núñez, M. E., 165, 168, 169(14), 170(14), 171(14), 181, 181(14), 184(14), 187
Nurnberger, J. I., Jr., 405
Nye, A. C., 393
Nyhlin, N., 570

O

Oakle, A., 449, 461(7)
Obendorf, M. S. W., 150, 155, 367, 402
Oberley, L. W., 569, 570(33), 586
Oberley, T., 569, 570(32)
Obermeier, A., 268, 269(27)
Obochi, M. O. K., 396
O'Brien, J. M., 198
Ochsner, M., 344, 345(5), 346(5), 384, 385(25), 387(25), 398(25)
O'Connor, J. E., 337, 340(24)
Oetjen, J., 16
O'Gara, J. F., 578, 579(13)
Ogilby, P. R., 40, 41, 43, 45, 47(10), 49, 66, 227
Ogino, N., 72, 73(30), 75(30)
Ohishi, N., 615
Ohkawa, H., 615
Ohki, S., 72, 73(30), 75(30)
Ohlweiler, D. F., 604, 609(5)
Ohno, S., 586
Ohta, N., 384
Ohya-Nishiguchi, H., 29
Oikawa, K., 29, 35(2), 36(2)
Oikawa, S., 156, 159(24), 331, 336

Okabe, K., 101
Okawa, K., 332
Okazaki, T., 234
Okochi, K., 208
Oku, T., 493
Olaizola-Horn, S., 17, 19(60), 121, 129(21), 130, 143(9), 154, 280, 298, 464
Olea, A. F., 39
Oleinick, N. L., 329, 343, 344(2), 345, 345(2), 346, 346(2), 348, 350, 351, 353(2, 7, 9), 354, 357
Oliver, C. N., 430, 431(30)
Oliveros, E., 45
Öllinger, K., 341
Olson, J. A., 495
Olsson, T., 570
Olvey, K. M., 311, 312(7), 315(7)
Omori, H., 113
O'Neil, J. J., 446
O'Neill, C. A., 510, 511, 512(17)
O'Neill, G., 68
O'Neill, P., 437
Op den Kamp, J. A., 604, 609(4)
Oppenheim, J. J., 554
Osada, K., 188
Osawa, T., 563, 569
Osborn, L., 553
Otani, H., 113
Oudard, S., 165
Ouellet, R., 150, 390
Overmyer, K., 525, 531(19), 532(19)
Owamoto, Y., 493
Oyama, M., 208
Ozawa, M., 302
Ozawa, N., 191, 193, 194, 194(10, 18)
Ozols, J., 67

P

Pacifici, R. E., 431(57), 434, 563
Packer, L., 393, 429, 430(22), 465, 466, 466(5), 478(17), 526, 536, 537, 538, 546
Page, F., 161
Pagels, N. R., 68
Pahl, H. L., 119, 274
Paillous, N., 121, 122(22), 127(22), 129
Pal, P., 385
Palade, G. E., 434
Palinski, W., 431(54), 434

Palmer, C. M., 362
Pamphilon, D. H., 212
Pancake, B. A., 212
Panthananickal, A., 68
Pap, E. H. W., 603
Papadopoulus, G., 467, 478(20)
Papamichail, M., 467, 478(20)
Pappas, S. P., 29, 36(6)
Paquette, B., 382
Parce, J. W., 336
Park, K.-K., 271
Park, S., 119, 125
Parker, A. W., 437
Parker, J. G., 66, 113, 388
Parker, R. A., 604, 609(5)
Parrish, J. A., 492
Parsons, P. G., 157
Parthasarathy, S., 431(54, 55), 434
Pascher, F., 13
Pasquet, S., 269
Pass, H. I., 342, 343(1), 344(1), 345(1), 346(1), 353(1)
Pathak, M. A., 481
Paulauskis, J. D., 556
Paulus-Thomas, J. E., 405
Paust, J., 230
Pavlov, Y. I., 161
Paweletz, N., 113
Pawliuk, R., 483
Paxton, J., 68
Payer, H.-D., 522, 523(15), 524(15), 525(15)
Payrat, J.-M., 212
Peak, J. G., 14, 155, 156, 158(30), 319, 417
Peak, M. J., 14, 155, 156, 158(30), 319, 417
Pearson, J. A., 258
Peipkorn, M., 492
Peker, S., 158, 159(47)
Pelech, S. L., 142, 356
Pell, E. J., 521, 527, 527(9), 531
Pelle, E., 612, 615, 616
Pellieux, C., 6, 11(19), 17, 131, 135, 135(20), 138(20), 139(20), 197, 199, 226
Pelosi, G., 69
Pendino, K. J., 553
Peng, Q., 141, 343, 344(4), 345(4), 346(4), 353(4), 377, 379(4), 381(4), 385(9), 386(4)
Peng, Y.-M., 498, 502(24)
Peng, Y.-S., 498, 502(24)
Pennathur, S., 431(48), 434

Peppelenbosch, M. P., 261
Perdrau, J. R., 16, 212
Pereboom, D., 396
Pereira, P., 416
Pereira-Smith, O. M., 593, 594(23)
Perez, P., 457, 464(16)
Perl, A., 607
Perlieux, C., 291
Perry, G., 431(46, 49), 434
Persad, S. D., 395
Peters, M., 230
Peters, M. J., 553, 557
Peterson, A. R., 191
Petit, P. X., 337, 340(24)
Petrussevska, R. T., 285, 439
Pettonilli, V., 329
Peyron, J. F., 274
Pfander, H., 499
Pflaum, M., 111, 121, 131, 145, 153, 156, 441, 442, 443, 444, 444(45)
Philips, L., 550
Phillips, D., 78, 377, 379, 379(2), 381(2), 383(2), 384, 392(2), 399(2)
Phipps, D. J., 325
Pianetti, P., 269
Picot, D., 68
Pidoux, M., 155, 157(10), 225, 291, 292(5)
Piechaczyk, M., 556
Pierceal, W. E., 417, 418, 418(2)
Pierlot, C., 3, 6, 7(18), 8(18), 10(18), 11(18, 19), 17, 131, 135, 135(20), 137, 138(20), 139(20), 197, 198, 199, 224, 226, 291
Piette, J., 14, 119, 121, 129, 129(20), 131, 157, 160, 160(37), 273
Ping, C., 234
Pinkerton, K., 552
Pintar, T. J., 89, 90(27), 96(27)
Piomelli, S., 137, 482, 483
Piret, B., 121, 131, 273
Pitts, W. J., 269
Pleogh, H., 278
Plettenberg, H., 301, 308
Plewig, G., 130
Plopper, C. G., 552
Plumb, M., 554
Podda, M., 466, 536
Podmore, I. D., 402, 403(18)
Podolski, D., 112
Poh-Fitzpatrick, M. B., 482, 485
Polefka, T. G., 283, 537

Polidoros, A. N., 115
Pollet, D., 158, 159(47)
Polter, V. I., 362
Polverelli, M., 153
Ponta, H., 271
Ponten, J. A., 365
Popovic, E., 457
Popovich, J., Jr., 552
Poppelmann, B., 319
Porter, G., 39, 78
Post, J. A., 603
Postaire, E., 496
Postlethwait, E. M., 505, 511, 547, 564
Poswig, A., 18, 130
Pot, D. A., 266
Potten, C. S., 157
Pottier, R. H., 381
Pouget, J.-P., 14, 15(40), 143, 151, 153
Poulsen, T. D., 49
Poupon, M. F., 165
Pourzand, C., 159, 295, 329, 331
Pouyet-Fesard, P., 483, 484(17)
Prabhu, V. S., 393
Prahalad, A. K., 144, 145
Prakash, I., 85
Prasad, L., 35
Prat, F., 48
Prefontaine, K. E., 555
Prenzel, K., 493
Prescott, A. L., 63
Prescott, S. M., 582
Pribnow, D., 258
Price, S., 155
Prieto, M. J. E., 107, 108(16)
Pross, D. C., 381
Proud, D., 552
Pryor, W. A., 194, 196, 505, 510(2), 512, 546,
    547, 564, 570, 571, 572, 572(6–11),
    575(12), 577(12), 582
Pulver, S. C., 169(18), 170, 171(18), 173(18),
    176(18), 184(18)
Punchard, N. A., 282
Purdue, P. E., 112
Puryear, J., 528
Pyle, A. M., 166, 173(12)

Q

Quail, P. H., 115

R

Raab, O., 376
Rabenstein, D. L., 538
Rabson, A., 16, 197, 198(12)
Radi, R., 431(41), 434
Radicella, J. P., 165
Radler-Pohl, A., 141, 258, 261, 268(14)
Rahmsdorf, H. J., 120, 141, 257, 258, 259, 261,
    261(9), 263(9), 268(9, 14), 269(9), 271,
    274
Raizenne, M. E., 552
Rajeshwar, K., 29
Rajini, P., 553
Rajski, S. R., 165, 168, 171(17)
Ralph, P., 119
Ramachandran, K. L., 177
Ramsden, J. M., 157
Randall, R. J., 89, 137
Rangel, P., 113, 114(52, 53), 118(52)
Rao, G., 401, 411(6)
Rao, M. V., 29
Rao, P. K., 607
Raoul, S., 14, 15(39), 144, 151, 152, 361,
    427, 436
Rapp, U., 141, 258
Raskin, I., 531, 535, 535(44)
Rasmussen, J. E., 431(48), 434
Ravanat, J.-L., 14, 15, 15(39, 40), 111, 143,
    144, 150, 150(6), 151, 151(5), 152, 153,
    361, 390, 411, 412, 412(56), 427, 436
Rawal, B., 212
Ray, A., 555
Razum, N., 381
Reardon, J. T., 158
Reddi, E., 198
Redemann, N., 271
Redmond, R. W., 43
Reed, G. A., 68
Reed, J. C., 329, 332
Reed, L. J., 202
Reed, M. W., 137, 141(32)
Reelfs, O., 159, 291, 295, 331
Reers, M., 338, 340(26)
Rees, J. L., 367, 371(5), 372(5), 376(5)
Régnier, C. H., 120
Rehemtulla, A., 319
Rehman, A., 401, 403, 404, 405(31), 407(31),
    408(31), 409(31), 410(31), 416, 416
    (25, 26)

Reid, T. M., 161
Reiners, J., Jr., 269
Reiss, J., 197
Remboutsik, E., 266
Remy, R., 553
Rensing-Ehl, A., 323
Rentier, B., 121, 131
Renz, H., 305
Requena, J. R., 429, 431(18)
Reszka, K., 30, 36(17), 418
Retèl, J., 409
Reupold-Popp, P., 525
Revzin, A., 288
Reznik, A. Z., 429, 430(22)
Rhijnsburger, E., 408
Ribaya-Mercado, J. D., 494
Ribeiro, D. T., 3, 14, 157, 160, 161(50)
Ricchelli, F., 111
Rice, E., 29, 36(11)
Rice-Evans, C., 78, 403, 416(25, 26), 494, 603
Richard, M.-J., 153
Richards, J. H., 546, 571
Richmond, R., 408
Richter, A. M., 381
Richter, C., 371, 466
Rickard, R. C., 152
Rider, J. R., 212
Rieber, P., 120
Riele, H. T., 158
Riendeau, D., 68
Rigaudy, J., 10, 50
Rihter, B. D., 348
Rijken, P. J., 603
Rippke, F., 466, 467(14), 475(14), 476(14), 477(14)
Rittenhouse-Diakun, K., 319
Rivett, A. J., 563
Rizzo, W. B., 112
Robberson, D. L., 16
Roberg, K., 341
Robert, C., 160(64), 162, 164(64), 362
Roberts, J. E., 482
Roberts, P., 208
Roberts, R. J., 168, 171(17)
Robinson, A., 164
Robinson, N., 446
Robinson, R., 16
Robson, J., 446
Rock, K. L., 278

Rodgers, M. A. J., 39, 50, 51(12), 61, 63, 86, 388, 389, 390(41), 391(41), 392, 392(41), 394, 395(41)
Rodriguez-Tome, P., 164
Roe, J.-H., 115
Roeder, R. G., 124
Roelandts, R., 490, 491(4)
Roelofsen, B., 604, 609(4)
Rogers, M. A. J., 227
Rogowska, J., 603
Roguet, R., 453, 457
Roldan-Arjona, T., 437
Romslo, I., 488
Ron, D., 287
Rong, X., 392
Rönnstrand, L., 268
Rorsman, C., 268
Ros, A. M., 491
Rose, S. D., 185
Rosebrough, N. J., 89, 137
Rosemann, D., 526
Rosen, G. M., 393
Rosen, J. E., 144, 145
Rosenfeld, M. E., 431(54), 434
Rosenkrantz, A. R., 465
Rosenstein, B. S., 111, 145, 336, 360
Rosenthal, I., 382, 392
Rosette, C., 111, 119, 258, 555
Rosner, B., 482
Ross, A. B., 25, 38, 39(2), 44, 45(2), 47(24), 132, 225
Ross, K. S., 410
Rossato, P., 196
Rossbroich, G., 43
Rossi, B., 274
Rossi, E., 398
Rossier, G., 159, 329, 331
Roth, G. J., 67
Roth, W. K., 207
Rothe, M., 120
Rothstein, L., 278
Rothwarf, D. M., 274
Rotig, A., 372
Rouault, T. A., 295
Rougee, M., 18, 54, 234, 513
Rougier, A., 448, 453, 457
Rounsley, S. D., 531
Rouse, J., 270
Röver, S., 230
Rowedder, W. K., 32

Roza, L., 144, 145(1), 156, 157, 158(28), 361, 362, 363, 442, 443, 444
Rozen, R., 405
Rozenmuller, E., 365
Ruan, K. H., 68
Rubsam, L. Z., 407
Rucker, N., 354, 356, 357
Rüdiger, H., 438
Rudolph, J. A., 365
Ruf, H. H., 68
Rünger, T. M., 156, 158
Rupec, R. A., 274
Rupee, R. A., 585
Russell, K. E., 237
Russell, R. M., 494
Russo, A., 85
Rustin, P., 372
Ruston, M., 381
Rutherford, A. W., 80
Ruzicka, T., 17(61), 18, 130, 137, 141, 141(33), 286, 296, 298, 300(2), 301, 303, 306(4), 308, 341, 367, 372(11), 373(11)
Ryter, S. W., 111, 293, 355, 357, 465
Rywkin, S., 197, 198(11), 400

S

Saarma, M., 525, 531(19), 532(19)
Sabbagh, F., 431(55), 434
Sacchi, N., 298, 556
Sachsenmaier, C., 141, 258, 261, 268(14)
Saetre, R., 538
Sagai, M., 552, 553(18)
Sagara, Y., 465
Sage, E., 158, 162
Saghafi, T. A., 45
Saha-Möller, C. R., 3, 427, 437
Saito, E., 14
Saito, I., 5, 7(16), 166, 168(21), 169(21), 170, 170(5), 183, 216, 217(4), 425, 426
Sakahira, H., 332
Sakai, H., 29, 36(13)
Sakaki, H., 29, 36(10)
Sakamoto, O., 219, 616
Sakanaka, T., 107, 108, 108(17)
Sakata, I., 384
Salet, C., 111
Salgo, M. G., 547, 570, 582
Salinas, J., 531, 532(41), 533(41), 534(41)

Salmon, M., 552, 553, 553(19), 554(19), 555, 555(20), 556(19), 557(19)
Salo, D. C., 563
Salokhiddinov, K. I., 223
Saltiel, A. R., 269
Saltzman, B. E., 507, 514
Salvatori, P. A., 69
Salvioli, S., 327
Sambrook, J., 180, 183(34), 221, 534
Samet, J. M., 571
Sampson, J., 494
Sampson-Johannes, A., 119
Samuelsson, B., 68, 73
Samuni, A., 428
Sancar, A., 158, 185
Sandberg, S., 488
Sandermann, H., 521, 522, 522(8), 523(15), 524, 524(15), 525, 525(12, 15), 526, 526(17), 527, 527(8, 10, 21), 528, 528(28, 29), 529, 529(28, 29), 530, 530(28, 29, 36)
Sandermann, H., Jr., 520
Sandy, M. S., 467, 468(19), 477(19)
Sanejouand, Y.-H., 121, 122(22), 127(22)
Sanghera, J. S., 142, 356
Sano, S., 341, 422
Sano, Y., 112, 222
Santanam, N., 431(55), 434
Santrock, J., 578, 579(13), 581, 582
Santus, R., 129, 130, 208, 227, 479, 480(3)
Saparbaev, M., 438
Sappey, C., 121, 131, 273
Sár, P. C., 85
Saran, M., 223, 395
Sarasin, A., 14, 153, 157, 158, 160(64), 161, 162, 164(64, 68), 362, 365
Sato, H., 208
Satoh, T., 262
Satriano, J. A., 120
Sauermann, G., 465, 475, 476(25), 477, 478(25)
Sauvaigo, S., 151, 153
Sawaki, Y., 217, 218(5)
Sawyer, D. T., 50
Sayre, L. M., 431(46, 49), 434, 563
Scaiano, J. C., 39
Scandalios, J. G., 115
Scannell, C., 552
Schaap, A. P., 50, 51(12, 13), 91
Schaefer, H., 445
Schaffner, W., 586

Scharffetter, K., 130
Scharffetter-Kochanek, K., 17, 18, 18(57–59), 111, 129, 130, 143, 154, 155(4), 225, 291, 294, 373, 493
Schauen, M., 130
Schaur, R. J., 562, 563(4), 564(4), 565(4)
Schausen, M., 129
Scheinman, R. I., 555
Schell, J., 531
Schellhorn, H. E., 362
Scherba, G., 212
Scherle, P. A., 269
Schietzel, M., 495, 497, 502(8)
Schieven, G. L., 111
Schieweck, K., 383
Schilderman, P. A. E. L., 408
Schlagnhaufer, C. D., 521, 527, 527(9), 531
Schlepegrell, R., 437
Schlessinger, J., 271
Schliess, F., 131, 141(21), 358
Schlondorff, D., 120
Schluter, G., 144
Schmaldienst, S., 465
Schmid, C., 494
Schmidt, H., 121, 129(21)
Schmidt, K. N., 119
Schmidt, R., 24, 37, 43, 45, 48(23), 66, 387
Schmitt, H., 17, 19(60), 130, 143(9), 154, 212, 280, 298, 464
Schmitz, A., 13
Schneider, J. E., 155
Schneider, R., 7
Schnurpfeil, G., 377, 379(6)
Schonberger, A., 427
Schönrock, U., 467
Schöpf, E., 296, 297, 298, 299(9), 300(2, 8, 9), 306, 371
Schothorst, A., 130, 294
Schraudner, M., 522, 524, 525(12), 526(17), 527
Schreck, R., 120, 553, 585, 586
Schreiber, E., 586
Schreiber, S. L., 278
Schreier, P. H., 528
Schreier, S., 108
Schrével, J., 208
Schroeder, J. L., 110
Schroeder, W. A., 112
Schroeter, D., 113
Schubert, D., 465

Schubert, R., 528
Schuitmaker, H. J., 393
Schulmann, W., 220
Schulte-Herbrüggen, T., 227
Schultz, G., 269
Schultz, P. G., 185
Schulz, W. A., 150, 155, 367, 402, 425
Schulze-Osthoff, K., 585
Schulze-Specking, A., 298, 299(11)
Schuster, G. B., 166, 169(19), 170, 425
Schwartz, S. E., 505
Schwarz, A., 17, 18, 18(59), 111, 130, 143, 154, 155(4), 291
Schwarz, T., 319
Sckiguchi, S., 197(14), 198
Scott, R. Q., 69, 70(20)
Scriver, C. R., 405
Scurlock, R. D., 40, 41, 45, 47(10), 227, 234
Sealy, R. C., 418
Seaman, C., 482
Seeberg, E., 437
Seeburg, P. H., 419
Seeds, M. C., 336
Seidman, J. G., 586
Seifen, 466, 475(15)
Seifried, E., 207
Seis, H., 388
Séité, S., 464
Sekiguchi, M., 111
Self, A. J., 262
Selke, E., 32
Selman, S. H., 398, 399
Seltzer, J., 552
Sempf, J. M., 569, 570(33)
Senden, H. C. M., 365
Separovic, D., 351
Sera, N., 155
Seragini, D. K., 362
Seraglia, R., 196
Seret, A., 123
Setlow, R. B., 331, 365
Sevanian, A., 76, 188, 191, 196
Shacter, E., 428, 430, 431, 431(35), 432, 434(31, 34), 436(27, 31)
Shaltiel, S., 430, 431(29, 30)
Shan, Z., 120
Shapiro, P. E., 418
Shapova, M., 377, 379(6)
Sharma, H. W., 418
Sharma, Y. K., 527, 531, 532(27), 535(44)

Sharman, W. M., 376
Sharp, P. A., 595
Sharpatyi, V. A., 227
Sharrocks, A. D., 131, 270
Shen, H. R., 111, 429
Sherr, C. J., 352
Sherry, B., 554
Sheu, C., 149, 150
Shevchenko, A., 120
Shewach, D. S., 407
Shibata, A., 107, 108(17)
Shibata, D., 369
Shibutani, S., 161, 362
Shigenaga, M. K., 145, 151(20), 158, 401, 402, 431(53), 434, 466
Shigeno, E. T., 431(53), 434
Shimizu, O., 42
Shimizu, Y., 69
Shimojo, N., 552, 553(18)
Shimokawa, T., 73
Shindo, 466
Shiozaki, H., 237
Shipman, J. M., 158
Shiraishi, M., 569
Shirasaki, H., 553, 555, 557
Shoffner, J. M., 367
Shook, F. C., 71, 86
Shopova, M., 392, 398(73)
Shore, G. C., 350
Shuin, T., 29, 36(10, 12)
Shuldiner, M., 120
Shuler, R. L., 553
Shulyupina, N. V., 362
Siebenlist, U., 119, 122, 125
Sieber, F., 86, 198
Siems, W. G., 563
Sies, H., 3, 4, 7, 13, 13(24), 14, 14(6, 8, 24), 16, 17, 17(61), 18, 18(57–59), 19(6, 60, 62), 36, 67, 69, 70(15, 16), 71, 75, 76, 76(37), 101, 105(1), 111, 112, 121, 129(21), 130, 131, 135, 135(19, 20), 137, 137(29), 138(19, 20), 139(20), 141(21, 33), 143, 143(9), 150, 154, 155, 155(4), 157, 198, 199(16), 218, 222, 223, 224, 225, 225(14), 226, 230, 273, 280, 283, 284, 291, 294, 298, 303, 304, 306(4), 341, 358, 367, 372(11), 373(11), 402, 425, 464, 465, 466(8), 478, 494, 495, 497, 498(14), 499, 499(11), 500(14), 502(8), 604, 609(6)
Signorini, N., 153

Silvester, J. A., 111, 430
Sima, P. D., 113, 505, 507(9), 510(9), 511, 512, 513, 515, 515(4, 5, 12), 516(12), 517(12), 518, 519(13), 547
Simon, E. H., 212
Simon, J. A., 365, 418
Simon, J. D., 154
Simons, J. W. I. M., 157, 158(34), 162(34), 361, 363(3)
Simpson, L., 554, 571
Simpson, M. C., 212
Singer, B., 401
Singh, A., 387, 391(36), 395(36)
Singh, H., 595
Singh, S., 274, 586
Siok, C. J., 67
Sistrom, W. R., 480, 481
Sitlani, A., 166, 173(12), 176
Skalkos, D., 398, 399
Slade, R., 546
Slaper, H., 364
Slight, J., 553
Sluis, K. B., 431(36), 432
Smiley, S. T., 338, 340(26)
Smith, C. D., 434
Smith, C. V., 428
Smith, I. C. P., 108
Smith, J. A., 586
Smith, J. C., 499
Smith, J. R., 593, 594(23)
Smith, L. L., 87, 89, 91(18, 28), 92(28), 188, 189, 190, 190(6), 196, 390
Smith, M. A., 431(46, 49), 434
Smith, T. W., 338, 340(26)
Smith, W. L., 68, 69(4), 73
Snell, M. E., 343
Snowden, P. T., 86
Soboll, S., 226
Soennichsen, N., 297
Sohal, R. S., 431(47), 432, 434
Sohmi, K., 107, 108(18)
Sokoloski, E. A., 563
Soliman, N., 283
Somerville, C. R., 531, 532(40)
Song, H. Y., 120
Sontag, Y., 365
Soong, N. W., 369
Sorensen, R., 377, 385(9)
Sosnovsky, G., 85
Sotelo, J. L., 508

Soussi, T., 162, 164(68)
Sowers, L. C., 416
Spanakos, G., 467, 478(20)
Spangler, C. J., 499
Spear, N., 409
Speizer, F. E., 551
Spence, H. M., 63
Spencer, B., 227
Spencer, J. P. E., 402, 415
Spengler, J. D., 551
Spetea, C., 78, 79(12), 80(12, 13)
Spikes, J. D., 50, 85, 111, 216, 382, 383, 385(19), 429
Spinelli, S., 14, 15(39), 144, 361, 436
Spöttl, R., 395
Squadrito, G. L., 547, 564, 570, 571, 572, 572(6–11), 575(12), 577(12), 582
Srinivasan, V. S., 112
Stäb, F., 465, 466, 467, 467(14), 475, 475(14), 476(14, 25), 477, 477(14), 478(25)
Stabinsky, Z., 176, 180(31)
Stables, G. I., 377
Stadler, R. H., 152, 411, 412(56)
Stadtman, E. R., 428, 429, 429(2, 5), 430, 431(17, 30, 35, 45), 432, 434, 436(5), 563, 564, 564(17), 565(5, 10, 17, 22), 566(5, 10), 567(5, 10), 570
Staehelin, T., 434
Stafforini, D. M., 582
Stahl, S. J., 431(42), 434
Stahl, W., 112, 137, 141(33), 230, 494, 495, 497, 499, 499(11), 500(14), 502(8), 604, 609(6)
Stanbro, W. D., 66
Standaert, R. F., 278
Standen, M. C., 227
Stanier, R. Y., 480, 481
Stankowski, L. F., Jr., 439
Stark, G. R., 349, 352(14)
Stary, A., 153, 160(64), 162, 164(64, 68), 362
Steck, A., 499
Steck, T. L., 88
Steele, G. D., Jr., 338, 340(26)
Steenken, S., 144, 150, 155, 166, 367, 402, 425, 427
Stege, H., 17(61), 18, 130, 286, 297, 301, 303, 306(4), 308, 328, 330(24), 341
Steigel, A., 230
Stein, R., 278
Steinbeck, M. J., 3
Steinkraus, V., 158, 159(47)

Stemp, E. D. A., 169(18), 170, 171(18), 173, 173(18), 176, 176(18), 184(18)
Stenhorst, F., 230
Stenn, S., 474
Stepanov, A. M., 508
Stephans, J. C., 185
Sterenborg, H. J. C. M., 364
Stern, R. S., 492
Sternberg, E., 381
Stetina, R., 153
Stevens, W. H., 66
Stobbe, C. C., 386
Stocker, R., 36, 218, 224, 226, 428, 431(12)
Stockton, G. W., 108
Stohs, S. J., 408
Stokes, R. P., 448
Stoltz, J.-F., 208
Stolze, K., 110, 111(6)
Stradley, D. A., 269
Straface, E., 614(6), 618
Straight, R. C., 50, 216
Stratton, S. P., 111
Stricklin, G. P., 17, 18(58), 143
Strong, M., 434
Struhl, K., 586
Stryker, M. H., 208
Stuhlmeier, K. M., 465
Stulborg, M., 552
Styring, S., 78
Su, B., 111, 259
Su, M. S., 131, 270
Sugimoto, H., 50
Sugiyama, H., 166, 170, 170(5), 183, 426
Suhr, O., 570
Sun, J., 552, 553(19), 554(19), 556(19), 557(19)
Sundquist, A. R., 135, 223, 226, 494, 495
Sundquist, F., 14
Suprunchuk, T., 234, 237
Susin, S. A., 321, 329(20), 332
Sutherland, B. M., 361
Sutherland, J. C., 361
Suwa, K., 91
Suzuki, A., 101, 102(3), 103(3–5), 104(3–5), 109(5), 112
Suzuki, Y. J., 429, 430(22)
Swipes, W., 197
Switala, J., 115
Swize, M. A., 527
Symons, M. C. R., 78
Synder, A., 381

Szeda, P. A., 563, 564(13), 565(13), 566(13), 567(13)
Szejda, P., 336
Szmigielski, S., 197
Szpakowski, M., 197
Szweda, L. I., 562, 563, 564, 564(9, 13), 565(5, 9, 10, 13, 22), 566(5, 9, 10, 13), 567(5, 9, 10, 13), 568(9)
Szweda, P. A., 562

T

Tabata, H., 217, 218(5)
Tada-Oikawa, S., 156, 159(24), 331, 336
Tajiri, H., 347
Tajiri, T., 111
Takagi, M., 197
Takahashi, M., 36, 107, 108(15)
Takahashi, N., 111
Takahura, F., 208
Takamura, Y., 237
Takayama, M., 166, 425
Takeda, H., 113
Takemura, T., 384
Takeshita, M., 161, 362
Takeuchi, T., 333, 408
Takigawa, M., 493
Talwar, H. S., 141
Tam, S. P., 283
Tamás, K., 110
Tamura, K., 113
Tan, O., 291
Tanaka, M., 570
Tanaka, T., 199
Tanew-Illitschew, A., 130, 156, 363
Tang, J. Y., 176, 180(31)
Tanielian, C., 24, 45, 111
Tanigaki, T., 101
Tao, J., 142, 356
Tappel, A. L., 429
Tarrant, A. W. S., 245
Tatano, T., 615
Tateishi, M., 191, 193, 194(10, 18)
Tatsuzawa, H., 36, 112, 199, 216, 217, 218(5), 222
Tauber, A. I., 59, 64(4)
Tauras, J. M., 141
Tavakkol, A., 283
Taylor, D. L., 603

Taylor, G. E., 521
Taylor, O. C., 521, 523(5)
Taylor, V. L., 227
Taylor, W. R., 349, 352(14)
Tedesco, A. C., 28
Teeve, V. E., 191, 195(12)
Telfer, A., 78
Tenev, T., 258, 259(8), 261(8), 263(8)
Teng, J. I., 87, 91(18)
Téoule, R., 155
Terao, J., 101, 102(3), 103(3–5), 104(3–5), 109(5), 112
Thayer, A. L., 50, 51(13)
Thepen, T., 296, 300, 300(2)
Thiele, J. J., 386, 397(33), 398(33), 536, 537, 538, 546
Thomas, C. E., 604, 609(5)
Thomas, D. P., 310, 311, 312(7), 314, 314(6), 315(7), 319(6)
Thomas, J. A., 429, 431(19, 43), 434
Thomas, J. P., 86, 389
Thomas, M., 234, 336
Thompson, C. B., 331
Thompson, K., 331, 365
Thornby, J. I., 191
Thornfeldt, C. R., 536
Thorpe, S. R., 429, 431(18)
Timmins, G. S., 111, 430
Timms, A., 110
Tindall, K. R., 439
Tindall, M., 367, 371(5), 372(5), 376(5)
Tingey, D. T., 521, 523(5)
Tobiasch, E., 257
Tobler, A. R., 323
Todd, C., 16
Todd, F. R. S., 212
Tokiwa, H., 155
Tokumura, A., 101, 102(3), 103(3–5), 104(3–5), 109(5), 112
Tokura, Y., 493
Toledano, M. B., 553
Toleutaev, B. N., 42, 60, 66(9), 388, 389, 513
Tomita, K., 30, 36(20)
Tomita-Yamagushi, M., 125
Tomizawa, H., 193, 194(18)
Tompkins, S. M., 119
Torii, Y., 569
Tortorella, D., 278
Tosteson, T., 551

Totpal, K., 593, 594(23)
Toullec, D., 269
Towbin, H., 434
Towers, G. M. N., 381
Towne, B., 410
Toyokuni, S., 563, 564(17), 565(17)
Tozer, T. N., 85
Traber, M. G., 466, 536, 537, 546
Traenckner, E. B., 119, 274
Traldi, P., 196
Tregubov, B. A., 508
Trelease, R. N., 115
Trent III, J. C., 154
Treshow, M., 520, 521(2)
Tritschler, H., 416
Tromp, J., 29, 35(3), 36(3)
Tronnier, H., 494, 495, 496, 496(16), 497, 502(8)
Tröscher, G., 45
Trost, M., 524, 526(17)
Trudel, L. J., 152
Trudle, L. J., 411, 412(56)
Trull, F. R., 45
Truscott, T. G., 78, 123, 133, 227, 389
Trzaskos, J. M., 269
Tsacmacidis, N., 131, 141(21), 358
Tsai, L., 428, 429(5), 436(5), 563, 564(9, 13), 565(5, 9, 13), 566(5, 9, 13), 567(5, 9, 13), 568(9)
Tsai, M., 434
Tsang, K., 536
Tsuchida, A., 166
Tsuchiya, J., 107, 108(15)
Tsuji, T., 301, 308
Tsuji, Y., 69
Tsurufuji, M., 554
Tsurufuji, S., 554
Tuche, Z., 153
Tulloch, A. P., 108
Tuomainen, J., 526, 527(21)
Turcotte, J., 162, 163(65), 362
Turesky, R. J., 152, 411, 412, 412(56)
Turmel, C., 385
Turner, R., 367, 371(5), 372(5), 376(5)
Turro, C., 166, 173(13)
Turro, N. J., 38, 39(2), 45(2), 166, 173(13)
Tutois, S., 483, 484(17)
Tyrell, R. M., 18, 86, 110, 111, 129, 130, 145, 154, 155, 155(7), 156, 156(5), 157(10, 26), 159, 161, 225, 290, 291, 292(5, 7), 293, 294, 295, 329, 331, 361, 363, 408, 427, 444, 465

U

U. S. Environmental Protection Agency, 546
Ubezio, P., 336
Uchida, K., 429, 431(17), 563, 564, 564(17), 565(5, 17, 22), 566(5), 567(5), 569, 570
Uchino, M., 570
Uemura, I., 108
Ueno, Y., 547
Ullrich, A., 259, 261(9), 263(9), 268, 268(9), 269, 269(9, 27), 271
Ullrich, O., 563
Ullrich, S. E., 460
Ullrich, V., 75, 76, 76(37)
Ultman, J. S., 564
Unsworth, M., 521, 523(6)
Untiedt, S., 466, 467(14), 475(14), 476(14), 477(14)
Uppu, R. M., 564, 571, 572, 572(6), 575(12), 577(12)
Urano, S., 107, 108, 108(17)
Urbach, F., 130
Ursini, F., 69, 196
Urumov, I. J., 397
Utsumi, K., 341
Utsumi, T., 341

V

Vadenas, E., 76
Vahala, J., 531
Vahnweg, G., 528
Valduga, G., 198
Valenti, P. C., 50, 51(13)
Valla, A., 227
Vallat, V. P., 302
Van Camp, W., 522, 525, 525(12), 527
van den Berg, J. J., 428, 431(12), 604, 609(4)
van den Bergh, H., 393
Vandenberghe, A., 283
van der Leun, J. C., 156, 158(31), 162, 331, 363, 364, 365, 366
van der Meer, J. B., 364
Vanderoef, R., 16, 197, 198(12)

van der Ouderaa, F. J., 67, 73
van der Schans, G. P., 156, 158(28), 363
van der Vliet, A., 434, 510, 511, 512(17)
van der Voorn, L., 262
van de Ven, J., 162
van de Vorst, A., 14, 78, 82, 121, 123, 129(20), 157, 160(37), 393, 398
van de Water, B., 283
van Dorp, D. A., 67
van Dranen, H. J., 162
van Dyk, D. E., 269
Vane, J. R., 73
van Hasselt, B. A., 450
van Heerden, J. A., 208
van Hoogevest, P., 383
Van Houten, B., 367
van Kranen, H. J., 365
van Kreijl, C. F., 162, 365
van Lier, J. E., 87, 150, 188, 191, 376, 382, 385(19), 387, 390, 390(34), 392(61), 395, 396, 397, 398
van Loon, A. P., 553
Van Montagu, M., 525, 527
van Steeg, H., 365
van Tilburg, M., 156, 158(31), 363
van Vloten, W. A., 156, 158(31), 363
van Weelden, H., 364, 480
van Zeeland, A. A., 157, 158(34), 162(34), 361, 363(3)
Vass, I., 77, 78, 79(12), 80, 80(12, 13), 82, 83, 84, 84(23), 110
Vassalli, P., 596
Vayssiere, J. L., 337
Velluz, L., 5
Velsor, L. W., 547, 564
Verdier, M. P., 464
Veregin, R. P., 29
Verma, I. M., 119
Vermeersch, G., 50, 51(17), 54(17), 227
Verneker, V. R. P., 29
Vichroy, T., 197
Vickers, P. J., 68
Vidigal, C. C. C., 70
Vigny, P., 144
Vile, G. F., 111, 129, 130, 156
Vile, G. T., 363
Villeneuve, L., 385
Vinogradova, O., 563, 564(13), 565(13), 566(13), 567(13)
Viola, A., 393, 395(77), 396, 396(77), 397(77)

Virgin, I., 78
Viviani, B., 260
Vogelsang, K., 286
Vogt, M., 585
Vogt, W., 428
von Laar, J., 494, 495, 497, 502(8)
von Röden, T., 271
von Sonntag, C., 401
von Tiedemann, A., 522, 535(11)
Voorhees, J. J., 141
Vowels, B. R., 319, 320(17)
Voyksner, R. B., 196
Vyas, G. N., 207, 212

W

Wachter, L., 177
Wachter, T., 158
Wagner, E. F., 271
Wagner, J. F., 36
Wagner, J. R., 151, 188, 218, 224, 390, 392(61)
Wagner, S. J., 197(14), 198
Wagnières, G., 393
Wagoner, M., 398, 399
Wahlländer, A., 226
Wahn, U., 305
Wainwright, M., 197, 385
Wakamatsu, C., 208
Wakayma, Y., 197
Walaschek, M., 111
Waleh, A., 173
Walker, P., 93
Wallace, D. C., 367, 370, 372(2)
Wallach, D. F. H., 88
Wallasch, C., 268
Wallenborn, E.-U., 185
Walsh, J. W. T., 247
Walten-Berger, J., 268
Walter, E., 7
Walter, P. B., 145, 151(20), 402
Walter, S., 298, 299(9), 300(9)
Wamer, W. G., 29, 36(14), 145
Wan, J. K. S., 237
Wan, Y., 141
Wang, D., 159, 233
Wang, K., 194, 196
Wang, L. H., 68
Wang, S. Y., 112, 294, 516
Wang, T. P., 112, 198

Wang, X., 332
Wang, Y., 111, 145, 226, 336
Wang, Z., 141, 283
Wanxue, Z., 234
Ware, J. H., 551
Warren, L. J., 490
Washko, P. W., 226
Wasserman, B. H., 5, 9
Watabe, T., 193, 194, 194(18)
Watanabe, J., 42
Watanabe, K., 554
Watkin, R. D., 295
Watson, J. J., 152
Watson, S. P., 269
Watson, Y., 208(13), 212
Wattre, P., 17, 135, 199
Watts, S. D., 396
Wayne, R. P., 389
Weatherall, J., 450
Webb, A. C., 550
Webb, R. B., 290
Weber, C., 466
Weber, M., 207
Weber, S. U., 536
Weeda, B., 123, 133
Wefers, H., 14
Wegener, A., 527, 528(28, 29), 529(28, 29), 530(28, 29)
Wegener-Strake, A., 529, 530(36)
Wehrly, K., 22, 102
Wei, H., 111, 145, 336
Wei, R. R., 29, 36(14), 145
Wei, Y.-H., 367, 372(3)
Weidenbach Gerbase, M., 552
Weil, R., 274
Weinfeld, M., 151
Weinmann, G. G., 552
Weinstein, G. D., 492
Weis, M., 467, 468(19), 477(19)
Weisel, C. P., 552
Weishaupt, K. R., 121, 387
Weiss, E. H., 553
Weiss, F. U., 268
Weiss, S. J., 59, 64(6), 111
Weitzner, G., 405
Weldon, D., 43, 49
Wellburn, A. R., 524, 525, 532(16)
Wellburn, F. A. M., 525
Welman, W. P., 38, 39(2), 45(2)
Wenck, H., 467

Wendel, A., 226
Wenk, J., 17, 18(59), 111, 129, 130, 143, 154, 155(4), 291
Werfel, T., 17(61), 18, 130, 301, 303, 305, 306(4), 328, 330(24), 341
Wermer, M., 569, 570(32)
Werner, T., 527, 528(28), 529(28), 530(28)
Wesselborg, S., 585
Wessels, J. M., 61
West, C. M. L., 379
West, M. S., 155
Wester, P. W., 162, 365
Westerhos, W., 450
Westerman, A., 365
Westmijze, E. J., 409
Westric, N. J., 112
Westwick, J. K., 262
Wever, R., 59, 64(6), 111
Whitacre, C. M., 350
White, W. H., 505
Whitehurst, C., 379
Whiteman, M., 402, 415(16)
Whiteside, S., 120, 258, 274
Whiteside, S. T., 274
Whitmarsh, A. J., 131, 270
Whitmore, E., 492
Whitten, D. G., 399
Widmer, U., 554
Wiebe, M. E., 208
Wiebusch, M., 495, 496(16)
Wiese, C., 522, 525(12)
Wijnhoven, S., 365
Wilcox, J. N., 431(55), 434
Wiles, D. M., 234, 237
Wilk, S., 119
Wilkenson, F., 391
Wilkes, G., 319
Wilkinson, F., 25, 38, 39, 39(2), 44, 45(2), 47(24), 52, 132, 225, 237
Will, O., 153, 444
Willekens, H., 525, 527
Williams, C. M., 309
Williams, E. H., 532, 533(46)
Williams, G. M., 144, 145
Williams, J. A., 430, 431, 431(35), 432, 434(31), 436(27, 31)
Williams, S. D., 183
Willis, J. I., 212
Wilson, A. D., 251
Wilson, B. B., 7

Wilson, C. B., 120
Wilson, G. L., 373
Wilson, M., 197, 198(9)
Wilson, T., 16
Wingerath, T., 495, 499, 499(11), 500(14), 501
Wingfield, P. T., 431(42), 434
Winter, M. A., 212
Winterbourn, C. C., 431(36), 432
Winyard, P. G., 405
Wirtz, K. W. A., 603
Wise, R. J., 483
Wiseman, H., 401
Witschi, H., 553
Witt, E., 466
Witztum, J. L., 431(54), 434
Wlaschek, M., 17, 18, 18(57–59), 129, 130, 143, 154, 155(4), 225, 291, 294, 493
Wlodawer, P., 68
Wodick, R., 496
Wogan, G. N., 160
Wöhrle, D., 377, 379(6)
Wolber, R., 465
Wolfe, J. T., 302
Wolfe, N. L., 50
Wolff, C., 24, 45
Wong, G., 142
Wong, J., 402
Wong, P. T., 152
Wong, S., 354, 355, 356, 357
Wong, V., 552
Wood, M. L., 161
Wood, S. G., 402
Woodall, A. A., 402
Woodford, T. A., 266
Woodhall, A. A., 145, 151(20)
Woodhead, A. D., 331
Woodhead, A. P., 365
Woog, J., 197
Woutersen, R. A., 480
Wright, J., 59, 64(4)
Wu, I. H., 111
Wu, R., 415, 552
Wyld, L., 137, 141(32)
Wyllie, A. H., 309, 325, 331

X

Xia, Y., 120
Xie, Z., 329

Xing, Y., 120
Xu, G., 563
Xu, Q., 132, 140(22)
Xue, L. Y., 329, 348, 350, 357

Y

Yabuki, M., 341
Yagi, K., 615
Yagi, Y., 547
Yakes, F. M., 367
Yalpani, N., 535
Yamada, M., 336
Yamada, Y., 582
Yamaguchi, K., 170
Yamaguchi, M., 30, 36(20)
Yamaizumi, Z., 427
Yamamoto, K., 166, 331, 333(4), 421, 422, 423, 423(14), 425, 426, 426(14)
Yamamoto, M., 166
Yamamoto, S., 72, 73(30), 75(30), 220
Yamamoto, Y., 29, 30, 35, 36, 36(18–20), 93, 219
Yamaoka, S., 274
Yamashita, N., 427
Yamashita, T., 570
Yamauchi, R., 547
Yamazaki, S., 191, 193, 194, 194(10, 18)
Yan, L. J., 431(47), 432, 434, 466
Yan, T., 586
Yang, C.-Y., 428
Yang, II.-X., 428
Yang, J., 332
Yang, J.-H., 367, 372(3)
Yang, L., 569, 570(32)
Yang, M., 403, 416(25), 428
Yang, S. H., 131, 270
Yano, K., 197
Yano, N., 115, 116(60)
Yano, S., 115, 116(60)
Ye, Y. Z., 434
Yen, B. T. S., 212
Yen, G. S. L., 212
Yen, H. C., 569, 570(32)
Yeo, H. C., 144, 145, 150(6), 151(20), 402
Yin, J., 29, 36(14)
Yin, Z.-H., 526, 527(21)
Yip, H., 431(53), 434
Yokoyama, H., 332

Yokoyama, I., 347
Yoo, E. K., 319, 320(17)
Yoritake, A., 570
Yoshida, S., 347
Yoshikawa, T., 191
Yoshikawa, Y., 35
Yoshioka, K., 259
Yoshioka, T., 29, 36(13)
Yoshioka, Y., 170
Young, A. R., 157
Young, D. B., 120
Young, P., 137
Young, P. R., 270
Young, R. H., 22, 102, 105(9), 106(9), 232
Yu, B. P., 563
Yu, K., 416
Yuasa, T., 586
Yurochko, A. D., 119

Z

Zacharide, C. O. C., 554
Zaidi, S. I., 345, 346, 353(7, 9), 391, 395(66)
Zaim, M. T., 345, 353(7, 8)
Zaklika, K. A., 50, 51(13)
Zamagishi, A., 69
Zamzami, N., 321, 329(20), 332, 337
Zandi, E., 274
Zanin, C., 337

Zastawny, T. H., 438
Zastrow, L., 475, 478(24)
Zeng, H., 385
Zepp, R. G., 50
Zhang, H. J., 586
Zhang, L. Y., 412, 416(59)
Zhang, X., 111, 145, 283, 336
Zhao, Q., 571, 581, 582
Zhao, W., 429, 431(19)
Zhou, S., 581, 582
Zhu, H., 120
Zhu, L., 434
Ziegler, A., 418
Ziereis, K., 50, 51(14)
Zigler, J. S., 429
Zimmerman, G. A., 582
Zimmerman, S., 460
Zinck, R., 261, 268(14)
Zinner, K., 70
Zinser, C., 530
Zinukou, S. U., 60, 66(9)
Zinukov, S. V., 42, 388, 389, 513
Zlabinger, G. J., 465
Zoeller, R. A., 112
Zoeteweij, J. P., 283
Zollner, H., 562, 563(4), 564(4), 565(4)
Zou, X., 547, 582
Zucker-Franklin, D., 212
Zweig, A., 238
Zwingermann, I., 408

Subject Index

A

N-Acetylcysteine, oxidative stress inhibition, 260–261

AGR, *see* α-Glucosyl rutin

ALA, *see* 5-Aminolevulinate

5-Aminolevulinate
mitogen-activated protein kinase induction in photodynamic therapy, 141–142
photodynamic therapy principles, 381–382
singlet oxygen generation, 137

Aminopyropheophorbide
colon carcinoma cell photosensitization
deuterium oxide effects, 123–124
irradiation conditions, 122–123
nuclear factor-κB activation
antioxidant effects, 128
deuterium oxide effects, 127, 129
electrophoretic mobility shift assay, 124–125, 127
irradiation conditions, 124
solution handling, 122
subcellular localization, 128

9,10-Anthracenedipropanoate, singlet oxygen trapping, 50–51

AP-2
skin end point for ultraviolet A exposure, 464
ultraviolet A-induced gene expression in keratinocytes, 280, 288, 290

Apoptosis
C11-BODIPY studies in fibroblasts, 607–608
caspase-3 role, 332
definition, 309–310
flow cytometry assay
cell requirements, 324–325
forward light scattering, 325, 329
instrumentation, 325
mitochondria transmembrane potential analysis, 326–328, 330
side light-scattering, 325, 329

HL-60 cell studies of ultraviolet A radiation
caspase-3 assay, 340–341
cell culture, 332
DNA analysis
gel electrophoresis of fragments, 333–334
ladder detection, 334–336
8-oxodeoxyguanosine determination, 333
flow cytometry assays
mitochondrial transmembrane potential, 336–340
peroxide detection, 336
irradiation conditions, 332–333
materials, 332
mechanisms of apoptosis, 341–342
irradiation of cells for studies
adherent cells, 315
buffers, 312–313
postexposure treatment
adherent cells, 315–316
suspended cells, 314–315
sources
ultraviolet, 313–314
visible light, 314
X-ray, 314
suspended cells, 314–315
temperature control, 313–314
vessels, 313
mitochondrial membrane potential role, 332
morphological features, 309
necrosis following apoptosis, 310–311
photodynamic therapy apoptosis
bcl2 role, 348–349
calcium signaling, 346–348
caspase activation
Bap31 fragments, 350–351
inhibitor studies, 349–350
cell cycle mediation, 351–353
ceramide generation, 351
early event in tumor shrinkage, 345–346

mechanisms, 346
nitric oxide role, 353
p21 role, 352, 354
p53 role, 349, 352
phospholipase C activation, 346–348
retinoblastoma protein role, 352–353
serine/threonine dephosphorylation, 348
preprogrammed cell death versus programmed cell death, assessment by translation inhibition with cycloheximide, 323–324
reactive oxygen species
induction, overview, 310, 316
identification of types in apoptosis induction
cyclosporin A inhibition of singlet oxygen, 316, 321–322, 328–329
deuterium oxide, 317–318
DNA damage induction, 320
hydroxyl radical generators, 319
sham controls, 317–318
singlet oxygen generators, 318–320
singlet oxygen quenchers, 317
superoxide generators, 319–320
recovery and handling of singlet oxygen-damaged cells
centrifugation, 311
pipetting, 312
suspending, 312
vortexing, 312
washing, 312
T cells in atopic dermatitis ultraviolet A therapy
cell isolation, 303, 305
deuterium oxide effects, 304
DNA laddering assay, 306
double-staining technique for detection, 308
immunofluorescence microscopy, 301
irradiation device and dosimetry, 304
irradiation studies, in vitro, 303–309
mitochondria transmembrane potential changes, 307
morphology, 300–301
phosphatidylserine assay, 307
singlet oxygen generator and quencher effects, 304–305
TUNEL assay, 306–308

ultraviolet A-1 radiation therapy, 300–302
ultraviolet B effects, 302
APP, see Aminopyropheophorbide
Ascorbic acid
C11-BODIPY, multiwell screening of topical effectiveness, 608
high-performance liquid chromatography analysis of ozone depletion in skin
equipment, 539
operation, 540
reagents, 539
sample preparation, 540
standards, 540
ozone-induced singlet oxygen in plants, chemiluminescence assay
ascorbic acid reaction, 512–513, 518–519
assay in leaves
carrier gas effects on emission intensity, 517–518, 520
cell wall emission, 516–517
gas volume per unit surface area effects, 519–520
instrumentation, 513–514
intensity of gas phase emission, 516
intercellular extract measurements, 515
kinetics, 519
principle, 513
spectral analysis, 517
reactants in generation, 514–515

B

Bcl2, photodynamic therapy, apoptosis role, 348–349
Bis(1-hydroxyheptyl)peroxide, synthesis, 578–579, 581
Bis(1-hydroxynonyl)peroxide, synthesis, 578–579, 581

C

C11-BODIPY
apoptosis studies in fibroblasts, 607–608

cell labeling, 606
confocal laser scanning microscopy and
    image processing, 606–607
excimer formation, 611–612
fluorescence spectra and properties, 604,
    606
multiwell screening of antioxidants
    ascorbate, 608
    numerical analysis, 610
    pitfalls, 611–612
    tocopherol, 608–609
photooxidation, 612
quenching, 611
reactive oxygen sensing, 603–604
Carbonyl group, see Protein oxidation
Carotenoids
    energy transfer in singlet oxygen quench-
        ing, 230
    erythropoietic protoporphyria, β-carotene
        treatment, 480–482, 484, 494
    high-performance liquid chromatography
        assay in skin
        chromatography conditions, 498–499
        identification of peaks, 499, 501
        quantitative analysis, 501–502
        sample preparation, 498
        standards, 499, 501
    metabolism, 495
    reflection photometry assay of skin
        β-carotene supplementation, 495
        data analysis, 496
        distribution, 497
        instrumentation, 495–496
        serum levels, 497
        validation, 497–498
    singlet oxygen scavenging rates in phos-
        pholipid membranes, 104–106,
        108–109
Caspase
    caspase-3
        apoptosis role, 332
        assay in irradiated HL-60 cells,
            340–341
    photodynamic therapy activation
        Bap31 fragments, 350–351
        inhibitor studies, 349–350
Catalase, singlet oxygen detection
    activity assays
        electrochemical assay, 116
        gel assay, 116

cell extract preparation, 116
    hanging drop system for in vitro modifi-
        cation, 117–118
    heme modification and electrophoretic
        mobility, 114, 118
    Neurospora crassa
        conidia system, 114
        heat shock, 115
        paraquat induction, 115
    purification for in vitro modification, 117
    rationale, 113
CD, see Circular dichroism
CHDDE, see 1,3-Cyclohexadiene-1,4-dietha-
    noate
Chemiluminescence, singlet oxygen
    dimol emission, 13, 71, 75–77, 222–223,
        388–389
    identification of singlet oxygen as emis-
        sion source, 66–67, 71
    monomol emission, 13, 71, 222–223,
        388–389
    near-infrared spectrometer
        assignment of 1270-nm emission to sin-
            glet oxygen, 66–67
        calibration, 63–64
        detector, 61
        electronics, 61–62
        eosinophil peroxidase assays
            kinetics, 64
            spectral analysis, 64–65
        injection port, 61
        interference filters, 60–61
        muon filter, 62
        schematic, 60
        sensitivity, 62–63
    ozone-induced singlet oxygen in plants
        ascorbic acid reaction, 512–513,
            518–519
        assay in leaves
            carrier gas effects on emission inten-
                sity, 517–518, 520
            cell wall emission, 516–517
            gas volume per unit surface area ef-
                fects, 519–520
            instrumentation, 513–514
            intensity of gas phase emission,
                516
            intercellular extract measurements,
                515
            kinetics, 519

principle, 513
  spectral analysis, 517
  reactants in generation, 514–515
photon generation, 59
plant assays, 78
prostaglandin endoperoxide H synthase,
    assay of arachidonic acid oxidation
  dimol emission, 71, 75–77
  filters, 70
  identification of singlet oxygen as emis-
    sion source, 71
  inhibitors of chemiluminescence, 74–75
  kinetics, 72–73
  oxygen requirement, 73
  peroxidase in singlet oxygen formation,
    75–76
  pH dependence, 73
  single photon-counting apparatus,
    69–70
  spectral analysis, 73–74
quencher assay
  biological quenchers, 224–226
  exogenous quenchers, 225
  generation of singlet oxygen, 222, 224
  instrumentation, 223–224
  principle, 223
time-resolved measurements, *see* Time-re-
    solved near-infrared phosphores-
    cence
Chlorpromazine, induced phototoxicity for
    sunscreen assay
  applications, 456
  cell viability assay, 45
  chlorpromazine dose–response, 455
  Episkin preparation, 454
  formulas for testing, 454–456
  incubation, 455
  interleukin-1α assay, 455
  principle, 453–454
  ultraviolet A radiation source, 454
Cholesterol
  oxidation products, 87, 188, 390–391
  ozonization
    high-performance liquid chromatog-
      raphy
      electrochemical detection, 195–196
      mass spectrometry detection, 196
    liposome cholesterol, 194–195
    products, 188–189
  singlet oxygen reporter

advantages, 87
  chemicals and reagents for assays, 88
  high-performance liquid chromatogra-
    phy with electrochemical detection
    of oxidation products
    erythrocyte membranes, 94–97
    operating conditions, 93–94
    photoxidized cells, 98–100
  iodometric assay, 90
  leukemia cell culture, 89
  lipid extraction, 90
  membrane preparation, 88–89
  photooxidation
    cells, 89–90
    membranes, 89
  thin-layer chromatography of erythro-
    cyte membrane oxidation prod-
    ucts, 91–93, 100
  ultraviolet A irradiation products
    high-performance liquid chromatog-
      raphy
      chemilumiscence detection, 192–194
      electrochemical detection, 192
    photoperoxidation reactions in cell-free
      systems, 191–192
    skin types and carcinogenicity, 190–191
CINC, *see* Cytokine-induced neutrophil
    chemoattractant
Circular dichroism, DNA assemblies with
    tethered photooxidant, 180
Coenzyme Q10, topical antioxidant screen-
    ing, 469 471, 474–475
1,3-Cyclohexadiene-1,4-diethanoate
  absorption, 51, 228
  disappearance pathways, 52–54
  nuclear magnetic resonance, 52
  peroxidation, 54–55
  physiochemical properties, 52
  synthesis of disodium salt, 52, 57–58
  trapped singlet oxygen measurement
    deuterium oxide assay, 56
    equations, 55–57
    instrumentation, 57
    ordinary water assay, 56–57
Cyclosporin A, inhibition of singlet oxygen-
    induced apoptosis, 316, 321–322,
    328–329
Cytokine-induced neutrophil chemoattrac-
    tant, ozone induction in rat lung
    dexamethasone inhibition, 560, 562

electrophoretic mobility shift assay, 557–558, 560
neutrophilia role, 554–555
Northern blot analysis, 556–557
reverse transcriptase polymerase chain reaction, 556

D

DanePy
  fluorescence measurements, 83–84
  leaf studies of singlet oxygen production, 83–85
  singlet oxygen reaction and fluorescence decay, 82
  stability, 82–83
Delayed-type hypersensitivity, *see* Skin, ultraviolet protection
Dexamethasone, inhibition of ozone-induced transcripts
  cytokine-induced neutrophil chemoattractant, 560, 562
  nuclear factor-κB, 560
DHPN, *see* N,N'-Di(2,3-dihydroxypropyl)-1,4-naphthalenedipropanamide
N,N'-Di(2,3-dihydroxypropyl)-1,4-naphthalenedipropanamide, *see* 1,4-Dimethylnaphthalene, singlet oxygen carrier derivatives
1,4-Dimethylnaphthalene, singlet oxygen carrier derivatives
  bacteria killing
    cell growth, 200
    N,N'-di(2,3-dihydroxypropyl)-1,4-naphthalenedipropanamide efficacy, 203, 208
    exposure conditions, 201
    overview, 199
    survival assays, 201–203
  cell culture exposure, 17–18
  DNA reactions
    deoxyguanosine, 14–15
    plasmid DNA, 15–16
  gene expression activation by 1,4-naphthalenedipropanoate, 18–19
  grafting sites for hydrophilic groups, 6–7
  history of development, 5–6
  hydrophilic functions insensitive to singlet oxygen, 6

peroxidation
  N,N'-di(2,3-dihydroxypropyl)-1,4-naphthalenedipropanamide, 12
  reactivity of compounds, 9–10
  singlet oxygen sources
    molybdate-catalyzed disproportionation of hydrogen peroxide, 9
    photooxidation, 9
    thermolysis rates, 10–11
  physiochemical properties, 7–8, 20
  singlet oxygen yield on thermolysis, 198
  stability, 5
  synthesis
    1-bromoethyl-4-chloromethylnaphthalene, 11
    N,N'-di(2,3-dihydroxypropyl)-1,4-naphthalenedipropanamide, 7, 11–12
    4-methyl-1-naphthalenepropanoate, 6–7
    4-methyl-N,N,N-trimethyl-1-naphthaleneethanaminium, 7
    1,4-naphthalenedimethanol, 7
    1,4-naphthalenedipropanoate, 7
  virus inactivation
    cultivation of virus, 200–201
    cytotoxicity assay, 202–203
    N,N'-di(2,3-dihydroxypropyl)-1,4-naphthalenedipropanamide efficacy, 204–207
    enveloped virus inactivation, 203–206
    exposure conditions, 201
    nonenveloped virus inactivation, 206
    overview, 16–17, 199
    survival assays, 201–202
9,10-Diphenylanthracene, singlet oxygen carrier, 5
1,3-Diphenylisobenzofuran, singlet oxygen probe, 22–24
DMN, *see* 1,4-Dimethylnapthalene
DNA
  1,4-dimethylnapthalene singlet oxygen carrier derivative reactions
    deoxyguanosine, 14–15
    plasmid DNA, 15–16
  gas chromatography–mass spectrometry, *see* Gas chromatography–mass spectrometry
  guanine, singlet oxygen oxidation chemistry, 149–151

HL-60 cell studies of ultraviolet A radiation, DNA analysis
  gel electrophoresis of fragments, 333–334
  ladder detection, 334–336
  8-oxodeoxyguanosine determination, 333
8-hydroxydeoxyguanosine
  assay in DNA
    acidic hydrolysis, 147–148
    cellular DNA, 151–153
    enzymatic digestion, 147
    extraction and isolation, 145–147
    gas chromatography–mass spectrometry, 148–149, 152
    high-performance liquid chromatography, 148, 152
    phosphorous-32 postlabeling, 152–153
  singlet oxygen role in formation, 144–145
long-range charge transport and oxidative damage
  assay using DNA assemblies with tethered photooxidant
    formamidopyrimidine-glycosylase digestion, 183
    gel electrophoresis, 183–184
    interduplex control reactions, 184
    photolysis, 182
    piperidine cleavage, 183
    solution conditions, 182
  construction of DNA assemblies with tethered photooxidant
    circular dichroism, 180
    complementary oligonucleotide synthesis, 180
    coupling of photooxidant, 175, 177–178
    design considerations, 170–171
    DNA synthesis, 176–177
    enzymatic ligation of long oligonucleotides, 180–182
    ethidium intercalators, 173, 175
    high-performance liquid chromatography, 178–179
    intercalation site determination, 171–173
    mass spectrometry, 179

rhodium intercalator as photooxidant, 166–168, 171, 173
    ruthenium complexes, 173
  photooxidant types, 168–171
  principle, 165–166
  stacking dependence, 168
  thymine dimer, oxidative repair
    characteristics of repair, 187
    cycloaddition reaction, 184–185
    DNA construct and analysis, 185, 187
    oxidative versus reductive repair, 185
mitochondrial DNA, see Mitochondrial DNA
mutation spectra
  rationale for study, 159, 164–165
  singlet-oxygen-induced DNA damage
    frequency of mutations, 160
    protection, 161
    transversion induction, 161
  ultraviolet A-induced damage
    aprt gene in Chinese hamster ovary cells, 163
    carcinogenicity, 161–162
    frequency of mutations, 160
    human embryonic kidney cells, 162–164
    p53, 163–164
    wavelength dependence, 162
radical cation formation potential of bases, 144
repair
  singlet-oxygen-induced DNA damage, 157–158
  ultraviolet A-induced damage, 158–159
singlet oxygen damage, overview, 154–155
transcriptional arrest as signal generator, 257–258
ultraviolet damage
  action spectra
    alkaline elution assay, 441–442, 444–445
    Chinese hamster ovary cell irradiation, 439–440, 442
    endonuclease analysis, 438–439, 444
    keratinocyte irradiation with sunlight, 440, 442
    overview, 360–361, 437–438

wavelength-dependent mechanisms, 442–444
cellular responses and repair, 363, 437
cyclobutane pyrimidine dimers, 360–361, 436
glutathione effects, 361
8-hydroxyguanine formation and effects, 361–362
overview of ultraviolet A damage, 155–157
sequence specificity of ultraviolet A damage with photosensitizer
  deuterium oxide effects, 422–423
  folic acid as photosensitizer, 422–423
  gel electrophoresis of DNA, 420–422
  hematoporphyrin as photosensitizer, 422–423
  irradiation conditions, 419–420
  menadione as photosensitizer, 422–423
  principle, 418
  radiolabeling of DNA, 419
  riboflavin as photosensitizer, 421–423
  sequencing of DNA, 420–421
  subcloning of *ras* fragments, 418
  type I versus type II photosensitization, sequence effects, 423, 425–427

E

EGF receptor, *see* Epidermal growth factor receptor
Electron paramagnetic resonance
  metal oxide oxidation product analysis, 31, 35
  oxygen radical detection in photodynamic therapy, 396–397
  photodynamic therapy singlet oxygen detection, 393
  TEMP, singlet oxygen detection in plants
    buffers, 78–79
    hydroxylamine conversion prevention, 79–80
    leaf infiltration, 81
    photoinhibitory treatment, 80–81
    reaction conditions, 79
    spectra, 79
    trapping, 78–79
  Tepy, singlet oxygen detection in plants
    leaf infiltration, 81
    trapping, 78–79
  topical antioxidant screening, 478
Electrophoretic mobility shift assay
  nuclear factor-κB
    aminopyropheophorbide, colon carcinoma cell photosensitization and gene activation, 124–125, 127
    binding reaction, 596–598
    cold competition assay, 598–599
    extract preparation, 593–595
    gel preparation, 595–596
    materials, 587
    oligonucleotide labeling, 589–591
    oxidant dose response and kinetics, 599
    oxidant treatment of cells
      adherent cells, 593
      suspended cells, 592
    ozone induction in rat lung, 557–558, 560
    sensitivity, 586
    stock solution preparation, 587–589
    supershift assay, 598–599
    troubleshooting
      background, 601–602
      high activation variability, 602
      protein content variability in extracts, 601
      quantitative analysis, 602
      smeared lanes, 599, 601
    tumor necrosis factor activation, 599
    ultraviolet A activation, 275–276
    Western blot analysis, 598
  ultraviolet A-induced transcription factor identification, 288–290, 585–586
EMSA, *see* Electrophoretic mobility shift assay
Eosinophil peroxidase, singlet oxygen chemiluminescence assays
  kinetics, 64
  spectral analysis, 64–65
Epidermal growth factor receptor
  inhibitors, 268
  radiation-induced signaling
    dephosphorylation assay in cell-free system
      membrane mix experiment, 264

membrane purification, 263
reaction conditions, 263–264
dephosphorylation assay *in vivo*, 265
dominant-negative mutants
gel electrophoresis analysis, 272
materials, 271–272
principles of study, 270–271
transfection, 272
overview, 258
phosphorylation assay with Western
blot analysis, 261–262
EPP, *see* Erythropoietic protoporphyria
EPR, *see* Electron paramagnetic resonance
Erythropoietic protoporphyria, *see also* Porphyrias
β-carotene treatment, 480–482, 484,
494
cysteine treatment, 482
gene defect, 479
gene therapy
bone marrow targeting, 482
fibroblasts, 483
mouse model, 483–484
peripheral blood stem cells, 483–484
laboratory testing, 482
mouse model, antioxidant effects,
479–480
singlet oxygen role, 479

F

Flow cytometry
apoptosis assays
cell requirements, 324–325
forward light scattering, 325, 329
instrumentation, 325
mitochondria transmembrane potential
analysis, 326–328, 330
side light-scattering, 325, 329
HL-60 cell studies of ultraviolet A radiation
mitochondrial transmembrane potential
DiOC$_6$, 337–338
JC-1, 338–340
peroxide detection, 336
keratinocytes, ultraviolet A-induced gene
expression analysis, 285

Fluorescent fatty acid, *see* C11-BODIPY
Fresh frozen plasma, *see* Virus inactivation

G

Gas chromatography–mass spectrometry
DNA analysis
acid hydrolysis, 411
applications, 401–403, 416–417
blood analysis
antioxidant addition during extraction, 408–409
blood preparation, 404
DNA extraction, 404–405, 410–411
materials, 403–404
oxygen exclusion during extraction,
408
ribonuclease digestion, 405, 407–408
storage of blood samples, 409–410,
416
derivatization, 411–412, 415
8-hydroxyguanine, 148–149, 152, 402
quantitative analysis, 415–416
running conditions, 415
metal oxide oxidation products, 31,
33–34
vitamin E ozonation analysis, 550–551
GC/MS, *see* Gas chromatography–mass
spectrometry
Gene therapy, erythropoietic protoporphyria
bone marrow targeting, 482
fibroblasts, 483
mouse model, 483–484
peripheral blood stem cells, 483–484
α-Glucosyl rutin, topical antioxidant screening, 469–471, 474, 476–477
Glutathione
DNA protection against ultraviolet radiation, 361
high-performance liquid chromatography
analysis of ozone depletion in skin
equipment, 541
operation, 541
reagents, 540
sample preparation, 542
standards, 541–542
oxidative stress inhibition, 260–261

H

High-performance liquid chromatography antioxidant depletion with ozone in murine skin, analysis
  advantages, 546
  ascorbate and urate analysis
    equipment, 539
    operation, 540
    reagents, 539
    sample preparation, 540
    standards, 540
  extraction, overview, 538
  glutathione analysis
    equipment, 541
    operation, 541
    reagents, 540
    sample preparation, 542
    standards, 541–542
  overview, 537
  ozone exposure, 537
  peak identification and quantification, 544–545
  tape stripping, 537–538
  troubleshooting, 545
  vitamin E analysis
    equipment, 543
    operation, 543
    reagents, 542
    sample preparation, 543–544
    standards, 543
cholesterol oxidation products
  electrochemical detection
    erythrocyte membranes, 94–97
    operating conditions, 93–94
    photoxidized cells, 98–100
  ozonization products
    electrochemical detection, 195–196
    mass spectrometry detection, 196
  ultraviolet A irradiation products
    chemilumiscence detection, 192–194
    electrochemical detection, 192
DNA assemblies with tethered photooxidant, 178–179
8-hydroxyguanosine separation and detection, 148, 152
squalene peroxide analysis, 616–617, 619–621
topical antioxidant analysis in tape-stripped samples, 618

HOG1, photodynamic therapy activation, 356–357
HPLC, see High-performance liquid chromatography
8-Hydroxyguanine, see DNA
4-Hydroxy-2-nonenal, see Lipid peroxidation

I

Immediate pigment darkening, see Skin, ultraviolet protection
Indigo, singlet oxygen quenching, 239–240
Ionizing radiation, oxidative stress, 260
Iron, role in singlet oxygen effects
  calcein assay, 295
  chelator studies, 294
  iron regulatory protein measurements, 295
  lipid peroxidation, 294, 296

K

Keratinocyte
  DNA damage from irradiation with sunlight, 440, 442
  mitochondrial DNA deletion, 376
  ultraviolet A-induced gene expression
    AP-2, 280, 288, 290
    cell culture, 281
    deuterium oxide studies, 284
    differential reverse transcriptase polymerase chain reaction, messenger RNA expression analysis, 285–286
    electrophoretic mobility shift assay for transcription factor identification, 288–290
    fluorescence-activated cell sorting, protein expression analysis, 285
    irradiation, 281–282
    materials, 284–285
    mimicing with singlet oxygen generators, 284
    reporter gene assays, 286–288
    singlet oxygen quencher effects, 282–283

L

Lipid peroxidation
 β scission pathways leading to aldehydes, 573
 Criegee ozonation, 572
 4-hydroxy-2-nonenal adduct detection
  applications, 563–564
  cytotoxicity, 562–564
  enzyme-linked immunosorbent assay, 569
  immunohistochemical analysis, 569–570
  polyclonal antibody preparation
   amino acid–HNE adduct antibody characterization, 565
   fluorescent lysine–HNE antibody preparation, 566–567
   principle, 564–565
   reductively-stabilized amino acid–HNE adduct antibody characterization, 565–566
  protein adducts, 563, 567
  reduction of protein adducts
   sodium borohydride, 567–568
   sodium cyanoborohydride, 568
  Western blot analysis, 568–569
 iron role, 294, 296
 signal transducers
  classes, 570–571
  formation, 571–574
 synthesis
  bis(1-hydroxyheptyl)peroxide, 578–579, 581
  bis(1-hydroxynonyl)peroxide, 578–579, 581
  glycerophospholipid hydroxyhydroperoxides, 574–575
  1-hydroxy-1-hydroperoxyalkanes, 575–576
  materials, 576
  1-palmitoyl-2-(1-hydroxy,1-hydroperoxynonanoyl)-*sn*-glycero-3-phosphocholine, 578
  1-palmitoyl-2-[3-(8-octanoyl)-5-(1-octyl)-1,2,4-trioxolane]-*sn*-glycero-3-phosphocholine, 577
  1-palmitoyl-2-(9-oxononanoyl)-*sn*-glycero-3-phosphocholine, 577–578

 phospholipid aldehydes, 574–575
 X-ray crystallography, 581–582
 skin equivalent assay
  Episkin preparation, 457
  K assay, 458
  sunscreen evaluation, 458–459
  ultraviolet A source, 457–458
 skin surface product analysis, *see* Skin, oxidation product measurement

M

α_2-Macroglobulin receptor/low-density lipoprotein receptor-related protein, photodynamic therapy response, 356
MAPK, *see* Mitogen-activated protein kinase
Mass spectrometry, *see also* Gas chromatography–mass spectrometry
 cholesterol ozonation product detection, 196
 DNA, long-range charge transport and oxidative damage, 179
Methylene blue, fresh frozen plasma treatment for virus inactivation
 absorption, 208
 photoproducts, 212
 principle, 208
 production, 212–216
 tolerability, 208, 212
 virus sensitivity, 208–209, 216
4-Methyl-1-naphthalenepropanoate, *see* 1,4-Dimethylnaphthalene, singlet oxygen carrier derivatives
3-(4′-Methyl-1′-naphthyl)propionic acid
 bacteria killing activity, 220–222
 dioxygenation of lipids in organic solvents, 218–220
 human umbilical vein endothelial cell toxicity, 220
 squalene susceptibility to oxidation, 219–220
 thermolysis, stoichiometry and kinetics, 217–218
4-Methyl-*N,N,N*-trimethyl-1-naphthaleneethanaminium, *see* 1,4-Dimethylnaphthalene, singlet oxygen carrier derivatives

Mitochondrial DNA
  genome, 367
  large-scale deletions in photoaging
    fibroblast cultures
      culture and medium, 374
      deuterium oxide effects, 375
      irradiation conditions, 374–375
      singlet oxygen generator effects, 375
      singlet oxygen quencher effects, 375
      ultraviolet A dose response of viability, 373–374
    keratinocytes, 376
    polymerase chain reaction detection
      advantages, 367
      DNA extraction, 368
      nested polymerase chain reaction, 369–370
      primers, 370
      principle, 368
      quantitative analysis, 371
      short-cycle polymerase chain reaction, 368–369
      tissue distribution of deletions, 372–373
  mutation frequency, 367
Mitochondrial transmembrane potential
  apoptosis role, 332
  flow cytometry assays, 326–328, 330, 336–340
  HL-60 cell studies of ultraviolet A radiation
    DiOC$_6$, 337–338
    JC-1, 338–340
  T cell apoptosis changes, 307
  topical antioxidant screening, 467–468, 473
Mitogen-activated protein kinase
  AP-1 activation, 586
  inhibitors, 269
  photodynamic therapy stimulation, 358
  singlet oxygen induction
    generation of singlet oxygen
      5-aminolevulinate, 137
      cellular accumulation of sources, 137–138
      chemical generation, 135–136
      rose bengal, 136–137
    photodynamic therapy effects, 141–142
    specificity of kinase activation in skin
      fibroblasts, 138–141, 143

ultraviolet A induction
  cell culture, 132
  deuterium oxide effects, 133
  immunoprecipitation and assay of kinases, 134–135
  irradiation conditions, 132
  overview, 131, 262
  singlet oxygen quencher effects, 132–138
  specificity of kinase activation, 135, 142
  Western blot analysis of phosphorylated kinases, 133–134
MNEA, see 4-Methyl-N,N,N-trimethyl-1-naphthaleneethanaminium
MNP, see 4-Methyl-1-naphthalenepropanoate

N

NAC, see N-Acetylcysteine
1,4-Naphthalenedimethanol, see 1,4-Dimethylnapthalene, singlet oxygen carrier derivatives
1,4-Naphthalenedipropanoate, see 1,4-Dimethylnapthalene, singlet oxygen carrier derivatives
NDMOL, see 1,4-Naphthalenedimethanol
NDP, see 1,4-Naphthalenedipropanoate
NEPO, see 3-(4′-Methyl-1′-naphthyl)propionic acid
NF-κB, see Nuclear factor-κB
Nitric oxide, photodynamic therapy
  apoptosis role, 353
p-Nitrosodimethylaniline, bleaching, 23, 25
Northern blot analysis
  Arabidopsis thaliana, ozone-sensitive mutants, 533–534
  cytokine-induced neutrophil chemoattractant induction by ozone, 556–557
Nuclear factor-κB
  activators, overview, 120–121, 273
  aminopyropheophorbide, colon carcinoma cell photosensitization and gene activation
    antioxidant effects, 128
    deuterium oxide effects, 127, 129
    electrophoretic mobility shift assay, 124–125, 127
    irradiation conditions, 124

cytokine gene regulation, 122
domains, 119
electrophoretic mobility shift assay
  binding reaction, 596–598
  cold competition assay, 598–599
  extract preparation, 593–595
  gel preparation, 595–596
  materials, 587
  oligonucleotide labeling, 589–591
  oxidant dose response and kinetics, 599
  oxidant treatment of cells
    adherent cells, 593
    suspended cells, 592
  sensitivity, 586
  stock solution preparation, 587–589
  supershift assay, 598–599
  troubleshooting
    background, 601–602
    high activation variability, 602
    protein content variability in ex-
      tracts, 601
    quantitative analysis, 602
    smeared lanes, 599, 601
  tumor necrosis factor activation, 599
  Western blot analysis, 598
inhibitors
  degradation, 120, 274, 277–278, 585
  kinase assay, 278–279
  phosphorylation, 273–274, 277–279,
    585
  types, 119
posttranslational processing, 119–120, 585
subcellular localization, 128
target genes, 273
ultraviolet A activation
  electrophoretic mobility shift assay,
    275–276
  extract preparations
    lysates for luciferase assay, 275
    nuclear extracts, 274–275
    whole cell extracts, 275
  inhibitor analysis
    glutathione S-transferase fusion, 279
    kinase assay, 278–279
    Western blot analysis of degradation
      and phosphorylation, 277–278
  luciferase reporter gene assay, 276–277
  nuclear translocation detection by indi-
    rect immunofluorescence, 276
ozone induction in rat lung

dexamethasone inhibition, 560
electrophoretic mobility shift assay,
  557–558, 560
inflammation role, 553–554
photodynamic therapy response, 355

O

8-Oxoguanine, see DNA
Oxygen
  singlet state, see Singlet oxygen
  triplet state, 3
Ozone
  antioxidant depletion in murine skin,
    high-performance liquid chromatog-
      raphy analysis
    advantages, 546
    ascorbate and urate analysis
      equipment, 539
      operation, 540
      reagents, 539
      sample preparation, 540
      standards, 540
    extraction, overview, 538
    glutathione analysis
      equipment, 541
      operation, 541
      reagents, 540
      sample preparation, 542
      standards, 541–542
    overview, 537
    ozone exposure, 537
    peak identification and quantification,
      544–545
    tape stripping, 537–538
    troubleshooting, 545
    vitamin E analysis
      equipment, 543
      operation, 543
      reagents, 542
      sample preparation, 543–544
      standards, 543
  β scission pathways leading to aldehydes,
    573
  biological targets, 505, 546–547
  Criegee ozonation, 572
  lipid peroxidation, 4-hydroxy-2-nonenal
    adduct detection
    applications, 563–564

cytotoxicity, 562–564
enzyme-linked immunosorbent assay, 569
immunohistochemical analysis, 569–570
polyclonal antibody preparation
 amino acid–HNE adduct antibody characterization, 565
 fluorescent lysine–HNE antibody preparation, 566–567
 principle, 564–565
 reductively-stabilized amino acid–HNE adduct antibody characterization, 565–566
protein adducts, 563, 567
reduction of protein adducts
 sodium borohydride, 567–568
 sodium cyanoborohydride, 568
Western blot analysis, 568–569
lung effects
cytokine-induced neutrophil chemoattractant induction
 dexamethasone inhibition, 560, 562
 electrophoretic mobility shift assay, 557–558, 560
 neutrophilia role, 554–555
 Northern blot analysis, 556–557
 reverse transcriptase polymerase chain reaction, 556
disease association, 551–552
inflammation, 552–554
nuclear factor-κB induction
 dexamethasone inhibition, 560
 electrophoretic mobility shift assay, 557–558, 560
 inflammation role, 553–554
penetration, 571–572
rat exposure, 555–556
plant response assays
Arabidopsis thaliana, ozone-sensitive mutant production
 advantages, 532
 gene isolation, 534
 genetic analysis, 533
 mutagenization and screening, 532
 Northern hybridization analysis, 533–534
 resources, 531
chloroplast effects, 521
disease association, 520–521
ethylene emission, 525–526

oxidative burst, 525
pathogen attack effects, 522
pathogen treatment
 Pseudomonas syringae, 535
 tobacco mosaic virus, 535
phytoalexin analysis, 526
phytotron exposure system, 522–524
plant dosing, 524–525
transcript induction
 complementary DNA libraries, 527
 poly(A) RNA isolation, 529–531
 RNA isolation from herbaceous plants, 527
 RNA isolation from trees, 528–529
pollutant levels, 551
reactive absorption
measurement
 antioxidant comparison, 510–511
 apparatus, 506–507
 assumptions, 508–509
 biomolecule concentration effects, 509–510
 data analysis, 507–509
 relative rate constants, 510
 respiratory system application, 510–512
modeling, 505
signal transducers
classes, 570–571
formation, 571–574
synthesis
 bis(1-hydroxyheptyl)peroxide, 578–579, 581
 bis(1-hydroxynonyl)peroxide, 578–579, 581
 glycerophospholipid hydroxyhydroperoxides, 574–575
 1-hydroxy-1-hydroperoxyalkanes, 575–576
 materials, 576
 1-palmitoyl-2-(1-hydroxy,1-hydroperoxynonanoyl)-sn-glycero-3-phosphocholine, 578
 1-palmitoyl-2-[3-(8-octanoyl)-5-(1-octyl)-1,2,4-trioxolane]-sn-glycero-3-phosphocholine, 577
 1-palmitoyl-2-(9-oxononanoyl)-sn-glycero-3-phosphocholine, 577–578
 phospholipid aldehydes, 574–575

X-ray crystallography, 581–582
singlet oxygen generation in plants
  ascorbic acid reaction, 512–513,
    518–519
  chemiluminescence assay in leaves
    carrier gas effects on emission intensity, 517–518, 520
    cell wall emission, 516–517
    gas volume per unit surface area effects, 519–520
    instrumentation, 513–514
    intensity of gas phase emission, 516
    intercellular extract measurements, 515
    kinetics, 519
    principle, 513
    spectral analysis, 517
    reactants in generation, 514–515
  stratum corneum effects, 536
  ultraviolet absorption, 359, 536
  uptake in plants, 521
  vitamin E reactions
    gas chromatography–mass spectrometry analysis, 550–551
    ozonation reaction, 549–550
    products and mechanisms, 547, 549

                    P

p21, photodynamic therapy apoptosis role, 352, 354
p53
  photodynamic therapy apoptosis role, 349, 352
  skin end point for ultraviolet A exposure, 464
  squamous cell carcinoma mutations, 355, 418
PACT, see Photodynamic antimicrobial chemotherapy
1-Palmitoyl-2-(1-hydroxy,1-hydroperoxynonanoyl)-sn-glycero-3-phosphocholine, synthesis, 578
1-Palmitoyl-2-[3-(8-octanoyl)-5-(1-octyl)-1,2,4-trioxolane]-sn-glycero-3-phosphocholine, synthesis, 577
1-Palmitoyl-2-(9-oxononanoyl)-sn-glycero-3-phosphocholine, synthesis, 577–578

PDGF receptor, see Platelet-derived growth factor receptor
PDT, see Photodynamic therapy
Peroxide, flow cytometry detection, 336
Persistent pigment darkening, see Skin, ultraviolet protection
PGHS, see Prostaglandin endoperoxide H synthase
Phospholipase C, photodynamic therapy activation, 346–348
Photodynamic antimicrobial chemotherapy, photosensitizers and mechanisms, 197–198
Photodynamic therapy, see also Ultraviolet radiation
  apoptosis role
    bcl2 role, 348–349
    calcium signaling, 346–348
    caspase activation
      Bap31 fragments, 350–351
      inhibitor studies, 349–350
    cell cycle mediation, 351–353
    ceramide generation, 351
    early event in tumor shrinkage, 345–346
    mechanisms, 346
    nitric oxide role, 353
    p21 role, 352, 354
    p53 role, 349, 352
    phospholipase C activation, 346–348
    retinoblastoma protein role, 352–353
    serine/threonine dephosphorylation, 348
  cell damage mechanisms, overview, 121–122
  clinical applications, 343–344
  dissipation of singlet state energy, 384
  early response gene activation, 355–356
  glucose-regulated proteins in response, 354–355
  heat shock protein induction, 357
  historical background, 343, 376
  HOG1 activation, 356–357
  hydrogen peroxide formation and detection, 394–396
  hydroxyl radical
    detection, 395, 397
    formation, 394
    reactivity, 394–395

α_2-macroglobulin receptor/low-density lipoprotein receptor-related protein response, 356
mitogen-activated protein kinase stimulation, 358
mode of action, 383–386
necrosis mechanism, 346
nuclear factor-κB response, 355
oxygen consumption rates, 386
photochemical process, 344, 383–384
photosensitizers, see also 5-Aminolevu-
linate; Aminopyropheophorbide
benzoporphyrin derivative monoacid
ring A, 381
electron-donating molecules, 394
hematoporphyrin derivative, 343
ideal criteria, 379, 385
lutetium texaphrin, 381
mono-L-aspartyl chlorin e6, 381
Photofrin II, 378–379, 386
Photofrin limitations, 344, 379
phthalocyanines, 382–383, 385
tetra(m-hydroxyphenyl) chlorin,
381
tin etiopurpurin, 381
tumor uptake, 377
principle, 342–343, 377–378
protein tyrosine phosphorylation,
357–358
singlet oxygen production
detection, 388–391, 393
deuterium oxide effects, 392–393
diffusion rate, 388
mechanism, 386–387
quencher studies, 391–392
stress-activated protein kinase activation,
356–357
superoxide anion
detection, 395–396
formation, 394
reactivity, 394
tumor response mechanism, 344–345
type I versus type II mechanisms
cell type dependence, 399–400
lipophilicity effects, 398
macromolecular binding effects,
398–399
oxygen availability effects, 397–398,
400
polarity effects, 398

Photosensitization, singlet oxygen production
cell photosensitization
linear dose–response, 26
photosensitizer efficiency comparisons,
27–28
rose bengal as photosensitizer, 26–27
subcellular localization of photosensitizers, 25–26
uniform illumination of cells, 26
energy transfer efficiency, 20–21
light sources, 21–22
metal oxides, see Titanium dioxide; Zinc
oxide
photon absorption rate, 20
photosensitizers, 21
principle, 20
solution photosensitization
1,3-diphenylisobenzofuran probe,
22–24
p-nitrosodimethylaniline bleaching, 23,
25
quantum yield measurement, 23, 24
substrate reaction modeling, 24–25
Photosystem II, singlet oxygen generation,
77–78, 80
PKC, see Protein kinase C
Platelet-derived growth factor receptor
inhibitors, 268–269
radiation-induced signaling
dephosphorylation assay in vivo, 265
overview, 258
PLC, see Phospholipase C
Polymerase chain reaction
mitochondrial DNA, large-scale deletions
in photoaging
advantages of detection, 367
DNA extraction, 368
nested polymerase chain reaction,
369–370
primers, 370
principle, 368
quantitative analysis, 371
short-cycle polymerase chain reaction,
368–369
RNA amplification, see Reverse transcriptase polymerase chain reaction
Porphyrias, see also Erythropoietic protoporphyria
clinical manifestations, 486–488

diagnosis, 486–487
heme synthesis, 485–486
inflammation, 487–488
phototherapy
  matrix metalloproteinase induction, 493
  melanin level manipulation, 489
  psoralen–ultraviolet A therapy
    action spectra, 491
    photosensitizers, 491
    regimens, 491–492
    risks, 492–493
    sources, 491
  ultraviolet B phototherapy
    dosing, 490–491
    risks, 490
    sources, 490
  ultraviolet penetration in skin, 489–490
phototoxicity, immediate versus delayed, 488–489
porphyrin photochemistry, 487
treatment, 485, 489
types and enzyme defects, 485–486
Prostaglandin endoperoxide H synthase
activities, 67–68
chemiluminescence assay of arachidonic acid oxidation
  dimol emission, 71, 75–77
  filters, 70
  identification of singlet oxygen as emission source, 71
  inhibitors of chemiluminescence, 74–75
  kinetics, 72–73
  oxygen requirement, 73
  peroxidase in singlet oxygen formation, 75–76
  pH dependence, 73
  single photon-counting apparatus, 69–70
  spectral analysis, 73–74
  suicide inactivation, 68–69
Protein kinase C, inhibitors, 269–270
Protein oxidation
  carbonyl group detection
    2,4-dinitrophenylhydrazine reaction, 430
    generation of modifications, 429–430
    tritiation, 430
    Western blot analysis
      advantages, 432

controls, 436
dinitrophenylhydrazine derivatization, 433
electroblotting, 434
immunoassay, 435
materials, 432–433, 435
principles, 431–432
sample preparation, 433
staining of total protein bands, 435
detection methods, overview, 429, 431
metal-catalyzed oxidation, 428
residues and modification types, 428–430
Protein tyrosine phosphatase, radiation-induced signaling
assays
  immunoprecipitation and counting, 267
  incubation conditions, 267
  principle, 266
  substrate labeling and purification, 266–267
targets, 258–259
Psoralen–ultraviolet A, porphyria therapy
  action spectra, 491
  photosensitizers, 491
  regimens, 491–492
  risks, 492–493
  sources, 491
PTP, see Protein tyrosine phosphatase
PUVA, see Psoralen–ultraviolet A

R

Radiometry, see Ultraviolet radiation
Ras
  DNA sequence specificity of ultraviolet A damage with photosensitizer
    deuterium oxide effects, 422–423
    folic acid as photosensitizer, 422–423
    gel electrophoresis of DNA, 420–422
    hematoporphyrin as photosensitizer, 422–423
    irradiation conditions, 419–420
    menadione as photosensitizer, 422–423
    principle, 418
    radiolabeling of DNA, 419
    riboflavin as photosensitizer, 421–423
    sequencing of DNA, 420–421
    subcloning of ras fragments, 418

type I versus type II photosensitization, sequence effects, 423,
425–427
skin cancer mutations, 418
Rb, *see* Retinoblastoma protein
Retinoblastoma protein, photodynamic therapy apoptosis role, 352–353
Reverse transcriptase polymerase chain reaction
cytokine expression analysis in atopic dermatitis therapy with ultraviolet A radiation
β-actin standard, 299–300
amplification cycles, 298–299
primers, 299
product purification and analysis,
299–300
RNA extraction, 298
cytokine-induced neutrophil chemoattractant, ozone induction in rat lung, 556
differential reverse transcriptase polymerase chain reaction in keratinocytes,
ultraviolet A-induced gene expression, 285–286
Rose bengal, photosensitization of cells,
26–27, 136–137
RT-PCR, *see* Reverse transcriptase polymerase chain reaction
RTC, *see* Tetrapotassium rubrene-2,3,8,9-
tetracarboxylate

S

SAPK, *see* Stress-activated protein kinase
Singlet oxygen
apoptosis induction, *see* Apoptosis
biological targets, 3–4
catalase detection, *see* Catalase, singlet
oxygen detection
chemiluminescence, *see* Chemiluminescence, singlet oxygen
cholesterol reporter, *see* Cholesterol
diffusion in air and detection, 112–113,
198
erythropoietic protoporphyria, *see* Erythropoietic protoporphyria
fluorescence assay, *see* DanePy

gene expression activation, *see also* Mitogen-activated protein kinase; Nuclear
factor-κB
1,4-naphthalenedipropanoate effects,
18–19
overview, 18–19, 111, 130–131
heat generation, 59–60
inorganic generators, 4
lifetimes and solvent effects, 3, 14, 25, 86,
121
peroxidase generation, 59
photosensitized production, *see* Photosensitization, singlet oxygen production
quantum yield components, 38–39
quenchers, *see* Singlet oxygen quencher
scavenging in phospholipid membranes
carotenoid scavenging rates, 104–106,
108–109
factors in rate measurement, 109–110
liposome preparation, 103
photosensitized singlet oxygen generation, 101–102
rate constant calculation, 102–103
α-tocopherol scavenging rates, 106–109
sources in biological systems, 3, 71, 85–
86, 110–111, 144
time-resolved detection
applications, 37
laser-induced optoacoustic spectroscopy, 43
thermal lensing, 42–43
time-resolved near-infrared phosphorescence
decay rate and quenching rate constant determination, 47–49
instrumentation, 40–42, 49
quantum yield measurement in homogeneous systems, 44–47
sample preparation and handling, 43
theory, 38–40
trace recording, 43–44
ultraviolet radiation production, *see* Ultraviolet radiation
water-soluble carriers, *see* 1,3-Cyclohexa-
diene-1,4-diethanoate; 1,4-Dimethyl-
naphthalene, singlet oxygen carrier
derivatives; 3-(4'-Methyl-1'-naph-
thyl) propionic acid
Singlet oxygen quencher, *see also specific
compounds*

amines, 232–233
amino acid-pyrrole *N*-conjugates, 228
assay with monomol emission
    biological quenchers, 224–226
    exogenous quenchers, 225
    generation of singlet oxygen, 222, 224
    instrumentation, 223–224
    principle, 223
indigoids, 239–240
keratinocytes, ultraviolet A-induced gene
    expression effects, 282–283
metal chelates, 234–235, 237–239
mitogen-activated protein kinase, ultravio-
    let A induction effects of quenchers,
    132–138
nonaromatic heterocycles, 239
olefins
    conjugated olefins and aromatic com-
        pounds, 228, 230, 232
    nonconjugated olefins, 227–228
phenols, 233–234
physical versus chemical quenching,
    226
rates of quenching, 227–228, 230, 232,
    237–238
sulfides, 232–233
Skin, oxidation product measurement
    ethanol-soluble sample preparation, 613
    high-performance liquid chromatography
        squalene peroxide analysis, 616–617,
            619–621
        topical antioxidant analysis in tape-
            stripped samples, 618, 621–622
    lipid hydroperoxides, colorimetric analy-
        sis, 615–616
    oxidation product generation, 619
    tape stripping, 613, 615
Skin, ultraviolet protection
    antioxidant defenses, 466
    chlorpromazine induced phototoxicity in
        skin equivalents, sunscreen assay
        applications, 456
        cell viability assay, 45
        chlorpromazine dose–response, 455
        Episkin preparation, 454
        formulas for testing, 454–456
        incubation, 455
        interleukin-1α assay, 455
        principle, 453–454
        ultraviolet A radiation source, 454

delayed-type hypersensitivity assay of im-
    munosuppression
    action spectra, 462, 464
    experimental groups, 460
    hypersensitivity test, 460, 462
    recall antigens, 460
    statistical analysis, 462
    sunscreen testing on human skin,
        460–462
    end points for ultraviolet A exposure,
        464
    immediate pigment darkening, ultraviolet
        A assay
        source definition and efficacy spec-
            trum, 451
        sunscreen evaluation, 452–453
        time course, 450
        ultraviolet A protection factor,
            451–452
    lipid peroxidation assay in skin equiva-
        lents
        Episkin preparation, 457
        K assay, 458
        sunscreen evaluation, 458–459
        ultraviolet A source, 457–458
    persistent pigment darkening, ultraviolet
        A exposure measurement, 449–450
    topical antioxidant screening
        overview, 466
        prescreening, *in vitro*
            cell culture, 467
            coenzyme Q10, 469–471
            α-glucosyl rutin, 469–471
            irradiation, 467
            mitochondria transmembrane poten-
                tial, 467–468
            phosphotyrosine assay, 467–468
            statistics, 468–469
            thiol assay, 467–468
            viability assays, 468
            vitamin E, 469
        ex vivo/in vitro screening
            biopsy and cell preparation, 473–474
            coenzyme Q10, 474–475
            α-glucosyl rutin, 474
            mitochondria transmembrane poten-
                tial, 473
            thiol assay, 473
        in vivo screening
            electron spin resonance, 478

erythema assay, 477
α-glucosyl rutin, 476–477
polymorphous light eruption prophylaxis, 476–477
ultraweak photon emission, 475–478
vitamin E, 476
transmission measurements of sunscreen
materials
advantages and limitations of assay,
448–449
carrier matrix, 446–447
light sources, 446–447
monochromatic protection factor calculation, 447–448
principle, 447
sun protection factor calculation,
447–448
ultraviolet penetration, 359, 366, 445,
489–490
Spectroradiometry, see Ultraviolet radiation
Squalene, peroxide analysis by high-performance liquid chromatography, 616–
617, 619–621
Stress-activated protein kinase
inhibitors, 270
photodynamic therapy activation,
356–357
Sun protection factor, see Skin, ultraviolet
protection
Sunscreen, see Skin, ultraviolet protection

T

T cell apoptosis, see Apoptosis
TEMP, see 2,2,6,6-Tetramethyl-4-piperi-
done
Tepy, singlet oxygen detection in plants
leaf infiltration, 81
trapping, 78–79
2,2,6,6-Tetramethyl-4-piperidone
photodynamic therapy singlet oxygen detection, 393
singlet oxygen detection in plants
buffers, 78–79
electron paramagnetic resonance spectra, 79
hydroxylamine conversion prevention,
79–80
leaf infiltration, 81

photoinhibitory treatment, 80–81
reaction conditions, 79
trapping, 78–79
Tetrapotassium rubrene-2,3,8,9-tetracarbox-
ylate, singlet oxygen trapping, 50
Thin-layer chromatography, cholesterol oxidation products, 91–93, 100
Time-resolved near-infrared phosphorescence, singlet oxygen detection
decay rate and quenching rate constant
determination, 47–49
instrumentation, 40–42, 49
quantum yield measurement in homogeneous systems, 44–47
sample preparation and handling, 43
theory, 38–40
trace recording, 43–44
Titanium dioxide
cosmetic applications, 29, 36
oxidation product analysis
butylated hydroxytoluene, 30–32,
35–36
electron paramagnetic resonance, 31,
35
gas chromatography/mass spectrometry, 31, 33–34
light source, 30
materials, 30
methyl oleate, 30–35
2,2,6,6-tetramethyl-4-piperidone, 30, 35
uric acid, 30–32
photochemical generation of active oxygen species, 29
TLC, see Thin-layer chromatography
α-Tocopherol, see Vitamin E

U

Ultraviolet radiation
apoptosis induction, see Apoptosis
atopic dermatitis therapy with ultraviolet
A radiation
cytokine expression analysis with reverse transcriptase polymerase
chain reaction
β-actin standard, 299–300
amplification cycles, 298–299
primers, 299

product purification and analysis, 299–300
RNA extraction, 298
irradiation technique, 297–298, 305–306
overview, 296–297, 302–303
T cell apoptosis
cell isolation, 303, 305
deuterium oxide effects, 304
DNA laddering assay, 306
double-staining technique for detection, 308
immunofluorescence microscopy, 301
irradiation device and dosimetry, 304
irradiation studies, *in vitro*, 303–309
mitochondria transmembrane potential changes, 307
morphology, 300–301
phosphatidylserine assay, 307
singlet oxygen generator and quencher effects, 304–305
TUNEL assay, 306–308
ultraviolet B effects, 302
carcinogenesis
action spectrum of murine squamous cell carcinoma, 364–355
mechanisms, 331, 360
p53 mutations, 355
papillomas, 365
tumor initiation and promotion, 364
cellular responses, 363–364
cholesterol oxidation, *see* Cholesterol
classification and radiation sources, 259–260, 280, 359, 417
DNA damage, *see* DNA; Mitochondrial DNA
dosimetry of ultraviolet A
cosine laws, 246
inverse square law, 246–247
objectives, 245
radiometric terms and units, 245–246
surface irradiation, 247
time of exposure, 246
volume irradiation, 247–248
genotoxicity, 153–154, 161–162, 360–362
histological analysis of skin effects, 612
keratinocyte, ultraviolet A-induced gene expression
AP-2, 280, 288, 290
cell culture, 281

deuterium oxide studies, 284
differential reverse transcriptase polymerase chain reaction, messenger RNA expression analysis, 285–286
electrophoretic mobility shift assay for transcription factor identification, 288–290
fluorescence-activated cell sorting, protein expression analysis, 285
irradiation, 281–282
materials, 284–285
mimicing with singlet oxygen generators, 284
reporter gene assays, 286–288
singlet oxygen quencher effects, 282–283
signal transduction induction, *see also* Epidermal growth factor receptor; Mitogen-activated protein kinase; Nuclear factor-κB; Platelet-derived growth factor receptor
DNA damage, transcriptional arrest as signal generator, 257–258
dominant-negative mutant studies, 270–272
kinase inhibitors, 267–270
protein tyrosine phosphatases as targets, 258–259
targets and sensors, overview, 256–257, 465–466
threshold hypothesis, 259
translational elongation, block as signal generator, 258
singlet oxygen role in ultraviolet A radiation effects
deuterium oxide studies, 292
iron role
calcein assay, 295
chelator studies, 294
iron regulatory protein measurements, 295
lipid peroxidation, 294, 296
mimicing with singlet oxygen generators, 293–294
modifiers of other active oxygen intermediates, 293
overview, 144, 290–291
singlet oxygen quencher effects, 292–293

skin protection studies, *see* Skin, ultraviolet protection
solar ultraviolet A levels, 446
spectroradiometry
 calibration of instruments, 250
 commercial instruments, 250–251
 detector, 250
 error sources, 250
 input optics, 249
 monochromator, 250
 spectral power distribution of sources, 248–249
T cell lymphoma therapy, 308
ultraviolet A radiometers
 angular response, 251, 253
 calibration, 253–255
 spectral sensitivity, 251
 stability, 255
Ultraweak photon emission, topical antioxidant screening, 475–478
UPE, *see* Ultraweak photon emission
Uric acid
 high-performance liquid chromatography analysis of ozone depletion in skin
 equipment, 539
 operation, 540
 reagents, 539
 sample preparation, 540
 standards, 540
 metal oxide reaction product analysis, 30–32

V

Virus inactivation
 1,4-dimethylnaphthalene singlet oxygen carrier derivatives
 cultivation of virus, 200–201
 cytotoxicity assay, 202–203
 N,N'-di(2,3-dihydroxypropyl)-1,4-naphthalenedipropanamide efficacy, 204–207
 enveloped virus inactivation, 203–206
 exposure conditions, 201
 nonenveloped virus inactivation, 206
 overview, 16–17, 199
 survival assays, 201–202
 fresh frozen plasma

photodynamic inactivation with methylene blue
 absorption, 208
 photoproducts, 212
 principle, 208
 production, 212–216
 tolerability, 208, 212
 virus sensitivity, 208–209, 216
 screening, 207
Vitamin E
 C11-BODIPY, multiwell screening of topical effectiveness, 608–609
 high-performance liquid chromatography analysis of ozone depletion in skin
 equipment, 543
 operation, 543
 reagents, 542
 sample preparation, 543–544
 standards, 543
 ozone reactions
 gas chromatography–mass spectrometry analysis, 550–551
 ozonation reaction, 549–550
 products and mechanisms, 547, 549
 singlet oxygen scavenging rates in phospholipid membranes, 106–109

W

Western blot analysis
 epidermal growth factor receptor phosphorylation assay, 261–262
 4-hydroxy-2-nonenal adduct detection, 568–569
 mitogen-activated protein kinase phosphorylation, 133–134
 nuclear factor-κB
 degradation and phosphorylation, 277–278
 electrophoretic mobility shift assays, 598
 protein carbonyl group detection
 advantages, 432
 controls, 436
 dinitrophenylhydrazine derivatization, 433
 electroblotting, 434
 immunoassay, 435
 materials, 432–433, 435

principles, 431–432
sample preparation, 433
staining of total protein bands, 435

Z

Zinc oxide
  cosmetic applications, 29, 36
  oxidation product analysis
    butylated hydroxytoluene, 30–32, 35–36
    electron paramagnetic resonance, 31, 35
    gas chromatography–mass spectrometry, 31, 33–34
    light source, 30
    materials, 30
    methyl oleate, 30–35
    2,2,6,6-tetramethyl-4-piperidone, 30, 35
    uric acid, 30–32
  photochemical generation of active oxygen species, 29

ISBN 0-12-182220-6

90038
9 780121 822200